Adhesive Joints
Formation, Characteristics,
and Testing

Adhesive Joints
Formation, Characteristics, and Testing

Edited by
K. L. Mittal
IBM Corporation
Hopewell Junction, New York

PLENUM PRESS • NEW YORK AND LONDON

Library of Congress Cataloging in Publication Data

International Symposium on Adhesive Joints: Formation, Characteristics, and Testing (1982: Kansas City, Mo.)
Adhesive joints.

"Proceedings of the International Symposium on Adhesive Joints: Formation, Characteristics, and Testing, held September 12-17, 1982, in Kansas City, Missouri" — T.p. verso.

Includes bibliographical references and index.

1. Adhesive joints—Congresses. I. Mittal, K. L., 1945- . II. Title.
TA492.A3I58 1982   668.3   84-2132

ISBN-13: 978-1-4612-9702-4   e-ISBN-13: 978-1-4613-2749-3
DOI: 10.1007/978-1-4613-2749-3

Proceedings of the International Symposium on Adhesive Joints: Formation,
Characteristics, and Testing, held as a part of the American Chemical Society meeting,
September 12-17, 1982, in Kansas City, Missouri

©1984 Plenum Press, New York
Softcover reprint of the hardcover 1st edition 1984
A Division of Plenum Publishing Corporation
233 Spring Street, New York, N.Y. 10013

All rights reserved

No part of this book may be reproduced, stored in a retrieval system, or transmitted,
in any form or by any means, electronic, mechanical, photocopying, microfilming,
recording, or otherwise, without written permission from the Publisher

PREFACE

This volume documents the proceedings of the International Symposium on Adhesive Joints: Formation, Characteristics and Testing held under the auspices of the Division of Polymer Materials:Science and Engineering of the American Chemical Society in Kansas City, MO, September 12-17, 1982.

There is a myriad of applications (ranging from aerospace to surgery) where adhesives are used to join different materials, and concomitantly the understanding of the behavior of adhesive joints becomes very important. There are many factors which can influence the behavior of adhesive joints, e.g., substrate preparation, interfacial aspects, joint design, mode of stress, external environment, etc., and in order to understand the joint behavior in a holistic manner, one must take due cognizance of all these germane factors. So this symposium was planned to address not only how to make acceptable bonds but their characterization, durability and testing were also accorded due consideration.

The symposium was organized with the following objectives in mind: to discuss the latest developments and activities anent adhesive joints, to provide a forum for cross-pollination of ideas, to bring together the researchers from various disciplines, and to highlight the areas which need intensified efforts. The interest and high tempo of research in adhesive joints is well manifested by the size of the program which contained 51 papers from many parts of the globe covering various ramifications of adhesive joints. So this was truly an international event and if the comments from the attendees are a measure of the success of a symposium, then it was more than successful and fulfilled very well the objectives of the symposium.

The present volume containing a total of 51 papers (921 pages) by 94 authors from 12 countries is divided into five parts. The topics covered include: substrate and interfacial aspects of adhesive bonding; determination of the locus of failure; factors influencing behavior of adhesive joints; evaluation, characterization and testing of adhesive joints; stress analysis and performance aspects, fracture aspects, and durability or stability behavior of adhesive joints. Apropos, the availability and widespread use of sophisticated surface analytical tools has proven a boon in adhesive joints research and in this volume are included many examples illustrating the usefulness of these techniques.

The purpose of a symposium is to present the state of knowledge of the topic under consideration and it can best be accomplished by a blend of overviews and original research papers. This is exactly what was done in this Symposium and as a result this volume contains a number of overview papers and unpublished original research contributions. It should be added here that all papers were properly reviewed as comments from the peers are a desideratum to maintain the standard of publications. As a result, most manuscripts were returned to the respective authors for suitable revisions. It should also be pointed out that this volume contains some papers which were not presented, and certain papers which were presented are not included for a variety of reasons. As for the discussion section, yours truly had hoped to include discussion at the end of each paper or group of allied papers, but the number of written questions received did not vindicate undertaking such endeavor. However, it should be recorded that there were many spontaneous and lively discussions both formally in the auditorium as well as in other (more relaxed) places.

Even a cursory look at the Table of Contents shows that many facets and ramifications of adhesive joints are accorded due coverage in this volume. It is hoped that this volume would be of interest to both the veteran (as a reference source for contemporary research and development activities) and to the tyro as a fountain of new ideas.

Acknowledgements: First of all I am thankful to the Division of Polymer Materials:Science and Engineering for sponsoring this event, and my thanks are due to H. R. Anderson, Jr. (IBM Corp.) for permitting me to organize this symposium and to Steve Milkovich (IBM Corp.) for his understanding and cooperation during the tenure of editing.

On a more personal note, I am thankful to my wife, Usha, for helping me in more ways than one, and to my children (Anita, Rajesh, Nisha and Seema) for rendering home environment conducive to work and for letting me use those hours which rightfully belonged to them. Special thanks are due to Phil Alvarez (Plenum Publishing Corp.) for his continued interest in this project. I would like to extend my appreciation to Barbara Mutino (Office Communications) for meeting various typing deadlines with a smile. Special thanks are due to the unsung heroes (reviewers) for their time and valuable comments. The generous support of the Petroleum Research Fund of the American Chemical Society towards travel

PREFACE

expenses of certain invited overseas speakers is gratefully acknowledged. Last, but not least, the cooperation and enthusiasm of the contributors is sincerely appreciated without which this book would not have seen the light of day.

K. L. Mittal
IBM Corporation
Hopewell Junction, New York 12533

CONTENTS

PART I. SUBSTRATE AND INTERFACIAL ASPECTS

The Role of Surface and Bulk Characterization in the
Evaluation of Adhesive Joints
    W. L. Baun 3

Practical Adhesion Measurement in Adhering Systems: A Phase
Boundary Sensitive Test
    A. A. Roche, M. J. Romand and F. Sidoroff 19

Effect of a Phenol Formaldehyde Resin on the Adhesion of a
Polysulfide Sealant
    R. Ramaswamy and P. Sasidharan Achary 31

Polysulfide-Polyurethane Interfacial Aspects
    A. M. Usmani 41

Epoxy Resin Wetting of E Glass Single Filaments as it
Relates to Shear Strength
    E. J. Berger and Y. Eckstein 51

PART II. FACTORS INFLUENCING BEHAVIOR OF ADHESIVE JOINTS

Influence of Surface Roughness on Mechanical Properties of
Joints
    Y. Gilibert and G. Verchery 69

The Effects of Surface Conditioning of AISI-304 Stainless
Steel on the Interfacial Properties of Alloy/Epoxy Composite
Structure; Adherend Surface Characterization Using X-ray
Emission Spectroscopy (LEEIXS)
    F. Gaillard, A. A. Roche and M. J. Romand 85

Surface Characterization of Anodic Oxides on Aluminium
Alloys by Means of Surface Potential Difference, Surface
Impedance and Surface Morphology
    A. Kwakernaak, R. Exalto and H. A. van Hoof 103

Investigation into the Effect of Surface Treatment on the
Wettability and the Bondability of Low Surface Energy
Materials
    J. P. Jeandrau 121

The Effect of Moisture on the Dimensional Stability of
Adhesively Bonded Joints
    E. J. Hughes, J. Boutilier, and J. L. Rutherford 137

The Dimensional Stability of Epoxy Adhesive Joints
    J. P. Sargent 151

Strength Characteristics of Mono and Multiple-Wire Steel to
Steel Joints Bonded with an Epoxy Adhesive
    R. W. Hylands 165

A Study of Adhesive Joints Between Aliphatic Polyamides and
Metals
    S. S. Pesetskii, V. E. Starzynskii, and
    S. V. Shcherbakov 195

"Honeymoon" Phenolic Fast-Setting Adhesives for Exterior-
Grade Finger Joints
    A. Pizzi 214

Ageing of Structural Film Adhesives - Changes in Chemical
and Physical Properties and the Effect on Joint Strength
    C. E. M. Morris, P. J. Pearce and R. G. Davidson 231

Structural Adhesives Based on Diallyl Phthalate
    A. M. Usmani 247

Possibilities of Integrating Surface Treatment of Bonding
Parts in the Adhesive Bonding Process
    K. Ruhsland 257

PART III: EVALUATION AND CHARACTERIZATION

Evaluation of Adhesive Test Methods
    G. P. Anderson, K. L. DeVries and G. Sharon 269

A Critical Appraisal of Dental Adhesion Testing
    W. J. O'Brien and S. T. Rasmussen 289

CONTENTS

Adhesive Joint Characterization by Ultrasonic Surface and
Interface Waves
    S. I. Rokhlin 307

Nondestructive Evaluation of Some Bonded Joints
    T. C. Ward, M. Sheridan and D. L. Kotzev 347

Ultrasonic Assessment of Cure Rate Effects in Bonded
Honeycomb Structures
    R. A. Pike and R. S. Williams 369

The Three-Point Bend Test for Adhesive Joints
    N. T. McDevitt and W. L. Baun 381

Accelerated Aging Procedures for Glue-Wood Bonds and Their
Use for In-Plant Quality Control
    R. B. Jathar 395

Peel-Strength and Energy Dissipation
    T. Igarashi 419

The Estimation of Adhesion in Filled Polymer Systems
    Yu.S. Lipatov, T. T. Todosiychuk, P. K. Tsarev and
    L. M. Sergeeva 433

PART IV. DURABILITY OR STABILITY ASPECTS

Adhesion and Durability of Metal/Polymer Bonds
    J. D. Venables 453

Failure Mechanisms in the Boundary Layer Zone of Metal/
Polymer Systems
    W. Brockmann, O.-D. Hennemann and H. Kollek 469

Joint Durability Studies with Abraded, Etched, Coated and
Anodized Aluminum Adherends
    J. D. Minford 485

Comparative Study of Aluminum Joint Strength and Durability
with Varying Thickness, Boehmite-Type Oxide Surfaces
    J. D. Minford 503

Epoxy Adhesion to Copper, Part II: Electrochemical
Pretreatment
    J. M. Park and J. P. Bell 523

Hydrothermal Stability of Titanium/Epoxy Adhesive Joints
    F. J. Boerio and R. G. Dillingham 541

A Study on Elastomer/Metal Bonds Applicable in Underwater
Sonar Systems
    R. Y. Ting 555

PART V. STRESS ANALYSIS AND PERFORMANCE ASPECTS

Life Prediction Methodology for Adhesively Bonded Joints
    J. Romanko, K. M. Liechti and W. G. Knauss 567

Stress Analysis of Adhesively Bonded Joints
    R. A. Kline 587

The Impact Strength of Adhesive Lap Joints
    J. A. Harris and R. D. Adams 611

The Performance of Adhesive-Bonded Thin-Gauge Sheet Metal
Structures with Particular Reference to Box-Section Beams
    A. Beevers and A. C. P. Kho 627

Cyclic Debonding of Adhesively Bonded Composites
    S. Mall, W. S. Johnson and R. A. Everett, Jr. 639

Effects of Low Cycle Loading on Shear Stressed Adhesive
Bondlines
    W. Althof 659

Effect of Scrim Cloth on Adhesively Bonded Joints
    E. C. Francis and D. Gutierrez-Lemini 679

PART VI. FRACTURE ASPECTS

Mechanical Measurement of Interatomic Bonding Energies at
Interfaces
    E. H. Andrews 689

Review of Continuum Mechanics Factors in Adhesive Fracture
    M. L. Williams 703

Fracture Energetics of Adhesive Joints
    L.-H. Lee 739

Fracture of Composite-Adhesive-Composite Systems
    E. J. Ripling, J. S. Santner and P. B. Crosley 755

Characterization of the Fracture Behavior of Adhesive Joints
    D. L. Hunston, A. J. Kinloch, S. J. Shaw and S. S. Wang 789

CONTENTS

Structural Precursors to Fracture in Adhesive Joints
    W. A. Jemian 809

Fracture Toughness of Elastomer Modified Epoxy Adhesives
    A. A. Donatelli, C. T. Mooney and J. C. Bolger 829

A Three-Dimensional Analysis of a Butt Joint with a Flaw
    R. S. Alwar and K. N. Ramachandran Nambisan 839

The Influence of Layer Thickness and Internal Stresses on the Bond Strength of Metal-to-Ceramic Joints
    W. Diem, G. Elssner, T. Suga and G. Petzow 855

Bond Strength Characterization of Metal-to-Ceramic and Adhesive Joints by Critical Energy Release Rates
    W. Diem, G. Elssner, T. Suga and G. Petzow 871

Interfacial Properties of Filled Epoxide Resins
    A. C. Moloney, H. H. Kausch and H. R. Stieger 883

About the Contributors 905

Index 923

Part I
Substrate and Interfacial Aspects

THE ROLE OF SURFACE AND BULK CHARACTERIZATION IN THE EVALUATION OF ADHESIVE JOINTS

W. L. Baun

Materials Laboratory, AFWAL/MLBM
Wright-Patterson AFB, Ohio 45433

Adhesive joints are often evaluated by a mechanical test followed by exposure of a similar joint to elevated temperature, humidity, and/or corrosive atmosphere. After this exposure to a deleterious atmosphere, the bonded joint is usually tested again using the same geometry and conditions. The original load to failure is usually taken as a measure of the quality of the joint and performance during the accelerated test determines how well the joint will hold up in service. In the past only these numerical values were noted and there was little diagnostic work on the failure surfaces. More recently, however, there has been greater emphasis on where and why a failure took place. In order to fully evaluate failure surfaces it is necessary to determine the locus of failure. Often this task requires modern methods of surface characterization, especially when the failure takes place along a weak boundary layer (WBL). These surface characterization probes use beams of ions, electrons or photons and include ISS, SIMS, AES, and XPS. Examples of the use of these techniques along with microscopy are shown for adhesive bonding research on aluminum, titanium and steel.

INTRODUCTION

Numerous technologies in addition to adhesive bonding rely on a system in which two materials are placed in intimate contact and must remain so for the life of the resulting part. Joining methods such as welding and soldering are examples of such processes. Metallic and polymeric coatings on metals to provide corrosion resistance, enhance appearance, or to take advantage of some property of the couple, provide other examples. Following fabrication, various mechanical tests are performed to determine how well one material adheres to the other. Tests are also performed under accelerated conditions to determine the durability of the bonded materials in long time service. Often the only information recorded from such tests is a numerical value of the force or energy necessary to cause the bonded structure to fail. Equally important is where the joint failed. Careful examination of the failure surfaces by the methods to be discussed here allows determination of the locus of failure. Good[1] says "There is great practical importance to the correct identification of the failure locus. It is obvious that the measures which must be taken to remedy an interfacial failure are different from those which must be taken to remedy "cohesive" failure in either bulk phase".

DISCUSSION

Where Can Failure Occur?

In a simple two component system such as shown in Figure 1[1] failure may take place in one or more of the five regions. That is, failure may propagate in either of the two bulk phases (1 and 5), at the interface (3) or in regions of A and B very near the interface (2 and 4). It has been the long-standing opinion of some workers[2] that most failures which are called interfacial (region 3) are in reality failures in a weak boundary layer (WBL) occurring very near the interface.

It has been shown earlier that it is not simple using visual or even microscopic examination of adhesive bonds to determine after testing whether an apparent adhesive failure occurred at the interface due to improper wetting or at some new interface, leaving behind a thin layer of adhesive on the adherend or oxide on the adhesive.[3] There is a resolution limitation for most scanning electron microscopes (SEMs) which makes very thin organic films difficult to detect, especially when the adhesive is a pure polymer containing no fillers of higher atomic number than the polymer to increase contrast. Optical and staining methods have been reported to determine the presence of adhesive films.

However, the optical technique uses the interference phenomenon, which is applicable only to fairly thick films, certainly not to films only a few molecules thick or boundary layers containing both adhesive and adherend components. Staining techniques are sensitive only to specific compounds present in the usually complex adhesive systems. Several investigative techniques on both sides of a joint failure are necessary to determine the locus of failure conclusively.

It is also important to determine the locus of failure of thin films subjected to many mechanical tests depending on peel, scratch, abrasion, and deceleration as reviewed by Mittal[4]. Generally, thin polymer films on metals present the same problems as encountered in adhesive bonding. On the other hand, evaporated metal films, by nature of their simpler composition and more ideal interfaces, present fewer difficulties. There are usually differences in color between the evaporated film and the substrate which make it easier to determine the mode of failure. However, when both metals are the same or nearly the same color or when a metal is deposited on a thin oxide film or on a polymer, there can be problems of interpretation of visual results.

In actual practice interfaces in adhesive bonding are not so simple as shown in Figure 1. Many more materials and interfacial regions exist in practical systems, as shown in Figure 2, adding considerably to the complexity of determining the exact locus of failure. To make it even more difficult, it has been suggested that these interfaces are not sharply separated but rather, are intermixed and diffuse. The term "interphase"[5] probably better describes these boundary areas. In order to determine precisely where a bonded system failed, it is necessary to use sensitive methods of analysis of chemistry and morphology on the failure surfaces. In Figure 2 it can be seen that cohesive failure may take place at A, B, C, or D and interfacial or near interfacial failure may take place in the regions marked 1, 2, and 3 at each interface.

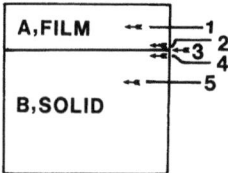

Figure 1. Model of the interface formed in a simple two-component system.

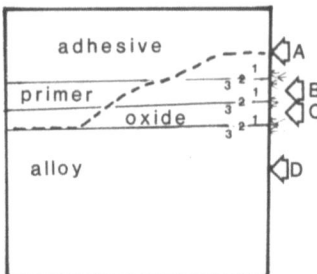

Figure 2. Model of the interfaces formed in a typical adhesive bonding system.

One of the most common failures, especially when the adhering phase is primarily organic, is depicted in Figure 3 and takes place along a layer of trapped air or other vapor. This type of weak boundary layer failure is a failure of the first kind, as described by Bikerman.[2a] This type of failure is especially prevalent in adhesive bonding where in addition to the trapped gases, it is also possible that the curing process causes gases to be released which cannot vent from the joint. Examples of such failures are seen in Figure 4, where the glass reinforced polymer Tefzel is shown from a failure specimen which had been bonded to steel. A significant area of voids exceeding 50% of the total area is seen in this scanning electron micrograph. Similar effects are seen in some adhesive bonded failure surfaces, especially where woven scrim cloth is used. Micrographs from such wedge specimen failure surfaces using the commercial adhesive FM 400 are seen in Figure 5. In this case, the voids were caused by improper application of temperature and pressure cycles during curing. Bascom and Cottington[6] show results in which a thin film of air is trapped and covers 50% of the interfacial area. Incomplete displacement was found to occur if the resin does not fully wet the adherend or if the resin does not become sufficiently fluid during the cure cycle. These authors found increases in bond strength of as much as 30% by proper vacuum processing to insure removal of air.

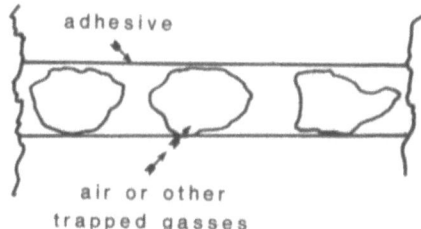

Figure 3. Failure of the first kind along region of trapped air or other vapor.

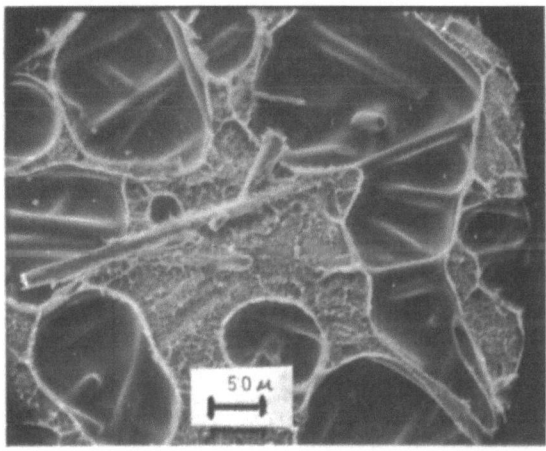

Figure 4. Failure surfaces from a steel lap shear specimen bonded with Tefzel.

What Characterization Methods are Applicable?

Each characterization method is based on an intrinsic property of the atoms and molecules on the surface. Such atomic and molecular surface properties are listed in Table I along with a partial list of techniques which use these properties. Park[6] has reviewed these methods and others and discusses the limitations as well as the advantages of each. He emphasizes in his critical review that in a field as new as surface characterization it is not surprising that much remains to be learned about the effective use of these tools. In the present paper we consider those techniques which show promise for determining the locus of failure when an adhesive bond breaks or a coating separates from a substrate. We have largely ignored

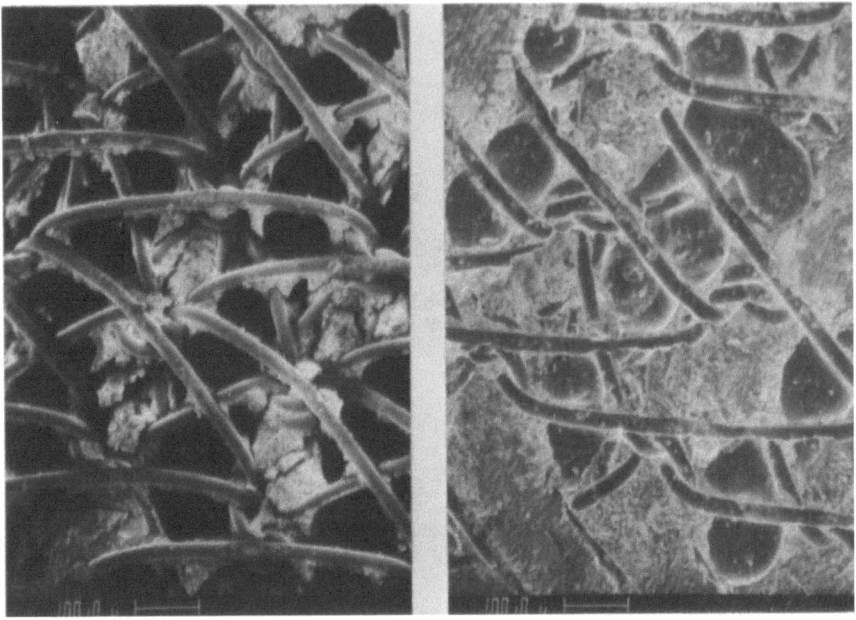

Figure 5. Wedge failure surfaces from titanium bonded with commercial FM 400 adhesive.

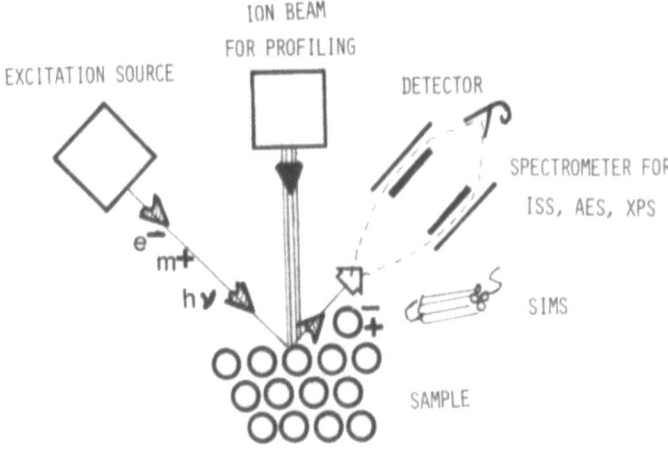

Figure 6. Representation of surface characterization methods.

techniques exploiting vibrational states and emphasized those using atomic properties because the latter appear more sensitive and easier to interpret. Figure 6 shows schematically how these surface methods are used to determine composition of the outer layers of material.

<u>Ion Scattering Spectrometry</u>. When an ion approaches a surface, it is most likely that the ion will become neutralized by electrons at the surface. Some ions do survive, however, and are back scattered from the surface after undergoing a simple elastic binary collision. The masses of the atoms at the surface can be determined by measuring the energy lost by the incoming ion beam. The ISS technique is only one which analyzes the first atomic layer at the surface. In each of the other techniques there is a finite sampling depth.

<u>Secondary Ion Mass Spectrometry</u>. The momentum transfer from the probing ion beam to the surface atom results in some atoms at the surface being dislodged. Although many of these sputtered species are neutral, some become ionized and can be analyzed by conventional mass spectrometry.

Table I. Properties of Surface Atoms and Molecules and Examples of Characterization Techniques Utilizing These Properties[7].

Properties	Technique
Atomic mass	Ion Scattering Spectrometry Secondary Ion Mass Spectrometry Thermal, Electron, Ion Desorption
Vibrational States	Infrared Absorption, Emission Raman Spectrometry Inelastic Electron Tunneling
Valence States and Core States	X-Ray Emission Auger Electron Spectroscopy Photoelectron Spectrometry Soft X-Ray Appearance Spectrometry
Crystallography	Electron Diffraction Field Ion Microscopy

<u>Auger Electron Spectrometry</u>. If a material is irradiated with an electron or photon beam of sufficient energy, an electron is ejected from an inner level, placing the atom in an excited state. When the atom reverts to the ground state, it may do so by emitting characteristic X-rays or by radiationless transitions, the so-called Auger transitions. One major disadvantage of AES for adhesive bonding research is that organic adhesives are very

unstable under the electron beam due to localized heating. Electrical charging is also a problem with adhesive surfaces. Naturally AES does not suffer such limitations for thin metal films. Information on chemical bonding is included in spectral details and is useful as a "fingerprint" but is extremely difficult to interpret from a theoretical point of view.

X-Ray Photoelectron Spectroscopy (Electron Spectroscopy for Chemical Analysis. In X-ray excited photoelectron spectroscopy a specimen is irradiated with a beam of X-rays denoted by $h\nu$ in Figure 6 (obtained by electron bombardment of, typically, aluminum or magnesium) exciting electrons from the target that are energy analyzed. X-ray photoelectron spectra contain chemical information similar to that mentioned for Auger spectra, but it is easier to interpret because it involves only one simple transition rather than several. As the chemical environment of an atom changes, the photoelectron spectrum undergoes changes in peak shape, position or intensity. Information such as this is invaluable in determining if chemical reactions take place in an adhesive joint following use in the field or after accelerated testing. In thin-film technology, changes in electron spectra allow the study of diffusion and alloying between thin films or thin-film/substrate combinations. Inert gas sputtering may also be used with XPS to provide elemental profile information.

What Do We Look For?

When we use these spectrochemical tools to determine species on the surface, we probably use them slightly differently each time, especially in adhesive bonding. Even when chemical and morphological information has been collected, interpretation may be difficult. Just how do we decide where failure has occurred? In the typical complex adhesive bonded system, we have several interfacial regions. Each of the materials coming together to form these interfaces has its own individual chemical signature. The substrate for instance usually contains alloying elements which vary in content between the surface and the bulk. In addition to alloying elements, surface treatments leave behind elements characteristic of each treatment. For instance the popular FPL (Forest Products Laboratory) etch for aluminum alloys consists of sulfuric acid and sodium dichromate in distilled water and leaves a detectable amount of chromium on the alloy surface. Primers often contain anions and cations which can be followed by spectrochemical methods. These additives (such as strontium chromate) are usually placed in the primer to provide corrosion protection in the coating. An example of using these elements as tracers are illustrated in Figure 7 where an aluminum alloy adherend which appeared to fail at the oxide/primer interface shows appreciable chromium. Analysis of the opposite side suggested a mixed mode failure near the suspected interface.

In addition to determining what elements exist on failure surfaces, it is most important to look at these surfaces using optical microscopy and SEM. Dwight[8] has used the SEM extensively, especially in all cohesive failures, to attempt to understand the mechanism by which failure occurs. Plastic and brittle failure mechanisms are easily differentiated on the polymer surface. Initial and final flaws and voids may be determined and are of importance in evaluating joint performance. An example of the use of the SEM was seen in Figures 4 and 5 where micrographs show the failure to be a weak boundary layer of the first kind, due to entrapped air or other vapor. Often we consider WBLs to occur only during bond processing but in actuality WBLs may be created by numerous processes including surface preparations to the original alloy or metal. An example of such chemistry changes is seen in Figure 8, the elemental profile prepared from AES data from a 2024 aluminum alloy treated with a sulfuric acid-chromate etch. As can be seen in this normalized profile, copper is concentrated in a band between the aluminum oxide and the metal. Similar results were shown by Sun and co-workers also using AES.[9] Another common weak boundary layer produced on aluminum alloys containing magnesium (most structural alloys) occurs due to concentration of MgO on the surface during heat treatment. ISS and SIMS spectra obtained from such a surface on 2024 Al are seen in Figure 10. Although initial bondability is often not affected by such an oxide layer, long time durability (especially under conditions of high humidity) is severely reduced.

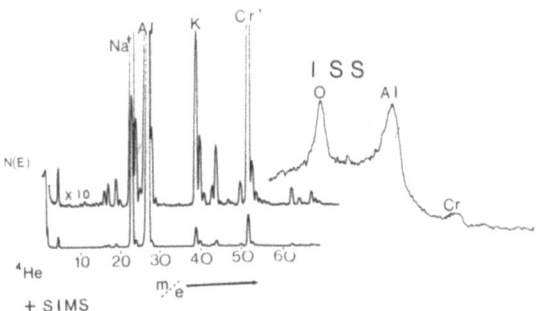

Figure 7. ISS and SIMS data from aluminum alloy failure surface.

Surface characterization techniques have been applied to a wide variety of failure surfaces obtained from several mechanical tests, such as lap shear, peel, crack opening and three-point bend.[11]

Failure surfaces from thin and thick-thin adherend wedge tests have been analyzed by the ISS-SIMS technique. The wedge test provides information about adherend surface preparation in that it is sensitive to different surface treatments and can discriminate between bonding processes that give good and poor service performance. The specimen consists of two thin adherends or a thick-thin configuration to concentrate stresses along the interface. A wedge is driven into the bondline and the position of the crack leading edge is determined microscopically. Then the specimen is subjected to various external stimuli such as changes in temperature and relative humidity. The propagation of the crack tip is followed with time. Sometimes when the wedge is driven into the bondline, separation of the specimen occurs over a portion of the bondline, first causing cohesive failure in the adhesive then apparent adhesive failure at the adhesive/adherend interface during testing at 71°C (160°F) and 95 percent relative humidity, followed by cohesive failure when the specimen is opened following the test. The adherend shows no indication of adhesive either visually or in the SEM, as seen in Figure 10 in the area "ABC". There are slight reflectivity differences across the surface. Perhaps staining techniques would be effective in outlining areas which contained thin films of adhesive when they exist. With extremely thin films, however, it is doubtful if staining would provide any information on the locus of failure where a primer is used and therefore several interfaces exist. An example of such a wedge test specimen is shown in Figure 10. Spectra (not shown) indicated failure in areas a and b near the oxide/adhesive interface. ISS/SIMS data shown here for area c along with depth profiles suggest that with time the failure locus changes to a failure either in the oxide or near the oxide/metal interface. The appearance of chromium (from the original etch) in area c and depth profiling data provide important clues to the exact location of the failure.

Some failures which could be classed as pure adhesive or cohesive have been examined by ISS/SIMS. Adherend surfaces which were obviously not wet by the adhesive showed no trace of the adhesive on the adherend. Often these surfaces were 'dirty" and showed a thin layer of contaminating elements on the adherend. This kind of failure probably should not be considered adhesive if proper bonding between the two surfaces never occurred. An example of this phenomenon is seen in Figure 11. ISS and SIMS spectra are shown for a low carbon steel surface following debonding of the thermoplastic Nylon 12 from the surface. The first ISS spectra show only weak peaks near potassium and iron.

The corresponding SIMS data show strong sodium and potassium peaks compared to iron. As sputtering continues, sodium and potassium decrease and iron increases. As was indicated earlier, this segregation of monovalent ions to the surface of adherends which have been heated is a common occurrence. When the temperature of adhesive bonding materials was raised to approximately the adhesive cure temperature (121 to 177°C) (250 to 350°F), the very mobile ions such as Li^+, Na^+, and K^+ were concentrated on the surface. Perhaps if the conditions for bond failure under water attack were those for hydrolysis, the diffusion of alkali ions to the surface would increase the osmotic potential and enhance the destructive ingress of water at the interface. Gledhill and Kinloch[12] have studied the iron-epoxy system and have found cohesive failure under dry conditions and adhesive failure under wet conditions. They show that substrate corrosion is not an operative mechanism in environmental failure but, rather, a post-failure mechanism.

Appendix 1 is a background bibliography of a few of the many publications which pertain to this subject. These examples illustrate test methods, problems, and surface characterization techniques related to the subject at hand.

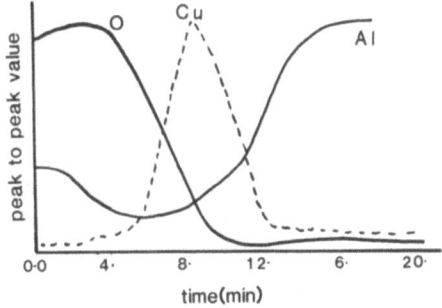

Figure 8. Elemental profile from AES data for 2024 aluminum alloy.

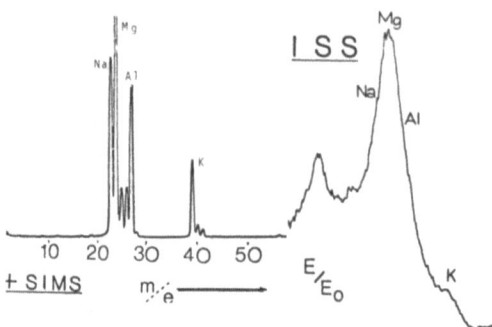

Figure 9. ISS and SIMS data from "As Received" 2024 aluminum alloy showing mill scale.

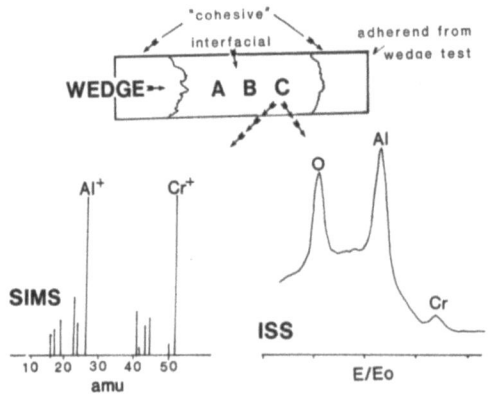

Figure 10. Failure propagations in wedge test specimen.

THE ROLE OF SURFACE AND BULK CHARACTERIZATION

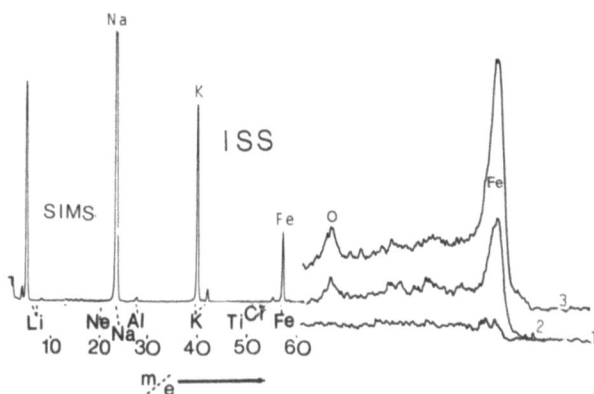

Figure 11. ISS and SIMS data for steel side of nylon/steel bonded lap shear specimen.

CONCLUSIONS

Spectrochemical techniques such as ISS, SIMS, AES, and XPS combined with microscopy can be usually used to gain a clear picture of where an adhesive joint has failed following testing or service. Unfortunately this information still does not always tell us the exact mechanism of failure. For instance, many failure surfaces, especially those exposed to high concentrations of water vapor, show elements of both adhesive and adherend. Such a result would suggest the presence of a weak boundary layer, but Good[1] shows that the transfer of some material from one phase to the other does not prove the existence of a WBL at or adjoining the interface before failure. He does not rule out the WBL, but states only that it is not a universal concept which explains all apparent interfacial failures. Regardless of these interpretive difficulties, chemical information about the surface certainly will help clarify the mechanism of failure in adhesive bonds.

APPENDIX I
PARTIAL BIBLIOGRAPHY RELATING TO LOCUS OF FAILURE AND ADHESIVE BONDING.

R. A. Gledhill and A. J. Kinloch, J. Adhesion 6, 315 (1974).
M. Gettings and A. J. Kinloch, J. Mater. Sci. 12, 2511 (1977).
W. J. Van Ooij, Surf. Sci. 68, 1 (1977).
L. H. Lee, Editor, "Adhesion Science and Technology", Plenum Press, New York, 1975.
W. D. Bascom and R. L. Cottington, J. Adhesion 4, 193 (1972).
W. L. Baun, N. T. McDevitt, and J. Solomon in "Surface Analysis Techniques for Metallurigcal Applications", pp. 86-101 ASTM STP 596, ASTM, 1976.
W. Bascom, C. Timmons, and R. L. Jones, J. Mater. Sci. 10, 1037 (1975).
W. L. Baun, J. Adhesion 7, 261 (1976).
W. L. Baun, in "Characterization of Metal and Polymer Surfaces" L.-H. Lee, Editor, p. 375, Academic Press, New York, 1977.
N. J. DeLollis and O. Montoya, J. Appl. Poly. Sci., 11 983 (1967).
J. M. Burkstrand, J. Appl. Phys. 52, 4795 (1981).
K. L. Mittal, Poly. Eng. Sci. 17, 467 (1977).
W. L. Baun, in "Surface Treatments of Plastics and Metals", D. M. Brewis, Ed., p. 45, Applied Science Pub., London, 1982.
H. E. Bair, S. Matsuoka, R. G. Vadimsky and T. T. Wang, J. Adhesion 3, 89 (1971).
W. L. Baun, Appl. Surf. Sci. 4, 291 (1980).
L. B. Sargent, ASLE Trans. 21 280 (1978).
A. J. Kinloch and N. R. Smart, J. Adhesion, 12, 28 (1981).
J. M. Burkstrand, J. Vac. Sci. Technol., 16 1072 (1979).
M. Gettings and A. J. Kinloch, Surf. Interface Anal., 1, 189 (1979).
F. Yamamoto, and S. Yamakawa, J. Appl. Poly. Sci., 25 2427 (1980).
W. L. Baun in "Industrial Applications of Surface Analysis" L. A. Casper and C. J. Powell, Editors, ACS Symposium Series No. 199, pp. 121-143 American Chemical Society, (1982).
K. L. Mittal, Pure Appl. Chem. 52, 1295 (1980).
W. L. Baun, in "Adhesion Aspects of Polymeric Coatings", K. L. Mittal, Editor, pp. 131-146, Plenum Press, 1983.

REFERENCES

1. R. J. Good, in, "Adhesion Measurement of Thin Films, Thick Films and Bulk Coatings", pp. 41-54, ASTM STP 640, K. L. Mittal, Ed., ASTM, Philadelphia, (1978).
2. J. J. Bikerman, ibid. pp. 30-40; and many earlier publications including J. J. Bikerman, "The Science of Adhesive Joints", 2nd Edition, Academic Press, New York, (1968).

3. W. L. Baun, in "Adhesion Measurement of Thin Films, Thick Films and Bulk Coatings", K. L. Mittal, Editor, pp. 41-53, ASTM, Philadelphia, PA, 1978.
4. K. L. Mittal, Electrocomponent Sci. Tech., $\underline{3}$, 21 (1976).
5. L. Sharpe, in "Recent Advances in Adhesion", L.-H. Lee, Ed., pp. 437-453, Gordon and Breach, New York, (1973).
6. W. D. Bascom and R. L. Cottington, J. Adhesion $\underline{4}$, 193 (1972).
7. R. L. Park, in "Surface Analysis Techniques for Metalurgical Applications", R. Carbonara and J. Cuthill, Eds. pp. 3-18, ASTM, Phila., PA, (1976).
8. D. W. Dwight, J. Colloid and Interface Sci. $\underline{59}$, 447 (1977).
9. T. S. Sun, J. M. Chen, J. D. Venables and R. Hopping, Appl. Surf. Sci. $\underline{1}$, 202 (1978).
10. A. J. Kinloch, H. E. Bishop and N. R. Smart, J. Adhesion $\underline{14}$, 105 (1982).
11. N. T. McDevitt and W. L. Baun, these proceedings, pp. 381-394
12. R. A. Gledhill and A. J. Kinloch, J. Adhesion $\underline{6}$, 315 (1972).

PRACTICAL ADHESION MEASUREMENT IN ADHERING SYSTEMS : A PHASE

BOUNDARY SENSITIVE TEST

A.A. Roche, M.J. Romand and F. Sidoroff[*]

Applied Chemistry Department (CNRS, ERA # 300)
University Claude Bernard - LYON I
69622 Villeurbanne Cedex, France

[*] Solid Mechanics Department
Ecole Centrale de LYON
69130 Ecully, France

       The strength of adhesive bonds is assessed by means of mechanical tests in which an increasing load is applied until failure occurs. This paper describes a three point flexure test using a single adherend/adhesive structure. Contrary to what occurs with usual tests such as single or double lap shear and peel tests, this configuration is sensitive to the adherend surface chemistry variations and adhesive cure conditions and thus allows evaluation of the properties of the adhesive bond joint interphase. Data are presented which demonstrate the flexure test sensitivity to the effects of surface prebonding treatments on interphasial mechanical properties, such as rigidity modulus. Photoelastic isochromatic fringes are recorded simultaneously with flexure test data as a means to monitor stress distribution, failure initiation and crack propagation. This test shows that the stresses are uniformly distributed along the specimen length and remain parallel to the interfacial region and insure failure totally along the adherend/adhesive interface.

INTRODUCTION

Industry is continuously seeking ways of producing durable and higher performance materials at lower costs. By way of example, adhesively-bonded structures are extensively developed and used in the aircraft and aerospace industries. Their advantages over mechanical fastening structures are well known, e.g., weight saving, aerodynamic and environmental resistance improvements and stress concentration reduction.

In single adherend/adhesive combinations - called adhering systems - various interfaces and interphases are created in addition to the bulk materials. Consequently, under stresses, failure can occur at any of the interfaces, interphases or bulk phases, and it is suggested that the bond strength in such systems be expressed as practical adhesion. For a detailed discussion of the concept of practical adhesion and how it relates to the interface/interphase between the adhesive (coating) and adherend, see ref. 1a, b, c. To improve bond joint performance, it is, then, very important to find out *"Where, how and why the bond fails"*.

Until recently, most adherend prebonding treatments (chemical etching, conversion, anodization, ...) have been developed mainly empirically, but today, many scientists are trying to establish why some prebonding processes yield better results than others. The present work shows firstly, how mechanical interphasial properties can be obtained using a flexure test and an appropriate specimen geometry ; and secondly, how analytical information from failed surfaces can be correlated with bond performances.

EXPERIMENTAL

Adherend sheets, prepared by machine process to provide identically sized strips (50 x 10 x .4 mm), are submitted, prior to bonding, to different surface treatments.

A single adherend-epoxy adhesive specimen is used. The specimen preparation fixture is shown in Figure 1. The adhesive forming mold (Figure 1 a) is made of RTV-501. It is held against the adherends by a clamping assembly (Figure 1 b and c). An exact amount of adhesive is applied to each adherend with a syringe (Figure 1 d). Care is taken to ensure that no bubbles evolve within the adhesive near the interfacial region. A similar specimen preparation fixture has been previously described [2,3].

Figure 2 is a schematic drawing of a specimen under applied load after debonding along the adherend/adhesive interface.

A three point flexure tester (INSTRON 1102 table model),

PRACTICAL ADHESION MEASUREMENT 21

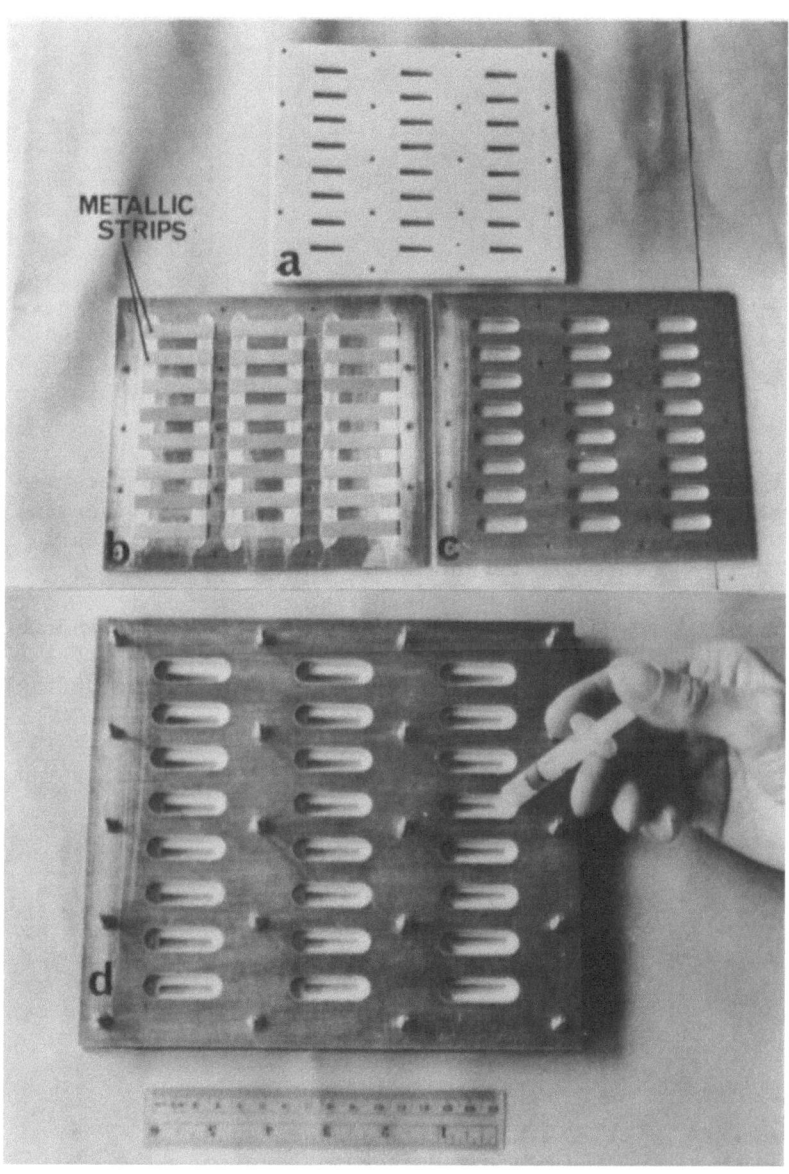

Figure 1. Specimen preparation fixture : a) forming mold, b) metallic holder, c) metallic retention plate, d) complete assembling.

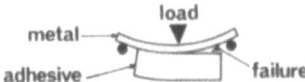

Figure 2. Illustration of the single adherend/adhesive test specimen under applied load.

fitted with a cross-head speed reducer and an appropriate load cell, is used. The cross-head displacement speed is typically 0.5 mm per minute and the loading sensitivity is as low as 100 N full scale. The bonded single adherend specimen is mounted in the tester behind a light polarizer (Figure 3).

For each tested specimen, the following can be observed simultaneously : - the recording of the load versus displacement
- the pattern modification of the photoelastic fringes through the adhesive bulk.

Figure 4 shows a typical load-displacement curve and the fringe pattern modification occurring during testing. The photograph numbers correspond to the various points along the load-displacement

Figure 3. Specimen mounted in the three point flexure test apparatus fitted with a light polarizer.

PRACTICAL ADHESION MEASUREMENT

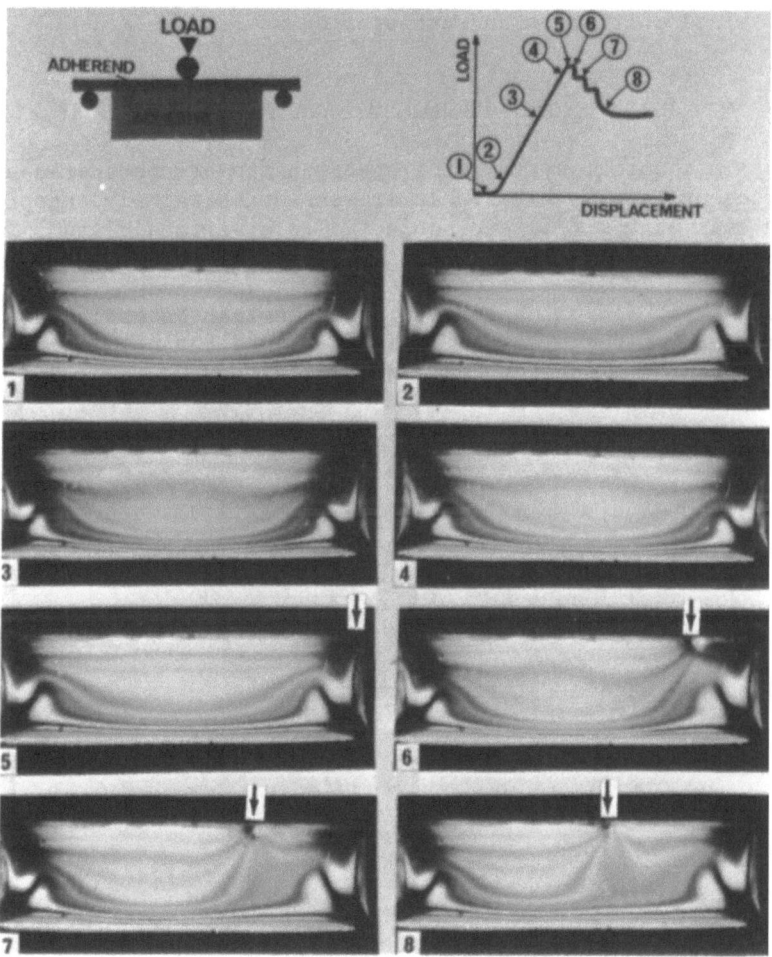

Figure 4. Photoelastic isochromatic fringes recorded during the testing of a single titanium adherend specimen.

curve at which they have been recorded. The failure front is pointed out by an arrow.

The following information can be derived from these experiments :
- the slope of the load-displacement curve
- the ultimate load before failure
- the stress distribution within the adhesive bulk
- the failure initiation and propagation.

SYSTEM MECHANICAL ANALYSIS

As a mechanical system, the specimen can be considered as an inhomogeneous beam with variable cross-section, simply supported at both ends ($x = \pm \ell/2$) and subjected to a vertical load P for $x = 0$ (Figure 5).

As a first approximation, this problem can be analyzed within the framework of the classical flexure theory for beams[4,5]. The bending moment **M** along the beam is given by

$$\begin{cases} \mathbf{M} = \frac{P}{2}(x - \frac{\ell}{2}) & \text{for} \quad 0 \leqslant x \leqslant \ell/2 \\ = -\frac{P}{2}(x + \frac{\ell}{2}) & \text{for} \quad -\ell/2 \leqslant x \leqslant 0 \end{cases}$$

and the elastic energy of the beam is

$$W = \int_{-\ell/2}^{+\ell/2} \frac{\mathbf{M}^2}{2H} dx = \int_{0}^{\ell/2} \frac{\mathbf{M}^2}{H} dx$$

where H is the bending stiffness of the beam.

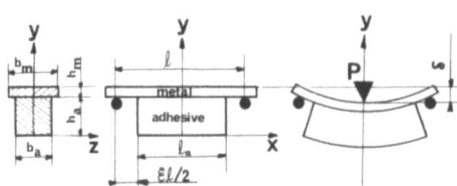

Figure 5. Specimen geometry and cross-section.

$$\begin{cases} H = H_m = E_m I_m = \dfrac{E_m b_m h_m^3}{12} & \text{for } \dfrac{\varepsilon \ell}{2} \leq |x| \leq \dfrac{\ell}{2} \\ H = H_S & \text{for } 0 \leq |x| \leq \dfrac{\varepsilon \ell}{2} \end{cases}$$

where $H_m = E_m I_m = (EI)_m$ is the usual bending stiffness for the rectangular metal sheet while H_S is the bending stiffness of the metal/interphase/adhesive composite system.

The elastic energy of the beam is then given by

$$W = \int_0^{(1-\varepsilon)\frac{\ell}{2}} \frac{P^2}{4 H_S} \left(x - \frac{\ell}{2}\right)^2 dx + \int_{(1-\varepsilon)\frac{\ell}{2}}^{\frac{\ell}{2}} \frac{P^2}{4 H_m} \left(x - \frac{\ell}{2}\right)^2 dx$$

$$= \frac{P^2 \ell^3}{96} \left[\frac{\varepsilon^3}{H_m} + \frac{(1-\varepsilon^3)}{H_S} \right]$$

According to Castigliano's theorem, the vertical displacement δ for $x = 0$ is

$$\delta = \frac{\partial W}{\partial P} = \frac{P \ell^3}{48} \left[\frac{\varepsilon^3}{H_m} + \frac{(1-\varepsilon^3)}{H_S} \right] = \frac{P}{k}$$

$$k = \frac{48 \, H_S}{\ell^3} \frac{1}{(1-\varepsilon^3) + \eta \varepsilon^3} \; ; \quad \eta = \frac{H_S}{H_m}$$

which shows that for a given geometry and for a given E_m the global stiffness k ($P = k\delta$) of the system only depends on H_S. This relation can be inverted to give

$$H_S = \frac{k \ell^3}{48} (1 - \varepsilon^3) \left[1 - \frac{k \ell^3 \varepsilon^3}{48 \, H_m} \right]^{-1}$$

or

$$\boxed{(EI)_S = \frac{P \ell^3}{48 \, \delta} (1 - \varepsilon^3) \left[1 - \frac{P \ell^3 \varepsilon^3}{48 \, \delta \, (EI)_m} \right]} \qquad (1)$$

It is quite clear that this approach is a very crude one since it neglects transverse shearing, edge effects and others and only gives a rough evaluation (in fact an excess value) for k. However it clearly shows that the slope of the load-displacement curve is directly related to $(EI)_S$ and, since all other parameters are kept constant, to the quality of the phase boundary layer. In fact, for all the different tested specimens, the whole geometry, the adherend size, the formulation, size and curing process of the adhesive are precisely controlled and kept constant. Under these conditions the interphase (metal/adhesive) properties vary with respect to the different possible reactions between adherend surface and adhesive during bonding. The above relationship is verified (see Figure 6) by considering experimental values of k for various preconditioned titanium adherend/adhesive systems (Table II) versus calculated $(EI)_S$ values according to Equation (1).

Starting from a similar analysis along the line of the

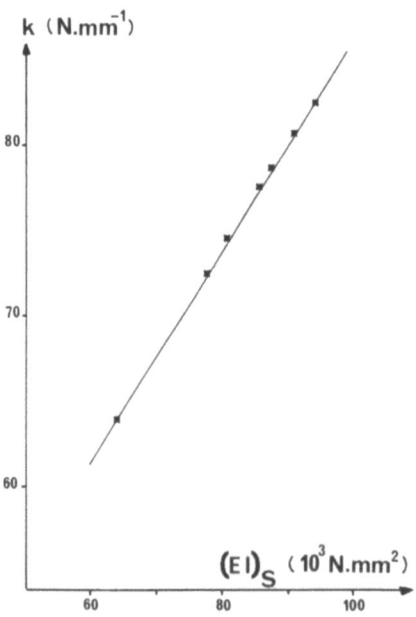

Figure 6. Experimental k values versus calculated $(EI)_S$ values.

classical flexure theory, an approximate value can be obtained for the value H_p of H_s in the ideal case of a perfectly adherend contact. Starting from Navier-Bernoulli assumption[4,5] (the cross-section remains plane and normal to the neutral line) the longitudinal strain ε can be written as

$$\varepsilon = K(y - y_0) \qquad K = \frac{1}{R}$$

where R is the curvature radius of the beam and y_0 the neutral line position (i.e. the line without length variation). The stress σ is then given by:

$$\sigma = E_a K (y - y_0) \qquad 0 \leqslant y \leqslant h_a \text{ (adhesive)}$$
$$= E_m K (y - y_0) \qquad h_a \leqslant y \leqslant h_a + h_m \text{ (metal)}$$

The value of y_0 is obtained from the condition

$$\iint_S \sigma \, dS = K \left\{ b_a E_a \int_0^{h_a} (y - y_0) \, dy + b_m E_m \int_{h_a}^{h_a + h_m} (y - y_0) \, dy \right\} = 0$$

which express the fact that there is no longitudinal force. The bending moment is then obtained as

$$\begin{cases} M = \iint_S \sigma y \, dS = H_p K \\ H_p = b_a E_a \int_0^{h_a} (y - y_0) y \, dy + b_m E_m \int_{h_a}^{h_a + h_m} (y - y_0) y \, dy \end{cases}$$

A somewhat lengthy calculation then leads for H_p to the following value[6]

$$H_p = \frac{E_a b_a h_a^3}{12} + \frac{E_m b_m h_m^3}{12} + \frac{E_a b_a h_a E_m b_m h_m}{E_a b_a h_a + E_m b_m h_m} \left(\frac{h_m}{2} + \frac{h_a}{2} \right)^2$$

which is found to be much higher than the values of H_s obtained from experiments. A reason for this may be the very high normal stresses induced in the very thin bonding layer and which cannot be analyzed by such a simple approach and which requires some kind of finite element analysis. Without such an analysis it is not possible to go beyond the qualitative discussion presented and to get quantitative information about the bonding layer (thickness, mechanical properties).

As for the photoelastic fringe pattern, it essentially describes the initial residual stresses and strains and is therefore related to the specimen preparation. Its evolution during the loading sequence qualitatively agrees with the above analysis.

RESULTS AND DISCUSSION

For a set of identical adherend/adhesive systems using the same sample geometry and size, a change in the slope of the load-displacement curve must be correlated with a change in the interfacial properties when the adherend has been subjected to various surface conditioning treatments.

Table I depicts a repeatability test. The rigidity modulus $(EI)_S$ and ultimate load values are given for a set of identically prepared specimens. Relative standard deviations for these values are shown to be 1 % and 10 %, respectively. Effects of the sample exposure to room atmosphere before testing are also shown. Such a statistical result takes into account reproducibilities of the mechanical tester, specimen sizing and preparation, and adherend surface state.

Table II gives system rigidity modulus $(EI)_S$ and ultimate load values for a set of titanium adherends submitted to various chemical etchings. A change in the ultimate load and rigidity modulus $(EI)_S$ values is observed for different titanium pretreatments. As seen previously, such changes cannot be explained in terms of the statistical error ; and therefore are characteristic of the different mechanical properties of the interphase created after titanium surface conditionings. Also listed in Table II are the respective adherend surface oxide thicknesses determined from a set of identically prepared specimens by Auger sputter profile analysis[7]. The optimum $(EI)_S$ value is shown for the thicker TiO_2-like oxide layer. The specimen subjected to treatment 4 has the thickest film layer but a lower $(EI)_S$. The relevant load-displacement curve shows that failure occurs in a slow, step-wise manner. In this case, the surface layer is found to be a mixture of titanium oxide and titanium fluoride TiF_4[8-9]. As already reported for Ti-6Al-4V alloy bonded with another adhesive[2] such specie appears to be responsible for this type of failure propagation process and interphase properties.

CONCLUSION

The results reported here show that a three point flexure test on a single metal adherend/adhesive bonded structure is sensitive to the adherend surface chemistry and topography, and therefore, to the material properties of the so-created interphase. According to the qualities of an "ideal test" as attributed by K.L. Mittal[1], the test described here fulfills more than half of the different requirements and therefore appears to be undoubtedly an important test.

Table I. Repeatability of the Ultimate Load and Rigidity Modulus for a Set of Fluoro-ammonium Etched Titanium Bonded to the CIBA Epoxy (LY 564) and Effect of the Exposure to Room Atmosphere before testing.

Exposure time (room atmosphere)	1 week						10 weeks	
Sample number	1	2	3	4	5	6	7	8
Ultimate Load (N)	18.6	21.4	21.7	21.7	17.7	17.7	19.1	17.6
			20 ± 10 %				18 ± 10 %*	
System rigidity modulus $(EI)_S (10^3 N.mm^2)$	129	127	127	130	127	130	94	94
			128 ± 1 %				94 ± 1 %*	

* Relative standard deviation based on the first six values

Table II. Effects of the Adherend Prebonding Treatments on the Values of Ultimate Load, Rigidity Modulus and Surface Oxide Thickness as Obtained by AES Sputter Profile Analysis.

Code	Treatment	Ultimate Load (N)	System Rigidity Modulus $(EI)_S$ $(10^3 N.mm^2)$	Adherend surface composition	Surface Oxide Thickness (nm)
1	Degrease	10.3	91	TiO_2	17
2	NaOH (Alkaline)	8.3	78	TiO_2	6
3	HNO_3/HF (Fluoro-nitric)	13.0	81	TiO_2	7
4	$Na_3PO_4/NaF/HF$ (Fluoro-phosphate)	10.1	87	TiF_4 + oxide	83
5	NH_4HF_2 (Fluoro-ammonium)	18.7	94	TiO_2	20
6	H_2SO_4/CrO_3 (Sulfo-chromium)	7.9	78	TiO_2	14
7	$HNO_3/HF/H_2O_2/NH_4HF_2$ (Fluoro-nitro-ammonium)	8.3	64	TiO_2	7
8	$NaOH/H_2O_2$ (65°C) (Hot alkaline)	7.2	86	TiO_2	16

ACKNOWLEDGMENT

Acknowledgments are made to W.L. Baun and J.S. Solomon for their helpful discussions.

REFERENCES

1a. K.L. Mittal, in "Adhesion Measurement of Thin Films, Thick Films and Bulk Coatings", K.L. Mittal, Editor, ASTM STP 640, pp. 5-17, American Society for Testing and Materials, Philadelphia, 1978.
1b. K.L. Mittal, Electrocomponent Sci., Technol., $\underline{3}$, 21 (1976).
1c. K.L. Mittal, Pure Appl. Chem., $\underline{52}$, 1295 (1980).
2. A.A. Roche, A.K. Behme and J.S. Solomon, Int. J. Adhesion and Adhesives, $\underline{2}$, No. 4, 249 (1982).
3. A.A. Roche, J.S. Solomon and M.J. Romand, in "Microscopic Aspects of Adhesion and Lubrication", J.M. Georges, Editor, Tribology Series 7, pp. 333-342, Elsevier, 1982.
4. S.P. Timoshenko, D.H. Young, "Theory of Structures", McGraw Hill, New York, 1965.
5. W. Flügge, "Handbook of Engineering Mechanics", McGraw Hill, New York, 1962
6. A.A. Roche, Thesis, Lyon, France, 1983.
7. A.A. Roche, Air Force Wright Aeronautical Laboratories, Dayton, OH, Technical Report AFWAL-TR-80-4004, 1980.
8. A.A. Roche, J.S. Solomon and W.L. Baun, Appl. Surf. Sci., $\underline{7}$, 83 (1981).
9. A.A. Roche, M. Charbonnier, F. Gaillard, M. Romand and R. Bador, Appl. Surf. Sci., $\underline{9}$, 227 (1981).

EFFECT OF A PHENOL FORMALDEHYDE RESIN OF THE ADHESION OF A
POLYSULFIDE SEALANT

R. Ramaswamy and P. Sasidharan Achary

Polymers and Special Chemicals Division
Vikram Sarabhai Space Centre
Trivandrum 695 022 (India)

The results obtained in a study to evaluate the effect of a phenol formaldehyde resin on the adhesion of a polysulphide sealant is presented. The adhesion of the sealant is improved significantly by the incorporation of phenol formaldehyde resin. This is explained on the basis of increased boundary layer strength and the results obtained are compared with the 'theory of attachment site'. Addition of phenol formaldehyde resin to the sealant caused an increase in cross-link density, ultimate tensile strength, stress at 100% elongation, hardness and a decrease in ultimate elongation and water absorption. The observed property changes are explained on the basis of polymer-filler interactions. Higher concentrations of phenol formaldehyde resin in the sealant caused a decrease in bond strength, cross-link density and mechanical properties. This is attributed to the lesser extent of cure of polysulphide resin.

INTRODUCTION

Polysulphide sealants which find major utility as elastic sealants in aircraft fuel tanks, aircraft runways, ship building and building construction are formulated from liquid polysulphide polymer, fillers, adhesion promoter, plasticizer and curing agent. The liquid polysulphide polymer generally used is mercaptan terminated having the structure,

$$HS-C_2H_4-O-CH_2-O-C_2H_4(-S-S-C_2H_4-O-CH_2-O-C_2H_4)_n - SH$$

with a proportion of thiol terminated branched chains. By the oxidation of thiol group, the liquid polymer is cured to a rubber.

In order to perform as a sealant, good adhesion (in this paper the term 'adhesion' refers to 'practical adhesion' [1,2]) of the sealant to the substrate is an essential requirement. However, polysulphide liquid polymer, in view of its low polar character, exhibits poor adhesion. In an earlier paper[3], the probable reasons for poor adhesion and possible mechanisms of adhesion development in a polysulphide sealant have been described. It was shown that by coreaction of polysulphide resin with epoxy resin before compounding into a sealant, adhesion of the sealant could be increased signficantly. Coreaction with epoxy resins results in introduction of polar hydroxyl groups in the polysulphide backbone resulting in increased adhesion.

In the present study, we have investigated the effect of a resol phenol formaldehyde resin (PF resin) on the adhesion of a polysulphide sealant composition to aluminium substrate. The substrate studied was B51 SWP aluminium. PF resin is chosen because of its high polarity and commercial use in adhesives and coatings. Also, it has been reported that the use of specific phenolic additives improve the adhesion of polysulphide sealants[4].

EXPERIMENTAL

Polysulphide sealants were prepared according to the formulation given in Table I. The constituents in the base component (Component A) are mercaptan terminated liquid polysulphide resin, PF resin and fillers. The constituents in the curing agent (component B) are dibutylphthalate which is a plasticizer, manganese dioxide and diphenyl guanidine which causes curing of the polysulphide resin.

Table I. Composition of the Sealant.

Component A
- Liquid polysulphide polymer   100 g
- Titanium dioxide   40 g
- Silica   10 g
- PF resin (resol)   varied from 0 to 5 g

Component B
- Manganese dioxide   12.5 g
- Dibutyl phthalate   4.0 g
- Diphenyl guanidine   2.5 g

The polysulphide resin used in this study was prepared in the laboratory by a conventional procedure[5] from dichloro diethyl formal and sodium polysulphide (Na_2S_x, x = 2 to 2.2) with 2 mole % trichloro propane as a cross-linking agent. The PF resin (resol) was also prepared in the laboratory from phenol and excess formaldehyde using sodium hydroxide as catalyst and was used as 70% solution in methyl ethyl ketone. Other ingredients used in the formulation are commercially available materials. A single batch of raw materials was used throughout the study to minimise variations in the ingredients.

Component A (resin portion) with various concentrations of PF resin was prepared by mixing the ingredients in a three roll mill. A large batch of Component B (curing agent) was prepared separately. Component A is a homogenous, viscous, Newtonian fluid white in colour with a yield value of 7.15 N m^{-2} (as indicated by Rheogram). Component B is a soft paste black in colour.

Appropriate amounts of Component A and Component B were mixed together, adhesion test specimens and slabs prepared as per procedure reported earlier[3]. The specimens and slabs were cured for 24 hours at room temperature followed by 24 hours at 70°C.

Adhesion of the sealant was measured as the force required to separate the sealant from the substrate (aluminium) in terms of shear and peel bond strengths. The force required to separate the sealant from the substrate was determined in INSTRON at a crosshead speed of 5 cm/min. The nature of failure was visually observed and recorded.

Stress-strain properties were measured in INSTRON at a crosshead speed of 5 cm/min. from the dumbell cut from the slabs. Ultimate tensile strength, stress at 100% elongation and ultimate elongation were calculated from the stress-strain curves obtained. Hardness was determined from the slabs by Durometer.

The cross-link density of the sealant was measured by swelling method. Cured samples of the sealant were immersed in toluene and they were allowed to swell to equilibrium for 24 hours. The swollen samples were compression tested at a cross-head speed of 0.5 cm/min. in INSTRON, keeping the sample immersed in toluene throughout the testing. The cross-link density was calculated using Equation (1)

$$\nu_e = \frac{2C_1}{RT} \quad (1)$$

where ν_e is the cross-link density in moles/cc, R is the gas constant, T is the absolute temperature and C_1 is the elastic constant defined by modified Mooney relation[6] which was determined[7] by using

Equation (2)

$$C_1 = \frac{h_o h_D^2 \sigma}{2(h_D^3 - h_s^3)} \qquad (2)$$

where h_o is the height of the unswollen sample, h_s is the height of the swollen sample and $h_D = h_s - \Delta h$ where Δh is the change in height for a compression load of σ, determined from the graph.

Water absorption was measured with specimens cut from slabs of constant thickness between 0.2mm to 0.3mm. Approximately 10 g. samples were cut, weighed accurately and immersed in 100 ml distilled water at 30°C. The weight changes were determined after blotting dry with filter paper.

The cure rate of the sealant was determined from the measurement of pot life using a gel timer.

RESULTS AND DISCUSSION

Adhesion of the polysulphide sealant is improved significantly

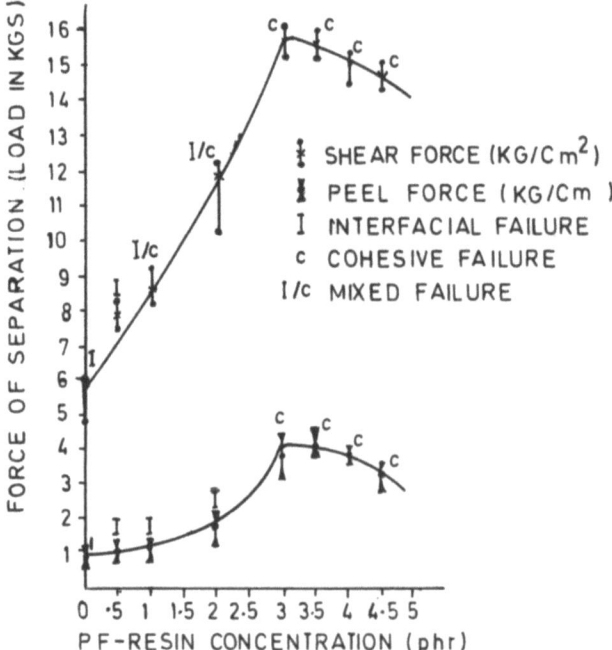

Figure 1. The effect of PF resin (Resol) concentration on the adhesion of the sealant.

by addition of PF resin as seen in Figure 1. The incorporation of highly polar PF resin into the polysulphide sealant probably helps to remove loosely bonded water layers (WBL)[8] present on the substrate surface by a process of solvation and diffusion and establish strong interfacial forces with the substrate. It is reported[9] that PF resin can establish strong interfacial forces with metal surfaces through dipole, ionic, covalent and co-ordinate interactions. The possible interfacial interactions between PF resin and metal surfaces are shown in Figure 2.

It is clear from Figure 1 that with the increase of PF resin concentration, both peel and shear bond strength increase accompanied by change in mode of failure from interfacial to cohesive. This result is comparable with the 'theory of attachment site' proposed by Lewis and Natarajan[10]. The theory proposes that as the degree of interfacial attachment increases, the joint strength also increases accompanied by a change to mixed mode of failure. At a certain saturation level of attachment site density, an optimum boundary strength is reached, where the boundary strength is equal to or greater than cohesive strength. Increasing the number of attachment sites beyond this point does not increase the joint strength as the strength of the joint has reached the so called cohesive plateau. Figure 1 indicates

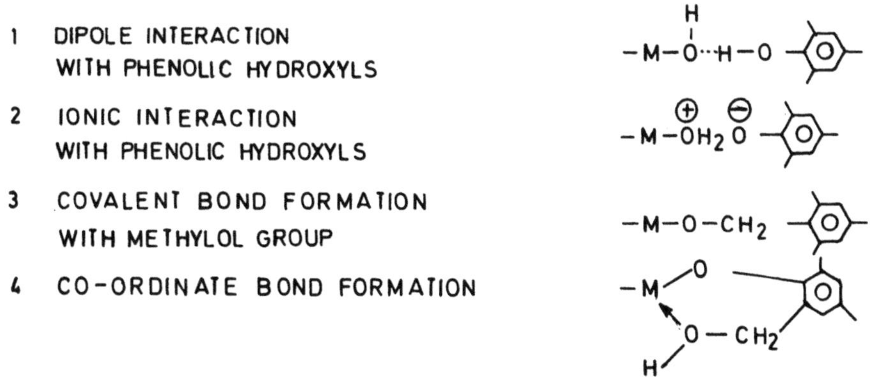

1   DIPOLE INTERACTION
    WITH PHENOLIC HYDROXYLS

2   IONIC INTERACTION
    WITH PHENOLIC HYDROXYLS

3   COVALENT BOND FORMATION
    WITH METHYLOL GROUP

4   CO-ORDINATE BOND FORMATION

Figure 2. Possible interfacial interactions between PF resin (resol) and metal surfaces.

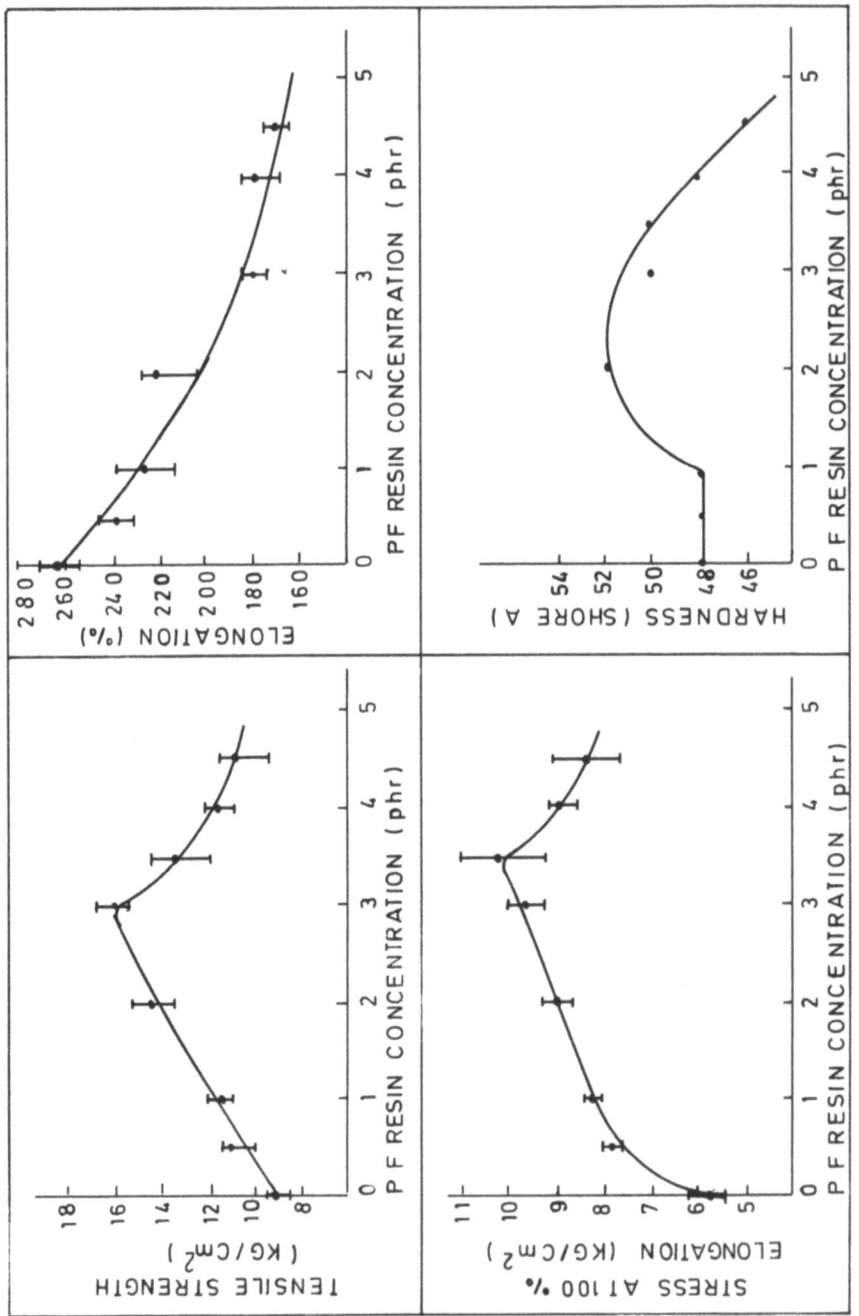

Figure 3. The effect of PF resin (resol) concentration on the mechanical properties of the sealant.

that the cohesive plateau is reached at 3 phr of PF resin. At and above this PF resin concentration, the boundary layer strength is greater than strength of the sealant and cohesive failure is observed. However, at higher levels of PF resin the peel and shear strengths decrease due to decrease in mechanical properties as seen in Figure 3.

The effect of PF resin on the cross-link density of the sealant is shown in Figure 4. We see that upto 3 phr the cross-link density increases and then decreases. Now the question arises, whether the increase in cross-link density is due to the additional cross-links formed in the polymer matrix or bonding with the filler surface. A separate study indicated that in the absence of filler, addition of PF resin reduces the cross-link density of the sealant (Table II). This suggests that addition of PF resin to the sealant produces additional cross-links with the filler surface due to the bonding between the polymer matrix and filler surface. The data obtained on the water absorption studies (discussed later) also supports this view. Thus the increase in ultimate tensile strength, stress at 100% elongation and hardness can be attributed

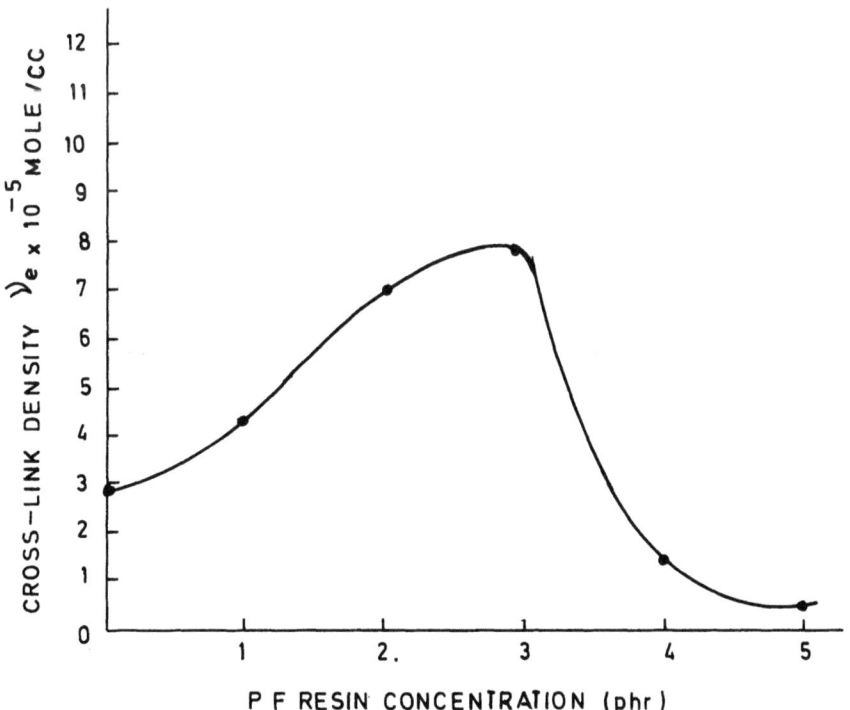

Figure 4. The effect of PF resin (resol) concentration on the cross-link density of the sealant.

Table II. Effect of PF resin (resol) concentration on the cross-link density of the polysulphide sealant (without fillers).

PF resin(resol) concentration (phr)	0	1	2	3
Cross-link density $\nu_e \times 10^{-5}$ mole/cc	14.5	12.6	10.3	5.6

to the increased interaction between the polymer matrix and filler surface. But the decrease in mechanical properties at higher levels of PF resin requires explanation of different sort. It is evident from the measurement of pot life of the sealant (Table III) that PF resin retards the cure of polysulphide sealant. The cure reaction is the simple oxidative coupling of terminal - SH groups of the polysulphide resin $2-SH \rightleftharpoons 2e + -S-S- + 2H^+$. Due to the acidity of phenol, PF resin retards the cure of polysulphide resin and can result in reduced extent of cure of polysulphide sealant. The decrease in cross-link density by the addition of PF resin (Table II) also indicates that PF resin inhibits the cure of polysulphide resin. Therefore, the decrease in shear strength, peel strength, mechanical properties and cross-link density at higher levels of PF resin can be attributed to the lesser extent of cure of polysulphide resin.

Figure 5 illustrates the effect of PF resin on the water absorption of the cured sealant samples. Contrary to the expectation, the water absorption of the cured sealant samples decrease with the increase of PF resin concentration. Water absorption of a polymer-filler system involves diffusion of water through the polymer and the adsorption of water at the polymer-filler interface. PF resin increases the polarity of the polymer system; therefore intensity of interaction between the sealant and water molecules increases which should result in the increased diffusion of water through the sealant. If the polymer and filler surfaces are not in intimate contact, water will displace the polymer easily from the filler surface and get

Table III. Effect of PF resin(resol) concentration on the pot life of the polysulphide sealant.

PF resin concentration (phr)	0	0.5	1	2	3	3.5	4	4.5
Pot life (minutes)	15	35	70	105	200	285	375	480

EFFECT OF A PHENOL FORMALDEHYDE RESIN

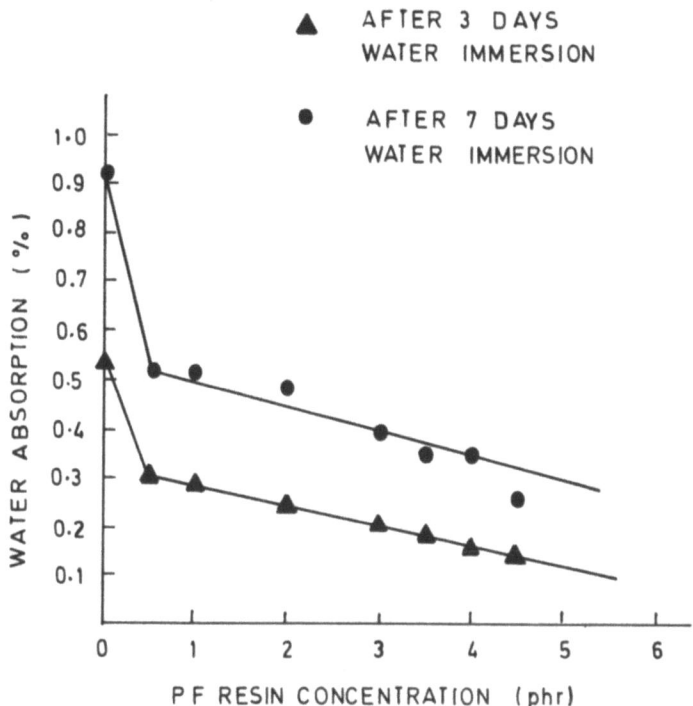

Figure 5. The effect of PF resin (resol) concentration on the water absorption of the sealant.

adsorbed around the filler particles. This will result in increased water absorption. If the polymer and filler surfaces are in intimate contact by chemical linkages, the appearance of water at the interface is delayed and the adsorption of water at the filler surface by the displacement of polymer is impossible until the breakage of chemical linkages occurs[11]. Therefore, the decrease of water absorption of the cured sealant samples with PF resin indicates that PF resin promotes the establishment of intimate contact when exposed to water probably by chemical interactions. Although there is not much direct evidence, it is assumed that PF resin establishes chemical interactions with high energy oxide surfaces[9]. It is also postulated that the very good durability characteristics of phenolic based adhesives is due to the presence of these interfacial chemical bonds[12]. Therefore, the reduced water absorption of the cured sealant with the increase of PF resin concentration can be due to the bonding between the polymer matrix and filler surface probably by chemical interactions which can show an increase in cross-link density[13].

In conclusion, PF resin improves adhesion of the polysulphide sealant. It also improves mechanical properties and water resistance of the polysulphide sealant. PF resin retards the cure of polysulphide resin and therefore extends the pot life of the sealant. Higher concentrations of PF resin in the sealant affects the extent of cure and therefore reduces the mechanical strength properties and hence bond strengths.

ACKNOWLEDGEMENTS

The authors thank Dr. K.V.C. Rao, Head, Chemicals Group for encouragement, Mr. S. Bhagwan, Propellant Engineering Division for determining cross-link densities and Mrs. B. Syamala Amma for assistance in the laboratory.

REFERENCES

1. K.L. Mittal, Polymer Engg. Sci.,17, 467 (1977).
2. K.L. Mittal in "Adhesion Measurement of Thin Films, Thick Films and Bulk Coatings" K.L. Mittal, Editor, ASTM STP 640, p.5, ASTM, Philadelphia, 1978.
3. R. Ramaswamy and P. Sasidharan Achary, J.Adhesion.,11, 305(1981).
4. J.R. Panek in "Handbook of Adhesives" I. Skeist, Editor, p.372, Van Nostrand Reinhold Company, New York, 1977.
5. M.B. Berenbaum in "Polyethers Part III, Polyalkylene Sulphides and Other Polythioethers" N.G. Gaylord, Editor, p.82, Interscience Publishers, New York, 1962.
6. L.R.G. Treloar in "The Physics of Rubber Elasticity" L.R.G. Treloar, Editor, p. 165, Clarendon Press, Oxford, 1975.
7. B.B. Moore and D.K. Thomas, RAE Technical Report No.66388, Ministry of Aviation, Farnborough,Hants (U.K.), 1966.
8. R.L. Patrick in "Treatise on Adhesion and Adhesives" R.L. Patrick, Editor, Vol.4, p.72, Marcel Dekker, New York, 1973.
9. J.C. Bolger and A.S. Michaels in "Interface Conversion for Polymer Coatings" P. Weiss and G.D. Cheever, Editors, p.3, Elsevier, New York, 1968.
10. A.F. Lewis and R.T. Natarajan in "Adhesion Science and Technology" L.H. Lee, Editor, Vol.9B, p.565, Plenum Press, New York, 1975.
11. W.C. Wake in "Fillers for Plastics" W.C. Wake, Editor, p.1. Ilffe Books, London, 1971.
12. A.J. Kinloch, J. Adhesion, 10, 193 (1979).
13. B.B. Boonstra in "Rubber Technology and Manufacture", C.M. Blow, Editor, p. 252, Butterworths, London, 1971.

POLYSULFIDE-POLYURETHANE INTERFACIAL ASPECTS

A. M. Usmani

The Research Institute
University of Petroleum & Minerals
Dhahran, Saudi Arabia

We studied polyurethane coating aging and determined polyurethane coating/polysulfide sealant interfacial integrity using scanning electron microscopy. A rapid microspecimen hand-pulled peel test was developed that will find application in designing future aircraft sealants. Dynamic mechanical analysis was used to follow polyurethane aging.

INTRODUCTION

A polyurethane coating is used on fuel tank interior surfaces prior to application of polysulfide sealant. The sealant/coating combination works well initially and there is apparently adequate adhesion of the polysulfide rubber to the polyurethane coating. Upon extended aircraft usage, however, leaks do develop. These leaks cannot be repaired by removing the old polysulfide rubber and then applying a fresh sealant because of insufficient adhesion.

The factors responsible for poor adhesion of the polysulfide sealant to "aged" polyurethane coating were determined by Usmani et al.[1] Using transmission and surface reflectance infrared spectroscopy, we monitored concentration changes in NCO groups of artificially aged polyurethane coatings. Heat treatment of polyurethane coating at 121°C for 48 hours produced a surface to which the polysulfide sealant would not adhere.

This paper describes sealant-coating interface examination by SEM and a hand-pull test on small specimens for rapid adhesion screening. It is hoped that these methods will find application in development of future generations of aircraft polysulfide sealant. Polyurethane reaction with polysulfide has also been treated.

BRIEF OVERVIEW OF POLYURETHANE AND POLYSULFIDE CHEMISTRY

Polyurethane

Polyurethanes are formed by condensation of an isocyanate and a polyol. The propagation involves the nucleophilic attack of the hydroxyl group on the isocyanate group to yield a substituted amide ester of carbonic acid, i.e., polyurethane. Much has been written about polyurethane chemistry. Excellent treatments of polyurethane chemistry and technology have been provided by Saunders and Frisch,[2] Dombrow,[3] and more recently by Usmani and Salyer.[4]

The chemical reactions involved in urethane polymerization are complex but reasonably well understood. Polyfunctional isocyanate reacts with polyfunctional polyol containing labile hydrogen to make the urethane structure. The formation of urethane groups is generally accepted to be the chain propagating reaction. If the polyol contains amine groups, a substituted urea becomes a principal product. Reactions that result in branching and crosslinking are the isocyanate-urethane reaction leading to allophanate linkages, and the isocyanate-urea reaction products which produce biuret linkages. These reactions are well known and are summarized in Table I.

Polysulfide

The highly reactive terminal mercaptans of liquid polysulfide polymers can be converted into high molecular weight products very easily at ambient temperatures by oxidation or by reaction with other active polymers. Polysulfides have excellent solvent resistance, predictable cure, good low-temperature performance, good weathering, and excellent adhesion to many substrates. These combination of properties have given polysulfide polymers a modest but significant position as specialty polymers since their commercialization in 1929.

Polysulfide sealants have found wide acceptance in the aircraft industry and although facing strong challenge by silicones and polyurethanes, continue to be the dominant segment of the elastomeric sealant market.

In a recent paper, Usmani has reviewed the chemistry, properties, compounding, processing and manufacturing, characterization and testing, applications, and future prospects of polysulfide

Table I. Polyurethane Coating Reactions.

$$R-N=C=O + R-OH \longrightarrow R-NH-\overset{\overset{O}{\|}}{C}-O-R$$
$$\text{Urethane}$$

$$R-N=C=O + H_2O \longrightarrow R-\overset{\overset{H}{|}}{N}-\overset{\overset{O}{\|}}{C}-OH \longrightarrow R-NH_2 + CO_2$$
$$\text{Unstable}$$

$$R-NH_2 + R-N=C=O \longrightarrow R-\overset{\overset{H}{|}}{N}-\overset{\overset{O}{\|}}{C}-\overset{\overset{H}{|}}{N}-R$$
$$\text{Substituted urea}$$

$$R-N=C=O + R-NH-\overset{\overset{O}{\|}}{C}-OR \longrightarrow R-NH-\overset{\overset{O}{\|}}{C}-\underset{\underset{R}{|}}{N}-\overset{\overset{O}{\|}}{C}-OR$$
$$\text{Allophanate}$$

$$R-N=C=O + R-NH-\overset{\overset{O}{\|}}{C}-NH-R \longrightarrow R-NH-\overset{\overset{O}{\|}}{C}-\underset{\underset{R}{|}}{N}-\overset{\overset{O}{\|}}{C}-NH-R$$
$$\text{Biuret}$$

sealants.[5] The author has also presented a generally accepted mechanism of cure which is as follows:

$$2-RSH + (O) \rightarrow -RSSR + H_2O.$$

CHEMICAL COMPOSITION OF COATING AND SEALANT

The polyurethane coating consisted of strontium chromate, titanium dioxide, phthalocyanine green, and talc dispersed in a polyester solution. The other component consisted of an aromatic diisocyanate. The NCO/OH ratio in the coating was 1.03/1.00. The coating was sprayed and allowed to cure for 14 days under ambient conditions to give a 3 mil thick film.

The chemical composition of the polysulfide sealant used in this work was as follows:

Sealant Base Component	%
Calcium carbonate	26.16
Titanium dioxide	3.10
Liquid polysulfide polymer	58.50
Volatile diluent	2.25

Accelerator Component	
Manganese dioxide	5.53
Processing oil	3.59
Volatile diluent	0.51

SEALANT-COATING INTERFACE SEM STUDY

Events occurring at the polysulfide-polyurethane interface are important in defining the cause of poor adhesion and ways to improve adhesion of sealant to the polyurethane surface. For this reason SEM was used to study the polysulfide-polyurethane interface on an aluminum aircraft alloy.

The method involved preparation of polyurethane coated minipanels (0.5 in. by 0.5 in.), followed by application of about 100 mil thick polysulfide sealant. The sealant was allowed to cure (two days under ambient conditions plus one day at 60°C). Then the minipanels were conditioned at room temperature for 4 hours followed by exposure to one of three different test environments.

- Two days at 60°C in air,
- Two days at 60°C in water, and
- Two days at 60°C in jet reference fuel (JRF).

After exposure, samples were examined for integrity at the interface or the extent of debonding (desorption).

In Figure 1 SEMs of sealant-coating interface, at 3,000X magnification, are shown. Most debonding was found with water. It can be also seen from Figure 1 that water has penetrated the sealant and that the morphology of the sealant has changed.

From this work we are led to believe that SEM study of the sealant-coating interface is useful for examination of the performance of both established and new sealants.

ARTIFICIAL AGING OF POLYURETHANE AND DEVELOPMENT OF HAND-PULLED PEEL TEST

An important aspect of our work was to artificially age poly-

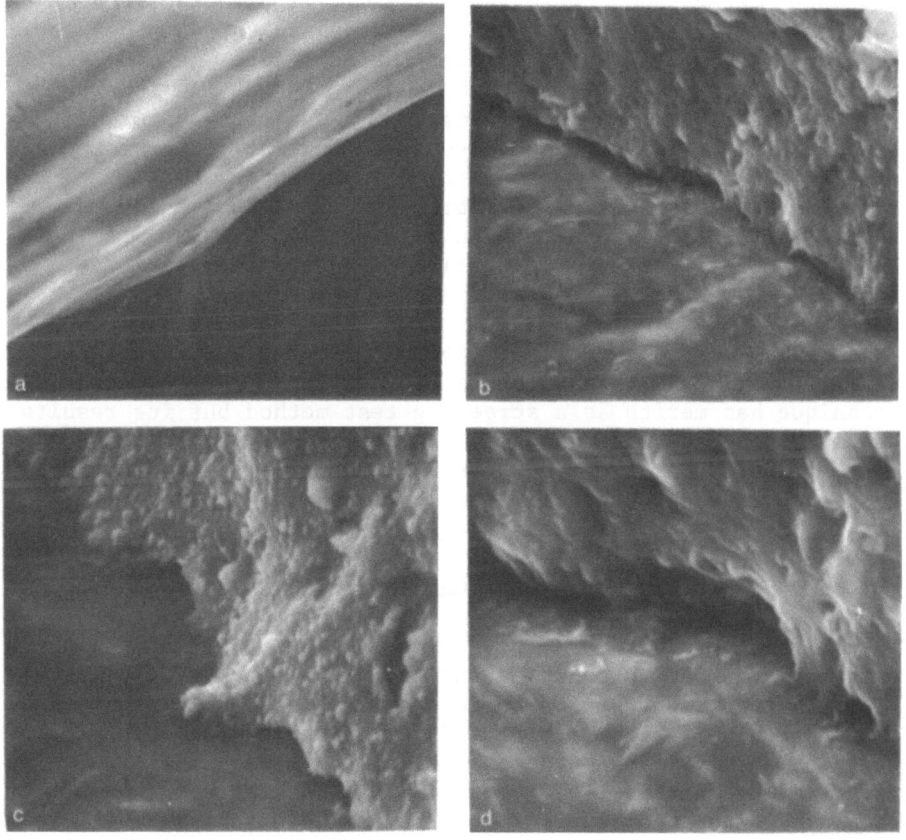

Figure 1. SEM of sealant-coating interface, at 3,000X magnification under conditions (a) new (b) 2 days at 60°C in air (c) 2 days at 60°C in JRF, and (d) 2 days at 60°C in water.

urethane coating to produce surfaces to which the polysulfide sealant would not stick. We were able to determine necessary conditions to artificially age the polyurethane coating (see Table II). After establishing the broad conditions, we proceeded to narrow down the conditions of aging at which limited or no adhesion with polysulfide will result. To narrow down the exposure time at 121°C at which no adhesion with polysulfide will result we developed a rapid screening method. Microspecimens (1 in. by 1 in.) of polyurethane coated panels were aged at 121°C in air for 1 through 10, 20, 30, 40, and 50 hours; the polysulfide sealant was then applied and cured. A strong but very fine fabric leader was embedded in the polysulfide (1" wide, ½" embedded and 1" outside to provide grip between thumb and index finger). A dual beam razor with 3/4" gap was used to indent the polysulfide sealant reaching the aluminum alloy substrate. The prepared test panel was held on the edges by the left hand. The fabric leader was pulled by the right hand. The same pull was used in all specimens although this is subject to operator error. Results of hand pulled peel tests are provided in Table III.

Table III indicates that exposure time greater than five hours at 121°C will result in a nonsticking surface. We feel that this technique has merits as a screening test method but its results should be checked with the standard peel test. Advantages of the test method are that it requires only about 10g of sealant and that the test can be quickly run.

DYNAMIC MECHANICAL ANALYSIS (DMA)

The polyurethane coating curing was studied by observing changes in damping properties and the glass transition temperature (T_g). DuPont 981 was used to record DMA spectra of polyurethane coatings on glass fabric (about 75 μm thick). The cure of polyurethane coating was found to be slow with steady increase in T_g with time over a three-week period. The coating was found to crosslink with an ultimate T_g of 80°C. The cure was accelerated by temperature and water immersion. In Figure 2, DMA of polyurethane coating cured for two-weeks at room temperature is shown. The film coated on glass fabric is then soaked in water for one-week and the resultant DMA is shown in Figure 3. Water soaking produces crosslinking and therefore increased T_g. This hardens the polyurethane surface and this is one of the principal reasons for insufficient adhesion of a polysulfide sealant to an aged polyurethane coating.

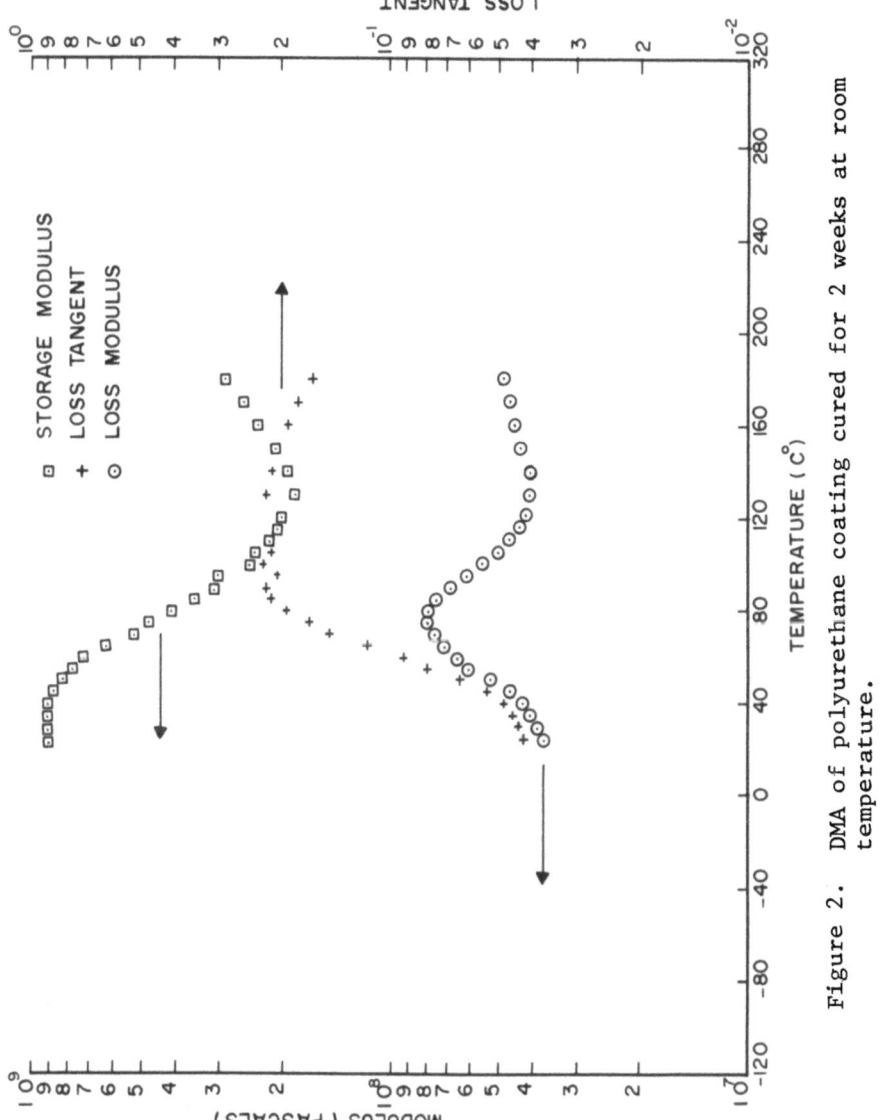

Figure 2. DMA of polyurethane coating cured for 2 weeks at room temperature.

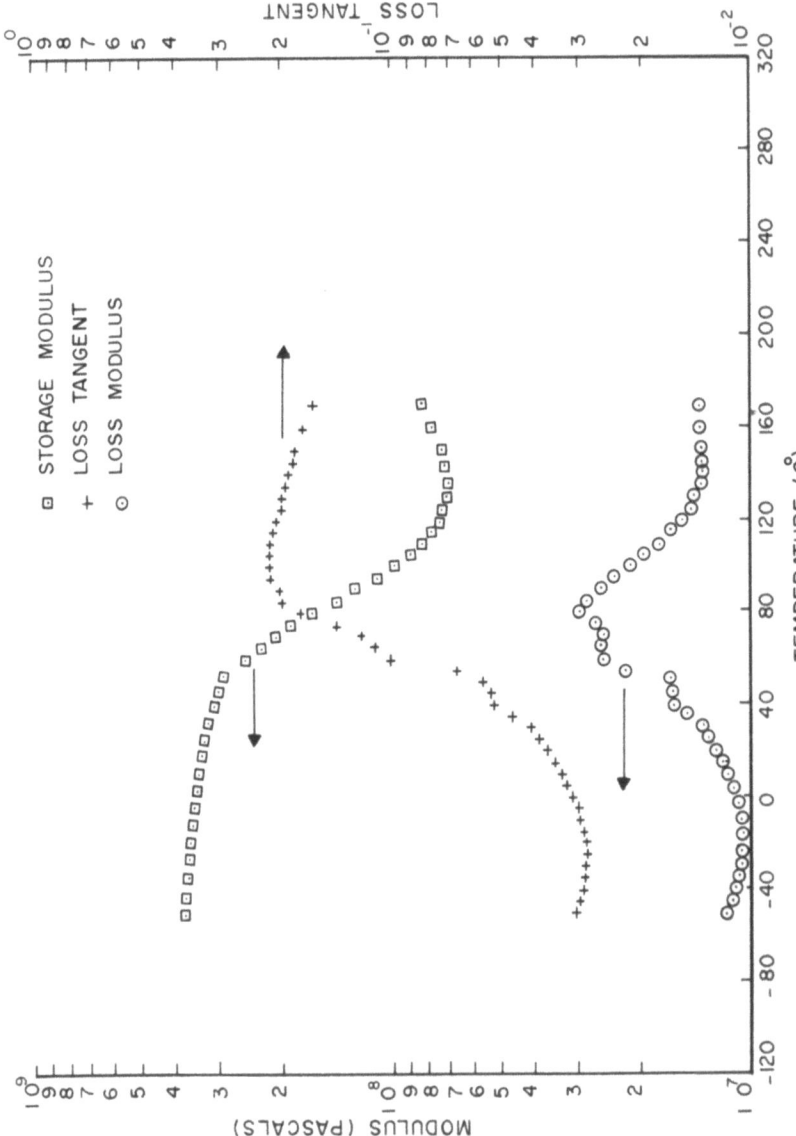

Figure 3. DMA of polyurethane coating cured for 2 weeks at room temperature followed by 1 week water soak.

Table II. Peel Tests on Polysulfide Sealants Applied over Artificially Aged Polyurethane Coating Surface.

Sample Treatment	Peel Strength, N/cm
5 days @ 60°C in air	104.2
12 days @ 60°C in air	86.0
5 days @ 60°C in water	106.2
2 days @ 121°C in air	0
5 days @ 121°C in air	0
5 days @ 60°C in air + 7 days @ 60°C in water	0
2 days @ 121°C in air + 2 days @ 60°C/95% R.H.	0

Table III. Microspecimen Hand-Pulled Peel Test Results.

Exposure Time at 121°C in air, hrs	Hand-Pull Peel Test Result
0	CF
1	~80% CF
2	~70% CF
3	~50% CF
4	~50% CF
5	~50% CF
6	AF
7	AF
8	AF
9	AF
10	AF
20	AF
30	AF
40	AF
50	AF

CF = Cohesive Failure
AF = Adhesive Failure

POLYURETHANE REACTION WITH POLYSULFIDE

The polyurethane reaction is slow in the coating because of the absence of tin or other catalyst compound. Newly coated polyurethane surfaces therefore contain residual NCO group. These groups can function as an internal coupling agent for the polysulfide sealant. The coupling reaction is shown below. The formation of thiourethane is responsible for excellent adhesion.

$$R-N=C=O + R'-SH \longrightarrow R-NH-\overset{\overset{O}{\|}}{C}-S-R'$$

Contained in   Polysulfide   Thiourethane
the cured   Sealant
Polyurethane

CONCLUSIONS

We have developed an accelerated aging method for polyurethane coatings. The hand-pull test that we developed on small specimens is a rapid adhesion screening method. This method, in conjunction with peel tests, can be utilized for developing new sealants.

ACKNOWLEDGEMENTS

The author appreciates help and support of Dr. Ali G. Maadhah and Mr. P. F. Cooke (both of the Research Institute) in preparation of this chapter. Technical help provided by D. E. Miller (University of Dayton) is acknowledged. Special thanks to Nihal Ahmad for typing this work.

REFERENCES

1. A. M. Usmani, R. P. Chartoff, W. M. Warner, J. M. Butler, I. O. Salyer, and D. E. Miller, Rubber Chem. Technol., 54, 1081 (1981).
2. J. H. Saunders and K. C. Frisch, "Polyurethanes: Chemistry and Technology," Vols. 1 and 2, Wiley-Interscience, New York, 1962 and 1964.
3. B. A. Dombrow, "Polyurethanes," Reinhold, New York, 1957.
4. A. M. Usmani and I. O. Salyer, Polym. Plast. Technol. Eng., 12, 61 (1979).
5. A. M. Usmani, Polym. Plast. Technol. Eng., 19, 185 (1982).

EPOXY RESIN WETTING OF E GLASS SINGLE FILAMENTS AS IT RELATES TO SHEAR STRENGTH

Elisabeth J. Berger and Yona Eckstein

Owens-Corning Fiberglas Corporation
Technical Center
Granville, Ohio 43023

The work of adhesion, W_a, of epoxy resin, water, and methylene iodide on single E-glass filaments was measured using a modified Wilhelmy apparatus. The resin used was a bisphenol A epoxy resin, and the fibers were coated with differing concentrations of silanes. Shear strengths were measured using a single filament adhesion test. It was found that increasing W_a of resin on the fibers correlates with increasing shear strength of dry and water-boiled composites. This result and the results of methylene iodide and water wetting are analyzed in terms of the geometric mean approximation for the work of adhesion, and good agreement is found.

INTRODUCTION

It has long been appreciated that wetting has a significant effect on adhesion in composites, and thus on composite properties.[1,2] However, the nature and extent of this effect have not been clearly defined. This work is a study of the correlation between glass/resin shear strength and wetting characteristics of the components of a composite system. No attempt is made to interpret our results in terms of the structure of the glass/resin interface.

Wetting experiments were done on single E-glass filaments, coated with differing concentrations of two silanes: γ-aminopropyl (triethoxy) silane and γ-methacryloxypropyl (trimethoxy) silane. The works of adhesion of epoxy resin, of methylene iodide and of water to these filaments were measured with a wetting balance. In the following, the results from these measurements and their correlation with the results from a single filament/resin interfacial shear test are presented and discussed.

EXPERIMENTAL

Single filaments of E-glass (10 micron diameter) were pulled from a one-hole bushing and coated from an applicator immediately under the bushing. The coatings were water or dilute aqueous silane solutions of concentrations from 0.2% to 2.5% by weight. The silanes used were Union Carbide A-1100 (amino silane) and A-174 (methacryl silane) dissolved at a pH of 3.5 (pH adjusted with acetic acid). A release agent, octyl(triethoxy)silane from Petrarch Systems Inc., was also used as a fiber coating in one experiment. The fibers were not rinsed.

The wetting balance used is similar to that described by Miller[3]. It consists of a Cahn 2000 electrobalance, an elevator and enclosure from Rame-Hart, and a Fisher strip chart recorder. The system is isolated from building vibrations by a Peabody air table. A fiber is glued to a small wire hook which is hung on the weighing arm of the electrobalance. The weight is tared off. The elevator platform is slowly raised until the beaker of DER331 epoxy resin (Dow Chemical Co.) or other wetting liquid resting on it touches the fiber. The elevator then continues rising for four millimeters at a rate of 0.15 microns per second. The contact of the resin with the fiber generates a force described as

$$mg = \pi d \gamma_{lv} \cos\theta \quad (1)$$

where m is the measured mass, g the acceleration due to gravity, d the fiber diameter, γ_{lv} the liquid surface tension, and θ the contact angle. The work of adhesion of the resin on the fiber can then be calculated, since

$$W_a = \gamma_{lv}(1+\cos\theta) \quad (2)$$

About ten fibers from each sample were measured and the results averaged.

Liquid surface tensions were determined using a Wilhelmy plate apparatus. The contact angle of resin against polypropylene was measured with a Rame-Hart NRL-100 contact angle goniometer in order to measure γ_{lv}^d.

The glass fiber/resin interfacial shear strength was determined using the method developed by Fraser et al.[5], and modified by Drzal et al.[6] In short, the technique consists of embedding a continuous glass monofilament in a resin. The dogbone-shaped specimen is then tensile loaded until the fragmentation of the embedded glass filament ceases. In order to analyze the results, the strain to failure of the resin must be four to five times larger than the ultimate elongation of the fiber. Since the fracture elongation of glass fiber is about 5%, the elongation at break of the matrix must be at least 25%. This requirement was satisfied by curing DER331, a diglycidyl ether of bis-phenol A, with the polybasic acid EMPOL 1052 (a mixture of 55% dibasic and 41% trimer fatty acids), using DMP-10 (dimethylaminoethylphenol) as an accelerator. The elongation at break of this resin varies between 35 and 55%.

The glass fibers were mounted on a stainless steel collecting fork by cementing their ends adjacent to the positioning notches. The single filaments were embedded in the resin using a silicone RTV-3112 resin mold. The mold consists of eight dogbone-shaped specimen cavities (1/8 inch wide x 1/16 inch deep x 1 inch long). Slots for the fibers are molded 1/32 inch deep through the ends of each specimen cavity. Since the distance between the slots in the mold is similar to the distance between neighboring notches on the steel fork, mounting and alignment of the fiber in the mold are readily accomplished.

The resin is injected into the mold with a disposable syringe to a level just above the height of the mold surface. The mold is then transferred to an oven and the resin cured. After curing, the mold is allowed to cool to room temperature. The specimens are easily removed by curling the silicone mold parallel to the fiber. In order to eliminate from consideration specimens with defects (e.g., broken fibers or improperly aligned fibers) each specimen is examined under a microscope prior to testing. After this initial inspection, those that passed are deformed on an Instron at jaw speed of 0.2 inch/min to just below their ultimate yield elongation.

Uniaxial tensile deformation of the specimens causes successive fragmentation of the fiber within the matrix. Theoretically, all lengths of fiber fragment should fall within the range of $\ell_c/2 \leq \ell \leq \ell_c$, where ℓ_c is the critical length. Assuming an even distribution of fragment lengths, the critical length can be calculated from

$$\ell_c = (4/3)\,\bar{\ell} \qquad (3)$$

where $\bar{\ell}$ is the average fragment length produced in a specimen after tensile deformation. The average interfacial shear strength, τ, is calculated from

$$\tau = (1/2)(\sigma_{Ff}d/\ell_c) \qquad (4)$$

where d denotes the fiber diameter, and σ_{Ff} is the average strength of the glass fragments.

After substituting Equation 4 into Equation 3, the formula becomes

$$\tau = (3/8)(\sigma_{Ff}d/\bar{\ell}) \qquad (5)$$

$\bar{\ell}$, the average length of glass fragments, can be estimated to a first approximation by dividing the gauge length of the specimen (one inch in our specimens) by the number of fragments produced upon tensile deformation. σ_{Ff} is calculated from the relationship

$$\sigma_{Ff} = \sigma_F \left(\frac{L}{\ell}\right)^{1/m} \qquad (6)$$

where σ_F denotes the mean strength of glass fiber of a known length L, and m is a statistical parameter describing the data scatter obtainable from a Weibull plot of the strength distribution of the fibers, measured prior to embedding the fibers in a resin.

The interfacial shear strength was measured in samples both before and after immersion in boiling water for six hours. These measurements are denoted dry strength and wet strength, respectively.

RESULTS AND DISCUSSION

The results of the glass/resin interfacial shear strength measurements on specimens coated with different concentrations of the silanes are given in Table 1. As demonstrated in Figures (1A) and (1B), the strength of the interface is a function of silane concentration, with the A-1100-coated fibers reaching a maximum dry strength at 1.0%, and the A-174-coated fibers giving the best dry strengths at the lowest concentration tested, 0.2%. For wet strength values, opposite trends prevailed for the two silanes. The wet strength of the A-1100-coated glass decreased with increasing silane concentration (Figure 3) while the A-174-coated samples increased in wet strength as the silane concentration increased.

The fiber coated with only water and with the release agent had shear strengths above most of the A1100- and A-174-coated fibers. Both systems may be viewed in terms of epoxy/glass adhesion, since the release agent coating is believed to be very nonuniform. It is known that epoxide groups are very reactive with water and adhere strongly to the glass surface[7]. McGarry[8], for example, reported that the epoxide/glass joint is about as strong as the cohesive strength of both. He observed that fracture itself never occurred at the interface. The large interfacial shear strength values reported for the water and release agent coated glass can thus be explained by the presence of strong glass/epoxy bondings. Apparently, even a small concentration of such bonding is sufficient to produce high interfacial shear strength.

Also listed in Table 1 are the measured works of adhesion of epoxy resin on glass filaments from the wetting experiments. As the work of adhesion increases, an increase in the interfacial shear strength of the dry specimens is observed. This effect is more easily observed when the dry strengths are plotted against the work of adhesion for the resin, W_a (Figure 2). The better molecular contact between the fiber and the uncured resin, which is suggested by the higher wetting, correlates with better interfacial adhesion in the cured composite. However, examination of Table 1 shows that no such correlation exists between the interfacial shear strength of the water-boiled samples and the work of adhesion of resin on the glass fibers.

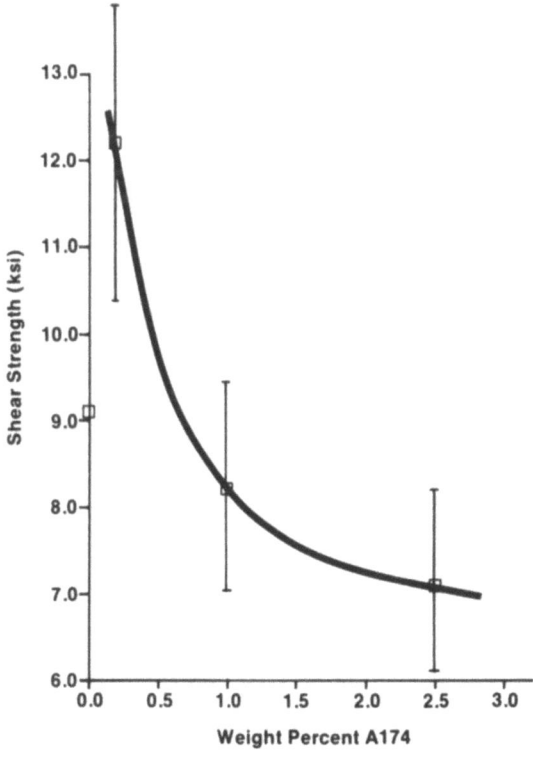

Figure 1A. Single Filament Interfacial Shear Strength vs. Concentration of A-174 on Glass Filaments.

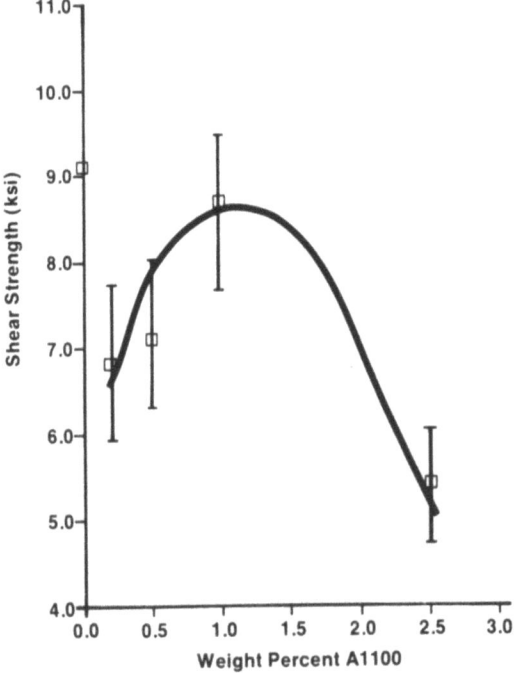

Figure 1B. Single Filament Interfacial Shear Strength vs. Concentration of A-1100 on Glass Filaments.

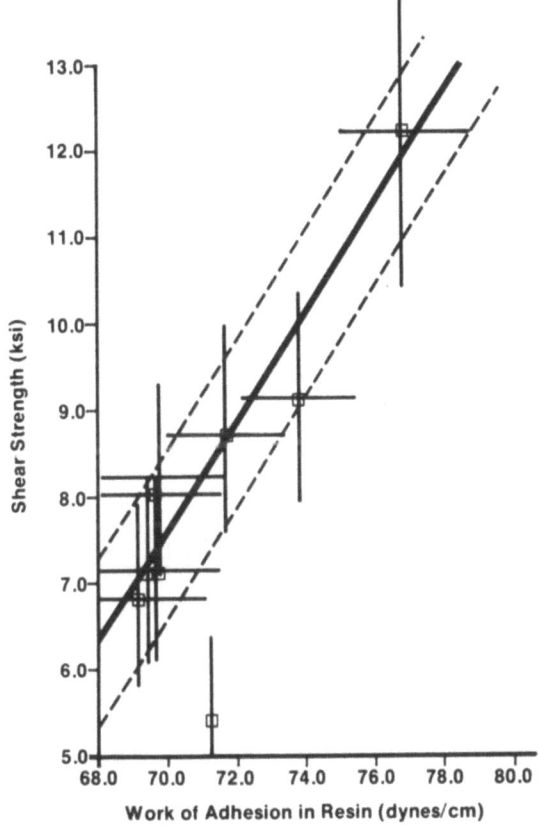

Figure 2. Single Filament Interfacial Shear Strength vs. Work of Adhesion of Epoxy Resin on Glass Filaments.

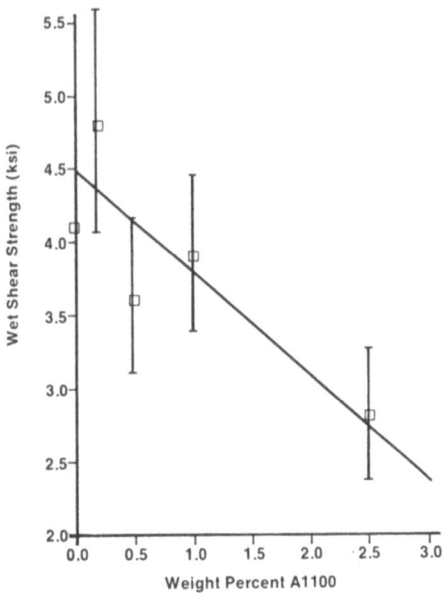

Figure 3. Single Filament Interfacial Shear Strength After Water Boil vs. Concentration of A-1100 on Glass Filaments.

Table I. Interfacial Adhesion and Wetting by Epoxy Resin, Water, and Methylene Iodide on Silanized Glass Fibers.

Coating	τ dry(a)	τ wet(a)	W_a in Epoxy Resin(b)	W_a in Water(b)	W_a in CH_2I_2(b)
A-1100(c)					
0.2%	6.8	4.8	69.2 ± 3	95.1 ± 6	68.9 ± 2
0.5%	7.1	3.6	69.5 ± 2	101.3 ± 9	64.3 ± 3
1.0	8.7	3.9	71.8 ± 2	108.6 ± 10	67.4 ± 5
2.5%	5.4	2.8	71.3 ± 2	119.3 ± 10	67.9 ± 1
A-174(c)					
0.2%	12.2	1.6	77.0 ± 2	111.1 ± 7	67.1 ± 2
1.0%	8.2	2.6	69.8 ± 2	114.1 ± 9	64.2 ± 1
2.5%	7.1	3.4	69.7 ± 2	111.5 ± 12	64.2 ± 1
Water	9.1	4.1	73.9 ± 2	103.3 ± 3	65.4 ± 2
Release agent					
1.0%	8.0		69.7 ± 3		

(a) Units are kpsi, errors are approximately 15%
(b) Units are ergs/cm², errors are 95% confidence intervals.
(c) Silane concentrations are weight percent in water.

In order to elucidate the glass/resin/water interactions, wetting experiments with water and methylene iodide were conducted on the glass fibers. These results are also shown in Table 1. The water wettability of the A-174-coated fibers appeared constant, but the A-1100-coated fibers showed a marked increase in hydrophobicity (lower water W_a) with decreasing silane concentrations. This corresponds to an increase in wet strength at lower A-1100 concentrations (Figure 3).

The degradation of the glass/resin interface during water boil may be seen as a competition between water and resin for the glass surface. Therefore, the larger the ratio of the work of adhesion of resin on glass to the work of adhesion of water on glass, the greater the wet strength should be. The trend of a general increase of wet strength with a greater ratio of resin wettability to water wettability is shown for our systems in Figure (4).

It is frequently assumed that the surface free energy of a substance can be divided into a polar and dispersion component (e.g. reference 8)

$$\gamma = \gamma^p + \gamma^d \qquad (7)$$

The γ^d is due to London dispersion forces, and the γ^p is due to dipole-dipole interactions and hydrogen bonding. One of the most prevalent theories of adhesion is that good adhesion is the result of a close match between the polar and dispersion components of the surface free energy of the adherend and the adhesive[9]. A difficulty with this theory is the lack of exact methods to measure the components of the surface free energy of solids. Several useful approximations have been suggested, however, including the geometric mean approximation[10], which describes the work of adhesion as

$$W_a = 2[(\gamma_l^d)^{\frac{1}{2}}(\gamma_s^d)^{\frac{1}{2}} + (\gamma_l^p)^{\frac{1}{2}}(\gamma_s^p)^{\frac{1}{2}}] \qquad (8)$$

Subscripts s and l refer to the solid and liquid, respectively. Dispersion components for liquids can be calculated from contact angle measurements against a nonpolar surface, such as polyethylene[11]. The polar component is then the total surface tension minus the dispersion component. If both sides of Equation 8 are divided by $2(\gamma_l^d)^{\frac{1}{2}}$ we have

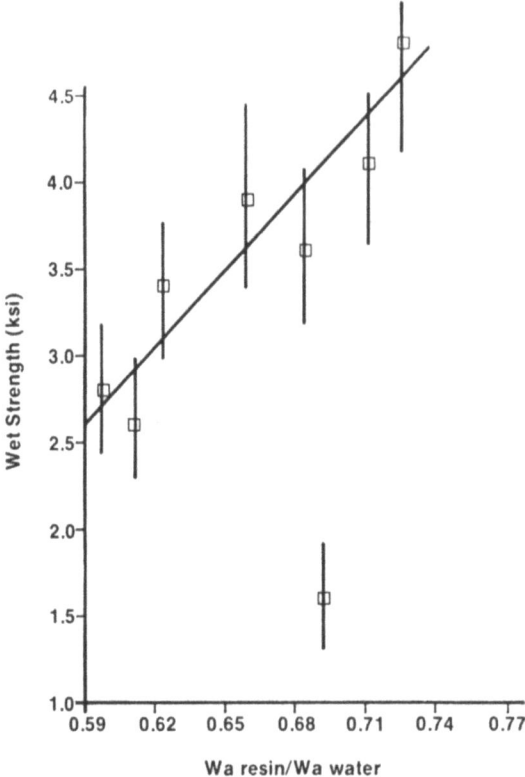

Figure 4. Single Filament Interfacial Shear Strength After Water Boil vs. The Ratio of Work of Adhesion of Epoxy Resin on Glass Filaments to the Work of Adhesion of Water on the Filaments.

$$\frac{W_a}{2(\gamma_1^d)^{1/2}} = (\gamma_s^d)^{1/2} + \frac{(\gamma_s^p)^{1/2}(\gamma_1^p)^{1/2}}{(\gamma_1^d)^{1/2}} \qquad (9)$$

Thus a solid with unknown surface characteristics can be wetted with two or more liquids with known surface energy components, and $W_a/2(\gamma_1^d)^{1/2}$ can be plotted against $(\gamma_1^p)^{1/2}/(\gamma_1^d)^{1/2}$. From Equation (9), the intercept is the square root of the dispersion component and the slope is the square root of the polar component.

Once the solid surface characteristics have been approximated, the adhesion can be discussed diagramatically[8]. The square root of the dispersion component is plotted against the square root of the polar component. A circle is then drawn through the coordinates of the adherend (phase 1) and the adhesive (phase 3) with radius R_o and center (K,H):

EPOXY RESIN WETTING AND SHEAR STRENGTH

$$R_o^2 = 1/4((\alpha_1 - \alpha_3)^2 + (\beta_1 - \beta_3)^2)$$

$$H = 1/2 (\alpha_1 + \alpha_3) \qquad (10)$$

$$K = 1/2 (\beta_1 + \beta_3)$$

where

$$\alpha = (\gamma^d)^{\frac{1}{2}}$$

$$\beta = (\gamma^p)^{\frac{1}{2}}$$

Sensitivity of this bond to another environment (phase 2; e.g., immersion in water) is then described by the Griffith fracture equation

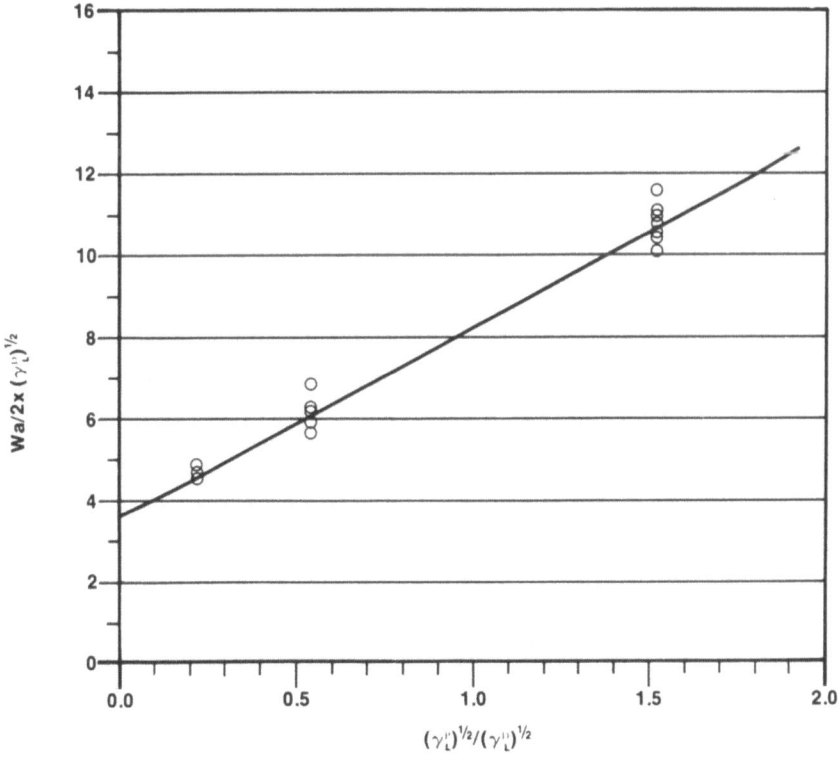

Figure 5. Plot for Surface Energy Calculation of Water-Coated Glass. Using the geometric mean approximation, the intercept is the square root of the dispersion component, and the slope is the square root of the polar component.

$$\gamma_G = R^2 - R_0^2 \tag{11}$$

in which

$$R^2 = (\alpha_2 - H)^2 + (\beta_2 - K)^2 \tag{12}$$

and γ_G is the critical fracture energy. If the immersion fluid has coordinates inside the circle, there will be spontaneous debonding, since $R < R_0$ giving a negative γ_G. For fluids outside the circle, the degree of sensitivity of the bond to the immersion fluid is described by the magnitude of γ_G. The farther the coordinates of immersion fluid are from the circle describing a particular bond, the less the bond will be affected by the fluid.

In this work, three wetting liquids were used to determine the surface characteristics of the fibers: water (γ_1^p = 50.8 ergs/cm², γ_1^d = 22.0 ergs/cm²), DER331 resin (γ_1^p = 10.8 ergs/cm², γ_1^d = 36.7 ergs/cm²), and methylene iodide (γ_1^p = 2.3 ergs/cm², γ_1^d = 48.5 ergs/cm²). A typical plot giving surface energy of a fiber is shown in Figure (5). The calculated polar and dispersion components of the fibers coated with A-1100 and A-174, and the Griffith fracture energy, are listed in Table II.

Table II. Square Roots of Polar and Dispersion Components of the Surface Free Eneregy of Silane-coated Fibers. Griffith Fracture Energy for Epoxy Composites in Water. Units are $(ergs/cm^2)^{\frac{1}{2}}$.

Coating	$(\gamma_s^d)^{\frac{1}{2}}$	$(\gamma_s^p)^{\frac{1}{2}}$	γ_G
A-1100(a)			
0.2%	3.68 ± 0.26	4.22 ± 0.17	9.89
0.5%	3.29 ± 0.29	4.88 ± 0.28	7.24
1.0%	3.60 ± 0.24	4.82 ± 0.27	7.74
2.5%	3.10 ± 0.37	5.87 ± 0.38	3.70
A-174(a)			
0.2%	3.60 ± 0.26	5.23 ± 0.25	6.24
1.0%	2.91 ± 0.40	5.80 ± 0.37	3.87
2.5%	2.95 ± 0.43	5.68 ± 0.40	4.34
Water	3.57 ± 0.12	4.76 ± 0.13	7.94

(a) Silane concentrations are weight percent in water.

Circles are drawn through the coordinates of the cured epoxy resin and coated fiber (Figure 6). The immersion fluid of interest is water, and it can be seen that the closer the circle is to the water coordinates, the lower the wet strength. If this distance, which corresponds to the Griffith fracture energy, is plotted against the wet strength, a correlation can be seen (Figure 7). As the Griffith fracture energy increases, more energy is needed to break the adhesive bond, and the wet strength increases.

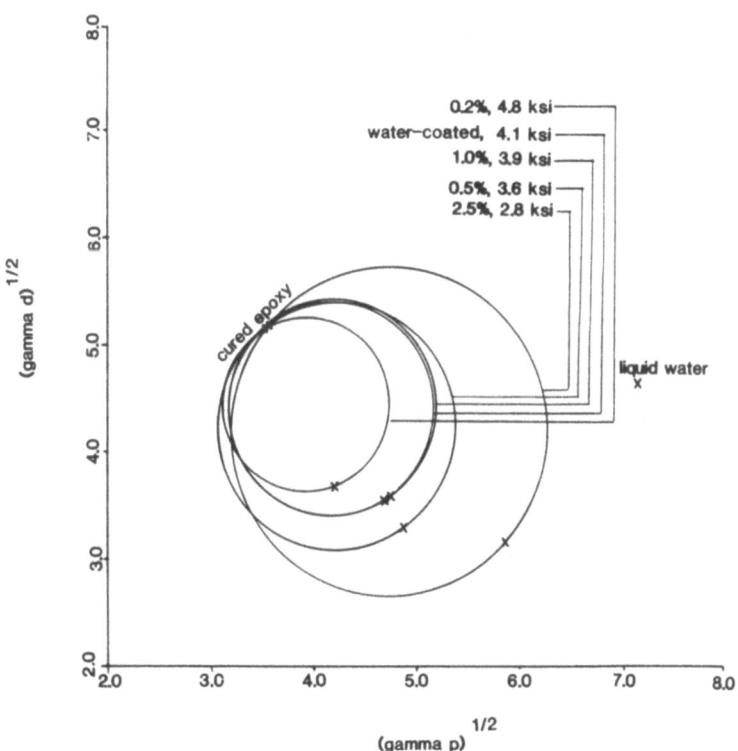

Figure 6. Square Root of the Dispersion Component of the Surface Free Energy vs. the Square Root of the Polar Component for Fiber Coated with A-1100 or Water, and for Liquid Water and Cured Epoxy Resin. Units are $(ergs/cm^2)^{\frac{1}{2}}$. The circles describe the bond between the fiber and the resin, with a smaller radius indicating a closer match between surface energy components of resin and fiber. Numbers associated with the circles are percent A-1100 and wet shear strength.

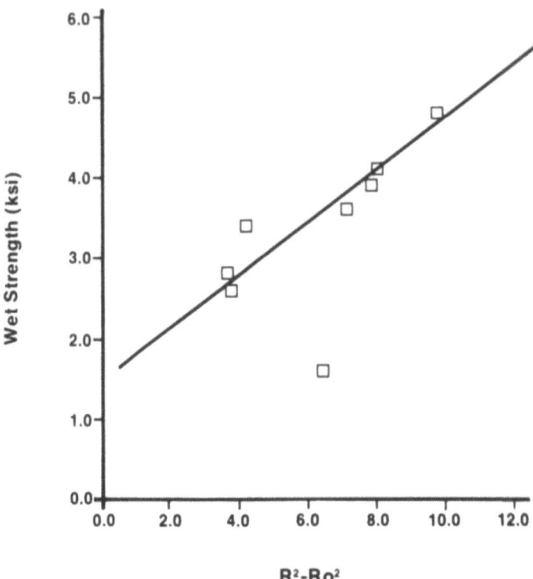

Figure 7. Single Filament Interfacial Shear Strength After Water-Boil vs. the Griffith Critical Fracture Energy.

It should be noted that in both Figure (4) and Figure (7) the 0.2% A-174 point does not fit the trend of the rest of the data. The reason for this is not known. We are confident, however, that this is a real phenomenon.

While it must be emphasized that γ_s^d and γ_s^p are approximations, this analysis works well for most of the systems studied. It demonstrates that while both components play a role in adhesion of coated E-glass fibers to epoxy resin, the polar component is the more important in determining resistance to debonding by water.

CONCLUSIONS

Interfacial shear strength in a glass fiber/epoxy resin composite is seen to be a function of the concentration of silane solution for γ-aminopropyl(triethoxy)silane and γ-methacryloxypropyl(trimethoxy)silane used as fiber coatings. Also, as wetting of the glass fiber surface by epoxy resin increases, interfacial adhesion in the composite increases, probably due to increased contact between the resin and the fiber. The ratio of wettability of the fibers by resin to wettability by water is shown to be an indicator of wet strength. Wet strength was also analyzed as a function of closeness of polar and dispersion components of the cured epoxy and the fibers, compared to their closeness to the water coordinates. On a plot of the square roots of the polar and the dispersion components, the distance from the liquid water coordinates to a circle describing the closeness of each fiber to the resin is seen to correlate to the wet strength. A smaller distance indicates a lower resistance of the glass/epoxy bond to water.

ACKNOWLEDGEMENTS

D. William Stroud and Sarah K. Wesson are gratefully acknowledged for help in data collection. Thanks are also due to Sheldon P. Wesson for helpful discussions.

REFERENCES

1. M. Yamamoto, S. Yamada, Y. Sakatani, M. Yaguchi and Y. Yamaguchi, Int. Conf. Carb. Fib., Compos. Appl. Pap., Plas. 21/1-21/6, 179, 1971.
2. W. D. Bascom, NRL Report 6140, Washington D.C., 1964.
3. B. Miller, in "Surface Characterization of Fibers and Textiles," M. J. Schick, Editor, Vol. 7, Part 2, Marcel Dekker, New York, 1977.
4. F. M. Fowkes, Ind & Eng Chem, $\underline{56}$ (12), 40 (1964).
5. W. A. Fraser, F. H. Achker, and A. T. DiBenedetto, Proc. Conf. Reinforced Plastics/Composites, SPI, Section 22A, 1975.
6. L. T. Drzal, M. J. Rich, J. D. Camping, and W. J. Park, Proc. Conf. Reinforced Plastics/Composites, SPI, Section 20C, 1980.
7. K. L. Loewenstein, in "Composite Materials," L. Halliday, Editor, Elsevier, p. 150, 1966.
8. F. J. McGarry, Proc. Conf. Reinforced Plastics/Composites, SPI, Section 11B, 1958.
9. D. H. Kaelble, J. Adhesion, $\underline{2}$, 66 (1970).
10. D. H. Kaelble, "Physical Chemistry of Adhesion," pp. 153-170, Wiley-Interscience, New York, 1971.
11. T. Smith, J. Adhesion, $\underline{11}$, 243 (1980).

Part II
Factors Influencing Behavior of Adhesive Joints

INFLUENCE OF SURFACE ROUGHNESS ON MECHANICAL PROPERTIES

OF JOINTS

Y. Gilibert ° and G. Verchery °°

° Ecole Nationale Supérieure
de Techniques Avancées
Centre de l'Yvette
91120 Palaiseau, France

°° Ecole Nationale Supérieure
des Mines
158, cours Fauriel
42023 St Etienne, France

Mechanical properties of joints are dependent on the geometry of surface defects caused during manufacture of the adherends. This paper presents experimental results on the influence of the roughness, due to grinding and sand blasting, on tensile properties of joints. We studied specimens made of low carbon steel rods, bonded by an epoxy resin containing mineral fillers. The surfaces were prepared with various roughness conditions : 2 grindings (coarse and fine), 8 grindings with sand blasting (combination of 2 grindings and 4 diameters of sand particles), and one shot blasting. The roughness parameters were measured, particularly the total depth and the average spacing of roughness. These measurements showed that the prepared surfaces had well defined and reproducible roughness. Electric strain gauges at specific points on the metallic parts were used ; these provided measurements of local strains (at ±1 µm/m) during the loading. The variation of these strains represents indirectly the joint behavior : the elastic limit, the appearance of micro-cracks and their propagation up to failure. Optical and scanning microscopic observations were used to check the results obtained by these mechanical measurements. The principal results of the present work are the following : with or without subsequent blasting, fine grinding produces better mechanical properties than coarse grinding ; sand blasting improves these properties when compared to shot blasting or pure grinding ; and finally, sand blasting gives the best behavior when the total depth of roughness is equal to the mean diameter of the dispersed particles in the resin.

INTRODUCTION

It is commonly observed that surface defects influence mechanical properties of joints : elastic limit, strength, and type of fracture[1-4]. In this work, we have studied, in particular, the effects of surface defects caused during machining of the adherends.

We realized from the beginning the necessity of precise controls during all the steps so as to obtain reproducible and dependable results. We have paid special attention to the following points :
- procedure for the pretreatment of the surfaces and the measurements of the resulting geometry of the adherend surfaces ;
- analysis of the adhesive characteristics, all the properties of which are not provided by the manufacturer ;
- design and manufacture of the test-specimens ;
- loading conditions ;
- selection of the characteristic mechanical parameters, and the procedure for measuring these parameters.

We used visual, optical and scanning microscopic observations to determine the aspect of fracture and to study the distribution of the fillers in the resin.

Most of the results of the present work can be found with more details in Reference 5, relevant chapters of which will be referred to in the following text.

PREPARATION OF THE SURFACE AND MEASUREMENT OF THE GEOMETRY

Properties of the Adherends

A low carbon steel (0.18 % carbon ; XC 18 French standard equivalent to SAE-AISI 1017) was used for the adherends of the specimens studied. The properties of the steel were controlled using mechanical measurements and microscopic observations. The Young's modulus (E_T = 207 700 MPa) and the Poisson's ratio were determined using tension test with strain gauges at a very low strain rate ($\dot{\varepsilon}$ about 10^{-6} s^{-1}) on steel specimens with 100 mm^2 square cross section and 250 mm length. The Brinnell hardness (HB = 210) test showed the uniform quality of the material, whereas microscopic observations indicated a ferritic microstructure.

Preparation of the Rough Adherends, Finishing and Blasting

A shaping machine with a traversing tool was used in the preparation of the rough adherends.

This was followed by milling and after which a grinding was effected with a horizontal surface grinding machine under different conditions : coarse or fine grain grinding wheels. For all these operations, machining conditions (speed, feed, depth of cut, lubrication, sharpening of the tools) were strictly controlled and are described in Reference 5, Chapter 4.

The finishing procedure for some of the specimens was completed with sand blasting under four conditions using different particle diameters (115, 169, 282 and 423 µm) or with shot blasting. The particles with a 115 µm diameter were made of high purity white alumina ($Al_2O_3 > 99.5\%$), whereas the particles with diameters of 423, 282 and 169 µm were of brown corundum, i.e. less pure alumina (composition : $Al_2O_3 > 94\%$; $TiO_2 < 2.5\%$; $SiO_2 < .5\%$). The blasting pressure used was 0.4 to 0.5 MPa. The sand jet was inclined at 60° from the specimen surface and displaced at a speed of 24 mm s^{-1}. The shot blasting was conducted with iron shots 1 mm in diameter.

In this paper, the various surface states are designated as follows :
RF - fine ground state ;   RFS - fine ground sand blasted state ;
RG - coarse ground state ; RGS - coarse ground sand blasted state ;
   RGG - coarse ground shot blasted state.
The numbers following RFS and RGS designate the particle mesh size.

Measurement of the Surface Roughness Parameters

The surface profiles were determined using probes with an air bearing of 25 µm and 750 µm travel distances. The specimens used had the same section as the adherends and were machined and treated at the same time under the same conditions. The surface profiles were measured, before and after blasting, along reference paths, the ends of which are marked with two notches remaining after blasting.

Data were recorded and processed by a computer to determine the various parameters of the surface defects describing the total profile, roughness and waviness. The calculations were done using a method developed by B. Scheffer of Renault, and J. Bielle and C. Berger of ENSAM[6,7].

We observed that the surface roughness parameters alter very quickly during the first few seconds of blasting and do not change after a while. We selected a time lapse of 30 seconds which ensured a stable state of these parameters.

Measurements showed that the surface roughness parameters are not dependent on the direction of measurement for blasted states, whereas condition as ground showed some anisotropy.

Table I presents values of the surface roughness parameters for the various surface states studied. These are the total depth of roughness, R_t ; the average depth of roughness, R ; the maximum depth of roughness, R_{max} ; the roughness levelling depth, R_p ; the arithmetical mean deviation from mean line of roughness, R_a ; and average spacing of roughness, A_R. In Figure 1, these parameters are plotted against the mean diameter d of sand particles.

Table I. Roughness Parameters of the Various Surface States. AVB is a trade name (Aluminium Oxide Vapor Blast) for an alumina with a purity over 99.5%.

Designation	d μm	ROUGHNESS (μm)					
		R_t	R	R_{max}	A_R	R_p	R_a
RF	—	2.73	0.82	2.73	40	0.44	0.29
RG	—	9.09	4.39	9.09	90	2.36	1.39
RGG	—	24.09	18.68	24.68	5650	9.00	5.20
RGS 60C	423	21.61	12.83	21.61	285	6.77	3.47
RGS 90C	282	7.20	4.96	7.20	185	2.77	1.40
RGS 150C	169	6.21	3.92	6.21	135	2.12	1.20
RGS 220AVB	115	9.01	4.73	9.01	155	2.51	1.41
RFS 60C	423	18.09	11.51	18.09	195	5.95	3.14
RFS 90C	282	14.42	8.06	14.42	100	4.34	2.54
RFS 150C	169	5.60	3.47	5.60	110	1.88	1.01
RFS 220AVB	115	3.64	2.27	3.64	105	1.25	0.69

From these data, it can be seen that the surface roughness parameters for sand blasted states depend on both the initial grinding and the mean diameter of the particles used for blasting. For all surface preparations, the maximum depth of roughness R_{max} is very close to the total depth R_t, which indicates that surfaces have no aberrant defects of roughness and are consequently fairly homogeneous ; for the fine ground state, the roughness parameters increase steadily with the particle diameter whereas for the coarse ground state these increase notably only for a diameter superior to 200 μm.

Other surface defect parameters were computed but proved of little importance as far as the mechanical properties of the joints are concerned. However, they confirmed that the prepared surfaces had well defined and reproducible parameters.

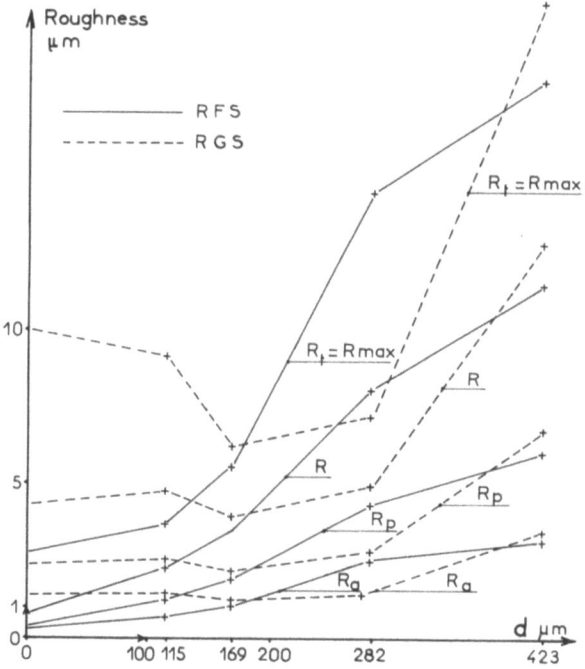

Figure 1. Roughness parameters versus sand particle diameter. The solid line refers to fine ground blasted states (RFS) ; the broken line refers to coarse ground blasted states (RGS).

ANALYSIS OF THE ADHESIVE

We used the commercial adhesive, Eponal 317, of Sonal, France, which is a two component system (an epoxy resin containing mineral fillers and a hardening agent). It was processed at room temperature ; at such temperature, about 20 minutes are available for use and polymerization rate reaches 90% within 2 hours.

We determined its mechanical, physical, structural and chemical properties. The elastic constants were measured at room temperature (20°C) using two methods. Tensile test using strain gauges and a very slow loading rate ($\dot{\varepsilon}$ about 10^{-6} s^{-1}) showed an elastic brittle behavior. Using traction test the Young's modulus was found to be 5800 MPa, the Poisson's ratio was 0.327 and the strength was 30 MPa ; whereas ultrasonic measurements gave 5940 MPa for the Young's modulus and 0.323 for the Poisson's ratio. The measured density was 1.433 kg dm^{-3}.

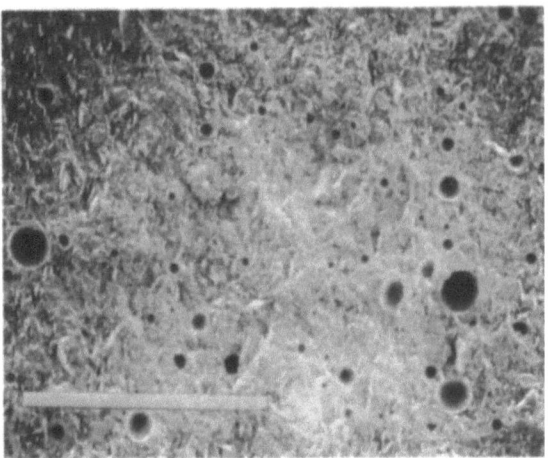

Figure 2. Fracture surface through the adhesive (scale bar 0.5 mm).

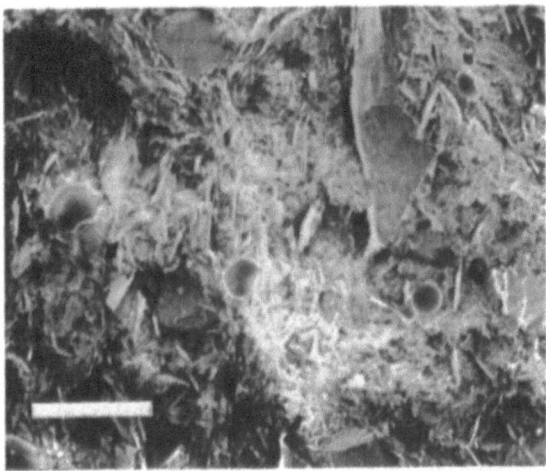

Figure 3. Fracture surface through the adhesive (scale bar 50 μm).

The resin was calcined at 1000°C, which produced a mineral residue weighing 40 % of its initial weight. A qualitative analysis of this residue was performed using two types of methods : chemical reactions and use of a X-ray spectrometer associated with the scanning electron microscope. It showed the presence of Mg, Fe and Si in significant quantities, barium sulphate, Ti, and traces of Al, Ca and various fluorides. The hardening agent was also calcined at 1000°C. The chemical reactions on the residue showed the presence of Ca, small quantities of Fe, carbonates and traces of Al.

Observations of the surfaces of the fractures in the adhesive using the scanning microscope revealed that the fillers are small particles with a prolate shape, a mean diameter of 5 µm, and a length up to 10 or 15 µm. Scanning electron micrographs (Figures 2, 3 and 9) showed that the fracture in the adhesive went through the mineral particles, with a smooth-looking surface, revealing a brittle behavior of the fillers. Cracks perpendiculars to the adherends were observed in the adhesive, and a brittle behavior of the adhesive is revealed. More details can be found in Reference 5, Chapter 3.

DESIGN AND MANUFACTURE OF THE TEST-SPECIMENS
LOADING CONDITIONS

We used specimens with two double-lap joints. One of these double-lap joints was used for strain measurements. The other joint, which was tightened transversely by a binding clip, ensured axial and uniform load in the adherends, while the clip eliminated any local bending. The extremities of the test-specimen were fixed on a universal testing machine by knee joints. In fact, this design of the specimen ensured symmetrical fractures in the tested part, and the clip decreased considerably the statistical deviation in the measurements, as shown in Reference 5, Chapter 2. Figure 4 shows a schematic view of the specimen.

The parts (10 mm x 4.5 mm cross-section for parts 2' ; 10 mm cross-section for parts 1') were produced at a ± 0.01 mm tolerance. The length of the overlap in the tested part of the specimen was 88 mm. A thickness of 0.5 mm was selected for the adhesive joint and controlled by gauged spacers. Preliminary tests, described in Reference 5, Chapter 11, showed that this thickness gives the best mechanical behavior for these dimensions of the metallic parts and this length of the joint.

The adhesive was applied at room temperature by hand with a spatula. A very thin layer was first spread on the adherends. Bubbles formed at the interface during this application vanished within a minute, after which a 1 mm adhesive layer was spread.

The various parts were positioned together into a frame. The excess adhesive was removed after the polymerization.

The test-specimens were loaded in tension at a slow rate (1000 N per minute) at room temperature (20°C).

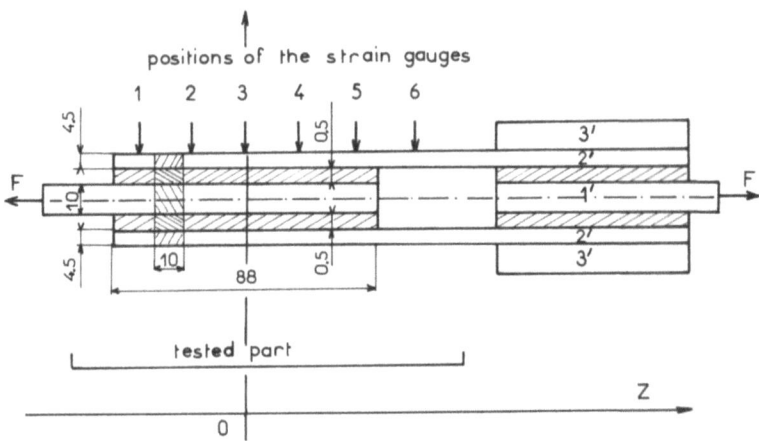

Figure 4. Schematic view of the test-specimen (dimensions in mm). A view of the cross-section is superimposed between gauges 1 and 2. 1' and 2' : adherends ; 3' : binding clip.

PRINCIPLE OF THE MECHANICAL MEASUREMENTS

Since mechanical quantities inside the joint cannot be obtained directly, we used electric strain-gauges placed at different points on external surfaces of the metallic adherends. Figure 4 shows the position of the gauges ; their distances from the middle of the joint are as follows :
- gauge 1 : −37   mm
- gauge 2 : −18.5 mm
- gauge 3 :   0
- gauge 4 : +18.5 mm
- gauge 5 : +37   mm

The positioning of the gauges is described in detail in References 5 (Chapter 5) and 8.

INFLUENCE OF SURFACE ROUGHNESS

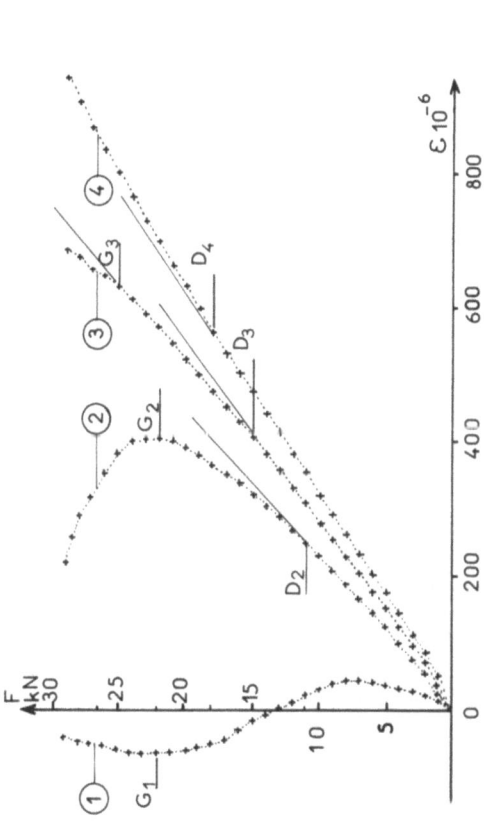

Figure 5. Strain versus tensile load for a ground sand blasted state (outputs of gauges 1-4). Non-linearity occurs over point D_2.

A sensitive and precise (±0.25%) bridge (VE 20 Micromeasurement) was used.

Figure 5 shows typical strain gauge recordings for a ground sand blasted state. Other recordings are given in Reference 5.

The gauges numbered 2, 3 and 4 show a linear relationship up to points D_2, D_3, D_4 respectively. The fact that gauge 2 is the first to deviate from linearity shows that initiation of micro-cracks takes place in the joint in area 2 under this gauge (at abscissa -18.5mm). When the load is increased, over the point D_2, micro-cracks continue to develop in area 2 and begin to appear in area 3 (point D_3), then 4 (point D_4). At G_2, micro-cracks in area 2 reach a sufficient size to propagate and later (point G_3), propagation begins in area 3. For a given load, this propagation is stable, but the speed of stable propagation increases when load increases. During this phase, area 2 can support less and less load. The absolute value of the slope of the load versus the strain of gauge 2 decreases. When it reaches zero, the propagation becomes unstable and fracture occurs very quickly.

Gauge 5 is in fact of little use as far as the analysis of the fracture process is concerned. Gauge 1 delivers a weaker output than the other gauges ; its values are not easy to analyze, especially at low loads. It can be seen that a compressive effect occurs about point D_2 and developed up to point G_2. Generally, non-zero values could be analyzed as the sign that the extreme end of the joint is not fractured until the final step.

Visual and microscopic observations of the fractured specimens have confirmed that in the central part of the joint a progressive shear fracture occurs inside the adhesive ; at the ends of the joint one can see either a shear fracture inside the adhesive or an adhesive failure between the adhesive and the metallic adherend along the upper or lower face, induced during the final phase of the fracture.

These facts differentiate the ground sand blasted state from the ground state, which showed predominantly metal-resin adhesive fractures.

The various stages for the ground sand blasted states, (i) elastic behavior, (ii) initiation of micro-cracks, (iii) coalescence of the micro-cracks, (iv) stable propagation of the cracks, and finally, (v) unstable propagation up to final fracture, proved to be absolutely reproducible (Reference 5, Chapter 8).

These experimental results can be partly explained by the various theoretical analyses[9-10], which have analyzed stress fields in adhesive joints. However, these theoretical analyses, which are based on continuum mechanics and use idealization of interfaces, cannot explain the significant differences observed using the above measurement procedure, and are presented below.

RESULTS

Here are presented the effects of roughness on the three typical mechanical parameters : elastic limit (crack initiation threshold), crack propagation threshold, and ultimate strength.

We found that roughness can be adequately represented by using the total depth of roughness R_t.

We observed a very small scattering in the measured values, which gives proof of the importance of our careful and well-defined process in preparation and testing. Deviation in the results is generally higher for the propagation threshold and the ultimate strength than for the elastic limit. Further, deviation in the ultimate strength is higher for non-blasted surfaces than for blasted surfaces (Reference 5, Chapter 8). Typical values, corresponding to the results presented in Figures 6 and 7, are, (i) 0.25 kN for the crack initiation threshold (about 12 kN), (ii) 0.5 kN for the crack propagation threshold (about 24 kN), and (iii) 0.5 kN for the ultimate strength (about 30 kN).

Mechanical Results

For the sand blasted states, we found that the mechanical parameters were influenced by the initial grinding and the roughness, which is related to the sand particle size.

The fine grinding is equivalent or better than the coarse grinding for the three mechanical parameters ; it gave better values for the elastic limit with the large particle diameters, and better values for the propagation threshold with all diameters.

For both types of grinding, we observed that the strain measurements given by the gauges at the external surface of the adherends were significantly smaller (for any position) and more uniform along the specimen, when the surfaces were prepared using particles with a diameter of 169 µm. Furthermore, this preparation shows the best values for the three mechanical parameters.

Figures 6 and 7 summarize the results. Almost all of the plotted points represent the mean value of four tests. In Reference 5, Chapter 9, one can find the curves representing F versus the other roughness parameters, R_a, R_p, R ; they have the same behavior as the curves in Figure 6.

Figure 6 shows that for fine grinding, with or without blasting, the three mechanical parameters reach a maximum value when the total depth of roughness equals 5.6 µm. It seems possible to associate this property with the fact that the total depth of roughness is approximately the size of the fillers in the resin.

No precise trends can be found in the case of coarse grinding.

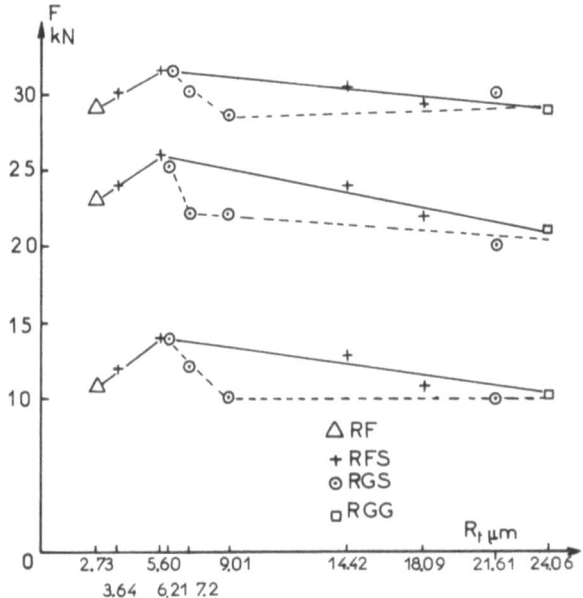

Figure 6. Mechanical parameters versus the total depth of roughness : from the bottom, (i) crack initiation threshold, (ii) crack propagation threshold, and (iii) ultimate strength. For fine grinding without blasting (RF), the values are in accordance with the extrapolation (with respect to the roughness) of those obtained for fine grinding with blasting (RFS).

Visual and Microscopic Observations of Fractures

Visual observations of the fractures show that for all the sand blasted states mostly shear fractures occur in the adhesive ; the shot blasted states show no such shear fracture.

INFLUENCE OF SURFACE ROUGHNESS

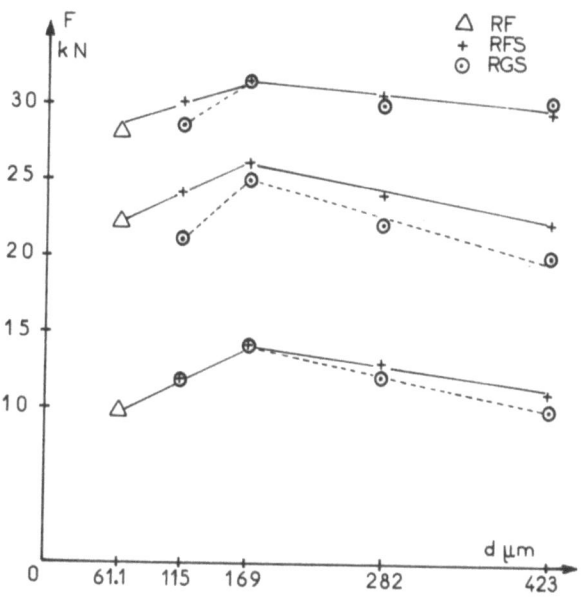

Figure 7. Mechanical parameters versus the mean diameter of the sand particles : from the bottom, (i) crack initiation threshold, (ii) crack propagation threshold, and (iii) ultimate strength. For fine grinding without blasting (RF), a fictitious diameter is considered, so as to retain the property shown in Figure 6, i.e. linear extrapolation from the nearest values for fine grinding with blasting (RFS).

We measured the length, L, of the shear fractures as they occurred in various states. Figure 8 is a plot of the mechanical parameters versus this length ; the corresponding curves are approximately linear.

Microscopic observations with use of a X-ray spectrometer were performed. For sand blasted states, they confirmed the predominance of shear fractures inside the joint. Furthermore, we observed that the mineral fillers tended to fill the cavities on the surface and, when prolate, to be inclined at 45° to the surface. Figure 9 shows a typical view. For the shot blasted states, the fracture surface revealed the metallic adherends with flakes of adhesive about 0.1 mm in diameter (Figure 10).

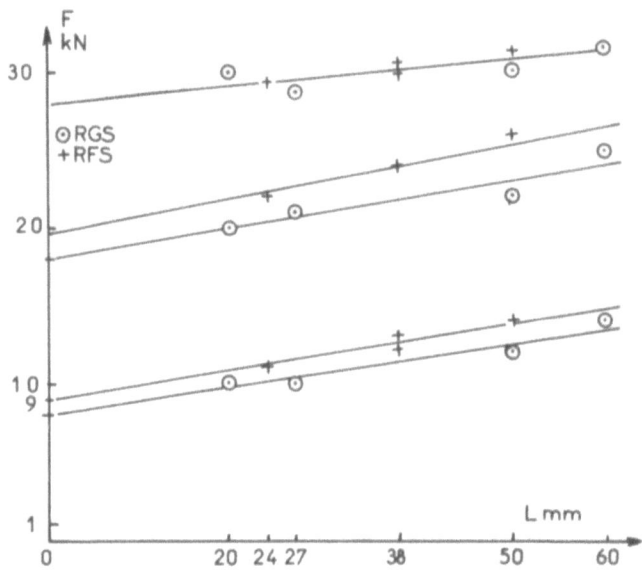

Figure 8. Mechanical parameters versus the length of shear fracture : from the bottom, (i) crack initiation threshold (two lines), (ii) crack propagation threshold (two lines), and (iii) ultimate strength (one line).

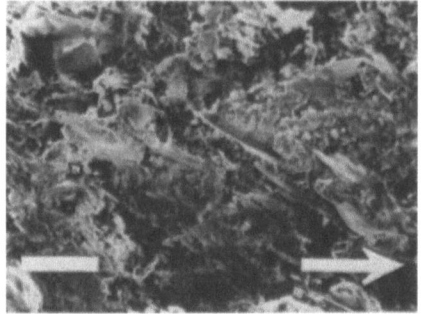

Figure 9. Fracture aspect for a sand blasted state ; the arrow is the direction of the surface of the adherend (scale bar 20 μm).

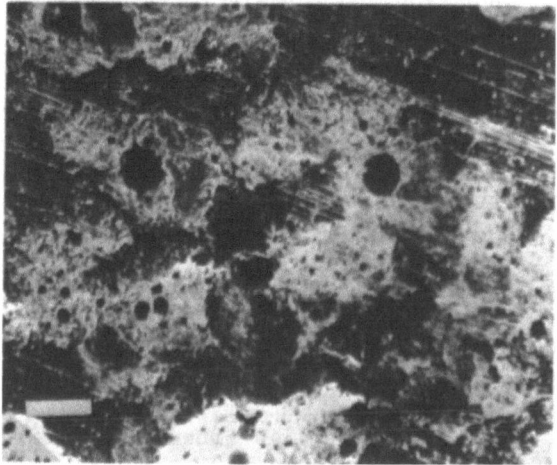

Figure 10. Fracture aspect for a shot blasted state ; the scale bar represents 0.2 mm.

CONCLUSION

Our experimental results show that the surface roughness has a significant influence on the mechanical properties (crack initiation threshold, crack propagation threshold, and ultimate strength) of joints. We have found the best conditions for joint design as follows :
- fine grinding is better than coarse grinding ;
- sand blasting improves mechanical properties ;
- sand particle diameter must be selected so as to give a total depth of roughness close to the size of the adhesive fillers.

REFERENCES

1. J.J. Bikerman, "The Science of Adhesive Joints," Academic Press, New York, 1968.
2. R. Houwink, "Adhesion and Adhesives, Vol. 2 : Applications," Elsevier, Amsterdam, 1967.
3. J. Corey McMillan, in "Bonded Joints and Preparation for Bonding," AGARD Lecture Series, No.102, Part 7, 1, 1979.
4. D. Hennemann and W. Brockmann, J. Adhesion, $\underline{12}$, 297 (1981).
5. Y. Gilibert, "Contribution à l'étude de l'adhésivité de matériaux collés par l'intermédiaire de résines époxydiques," Doctoral Dissertation, Université de Reims, France, 1978.
6. J. Bielle, Revue Arts et Métiers (Paris), No.3, 44 (1973).
7. J. Bielle, "Etat géométrique des surfaces techniques," lecture notes, ENSAM, Châlons sur Marne, France, 1981.
8. Y. Gilibert, H. Miller and G. Gautron, Annales de l'Université et de l'ARERS, Reims, France, $\underline{14}$, 17 (1976).
9. M. Goland and E. Reissner, J. Appl. Mech., $\underline{11}$, A17 (1944).
10. D. Volkersen, Construction Métallique (Paris), No.4, 3 (1965).
11. R.D. Adams and N.A. Peppiatt, J. Strain Analysis, $\underline{8}$, No.2, 134 (1973).
12. Y. Gilibert and A. Rigolot, J. Mécanique Appliquée, $\underline{3}$, No.3, 341 (1979).

THE EFFECTS OF SURFACE CONDITIONING OF AISI-304 STAINLESS STEEL ON THE INTERFACIAL PROPERTIES OF ALLOY/EPOXY COMPOSITE STRUCTURE; ADHEREND SURFACE CHARACTERIZATION USING X-RAY EMISSION SPECTROSCOPY (LEEIXS)

F. Gaillard, A. A. Roche and M. J. Romand

Applied Chemistry Department (CNRS, ERA # 300)
University Claude Bernard - LYON I
69622 Villeurbanne Cedex, France

AISI-304 stainless steel adherend-epoxy bonded specimens were subjected to a three point flexure test. This test used single interface type specimens. The measurements were conducted in order to: (i) investigate the effects of adherend surface reproducibility and metal surface conditioning on joint strength, (ii) determine the sensitivity and repeatability of the test, (iii) gain a better idea of the range of applicability of the test.

The results presented emphasize that mechanical properties such as modulus of rigidity and ultimate load values of the metal/epoxy interface are highly dependent upon adherend surface treatments (chemical etching, electropolishing, cathodic reduction, thermal treatment). In order to understand the chemical factors affecting adhesive strength, a near-surface sensitive technique (Low Energy Electron-Induced X-ray Spectroscopy (LEEIXS)) has been used and effects of surface conditioning of the stainless steel on the surface layer composition and thickness were studied.

INTRODUCTION

Different standard mechanical tests are commonly used to evaluate adhesive bonded joint strength or quality. These tests may be categorized in three major groups i.e. tension, shear or peel, even though various geometries have been adapted from these configurations.

In order to investigate the influence of the adherend preparation processes on the strength and the mechanical properties of the adhesive bonded joint interfacial region, it has been shown recently[1,2] that a three point flexure test applied to a single interface type specimen can be employed advantageously. The objective of the present work is to gain a better idea of the range of applicability of the test and therefore to define ways and means to achieve fully reliable bonded joints. The relevant study deals with AISI-304 stainless steel adherend/epoxy bonded specimens. In addition LEEIXS (Low Energy Electron-Induced X-ray Spectroscopy)[3], a near-surface sensitive technique, was used in order to understand the chemical factors which affect the adhesive strength.

I. INSTRUMENTATION

I.1 Three point flexure test

A table model 1102 INSTRON tester fitted with a cross head speed reducer and a 500 N load cell was used for testing. The cross head speed was typically 0.5 mm.mn^{-1} with a loading sensitivity down to 100 N full scale.

The specimen preparation fixture has been described previously [1,2,4]. A Plexiglass holder was used as a form for the RTV 501 silicone forming mold. Rectangular Plexiglass strips (25 × 5 × 10 mm) were prepared simultaneously by a machining process to an exactly identical size. In this manner, the area of all mold cavities was identical. By applying an exact amount of adhesive (0.5 cm^3) to each adherend with a syringe, bonded adhesive size and volume were then precisely controlled.

Industrial AISI-304 stainless steel foils 0.47 mm thick were milled simultaneously to an identical size (10 × 50 mm).

1.2 LEEIXS instrument

A detailed description of the LEEIXS instrument has been given elsewhere[3,5]. We shall only mention here the basic principles of such an apparatus, which is shown schematically in figure 1. This equipment consists of a wavelength-dispersive X-ray spectrometer equipped with a cold cathode tube (CCT). This gas discharge tube is an open window device which operates directly in the spectrometer primary vacuum. Normal operating reduced pressure is 1-10 Pa. The CCT is used to bombard the sample surface with a quasi monoenergetic electron beam which may be selected in the range of 0.5-5 keV.

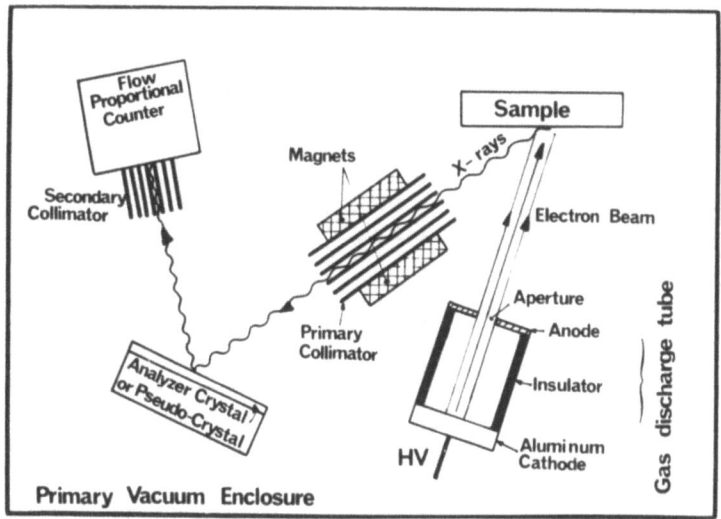

Figure 1. Schematic drawing of the Low-Energy Electron Induced X-ray Spectrometer (LEEIXS).

The beam spot size at the sample surface is limited by the anode aperture. Typically, the spot diameter is 1 cm. During experimentation, for a given acceleration voltage, the selected electron current is stabilized by an automatic pressure regulator in the range of 0 to 2 mA. Both voltage and current are digitally controlled. The soft X-ray spectrometer employed in the present measurements uses flat analyzing crystals or pseudo-crystals. The detector is a flow proportional counter with an ultra-thin (0.5 µm aluminized polypropylene) window and uses, at atmospheric pressure, an argon-methane (P-10) gas flow. The bombarding electron beam incidence angle is 70° and the take-off angle of X-rays is 35°.

The effective depth of analysis of LEEIXS is determined by the material thickness at which characteristic X-rays are produced within the sample. Obviously, this depends upon the target material, the electron energy and the electron beam incidence angle.

From our previous results and their comparison with Rutherford Backscattering Spectroscopy data[6,7], it appears that the average excitation depth (R_e) at which X-rays are produced can be formulated as follow :

$$R_e \, (\mathring{A}) = 250 \, (\bar{A}/\rho(\bar{Z})^{n/2}) \, (E_o^n - E_x^n),$$

Table 1. Excitation Depth for OK_α, CrL_α and FeL_α Radiations.

E(keV)	1.5			4.5		
Radiation	OK_α	CrL_α	FeL_α	OK_α	CrL_α	FeL_α
R_e (nm) AISI-304 matrix	-	13	11.5	-	136	134.5
R_e (nm) Fe_2O_3 matrix	21	21	19	203	202	200

where $n = 1.2/(1 - 0.29 \log \bar{Z})$, E_0 is the electron energy in keV, E_x the excitation energy in keV of the concerned emission process. $\bar{A}$, $\bar{Z}$ and ρ are respectively the atomic weight, the atomic number and the density of the material under investigation. This expression for the depth of excitation has been deduced from the Feldman relationship[8] giving the practical maximum range of 1-10 keV electrons in solids. This semi-empirical relationship takes into account the electron backscattering effects and the energy loss of electrons during their path in the target material.

The calculated excitation depths for OK_α, CrL_α and FeL_α radiations are given in table I for 1.5 and 4.5 keV electron energies.

I.3 Cathodic reduction device

The oxide layers formed at the sample surfaces can be reduced using a standard amperostatic circuit[9] shown in figure 2.

Reduction takes place in a medium of concentrated H_2SO_4 and the reduction current density can be fixed in the range from 0.6 to 13 $\mu A.cm^{-2}$. Under reduction, the sample potential varies from oxide to metal/medium potential. The amount of charge necessary for this potential variation can be related to the coulombic thickness of the reducible oxide layer[9,10,11].

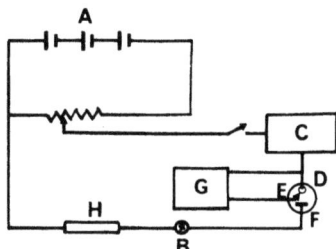

Figure 2. Schematic amperostatic circuit diagram : A, 75 V battery; B, ammeter ; C, current integrator ; D, sample ; E, reference electrode ; F, auxiliary electrode (Pt) ; G, E = f(t) recorder ; H, 2 MΩ resistance.

II. STATISTICAL STUDY

Different adherend surface conditioning treatments were used as depicted in Table II.

Firstly, it is necessary to clarify that the different chemical conditionings do not affect the modulus of rigidity of the alloy $(EI)_M$ and therefore the substrate mechanical properties. For this purpose, unbonded metallic strips were submitted to different cleanings or etchings (degreasing, electropolishing, concentrated and diluted sulfuric acid oxidations) and then subjected to the three point flexure test. The modulus of rigidity of the metallic strip $(EI)_M$ is deduced from the slope P/δ of the load/displacement curve in the region of elastic deformation by :

$$(EI)_M = \frac{P \ell^3}{48 \delta}$$

where ℓ is the distance between the two supports. The values of $(EI)_M$, shown in table III, exhibit a relative standard deviation value of ± 1.6 %. Such a statistical result takes into account reproducibilities of the mechanical tester, specimen sizing and adherend surface state.

Table II. Different Treatments for AISI-304.

Code	Treatment
A	Sample degreased in acetone for ten minutes.
B	Sample degreased as in A, submerged in 1 % HCl, 2 % HF, 10 % HNO_3 at 80°C for one minute, rinsed with deionized water for two minutes.
C	Sample degreased as in A, submerged in 5 % H_2SO_4 at 60°C for 15 minutes, rinsed as in B.
D	Sample degreased as in A, electropolished in a CH_3CO_2H, $HClO_4$ 3.2 : 1 solution, rinsed as in B.
E	Sample cathodically reduced in concentrated H_2SO_4 till dissolution of any oxide layer.
F	Sample degreased as in A, oxidized at 300°C at air for one hour, air cooled.
G	Sample degreased as in A, submerged in boiling concentrated H_2SO_4 for 15 minutes, rinsed as in B.

Secondly, it is necessary to emphasize the test reproducibility for the composite materials prepared. After bonding, sets of identically prepared specimens (identical surface preparation, epoxy formulation, curing cycle ...) were tested. The results are tabulated in table IV.

For panels oxidized by means of treatment C, relative standard deviation σ % of the modulus of rigidity of the system $(EI)_s^4$ was less than 1 %, while this deviation is less than 7 % for the values of the ultimate load (F_{max}). Such a statistical result includes reproducibilities of the tester, specimen sizing, adherend surface state and bonding processing.

Table III. The Effects of Surface Conditioning on the Modulus of Rigidity of the Metal.

Treatment	A	B	C	D
$(EI)_M$ (10^3 N . mm^2)	16.5	17.1	17.0	17.0
	$(EI)_M \pm \sigma$ % = 16.9 ± 1.6 %			

Table IV. The Effects of Surface Conditioning on the Mechanical Properties of the System.

Treatment	A		C		D		G	
Sample number	$(EI)_S$ $(N.mm^2)$	F_{max} (N)	$(EI)_S$ $(N.mm^2)$	F_{max} (N)	$(EI)_S$ $(N.mm^2)$	F_{max} (N)	$(EI)_S$ $(N.mm^2)$	F_{max} (N)
1	122.10^3	31	172.10^3	33	104.10^3	19	128.10^3	30
2	86.10^3	12	174.10^3	38	97.10^3	18	128.10^3	31
3	121.10^3	9	174.10^3	33	112.10^3	18	124.10^3	33
4	85.10^3	9	174.10^3	34	104.10^3	20	150.10^3	30
5			172.10^3	32			145.10^3	35
6			174.10^3	37			133.10^3	29
Mean values $\pm \sigma \%$	103.10^3 $\pm 20\%$	15 $\pm 67\%$	173.10^3 $\pm 0.9\%$	35 $\pm 7\%$	104.10^3 $\pm 6\%$	19 $\pm 4\%$	135.10^3 $\pm 8\%$	32 $\pm 8\%$

$$\overline{(EI)}_S \pm \sigma \% = 129 \pm 26 \%$$
$$\overline{F}_{max} \pm \sigma \% = 25 \pm 38 \%$$

Table IV also shows the $(EI)_S$ and F_{max} values for different series of specimens subjected to various conditioning treatments. $\overline{(EI)}_S$ and $\overline{F}_{max}$ are respectively the average of the various $(EI)_S$ and F_{max} values obtained for each surface conditioning. Relative standard deviation of $\overline{(EI)}_S$ was found to be 26 % while $\overline{F}_{max}$ relative standard deviation was 38 %. Such variations cannot be explained by statistical error (i.e. reproducibilities of apparatus, specimen sizing, adherend surface state) and are therefore characteristic of the properties of the created interphase.

III. INFLUENCE OF ADHEREND CONDITIONING

A set of eight samples was identically electropolished (treatment D) ; four of them were then submitted to cathodic reduction (treatment D + E). The whole set was immediately bonded and submitted to the three point flexure test. For panels bonded after a cathodic reduction, there is a considerable decrease in the values of F_{max} whereas the modulus of rigidity of the system remained constant (see table V). The cathodic reduction curves are shown in figure 3. These curves were obtained by reducing the surface oxide layer at constant current density ($j = 5$ µA.cm^{-2}) in a concentrated sulfuric acid (18 N) bath[12]. The amount of charge Q (mC.cm^{-2}) used for this reduction has been plotted against the sample potential relative to the standard calomel electrode (S.C.E.). The Q

Table V. The Effects of Cathodic Reduction on the Mechanical Properties of the System.

Treatment	D		D + E	
Sample Number	$(EI)_S$ (10^3 N.mm^2)	F_{max} (N)	$(EI)_S$ (10^3 N.mm^2)	F_{max} (N)
1	104	19	127	7
2	97	18	111	7
3	112	18	106	8
4	104	20	103	11
Mean values $\pm \sigma$ %	104 $\pm$ 6 %	19 $\pm$ 4 %	111 $\pm$ 10 %	8 $\pm$ 25 %

value necessary to reach the metal potential is then representative of the apparent coulombic thickness of the oxide layer[13].

Figure 3 shows the amount of charge Q necessary for the reduction of (a) the AISI-304 natural oxide and (b) the AISI-304 surface layer formed during cathodic reduction and / or immediately after.

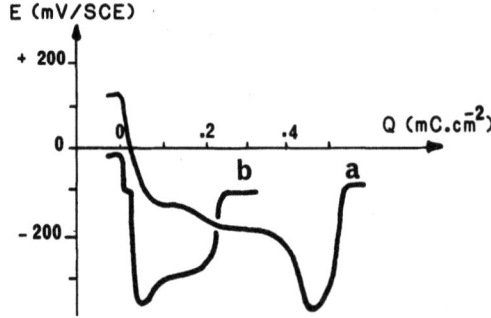

Figure 3. Cathodic reduction curves.

EFFECTS OF SURFACE CONDITIONING OF STAINLESS STEEL

In this last case the metallic surface is covered by a very thin layer. The subtle metal potential difference exhibited by curves (a) and (b) may be due to differences in surface state and/or contaminant deposition on the electrodes. Inversely, treatment by sulfuric acide (C) provided a thick oxide layer (table VI) and gave the highest F_{max} values.

It is also worthy of note that $(EI)_S$ and F_{max} values and their corresponding σ % are strongly dependent upon the preconditioning steps. As can be seen in table VII, treatment C provided F_{max} and $(EI)_S$ values higher than those obtained after treatment B. If treatment C was performed after treatment B (treatment B + C), corresponding values were lower than those obtained for treatment C alone.

Table VI. The Effects of Surface Conditioning on Oxide Thickness.

	OK_α INTENSITY (a.u.)*	OXIDE LAYER THICKNESS (nm)**
Fe single crystal after long time air exposure	5690	13
Etched AISI-304 (B)	2210	5
Electropolished AISI-304 (D)	3010	7
Etched AISI-304 followed by sulfuric acid treatment (B + C)	9090	21
Sulfuric acid treated AISI-304 (C)	10140	23
Thermally oxidized AISI-304 (350°C, 1 hour)	10440	24
Thermally oxidized AISI-304 (500°C, 1 hour)	21610	49

* LEEIXS measurements performed with 4.5 keV electron beam and a 0.5 mA.cm^{-2} electron current density.

** Thickness calculated using Feldman's relationship, assuming that the surface oxide layer formed after each treatment is an homogeneous Fe_3O_4 film (density = 5.18).

Table VII. The Effects of Treatments B, C and B + C on the Mechanical Properties of the System.

Treatment	B		C		B + C	
Sample number	$(EI)_S$ (10^3 N.mm^2)	F_{max} (N)	$(EI)_S$ (10^3 N.mm^2)	F_{max} (N)	$(EI)_S$ (10^3 N.mm^2)	F_{max} (N)
1	151	17	172	33	117	32
2	130	19	174	38	119	23
3	152	16	174	33	126	26
4	160	16	174	34	126	30
5			172	32	104	26
6			174	37	116	29
Mean values ± σ %	148 ± 9 %	17 ± 9 %	173 ± 1 %	35 ± 7 %	118 ± 7 %	28 ± 9 %

The standard deviations in $(EI)_S$ and F_{max} for (B + C) or (B) treated metallic strips were similar. This emphasizes that the material surface keeps remanent traces of each treatment step.

Figure 4 shows the LEEIXS spectra of (B), (C) and (B + C) treated specimens. These spectra were obtained by bombarding the sample surface with 1.5 keV electrons, the electron current density being 0.5 mA.cm^{-2}. Under these conditions, the depth probed is less than 25 nm. X-ray dispersion is provided by a TℓAP single crystal analyser. These spectra show that surface enrichment of nickel is produced by the 5 % H_2SO_4 oxidation process (C). This enrichment is revealed by the lower ratio of intensities FeL_α/NiL_α exhibited by spectra C and B + C compared to this observed in the case of spectrum B.

It is also noteworthy that the fluorine impurity at the B etched sample surface, remained within the surface oxide layer even after the sulfuric oxidation (B + C).

Both the mechanical and the spectroscopic studies emphasize the effect of fluorine contaminant incorporated within the oxide layer. This impurity is probably responsible for the poor adhesion (lower F_{max} and $(EI)_S$ values), as has been previously reported for titanium and titanium alloys[5,14,15].

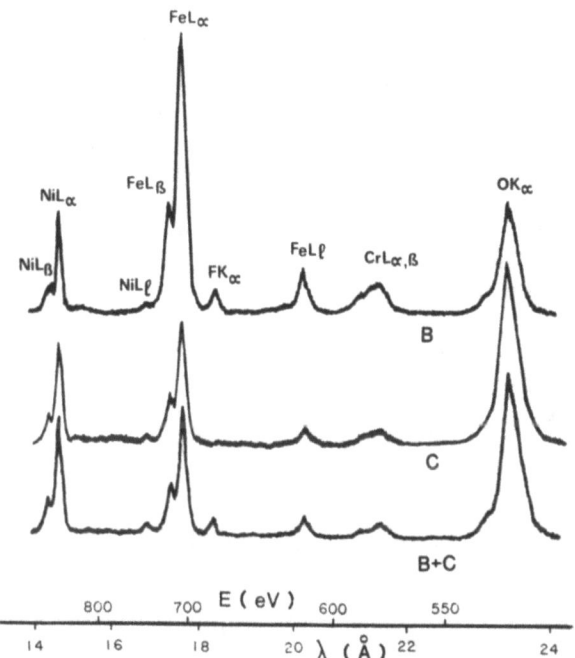

Figure 4. LEEIXS spectra of AISI-304 after treatments (B), (C), (B + C).

IV. INFLUENCE OF EXPOSURE TIME TO AIR BEFORE BONDING

In order to investigate surface ageing effects on bonded specimen performances, the treated panels were submitted to room atmosphere exposure before bonding. After the desired ageing, aged and identically treated unaged samples were bonded at the same time and tested. The results are shown in table VIII.

In order to understand further these results it is necessary to clarify what are the physical meanings of $(EI)_S$ and F_{max} parameters. Account been taken of the sample preparation processing, an increase (or decrease) in the value of $(EI)_{System}$ is associated with an increase (or decrease) in the value of $(EI)_{Interphase}$. For the materials investigated in this study $(EI)_{Oxide}$ is higher than the corresponding $(EI)_{Metal}$. Consequently, formation of an oxide of high rigidity induces an increase in $(EI)_{Interphase}$ value and therefore an increase in $(EI)_{System}$[4]. F_{max} values depend firstly upon the viscoelastic properties of the so-created interphase (i.e. upon the presence or not of a weak boundary layer) and secondly upon diffusion of hydroxide groups[16].

In the case of C-treated specimens, $(EI)_S$ and F_{max} values decrease and their respective σ % increase drastically. These

Table VIII. Adherend Ageing Effects on the Mechanical Properties of the System.

Treatment	C				G			
Ageing	no		one week		no		24 hours	
Sample number	$(EI)_S$ $(N.mm^2)$	F_{max} (N)	$(EI)_S$ $(N.mm^2)$	F_{max} (N)	$(EI)_S$ $(N.mm^2)$	F_{max} (N)	$(EI)_S$ $(N.mm^2)$	F_{max} (N)
1	173.10³	33	185.10³	24	128.10³	30	100.10³	45
2	174.10³	38	144.10³	39	128.10³	31	131.10³	44
3	174.10³	33	151.10³	25	124.10³	33	150.10³	26
4	174.10³	34	137.10³	14	150.10³	30	200.10³	44
5	172.10³	32	124.10³	19	145.10³	35	165.10³	40
6	174.10³	37			133.10³	29	145.10³	36
7							157.10³	34
Mean values ± σ %	173.10³ ± 1 %	35 ± 7 %	148.10³ ± 16 %	24 ± 40 %	135.10³ ± 8 %	32 ± 8 %	150.10³ ± 20 %	39 ± 18 %

results are associated with changes in the properties of the oxide layer which is well known to decay when exposed to air and to lose its protective properties[10].

In the case of G-treated specimens $(EI)_S$ and F_{max} values increased with the surface ageing. This is in agreement with electrochemical data which report that stainless steel treated with boiling sulfuric acid undergo an increase in passivity after a 24-h exposure to air[17].

In these conditions the oxide so-produced improves the adhesion properties of the system.

V. INFLUENCE OF AGEING BONDED SPECIMENS

In order to investigate the ageing effects on bonded structures, single interface type specimens were submitted to a three weeks exposure in ambient atmosphere. Although such an ageing may be considered as a short time, it has been reported that it is long enough to modify rheological properties of bonded joints and bulk adhesive[18,19]. After ageing, identically prepared unaged samples and aged samples were tested at the same time. Table IX shows the ageing effects on the modulus of rigidity of the system and on the maximum load values. The results concern adherend sheets submitted either to chemical or thermal oxidation processes.

In the case of C-treated specimens, two opposite phenomena can be observed. The decrease in $(EI)_S$ reflects a decay of the oxide layer with exposure time in air. This correlates with previous results (table VIII). This oxide layer degradation would be expected to lead to a decrease in F_{max}. In fact, experimental F_{max} values are nearly constant with, or without, ageing. This can be due to the diffusion of water and/or oxygen into the adhesive volume and/or into the interfacial region[16,18]. The diffusion process of water may relieve internal stresses and/or transform some surface oxide to hydroxide and so may create a weak boundary layer. In both cases, adhesion is impaired. Diffusion process of oxygen through the polymer may modify its mechanical properties and so enhances adhesion. The combination of these two phenomena may allow to explain why the values of F_{max} remain constant.

It is well known that treatment F produces a homogeneous oxide layer, stable to air exposure. The oxide layer produced thermally has a low permeability to water. In this case, a constant value of $(EI)_S$ is predictable and was observed. Since the oxide layer is unaffected by exposure to air, diffusion phenomena occur mostly at oxide/adhesive interface or through the adhesive to the interface. Then a F_{max} increase can be expected if oxygen diffusion is prevailing. This was experimentally observed.

Table IX. Ageing Effects on the Mechanical Properties of the System

Treatment	C				F			
Ageing	no		three weeks		no		three weeks	
Sample number	$(EI)_S$ $(N.mm^2)$	F_{max} (N)	$(EI)_S$ $(N.mm^2)$	F_{max} (N)	$(EI)_S$ $(N.mm^2)$	F_{max} (N)	$(EI)_S$ $(N.mm^2)$	F_{max} (N)
1	173.10^3	34	146.10^3	26	108.10^3	20	160.10^3	43
2	176.10^3	32	139.10^3	39	115.10^3	24	127.10^3	41
3	175.10^3	38	148.10^3	40	135.10^3	19	153.10^3	56
4	170.10^3	31	165.10^3	37	154.10^3	29	165.10^3	31
5	169.10^3	37	162.10^3	36	164.10^3	32	124.10^3	30
6	174.10^3	35	152.10^3	34	131.10^3	27	148.10^3	30
7	173.10^3	41	148.10^3	41	155.10^3	21	143.10^3	40
8	172.10^3	31	143.10^3	53	145.10^3	21	108.10^3	25
Mean values $\pm \sigma$ %	177.10^3 $\pm 1\%$	35 $\pm 10\%$	150.10^3 $\pm 6\%$	38 $\pm 20\%$	138.10^3 $\pm 14\%$	24 $\pm 21\%$	141.10^3 $\pm 14\%$	37 $\pm 27\%$

VI. INFLUENCE OF THERMAL TREATMENTS

a) Influence of the duration of thermal treatments

The degreased adherend sheets were oxidized in air for various durations at 300 and 600°C. CrL_α, FeL_α and OK_α radiation intensities were mesured using LEEIXS. These data are plotted with respect to the duration of the thermal treatment as shown in figure 5.

The OK_α intensity can be related to the thickness of the oxide layer. Using a 4.5 keV electron beam, the depth probed is higher than 150 nm. This value is considerably greater than studied oxide thicknesses.

For an identical duration of thermal treatment, the surface layer formed at 600°C was found to be thicker than that formed at 300°C.

The CrL_α radiation intensity is representative of the outer 21 nm of the oxide layer (electron energy : 1.5 keV). For 300°C treated samples a regular decrease in the intensity of the CrL_α intensity was observed when the duration of treatment was increased Conversely at 600°C, samples treated for more than 15 minutes exhibited an enrichment of surface chromium. This is supported by the uniform increase in the ratio of CrL_α/FeL_α intensities.

EFFECTS OF SURFACE CONDITIONING OF STAINLESS STEEL

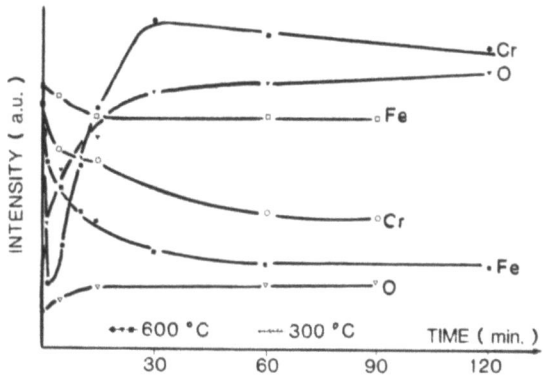

Figure 5. CrL_α, FeL_α, OK_α radiation intensities versus duration of thermal treatment.

Panels oxidized in air at 600°C for various lengths of time, were bonded and tested. This temperature was chosen to determine the effects of chromium diffusion within the oxide layer. The results of the mechanical test are given in table X.

Figure 6 compiles CrL_α, OK_α radiation intensities, F_{max} and $(EI)_S$ values versus the duration of treatment. The highest F_{max} values have been found to correspond to treatment durations providing chromium depleted surfaces. Cohesive failure of the oxide was observed for specimens treated for more than 15 minutes at 600°C. This leads to the hypothesis of failure below the chromium-enriched surface layer. In addition a value increase in $(EI)_S$ can be observed up to 30 minutes treatment. For longer durations a decrease of this parameter occured. This fact can be related to the apparent

Table X. Thermal Exposure Time Effects on the Mechanical Properties of the System.

Treatment duration (min.)	0	1	5	15	30	60	120
$(EI)_S$ (10^3 N.mm^2)	129	140	152	153	155	134	114
F_{max} (N)	15	21	28	29	21	23	23

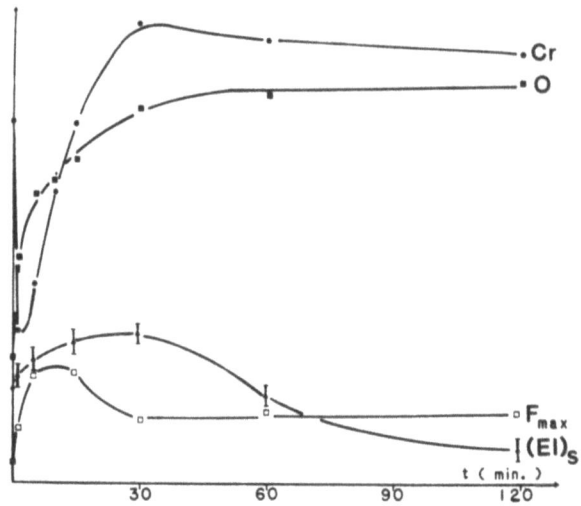

Figure 6. OK_α, CrL_α radiation intensities, $(EI)_S$ and F_{max} values versus duration of thermal treatment.

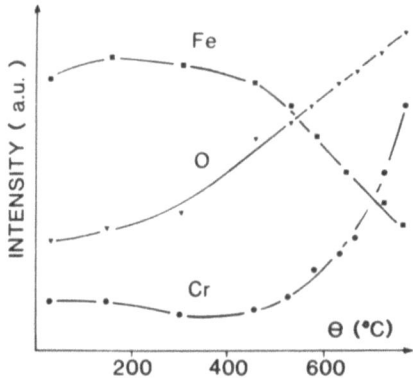

Figure 7. OK_α, FeL_α and CrL_α radiation intensities versus treatment temperature.

inhomogeneity of the oxide layer which can be easily observed for treatments of long duration.

b) Influence of thermal treatment temperature.

Degreased adherend sheets were air-oxidized at various temperatures for half an hour. The choice of this duration was deduced from previous experiments results (figure 5). In figure 7, the LEEIXS data are plotted against the air oxidation temperature ; the relevant experiments were performed in the same conditions as described previously.

The intensity of the OK_α radiation and therefore the oxide thickness increased with temperature of treatment up to 750°C. An enrichment of surface chromium was observed for $\theta \geqslant 450°C$. This fact is indicated both by the continuous increase of the intensity of the CrL_α radiation an by the decrease in the FeL_α radiation intensity. These data are in good agreement with the transformation of the oxide layer structure at high temperatures (over 400°C)[20].

Sheets treated at different temperatures were bonded and tested, the results are tabulated in table XI.

In the temperature range studied, F_{max} increased continuously while $(EI)_S$ increased and then remained constant. A delamination along the oxide/adhesive interface was observed for specimens oxidized up to 450°C, whereas cohesive failure, within the oxide layer, was observed for higher temperatures. The failure locus changes occur at temperatures at which chromium diffusion is observed. This last phenomenon seems to be responsible for the localization of failure within the oxide layer.

Table XI. Treatment Temperature Effects on the Mechanical Properties of the System.

Treatment temperature (°C)	20	150	300	450	600	750
$(EI)_S$ (10^3 N.mm^2)	104	122	160	157	157	161
F_{max} (N)	15	18	22	23	24	30

CONCLUSION

The test used in this work undoubtedly appears to be a promising technique for measuring the practical adhesion in adhering systems. Surface analytical techniques such as LEEIXS, combined with an appropriate mechanical test provide chemical information on the adherend surfaces prior to bonding or after delamination. and therefore can allow a better understanding of the chemical parameters which influence practical adhesion.

REFERENCES

1. A.A. Roche, J.S. Solomon and M.J. Romand, in "Microscopic Aspects of Adhesion and Lubrication", J.M. Georges, Editor, Tribology Series 7, pp. 333-342, Elsevier, 1982.
2. A.A. Roche, A.K. Behme and J.S. Solomon, Inter. J. Adhesion Adhesives, 2, No. 4, 249 (1982).
3. R. Bador, M. Romand, M. Charbonnier and A. Roche, Adv. in X-ray Analysis, 24, 351 (1981).
4. A.A. Roche, M.J. Romand and F. Sidoroff, these proceedings, pp.19-30.
5. A.A. Roche, M. Charbonnier, F. Gaillard, M. Romand and R. Bador, Appl. Surf. Sci., 9, 227 (1981).
6. A. Roche, A. Cachard, R. Bador, F. Buiguez, M. Charbonnier and M. Romand, J. Microsc. Spectrosc. Electron., 2, 627 (1977).
7. R. Bador, Thesis, Lyon, 1980.
8. C.F. Feldman, Phys. Rev., 117, 455 (1960).
9. R.P. Frankenthal, J. Electrochem. Soc., 116, 580 (1969).
10. P. Berge, Etude de la Stabilité de la Passivité des Aciers Inoxydables en Solution Sulfurique, Publications Scientifiques et Techniques du Ministère de l'Air, p. 37, Paris, 1961.
11. M. Nagayama and M. Cohen, J. Electrochem. Soc., 110, 670 (1963).
12. P. Berge, C. R. Acad. Sci., Paris, 245, 1239 (1957).
13. M. da Cunha Belo, B. Rondot, F..Pons, J. le Héricy and J.P. Langeron, J. Electrochem. Soc., 124, 1317 (1977).
14. C. Hamilton, Appl. Polymer Symp., 19, 105 (1972).
15. P.M. Stifel, New Ind. Applications for Advanced Materials Techn., 19th National SAMPE Symp. Exhibition, 75 (1974).
16. D.M. Brewis, J. Comyn, B.C. Cope and A.C. Moloney, Polymer, 21, 1477 (1980).
17. J. Harwood and Schulman, Corrosion, 203 (1949).
18. J. Schultz and K.C. Sehgal, Proc. 13ème Congrès FATIPEC, EREC Ed., Cannes, p. 575, 1976.
19. A. Roche, Thesis, Lyon, 1983.
20. S. Storp and R. Holm, Surf. Sci., 68, 10 (1977).

SURFACE CHARACTERIZATION OF ANODIC OXIDES ON ALUMINIUM ALLOYS BY MEANS OF SURFACE POTENTIAL DIFFERENCE, SURFACE IMPEDANCE AND SURFACE MORPHOLOGY

A. Kwakernaak, R. Exalto and H.A. van Hoof

Fokker B.V., Technological Centre
P.O. Box 7600, 1117 ZJ Schiphol-Oost
The Netherlands

To obtain acceptable adhesion and durability of adhesive bonded structures, anodizing of aluminium alloys in chromic acid or phosphoric acid is commonly used in the aircraft industry. In order to find a non-destructive testing technique to inspect pretreated parts and to gain further understanding, we used some simple techniques to characterize surfaces.
Using Transmission Electron Microscopy (TEM) as a tool to characterize surface morphology showed that porous oxides gave the best adhesion quality. The Surface Potential Difference (SPD) measurement was found to be very sensitive to adsorbed dipole layers on surfaces. Surface Impedance (Z) measurements were introduced for determination of the sealing quality of anodic oxides, but can also be used to characterize non-sealed oxide layers. On the basis of experimental data obtained, the effects of several variables in the anodizing process on SPD and Z measurements and TEM morphology are discussed. Also the effects of ageing of anodic oxides by dipping in ambient temperature water are shown. None of the techniques discussed alone can fully characterize the anodic oxide; the methods used are supplementary to each other. SPD, Z, and surface morphology can be described as functions of all influencing parameters. With the use of such functions characterization nomograms of anodizing processes can be determined. Finally a new further refined impedance analysis is proposed as a technique to characterize more fully the anodic layer.

INTRODUCTION

Adhesive bonded aluminium structures have been used in the aircraft industry to an increasing extent for the last thirty years. During these years more and more knowledge and understanding of adhesive bonding has been gained by operational experience and experimental research work. Most of the problems encountered were at the interface between the metal surface and the adhesive and it was recognized soon that, in order to obtain optimal adhesion and durability, the surface treatment of the aluminium alloy was of utmost importance. Nowadays, it is widely accepted that adhesion as such can be attained by etching processes, such as the sodium dichromate-sulphuric acid etch or the chromic acid-sulphuric acid etch[1]. However, to obtain a more durable bonded structure a thicker oxide layer is needed. The chromic acid anodizing process without hot water sealing, as used in Europe since 1953, and the phosphoric acid anodizing, as used in the U.S.A. since 1974, are the two most commonly used processes in the aircraft industry.

Since the first introduction of adhesive bonding as a production process, much emphasis has been placed on the quality control aspects of bonding. In addition to inspecting the cohesion quality by nondestructive techniques, close control of the pretreatment variables has always been carried out. To ensure proper etching and anodizing of components, separate samples are pretreated together with the production components. After bonding, these samples are tested destructively and if their peel strength is sufficient the pretreatment is accepted. In order to support the empirical approach of the early days, and to find a non-destructive technique to inspect pretreated parts for surface quality , new ways to characterize surface quality have been explored. Today, several techniques for surface characterization are available. The methods range from simple physical measurements, such as Surface Potential Difference or Voltapotential, to sophisticated methods for surface analysis in ultra high vacuum such as X-ray Photo-electron Spectroscopy (XPS or ESCA). In this paper attention will be focussed on the three techniques which were introduced at Fokker in the sixties, viz:
- Surface morphology by electron microscopy
- Surface potential difference
- Surface impedance.

Morphology of Anodic Oxides

Transmission Electron Microscopy (TEM) has earlier been used to study the structure of anodic oxides[2,3]. Early TEM work carried out in cooperation between N.L.R., Fokker and T.N.O.[4] gave an indication of the importance of the oxide morphology on adhesion. From studies with variations in e.g. electrolytes,

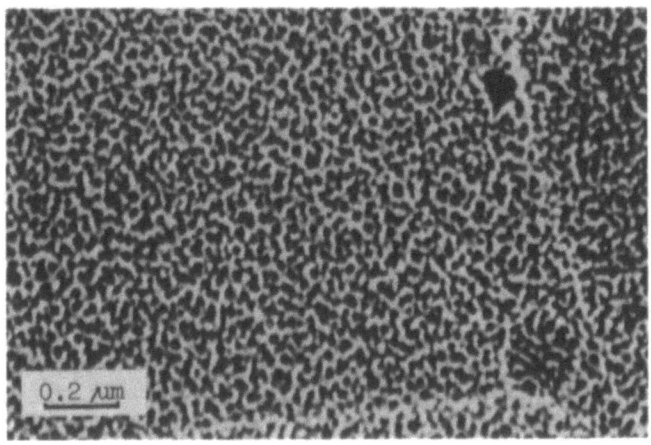

Figure 1. Scanning Electron Micrograph of CrO_3 anodized surface.

solution temperatures and treatment times, correlations were found between oxide morphology and the adhesion quality as measured by the peel test[1,5]. New high-resolution scanning electron microscopes provide even more detailed information on surface morphology. Using stereopairs Venables et al.[6] were able to develop oxide structure models. From both the earlier TEM work[7] and from micrographs using the same technique (Figure 1), we found an open pore structure for the European chromic acid anodizing process with a pore size of about 400 Å, and an average pore wall thickness of about 150 Å (Figure 2), which is quite different from

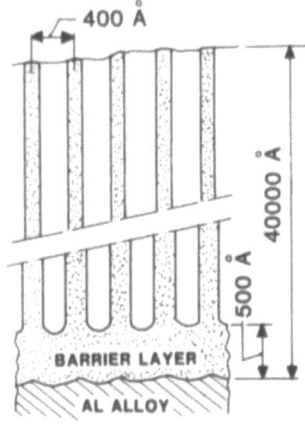

Figure 2. Schematic representation of CrO_3 anodic oxide.

the model proposed by Venables et al. We varied the anodizing parameters, but could not reproduce their results.

Surface Potential Difference of Anodic Oxides

The Surface Potential Difference (SPD) or Voltapotential is the difference in workfunction between a reference surface and the sample of interest. The extreme sensitivity of this technique to changes in the oxide and to the presence of contamination had been reported about 20 years ago[8]. The SPD can be measured with the vibrating capacitor technique known as the Kelvin method[9]. Based on this principle the Fokker Contamination Tester was developed[10]. With specimens etched in chromic acid-sulphuric acid solutions, correlations were found between SPD and bondability[1,10].

With anodic oxides the situation seems to be more complicated. Effects due to adsorption and desorption of water were monitored with SPD[10] and can be explained in terms of oriented dipoles or a residual space charge[11,12] in the barrier layer.

This effect is also illustrated in Figure 3, where experimental data on different anodic oxides (barrier layer and porous oxides) are shown. The effect of adsorption of methanol on the SPD change (Δ SPD) was measured and is plotted against the SPD measured after anodizing. The same effect was also measured for two

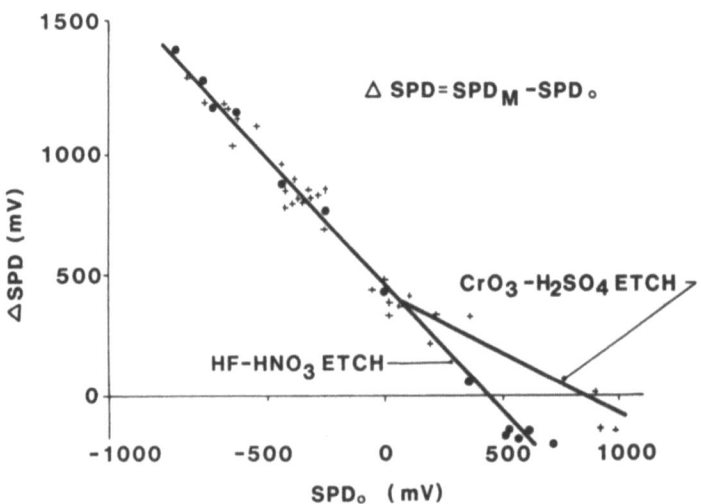

Figure 3. The effect of adsorption of methanol on the SPD of anodic oxides (● barrier layer oxides, + porous oxides).

etching solutions and is shown in the same diagram. It seems that when the residual space charge is large, it controls the adsorption. The influence of the surface morphology is seen only when the residual space charge is small (SPD $\gtrsim$ 0). This shows the sensitivity of the SPD measurement for adsorbed dipole layers on substrate surfaces, while it can also be used to detect organic contaminants[13] and changes in anodic oxides due to ageing.

Surface Impedance Measurements of Anodic Oxides

The surface impedance (Z) measurement of anodic coatings was introduced as a simple method to measure the quality of sealing[14] and nowadays is commonly used in industry[15,16]. The measurement is carried out with a test cell with a well-defined measuring area using 3.5% sodium chloride or 5% potassium sulphate as an electrolyte. The impedance or admittance of the oxide layer is generally measured at a frequency of 1000 Hz. Some instruments are also able to measure the dissipation or loss factor. The method was used in our laboratory to measure all kinds of oxide layers. The effect of different alkaline cleaning solutions prior to etching could be measured, and thus we could discriminate the aggressiveness of the alkaline cleaners[17].

THE INFLUENCE OF PARAMETER VARIATIONS ON SPD, Z AND MORPHOLOGY OF ANODIC OXIDES

Variations in aluminium alloy, pretreatment, anodizing process and post-anodizing treatment are considered to be of great importance for the oxide film properties. In order to determine the effect of some of these parameters on oxide layer properties, several experiments were carried out. SPD was measured using the Fokker Contamination Tester, which has a galvanic gold layer as a reference surface. The impedance was measured with a Radiometer vector impedance bridge at 1 kHz on an area of 12.9 mm^2 with a 5% NaCl solution as electrolyte. The structure of the anodic oxide was determined with transmission electron microscopy (TEM) which provides information about the porosity of the anodic oxide and by measuring the oxide layer mass (w), by stripping in a selective oxide stripping solution (CrO_3 20 g/dm^3, H_3PO_4 50 g/dm^3, 100°C, 5 min.).

The Effects of Pretreatment, Anodizing Temperature and Anodizing Solution

1100 aluminium alloy (99.0% Al) was subjected to the following treatments before chromic acid anodizing:
- solvent cleaning with methyl ethyl ketone,

- alkaline cleaning in Turco 4215 S for 15 minutes,
- alkaline cleaning followed by etching in a chromic acid-sulphuric acid solution.

Chromic acid anodizing was carried out in a fresh solution:

- Total chromic acid present as Cr^{6+}: 44 g/dm^3 CrO_3

and in an aged solution:

- Chromic acid present as Cr^{6+}: 40 g/dm^3 CrO_3;
  total present as Cr^{3+} and Cr^{6+}: equivalent to 81 g/dm^3 CrO_3.

Anodizing was carried out at temperatures of 30, 40 and 50°C, with voltage program for 40 minutes according to the Fokker standard process[7]. The results of SPD, Z and surface morphology are shown in Table I.

The SPD results show the interaction between pretreatment and anodizing temperature. The effect of the aged anodizing solution on SPD disappears at higher temperatures.

Table I. The SPD, Z and Surface Morphology After CrO_3 Anodizing of 1100 Aluminium Alloy.

		SOLVENT CLEANING		ALKALINE CLEANING		CrO_3-H_2SO_4 ETCHING	
ANODIZING TEMPERATURE	30°C						
	SPD (mV)	-155	-255	-220	-445	-385	-640
	Z (KΩ)	5.35	5.06	5.33	4.98	5.50	5.10
	40°C						
	SPD (mV)	-485	-370	-415	-345	-390	-370
	Z (KΩ)	4.78	3.53	4.81	3.63	5.00	3.70
	50°C						
	SPD (mV)	-310	-320	-270	-270	-215	-230
	Z (KΩ)	4.61	4.53	4.63	4.38	4.57	4.42
0.5 μm		FRESH	AGED	FRESH	AGED	FRESH	AGED
		ANODIZING SOLUTION					

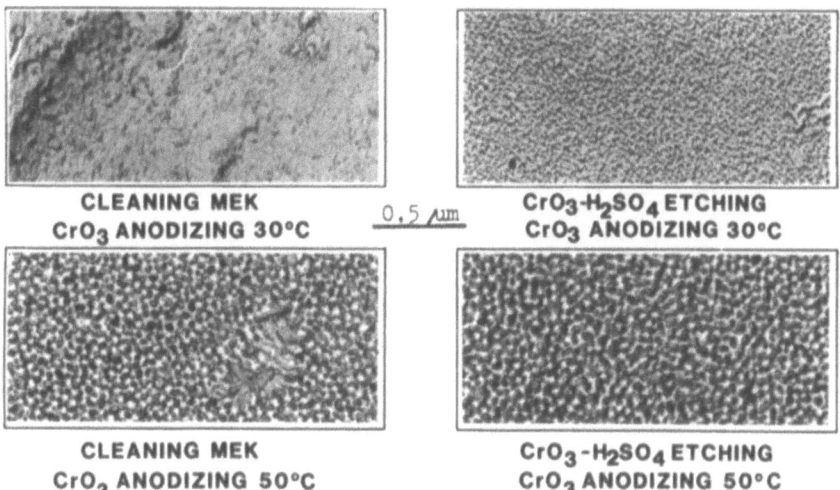

Figure 4. The effect of pretreatment and anodizing temperature on surface morphology (part of Table I).

Z decreases with increasing temperature and is not significantly influenced by the pretreatment. The surface morphology shows the effect of temperature and pretreatment and is shown in more detail in Figure 4.

The Effect of Ageing by Water Soaking

Specimens of 2024-T3 Alclad (Clad layer: 99.5% Al) chromic acid anodized were stored for different times up to 30 days in deionized water at room temperature. After removing from the tank, the specimens were dried at 60°C for 1 hour and SPD, Z, and the mass increase were measured. The results are shown in Figure 5. The SPD shows a rapid change to positive values after storage for a short time and a change in behaviour after 150 hours of exposure. The mass increases linearly with time. Z stays constant until 30 days when the specimen was visually corroded, which resulted in a lower impedance. The surface morphology after 70 hours did not show any change, but after 210 hours, the structure is changed (Figure 6).

As a reference, the morphology, SPD and Z are shown for a specimen after dipping for 15 minutes in 100°C water, which shows clearly a sealing effect. Figure 7 shows the results of Infrared Spectroscopy, using specular reflection, of chromic acid anodized surfaces.

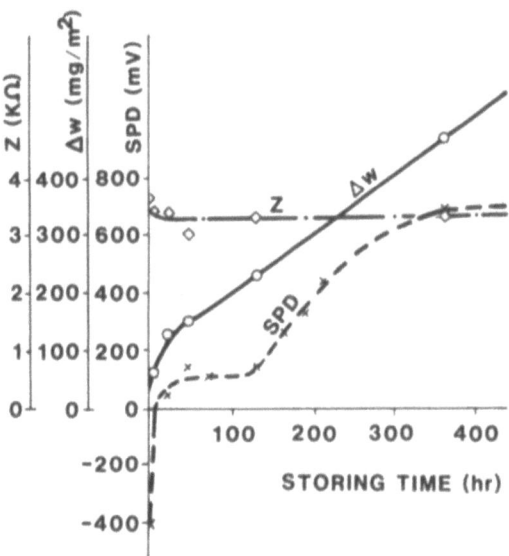

Figure 5. The influence of storing CrO₃ anodic oxides in 25°C water.

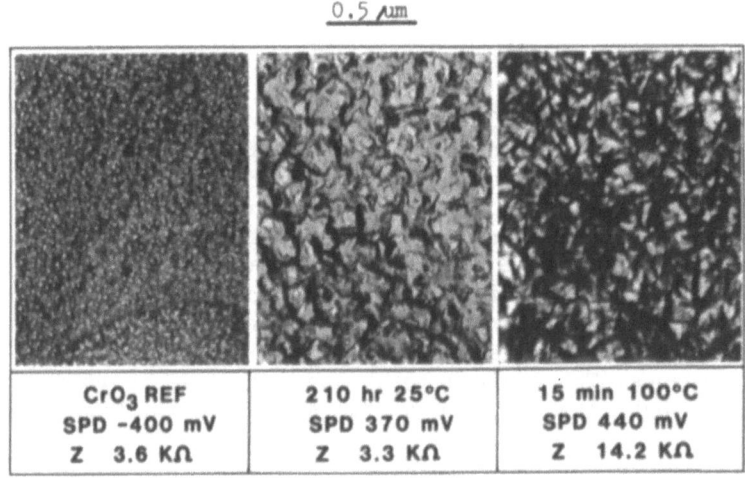

Figure 6. Electron micrographs of CrO₃ anodized surfaces after storing in water.

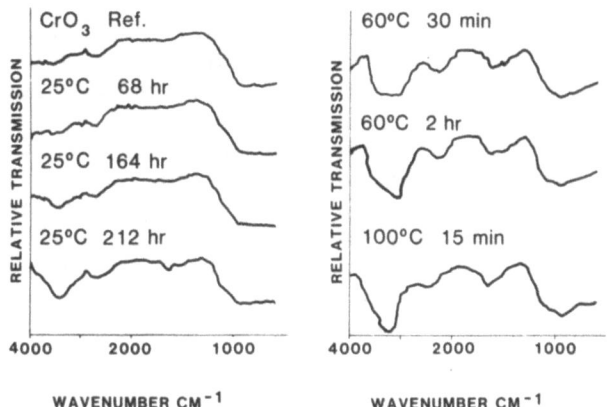

Figure 7. Reflection IR spectra of CrO_3 anodic oxides after storing in water.

The change in SPD after 150 hours corresponds with the appearance of a peak at 3400 cm^{-1}. The sealed surface morphology which appears after 15 minutes in 100°C water or after 30 minutes in 60°C water, results in a different peak in the IR spectrum at 3100-3600 cm^{-1}.

The Effects of Anodizing Time and Voltage

Chromic acid anodizing at a temperature of 40°C in a solution of 40 g/dm^3 CrO_3 was carried out with 2024-T3 Alclad specimens. The anodizing time and voltage were varied, viz. 5 and 15 minutes and 20 and 40 Volts. The results are shown in Table II. Phosphoric acid anodizing was carried out with specimens of the same alloy at a temperature of 25°C in a solution of 100 g/dm^3 H_3PO_4. Also anodizing time and voltage were varied, viz. 5 and 25 minutes and 10, 25 and 40 Volts. The results are shown in Table III.

Next to the SPD, Z and w values in Tables II and III, climbing drum peel test results using test specimens with a width of 25 mm bonded with a vinyl phenolic adhesive (Redux 775 L/P) are given. The electron micrographs (TEM) shown were taken using two different specimen preparation techniques; on the left-hand side by the carbon replica technique and on the right-hand side by the oxide transmission technique. The results of both the chromic acid and the phosphoric acid anodizing show generally a similar behaviour. The pore size increases with increase in voltage and the thickness of the porous layer (or layer mass) increases with time and voltage. The SPD decreases and Z increases with increase in voltage. At higher voltage Z decreases with increase in time. Only the effect of time on SPD shows a difference between chromic acid and phosphoric acid anodizing. With phosphoric acid the SPD did not show any effect, but with chromic acid the SPD increases with increase in time.

Table II. CrO_3 Anodizing of 2024-T3 Alclad.

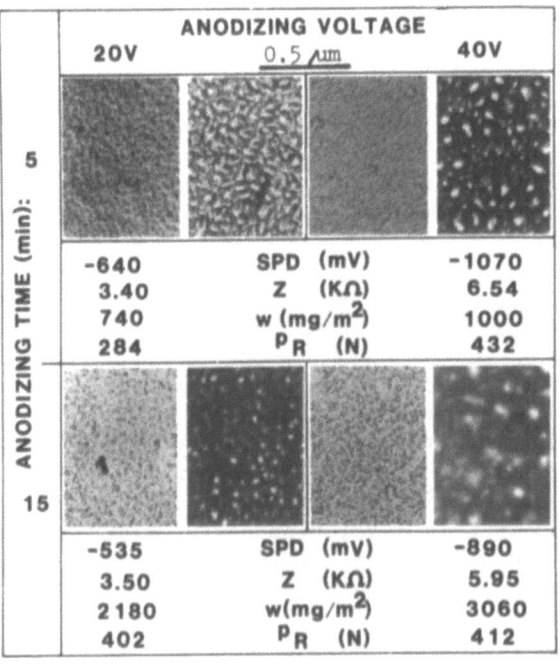

SURFACE CHARACTERIZATION OF ANODIC OXIDES

Table III H_3PO_4 Anodizing of 2024-T3 Alclad.

ANODIZING TIME (min)	ANODIZING VOLTAGE	10V	25V	40V
5	SPD (mV)	280	60	-380
	Z (KΩ)	1.69	3.47	5.13
	w(mg/m^2)	360	590	830
	P_R (N)	263	161	87
25	SPD (mV)	270	-150	-390
	Z (KΩ)	1.62	3.55	4.25
	w(mg/m^2)	650	1380	2050
	P_R (N)	370	263	159

0.5 μm

REPRESENTATION OF OXIDE FILM PROPERTIES BY CHARACTERIZATION NOMOGRAMS

From these experiments, it can be seen that several parameters largely influence the quality and structure of the anodic oxide. Next to the effects of pretreatment, anodizing voltage, time, temperature and oxide ageing by water adsorption, there are other influences such as the alloy composition, impurities in solutions, rinsing temperature and time, which will lead to changes in SPD, Z, layer mass and surface morphology. None of the discussed techniques alone can fully characterize the anodic oxide. The measuring methods are supplementary to each other.

To find the relationship between a measured quantity y and the investigated processing parameters, the method of regression analysis was used[18,19]. In general the dependent variable y is a linear function of m independent variables x_j which can be written as:

$$y_i = b_0 + b_1 x_{i1} + b_2 x_{i2} \ldots b_m x_{im} + e_i,$$

where the index i identifies a particular set of processing parameters, and e_i is the experimental error. The regression coefficients $b_0, b_1, \ldots b_m$ can be determined by least squares fitting. Also the covariance matrix can be developed, which shows the dependency of one parameter on the others and gives the standard error for each of the regression coefficients.

SPD, Z, oxide film mass, and surface morphology were determined in a chromic acid (50 g/dm^3 CrO$_3$) anodizing experiment on 2024-T3 Alclad material. The following parameters were varied: process time, 5, 25 and 45 minutes; voltage, 10, 25 and 40 Volts; and solution temperature, 30, 40 and 50°C.

With the measured data the following regression equations were deduced:

SPD = -590 - 13 t' - 521 V' + 36 T' + 42 t'V' + 79 V'T' (mV)

Z   = 3.28 - 0.31 t' + 1.69 V' - 0.10 T' - 0.31 t'V' (kΩ)

w   = 4174 + 3138 t' + 1876 V' + 2000 T' + 1547 t'V' + 1366 t'T' + 1139 V'T' + 929 t'V'T' (mg/m^2)

The parameters t', V', T' represent, respectively, the process time, anodizing voltage and solution temperature and are expressed as dimensionless parameters, where $-1 \leq$ t', V', T' ≤ 1.
The values of the multiple correlation coefficient (r) as well as the standard error of estimate (s) are:

r_{SPD} = 0,973   ; s_{SPD} = 115 mV

r_Z = 0,973   ; s_Z = 0,37 kΩ

r_w = 0,994   ; s_w = 490 mg/m^2

To allow a graphical characterization, nomograms can be developed with the regression equations, showing for a fixed value of one parameter the effect of the other parameters on SPD, Z and w. Figure 8 shows the characterization nomogram for the CrO$_3$ acid anodizing process at 40 V, showing the effects of time and temperature. For this nomogram, the change in surface morphology is also given by the electron micrographs in figure 9.

Such a characterization nomogram, developed for a certain production process, would allow the detection of processing deviations.

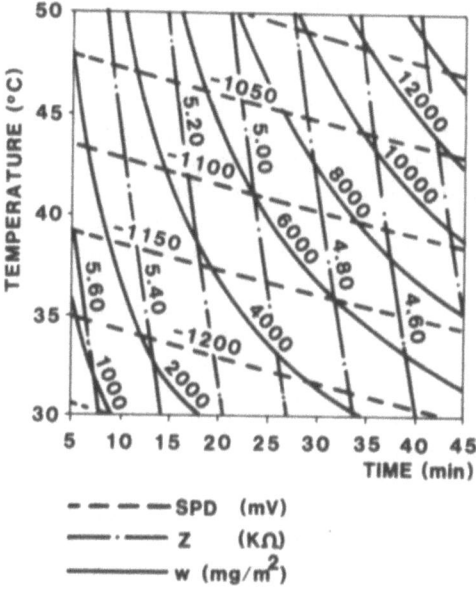

Figure 8. Characterization nomogram for CrO_3 (50 g/dm^3) anodizing at 40 V.

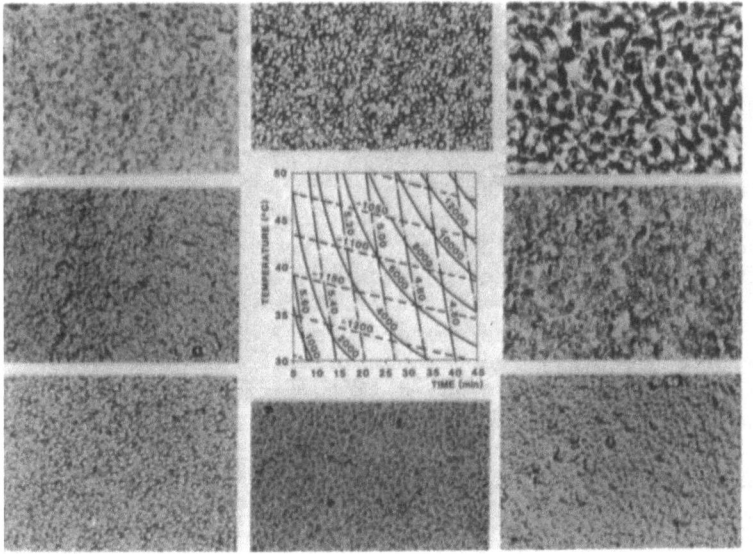

Figure 9. Electron micrographs of the characterization nomogram for CrO_3 anodizing at 40 V.

FREQUENCY ANALYSIS OF THE SURFACE IMPEDANCE

The surface impedance measurements mentioned before were all performed at a fixed frequency of 1 kHz. For a simple homogeneous oxide layer the capacitance and loss factor can be obtained from the magnitude and phase, respectively. These represent approximately the barrier layer of the porous oxides. Most oxide layers occurring in practice are far from being homogeneous. The porous layer is much thicker than the barrier layer (Figure 2), but has a high conductivity due to the penetrating electrolyte. The layers need not be homogeneous in a direction perpendicular to the surface nor in directions along[20] the surface. Nevertheless, the electrical properties of such complex systems for small AC signals can always be represented by an equivalent circuit consisting of simple components, like resistors and capacitors. The validity of such models can only be checked by measuring the frequency dependence of the surface impedance[21]. The range of frequencies should be as wide as possible. A computer program was developed to fit a complex equivalent circuit to the measured frequency response. Each component is switched in parallel or in series with the circuit formed by the preceding components. The values of some or all the components are considered to be adjustable parameters in a non-linear least squares fit. As the derivative of the complex impedance with respect to the parameters is known for each component, a local linear approximation can be made. The resulting linear least squares problem is then solved with singular value decomposition[22].

The number of iterations required depends strongly on the complexity of the model and on the ratio between the largest and smallest singular value. After convergence the program calculates the covariance matrix of the parameters. This is used mainly to evaluate the significance of changes in the parameters for different surfaces. In addition to normal resistors and capacitors a very useful component is the constant phase angle element[23] defined by its admittance: $Y = Q(i\omega)^P$. Here Q and P are both considered to be adjustable parameters. For P close to 1, this is a capacitor with a frequency-independent loss factor. Sometimes inductive elements are also needed[24]. Both infinite and finite Warburg impedances can also be treated.

Preliminary results on both CrO_3 and H_3PO_4 anodized surfaces indicate that indeed a good fit with correlations of the order of 95% can be attained over the frequency range $1-10^5$ Hz. The fit for the CrO_3 anodizing process, shown in Figure 10, is dominated by the capacitance of the barrier layer. After sealing in boiling water (Figure 11), the effect of the barrier layer remains but, for higher frequencies, is exceeded by the increased impedance of the sealed porous layer.

SURFACE CHARACTERIZATION OF ANODIC OXIDES   117

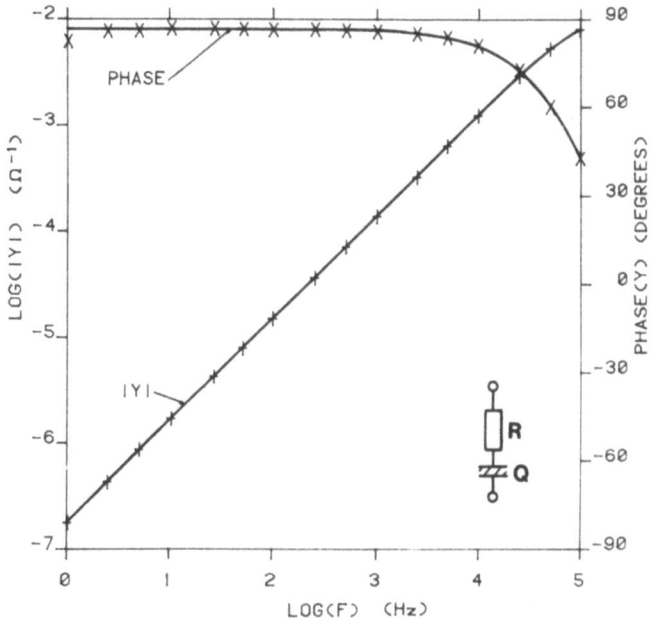

Figure 10. Surface admittance (measured values and calculated fit) of CrO_3 anodic oxide.

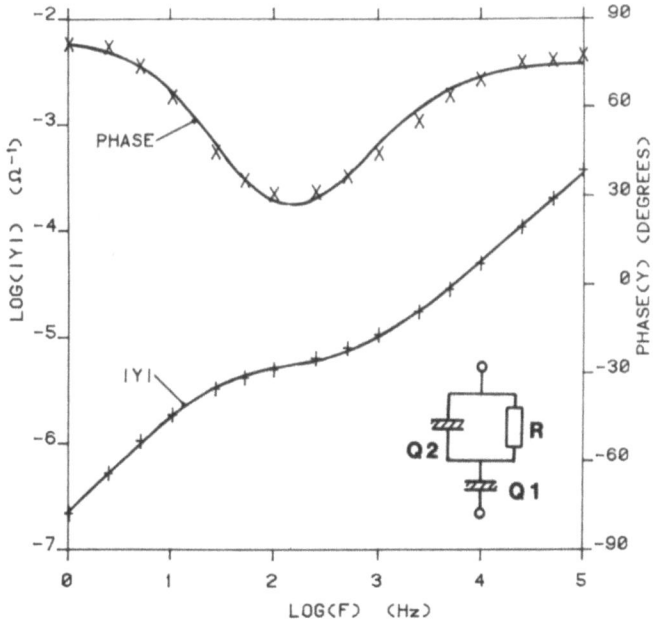

Figure 11. Surface admittance (measured values and calculated fit) of sealed CrO_3 anodic oxide.

CONCLUSIONS

The measurement of SPD, Z, w, and surface morphology can provide information about the formation history of oxide layers. The surface impedance is largely influenced by the anodizing voltage and slightly by the process time. The surface potential difference depends mainly on the anodizing voltage and to a lesser extent on the time and temperature. It also provides information about rinsing and ageing after anodizing. The oxide layer mass depends strongly on process time, temperature and anodizing voltage. The surface morphology is largely influenced by temperature and less by voltage and pretreatment.

In order to obtain a more accurate view of the effect of different parameters, it is possible to find equations describing the relationship between the measured quantities and the influencing parameters by regression analysis. With these equations characterization nomograms can be developed, representing the effect of process variations for a certain anodizing process.

Some important structural elements of the anodic oxide can be described by an electrical analogue. This model can be developed by fitting the equivalent circuit to the measured surface impedance of the oxide over a wide range of frequencies.

REFERENCES

1. P.F.A. Bijlmer and R.J. Schliekelmann, SAMPE Quarterly 5, 13 (1973).
2. F. Keller, M.S. Hunter and D.L. Robinson, J. Electrochem. Soc. 100, 411 (1953).
3. H. Ginsberg and K. Wefers, Metall. 17, 202 (1963).
4. A. Hartman, N.L.R. Report TN M.2091 National Aerospace Laboratory the Netherlands, Amsterdam (1961).
5. P.F.A. Bijlmer, J. Adhesion 5, 319 (1973).
6. J.D. Venables, D.K. McNamara, J.M. Chen, T.S. Sun and R.L. Hopping, Appl. Surface Sci. 3, 88 (1979).
7. P.F.A. Bijlmer, Metal Finishing 70, 30 (April 1972).
8. A. Matting and K. Ulmer, Kautschuk und Gummi. Kunststoffe 16, 280, 387 (1963).
9. W. Thomson (Lord Kelvin), Phil. Mag. 5, 82 (1898).
10. P.F.A. Bijlmer in "Surface Contamination: Genesis, Detection and Control", K.L. Mittal, Editor, Vol.2, pp. 723-748, Plenum Press, New York, 1979.
11. C.E. Michelson, J. Electrochem. Soc. 115, 213 (1968).
12. A.T. Fromhold, Jr. and J. Kruger, J. Electrochem. Soc. 120, 722 (1973).
13. T. Smith and G. Lindberg, Surface Technol. 9, 1 (1979).

14. E.T. Englehart and D.J. George, Materials Protection 3, 24 (Nov. 1964).
15. H. Birtel and W. Leute, Aluminium 43, 93 (1967).
16. A. Ott, Aluminium 52, 491 (1976).
17. R.J. Schliekelmann, Non Destructive Testing 5, 79 (1972).
18. L.M. Rose, "The Application of Mathematical Modelling to Process Development and Design", Applied Science Publishers Ltd., London, 1974.
19. O.L. Davies, "The Design and Analysis of Industrial Experiments", Oliver and Boyd, London, 1967.
20. R. Leek and N.A. Hampson, Surface Technol. 7, 151 (1978).
21. J. Ross MacDonald and J.A. Garber, J. Electrochem. Soc. 124, 1022 (1977).
22. G.H. Golub and C. Reinsch in "Handbook for Automatic Computation, Vol. II: Linear Algebra", J.H. Wilkinson and C. Reinsch, Editors, pp. 134-151, Springer Verlag, 1971.
23. K.S. Cole and R.H. Cole, J. Chem. Phys. 9, 341 (1941).
24. H.J. de Wit, C. Wijenberg and C. Crevecoeur, J. Electrochem. Soc. 126, 779 (1979).

INVESTIGATION INTO THE EFFECT OF SURFACE TREATMENT ON THE WETTABILITY AND THE BONDABILITY OF LOW SURFACE ENERGY MATERIALS

J.P. Jeandrau

Etablissement Technique Central de l'Armement
Centre Mécanique-Chimie-Matériaux
16 bis, avenue Prieur de la Côte d'Or
94114 Arcueil Cédex, France

An experimental effort has been undertaken to examine the effect of surface treatment on various low surface energy thermoplastic materials to promote wettability and bondability of these substrates. Changes in wettability were followed by contact angle measurements. These measurements were correlated with the bondability of treated surfaces using a usual two-part epoxy adhesive. Air-plasma cleaning is a more effective treatment than the usual chemical-acid etching in increasing the surface energy of polymers to make them more wettable and bondable, except for polypropylene and for polytetrafluoroethylene. This work has enabled several satisfactory solutions for bonding these substrates by selecting suitable surface treatments and adhesives for each.

INTRODUCTION

The continuing increase in the use of thermoplastic materials has led to investigate the different means of joining them together. The use of adhesives offers many advantages when compared to other more conventional methods such as welding, riveting and bolting. A necessary, though sometimes insufficient, requirement for developing strong adhesive joints is the establishment of intimate molecular contact at the interface. Many theories or the mechanisms of adhesion are found in the literature[1,2,3,4]. For adhesive joints, the adsorption and wetting theory is the most generally accepted one. Wetting may be quantitatively defined by reference to a liquid drop resting in equilibrium on a solid surface.

The main difficulty for bonding thermoplastic materials is due to their low surface free energies. Zisman and co-workers[5,6] established that, for low energy solids and a series of liquids, a rectilinear relationship frequently existed between the cosine of the contact angle, $\cos \theta$, and the surface tension of the wetting liquid, γ_{LV}. He defined the critical surface tension of wetting, γ_c, as the value to which γ_{LV} tends as $\cos \theta$ approaches unity, i.e. as θ approaches zero degree. Then he used the critical surface tension to characterize and compare the wettability behavior of low-energy surfaces.

Sharpe and Schonhorn[7] proposed that the single most important factor influencing adhesive joint strength is the ability of the adhesive to spread spontaneously on the substrate. They developed a criterion for the case when the adhesive will spontaneously spread on the substrate and, using Zisman's critical surface tension concepts, proposed that the γ_c of the adhesive must be less than or equal to that of the substrate. When we look at the values of γ_c given in Table I, it becomes easy to understand why thermoplastic materials such as polyethylene and polytetrafluoroethylene are difficult to bond with usual adhesives.

However, other ways to characterize surfaces of low-energy materials exist and are reported in the literature[8,9]. In these two articles, Mittal has reviewed and discussed the relationship between γ_c or other surface energetic parameters and joint strength.

The purpose of this work was to investigate different surface treatments for increasing the wettability of seven difficult-to-bond substrates. Various surface treatments of polymers are reported in the literature[10]. These surface treatments include the usual chemical acid-etching and also air-plasma cleaning. This last surface treatment had been reported by Schonhorn and Hansen[11] to be highly effective for the surface preparation of low surface energy polymers to enhance adhesive bonding. Essentially, the technique consists of exposing the polymer surface to a gas plasma at a reduced pressure generated by an electrodeless glow discharge.

Table I. Values of Critical Surface Tensions of Polymers at Room Temperature.

Solid Surface	Critical Surface Tension * γ_c (m N m^{-1})
Polytetrafluoroethylene	18.5
Silicone rubber	22.0
Polyurethane	29.0
Polyethylene	31.0
Polystyrene	32.8
Polychloroprene	38.0
Polymethylmethacrylate	39.0
Nylon 11	42.0
Nylon 6.6	42.5
Phenol-resorcinol adhesive	52.0
Urea-formaldehyde resin	61.0

* Values from reference [3]

The change in surface energy resulting from surface treatment was characterized by the change in the contact angle of water with the surface of the material being evaluated.

The increase in the bondability of the substrates was characterized by the shear strength resulting from treated specimens bonded with usual adhesives.

The main object of this work was to select the most suitable surface treatments and adhesives required for developing strong adhesive joints.

EXPERIMENTAL

Substrates

Seven semi-crystalline thermoplastic substrates were selected because of their low energy surfaces:

Polytetrafluoroethylene	PTFE	(TEFLON, Dupont de Nemours)
Low density polyethylene	LDPE	(ARCODUR, Arcotec)
High density polyethylene	HDPE	(6M50/50, Hoechst)
Polypropylene	PP	(PPH 2210, Hoechst)
Polyoxymethylene	POM	(DELRIN 15, Dupont de Nemours)
Polyamide 6.6	PA 6.6	(ULTRAMIDEA, BASF)
Polyamide 11	PA 11	(BMNO RILSAN, Ato-Chimie)

Surface Treatments

The following surface treatments were selected to improve the wettability of each substrate:

<u>No treatment.</u> Samples were degreased by placing them in absolute ethanol in an ultrasonic cleaner for 15 minutes, then dried at 80°C for 30 minutes.

<u>Chromic sulfuric etching.</u> After the first operation of degreasing, samples were immersed in etching solution for one hour at room temperature. The samples were thoroughly rinsed in deionized water and dried at 80°C for 30 minutes.

The etching solution was composed of:

- 2000 ml sulfuric acid (1.84 sp. gr.)
- 184 g potassium dichromate
- 294 ml deionized water.

<u>Naphthalene-sodium etching.</u> This surface treatment is specific and recommended for polytetrafluoroethylene. After degreasing in pure ethanol, samples were immersed for 48 hr at room temperature in the following etching solution:

- 500 ml tetrahydrofuran
- 128 g naphthalene
- 24 g sodium

Then samples were thoroughly rinsed in tetrahydrofuran and in deionized water and dried at 80°C for 30 minutes.

Air-plasma cleaning. The equipment used for generating the gas plasmas consisted of a vacuum system capable of achieving 10^{-4} to 10^{-5} torr pressure and a tubular reactor (200 mm high and 400 mm in diameter). A high voltage was applied across the cathode (a 200 mm long and 30 mm in diameter aluminium cylinder) and the top and bottom of the reactor. After degreasing in absolute ethanol, samples were placed in the apparatus, at a 200 mm constant distance from the cathode, and continuously exposed to the gas plasma. During the 15-minute air-plasma treatment the power intensity and gas pressure were nominally 50 W and 5.10^{-2} torr After the air-plasma treatment was completed, the samples were immediately bonded (15 minutes maximum after treatment). Some work was done to optimize the plasma treatment process, and the effect of different parameters on wettability and bondability were evaluated: plasma atmosphere, exposure time, power intensity, gas pressure and the elapsed time between the end of the treatment and the bonding operation. The operating conditions given above were selected with respect to these preliminary results on polyethylene substrate.

Measurement of Contact Angles

Distilled water was used to determine the contact angles.

Constant volume droplets of 0.5 µl were delivered from the stainless steel tip of a 5 µl syringe onto the substrate. The contact angle of the droplet was taken one minute after deposition using a KERNCO goniometer. This interval was sufficient for the droplet to attain an equilibrium contact angle θ, yet short enough for negligible evaporation loss. Each contact angle quoted is the mean of at least four measurements on droplets placed on the sample.

Evaluation of Adhesive-Bonded Joints

Bond strengths were determined using 10 mm thick thermoplastic substrate shear specimens bonded as shown in Figure 1. The bonded area was 625 mm².

Each shear bond strength quoted is the mean of at least ten specimens tested.

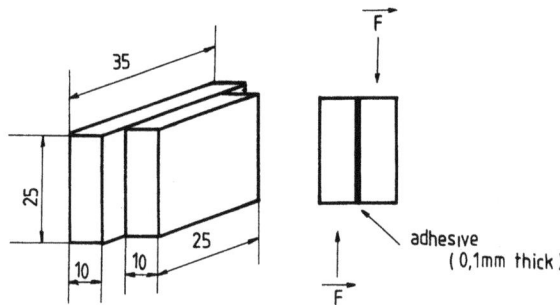

Figure 1. Single lap shear test specimen.

Adhesives

Four structural adhesives were selected for bonding different substrates.

- EC 2216 B/A is a two-part epoxy adhesive curing at room temperature (3M Company)

- AW 116/HV 953U is a two-part epoxy adhesive curing at room temperature (CIBA GEIGY)

- UK 8212/UK 5400 is a two-part polyurethane adhesive curing at room temperature (HENKEL)

- Multibond 329 is a toughened acrylic adhesive curing at room temperature with the initiator No. 738 (LOCTITE)

RESULTS AND DISCUSSION

Table II summarizes the effect of various surface treatments on contact angles and shear strengths of adhesive bonded joints for the seven low-energy thermoplastic substrates investigated.

The measured contact angles of untreated substrates are in good agreement with those found in the literature (94° for low density polyethylene[12], 108° for polytetrafluoroethylene[12] and 90° for polypropylene[13]. Three important conclusions may be drawn from the test results:

1) For untreated surfaces, the shear strengths increase as the contact angles decrease (i.e. as the substrate surface energies increase). It clearly shows why untreated polyamides give stronger adhesive bonded joints than polyolefins which have lower surface energies (when the substrates are bonded with the same adhesive under the same conditions).

2) The second fact observed is that the usual chromic sulfuric etching surface treatment leads to a weak improvement of the wettability and bondability of low energy surface thermoplastic materials. With this surface treatment, shear strengths are increased by only a factor of two or three in the best cases.

3) The third conclusion is that the air-plasma cleaning is very effective in increasing the surface energy of polymers (contact angles decrease and tend to zero degree) to make them more wettable except for polytetrafluoroethylene. In five cases, wettability and bondability are enhanced by air-plasma treatment. For these five substrates, contact angles decrease from more than 70 degrees to less than 17 degrees and shear strengths increase from very low values to more than 7 MPa. So air-plasma treatment seems to be a very effective treatment to improve wettability and bondability of the five following thermoplastic materials: polyethylene (low density and high density), polyoxymethylene and polyamides (6.6 and 11). However, exceptions are observed for polytetrafluoroethylene and polypropylene. For the first (PTFE), air-plasma treatment improves neither wettability (contact angles increasing) nor bondability (shear strength staying constant and very low, i.e. 1.1 MPa). The only efficient surface treatment to improve wettability and bondability of polytetrafluoroethylene is the naphthalene sodium etching. For polypropylene, the air-plasma treatment increases the wettability of the substrate but surprisingly the bondability is not improved.

The understanding of the mechanism of plasma surface treatment in enhancing the adhesive properties of polymer surfaces is still a controversial subject[13,14,15,16,17,18]. The objective of our investigation was to demonstrate a trend in improvement in the bond strength between thermoplastic substrates. As plasma treatments are reported to be efficient surface treatments for polymer surfaces, we have simply used this technique without attempting to

Table II. Effect of Surface Treatment on Wettability (Contact Angles) and Bond Shear Strength of Low Energy Thermoplastic Substrates.

Thermoplastic Substrate	Surface Treatment (a)	Contact Angle	Shear Strength (b) (MPa)	Mode of failure (c)
Polytetrafluoroethylene PTFE	NT	107°	1.1	A
	APC	120°	1.1	A
	NSE	0°	6.3	C*
Low density polyethylene LDPE	NT	93°	1.6	A
	CSE	74.5°	3.4	A
	APC	0°	8.3	C
High density polyethylene HDPE	NT	86°	1.6	A
	CSE	71°	3.5	A
	APC	0°	8.3	C
Polypropylene PP	NT	92°	1.3	A
	CSE	90°	3.6	A
	APC	10°	1.8	A
Polyoxymethylene POM	NT	82°	2.7	A
	CSE	89°	4.0	A
	APC	17°	8.9	C
Polyamide 6.6 PA 6.6	NT	76°	5.2	A
	APC	0°	10.8	C
Polyamide 11 PA 11	NT	71°	4.1	A
	APC	0°	7.6	C

a) NT: No treatment; CSE: chromic-sulfuric etching
APC: air-plasma cleaning; NSE: naphthalene sodium etching

b) Shear strengths were determined on ten test specimens bonded with the EC 2216 B/A two-part epoxy adhesive.

c) A: interfacial failure
C: cohesive failure in the adhesive
C*: the failure occurs in the substrate in the surface region.

The locus of failure was determined by unaided eye. However, we have to be careful because many times, a failure may look interfacial by visual inspection, but the sophisticated analysis (eg, by ESCA) shows that a thin layer of adhesive is left behind on the substrate. The question of locus of failure and its precise determination by surface analytical techniques has been widely discussed in the literature[19,20,21].

contribute to a better understanding of the physical or chemical modification of the substrate surfaces. The fact that this surface treatment did not improve the bondability of polypropylene and polytetrafluoroethylene under our experimental conditions does not mean that it is inefficient on these substrates under all conditions. As a matter of fact, Tira[13] reported that polypropylene samples treated in an oxygen plasma at 100 W for 30 minutes could be stored after treatment over a period of two weeks and then successfully bonded with an RTV silicone.

The shear strength results in Table III confirm the inefficiency of plasma treatment to improve the bondability of polytetrafluoroethylene by adhesives from different chemical families and of different rigidities. These results confirm also

Table III. Shear Strength of Polytetrafluoroethylene Bonded to Polytetrafluoroethylene.

Adhesive	Surface Treatment (a)	Shear Strength (b) (MPa)	Mode of Failure (c)
EC 2216 A/B	NT	1.1	A
Epoxy	APC	1.1	A
	NSE	6.3	C*
UK 8812/UK 5400	NT	1.1	A
Polyurethane	APC	1	A
	NSE	8.4	C*
Multibond 329	NT	0.7	A
Acrylic	APC	1.5	A
	NSE	5.4	C*
AW 116/HV 953U Epoxy	APC	1.5	A

a) NT: No treatment; CSE: chromic-sulfuric etching
   APC: air-plasma cleaning; NSE: naphthalene sodium etching

b) Shear strengths were determined on ten test specimens

c) A: interfacial failure
   C: cohesive failure in the adhesive
   C*: the failure occurs in the substrate in the surface region.

Table IV. Shear Strength of Low Density Polyethylene Bonded to Low Density Polyethylene.

Adhesive	Surface Treatment (a)	Shear Strength (b) (MPa)	Mode of Failure (c)
EC 2216 A/B	NT	1.6	A
Epoxy	CSE	3.4	A
	APC	8.3	C
UK 8212/UK 5400	NT	1.8	A
Polyurethane	APC	7.5	C
Multibond 329	NT	1.0	A
Acrylic	APC	3.2	A
AW 116/HV 953U	CSE	6.0	A 50% C 50%
Epoxy	APC	9.0	C

Table V. Shear Strength of High Density Polyethylene Bonded to High Density Polyethylene

Adhesive	Surface Treatment (a)	Shear Strength (b) (MPa)	Mode of Failure (c)
EC 2216 A/B	NT	1.6	A
Epoxy	CSE	3.5	A
	APC	8.3	C
UK 8212/UK 5400	NT	1.8	A
Polyurethane	APC	9.3	C
Multibond 329	NT	1.2	A
Acrylic	APC	3.1	A
AW 116/HV 953U	NT	1.7	A
Epoxy	APC	8.4	C

a) NT: No treatment; CSE: chromic-sulfuric etching
   APC: air-plasma cleaning; NSE: naphthalene sodium etching

b) Shear strengths were determined on ten test specimens

c) A: interfacial failure
   C: cohesive failure in the adhesive

that the naphthalene sodium etching is the only efficient surface treatment for bonding polytetrafuoroethylene. This last surface treatment leads to shear strengths greater than 5 MPa with the different adhesives used. All the failures occur in the polymer surface region because of the decrease in the cohesive strength of the surface layer due to the chemical etching.

Tables IV and V summarize the effect of changing various parameters (adhesives and surface treatments) on polyethylene shear strengths (low density and high density respectively). These results show and confirm that air-plasma cleaning is the only efficient surface treatment for bonding polyethylene. Usual adhesives (epoxy and polyurethane) lead to cohesive failures (shear strength greater than 7.5 MPa) on air-plasma treated samples. However, the acrylic adhesive does not adhere well either on low density polyethylene or on high density polyethylene even after air-plasma treatment.

Table VI summarizes the shear strength results of bonded polypropylene samples. This table shows that all failures are interfacial in nature irrespective of the adhesive used or surface treatment employed. These results confirm that the surface treatments investigated on polypropylene do not improve the bondability of polypropylene substrate.

Table VII summarizes the shear strength results of bonded polyoxymethylene samples. These results confirm the efficiency of air-plasma cleaning to improve wettability and bondability of polyoxymethylene. Cohesive failures occur on air-plasma treated samples bonded with epoxy and acrylic adhesives. Interfacial failures at a lower strength level ($\simeq 6$ MPa) occur when a polyurethane adhesive is used on the treated surfaces.

Tables VIII and IX summarize the results of adhesive bonded polyamide shear strengths. These results show and confirm that strong adhesive bonded joints (shear strengths higher than 7 MPa) occur when epoxy and polyurethane adhesives are used on plasma treated polyamide surfaces. However, the acrylic adhesive leads to interfacial failures in every case. The shear strength is not affected by the surface treatment when samples are bonded with Multibond 329.

Table VI. Shear Strength of Polypropylene Bonded to Polypropylene.

Adhesive	Surface Treatment (a)	Shear Strength (b) (MPa)	Mode of Failure (c)
EC 2216 A/B Epoxy	NT	1.3	A
	CSE	2.8	A
	APC	1.8	A
UK 8212/UK 5400 Polyurethane	NT	1.0	A
	APC	2.7	A
Multibond 329 Acrylic	NT	0.8	A
	APC	1.8	A
AW 116/HV 953U Epoxy	APC	3.8	A

Table VII. Shear Strength of Polyoxymethylene Bonded to Polyoxymethylene.

Adhesive	Surface Treatment (a)	Shear Strength (b) (MPa)	Mode of Failure (c)
EC 2216 A/B Epoxy	NT	2.7	A
	CSE	4.0	A
	APC	8.9	C
UK 8212/UK 5400 Polyurethane	NT	2.9	A
	APC	4.8	A
Multibond 329 Acrylic	NT	1.9	A
	APC	6.1	C
AW 116/HV 953U Epoxy	APC	10.1	C

a) NT: No treatment; CSE: chromic-sulfuric etching
   APC: air-plasma cleaning; NSE: naphthalene sodium etching

b) Shear strengths were determined on ten test specimens

c) A: interfacial failure
   C: cohesive failure in the adhesive

Table VIII. Shear Strength of Polyamide 6.6 Bonded to Polyamide 6.6.

Adhesive	Surface Treatment (a)	Shear Strength (b) (MPa)	Mode of Failure (c)
EC 2216 A/B Epoxy	NT	5.2	A
	APC	10.8	C
UK 8212/UK 5400 Polyurethane	NT	3.0	A
	APC	8.8	C
Multibond 329 Acrylic	NT	6.0	A
	APC	6.3	A
AW 116/HV 953U Epoxy	NT	5.0	A
	APC	10.2	C

Table IX. Shear Strength of Polyamide 11 Bonded to Polyamide 11.

Adhesive	Surface Treatment (a)	Shear Strength (b) (MPa)	Mode of Failure (c)
EC 2216 A/B Epoxy	NT	4.1	A
	APC	7.6	C
UK 8212/UK 5400 Polyurethane	NT	4.6	A
	APC	10.4	C
Multibond 329 Acrylic	NT	2.9	A
	APC	3.4	A
AW 116/HV 953U Epoxy	NT	3.7	A
	APC	9.4	C

a) NT: No treatment; CSE: chromic-sulfuric etching
   APC: air-plasma cleaning; NSE: naphthalene sodium etching

b) Shear strengths were determined on ten test specimens

c) A: interfacial failure
   C: cohesive failure in the adhesive

Table X gives the most suitable surface treatments and adhesives required for developing strong adhesive joints on low energy thermoplastic materials bonded together. These results show that the difficult-to-bond substrates can be strongly bonded with usual adhesives when suitable prior surface treatment is carried out on the surface.

Table X. Suitable Surface Treatments and Adhesive Required for Developing Strong Adhesive Joints on Low Energy Thermoplastic Materials (Shear Strength > 7 MPa).

Thermoplastic Material	Suitable Surface Treatment	Suitable adhesive	Shear Strength (MPa)
PTFE	NSE	UK 8212/UK 5400 (polyurethane)	8.4
LDPE	APC	EC 2216 A/B (epoxy)	8.3
	APC	UK 8212/UK 5400 (polyurethane)	7.5
	APC	AW 116/HV 953U (epoxy)	9.0
HDPE	APC	EC 2216 A/B (epoxy)	8.3
	APC	UK 8212/UK 5400 (polyurethane)	9.3
	APC	AW 116/HV 953U (epoxy)	8.4
POM	APC	EC 2216 A/B (epoxy)	8.9
	APC	AW 116/HV 953U	10.1
PA 6.6	APC	EC 2216 A/B (epoxy)	10.8
	APC	UK 8212/UK 5400 (polyurethane)	8.8
	APC	AW 116/HV 953U	10.2
PA 11	APC	EC 2216 A/B (epoxy)	7.6
	APC	UK 8212/UK 5400 (polyurethane)	10.4
	APC	AW 116/HV 953U	9.4
POM*/PA6.6**	APC* APC**	AW 116/HV 953U (epoxy)	8.2
LDPE*/PA 11**	APC* APC**	UK 8212/UK 5400 (polyurethane)	8.4
PTFE*/PA 6.6**	NSE* APC**	UK 8212/UK 5400 (polyurethane)	8.0

CONCLUSION

Air-plasma treatment of surfaces is a versatile technique for improving the wettability and bondability of low surface energy thermoplastic materials. Five thermoplastic materials (LDPE, HDPE, POM, PA 6.6 and PA 11) could be treated for fifteen minutes at 50 W power successfully to render them wettable and bondable. Once the surface is treated in an activated gas plasma, it should be bonded within thirty minutes. Epoxy and polyurethane adhesives are suitable for developing strong adhesive joints (shear strength higher than 7 MPa) on these five air-plasma treated polymers).

Naphthalene-sodium etching is the only efficient surface treatment for improving the wettability and bondability of polytetrafluoroethylene. The polyurethane adhesive UK 8212/UK 5400 is a suitable adhesive for developing high shear strength (higher than 8 MPa) on thoroughly treated polytetrafluoroethylene to itself or to polyamide 6.6. No adhesive and (or) surface treatment among those studied in this work was suitable for developing strong adhesive bonded joints on polypropylene.

For a better understanding of the inefficiency of air-plasma cleaning for improving the bondability of polypropylene and polytetrafluoroethylene, additional work should be done to optimize the plasma cleaning process. A study of the chemical and physical effects of plasma treatment of polymer surfaces by modern surface analysis methods (ECSA or Auger Spectroscopies) will be undertaken in our laboratories.

REFERENCES

1. R. Houwink and G. Salomon, Editors, "Adhesion and Adhesives", Vol. 1 and 2, Elsevier, Amsterdam, 1965.
2. D. H. Kaelble, "Physical Chemistry of Adhesion", John Wiley and Sons, (Interscience), New York, 1971.
3. I. Skeist, Editor, "Handbook of Adhesives", Von Nostrand, New York, 1977.
4. A. J. Kinloch, J. Materials Sci., $\underline{15}$, 2141 (1980).
5. H. W. Fox and W. A. Zisman, J. Colloid Sci., $\underline{7}$, 428 (1952).
6. W. A. Zisman, in "Contact Angle, Wettability and Adhesion", ACS Adv. Chem. Ser., No. 43, p. 1 - American Chemical Society, Washington, D. C. 1964.
7. L. H. Sharpe and H. Schonhorn, ibid, p. 189.
8. K. L. Mittal, in "Adhesion Science and Technology", L. H. Lee, Editor, Vol. A, pp. 129-168, Plenum Press, New York, 1975.
9. K. L. Mittal, Polymer Eng. Sci., $\underline{17}$, 467 (1977).
10. J. A. Koutsky, in "Surface Contamination: Genesis, Detection and Control", K. L. Mittal, Editor, Vol. 1, pp. 351-357, Plenum Press, New York, 1979.

11. H. Schonhorn and R. H. Hansen, J. Polymer. Sci., **84**, 203 (1966).
12. D. K. Owens and R. C. Wendt, J. Appl. Polymer. Sci., **13**, 1741 (1969).
13. J. S. Tira, "Gas Plasma Treatment to Improve the Bondability fo an RTV Silicone to Foamed Polypropylene", Department of Energy, Contract Number DE-AC04-76-DP00613, BDPX-613-2224 (Rev), July 1979.
14. N. J. De Lollis, "Activated Gas Treatment of Silicone Surfaces", Sandia Laboratories, Albuquerque, New Mexico, report SAND 77 - 1256C, 1977.
15. D. Briggs, D. G. Rance, C. R. Kendall and A. R. Blythe, Polymer, **21**, 895 (1980).
16. M. R. Wertheimer and H. P. Schreiber, J. Appl. Polymer. Sci., **26**, 2987 (1981).
17. T. Riley, T. C. Mahuson and K. Seibert, "Investigation into the effect of plasma pretreatment on the adhesion of Parylene to various substrates", National Aeronautics and Space Administration, Lewis Research Center, Contract 8110-72-0435, pp. 95-114, 1981.
18. G. L. Flowers and J. L. Montague, "Evaluation of Cleaning Procedures and Bonding Materials for TPX Materials", Department of Energy, Washington, D. C. MHSMP-8140, Contract AC04-76DP00487, pp. 1-9, 1981.
19. W. L. Baun, in "Adhesion Measurement of Thin Films, Thick Films and Bulk Coatings", K. L. Mittal, Editor, pp. 41-53, American Society for Testing and Materials, Philadelphia, PA, 1980.
20. K. L. Mittal, Pure Appl. Chem., **52**, 1295 (1980).
21. W. L. Baun, in "Adhesion Aspects of Polymeric Coatings", K. L. Mittal, Editor, pp. 131-146, Plenum Press, New York, 1983.

THE EFFECT OF MOISTURE ON THE DIMENSIONAL STABILITY OF ADHESIVELY
BONDED JOINTS

E.J. Hughes, J. Boutilier, and J.L. Rutherford

Kearfott
A Division of the Singer Co.
Little Falls, NJ 07424

Torsion shear creep tests were made on two
bisphenol A-type epoxy adhesives as a function of
temperature, humidity, and applied stress. The
effects of varying the filler content of the two systems
were also determined. The relationship between creep rate
and stress was logarithmic. Although the presence of
moisture had a much greater effect in the unfilled
epoxy systems than in the filled systems, the
relationship between humidity effect and filler
content was not a simple one. The results are discussed
in terms of the residual stresses generated by
temperature changes and by moisture absorption.
The non-monotonic nature of the creep rate-stress
relationship previously shown to exist in a number
of filled epoxy systems is explained in terms of the
combined effects of thermal and moisture-induced stresses.

INTRODUCTION

The use of adhesive bonding in many critical inertial guidance applications where instrument stability is of paramount importance has given rise to concerns over the effects of environmental factors on the stability of adhesive bonds. Of particular interest is the generation of stresses in bonded adhesives as a result of moisture absorption[1-4] and differences in the coefficients of thermal expansion of the adhesive and adherends. One measure of the dimensional stability is the change in creep behavior which occurs as a result of either moisture absorption or thermally-induced stresses.

Previous work[5] has shown the effect of thermally-induced stresses on creep behavior. The work described in this paper demonstrates the effect of moisture absorption on creep behavior and an effort is made to describe the creep behavior of bonded joints in terms of the combined effect of thermally and moisture-induced residual stresses.

EXPERIMENTAL METHODS AND RESULTS

All creep tests were made using napkin-ring type torsion-shear specimens. The adherends were made of Ti-6Al-4V alloy.[6,7] Bondline thicknesses of about 0.003 inch were used for all specimens. The adhesives studied are listed in Table I, along with their cure cycles. Creep rates refer to the steady-state portion of the creep curve (strain vs time).

Unfilled Epoxy Systems

The effects of moisture absorption on the creep rates of two bisphenol-A type epoxy adhesives were measured as a function of the humidity level of the test environment. The bonded specimens were conditioned for 100 hours at the test humidity level and temperature prior to making creep tests as a function of applied stress. Typical results for tests made at $120°F$ are shown in Figures 1 and 2. At any given applied stress the creep rate increases with increasing humidity of the test environment.

Filled Epoxy Systems

To improve dimensional stability, most gyro-grade adhesives usually have fillers added. The Epotek epoxy system normally contains aluminum oxide as a filler, while the LCA4 system contains calcium carbonate. To determine the effects of these fillers on the susceptibility of the two systems to moisture absorption, creep tests were made on bonded epoxy adhesives which contained different amounts of filler. The results are summarized in Table II and plotted in Figures 3 and 4. In the

Table I – Bisphenol-A-type Epoxy Adhesives.

Adhesive	Hardener	Cure Cycle	Filler Material	Manufacturer
Epotek H-74	amine	116°F for 50 min / 212°F for 20 min	70 wt % Al_2O_3 44 µm size	Epoxy Technology, Inc.
Epotek H-74	amine	116°F for 50 min / 212°F for 20 min	35 wt % Al_2O_3 44 µm size	Epoxy Technology, Inc.
Epotek H-74	amine	116°F for 50 min / 212°F for 20 min	unfilled	Epoxy Technology, Inc.
LCA4	amine	212°F for 120 min	70 wt % $CaCO_3$ 14 µm size	Bacon Industries
LCA4	amine	212°F for 120 min	unfilled	Bacon Industries

Table II - Effect of Humidity and Filler Content on the Creep of Epoxy Adhesives.

Epoxy	Filler Content (percent by weight)	Relative Humidity (percent)	$d\frac{(\log \dot{\gamma})}{d\tau}$ x 10^{-4} $(sec^{-1} \cdot psi^{-1})$
Epotek H-74	0	55	8.62
	0	70	14.08
Epotek H-74	35	55	4.57
	35	70	5.78
Epotek H-74	70	55	3.28
	70	70	4.22
LCA4	0	40	2.51
	0	60	3.25
LCA4	70	40	2.37
	70	60	7.07

EFFECT OF MOISTURE ON DIMENSIONAL STABILITY

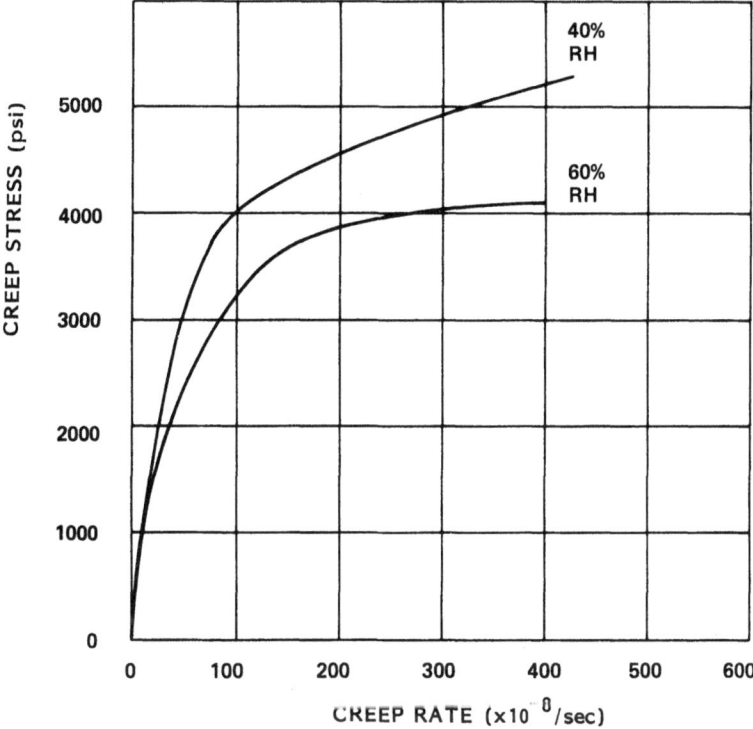

Figure 1 - Effect of Humidity on Torsion-Shear Creep of Unfilled LCA4 at 120°F.

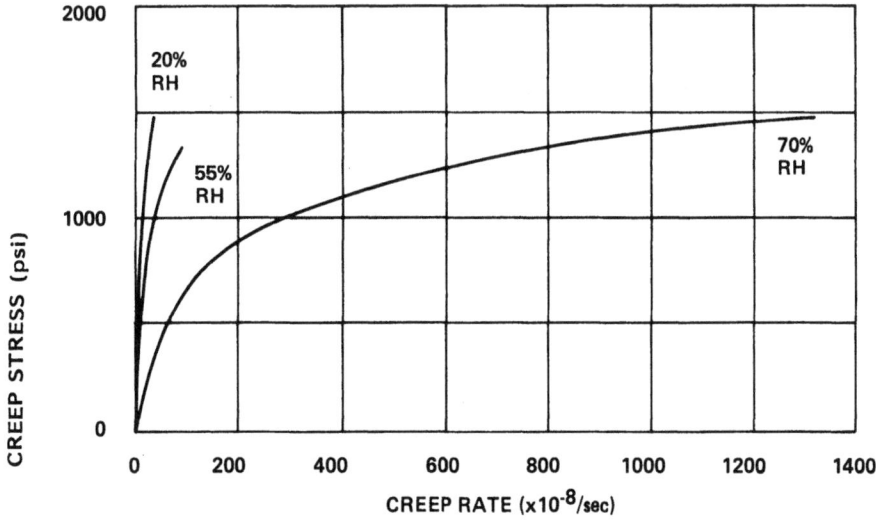

Figure 2 - Effect of Humidity on Torsion-Shear Creep of Unfilled Epotek H-74 at 120°F.

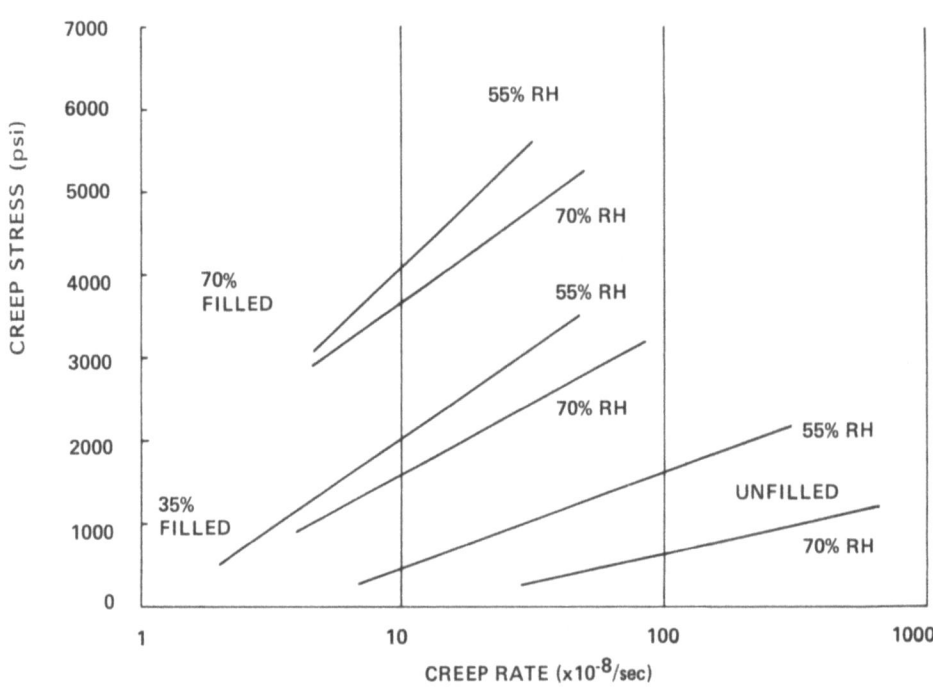

Figure 3 - Effect of Humidity on Creep of Epotek H-74 Systems.

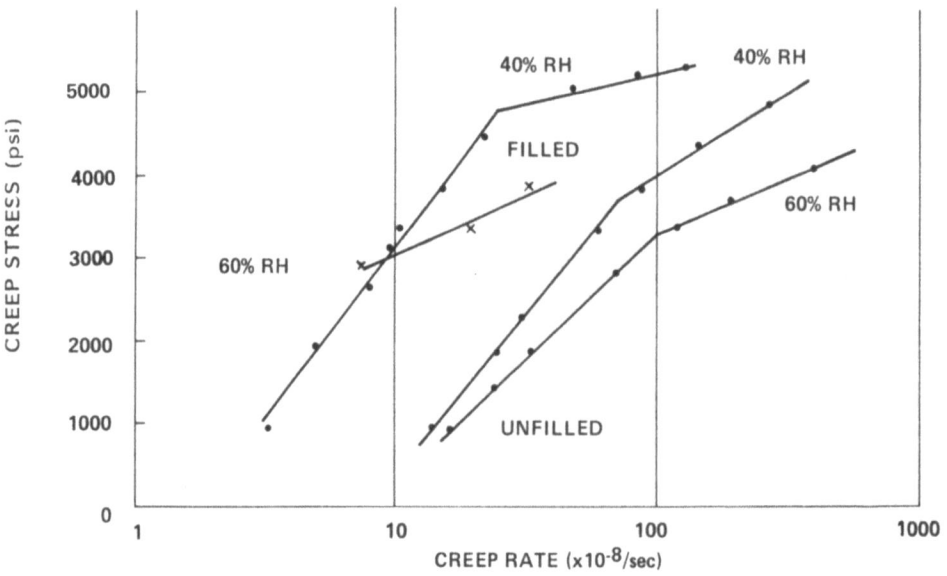

Figure 4 - Effect of Humidity on Creep of LCA4.

Epotek H-74 epoxies, changing the humidity level of the test environment was seen to have a much greater effect in the unfilled epoxy than in the two filled systems. The data for the filled LCA4 at the higher humidity level were anomalous. As shown in Figure 4, there was a distinct change in slope of the log (creep rate) versus creep stress curve at the higher stress levels in three of the four specimens tested. At the lower stress levels the slopes for these three specimens were similar.

Effects of Temperature

Creep tests were made with a filled LCA-4 adhesive at three temperatures ($90°$, $130°$, and $150°F$) using the same applied stress (3100 psi) and relative humidity (43 percent). The specimens were conditioned for 100 hours at the test temperature and humidity level before the creep tests were made. Typical creep curves are shown in Figure 5. The creep rates ($\dot{\gamma}$) obey an Arrhenius-type relationship as shown in Figure 6:

$$\dot{\gamma} = K\, e^{-Q/RT} \tag{1}$$

where

Q = activation energy for creep

R = gas constant

T = absolute temperature

K = constant

DISCUSSION

Effects of Filler Content on Moisture Absorption

Although the creep properties are dependent on the humidity of the test environment, the magnitude of the effect is not directly related to the amount of filler present. For example, in the Epotek systems the volume of epoxy resin in the bonds is in the ratio 100:65:30 for the unfilled, 35 percent, and 70 percent filled specimens, respectively. However, at constant humidity and creep stress level, the changes in creep rate are measured more in orders of magnitude than in these simple ratios. From Figures 3 and 4, the creep behavior can be described by an equation of the form

$$\log \dot{\gamma} = A\tau + B \tag{2}$$

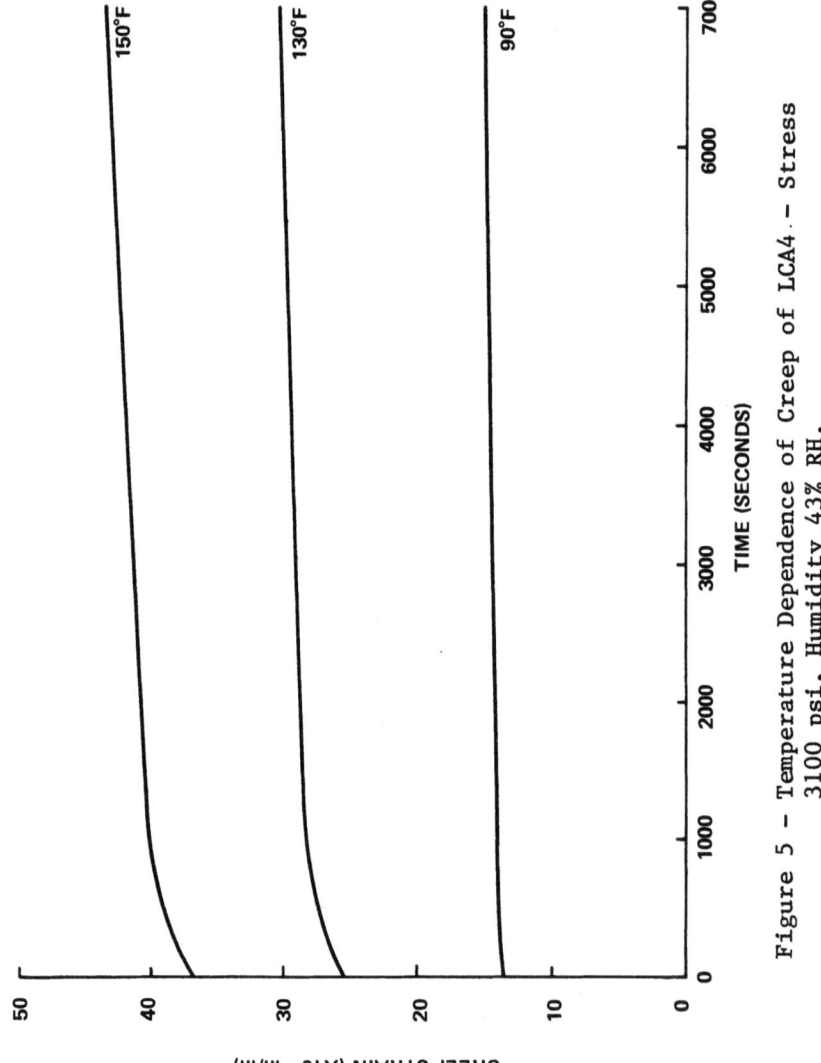

Figure 5 - Temperature Dependence of Creep of LCA4 - Stress 3100 psi, Humidity 43% RH.

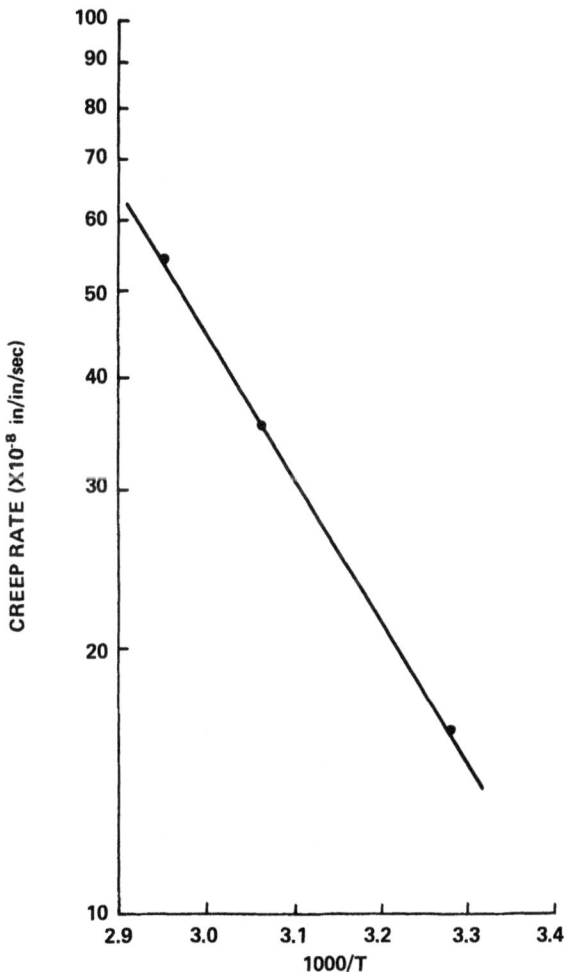

Figure 6 - Logarithm of Creep Rate vs 1/T for LCA4.

where

$\dot{\gamma}$ = creep rate

τ = creep stress

A, B = constants

The slopes of the log (creep rate) vs creep stress plots are given in Table II. It appears that the presence of a filler reduces the susceptibility of the epoxy adhesives to effects due to moisture absorption. This is not surprising since adding filler to the system reduces the amount of moisture absorbing epoxy in the adhesive bond. What is somewhat more surprising is that after an initial relatively large change in the creep behavior of Epotek H-74 on the addition of 35 percent by weight of filler, further increases in the filler content had a much smaller effect on the subsequent creep properties.

Effects of Temperature on Moisture Absorption

The creep stress obeyed an Arrhenius-type relationship as shown in Figure 6. The activation energy for creep is calculated to be 7250 cals/mole which is about three times higher than that for nylon epoxide adhesives measured at the same temperature but at a lower stress level.[8] The high activation energy for creep is consistent with the high strength, high modulus properties of filled gyro-grade adhesives.

Combined Effects of Thermal and Moisture-Induced Stress

The generation of residual stresses in adhesively bonded joints can arise from: differences in thermal expansion coefficients between the adhesive and adherend[5,9,10] and from moisture absorption by the adhesive.[11] The presence of internal or residual stresses in any material is always a potential source of dimensional instability. When external stresses are applied to the material during normal operation, their effects can be greatly modified by the presence of existing internal stresses. Previous work on several filled epoxy adhesive systems[5] has shown that the relationship between creep stress and creep rate is not monotonic. The type of relationship observed is shown schematically in Figure 7. This unusual relationship was interpreted in terms of the thermally-induced residual stresses which exist in bonded joints and which are relieved as a result of plastic flow during creep -- accounting for the initial decrease in creep rate with increasing stress. A finite element analysis of the residual stresses in the bonds[5] showed that although they were significant, their magnitude was too low to fully account for the observed behavior. In that treatment, no effort was made to include the effects of moisture-induced stresses.

These latter stresses can be determined in a manner similar to that of the thermally-induced stresses; provided that the humidity and temperature of the test environment is known together with the coefficient of moisture absorption of the epoxy at the test temperature. The latter quantity, which is equivalent to the coefficient of thermal expansion, is the change in length per unit length per percent relative humidity of the test environment. An apparatus has been developed by Singer-Kearfott[1] which enables this property to be measured as a function of temperature. Typical values for the LCA4 and Epotek H-74 systems are given in Table III. To determine whether the combined effects of thermal and moisture-induced residual stresses can account for the non-monotonic slope of the creep relationship shown in Figure 7, a finite element analysis was performed (using an isoparametric method) to determine the residual stresses arising from the combined effects of the two phenomena. The average values of the radial (σ_x), longitudinal (σ_y), and tangential (σ_z) components of the residual stress derived from the analysis are shown in Table IV for filled Epotek H-74. Included in Table IV are the components of the residual stress arising from the temperature difference between the cure and the test temperatures. The stresses were determined for a typical 0.003 inch bond line specimen at a test temperature of 120°F and relative humidity of 23 percent. The combined effects of the applied and the internal stresses can be represented by the octahedral shear stress, τ_{oct}, where

$$\tau_{oct} = 1/3 \ (\sigma_x - \sigma_y) + (\sigma_y - \sigma_z) + (\sigma_z - \sigma_x)^2 \qquad (3)$$
$$+ 6(\tau_{xy}^2 + \tau_{yz}^2 + \tau_{zx}^2)^{\frac{1}{2}}$$

If the non-monotonic shape of the creep rate curve in Figure 7 is due to residual stresses, τ_{oct} calculated at the threshold stress (τ_T in Figure 7) should be equal to the value of τ_{oct} calculated from that value of the applied stress above the knee of the curve which results in the same creep rate as the threshold stress (τ_A in Figure 7). At the threshold stress, both the residual and the applied stresses are operative, whereas at τ_A it is assumed the residual stresses have been relieved by plastic flow and only the applied stress is operative. In Equation 3, τ_{xy}, τ_{zx}, and the σ terms are contributed by the residual stresses while the τ_{yz} term is the applied stress. In napkin-ring type specimens τ_{xy} and τ_{zx} are either zero or close to zero and can be ignored. By substituting the appropriate stress values in Equation 3, the octahedral shear stress can be determined at τ_T and τ_A. The results are shown in Table V where the two values of τ_{oct} derived from the analysis are in reasonable agreement, lending support to the hypothesis that

Table III - Values of the Coefficient of Moisture Absorption for Epotek H-74 and LCA4.

Epoxy Identification	Test Temperature (°F)	Coefficient of Moisture Exp. (in/in/% RH x 10^{-6})
LCA4 (filled)	160	18.22
	140	14.11
	120	8.55
Epotek H-74 (filled)	160	51.9
	140	48.88
	120	45.0

Table IV - Average Values of the Residual Stresses in an Epotek H-74 Adhesive Bond Derived from the Finite Element Analysis. Humidity 23% RH, Temperature 120°F.

Adhesive	σ_x lb/in^2	σ_y lb/in^2	σ_z lb/in^2
Combined Thermal and Moisture-Induced Stresses	2385	-14	2327
Thermally-Induced Stress	1174	-11	1214

Table V - Comparison of Octahedral Shear Stresses at τ_T and τ_A in Epotek H-74.

Threshold Stress τ_T (psi)	Applied Stress τ_A (psi)	Octahedral Shear Stress at τ_T (psi)	Octahedral Shear Stress at τ_A (psi)
950	2660	1360	2170

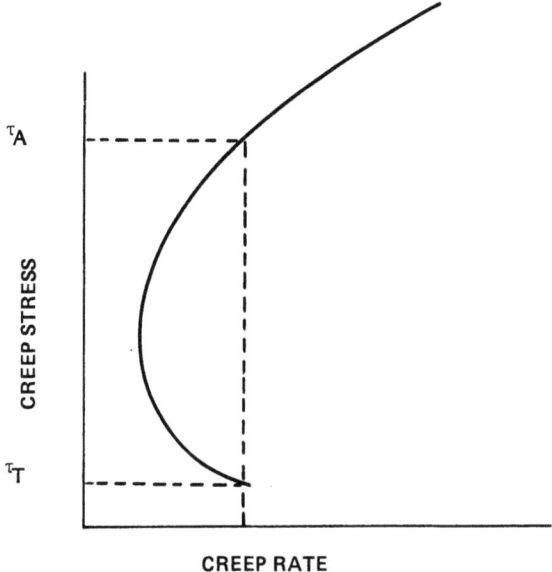

Figure 7 - Schematic of Creep Rate-Creep Stress Relationship in Filled Epoxy Adhesive Systems.

the non-monotonic slope of the creep curves is due to residual stresses. The values of τ_T and τ_A shown in Table V were obtained from Reference 5. In this analysis no consideration was given to other sources of residual stresses such as shrinkage stresses during cure. Inclusion of these additional stresses would result in even closer agreement in the values of τ_{oct} calculated at τ_A and τ_T.

CONCLUSIONS

1. The torsion-shear creep of filled and unfilled epoxy adhesives is increased by the presence of moisture.

2. Unfilled systems have a greater susceptibility to moisture-induced creep effects than do filled systems.

3. In the filled systems examined, increasing the amount of filler had a minimal effect on the creep behavior in humid environments.

4. The activation energy for creep at approximately 3000 psi stress was comparatively high, probably reflecting a high degree of cross linking and the innate stability of these epoxy systems.

5. The non-monotonic creep behavior of this type of epoxy adhesive system can be accounted for by considering the combined effects of thermally-induced, moisture-induced, and externally applied stresses.

ACKNOWLEDGEMENTS

The authors would like to express their appreciation to LaVerne Dunham who performed much of the creep testing described in this paper. We are also indebted to Joseph Rygelis who performed the finite element analyses. Finally, we would like to thank Professor Norman Brown, University of Pennsylvania, Philadelphia, Pennsylvania, for his many helpful ideas and suggestions during the course of this work.

REFERENCES

1. J. Boutilier, E.J. Hughes, and J.L. Rutherford, in "Proc. 26th National SAMPE Symposium," Los Angeles, 781 (1981).
2. Y. Weitsman, "Residual Stresses in Adhesive Joints," Materials Lab., Wright Patterson AF Base, Final Rep., AFWAL-TR-81-4121 (1981).
3. J.P. Sargent and K.H.G. Ashbee, J. Adhesion, $\underline{80}$, 175 (1980).
4. W. Althof, in "Proc. 11th National SAMPE Tech. Conf.," 309, Boston 1979.
5. E.J. Hughes and J.L. Rutherford, Materials Sci. Eng., $\underline{44}$, 57 (1980).
6. E.J. Hughes, J.L. Rutherford and F.C. Bossler, Rev. Sci. Instrum., $\underline{39}$ (5), 666 (1968).
7. F.C. Bossler, M.C. Franzblau and J.L. Rutherford, J. Phys. E, Ser. 2, $\underline{1}$, 820 (1968).
8. H. Shen and J.L. Rutherford, Materials Sci. Eng., $\underline{9}$, 323 (1972).
9. H. Dannenburg, Soc. Plast. Eng. J, $\underline{21}$ (7), 669 (1965).
10. W.T. Chen and C.W. Nelson, IBM J. Res. Develop. $\underline{23}$ (2), 179 (1979).
11. S. Gazit, J. Appl. Polymer Sci., $\underline{22}$, 3547 (1978).

THE DIMENSIONAL STABILITY OF EPOXY ADHESIVE JOINTS

J.P.Sargent

H.H.Wills Physics Laboratory
University of Bristol
Royal Fort, Tyndall Avenue
Bristol BS8 1TL, England

An interferometric technique which employs Moiré patterns has been used to study the deformation that occurs when an adhesive joint undergoes water uptake. It has been demonstrated that very large swelling stresses can be generated as a result of the inhomogeneous swelling. The magnitude and position of these stresses can be obtained by application of equations developed from beam and elasticity theory. A consequence of this analysis is the prediction of damped normal displacement waves in the cover slip in regions of negligible water concentration. This is observed experimentally. Measurements are reported on the irreversibility of dimensional changes when an adhesive joint is subjected to uptake and subsequent removal of the water responsible for swelling - on the assumption that the epoxy based adhesive is free from shear stresses during the cure - and on the influence of dissolved inorganic salts upon the swelling behaviour.

INTRODUCTION

The swelling inhomogeneity that occurs when a resin adhesive absorbs water may be conveniently demonstrated by making model joints consisting of an adhesive layer sandwiched between a rigid substrate and a flexible microscope cover slip. If such a joint is used as one of the components of an interferometer, the resulting interference pattern may be analysed to give information about the deformation of the cover slip and hence of the underlying adhesive. The experimental technique, reported by Sargent and Ashbee[1] makes use of photographs of the interference pattern to generate Moiré patterns in order to precisely follow the development of swelling in the adhesive layer.

EXPERIMENTAL METHOD

Figure 1a is a schematic diagram of an adhesive joint manufactured from a microscope cover slip as one adherend and a rigid block of glass as the other. Figure 1b shows a schematic representation of the apparatus and optical path. The adhesive joint is mounted in close proximity to an optical flat and when illuminated with monochromatic light a complex pattern of interference fringes is formed. These fringes precisely reflect the topography of the cover slip and therefore of the underlying adhesive. Should the shape of the cover slip change, as happens when the resin swells during water uptake, the pattern of interference fringes changes. If images of the interference pattern photographed during swelling are superimposed onto an image obtained at the start of the experiment, a Moiré pattern is generated, the development of which faithfully follows any changes in shape of the cover slip.

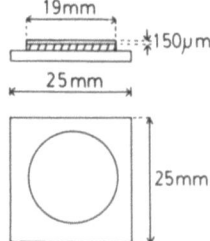

Figure 1a. Schematic diagram showing the test specimen.

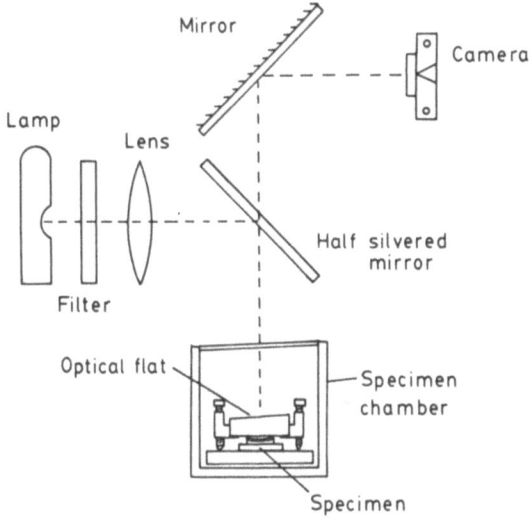

Figure 1b. Schematic of the optical path.

Application of the Moiré technique to these interference patterns is valid so long as there is no change in sign of the gradient of the surface, at the point of interest, between the initial photograph and any subsequent photographs.

Specimens usually took the form of 19mm diameter circular glass cover slips bonded to a rigid block of glass, some specimens however, were manufactured using cover slips which were square in outline, some in which the glass substrate was replaced by anodised titanium, and some in which all the glass components were replaced by mica.

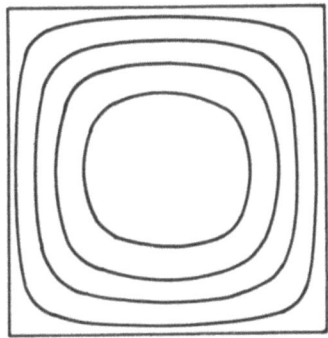

Figure 2a. Predicted water concentration contours.

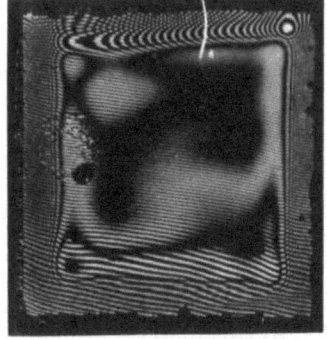

Figure 2b. Moiré contours of displacement.

RESULTS

1. Swelling Behaviour[1,2,3]

Figure 2a shows the predicted water concentration contours obtained by solution of Fick's diffusion laws for a specimen having a square outline. Figure 2b is a Moiré pattern photograph for an FM1000 specimen immersed in distilled water for 4 hours at 60C. The Moiré finges are the circumferential lines running around the rim of the specimen and which are superimposed upon the interference pattern. For small times, when both the displacement and penetration distance of the advancing water front is small, then the correlation between the experimentally derived displacements and the water concentration contours is good. Since the Moiré fringes are displacement contours an additional effect is apparent at longer times of immersion, namely the presence of a negative displacement ahead of the swelling front. This is seen as closed Moiré loops towards the centre in Figure 2b. This negative displacement is a consequence of the cover slip compensating for the swelling around the rim and as a result a complicated normal stress system is introduced into the adhesive of a magnitude which eventually leads to fracture of the cover slip.

Equations derived by Love[4] show that the normal displacement (w) of a thin plate rigidly supported at its edge and subjected to a normal pressure p is given by $p = -D\partial^4 w/\partial x^4$, where D is the flexural rigidity. Using this equation in conjunction with the equations of linear elasticity and beam theory to describe the deformation of the cover slip, it is possible to predict the behaviour of an adhesive joint when it undergoes water uptake. An important consequence of this analysis[3] was the prediction of damped waves of normal displacement in the region of negligible water concentration. The predicted waves are sketched in Figure 3a.

Figure 3b shows the results of a microdensitometer scan across a radius of the interference pattern for the specimen shown in Figure 2b. The specimen had undergone 4 hours exposure to distilled water at 60C. The initial depression and first wave are clearly resolved. The height of the first peak is approximately 1/26 of the initial depression (∇), which is in good agreement with the theoretical prediction of Figure 3a.

The normal displacement may be differentiated graphically in order to obtain the normal stress acting across the cover slip. A more satisfactory method, since it incurs less error upon differentiation, is to obtain a functional form for the normal displacement. Analysis is reported in reference 3 which enables this to be achieved, and Figure 4a shows a plot of $W^{IV}(x)$ (the

DIMENSIONAL STABILITY OF EPOXY ADHESIVE JOINTS 155

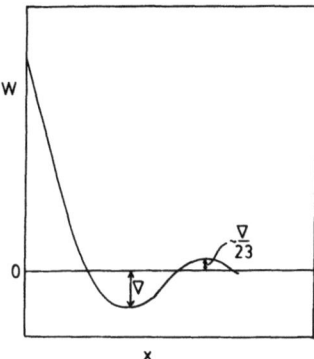

Figure 3a. The predicted damped normal displacement waves.

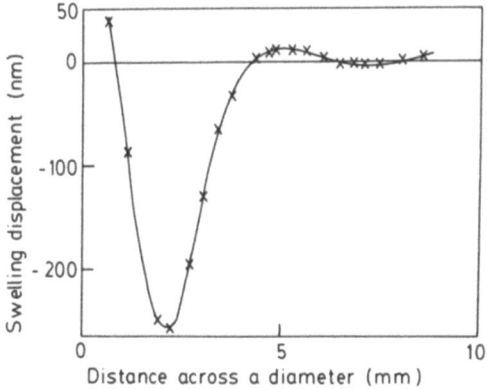

Figure 3b. The normal displacement for the specimen from Figure 2b showing the damped normal displacement waves.

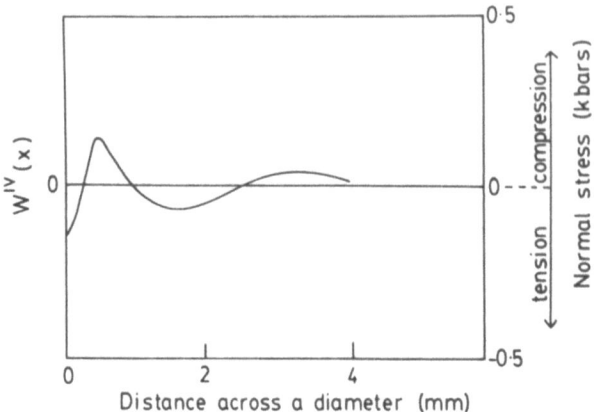

Figure 4a. The form of $W^{IV}(x)$ together with the normal stress distribution.

fourth differential) obtained using this technique. Physically the form of this may be understood by considering the time dependence of the pressures exerted on the cover slip by the waterfront as it progresses into the resin. This is shown in Figure 4b.

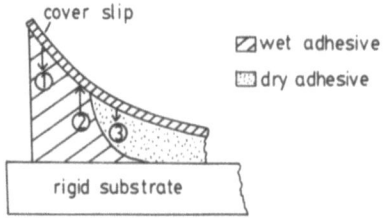

Figure 4b. A schematic explanation for the form of the $W^{IV}(x)$ curve in Figure 4a. In regions 1 and 3 the cover-slip is pulled down by dry and saturated regions of the resin. This counteracts the upward stress exerted in region 2 due to the pressure of the waterfront.

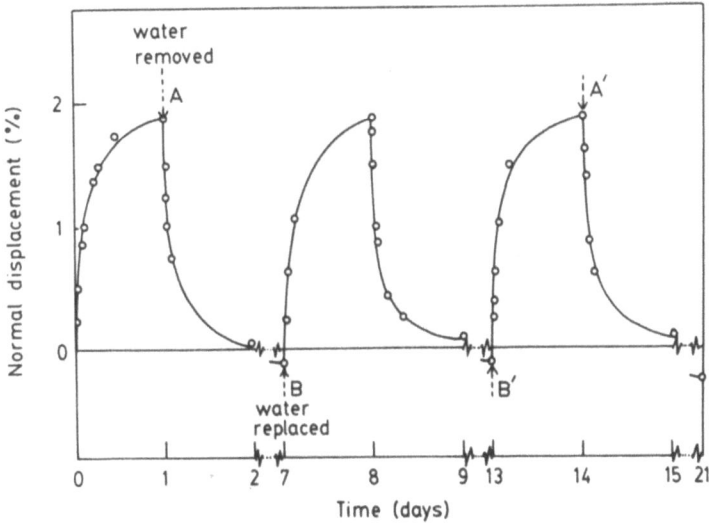

Figure 5. Repeated cycles of water uptake and expulsion.

2. The Irreversibility of Dimensional Change[5]

The Moiré technique may also be used to observe the behaviour of an adhesive joint when it is subjected to alternate periods of water uptake and expulsion. Figure 5 is a plot of the percentage displacement of a point immediately adjacent to the edge of an adhesive joint, and extended to cover several cycles of water uptake and expulsion. Creation of Moiré patterns from images photographed at identical amounts of swelling, and which are separated by one or more water uptake/expulsion cycles permit immediate detection of any irreversible changes in the adhesive film dimensions. Figure 6a is a Moiré pattern formed between photographs taken at the points referred to as A and A' in Figure 5 (representing states of maximum water uptake). Figure 6b is another pattern corresponding to B and B' in Figure 5 (representing states of minimum water uptake). Although the absence of any circumferential Moiré patterns in either of the Figures 6a or 6b demonstrates that repeated cycling has so far not led to any overall irreversible changes in the displacement field, small changes are evident. One such change is indicated by an arrow in Figure 6b and shows the appearance of a small Moiré loop indicating that a permanent irreversible local swelling displacement has occurred. Longer immersion times eventually lead to a complex debonding pattern. However, it was found that substitution of adhesive FM 1000 in place of FM 73M, led to a well defined debonding pattern in which the development of the debonding edge could be carefully monitored.

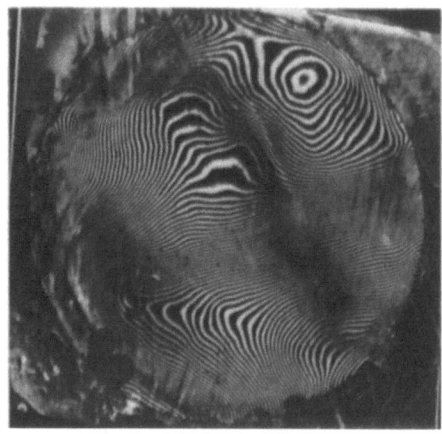

Figure 6a. Moiré image formed between photographs taken at points indicated as A and A' in Figure 5.

3. The Influence of Dissolved Inorganic Salts

When an adhesive joint manufactured from an anodised titanium/epoxy adhesive/glass cover slip sandwich is exposed to a solution that contains sodium or potassium ions, it shows a marked reduction in the magnitude of the swelling relative to that when exposed to distilled water[6]. Figure 7 shows the normal

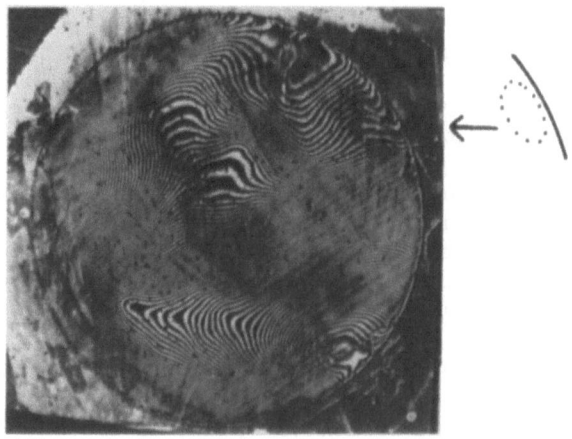

Figure 6b. Moiré image formed between photographs taken at points indicated as B and B' in Figure 5.

displacement across a diameter for three specimens which were immersed in distilled water and saturated solutions of NaCl and KCl solution for 71 hours at 81C. The data presented in Figure 7 corroborates earlier evidence (Ashbee and Wyatt[7], Farrar and Ashbee[8]) that the cured resins behave as semipermeable membranes. If for each of the immersion media the diffusion species in the adhesive is pure water, then when it reaches either the oxide surface, the glass surface, or impurities present in the adhesive, water solubles will be dissolved with a consequent decrease in chemical potential. To compensate for this and hence maintain chemical equilibrium between pockets of aqueous solution and the water or water solution to which the joint is exposed, the pressure of the liquid increases, thereby causing the pockets to expand and contribute to the swelling. When, instead of pure water, the aqueous environment is a solution that contains sodium or potassium ions, the difference in chemical potential and hence the magnitude of the osmotic pressure and the magnitude of the swelling, are all diminished.

Experiments in which the rigid titanium substrate was replaced by its glass equivalent showed that immersion of the adhesive joint in saturated NaCl or KCl solution had a similar mitigating effect upon the swelling. Examination of the specimen

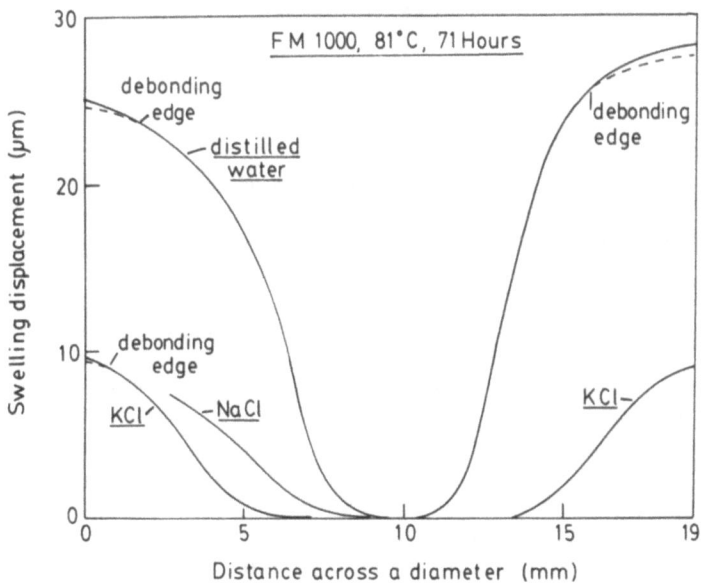

Figure 7. Displacement normal to 3 adhesive joints after immersion in distilled water and saturated solutions of NaCl and KCl respectively.

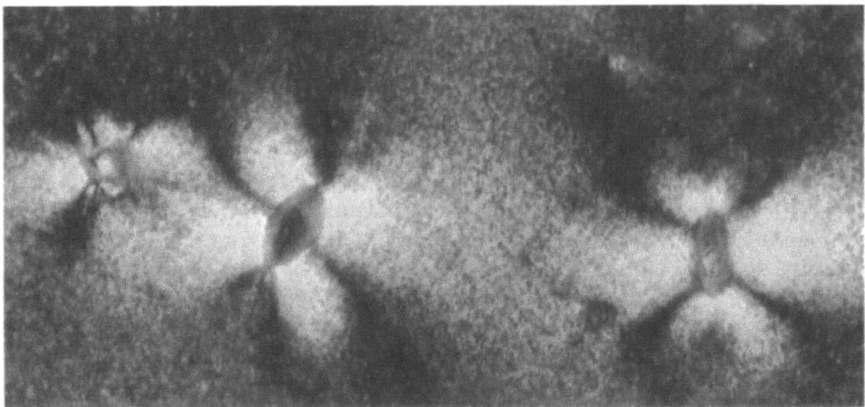

Figure 8. Pressure filled cavities observed between crossed polarizers for adhesive FM 1000 after immersion in distilled water.

with a transmission polarising optical microscope revealed the presence of numerous pressure pockets contained within the adhesive layer[9], (Figure 8). These showed patterns of stress birefringence consistent with their having a large internal pressure. Comparison between two specimens, one of which was immmersed in distilled water and the other immersed in a saturated solution of NaCl, showed that there was a reduction in the number of cavities per unit volume for the NaCl immersed specimen.

4. Curing Stresses[9]

Non-uniform changes in glue-line thickness during cure may also be measured using the Moiré technique by monitoring the normal displacements of the deformation of the cover slip in an adhesive butt joint as the underlying adhesive cures.

The accelerated cure reactions that take place during the elevated temperature curing recommended by manufacturers for proprietary resin-based adhesive films promote cross-linking and thereby give rise to shrinkage. There are no superimposed dimensional changes attributable to such processes as chain-scissioning, as could be the case in polyesters that contain diffused water, or to the release of volatiles, as might be the consequence of condensation reactions in polyamides for example. The shrinkage is assumed to be homogeneous because any tendency to create shear stress is thought to be relieved by viscoelastic flow. This assumption is based on the premise that before it gels, the fluid resin behaves in a Newtonian fashion and, in particular, is unable to support shear stress.

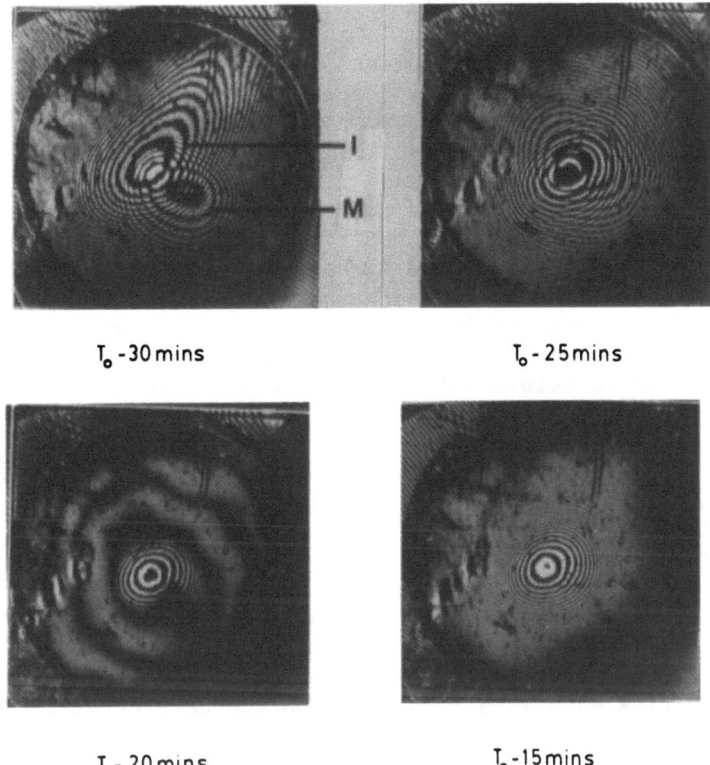

Figure 9. Moiré fringes (M) generated by superimposition of interference patterns (I) in order to reveal the deformation of the cover slip after reaching the cure temperature (120C). The cover slip was 150 μm thick.

Each specimen was mounted in a specimen chamber so that the free surface of the cover slip was in close proximity to an optical flat. This assembly was then mounted on an optical bench. In order that a uniform temperature distribution was maintained across the specimen it was mounted on a thick disc of copper, this in turn was held in good thermal contact with an aluminium specimen chamber. Figure 9 shows a sequence of Moiré patterns taken during the cure of a specimen manufactured from Redux 312/5 (manufactured by Ciba-Geigy). Figure 10 shows the normal displacement at different points across a diameter of the same specimen after 5 minutes at the cure temperature, where $t_o = 30$ minutes is the total time at the cure temperature. The curve fits the equation

$$w = .073 (x - .0095)^2 \tag{1}$$

where w is the normal displacement
x is the distance measured from the edge of the specimen

Since the data are described by a parabola, the fourth differential $\partial^4 w/\partial x^4$ and hence the normal stress σ_{zz} are evidently zero. Hence it is concluded that the deformation is caused by radial stresses transmitted from the resin to the cover slip and not by stresses created normal to the joint. To check that curing really does cause the parallel sided disc of resin to transform into a concave lens shape, a specimen was manufactured using cover slips for both adherends. As expected, curing caused this sandwich to deform into a double concave lens as might have been produced by the application of edge tractions.

Figure 11 shows a sequence of Moiré patterns for the specimen from Figure 9 on cooling from the cure temperature. In addition to a small increase in the concave deformation during cooling (when the joint presumably behaves as an elastic sandwich with contraction in the resin exceeding that in the glass cover slip), the Moiré patterns reveal a generally complex pattern of deformation, as shown by small loops and perturbations on the circumferential fringes. These may be associated with local inhomogeneities such as air bubbles and the carrier cloth present in the adhesive.

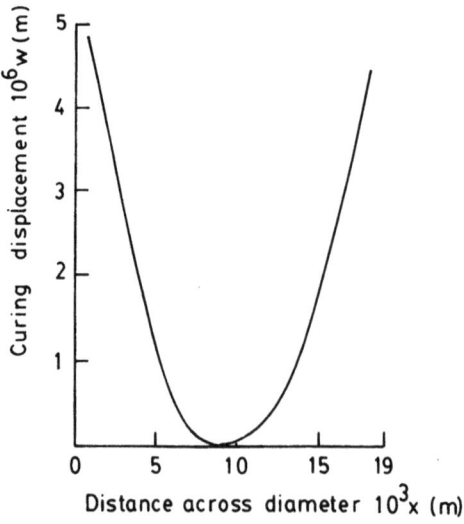

Figure 10. Normal displacement across one diameter of the cover slip after 5 minutes at the cure temperature i.e.(T_o-25mins).

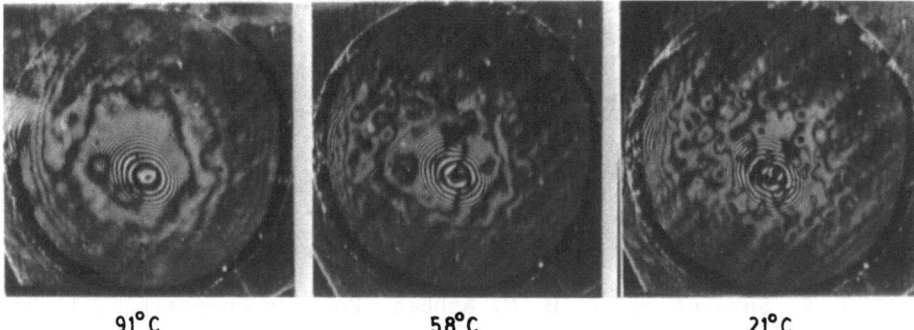

Figure 11. Moiré fringes generated by superimposition of interference patterns in order to reveal the deformation of the cover slip on cooling down from the cure temperature to room temperature. The cover slip was 150 μm thick.

CONCLUSIONS

1. Using an optical interferometric technique it has been demonstrated that the swelling of a model epoxy adhesive joint exposed to aqueous environments is strongly inhomogeneous, and that the stresses generated can be sufficient to fracture one of the adherends.

2. The application of beam and elasticity theory to the swelling of the adhesive joint has resulted in the prediction of a negative displacement of the cover slip and damped normal displacement waves. Both have been observed experimentally.

3. An adhesive film subjected to alternate periods of immersion in distilled water and exposure to dry air, both at 62°C undergoes reversible changes in linear dimensions when the period of water immersion is less than 1 day, but that irreversible changes occur after longer times.

4. Dissolved salts in the immersion water lead to reductions in the swelling displacement.

5. An optical interference method has been used to look for non-uniform shrinkage during the curing of an epoxy based adhesive film. Inhomogeneities have been measured, the magnitudes of which indicate that stresses of the order of 10^5 Pa are established and maintained throughout curing.

ACKNOWLEDGEMENTS

The author is grateful to Dr. K. H. G. Ashbee for his helpful comments in writing this paper.

REFERENCES

1. J. P. Sargent and K. H. G. Ashbee, J. Adhesion, $\underline{11}$, 175 (1980).
2. J. P. Sargent and K. H. G. Ashbee, Polymer Composites, $\underline{1}$, 93 (1980).
3. D. E. Jesson and J. P. Sargent, J. Adhesion, $\underline{14}$, 119 (1982).
4. A. E. H. Love, "A Treatise on the Mathematical Theory of Elasticity", Dover Publications, Inc., New York, 1944.
5. J. P. Sargent and K. H. G. Ashbee, Polymer, $\underline{23}$, 327 (1982).
6. J. P. Sargent and K. H. G. Ashbee, J. Phys. D, $\underline{14}$, 1933 (1981).
7. K. H. G. Ashbee and R. C. Wyatt, Proc. Roy. Soc., Lon., $\underline{A312}$, 553 (1969).
8. N. R. Farrar and K. H. G. Ashbee, J. Phys. D., $\underline{11}$, 1009 (1978).
9. J. P. Sargent and K. H. G. Ashbee, To be published.
10. J. P. Sargent and K. H. G. Ashbee, I&EC Product R&D, $\underline{21}$, 302 (1982).

STRENGTH CHARACTERISTICS OF MONO AND MULTIPLE-WIRE STEEL TO STEEL JOINTS BONDED WITH AN EPOXY ADHESIVE

R. W. Hylands

Portadown College of Further Education
Lurgan Road, Portadown
Craigavon BT63 5BL, U.K.
*Formerly of Ulster Polytechnic

Strength characteristics of mono-wire and multiple-wire specimens were investigated as support work for a research project concerned with the pre-stressing (energising) of structural materials.[1,2,3,4,5] The mono-wire joints were composed of cylindrical steel to steel interfaces, the adhesive used being Araldite AY 103 with Hardener HY 991. Compression and tension joints were concentrically constructed and after curing, axially loaded to failure. Nine glue line thicknesses and five bond lengths were selected ranging from 0.001 in to 0.100 in (0.0254 mm to 2.54 mm) and 0.50 in to 2.50 in (12.7 mm to 63.5 mm) respectively. Each test was performed five times. The strongest joints were obtained with the thinnest glue lines and increases up to 0.060 in (1.524 mm) diminished the strength by approximately 32% for both compression and tension specimens. For glue line thicknesses from 0.60 in to 0.100 in (1.524 mm to 2.54 mm) the joint strength was almost constant. Test results displayed a coefficient of variation for the joint strength of between 6.89% and 13.41%. A relationship was found among mean breaking force, glue line thickness and bond length. The multiple-wire joints consisted of Bridon prestress wire manufactured by British Ropes Limited[6] of 0.276 in (7 mm) diameter 'patented' plain cold drawn and produced to BS 2691[7]. Four of these were placed inside a standard square steel tube together with a central 0.1875 in (4.76 mm) diameter wire which prevent the four larger wires from undue warping or twisting during use. Five bond lengths were selected for compression and tension joints ranging from 0.50 to 2.50 in (12.7 mm to 63.5 mm) respectively and after curing all joints were axially loaded to failure. Each test was performed five times. Test results displayed a coefficient of variation for the joint strength of between 5.54% to 12.47%. In general, compression specimens were slightly stronger than the corresponding tension ones for all adhesive tests.

INTRODUCTION

The main research project compared the bonding of the cables of a poststressed timber beam with a similar unbonded beam. The cable for the poststressed timber beams consisted of the multiple-wire and standard steel tube arrangement shown in Figure 1.

Preparatory to this main area of research a suitable adhesive was selected and its strength properties determined. The chosen adhesive required had to possess the following essential characteristics: ability to bond to a variety of different materials, adhesion, cohesion, low shrinkage during cure, set 100% solid, durability, resistant to moisture, resistant to dilute acids alkalies and many solvents, ability to be flexible, low creep, possibility of room temperature cure, low viscosity, ability to solidify for a range of glue line thicknesses and an assembly period of about four hours. The epoxy resin selected for the research project satisfied all of these characteristics.

The strength properties of this adhesive had to be determined to establish strength behaviour of joints with varying glue line thicknesses. This information was necessary as the glue line thickness within the cross-section of the duct is variable. For these tests mono-wire cylindrical joints were used and the range of glue line thickness chosen to be compatible with those within the duct. Tests were also required to be carried out using a multiple-wire arrangment similar to the one which was used in the poststressed beam. Consequently, joints were constructed using steel to steel interfaces; these results were a useful prelude to the design and analysis of the poststressed beam with bonded cables.

For all adhesive tests both tension and compression bonded joints were used.

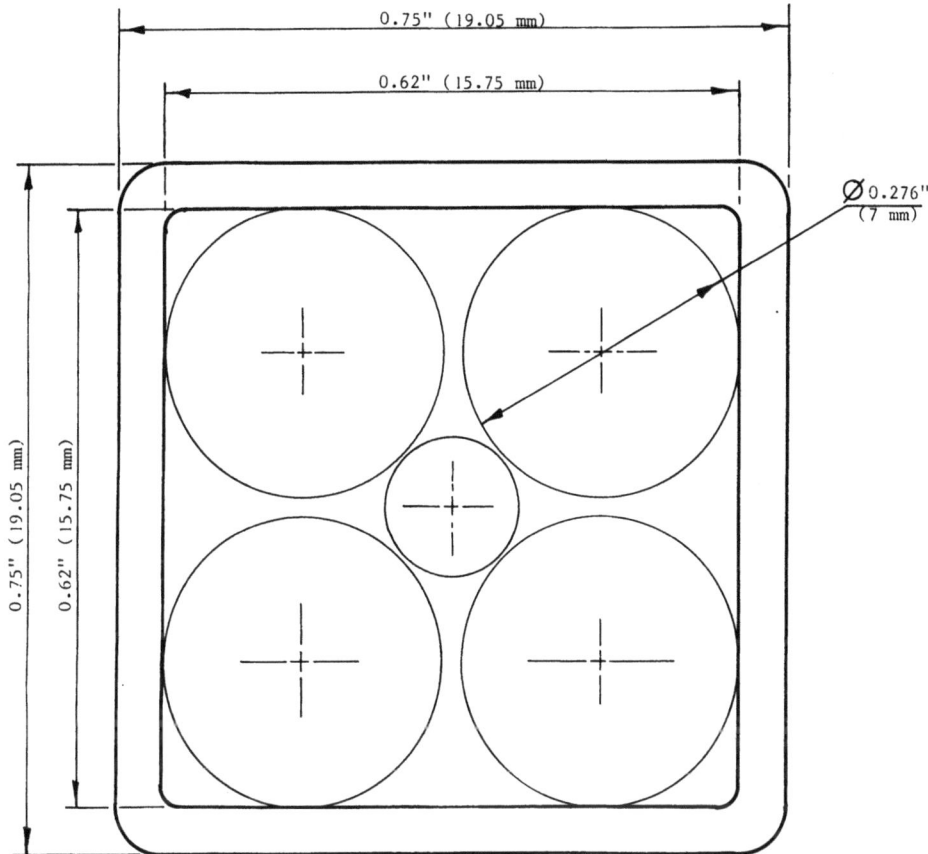

Figure 1. Cross-section of multiple-wire arrangement for post-stressed timber beams.

Mild Steel Specimens used in Mono-Wire Tests

The glued joint which was constructed consisted essentially of a steel rod placed concentrically inside a hole which was drilled along the centroidal axis of a circular steel bar. The steel rod had part of its length geometrically similar to the 0.276 in (7 mm) diameter prestress wire and had a polished bright surface finish. A hole was drilled and reamed at the centroid of the bar cross-section along the length of the specimen. These specimens were used to enable accurate control of the bond length and glue line thickness of each joint. Nine different hole diameters were selected to give glue line thicknesses ranging from approximately 0.001 in (0.0254 mm) to 0.100 in (2.54 mm) and five different bond lengths, viz., 0.5 in (12.7 mm), 1.0 in (25.4 mm), 1.5 in (38.1 mm), 2.0 in (50.8 mm) and 2.5 in (63.5 mm) used to increase test data. Figure 2 shows the test specimens.

Design of Jigs for Alignment of Mono-Wire Joints

It was necessary to design jigs which would enable both proper alignment and a concentric joint of uniform thickness to be maintained. Two types of jigs were designed and manufactured to accommodate compression and tension specimens. Five complete sets of jigs for each type of test were made so that the testing programme period could be shortened. Figure 3 shows typical alignment jigs for compression and tension test specimens.

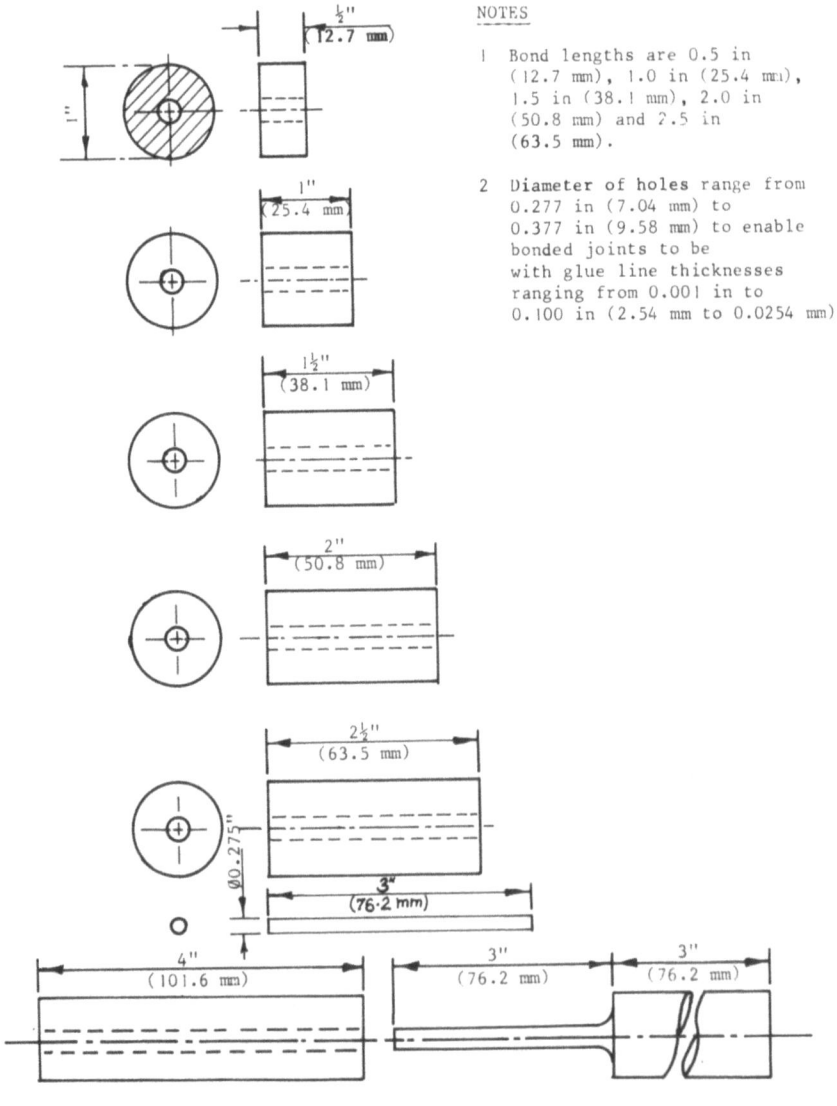

NOTES

1. Bond lengths are 0.5 in (12.7 mm), 1.0 in (25.4 mm), 1.5 in (38.1 mm), 2.0 in (50.8 mm) and 2.5 in (63.5 mm).

2. Diameter of holes range from 0.277 in (7.04 mm) to 0.377 in (9.58 mm) to enable bonded joints to be with glue line thicknesses ranging from 0.001 in to 0.100 in (2.54 mm to 0.0254 mm).

Figure 2. Compression and tension specimens for adhesive tests.

STRENGTH CHARACTERISTICS OF STEEL TO STEEL JOINTS

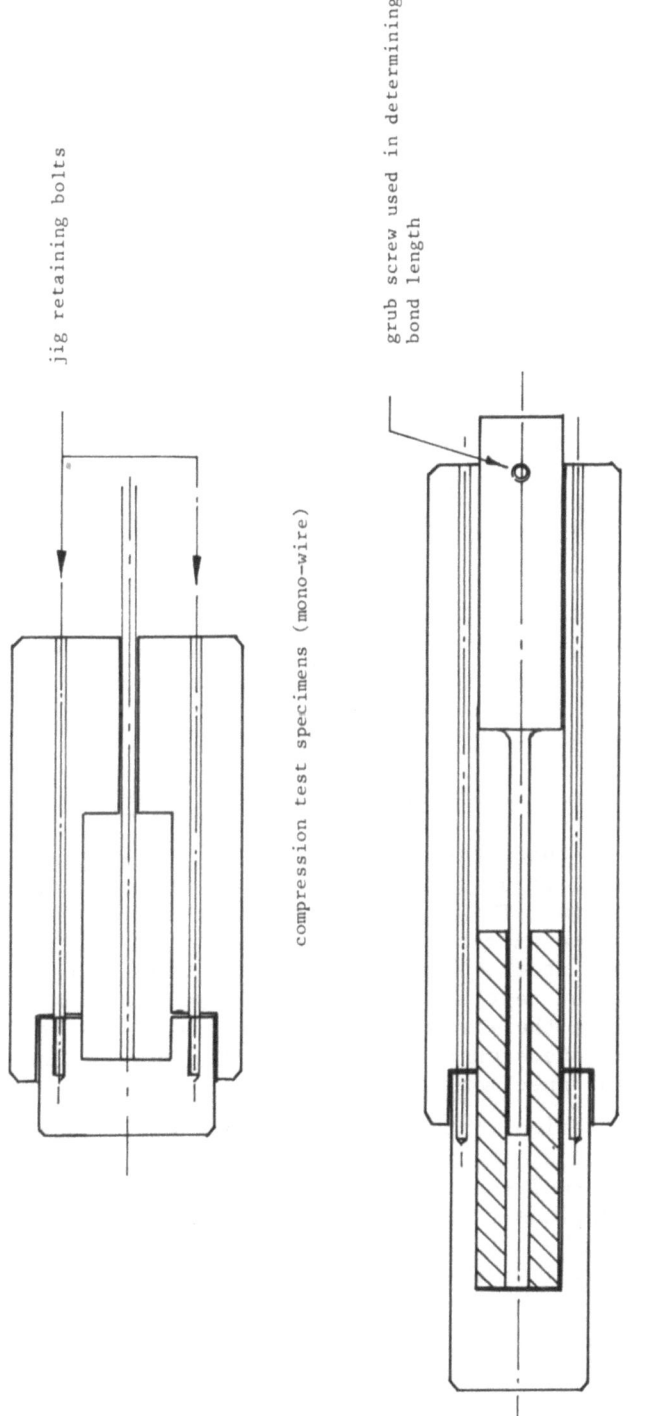

Figure 3. Typical alignment jigs for compression and tension test specimens.

To ensure easy separation of the jigs after the adhesive had been cured, a film of grease was smeared on the portion of the test specimens adjacent to the jigs. Great care was taken to ensure that no grease came into contact with the surfaces being bonded or indeed with the adhesive. When assembled, both parts of the jig and joint specimen were held firmly in position using hexagon screws as indicated in Figure 3.

The jigs for the compression joint specimens were made to receive the complete range of bond lengths, the bond length being determined in each case by the length of circular steel bar with the concentric hole along its length. The length of these bars was cut to agree with the correct length of bond.

For the compression tests a special gripping device was designed and manufactured to fit the top jaws of the Avery testing machine and also the steel specimen which was to be tested. A circular steel block was made to receive the lower portion of the jointed specimen being tested and to maintain an axially applied load during testing.

The jigs for the tension joint specimens were so designed to facilitate the assembly of the five different bond lengths. This was achieved by means of a small grub screw attached to the female portion of the jig which allowed the joint specimen to be positioned at the desired bond length.

Steel Specimens used in Multiple-Wire Tests

The purpose of these tests was to examine the strength characteristics of a group of wires which were bonded to a standard steel sheath. The glue line thickness varied from 0.001 in (0.0254 mm) to approximately 0.100 in (2.54 mm). For the tension tests standard steel tubing and prestress wires were used. Five complete sets of test specimens were designed and manufactured to accommodate the five selected bond lengths. This allowed twenty-five tension tests to be carried out on these multiple-wire joint specimens.

For the compression tests the steel sheath was embedded in a concrete cube surround, the multiple-wire arrangement being then placed inside the steel sheath and bonded at the required bond length. It was previously calculated that the joint would fail before slip would take place between the steel sheath and concrete cube surround. Twenty-five concrete cube surrounds with embedded steel ducts and five sets of multiple-wire specimens were constructed to facilitate a total of twenty-five compression tests to be performed. Figure 4 shows the multiple-wire test specimens.

STRENGTH CHARACTERISTICS OF STEEL TO STEEL JOINTS

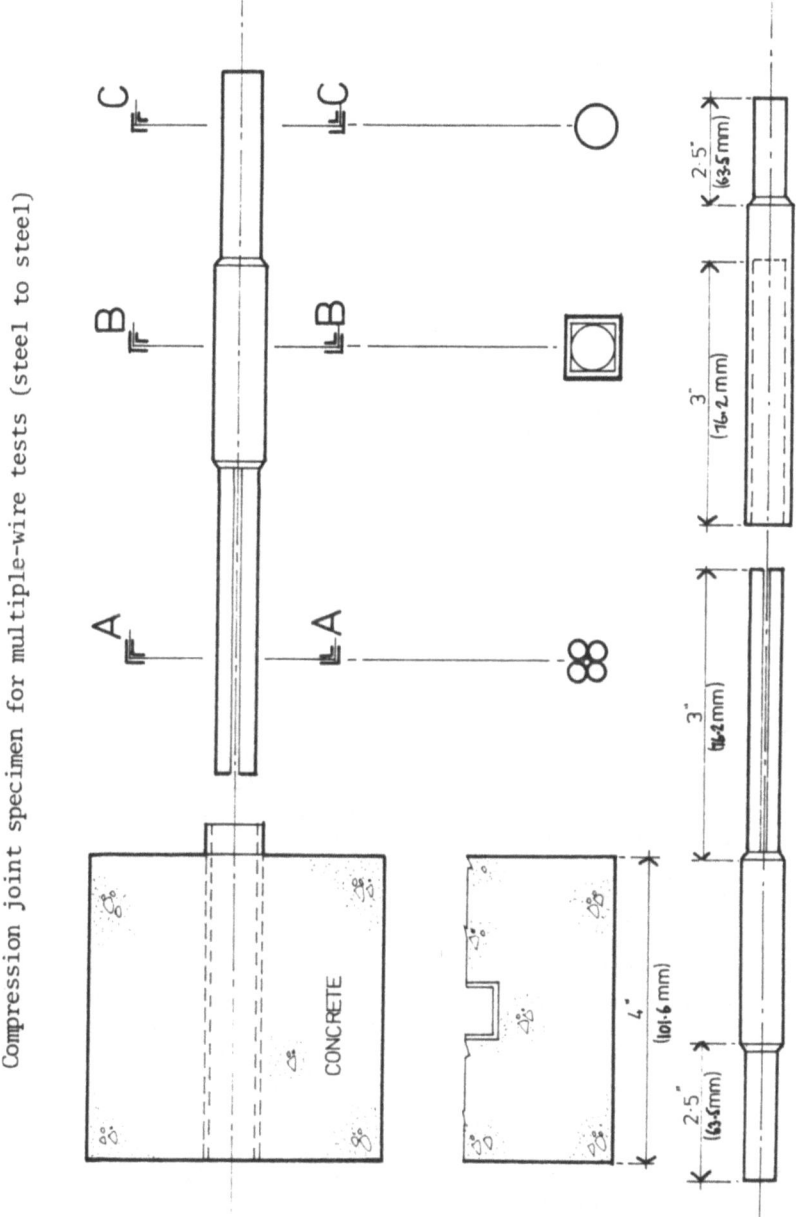

Figure 4. Tension joint specimen for multiple-wire tests (steel to steel).

Requirements of Testing Apparatus

The main objects of this part of research was to determine:

(a) Strength characteristics of a selected adhesive for the mono-wire and multiple-wire tests (compression and tension), steel to steel.

(b) How the strength of a joint varies with glue line thickness.

(c) How the strength varies with bond length.

(d) The mode of failure for the adhesive tests.

The testing machine satisfied the following requirements:

(a) Accuracy - The test machine shall be verified according to the requirements of Grade A, BS 1610, 'Verification of testing machines', Part I, 'Methods of load verification requirements for elastic proving devices and verification of machines for tension and compression testing'.[8]

(b) Rate of loading - with the test piece in position and under load. The rate of increase in load shall be between 300 and 600 lbf/minute, according to BS 1203 : 1963 Appendix B.[9]

(c) Type of jaw assembly - The jaws shall grip the test piece with a wedge action. Each pair of jaws shall be attached by loose-fitting pin joints which in turn are fitted by ball-and-socket joint to the straining heads. In a machine mounted horizontally the dead weight of the jaw assembly shall be carried by freely moving cross-members. This shall not affect the freedom of rotation of the jaws.

Alternatively, one of the pair of jaws shall be attached rigidly to the tensile machine and the other jaws shall be suspended from a ball-and-socket joint allowing sufficient sidesway movement of the jaw to permit self-alignment of this jaw while the test pieces are being pulled.

(d) Axial load for compression testing - The machine will require adaptability for the addition of suitably designed portions of apparatus to facilitate proper compression tests to be executed.

STRENGTH CHARACTERISTICS OF STEEL TO STEEL JOINTS

(e) Be capable of adjustment to operate with a scale not exceeding 20 kN.

(f) Be capable of producing a graphical recording of load against change in length to enable observation of the mode of failure for a particular test.

The machine used for the adhesive tests was the Avery-Denison Universal Testing Machine.

The number of tests to be repeated for each joint specimen was another important consideration. BS 1203 : 1963, Appendix B[9] specifies that five tests should be carried out for each specimen and the mean of the results reported for the failing load. It was also decided to consider five different bond lengths so that such things as edge effects, Poisson's effect, etc, could be considered.

Test Procedure

The following is a generalised procedure for carrying out adhesive tests.

All steel specimens were thoroughly degreased using carbon tetrachloride on a fluff-free cloth. A close inspection of the surfaces to be bonded was carried out to check the presence of any undesirable foreign particles that would interfere with the glue line of the joint. The surfaces were also checked to ensure that they were free from damage.

The adhesive was prepared in accordance with the manufacturers instructions and due regard given to handling precautions for Araldite epoxy resin materials.[10] Araldite AY 103, a plasticised liquid epoxy resin was mixed with Hardener HY 991, a light-brown liquid in the recommended mix proportions of 100 parts Araldite to 40 parts Hardener by weight or 100 parts Araldite to 50 parts Hardener by volume. The adhesive was prepared at room temperature, ie 20°C (68°F) using volumetric proportioning and was thoroughly stirred to give a uniform liquid of low viscosity. The mixed adhesive was then poured into the female specimen and the male portion inserted to form the required joint. The excess glue displaced was removed from the joint.

The jointed samples were placed in the Standards Laboratory and left for twenty-four hours to cure. This laboratory had a controlled room temperature of 20°C ($\pm$ 1°C) ie 68°F ($\pm$ 1.8°F); a certificate having been issued to this effect by the British Standards Institution.

After curing, each joint specimen was carefully assembled between the jaws of the Avery-Denison Universal Testing Machine. The specimens had been identified by letter or number and were tested at approximately ten minute intervals to agree with the twenty-four hour curing period. An axial load was applied to the jointed specimen, the rate of increase in load being kept within the limits of 300 and 600 lbf/minute as stated in BS 1203 : 1963 Appendix B.[9] The rate of loading was controlled using the load pacer on the testing machine. Careful records were kept of the quality of the joint specimen, mode of failure, etc, for each test sample.

After testing the specimens were soaked in chloroform and placed in a fume cupboard for twenty-four hours to break down the epoxy resin on the specimen. The specimens were then spotlessly cleaned and stored in methylated spirit preparatory to further testing.

Mono-wire Tests

The steel specimens to be jointed were designed to give nine glue line thicknesses: 0.001 in (0.0254 mm), 0.004 in (0.1016 mm), 0.012 in (0.3048 mm), 0.020 in (0.508 mm), 0.035 in (0.889 mm), 0.040 in (1.016 mm), 0.060 in (1.524 mm), 0.080 in (2.032 mm) and 0.100 in (2.54 mm).

Provision was also made to test each glue line thickness over five different bond lengths, viz; 0.5 in (12.7 mm), 1.0 in (25.4 mm), 1.50 in (38.1 mm), 2.0 in (50.8 mm) and 2.5 in (63.5 mm). The order of carrying out the tests on the specimens was as follows: Day No 1, all 0.5 in(12.7 mm) bond lengths (nine tension and nine compression tests), Day No 2, all 1.0 in (25.4 mm) bond lengths, etc. This testing procedure was adopted to retain consistency within each group of results.

In the compression tests, the male specimens were designed to ensure that there would be no buckling of the member when the joint was being loaded to failure.

The length of specimen between the upper jaws of the testing machine and the glued joint was calculated to be 1.875 in (47.63 mm), this value being the greatest possible load anticipated, viz 6000 lbf (26.7 kN). In the tension specimen the length of member was not a ruling factor.

The specimens whose glue line thickness was 0.001 in (0.0254 mm) provided some difficulty in ensuring that the adhesive completely covered all parts of the joint. As previously stated alignment jigs were used to assemble the concentrically jointed specimens. A film of grease was smeared around the surfaces of the parts of the jointed specimen not being bonded and also on the adjacent surfaces of the alignment jigs. This facilitated easy removal of the jointed specimen from the steel jigs after the twenty-four hour curing period.

The analysis of tests results show an approximate spread of 12% between the lowest and highest readings after five repeated tests for a particular joint. This spread is satisfactory for adhesive tests.[11] In general, it was found that the tension joint specimens were slightly weaker than the corresponding compression joint specimens. It is believed that this is because of the Poisson's ratio effect when the jointed specimen is under axial load. The compression male member is obviously subjected to a small lateral increase in dimension whereas with the tension arrangement the member is subjected to a small contraction. This is a possible reason for both types of tests having differing results. Compression and tension specimens failed with a marked explosive bang with a well defined reading being recorded on the testing machine. The results of 450 tests using the mono-wire arrangement revealed information regarding varying glue line thickness and bond length for both compression and tension test specimens.

Graphs of <u>mean breaking force</u> versus <u>glue line thickness</u> were plotted for both compression and tension experiments using a mean of five readings for each point plotted and are as shown in Figures 5 and 6. Additional graphs of <u>mean breaking force</u> versus <u>bond length</u> were plotted to check if there was a linear relationship between breaking force and each equal increment of bond length. An approximate linear relationship was found to exist for test specimens having a similar glue line thickness. Figures 7 and 8 show graphs of <u>mean breaking force</u> versus <u>bond length</u> with all results plotted.

These give a pictorial representation of the spread of results for each bond length. Calculations for the coefficient of variation indicate a spread of values ranging from 6.89% to 13.41%.

From the graphs of <u>mean breaking force</u> versus <u>glue line thickness</u>, Figures 5 and 6, it can be seen that the strongest joint is obtained with a thin glue line. As the glue line becomes thicker, up to about 0.060 in (1.524 mm), the joint strength diminishes for both compression and tension tests, and for glue line thicknesses in the range 0.060 in (1.524 mm) to 0.100 in (2.54 mm) the joint strength is approximately constant.

A probable reason for the diminution in strength with the thicker glue line joints is the slight shrinkage of the adhesive from the interfaces during curing. The shrinkage property for the adhesive under consideration is approximately 1.5%. Self straining in the adhesive is more likely during shrinkage in a thick glue

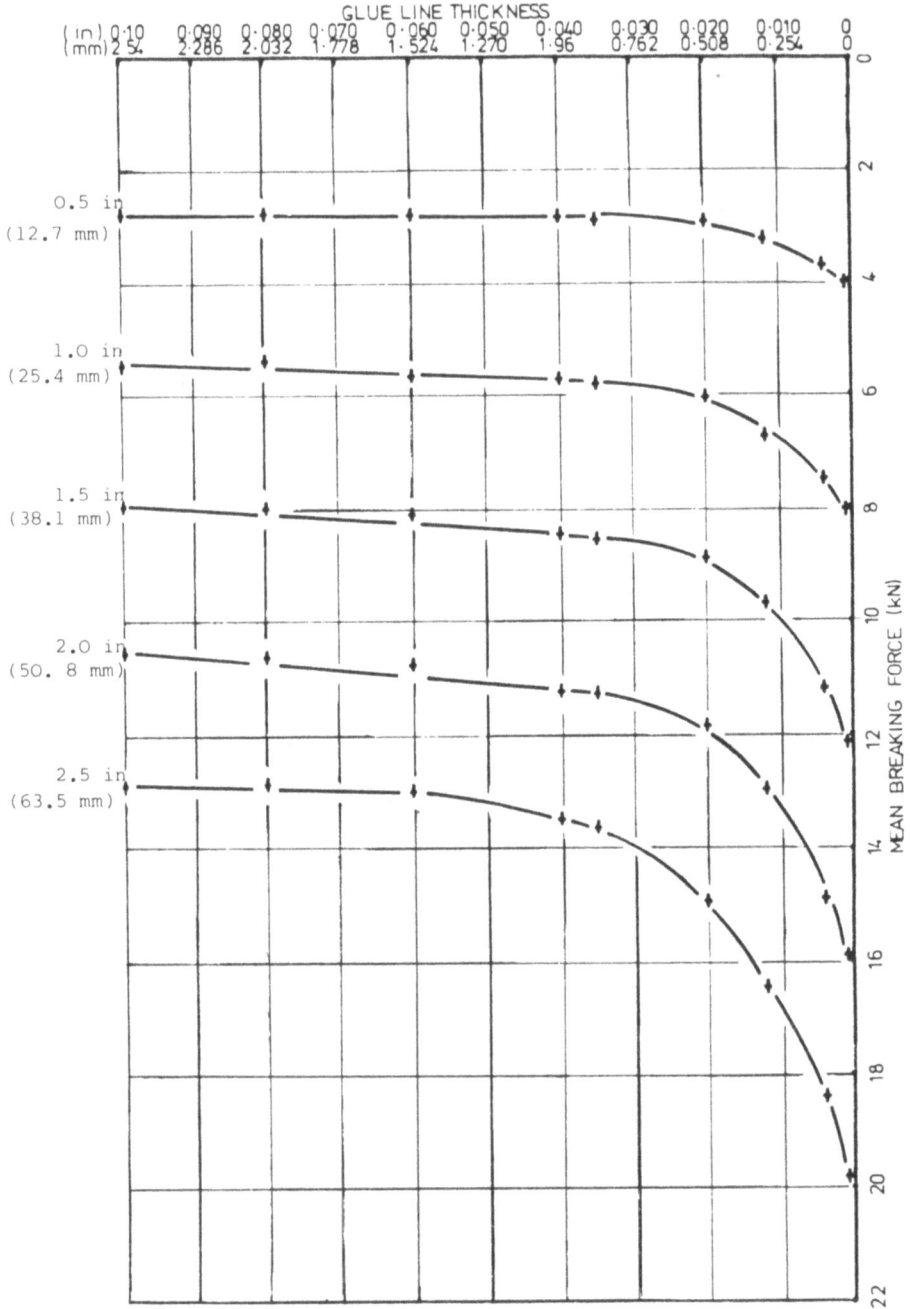

Figure 5. Mean breaking force versus glue line thickness for mono-wire tests on compression specimens, steel to steel. (5 bond lengths)

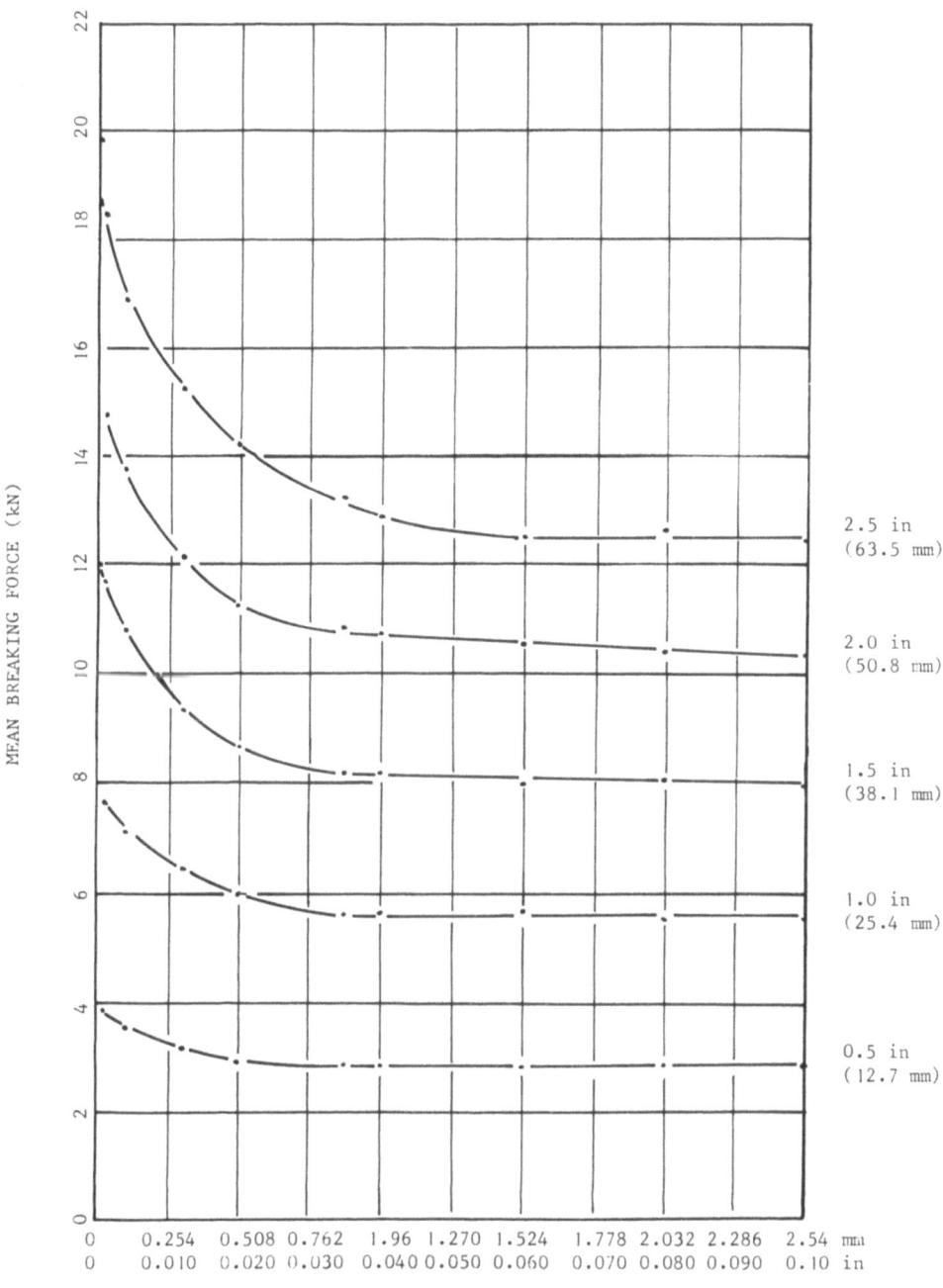

Figure 6. Mean breaking force versus glue line thickness for mono-wire tests on tension specimens, steel to steel. (5 bond lengths)

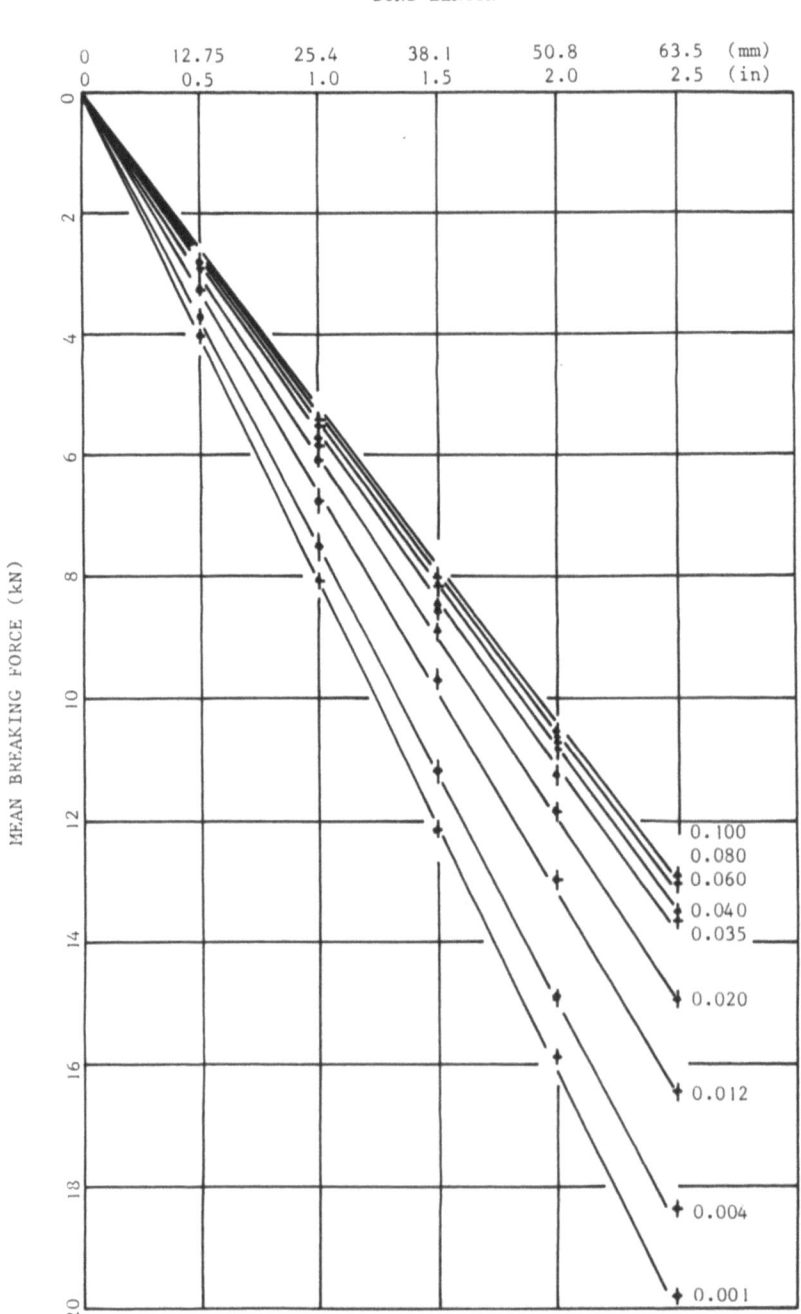

Figure 7. Mean breaking force versus bond length for mono-wire tests on compression specimens, steel to steel.

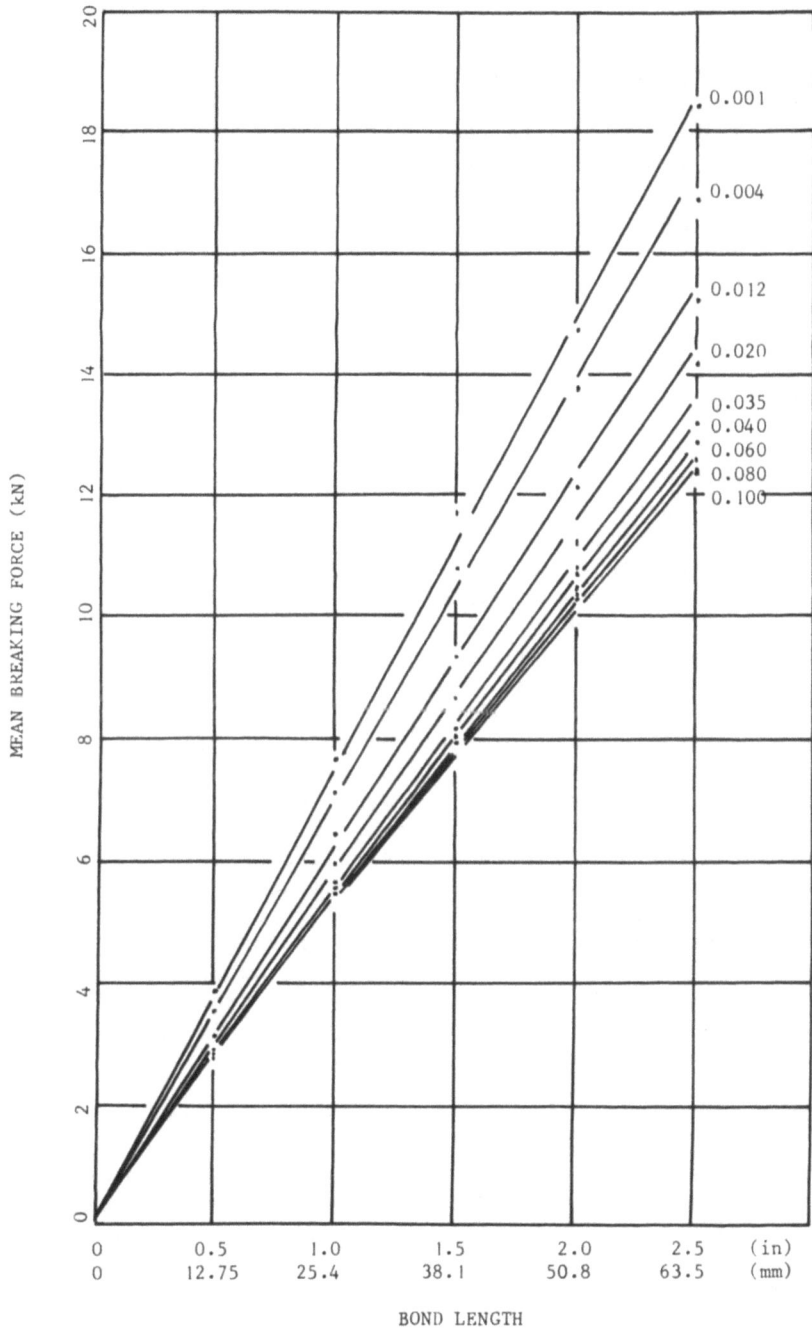

Figure 8. Mean breaking force versus bond length for mono-wire tests on tension specimens, steel to steel.

line than a thin one : in a very thin one it would be negligible. From test results it would appear that the shrinkage of adhesive during curing period affects the joint strength to varying degrees up to a glue line thickness of approximately 0.060 in (1.524 mm); beyond this joint thickness the adhesive shrinkage does not significantly diminish the strength of the jointed steel specimen.

From the graphs of <u>mean breaking force</u> versus <u>bond length</u>, Figures 7 and 8, it is seen that there is an approximate linear relationship in both compression and tension tests. There is however a tendency for the joint strength to deviate from this linear relationship and diminish slightly in strength as the bond length increases; it is believed that if the bond length range was substantially extended this linear relationship would not necessarily prevail.

A comparison of compression and tension test results for <u>mean breaking force</u> versus <u>glue line thickness</u> is shown in Figure 9 and it may be generally seen that the compression tests are slightly stronger than the corresponding tension tests. The following may provide a possible explanation. When the 0.276 in (7 mm) diameter steel bar is loaded there is a change in its lateral dimension which would tend to be more significant in the higher load range and almost negligible in the lower load range. In both compression and tension tests the shrinkage of the adhesive during curing and Poisson's effect must contribute to the joint strength of corresponding test specimens. It is reasonable to argue that the shrinkage is the same for both compression and tension tests and hence the difference in joint strength in the corresponding tests is caused by the Poisson's effect alone. The Poisson's effect in the compression tests would tend to make the joint stronger because of the additional frictional force developed by the increased lateral spread of the 0.276 in (7 mm) diameter bar. In the tension tests the bar would contract a small amount laterally and tend to pull away from the joint interface thus reducing the strength of the joint. The graphs in Figure 9 support this reasoning.

From the results of the test series a relationship was derived between the three variables mean breaking force (Y) kN, glue line thickness (X_1) mm and bond length (X_2) mm, a separate equation being determined for compression and tension test situations. Using a computer program obtained from the University of Kansas,[12] a facility exists to express a dependent variable Y in terms of two independent variables X_1 and X_2 using a polynominal equation up to the power of 7. However, a polynominal equation of degree 4 gave sufficient accuracy for adhesive test results, viz,

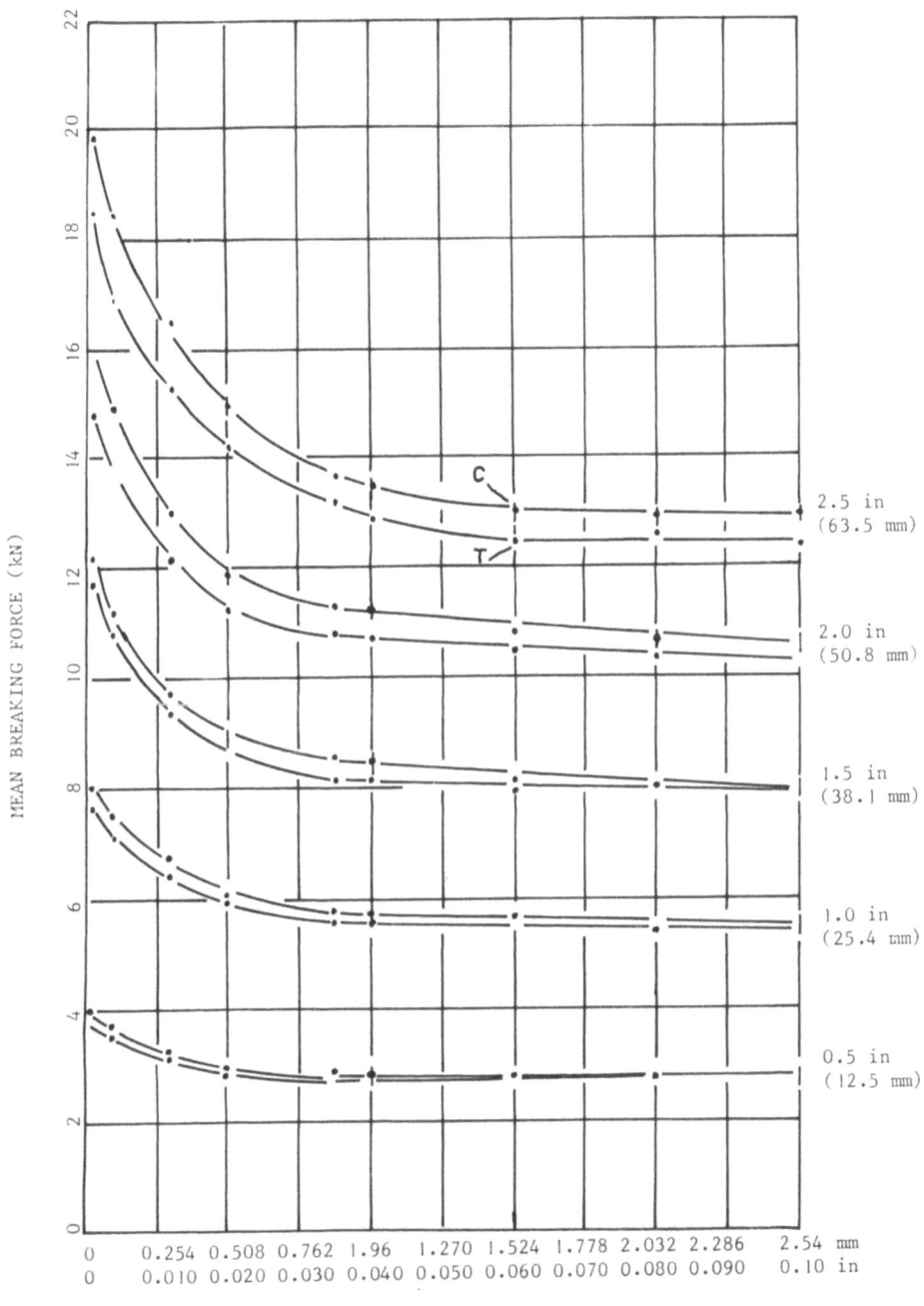

Figure 9. Mean breaking force versus glue line thickness for mono-wire tests on compression and tension specimens, steel to steel. (5 bond lengths)

$$Y = C_1 + C_2 X_1 + C_3 X_2 + C_4 X_1^2 + C_5 X_1 X_2 + C_6 X_2^2 + C_7 X_1^3 + C_8 X_1^2 X_2$$
$$+ C_9 X_1 X_2^2 + C_{10} X_2^3 + C_{11} X_1^4 + C_{12} X_1^3 X_2 + C_{13} X_1 X_2^3 + C_{14} X_2^4 + C_{15} X_1^2 X_2^2$$

This program is an outgrowth of research in geological trend analysis. The method consists of expanding the desired linear regression into a matrix of normal equations, which is then solved by inversion, giving coefficients of the regression. A simple Gaussian elimination is used for inverting the matrix.

The coefficients of the equations for compression and tension specimens are as follows:

Compression joints (maximum error = 3%)

$C_1 = -0.8156$ $C_6 = -0.7032 \times 10^{-2}$ $C_{11} = 0.7864$

$C_2 = -3.5370$ $C_7 = -4.1023$ $C_{12} = -0.2121 \times 10^{-1}$

$C_3 = 0.4710$ $C_8 = 0.1010$ $C_{13} = -0.7808 \times 10^{-5}$

$C_4 = 6.8191$ $C_9 = 0.3828 \times 10^{-3}$ $C_{14} = -0.7563 \times 10^{-6}$

$C_5 = -0.1873$ $C_{10} = 0.1236 \times 10^{-3}$ $C_{15} = 0.1673 \times 10^{-3}$

Tension joints (maximum error = 6%)

$C_1 = 0.4274 \times 10^{-1}$ $C_6 = 0.6253 \times 10^{-3}$ $C_{11} = 0.6724$

$C_2 = -2.7993$ $C_7 = -3.5703$ $C_{12} = -0.1933 \times 10^{-1}$

$C_3 = 0.3185$ $C_8 = 0.1060$ $C_{13} = 0.7919 \times 10^{-5}$

$C_4 = 5.9071$ $C_9 = 0.1014 \times 10^{-2}$ $C_{14} = 0.3780 \times 10^{-6}$

$C_5 = -0.2065$ $C_{10} = -0.4092 \times 10^{-4}$ $C_{15} = -0.7295 \times 10^{-4}$

Multiple-Wire Tests

The steel specimens to be jointed were designed, constructed and tested for five different bond lengths, viz; 0.5 in (12.7 mm), 1.0 in (25.4 mm), 1.50 in (38.1 mm), 2.0 in (50.8 mm) and 2.5 in (63.5 mm). The glue line thicknesses within the sheath varied from 0.001 in (0.0254 mm) to approximately 0.100 in (2.54 mm). To retain consistency within each group, the order of testing was as follows: Day No 1, all 0.5 in (12.7 mm) bond lengths (five tension and five compression), Day No 2, all 1.0 in (25.4 mm) bond lengths, etc.

STRENGTH CHARACTERISTICS OF STEEL TO STEEL JOINTS

The main problem with multiple-wire tests was to ensure that all specimens were spotlessly clean particularly when being re-used for subsequent tests. The top portion of the jointed specimen, which is common to both tension and compression arrangements, was steeped in chloroform for twenty-four hours to destroy the adhesive, cleaned with a cloth and then stored in methylated spirit for a further twenty-four hours, removed and placed in acetone for half an hour and allowed to dry preparatory to further testing.

The lower portion of the tension jointed specimen was treated in a similar manner. With the lower portion of the compression specimen it was slightly more difficult because of the concrete surround. However, the steel sheath which was embedded within the concrete was thoroughly cleaned prior to constructing this portion of the specimen. During curing of the concrete surround the ends of embedded steel sheath were 'stopped' with plasticene to prevent unwanted materials from entering the previously cleaned duct. Prior to constructing the adhesive joint, the plasticene was removed and a fluff-free cloth containing carbon tetrachloride was dragged through the duct to retain inner cleanliness. This lower portion of the compression test specimen was not re-useable after testing.

There was a difference in mode of failure with tension and compressive specimens. The tension specimens failed with a marked explosive bang with a well defined reading being recorded by the testing machine, whereas with the compression specimens there was little or no noise at the point of failure although the testing machine recorded a definite failure load.

For all tests the jigs and general arrangement for testing proved to be satisfactory.

The results of 50 tests using the multiple-wire arrangement give information with regard to the variation of breaking force for differing specimen bond lengths for both compression and tension test specimens.

Graphs of _mean breaking force_ versus _bond length_ were plotted to check if there was a linear relationship between breaking force and each equal increment of bond length. An approximate linear relationship was found to exist for both compression and tension situations as shown in Figures 10 and 11. Each point on the graphs was an average of five similar experiments. Additional graphs of _breaking force_ versus _bond length_ were plotted to give a pictorial representation of the spread of results for each bond length. These are shown in Figures 12 and 13.

Calculations for the coefficient of variation indicate a spread of values ranging from 5.54% to 12.47%

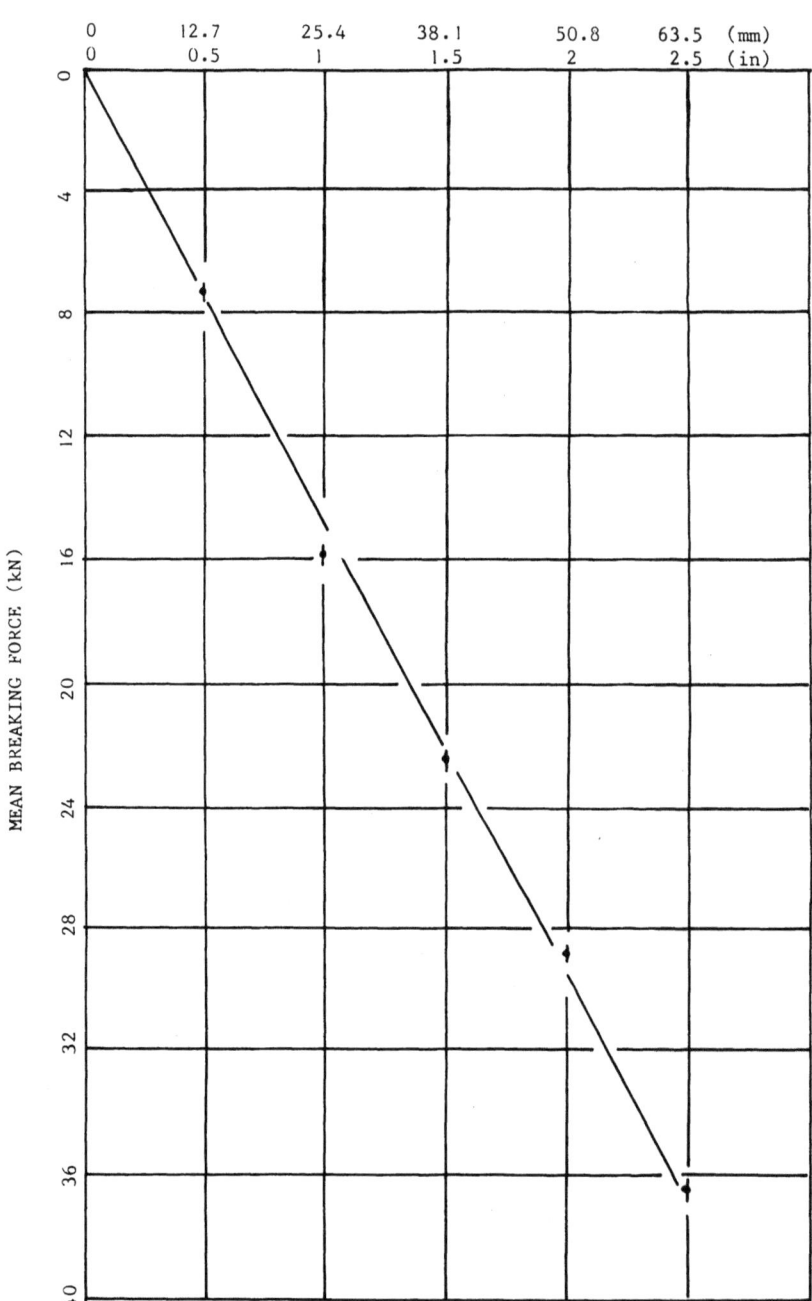

Figure 10. Multiple-wire tests (compression specimens) steel to steel.

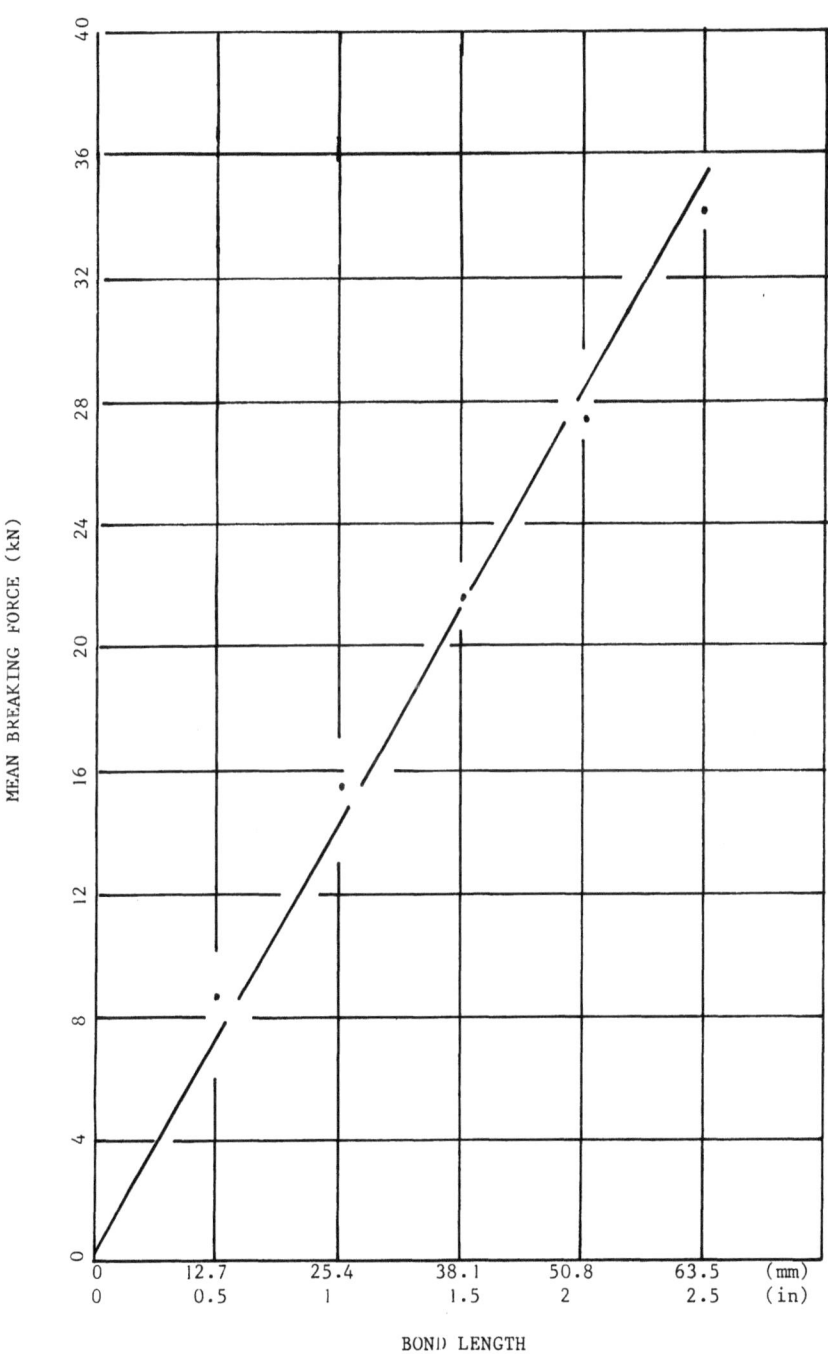

Figure 11. Multiple-wire tests (tension specimens) steel to steel.

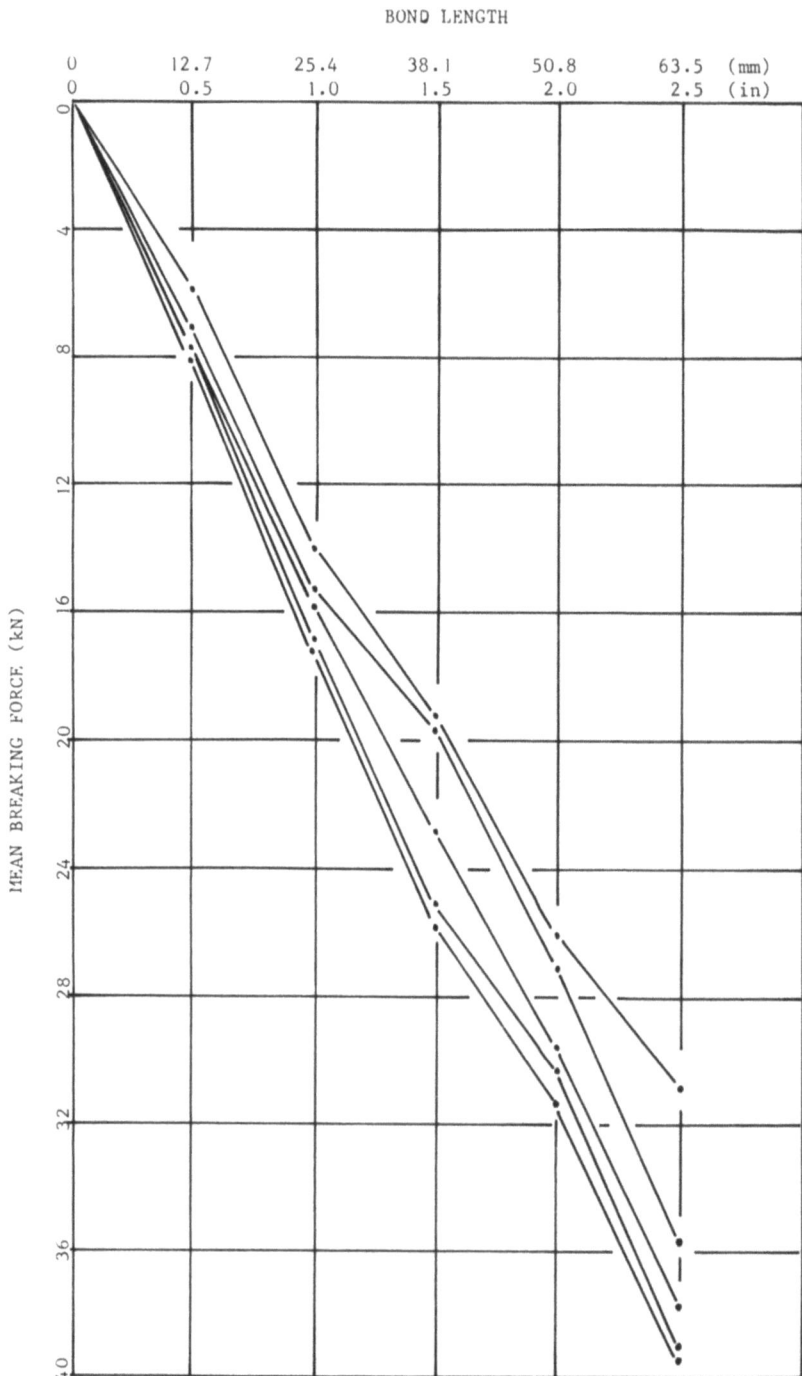

Figure 12. Multiple-wire tests (compression specimens) steel to steel.

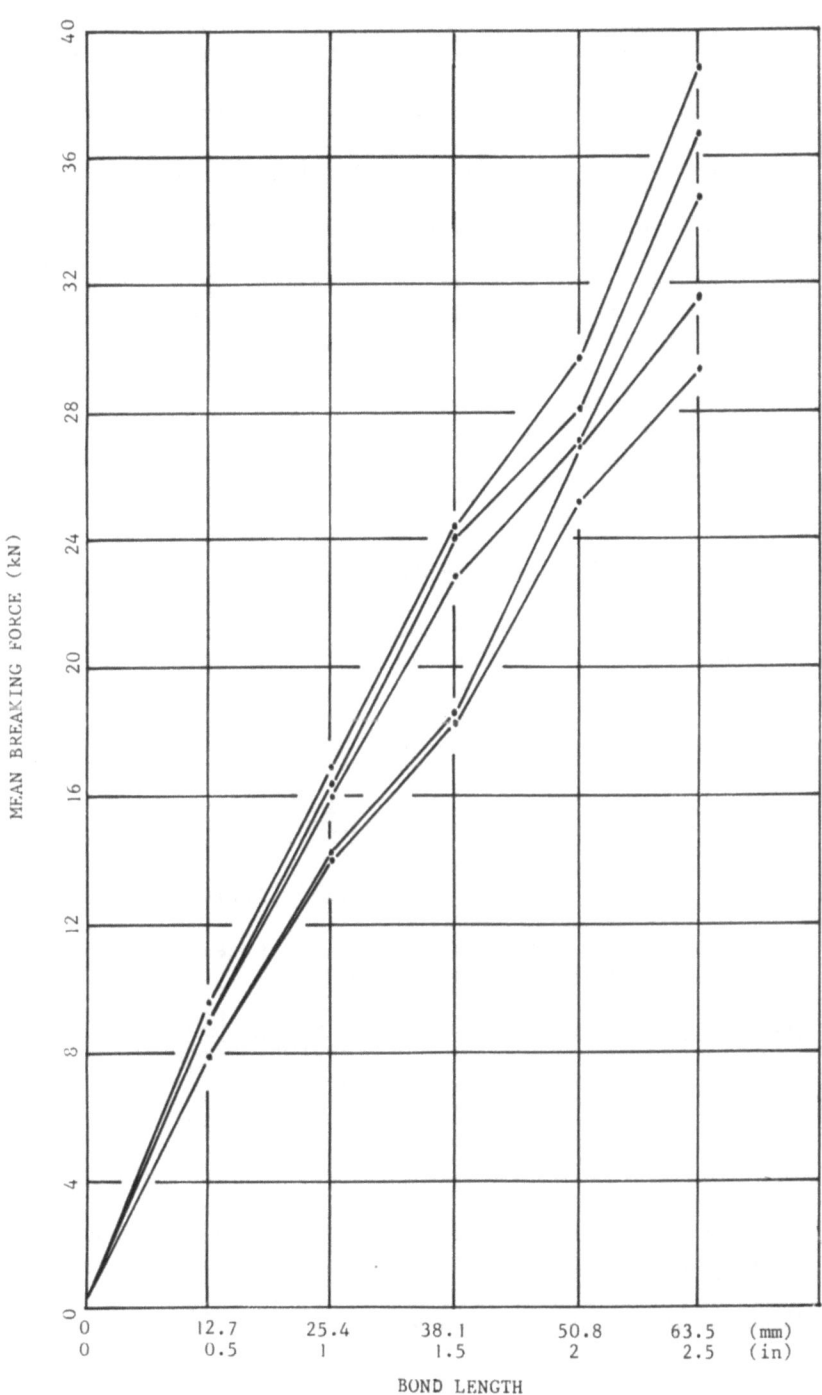

Figure 13. Multiple-wire tests (tension specimens) steel to steel.

From the graphs of <u>mean breaking force</u> versus <u>bond length</u>, Figure 14, it can be seen that the compression specimens are slightly stronger than the tension specimens particularly in the higher load range. This is probably caused by the very small increase in lateral dimension of the multiple-wire group which gives rise to a small additional force which thus increases the breaking load. The glue line thickness in these tests varies from 0.001 in (0.0254 mm) to about 0.100 in (2.54 mm) within the cross-section of the multiple-wire arrangement. The multiple-wire specimens generally were found to fail between the outside periphery of the wire group and the inside surface of the steel sheath; it is therefore reasonably simple to evaluate the approximate surface area of the glue line at failure. Consequently, if we consider the mean shear bond stress of the adhesive for glue line thicknesses > 0.060 in (1.524 mm) from the mono-wire test results and compare with the mean shear bond stress at failure of the multiple-wire tests, it is found that the values obtained are in reasonable agreement.

Consider the following examples:

Compression specimens (mono-wire) 2.5 in (63.5 mm) bond length, glue line thicknesses > 0.060 in (1.524 mm) see Figure 5.

Mean shear stress of adhesive = $\dfrac{\text{mean breaking force}}{\text{shear area}}$

$$= \dfrac{1.29 \times 2240}{\pi \times 0.276 \times 2.5}$$

$$= 1333 \text{ lbf/in}^2 \ (9.22 \text{ N/mm}^2)$$

Compression specimens (multiple-wire) steel to steel 2.5 in (63.5 mm) bond length see Figure 10.

Mean shear stress of adhesive = $\dfrac{\text{mean breaking force}}{\text{approx. shear area}}$

$$= \dfrac{3.55 \times 2240}{2.4984 \times 2.5}$$

$$= 1273 \text{ lbf/in}^2 \ (8.78 \text{ N/mm}^2)$$

Tension specimens (mono-wire) 2.5 in (63.5 mm) bond length, glue line thicknesses > 0.060 in (1.524 mm) see Figure 6.

Mean shear stress of adhesive = $\dfrac{1.25 \times 2240}{\pi \times 0.276 \times 2.5}$

$$= 1292 \text{ lbf/in}^2 \ (8.91 \text{ N/mm}^2)$$

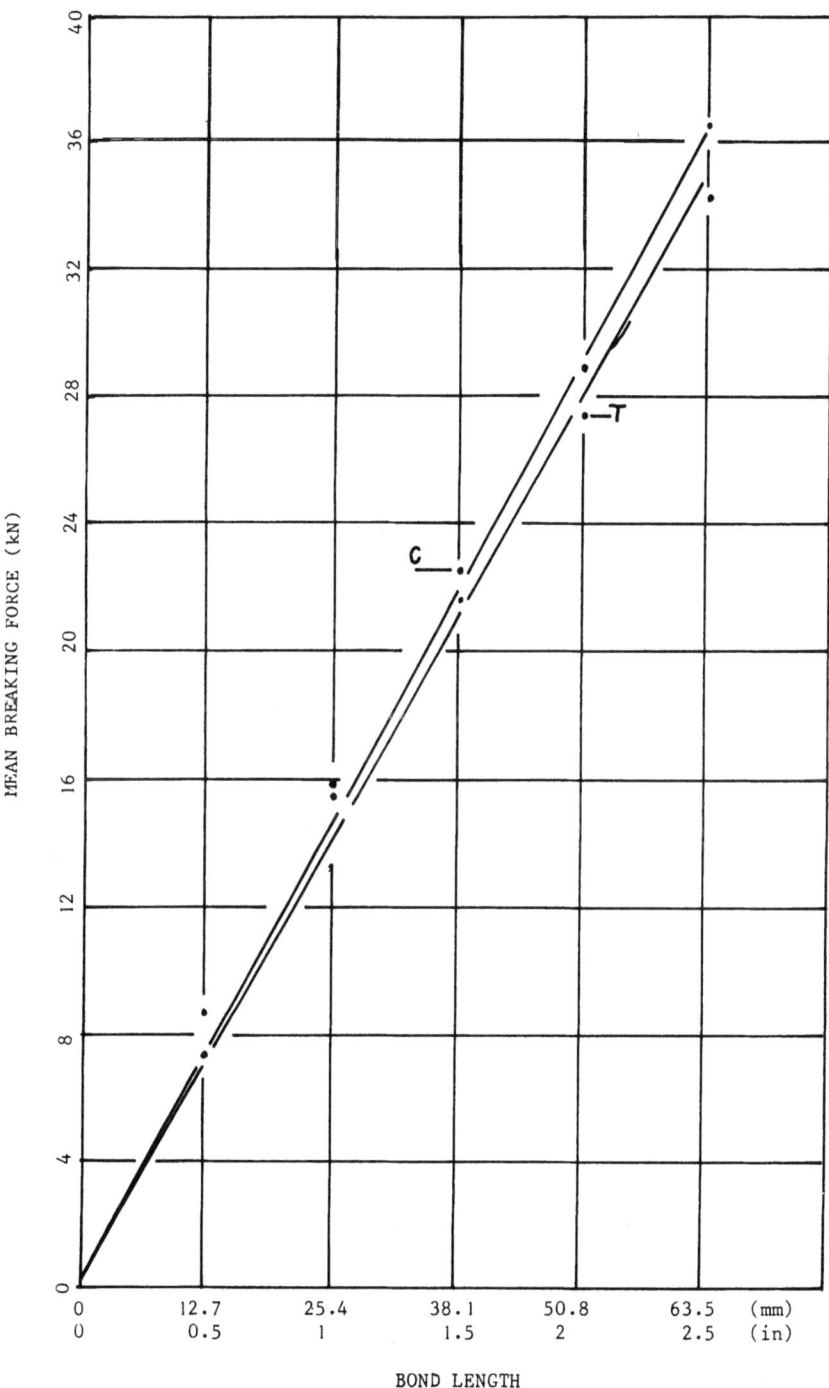

Figure 14. Multiple-wire tests, steel to steel (compression and tension tests).

Tension specimens (multiple-wires) steel to steel 2.5 in (63.5 mm) bond length see Figure 11.

$$\text{Mean shear stress of adhesive} = \frac{3.42 \times 2240}{2.4984 \times 2.5}$$

$$= 1227 \text{ lbf/in}^2 \ (8.46 \text{ N/mm}^2)$$

As a result of this comparison one may suggest that it would be feasible to use the results of mono-wire tests for glue line thicknesses beyond say 0.060 in (1.524 mm) to determine a mean shear bond stress which could be used in a situation involving a bonded structural element whose glue line thickness is variable.

The foregoing comparisons indicate a reasonable degree of consistency with the general adhesive testing.

Figure 15 shows a typical breaking line which occurred with compression and tension specimens at failure.

CONCLUSIONS

The following conclusions are indicated from results of tests employing an epoxy resin adhesive (Araldite AY 103 with Hardener HY 991).

(i) The strongest joints are obtained using thin glue lines.

(ii) The strength of a bonded joint diminishes as the glue line increases in thickness up to about 0.060 in (1.524 mm); the joint strength being approximately constant for thicker glue lines in the range 0.060 in (1.524 mm) to 0.100 in (2.540 mm).

(iii) Jointed specimens subjected to compression loads tend to be slightly stronger than corresponding joints subjected to tension, due to Poisson's ratio effect when load is applied to the joint.

(iv) A relationship does exist between mean breaking force, glue line thickness and bond length for the mono-wire joints; a separate equation being necessary for compression and tension joints.

(v) The strengths of mono-wire bonded joints for glue line thicknesses beyond 0.060 in (1.524 mm) were in reasonable agreement with the strengths of multiple-wire joints.

STRENGTH CHARACTERISTICS OF STEEL TO STEEL JOINTS

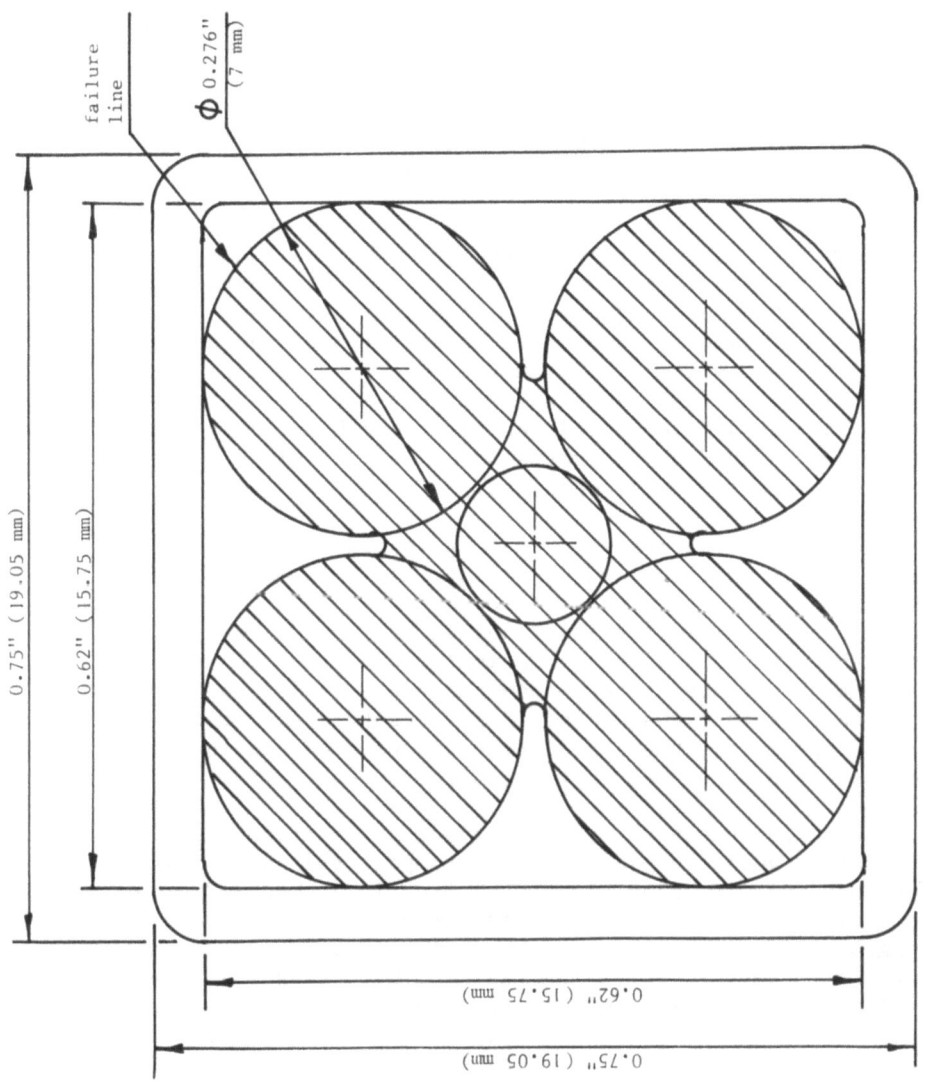

Figure 15. Typical breaking line for multiple-wire specimens.

RECOMMENDATIONS FOR FURTHER RESEARCH

The range of glue line thicknesses in this research project has been confined to 0.001 in (0.0254 mm) to 0.100 in (2.54 mm). From test results associated with the mono-wire arrangement it would appear that the ultimate strength of a bonded joint is approximately constant for glue line thicknesses between 0.060 in (1.524 mm) and 0.100 in (2.54 mm); however, it would be a useful exercise to extend this range to check if this constant relationship still exists for thicknesses beyond 0.100 in (2.54 mm).

There is also lack of information regarding the behaviour of joints having a wide variety of bond lengths and consequently more work requires to be done to obtain strength characteristics of the adhesive when subjected to an extended range of bond lengths. For example, results of tests for bond lengths varying from 100 mm to say 1 m in 100 mm increments of bond length would be invaluable. From these results graphs may be plotted and an appropriate law developed related to strength and bond length which could be used by designers whose work involves a detailed knowledge of epoxy resin adhesives.

The information derived relating to mono-wire bonded joint situations was concerned with concentrically formed joints giving a uniform glue line thickness throughout the joint. Test results data related to a similar joint having variable glue line thickness within the mono-wire arrangement would prove beneficial. Graphs may be plotted to show how the adhesive strength varies with glue line thickness under these eccentric conditions using a single parameter; these results could be compared in detail with test results of the multiple-wire joints.

A detailed study as to the surface preparation and finish of the adherend would be useful. For example, polished bright surfaces compared with sand blasted chemically etched and other surface finishes could be compared from the point of view of strength variation. To ensure consistency it is suggested that the tests be carried out using the mono-wire arrangement and a glue line thickness kept within the constant portion of the graphs of Figures 5 and 6; the only variable would then be surface preparation and finish. From the results of this work the designer could be given positive direction as to the choice and standard of finish for a particular practical situation.

All of the tests reported are associated with one epoxy resin adhesive. The author believes that similar tests should be carried out on a number of different epoxy resins and a comparison made of their structural inter-relationships.

REFERENCES

1. E. H. Sidwell, Civil Eng. 64, No. 751, 120 (Feb. 1969)
2. S. R. Port and E. H. Sidwell, Civil Eng. 64, No. 752, 226 (March 1969)
3. E. H. Sidwell and B. C. Carroll, Civil Eng. 66, Nos 774, 775, 47 (Jan/Feb. 1971).
4. E. H. Sidwell and F. F. Keatley, Civil Eng. Oct. 1974, p.28; Jan. 1975, p.50.
5. R. W. Hylands, "The bonding of the cables of a poststressed timber beam and a comparison of its performance with a similar unbonded beam". PhD Thesis Council for National Academic Awards, June 1978.
6. British Ropes Limited, Doncaster, England. "Bridon wire and strand for prestressed concrete (1962)"
7. BS 2691 (1969) "Specification for steel wire for prestressed concrete". British Standards Institution.
8. BS 1610 Grade A "Verification of testing machines, Part 1. Methods of load verification of machines for tension and compression testing". British Standards Institution.
9. BS 1203 (1963) "Synthetic resin adhesives for plywood". British Standards Institution.
10. CIBA-GEIGY (UK) Limited, Plastics Division, Duxford, Cambridge. "Handling precautions for aralidite epoxy resin materials". July 1974 Sheet No. M37c.
11. R. Houwink and G. Salomon, "Adhesion and Adhesives" Vol 2 - Applications. Elsevier Publishing Company. Amsterdam 1967.
12. The University of Kansas State Geological Survey. "Computer contributions No. 28 KWIKR8 - A FORTRAN IV program for multiple regression and geological trend analysis".

A STUDY OF ADHESIVE JOINTS BETWEEN ALIPHATIC POLYAMIDES AND METALS

S. S. Pesetskii, V. E. Starzynskii, and S. V. Shcherbakov
Institute of Mechanics of Metal-Polymer Systems
BSSR Academy of Sciences, Gomel, USSR

Injection-moulded adhesive joints of aliphatic polyamides and metals were studied and showed that intermediate thin adhesive films applied from polymer or oligomer solutions onto metals is an efficient means of monitoring the bond strength. Contact events in a metal/thin polymeric film/polyamide system have been investigated and the results are described here. The thickness of intermediate films whose optimum values range from a few tenths to 1 - 2 μm were found to be a most important parameter that determines the adhesive strength. The bond strength improvement is most obvious during a thermal contact between the intermediate film and metal in air. Adhesive bonds are formed between the metal and the film which are resistant to dynamic stresses and liquids. The bond between polyamide and the intermediate film is usually formed owing to molecular interactions of polar groups in macromolecules at the interface. The joints were found to fail cohesively through the intermediate film. The improved adhesion to metals and cohesive strength of thin intermediate films are believed to result from the fact that thermo-oxidative degradation is inhibited in these during the thermal contact with metal.

INTRODUCTION

Aliphatic polyamides contain adhesively active amide groups, yet their adhesion to metals is not strong enough. Moreover, the bond is unstable, it may change spontaneously during the storage of joint structures, or under alternate exposure to liquids and atmospheric factors, or in the course of their thermal treatment in air, etc. [1-5]. The adhesive strength of joints is also dependent on the way they are formed. If the process of joint making allows only a momentary contact between the molten polyamide and a metallic surface, the joints have low adhesive strength. This is particularly characteristic of the injection-moulded adhesive joints[6]. To improve their adhesive strength, it is advisable to introduce intermediate adhesive layers (IL) such as thin polymeric films that can be applied onto metallic surfaces prior to coating with polyamide [7-10]. However, the properties of thin ILs and characteristics of adhesive contacts with ILs have not been studied in detail.

This paper describes the results of studies of adhesive joints between aliphatic polyamides and metals which contain ILs applied from polymeric solutions.

EXPERIMENTAL

Adhesive joints of polyamides P6 (poly-ε-caprolactam), P66 (poly(hexamethylene adipamide)), P610 (poly(hexamethylene sebacamide)) with steel and aluminum have been studied. Two types of specimens were used, namely, unilateral and bilateral adhesive joints. The first type was represented by polyamide coatings (120mm x 12mm x 1mm in size) on aluminum foil (A99 grade, 100 μm in thickness). The second type consisted of two steel rods (Steel 45, 'C o.45%) and an adhesive polyamide layer (2 mm thick) to form a butt joint. The specimens were injection-moulded following the procedure published earlier [11].

For ILs, polyamide P548 (ε-caprolactam(44%)-hexamethylene adipamide(37%)-hexamethylene sebacamide(19%) copolymer) or polyvinyl butyral (PVB) were dissolved in organic solvents and applied onto the metals from solutions. The succession of the steps was as follows. The surface of the metallic substrate was placed horizontally and the solution applied onto it. Then the solvent was evaporated at 30 - 100°C. The ILs thickness was calculated as well as determined by interferometry[7].

The adhesive strength was determined by the normal separation of rods and peeling-off the foil from the coating[11]. The atmospheric resistance of the joints was tested using an accelerated method [10]. According to this method, changes in the adhesive strength with time of exposure to atmospheric conditions have been determined. One cycle of the accelerated test consists of the exposure of the specimens to 100% RH at 55°C for 3 hours along with IR(0.8 kWt) and UV (1.0 kWt)-irradiation with subsequent exposure to the temperatures -20 up to -25°C for 3 hours. The test cycle was repeated several times. The adhesive bond resistance to dynamic loading was determined from the breaking contact load as measured for polyamide-coated steel rollers paired with uncoated ones. The tests were carried out at the speed of 7 rev.s^{-1}, the load was increased stepwise by 10 kN/m after every $2.5 \cdot 10^4$ cycles.

IR-spectral analysis was performed on UR-20 spectrophotometer fitted with an attachment for a multiple frustrated total reflection (MFTR) (MFTR unit, KRS-5, number of reflections, 14; angle of incidence, 45°). The kinetics of degradation was studied by thermogravimetry (Derivatograph OD-102) using a known procedure [12] based on the analysis of the thermograms recorded at various heating rates.

Effect of Intermediate Layer Thickness on Adhesive Strength

The IL thickness (δ) was found to be a most important parameter that controls the adhesive strength. The adhesive strength (σ) is described as a function of δ and it is obvious that a maximum in σ is reached for δ values ranging from a few tenths to 1-2 μm (Figure 1 a and b).

The maximum adhesive strength is 2 - 3 times higher in the presence of IL than without it. Normal separation tests showed that it reaches the value of the tensile strength (Figure 1 a). Joints with thin ILs appeared resistant towards the operating conditions, namely, atmospheric factors (Figure 2), liquids[6], and dynamic contact loadings (dynamic tests showed that the plastic deformation of the polymeric coating, and not the breaking of adhesive bonds, cause the failure of the adhesive joints (Table 1).

The adhesive strength of joints with ILs of optimum thickness is in fact not influenced by the dissolving solvent (Table II) or the solution concentration.

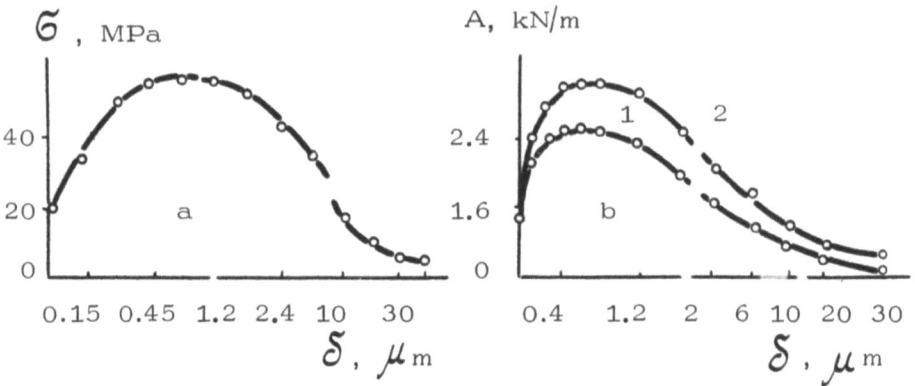

Figure 1. Variation of adhesive strength in (a) P6/Steel and (b) P6/Aluminum joints with thickness of intermediate layers (a, b - 1, P548) and (b - 2, PVB).

Table I. Load Bearing Capacity of Polyamide Coatings (2 mm thick) Bonded to Steel 45 Rollers.

Coat Material	Mode of Coat Bonding to Metal	Breaking Contact Stress, kN/m	Mode of Coat Failure
P6, P66, P610	Without IL	30–40	Separation from metal
P6, P66, P610	With IL of P548 $\delta = 1\,\mu m$	130–140	Plastic deformation

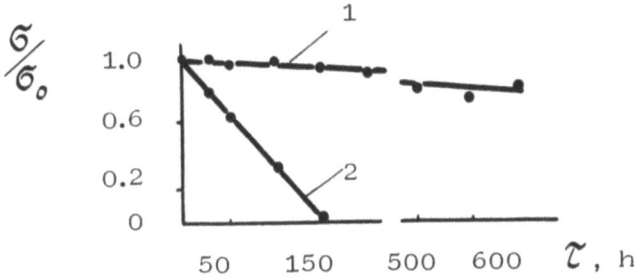

Figure 2. Variation of relative adhesive strength in P6/Steel joints (1) with P548 layer and (2) without, as a function of time of exposure. (σ_0) initial adhesive strength; (σ) adhesive strength after atmospheric action.

Table II. Effect of Dissolving Solvent Used on Adhesive Strength of P6/Steel Joints (IL of P548, δ = 0.8 μm; conditions of IL's thermal contact with substrate : temperature, 320°C; time, 0.9 ks).

Solvent	Solubility Parameter $(J/m^3)^{0.5} \cdot 10^{-3}$	Adhesive Strength, MPa
Aniline	22.44	60±2.3
N-Propanol	24.28	57±2.7
Ethanol	26.11	58±2.4
O-Cresol	27.13	57±3.0
Phenol	29.58	59±3.4
Resorcinol	32.44	58±3.2

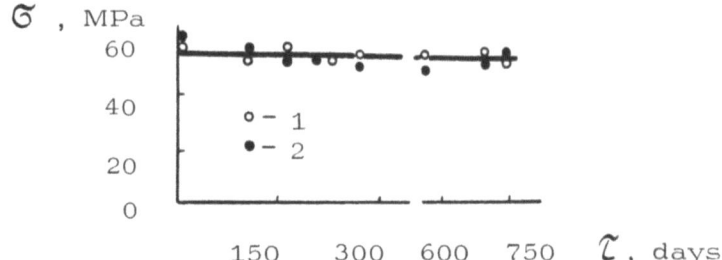

Figure 3. Effect of steel storage period (1) without IL and (2) with P548 layer on the adhesive strength of joints with P6.

Ultimate strengths similar to that for polyamides were achieved for substrates with different roughnesses. Therefore, it is believed that the adhesive properties of ILs are not affected by the time of film formation from a solution.

It is obvious from Figure 3 that the storage of metallic substrates in air at about 25°C and 70% RH, i. e. laboratory ambient conditions, for over 2 years did not affect the adhesive strength significantly. It is clear, therefore, that IL's adhesion to metals cannot result from specific interactions between ILs and the newly formed metallic surfaces.

Atmospheric factors, temperatures, and periods of IL formation on metallic surfaces are most important for adhesive strengths. The improved adhesion resulting from thin IL is most vividly manifested during thermal treatment, i.e. thermal contact (TC) between IL and metal in air. When an inert gas was used instead of air, the improvement was not so obvious (Table III). Hence, the improved adhesiveness of thin polymeric films may be expected to result from their kinetic variations and, probably, a different mechanism of thermal oxidation.

To verify this idea, the effect of time-temperature conditions of thermal contact on the adhesive strength was studied between IL and metal in air, as well as the kinetics of thermal oxidative degradation of ILs of different thicknesses on substrates.

Table III. Effect of Thermal Treatment Conditions at 280°C on Adhesive Strength of P6/Aluminum Joints.

IL Material	Thermal Contact (TC) Environment	IL's Thickness, μm	Adhesive Strength in kN/m, Against TC Period in ks			Mode of Joint Failure
			0.6	1.2	1.8	
P548	Air	1	2.20	1.94	1.86	Failed cohesively through ILs
		30	0.25	0.25	0.30	
	Argon	1	1.10	1.10	1.00	ILs separated from metal
		30	1.60	1.50	1.50	
PVB	Air	1	1.80	2.25	2.40	Failed cohesively through ILs
		30	0.90	0.80	0.75	
	Argon	1	1.30	1.30	1.25	ILs separated from polymer
		30	1.40	1.35	1.35	

Properties of ILs and Metal as Influenced by the Conditions of Their Thermal Treatment

The mode of kinetic dependence of the adhesive strength on the thermal contact (TC) temperature changes with IL thickness (Figures 4, 5). The adhesive strength of joints formed with thick ILs (δ = 30 μm) either decreases (from 1.7 - 1.3 kN/m up to 0.4 - 0.1 kN/m, Figure 4 b) on increasing the TC time (τ) or depends, in a complicated way, upon τ remaining in general low (0.4 - 0.9 kN/m, Figure 4 d). With thin ILs (δ = 1 μm) the variation of the adhesive strength depends on the temperature (Figure 4 a, c).

The mode of joint failure is also influenced by TC conditions. At the initial stages of TC ($\tau < 0.6$ ks) with ILs of δ = 30 μm, as well as over the range of minimum adhesive strengths with ILs of δ = 1 μm (Figure 4 a, c), the joints failed (as MFTR indicated) at the IL/metal interface. In other cases (irrespective of δ) a cohesive failure in IL's material was observed (cohesive mode of IL's failure was observed visually and recorded by MFTR technique (Figure 5).

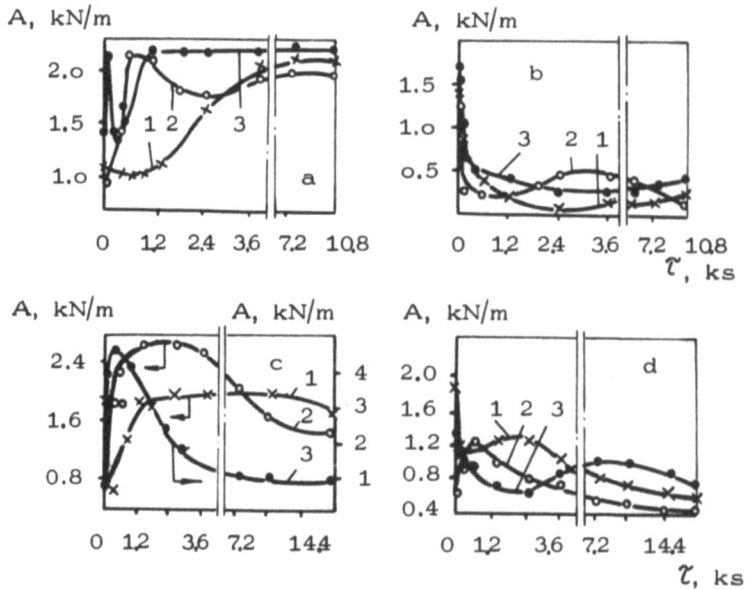

Figure 4. Variation of adhesive strength with period of TC in air between P548 (a, b) and PVB (c, d) on aluminum and P6 at: (1) 240, (2) 280, (3) 340°C. IL's thickness: a, c - 1 μm; b, d - 30 μm.

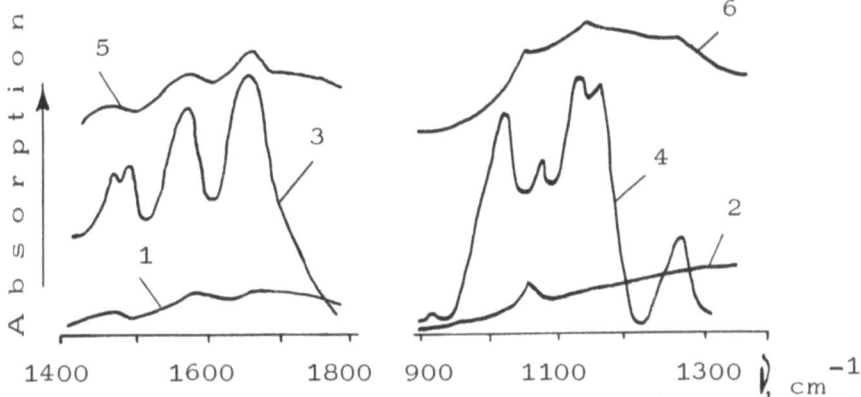

Figure 5. IR-spectra obtained by MFTR for aluminum foil-substrate (1, 2), P548 (3) and PVB (4) films 30 μm thick along with spectra of aluminum foil after P6/aluminum joint had failed (5, 6) that incorporated ILs (δ = 1 μm) of P548 (5) or PVB (6); TC conditions between ILs and substrate: temperature, 240°C; time, 3.6 ks.

Therefore, the differences in the adhesive strengths of joints with thin and thick films result from IL's thickness effect on the cohesive strength of IL's material : cohesive strengths of thin films formed in contact with metals are much higher than those of thick films.

High values of adhesive strengths with thin ILs can be achieved under severe conditions of thermal contact when the temperature is 100 - 160°C higher than the temperature of the viscous flow of IL's material, and the time of thermal contact was several hours (Figure 4 a,c, curves 2 and 3). Under such conditions blocks and bulky polymeric films undergo severe degradation. Hence, the resistance of thin films to thermal oxidative degradation should increase.

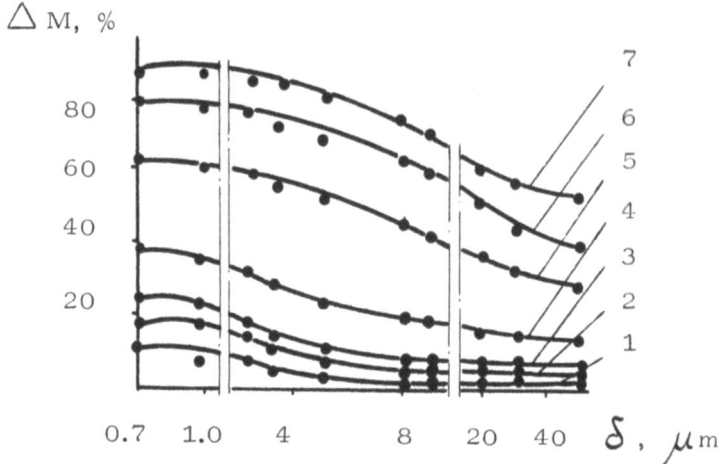

Figure 6. Weight loss with thickness of P548 films on aluminum: heating rate 3.07°C/min ; temperature of film degradation : (1) 150, (2) 200, (3) 250, (4) 300, (5) 350, (6) 400, (7) 450°C.

The kinetic curves in Figure 6 plotted based on thermogravimetric data and describing the degradation of P548 films indicate that volatile products can be easily removed from films of 2 - 3 μm thick, particularly at an early stage of degradation (curves 1 - 3). The differences are not significant in the kinetics of weight loss for films of 20 - 30 μm and thicker.

Calculations showed[6] the effective activation energy E^* of the thermal oxidative degradation of P548 thin films to be higher than that of the films 30 μm thick during the process of degradation (Table IV). With low percentage of conversion (20 - 40 %) the difference in values reached about 50 %.

Similar results were obtained with PVB films. For the conversion range of 10 to 60%, E^* for films of $\delta = 1 \mu$m exceeded 1.3 times E^* for films of $\delta = 30 \mu$m (Figure 7).

Table IV. Effective Activation Energy of P548 films on Aluminum Thermal Oxidative Degradation.

Film thickness, μm	E* in kJ/mole depending on conversion, %				
	20	30	40	80	90
1	116.9	116.9	116.9	156.3	175.1
30	76.2	76.2	76.2	149.0	141.2

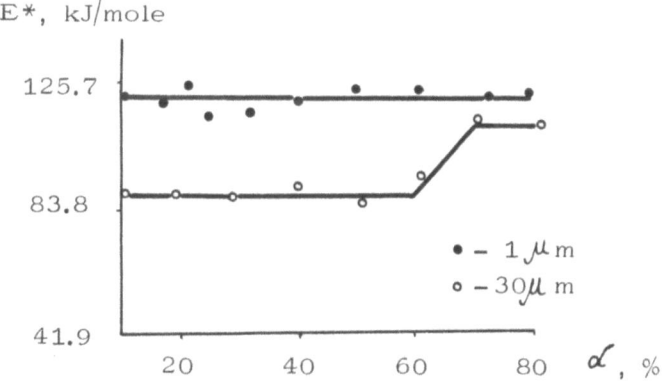

Figure 7. Variations in the effective activation energy during thermal oxidative degradation of PVB films (1 and 30 μm in thickness) on aluminum. α , conversion percentage (weight of the decomposed polymer in % with respect to the initial material).

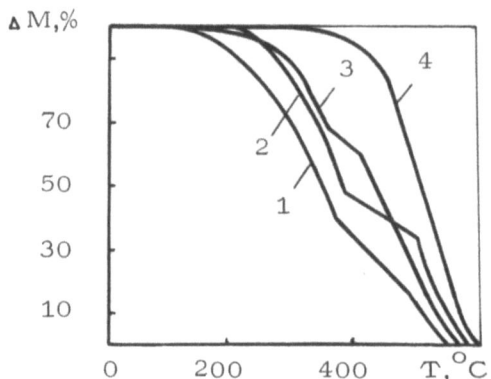

Figure 8. TG-curves of P548 films (1, 2 - 1 μm thick), (3, 4 - 30 μm thick) on aluminum. (1, 3) original films; (2, 4) films after thermal treatment on aluminum at 340°C for 10.8 ks. Heating rate 3.07°C/min.

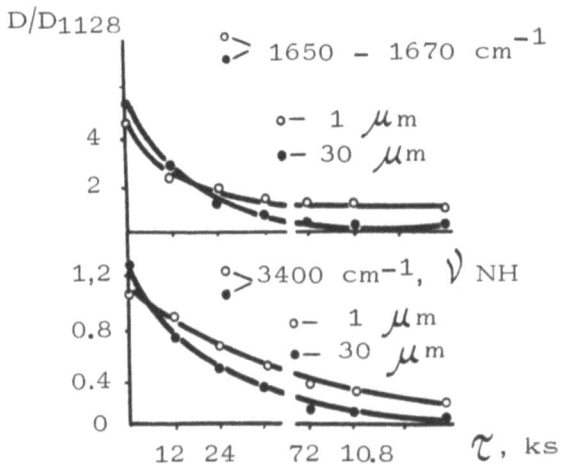

Figure 9. Variation of relative optical density of absorption bands for P548 films after their thermal treatment on aluminum foil at 240°C. D/D_{1128} is the ratio of optical density (D) of the absorption band under discussion to the optical density of the absorption band at 1128 cm^{-1}.

Keeping in mind the direct correlation between E^* measured during dynamic heating and the thermal stability of polymers under isothermal conditions, one may expect the thin films to increase their resistance to the thermal oxidative degradation under the thermal conditions as well. This expectation is also supported by TGA and IR-spectroscopy (Figures 8 and 9).

Figure 8 shows that thick films undergo more severe changes than thin ones in the course of thermal contact with metals. A film of 30 μm thick after thermal treatment at 340°C for 10.8 ks gave a TG-curve that describes a typical single-stage process of decomposition (Figure 8, curve 4) which differs radically from the curve for the original film (Figure 8, curve 3). TG-curves for films of $\delta = 1$ μm are almost the same qualitatively before the thermal contact and after that (Figure 8, curves 1 and 2).

Figure 9 describes the variations in IR-spectra of P548 films taken with MFTR in the course of TC with aluminum at 240°C. It follows, therefore, that with thin films the band related to CONH group concentration (amide I, ν NH) retains, or is quite intense after a long TC with the metal (τ = 3.6 up to 18.0 ks). Hence, the amide groups within thin films fail more easily than those within thick films under the conditions of the thermal oxidative degradation.

One can conclude from the above that the inhibition of thermal oxidative degradation of macromolecules in thin ILs during thermal contact increases their cohesive strength. The easy removal of low molecular weight products of degradation from thin films after their TC with metal also increases the cohesive strength and adhesion of these to the metal [15].

On the Formation of Contacts between Metal, IL and Polyamide

Qualitative information on the type of bonds formed during polymer/metal contact may be obtained by studying the resistance of joints to liquids. After TC with the substrate in air, thin films would not be removed from the surface with a solvent which may be associated with polymer cross linking and their chemisorption on the metal. That the adhesive IL/metal bond was not completely broken after a prolong exposure (dozens hours) in boiling water also indicates a probable chemisorption interaction.

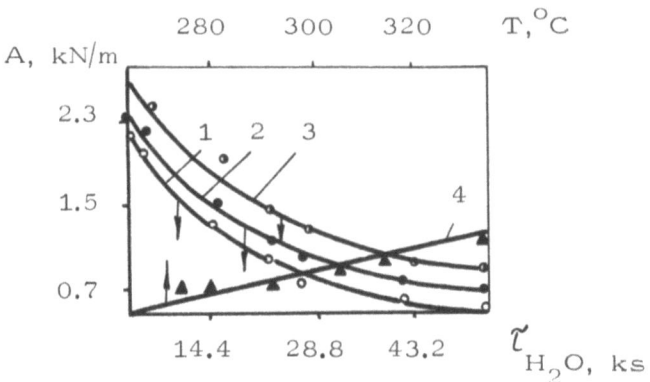

Figure 10. Variation of adhesive strength in P6/aluminum joints containing ILs of P548 ($\delta = 1 \mu m$) with time of their exposure to water (τ_{H_2O}) at 100°C. Conditions of IL's thermal contact with substrate when joints were formed: temperature, 280°C; time of contact, 0.6 ks (1, 4); 2.4 ks (2); 7.2 ks (3); (4) residual adhesive strength versus TC temperature after 60 ks of joints' exposure to water.

After water exposure, the residual strength of joints is dependent on the time-temperature regimes of TC between IL and metal. The higher the TC temperature, the stronger the residual strength (Figure 10, curve 4). At a fixed temperature the residual strength increases with TC period (Figure 10, curves 1 - 3).

After the joints were exposed to water, the amount of IL material retained on the substrate after separation decreased. Figure 11 shows that ILs treated at 280°C for 0.6 ks on the substrate were separated completely after a prolong exposure (10 - 15 h) to boiling water. The adhesive strength of joints formed on substartes which had been separated from coatings exposed to water for 10 - 15 h (curve 1) equals to the adhesive strength of the joints that were free of ILs (dashed curve). With ILs, the bond with metal did not fail completely in water after a prolong (7.2 ks) thermal contact at 280°C. The residual IL on

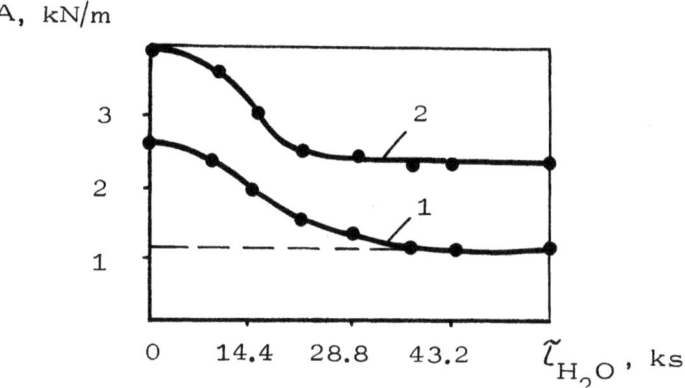

Figure 11. Variation of adhesive strength in P6/aluminum joints formed after coating was separated from aluminum against the time of specimen exposure to water ($\tilde{\tau}_{H_2O}$) at 100°C. IL was of P548 with $\delta = 1 \mu m$. TC conditions: temperature, 280°C; time of exposure : (1) 0.6 ks; (2) 7.2 ks. Dashed line, joints free of ILs.

the substrate could be seen with a naked eye and recorded with IR-spectroscopy (MFTR). When these were reused as ILs, high adhesive strengths were recorded (Figure 11, curve 2).

Thus, under proper conditions of thermal contact for ILs and the metal, adhesive bonds are formed that do not fail in water completely. Similar results have been obtained by the authors when adhesive joints were investigated after exposure to ethanol.

A polyamide/IL contact may be the result of physical adsorption, diffusion, or interdiffusion of macromolecules along with chemical bonds formed between IL and polyamide.

Polyamide/IL contacts are usually formed when polyamide is in a viscous-flow state. Since the accepted technology of specimen production allows a short period of contact (below 10 - 20 s) between the molten polyamide and IL, and because IL undergoes cross linking after which the molecular (segmental) mobility is restricted in it

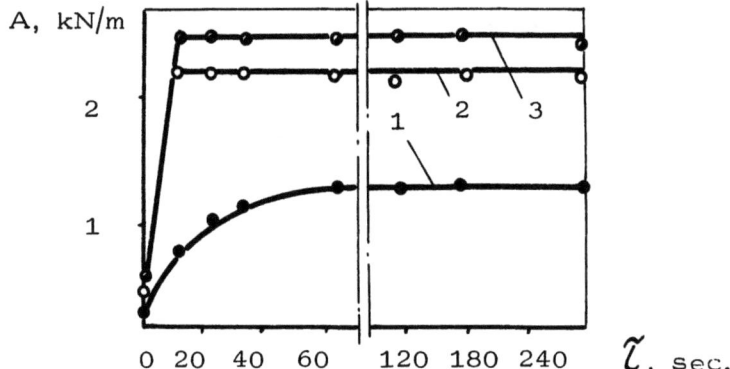

Figure 12. Adhesive strength of P6/aluminum joints against the period of P6 contact with PVB intermediate layer ($\delta = 1\,\mu m$) at 225°C. (1) heat untreated IL; (2, 3) heat treated IL at 280°C in air for 0.9 ks and 7.2 ks, respectively.

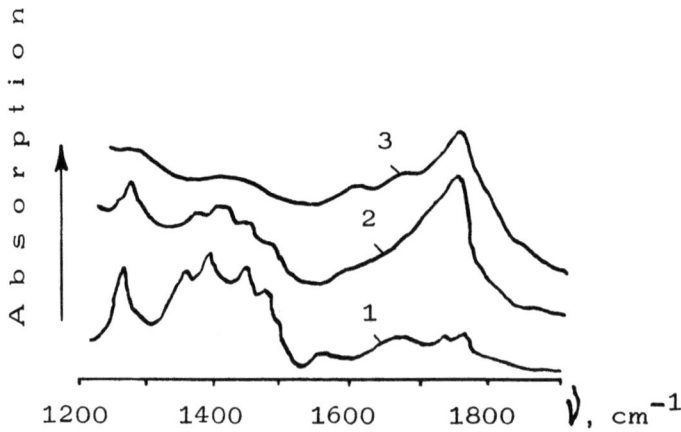

Figure 13. IR-spectra (MFTR) of PVB films ($\delta = 1\,\mu m$) on aluminum foil. (1) heat untreated film; (2, 3) heat treated film at 280°C for 0.6 and 3.6 ks, respectively.

during thermal contact with metals [7,8], the diffusion mechanism of contact formation seems almost unlikely[16,17]. With short-time contacts of thermoplastics, chemical reactions seem unlikely between these. Hence, it is believed that the physical intermolecular relations make a major contribution to the bond development between IL and metal. This idea is supported by the results of kinetics study of the adhesive contacts when PVB was used for ILs (Figure 12).

The joints produced with heat untreated PVB intermediate layers possess low adhesive strengths after instantaneous contact and fail at the P6/PVB interface. Their strengths improve gradually with the time of contact (Figure 12, curve 1). Heat treated ILs give adhesive strengths similar to the ultimate magnitude after a short contact period (Figure 12, curves 2 and 3). The heat treated ILs in air provided a higher concentration of oxygen-containing polar groups (carbonyl groups in particular) in PVB surface layers which is shown through an increased IR-absorption at 1730^{-1} cm (Figure 13). Thus, the surface oxidation of intermediate layers (ILs) improves the bonding capacity in IL/polyamide systems.

It may be stated from the above that intermolecular relations of macromolecular polar groups at the interface play a major role in the development of contact between the coating and intermediate layer.

CONCLUSION

The application of intermediate adhesive layers in the form of thin films (applied from polymeric solutions) onto metals appears an efficient means of controlling the adhesive strength of aliphatic polyamides/metal joints. One of the major parameters that determine the adhesive strength was found to be the IL thickness, the optimum value of which ranges from a few tenths to $1 - 2\,\mu$m. The improvement in adhesive strength is most obvious during thermal contact of thin IL with a metal in air. An adhesive bond develops between the metal and IL which is resistant to dynamic loadings and liquids. Polyamide/IL bonds result usually from intermolecular interactions of macromolecular polar groups at the interface.

These investigations reveal that a cohesive mode of joint failure through IL material is dominant. It is suggested that the inhibition of thermal oxidative degradation in thin ILs under thermal contact with metal improves their adhesion to the metal, and also the cohesive strength.

REFERENCES

1. S.S. Pesetskii, N.I. Egorenkov, and S.V. Shcherbakov, in "Composite Polymeric Materials", Issue II, p. 39, Naukova Dumka, Kiev, 1981.
2. S.S. Pesetskii, N.I. Egorenkov, and S.V. Shcherbakov, Colloid Journal, No. 5, 992 (1981).
3. N.I. Egorenkov, V.N. Mlynskii, and E.N. Sokolov, Mechanics of Polymers J., No. 1, 40 (1978).
4. V.N. Mlynskii, N.I. Egorenkov, and V.A. Belyi, Physico-Chem. Mechanics of Materials J., No. 2, 83 (1975).
5. N.I. Egorenkov, A.I. Isterin, and V.L. Potekha, Mechanics of Polymers J., No. 1, 170 (1976).
6. S.S. Pesetskii, "Study of Moulded Polyamide/Metal Adhesive Joints". Dissert. Thesis, Riga, 1980.
7. S.V. Shcherbakov, V.E. Starzhynskii, and S.S. Pesetskii, BSSR Academy of Sciences Reports, $\underline{20}$, No.12, 1086 (1976).
8. S.S. Pesetskii and S.V. Shcherbakov, BSSR Academy of Sciences Trans., Ser. Phys. & Techn. Sci., No. 1, 45 (1979).
9. W.A. Belyj, S.W. Scherbakow, und S.S. Pessezkij, Plaste und Kautschuk, $\underline{27}$, No. 11, 642 (1980).
10. S.V. Shcherbakov, S.A. Osipenko, and S.S. Pesetskii, BSSR Academy of Sciences Trans., Ser. Phys. & Tech. Sci., No. 1, 127 (1974).
11. S.V. Shcherbakov, S.S. Pesetskii, S.A. Osipenko, and O.N. Kurganova, BSSR Academy of Sciences Trans., Ser. Phys. & Techn. Sci., No. 3, 121 (1976).
12. H.L. Fridman, in Polymer Preprints, $\underline{4}$, No.2, 662 (1966). J. Polymer Sci., $\underline{6}$, 183 (1964).
13. M.B. Neuman, Editor, "Aging and Stabilization of Polymers", Nauka, Moscow, 1964.
14. E.T. Gevorkyan and L.V. Baranova, High Molecular Weight Compounds J., $\underline{196}$, No. 10, 758 (1977).
15. V.P. Karlivan, et al., Plastics J., No. 11, 46 (1976).
16. S.W. Stscherbakow, Herstellung verstarkter metallpolymerer Erzeugnisse durch Spritzguß. Sammelband "Verstarkte Plaste-80", S. K.9.1-K.9.8, Dresden, DDR, (1980).
17. A.A. Berlin and V.E. Basin, "Fundamentals of Polymer Adhesion", Khimia, Moscow, 1974.

"HONEYMOON" PHENOLIC FAST-SETTING ADHESIVES FOR EXTERIOR-GRADE FINGER JOINTS

A. Pizzi

National Timber Research Institute
Council for Scientific and Industrial Research
P.O. Box 395, Pretoria 0001, Republic of South Africa

All fast-setting adhesive systems tested proved satisfactory for accelerated finger joints production as they show development of strength at ambient temperature rapid enough to allow finger joint machining five to 30 minutes after manufacture, according to the resin system used. The results of accelerated weathering tests showed all the adhesive systems to be weather- and boil-proof. The main advantages obtained were:

(i.) A considerable decrease in the cost of phenolic adhesives for fingerjointing obtained by the use of tannin extracts, and

(ii) improvement in production flow obtained by the decrease of the time delay between the manufacture of the finger joints and their machining or dispatch from an overnight period to approximately ten to 30 minutes at ambient temperature according to the adhesive system used.

INTRODUCTION

Finger joints are commonly used in South Africa to produce long boards from short-length timber. Such joints are acceptable in structural timber and in laminations for glulam (Table I). The adhesives normally used for finger jointing are melamine/urea/formaldehyde and phenol/resorcinol/formaldehyde which require lengthy periods or temperature higher than ambient to set. There is, therefore, a one-day delay between fingerjointing and further processing and despatch which is inconvenient and interferes with production flow.

Separate application adhesives capable of setting faster than conventional adhesives were developed in the U.S.A.[10,11] and in other countries[7,12] to glue large components where presses were impractical[1-4]. Kreibich[4] describes these "honeymoon" systems as follows: "Component A is a slow-reacting resin with a reactive hardener. Component B is a fast-reacting resin with a slow-reacting hardener. When A and B are mated, the reactive parts of the components react within minutes to form a joint which can be handled and processed further. Full curing of the slow-reacting part of the system takes place with time." (Figure 1).

In South Africa the basic "honeymoon" gluing system was considerably modified by: (i) the use of tannin-based rather than synthetic phenols for both components A and B; (ii) the elimination of the hardener from component B, and (iii) the elimination of the expensive m-aminophenol from component B.

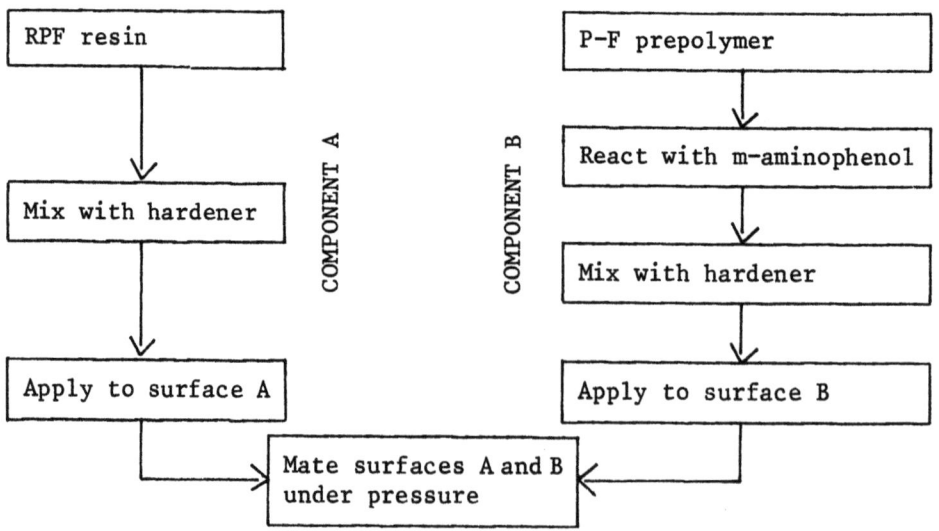

Figure 1. Flow diagram of fast-set adhesives according to Kreibich[4].

Table I. Finger joint dimensions.

The dimensions in a finger joint profile should be as follows (see Figure of Finger Joint Profile, below):

1. The finger length, L, should be at least 7.5 mm.
2. The pitch should not exceed the appropriate of the following maxima:

Finger length, L, mm	Pitch, p, mm, max.
7.5	2.5
Over 7.5 up to and including 10	$\frac{L}{2.5}$
Over 10 up to and including 24	$\frac{L}{3.0}$
Over 24 up to and including 35	$\frac{L}{3.5}$
Over 35	$\frac{L}{4.0}$

3. The weakening ration v=b/p, where b=width of slot-base should not exceed the appropriate of the following values:

Finger length, L, mm	Weakening ratio, v, max.
7.5	0.08
Over 7.5 up to and including 35	0.16
Over 35	0.20

NOTE: The maximum slot-base width can be calculated, the relationship b=vxp being used.

4. The finger-tip thickness should be within the limits 1.1b-1.2b.
5. The profile dimensions should be such that the clearance, c, in the glued joint (after assembly), is not greater than 0.08L of 4.00 mm, whichever is smaller.
   NOTE: The presence of some clearance is desirable in joints of any finger length.

Figure Joint Profile

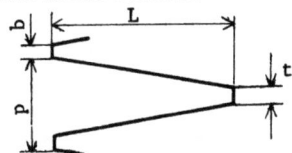

L = Finger length
b = Slot-base width
p = Pitch
t = Thickness of finger-tip

The following end pressures are recommended:

| Finger length, L, mm | End Pressure, MPa, min. | |
	Dense hardwood[1]	Softwoods and other hardwoods
7.5 up to and including 10	15.3	7.8
Over 10 up to and including 21	12.75	5.8
Over 21 up to and including 31	10.2	4.0
Over 31 up to and including 41	7.65	2.0
Over 41	5.1	1.0

1) Of density, at a moisture content of 12%, greater than 600 kg/m^3

This allowed the economical use of these types of adhesives. South Africa is the first country to have used industrially and extensively these types of adhesives for fingerjointing.

In the original U.S.A. adhesive, component A is a phenol/resorcinol/formaldehyde cold-setting adhesive to which paraformaldehyde hardener has been added. Coconut shell and wood flour fillers may be added or omitted from the glue mix. Component B is instead a phenol/m-aminophenol/formaldehyde resin. This system has been proved to be successful at laboratory levels. While finger joints manufactured with this adhesive proved to be excellent, both in speed of curing and durability, a few problems were encountered in the availability, ease of supply and high price of m-aminophenol chemical.

This study describes the development of alternative fast-setting adhesives not containing m-aminophenol.

The adhesives used for components A and B of the "honeymoon" systems are shown in Table II.

All the combinations of components A and B were used to manufacture finger joints. Particular attention was paid to, and extensive tests were done with, the adhesive systems A1/B1, A1/B2 and A1/B3.

To better understand the function of the different A and B components of the different adhesive systems presented, it is necessary to give some background about the resins they are composed of.

Table II. Adhesive Components (A and B) used for "honeymoon" systems.

Component A	Component B
A1. Commercial phenol/resorcinol formaldehyde cold-setting laminating adhesive + hardener + fillers (pH=8,0)[1]	B1. Pine tannin extract; no hardener; pH adjusted to 12.4
	B2. Wattle tannin extract[2]; no hardener; pH adjusted to 12.6
	B3. Commercial phenol/resorcinol/formaldehyde resin[2]; no hardener; pH adjusted to 11.4

1) Cascophen RS12 supplied by General Chemical Corporation Ltd, Johannesburg, South Africa.
2) Mimosa extract supplied by Natal Tannin Extract Co. Ltd, Pietermaritzburg, South Africa

Components A

A1. Phenol/resorcinol/formaldehyde laminating adhesives. This is one of the many resorcinol/terminated phenolic resins available on the market for cold-setting laminating applications. All these classes of adhesives are weather- and boil-proof.

Components B

B1 and B2. Pine bark tannin extract and wattle bark tannin extract. The phenolic nature and consequent behaviour of pine and wattle tannins is evident by the schematic structure of their main monomeric constituents.

Wattle Tannin   Pine Tannin

These schematic wattle and pine tannin structures represent the two major types of flavonoid tannins known, wattle tannins being composed of mainly resorcinolic A-rings and pine tannins by phloroglucinolic A-rings[5-8]. The pyrogallic and catecholic B-rings do not take part in the reaction with formaldehyde to any noticeable extent, up to pH=+10, due to their low nucleophilic character[5-8]. A further difference between these two tannins is the mode of linkage of the flavonoid units 4,6 in the case of wattle tannins and 4,8 in the case of pine tannins.

The stronger nucleophilicity of the pine tannins phloroglucinolic A-rings causes pine tannin to be faster reacting, with formaldehyde, at comparable pH's, than the resorcinolic A-rings of the wattle tannins. As a consequence pine tannin extract, with the pH adjusted to 12.4 was a more logical choice as a component of the "honeymoon" system as it should allow faster curing, hence shorter time delay before finger joint machining, than wattle tannin extract.

B3. Phenol/resorcinol/formaldehyde without any hardener and with pH adjusted to 11.4.

The bark extracts of the black wattle tree (*Acacia mearnsii*, formerly *mollissima*), commercially available and containing 70 %

to 80 % tannins, and of the pine species *Pinus patula*, pilot plant produced by ambient temperature countercurrent methanol extraction were used for this study.

The influence of the following parameters on the performance of the different adhesive systems and of the finished finger joints were evaluated: (i) wood density, (ii) wood species, and (iii) tolerance to quantity variations, namely variations of the component A : component B ratio of 2 : 1; 1 : 1 and 1 : 2 were investigated.

To show that the "honeymoon" systems curing speeds are considerably faster than conventional adhesive systems, finger joints were also prepared by gluing both finger joint profiles with a commercial phenol/resorcinol/formaldehyde resin + hardener and with a commercial wattle/resorcinol/formaldehyde resin + hardener of unadjusted pH's. Thus, the components A1 and A2 were used to glue both sides of the finger joints in order to evaluate the rate of strength increase on curing of traditional fingerjointing adhesives.

EXPERIMENTAL

Adhesive Resins Preparation

The phenol/resorcinol/formaldehyde resin used was a commercial product. However, any of the numerous phenol/resorcinol/formaldehyde resins commercially available can be used with similar results.

Glue Mixes

(All pH's adjusted with NaOH 30 % to 40 % solution.)

Component	Parts by weight
A1. PRF liquid resin (52 % solids)	100
hardener + fillers, powder	24 (20 paraformaldehyde)
pH	8.0
pot-life	1.5 to 3.0 hours at 25 °C
B1. Pine extract solution (50 % solids)	100
pH	12.4
pot-life	Indefinite as no hardener is added
B2. Wattle extract solution (50 % to 55 % solids)	100
pH	12.6
pot-life	indefinite as no hardener is added

B3. PRF liquid resin   100
    pH   11.4
    pot-life   indefinite as no hardener
                                    is added

Fingerjointing

Boards of South African pine (*Pinus patula, Pinus radiata, Pinus pinaster, Pinus taeda* and *Pinus elliotii*) of dimensions 500 mm x 120 mm x 37 mm and of *Eucalyptus grandis* of dimensions 500 mm x 75 mm x 25 mm were mechanically fingerjointed, using the various adhesive systems presented. All the adhesive systems were used to produce finger joints to be tested 30 minutes, one hour, two hours, four hours, 24 hours and seven days after gluing to evaluate the rate of strength increase with time.

The experimental series of finger joints which were manufactured were the following:

General. Finger joints were prepared using medium density pine (approximate density range 0.450 g/cm³ to 0.550 g/cm³), using the adhesive systems presented.

Effect of wood density. Finger joints were prepared using high density pine (approximate density range 0.650 g/cm³ to 0.750 g/cm³), using the A/B adhesive systems A1/B2 and A1/B3.

Tolerance to quantitative and viscosity variations.
As it is practically impossible, under factory production conditions, to have identical quantities of adhesive on both the two profiles of each finger joint, it was deemed necessary to investigate the tolerance of the adhesive systems to variations of quantities. This was achieved by varying the glue mix viscosity of the two adhesive components in order to have approximate proportions by weight of A : B of 2 : 1; 1 : 1 and 1 : 2. The viscosities used were as follows:

Component A	Component B	Approximate mass
3.8 Pa.s [1]	2.0 Pa.s	(2 : 1)
1.9 Pa.s	1.9 Pa.s	(1 : 1)
1.9 Pa.s	3.5 Pa.s	(1 : 2)

1) 1 Pa.s = 1 000 centipoises

The A/B adhesive systems used, were 1/2 and 1/3. Gel times at 60 °C of mixtures of different mass ratios of component A :

component B, namely 25 : 75; 33 : 67; 50 : 50; 75 : 25 and 100 : 0 were also measured. The gel times results for the A/B adhesive system A1/B1 are shown in Table III.

Control. Series of finger joints in which both profiles of each finger joint were glued with component A1 were also prepared. This was deemed necessary to compare the rate of strength increase of the "honeymoon" adhesive systems with that of more traditional cold-setting adhesives for finger joints.

Testing

The finger joints prepared were tested to failure in center loading, three points, bending, 30 minutes, one hour, four hours, 24 hours and seven days after gluing and assembling. Breaking strength in bending for a span of 914.4 mm as well as percentage wood failure of the broken finger joints was assessed according to the South African Bureau of Standards SABS 096-1976 specification for structural finger joints[9]. The SABS specification requires a minimum of 20 MPa in bending. Several samples for each time span were tested. For all the finger joints series prepared with the different adhesive systems, accelerated weathering tests according to the South African Bureau of Standards SABS 096-1976 specification[9] for structural finger joints were carried out seven days after gluing and assembling. The accelerated weathering tests consisted of testing in tension, with a universal testing machine, finger joint samples of 250 mm x 25 mm x 3 mm, dry, after 24 hours cold water soaking and after six hours boiling. The SABS 096-1976 requirements for exterior grade structural finger joints after 24 hours cold water soaking and/or six hours boiling are the following:

Minimum failing load (Newtons)	Wood failure (%)
1 400	90-100
1 700	70-89
2 100	50-69
2 500	30-49
2 800	10-29

The average shear strength and percentage wood failure results obtained for the different adhesive systems and finger joint series are shown in Table IV.

Table III. Gel time at 60 °C of adhesive A1/B2 containing different ratios of components A and B.

A Parts by Mass	B Parts by Mass	Mean Gel Time Seconds
25	75	50.0
33	67	39.3
50	50	37.3
67	33	39.3
75	25	45.6
100	0	320.0

Statistical analysis

As the total experiment involved nearly 1 000 finger joints, the large sample allowed meaningful statistical analysis to be applied to the results. In this type of experiment the most useful information concerns the relationship between the strength of a joint and the period of time after gluing and assembling. The strength versus time curves obtained are of considerable importance as they show how long after gluing and assembling the finger joint has enough strength to be machined or dispatched without damages. This relationship can normally be expressed in some form of regression equation or graphic curve.

The analysis of the experimental results was looked at according to two different approaches. In the first method one adhesive system was selected and simple regression analysis was used to establish the relationship of strength as a function of the density of the timber. Several regressions were performed where each equation corresponded to a unique gluing time, viz. for 30 minutes, one hour, two hours, four hours, 24 hours and seven days. The strength versus timber density graphs were plotted for the 30 minutes to four hours period as this was the more interesting time span. A certain density was then selected and the corresponding strength value was determined from the strength versus density curves for each time span. These strength versus time span values were then plotted. This procedure was repeated for the rest of the adhesive systems.

In the second and more comprehensive approach it was decided to use multiple regression analysis to obtain, for a certain adhesive system, an equation which would describe strength in terms of the other variables. The independent variables considered were the following:

Table IV. Average Shear Strength and Wood Failure Results of S.A. Pine and E. saligna Finger Joints Prepared using A1/A1; A1/B1 and A1/B3 Adhesive Systems.

Adhesive System	S.A. Pine							E. saligna						
	Dry		24h cold soak		6h boil		Average density (g/cm³)	Dry		24h cold soak		6h boil		Average density (g/cm³)
	Strength, tension, (N)	Wood failure, (%)	Strength, tension, (N)	Wood failure, (%)	Strength, tension, (N)	Wood failure, (%)		Strength, tension, (N)	Wood failure, (%)	Strength, tension, (N)	Wood failure, (%)	Strength, tension, (N)	Wood failure, (%)	
A1/A1	3 720	96	2 132	76	2 170	87	0.500	–	–	–	–	–	–	–
A1/B1	5 252	90	3 474	85	3 558	89	0.512	–	–	–	–	–	–	–
A1/B3	4 060	94	2 658	94	2 824	92	0.531	3 985	88	4 182	78	3 538	80	0.481

1 - curing temperature in °C
2 - curing time in hours
3 - percentage wood failure
4 - wood density in kg/m³

The dependent variable was the modulus of rupture (MOR) expressed in MPa. As the points obtained appeared to follow a logarithmic law for the curing time, it was decided to use both the curing time and its natural logarithm in order to obtain more elastic and pliable equations able to better describe the phenomenon. The multiple regression analysis was performed with forced variables and not-forced variables. The multiple regression equations with the relative coefficients for the statistically significant independent variables and the percentages of the phenomenon explained for one of the adhesive systems for which there were enough data to perform a multiple regression analysis are shown in Table V.

Table V. Example of dependence, by non-linear multi-regression analysis, of MOR increase from curing time, temperature, percentage wood failure and wood density for S A Pine finger joints glued with a fast-setting adhesive system.

Adhesive system A1/B1

- Unforced variables:

$$\text{MOR} = 2.773 \text{ temperature} + 5.285 \ln \text{time} - 48.308$$

Percentage of phenomenon explained = 79.00 %
Level of significance of variables:
- curing temperature   = 100.00 %
- ln of curing time   = 100.00 %

- Forced variables:

$$\text{MOR} = 2.119 \text{ temperature} + 0.047170 \text{ time} + 0.075485 \text{ wood failure} + 0.029009 \text{ density} + 4.026 \ln \text{time} - 50.583$$

Percentage of phenomenon explained = 80.95 %
Level of significance of variables:
- curing temperature   = 98.97 %
- curing time   = 75.58 %
- percentage wood failure = 88.79 %
- wood density   = 93.24 %
- ln of curing time   = 100.00 %

INDUSTRIAL APPLICATION

In Tables VI and VII are shown the testing results of finger joints manufactured in a three-day production factory trial using the adhesive systems A1/B2 and A1/B3. To investigate if planing of the finger joints 30 minutes after manufacture would damage the glue line and decrease its strength, finger joints were tested rough, as manufactured, and after planing. From the bending and tensile strength results it appeared that planing of the joints 30 minutes after manufacture has no bearing on the strength of the joint. The joints were tested 30 minutes, one hour, two hours, 24 hours and seven days after manufacture. All the components A and B of the adhesive system used had been industrially manufactured. The viscosities of the adhesive systems used, were the following: A1/B2 = 3 400/3 200 cps; A1/B3 = 3 800/3 700 cps. As a consequence of this industrial production trial, the South African Bureau of Standards accepted the new adhesive systems as suitable for exterior structural grade finger joints of timber of grades 4, 6 and 8. Timber of grades 4 and 6 constitute $\pm$99 % of the timber finger-jointed in the Republic of South Africa.

DISCUSSION

The results shown in the figures and tables indicate that all the adhesive systems tested perform sufficiently well to allow fast binding of finger joints. The interesting feature of all the adhesives tested is that notwithstanding the differences in curing and setting rates among the different systems, they are fast enough in developing sufficient strength to allow short delay between assembly and machining or dispatching. The systems A1/B1 and A1/B2 are of considerable economical importance as a proper manufactured adhesive is used only on the "A" profile of the finger joint while the "B" profile is coated with a considerably cheaper and easily available tannin extract.

The curves in Figure 2 show that phenol/resorcinol/formaldehyde (PRF) adhesives are very well suited to the function of components A and B, and that tannin extract are as suitable to the function of component B as PRF resins. Very little difference exists in the results obtained with the A1/B1 and A1/B3 resin systems.

The curves in Figure 2 clearly indicate that the rate of strength development in finger joints glued with conventional adhesives are much slower than for joints glued with the various "honeymoon" adhesives. The 20 MPa level is reached, for conventional phenolic cold-setting adhesives, in 100 minutes for the A1/A1 system, hence in periods considerably longer than with "honeymoon" systems.

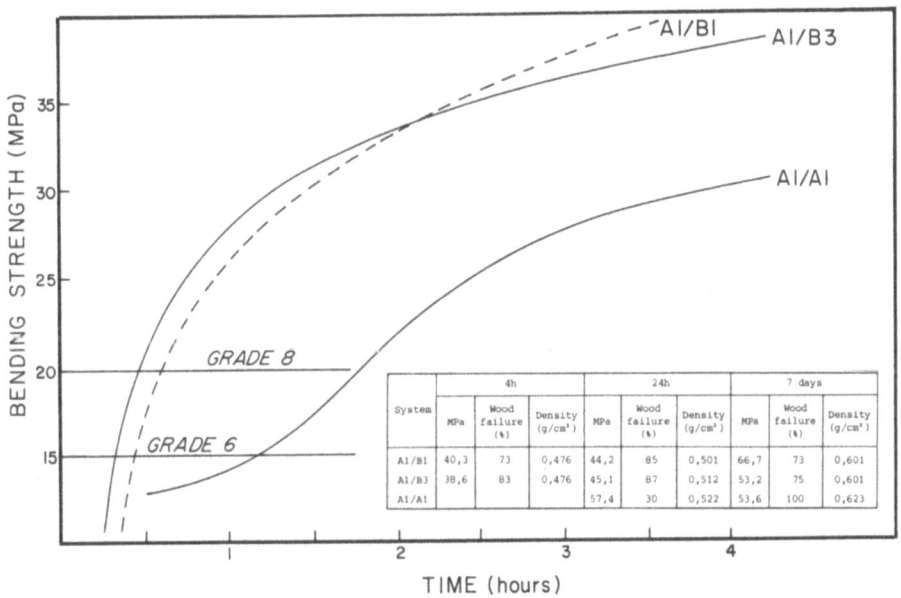

Figure 2. Development of finger joint bending strength as a function of time. Systems A1/A1; A1/B1; A1/B3 = S.A. Pine.

Table VI. Development of Bending Strength versus Time; Industrial Production Finger Joint Trial.

Curing time (hours)	Treatment	Mean Values[1]	A1/B2	A1/B3
		Ambient Temperature °C	21	24
0.5	Planed	MOR (MPa)	8.6	9.6
		Density (kg/m³)	538	555
		Wood failure (%)	2	1
	Rough	MOR (MPa)	8.5	12.1
		Density (kg/m³)	527	531
		Wood failure	1	0
1.0	Planed	MOR (MPa)	12.5	11.5
		Density (kg/m³)	617	612
		Wood failure (%)	2	5
	Rough	MOR (MPa)	12.9	13.4
		Density (kg/m³)	598	584
		Wood failure (%)	2	2
2.0	Planed	MOR (MPa)	18.6	16.8
		Density (kg/m³)	505	501
		Wood failure (%)	26	18
	Rough	MOR (MPa)	17.1	16.2
		Density (kg/m³)	510	480
		Wood failure (%)	19	18
24.0	Planed	MOR (MPa)	35.4	34.2
		Density (kg/m³)	527	640
		Wood failure (%)	100	70
	Rough	MOR (MPa)	35.1	33.0
		Density (kg/m³)	512	633
		Wood failure (%)	91	47

1) Mean taken of five specimens

Table VII. Accelerated Weathering Test of Finger Joints Manufactured Industrially. Tension Test of 250 mm x 25 mm x 5 mm Specimens.

Adhesive system		Dry Failing load (N)	Dry Wood failure (%)	24h Cold Water Soak Failing load (N)	24h Cold Water Soak Wood failure (%)	6h Boiling Water Failing load (N)	6h Boiling Water Wood failure (%)
A1/B3	Planed	5 741	97	3 467	25	3 946	47
	Rough	5 846	97	3 833	34	4 247	54

Each value presented in this table is the mean taken of 25 specimens.
Thanks for this table are due to the South African Bureau of Standards.

The results of the accelerated weathering tests showed that all the finger joints produced with the various "honeymoon" adhesive systems satisfy with ease the requirements for weather- and boil-proof fingerjointed structural timber.

Interesting is also to compare the equations correlating the modulus of rupture (MOR) in bending with other independent variables obtained by non-linear multiple regression analysis for adhesive systems[13]. This comparison gives an insight on the differences in intrinsic characteristics of the various resins which are related to their molecular structure. The independent variables which mostly contribute to the MOR values are those appearing in the unforced variables equations[13]. The MOR of the conventional control adhesive, which is quite slow-curing, depends mainly on the length of time the adhesive is left to cure[13]. The MOR of the adhesive system containing the very reactive m-aminophenol is instead quite independent of the curing time as high MOR's are achieved very quickly after assembly of the joint. The latter adhesive system is though strongly dependent on the density of the wood used (both the percentage wood failure and the wood density variables appear in the equation). Such a behaviour has often been noticed in industrial practice and it is caused by the rapid thickening and cure, on assembly, of the m-aminophenol-fortified B1 component, the rate of setting and molecular size increase of which causes so fast a thickening to allow only limited penetration of the adhesive in the wood grain and adhesion only to the end-grain of the more porous, lower to medium densities, wood species.

CONCLUSIONS

Pine and eucalyptus finger joints manufactured with the new adhesive systems presented develop sufficient strength to allow further processing within a very short period (five to 30 minutes at 25 °C to 28 °C).

The "honeymoon" adhesive systems tested are capable of such fast curing that pine and eucalyptus finger joints comply with the requirements of the South African Bureau of Standards SABS 096-1976 specification for the manufacture of fingerjointed structural timber for bending strength (MOR) within 15 to 45 minutes, and for weather- and boil-proof tensile strength within one week from assembly.

The main advantages of these adhesives are economical as:
(i) the delay between the manufacture of fingerjointed timber and further processing or dispatch is decreased from an overnight period to anything between five and 30 minutes according to the adhesive system used. This eliminates the accumulation of a full-day production stock and therefore improves production flow. This is

achieved without altering the existing equipment and method of production, and can therefore be easily implemented; (ii) components B1 and B2 are simple tannin extracts with adjusted pH. B2 is commercially available at the ridiculously low price of only 13 % to 15 % of commercial phenol/resorcinol/formaldehyde adhesives mass/mass and solids/solids. The coupling of the A2 component, commercial phenol/resorcinol/formaldehyde adhesive, with the B1 and B2 components also considerably decrease costs in adhesives for fingerjointing. The A1/B3 system, all synthetic, disprove, once and for all, the need of addition of fast-reacting and expensive components, such as m-aminophenol, to fast-setting finger joint adhesives as claimed by previous authors[1,4,11,12]. Unnecessary is also the preliminary warming of the timber while using conventional phenol/resorcinol/formaldehyde adhesives as now sometimes advocated in North America; (iii) increased independence from oil-derived synthetic adhesives and from a scarce, expensive and not easily available raw material such as m-aminophenol, and (iv) use of only one adhesive for beam laminating and fingerjointing; the preparation of the finger joint adhesive systems requires only the addition of one component, component B, to the usual beam-laminating adhesives.

Disadvantages of these systems are: (i) preparation of the adhesive is complicated by the fact that it consists of more ingredients than conventional adhesives, namely two resins and one hardener against the one resin and one hardener of conventional systems; (ii) complete coverage of both profiled faces is essential, and (iii) mis-matching of profiled faces after adhesive application would lead to slow curing or no curing at all.

These disadvantages though have been found to be minimal in industrial production practice. Sixty per cent of the South African fingerjointing industry is now using one of these fast-setting adhesives on a regular production basis.

REFERENCES

1. G. F. Baxter and R. E. Kreibich, For. Prod. J., 23, (1), 17 (1973).
2. W. Caster, For. Prod. J., 23, (1), 26 (1973).
3. H. Ericsson, Papper och Trä, (1), 19 (1975).
4. R. E. Kreibich, Adhesives Age, (17), 26 (1974).
5. A. Pizzi, J. Appl. Polym. Sci., 23, 2777 (1979).
6. A. Pizzi, J. Appl. Polym. Sci., 24, 1257 (1979).
7. A. Pizzi and D. G. Roux, J. Appl. Polym. Sci., 22, 1945 (1978).
8. A. Pizzi and H. O. Scharfetter, J. Appl. Polym. Sci., 22, 1745 (1978).
9. "South African Bureau of Standards SABS 096- 1976", Specification for the manufacture of fingerjointed structural timber,

Pretoria, South Africa, 18 pp.
10. G. T. Tiedeman and M. R. Sanclemente, J. Appl. Polym. Sci., 17, 1813 (1973).
11. G. T. Tiedeman, M. R. Sanclemente and H. A. Smith, J. Appl. Polym. Sci., 17, 1819 (1973).
12. P. K. van der Westhuizen, A. Pizzi and H. O. Scharfetter, Wood Southern Africa, 3, 7 (May 1978). Paper presented at IUFRO Wood Gluing Working Party S5.04, Mèrida, Venezuela, October 1977.
13. A. Pizzi, D. du T. Rossouw, W. E. Knuffel and M. Singmin, Holzforschung und Holzverwertung, 32, (6), 140 (1980).

AGEING OF STRUCTURAL FILM ADHESIVES - CHANGES IN CHEMICAL AND PHYSICAL PROPERTIES AND THE EFFECT ON JOINT STRENGTH

C.E.M. Morris, P.J. Pearce and R.G. Davidson

Materials Research Laboratories
Department of Defence Support
Ascot Vale, Vic., 3032, Australia

Epoxy-based film adhesives are extensively used in structural aircraft applications but although the one part nature of these materials has many advantages in terms of ease of use, the short shelf-life can be a serious disadvantage, especially when the material spends lengthy times in transit between manufacturer and user. Studies on a number of epoxy and nitrile-epoxy adhesives have shown that slow cure, hydrolysis of the resin and specific interactions between components can occur during storage which result in modification of various chemical and physical properties of the uncured adhesives. The relative importance of these reactions depends on the adhesive composition. These modifications are reflected in changes in the strength of joints made with aged adhesives. This paper presents examples of these effects drawn from the results of a number of ageing studies.

INTRODUCTION

Many of the adhesives used for structural applications in aircraft are in the form of one part films of very limited shelf life, even when stored at low temperatures. The increasing use of bonding in critical areas, including primary aircraft structure, has focussed attention on the nature of the reactions which occur during storage, chemical methods of assessing the extent of these reactions and the effect of such reactions on the joint strength of the cured material. These considerations are of particular concern in Australia since all adhesives of this type are imported and must be transported long distances from the point of manufacture. In addition, the usage of particular adhesives is quite small so that means of extending the usable life are of interest.

Considerable efforts by many workers have been devoted to the establishment of chemical methods of characterising epoxy-based adhesives and composite prepregs for the purposes of quality control.[1-4] Some attention has also been given to the (adverse) effects of a high humidity atmosphere on the properties and cure characteristics of adhesives and prepregs.[5-7] While some consideration has been given to certain aspects of the ageing of these materials,[8,9] little attention seems to have been paid to an overall study of the changes in the chemical and physical properties of adhesives during storage on the one hand and the strength of joints made with aged adhesives on the other.

This paper summarises the results of a number of ageing studies on two classes of film adhesives: 177°C curing, epoxy systems and 121°C curing, nitrile-epoxy systems.[10-13] Several types of reactions have been identified and their effects on joint strength considered.

EXPERIMENTAL

Examples of two classes of film adhesives have been studied: (a) epoxy-based, 177°C - curing systems, designed for metal-to-metal bonding in aerospace applications, for service up to about 200°C and (b) nitrile rubber modified epoxy systems, 120°C - curing, designed for aerospace metal-to-metal bonding for service up to about 100°C.

For ageing, adhesive specimens were stored in polyethylene bags at about 23°C. Samples were removed at intervals for testing or, if necessary, stored at -18°C until required.

For high pressure liquid chromatography (HPLC) a modified

Varian 8500 instrument was used with an Altex 153 UV detector operating at either 254 or 280 nm. The equipment was operated in the reverse phase, gradient elution mode with a Du Pont Zorbax column and the combination of tetrahydrofuran (THF) and water as the solvent. For gel permeation chromatography (GPC) essentially the same equipment was employed with a Waters model R401 refractive index detector in series with the UV detector, and using four μStyragel columns and THF as the solvent.[11,14]

Materials associated with selected peaks in HPLC and GPC traces were collected and identified by infrared analysis using a Perkin Elmer 580B, ratio-recording, double beam spectrophotometer, interfaced with a model 3500 data handling system. Procedural details are given elsewhere.[12]

Epoxide content was determined by non-aqueous titration, as previously described.[14]

Adhesive flow was assessed by measurements on a disc of adhesive cured between release sheets in a heated platen press using the cure cycle recommended for the adhesive.

Single overlap joints were made with 1.6 mm 2024-T3 Alclad aluminium sheet for which the surface preparation was a vapour degrease with 1,1,1-trichloroethane followed by a chromic acid etch. No primer was used. Joints were made in a heated platen press using the cure cycle recommended for the adhesive. Joint geometry and test methods were in accord with standard procedures.[15] The testing was accomplished with a model TT-C-L Instron tensile testing machine. Results are reported as the average of 6 replicates.

Bonded honeycomb panels were made using 9.5 mm cell size aluminium honeycomb 12.5 mm thick with 0.13 mm 2024-T3 face sheets. These panels (75 mm wide) were tested by climbing drum peel.

Dynamic mechanical properties of cured adhesive samples were assessed with a Rheovibron DDV-II-C direct reading dynamic viscoelastometer, operating in the tensile mode at 110 Hz and a heating rate of about 2°C/min. Samples for examination were cured in batches in the press, using a polyester template, to ensure equivalent thermal history.

REACTIONS OCCURRING DURING AGEING

Slow Cure

The adhesives in this study all contain latent curing

systems (dicyandiamide, dicyandiamide/Monuron or the 2,4-tolylene di-isocyanate/dimethylamine adduct). Room temperature (about 23°C) is far below the recommended cure temperature for these systems (177°C for the former and 121°C for the latter two). In addition, these curing agents are only very slightly soluble in epoxy resins below about 80°C. In spite of this, over a period of weeks at ambient temperature some polymerisation (chain extension and crosslinking) can occur.

Polymerization can be expected to result in a reduction in the epoxide content (increase in epoxy equivalent weight, EEW) and a reduction in the solubility of the adhesive in organic solvents. Where a substantial increase in molecular weight occurs, reduced flow of the adhesive during cure would be expected.

Hydrolysis

Certain types of epoxy resins, notably those containing the N-glycidyl aromatic amine moiety, are susceptible to hydrolysis.[14] Resins of this type are frequently incorporated in adhesives for high temperature applications. Resins based on the diglycidyl ether of bisphenol A (DGEBA) are regarded as stable to hydrolysis. However, recent evidence indicates that in certain adhesives based on DGEBA resins some hydrolysis of the resin occurs during ageing[13] (Figure 1).

Hydrolysis of the resin leads to a reduction in epoxide content and increase in polarity. The latter effect may modify the moisture uptake properties of the resin.

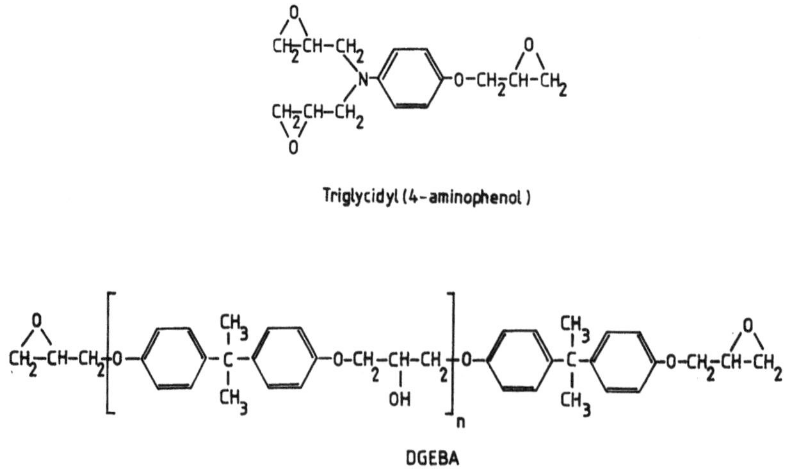

Triglycidyl (4-aminophenol)

DGEBA

Figure 1. Epoxy resins.

Interactions Between Components (Excluding Cure)

Adhesives consist of a number of components and there is the possibility of other reactions occurring during storage. Nitrile modified epoxy systems contain carboxy-terminated, low molecular weight, butadiene acrylonitrile (CTBN) or polybutadiene (CTB) polymers; or higher molecular weight, carboxylated nitrile rubber; or a combination of these (Figure 2). A reaction can occur between the carboxyl and epoxide groups. In some nitrile-epoxy adhesives these groups are prereacted (that is, deliberately reacted as part of the manufacturing process) whereas in other systems they are not. In the latter case this reaction can proceed slowly during storage of the adhesive.

Occurrence of the carboxyl-epoxide reaction during storage results in a reduction in the epoxide content. In addition, depending on the rubber molecular weight, a decrease in adhesive solubility and in adhesive flow during cure may occur.

EFFECT ON JOINT STRENGTH

The effect of the reduction in the epoxide content during ageing on the properties of the subsequently cured adhesive depends chiefly on modification of the crosslink density. If the epoxide loss results from the normal cure reactions proceeding very slowly, the chemical structure of the cured adhesive may not differ greatly from that achieved from a normal cure of fresh adhesive. It would be anticipated that the joint strength would then be little affected. On the other hand,

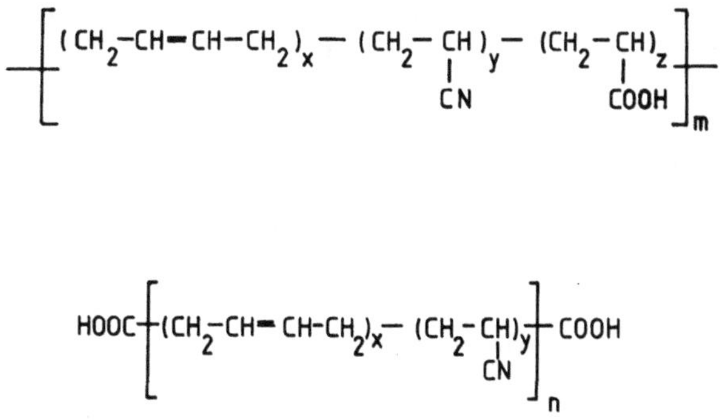

$$\left[(CH_2-CH=CH-CH_2)_x - (CH_2-\underset{CN}{CH})_y - (CH_2-\underset{COOH}{CH})_z \right]_m$$

$$HOOC \left[(CH_2-CH=CH-CH_2)_x - (CH_2-\underset{CN}{CH})_y \right]_n COOH$$

Figure 2. Nitrile rubbers.

epoxide loss from non-cure processes potentially can lead to a
cured resin of lower crosslink density. This can be expected to
result in reduced joint strength, especially at high
temperatures.

A reduction in flow during cure would be most significant
in the bonding of honeycomb panels, where good flow is needed
for adequate filleting and thus the formation of strong bonds.
In the case of overlap joints, the pressure applied during the
cure cycle may be sufficient to overcome deficiencies in
substrate wetting ability of an adhesive of moderately
diminished flow. For systems of greatly altered flow, or in
circumstances where only light pressures are employed during
cure, the applied pressures may be insufficient for this purpose
and defective joints result.

AGEING STUDIES

Examples of the above reactions and their effect on joint
strength for two types of epoxy adhesives are provided from the
results of extensive ageing trials conducted at, or slightly
above, ambient temperature.

High Temperature Curing, Epoxy Adhesive

A study was conducted on two 177°C curing, epoxy adhesives
designed for service up to about 200°C. These systems contained
triglycidyl-(4-aminophenol) as well as other, glycidyl ether
epoxy resins.[10,12] Results obtained from just one of these
adhesives is described in this paper.

The HPLC analysis of fresh and aged samples is shown in
Figure 3. As the extinction coefficient of triglycidyl-
(4-aminophenol) at 254 nm is much larger than that of the other
components, the chromatograms are dominated by the peaks due to
this compound, its oligomers and derivatives. Fractions
corresponding to the peaks designated 1, 2 and 3 were collected
and examined by IR. The spectra were consistent with
triglycidyl-(4-aminophenol) having 1, 2 and 3, respectively,
epoxide groups hydrolysed.[12] An indication of the rate of
hydrolysis is provided in Figure 4 which shows the increase in
peak heights for the material with 2 and 3 epoxide groups
hydrolysed.

Table I summarises the changes in epoxide content,
solubility in chlorobenzene, and flow during cure.[12] These
quantities altered slowly over the first 16-20 weeks of the
investigation and thereafter changed rather more rapidly.

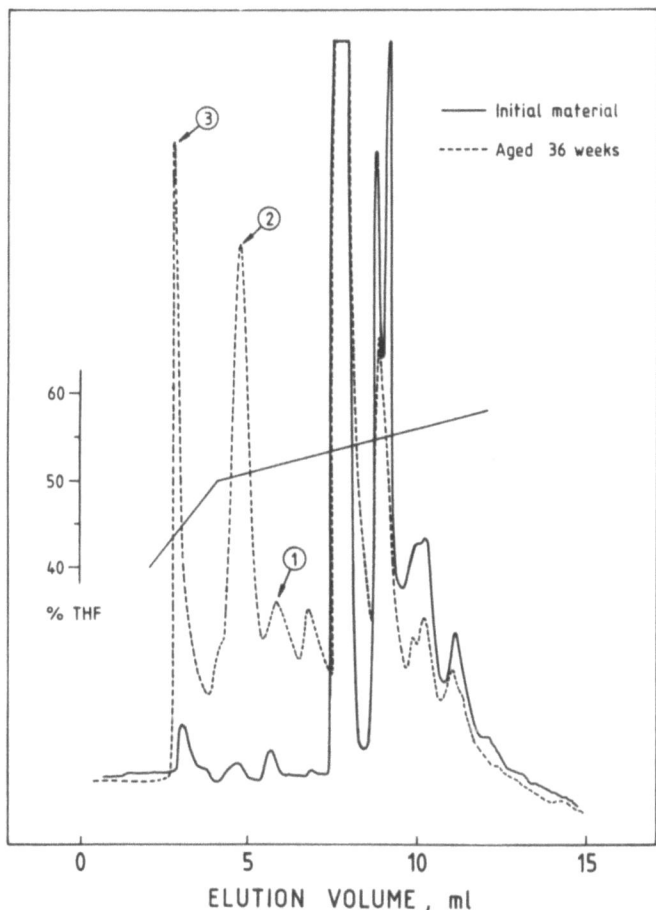

Figure 3. HPLC analysis of fresh and aged epoxy adhesive. Material from labelled peaks was collected and analysed by IR.

Table 1. Change in Epoxide Content, Solubility, and Flow during RT Ageing of a High Temperature Curing Epoxy Adhesive

Ageing Time weeks	EEW g/eq.	Chlorobenzene Solubility wt %	Flow %
0	229	66	370
10	245	65	410
20	284	60	300
30	474	38	0

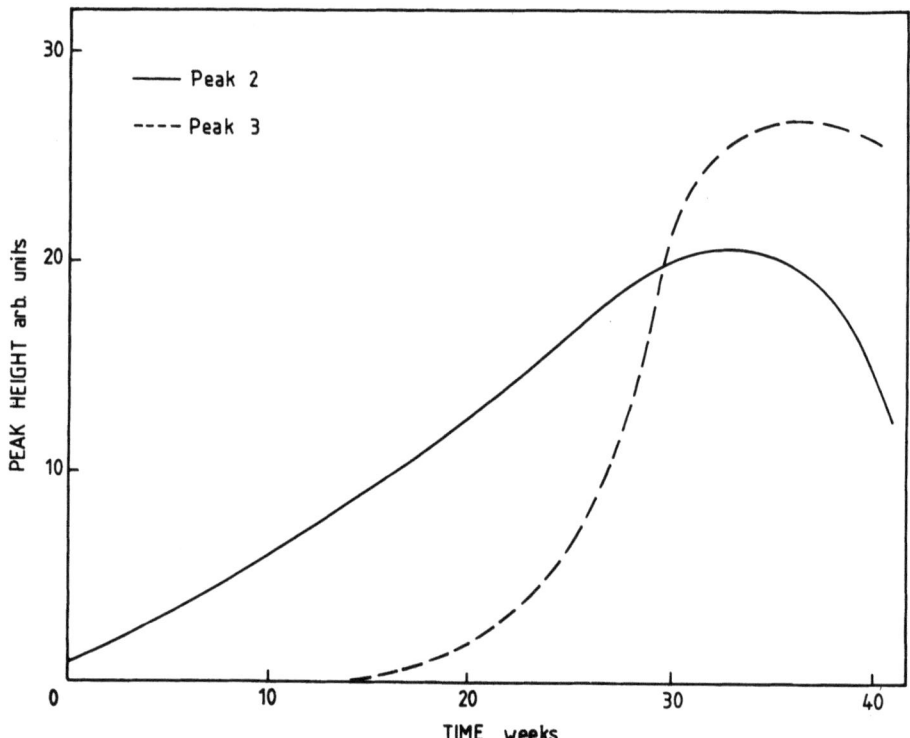

Figure 4. Change in peak height for peaks 2 and 3 in Figure 3.

Results of dynamic mechanical analysis of cured, aged samples are shown in Figure 5. By 16 weeks substantial changes had occurred, especially in the region above 120°C.

Tensile-shear strength of lap joints, tested at ambient temperature, 132°C and 177°C, are given in Figure 6. High temperature joint strength was substantially reduced by ageing long before any effect was discernible for joints tested at ambient temperatures. The increased strength observed after several weeks ageing in the 132°C and ambient cases is attributable to increased flexibility (lower crosslink density) of the adhesive.[12]

These results are consistent with a combination of hydrolysis and slow cure. For times up to about 15 weeks ageing, hydrolysis effects predominate, as indicated by the HPLC

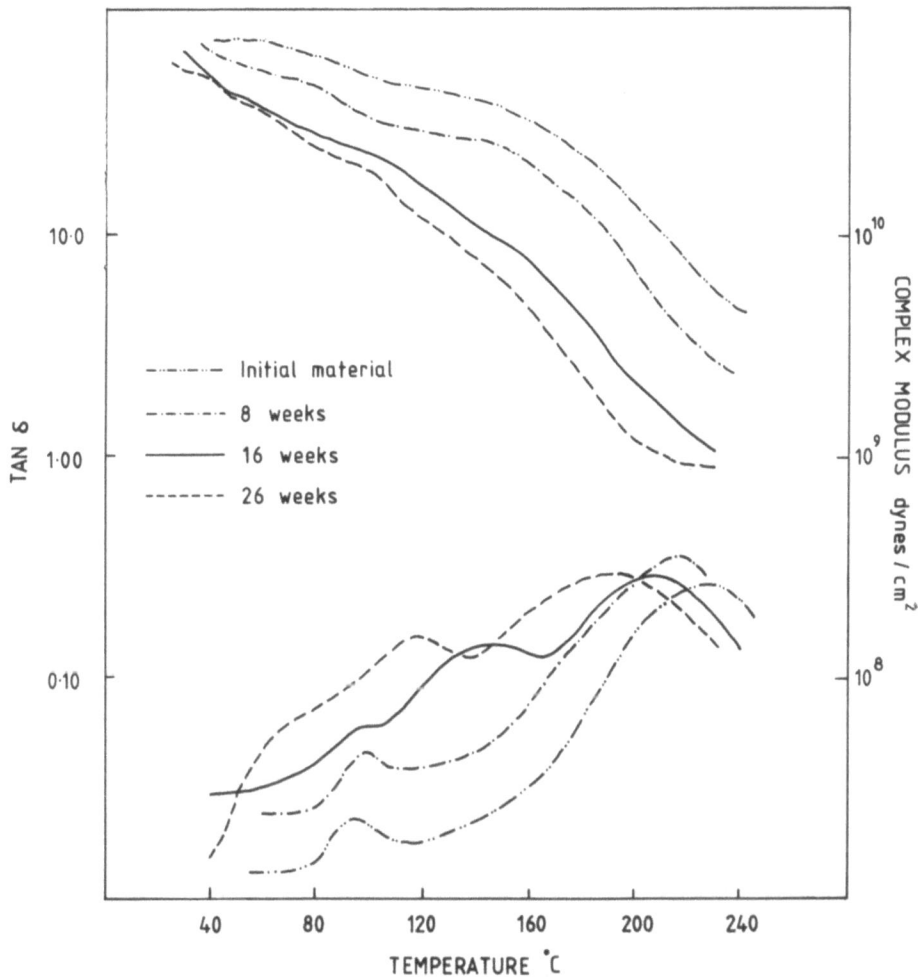

Figure 5. Dynamic mechanical analysis of aged epoxy adhesive. Upper curves - complex modulus; lower curves - tan δ.

analysis and the increase in EEW but only minor changes in the solubility and flow. Loss of epoxide content by hydrolysis leaves the way open for a reduction in crosslink density, and thus a reduction in the high temperature performance, of the cured system. This is reflected in the reduced joint strength at high temperatures. For ageing times longer than 15 weeks the effects of slow cure (chain extension) are more evident, namely reduced solubility and flow and a large fall in epoxide content. These effects are ultimately reflected in joints of very poor strength.

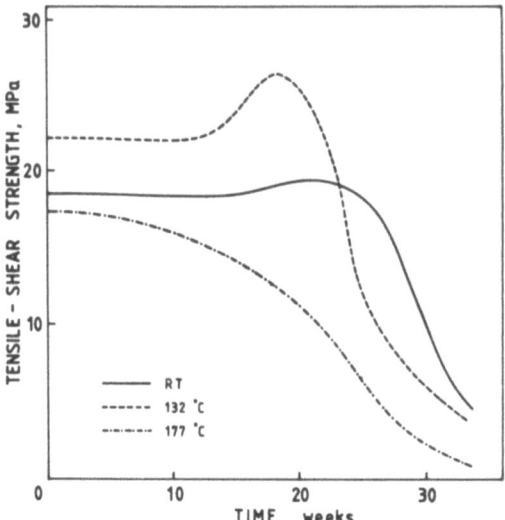

Figure 6. Tensile-shear strength of Al-Al lap joints made with aged adhesive and tested at the temperatures indicated.

Nitrile-Epoxy Adhesives

Two 121°C curing systems were studied, both of which are based on DGEBA epoxy resins. One system, designated Adhesive A, contains a carboxylated nitrile rubber of molecular weight about 30,000. The nitrile rubber and epoxy resin have not been prereacted.[16] The other system (Adhesive B) contains the same, or a similar, rubber and some carboxy-terminated polybutadiene of molecular weight about 4,000. The rubber and epoxy have been prereacted.[16]

Tables II and III show EEW and solubility data for the two systems.[13] A substantial and rapid change is obvious in Adhesive A which is not evident in Adhesive B. More gradual changes in these properties are seen in both adhesives over longer periods of time.

GPC studies on the ageing of Adhesive A at 40°C, carried out using both refractive index (RI) and ultraviolet (UV) detectors, showed that a carboxyl-epoxide reaction occurred[11] (Figure 7). This is indicated by the increased UV absorbance in the high molecular weight region (nitrile rubber) and, on further ageing, the reduction in solubility of this high molecular weight material. The corresponding GPC analysis of fresh Adhesive B showed that the UV and RI traces were very similar at the high molecular weight end indicating that the carboxyl-epoxide reaction had already taken place[16].

Table II. Change in Epoxide Content during RT Ageing of Two Nitrile-Epoxy Adhesives.

Ageing Time weeks	EEW g/eq. Adhesive A	Adhesive B
0	425	442
5	516	446
10	535	526
15	572	676
20	598	817

Table III. Change in Solubility during RT Ageing of Two Nitrile-Epoxy Adhesives.

Ageing time weeks	Solubility in Chlorobenzene wt % Adhesive A	Adhesive B
0	84	88
5	60	88
10	57	80
15	54	63
20	50	52

The flow of Adhesive A dropped sharply over a short ageing time while that for Adhesive B fell more slowly (Figure 8).[11,13] These changes were mirrored in the peel strength of bonded honeycomb panels[13] (Figure 8).

HPLC/IR analysis indicated slow hydrolysis of the resin in both adhesives.[13] Figure 9 shows the analysis for Adhesive B. Material giving rise to the peak designated B gave a spectrum consistent with the half hydrolysed monomeric epoxy compound. The spectrum of A was consistent with the fully hydrolysed monomer while those of C and D corresponded to the partly hydrolysed n = 1 and n = 2 oligomers respectively.[13] These reactions occur slowly in the initial stages of ageing and more rapidly in the later stages, and are rather faster in Adhesive B than in Adhesive A.[13]

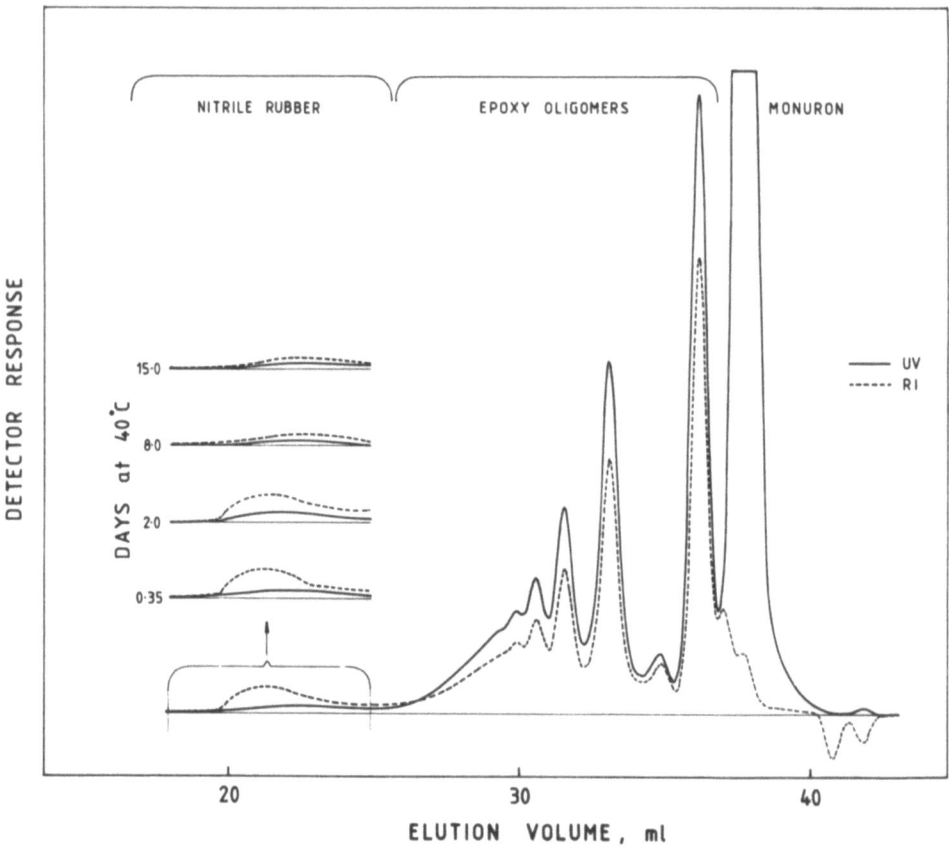

Figure 7. GPC analysis of a nitrile-epoxy adhesive (Adhesive A) aged at 40°C.

Tensile-shear strength of lap joints tested at ambient temperature and at 100°C showed only small changes on ageing[13] (Figure 10). Joints made with adhesive aged for more than 15 weeks were more significantly reduced in strength when tested at elevated temperatures. Dynamic mechanical analysis indicated a downwards shift of about 15°C in the T_g of cured samples over the time of the ageing trial.

These results are consistent with the following interpretation. In Adhesive A a carboxyl-epoxide reaction occurs over the first two weeks of ageing which results in the formation of a small amount of gel incorporating the nitrile rubber. The presence of this gel is responsible for the rapid fall in adhesive solubility and flow during cure.[10,13] The latter, in turn, results in poor peel strength of bonded honeycomb panels. No equivalent reaction occurs in Adhesive B.

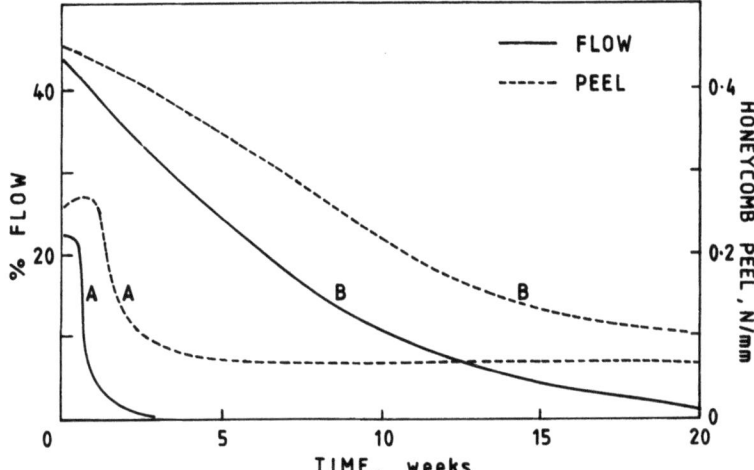

Figure 8. % flow and climbing drum peel strength of aged samples of two nitrile-epoxy adhesives.

Over longer ageing times, slow cure and hydrolysis reactions take place in both systems.[13] Slow cure results in material of reduced flow. The pressure applied during the formation of lap joints is evidently sufficient to overcome the effect of reduced flow and the tensile-shear strength of such joints is only slightly affected for ageing up to about 15 weeks. In the case of bonded honeycomb panels the different configuration prevents the applied pressure from compensating for the effects of reduced flow and the peel strength of such joints is severely reduced after even 3 weeks ageing. As indicated above, hydrolysis can lead to the formation of cured material of lower performance, especially at elevated temperatures. This effect is apparent in these systems after prolonged ageing.

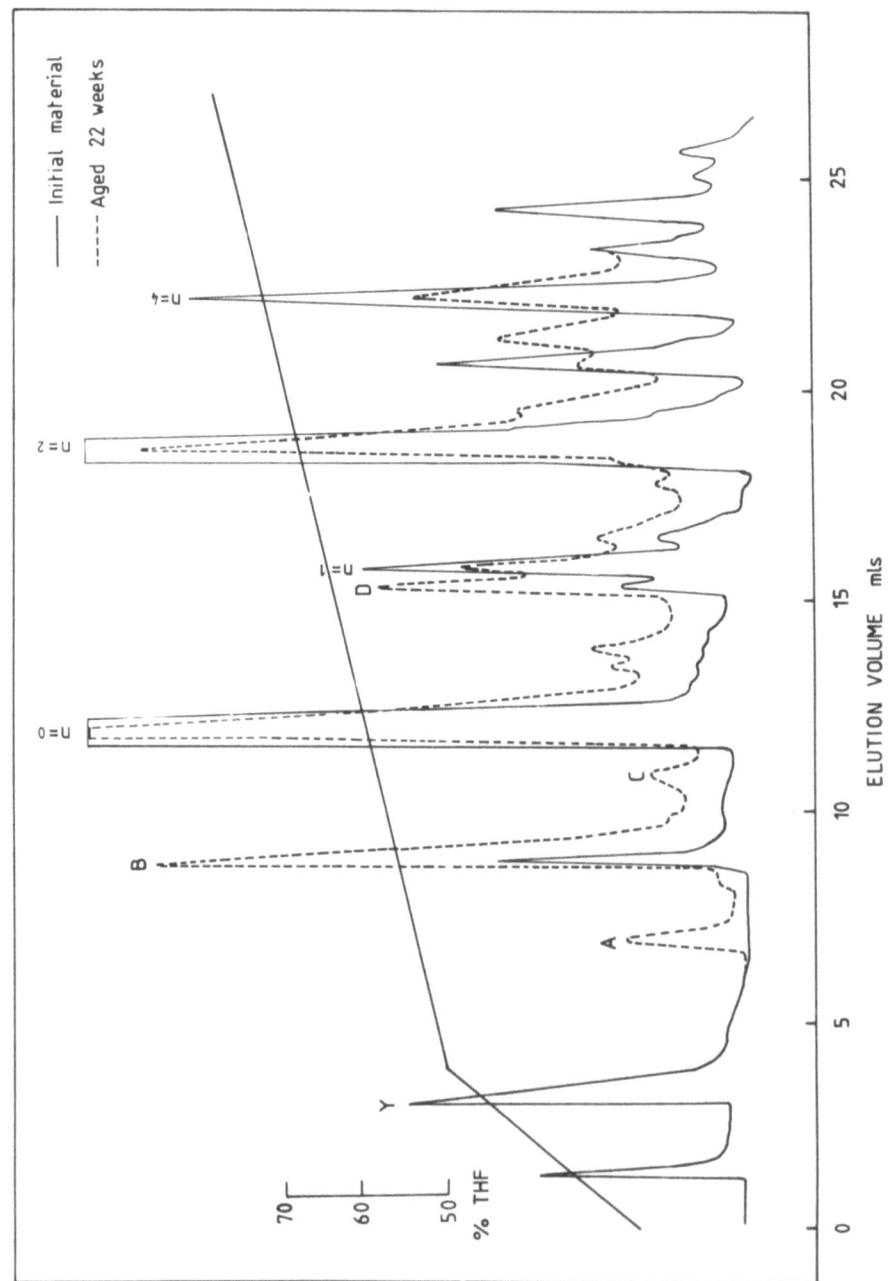

Figure 9. HPLC analysis of fresh and aged nitrile-epoxy adhesive (Adhesive B). Peaks labelled n are due to epoxy oligomers; A, B, C and D were analysed by IR; Y is due to the curing agent.

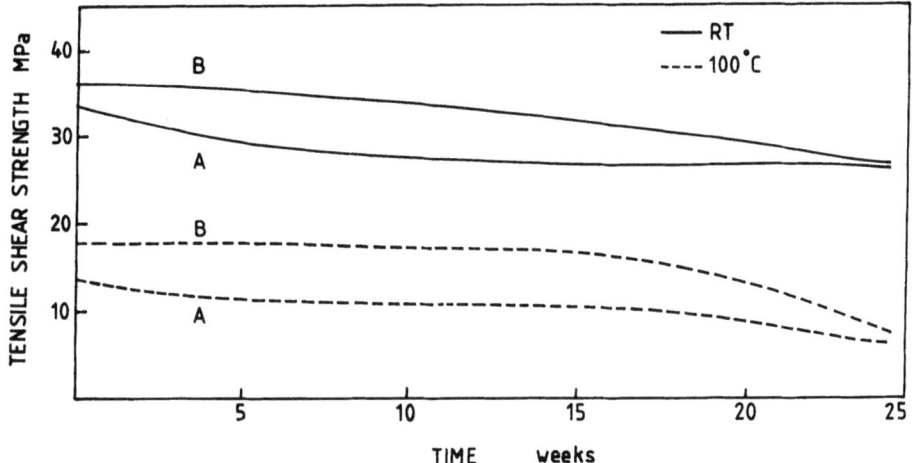

Figure 10. Tensile-shear strength of Al-Al lap joints made with aged samples of two nitrile-epoxy adhesives. Test temperatures as indicated.

REFERENCES

1. J. F. Carpenter and T. T. Bartels, SAMPE Quart., 7 (1), 1 (1976).
2. C. A. May, D. K. Hadad and C. E. Browning, Polym. Eng. Sci., 19, 545 (1979).
3. J. E. Twichell, J. O. Walker and J. B. Maynard, J. Chromatogr. Sci., 17, 259 (1979).
4. G. L. Hagnauer, Polym. Composites, 1, 81 (1980).
5. D. L. Paradis, in "Structural Adhesives and Bonding", p. 110, Technical Associates Conference, El Segundo, CA, 1979.
6. G. W. Lawless, in "Proc. 24th Natl. SAMPE Symp.", Vol. 2, p. 979, 1979. Available from SAMPE, Azusa, CA.
7. R. A. Pike, F. P. Lamm and J. P. Pinto, J. Adhesion, 13, 229 (1982).
8. D. J. Crabtree, in "Resins for Aerospace", C. A. May, Editor, p. 459, American Chemical Society, Washington, D.C., 1980.
9. R. A. Pike, F. P. Lamm and J. P. Pinto, J. Adhesion, 12, 143 (1981).
10. C. E. M. Morris, A. G. Moritz and R. G. Davidson, in "Adhesion and Adsorption of Polymers", L.-H. Lee, Editor, Part A, p. 313, Plenum Press, New York, 1980.

11. J. J. Pitt, P. J. Pearce, T. W. Rosewarne, R. G. Davidson, B. C. Ennis and C. E. M. Morris, J. Macromol. Sci., Chem., A17, 227 (1982).
12. P. J. Pearce, R. G. Davidson and C. E. M. Morris, J. Appl. Polym. Sci., 27, 4501 (1982).
13. P. J. Pearce, R. G. Davidson and C. E. M. Morris, J. Appl. Polym. Sci., 28, 283 (1983).
14. P. J. Pearce, R. G. Davidson and C. E. M. Morris, J. Appl. Polym. Sci., 26, 2363 (1981).
15. "Adhesives, Heat Resistant, Airframe Structural, Metal-to-Metal", U. S. Federal Aviation Specification MMM-A-132.
16. C. E. M. Morris, P. J. Pearce and R. G. Davidson, J. Adhesion, 15, 1 (1982).

STRUCTURAL ADHESIVES BASED ON DIALLYL PHTHALATE

A. M. Usmani

The Research Institute
University of Petroleum & Minerals
Dhahran, Saudi Arabia

This paper describes diallyl phthalate resins as a candidate replacement for the moisture sensitive epoxies used in formulating structural adhesives. Polymerized diallyl phthalate had most of the performance properties desired but was excessively brittle. Flexibilizing monomers were evaluated. Some promising copolymers were defined. Diallyl phthalate/epoxy interpenetrating polymer networks were also explored.

INTRODUCTION

Surfaces are involved in many physical processes. Chemical interaction across interfaces are studied as "pure" science or as a way to investigate applied problems, e.g., adhesion.[1] Conversely, adhesion can be studied as a basic phenomenon or as a way to investigate surface interactions.

Use of adhesively bonded structures will become more prevalent in Air Force and Naval systems, e.g., satellites, missiles, aircraft, weapons, and ships in the present decade and beyond. Growth in the usage of adhesively bonded structures in the future will result primarily from improved structural integrity of such systems.

Epoxy type, 125°C curing adhesives which represents the current state-of-the-art in structural adhesive bonding have several shortcomings. Their most serious deficiency is the loss of mechanical properties after exposure to elevated-temperature, high-humidity environments. Another problem with the currently available adhesives is moisture sensitivity during layup of bonded parts. Absorbed moisture in the uncured adhesive can dramatically alter curing behavior and final properties. There is a pressing need for low-moisture-absorbing structural adhesives for use on aircraft and portable tactical shelters. What is required is a chemically well-balanced adhesive resin (aromatic or cycloaliphatic groups for thermal and mechanical stability, alkyl groups for moisture resistance, few ether linkages and hydroxyl groups for adhesion and multifunctional reactive sites for curing) that will provide prolonged retention of high mechanical properties when exposed to elevated temperature in a humid environment.

This paper describes the exploratory and development work that was performed on flexibilized diallyl phthalate (DAP) and DAP/epoxy interpenetrating polymer network adhesive resins. Optimization of the composition and formulation, evaluation, and product scale-up studies are still needed before the material will be ready for commercial application. Schematic of DAP adhesive optimization has been provided, however.

FORM AND COMPOSITION OF STRUCTURAL ADHESIVES

Structural adhesives are generally classified by the physical form in which they are used. Both solids and spreadable pastes are commonly employed.[2] Examples are:

1. Solid adhesives

    1.1 Film (unsupported)
    1.2 Tape (supported)
    1.3 Solid powders and melt-applied crosslinking types
    1.4 Solvent-based

2. Solventless and high solids pastes

    2.1 One-component, long shelf life
    2.2 Two-component, short shelf life

Structural adhesives based on epoxies and phenolics are most popular. Other resins used to make structural adhesive formulations are epoxy-phenolic, nylon-epoxy, nitrile modified epoxy, and polyimide (for tape and film adhesives) and one-component heat-cured epoxy, two-component room-temperature curing epoxy, and amine-cured polyisocyanate (for paste and liquid adhesives).

EXPLORATORY WORK ON DAP ADHESIVE

Requirements of Metal Bonding Adhesives

The adhesives should cure fully at 125°C, have good adhesion and bonding properties, have low moisture absorption, provide excellent bonded strength over a wide temperature range, and retain mechanical properties when exposed to elevated temperature and high humidity for prolonged periods.

Property-Moiety Relationship in a Resin

The properties obtained from specific chemical groups in a resin are shown in Table 1. For low moisture absorption, a resin should contain few hydroxyl, amide, and ether linkages and many aromatic and alkyl groups. For bonding properties it is important to have some hydroxyl, carboxyl, or nitrogen in the resin. A judicious choice of the various groups in a resin can lead to an adhesive meeting all the requirements outlined in the earlier paragraph.

Creating an interpenetrating polymer network (IPN) is a useful approach for modifying a polymer's architectural design. An IPN produces permanent entanglement that restricts the motions of segments and thus mimics covalently bonded crosslinks. Since property maxima and minima are obtained in IPN's, a valuable development tool is available to develop low moisture absorbing adhesives that will provide high retention of mechanical properties under elevated heat/high humidity conditions.

Table I. Properties versus Groups in a Resin.

Group	Property Obtained
Aromatic	Thermal and mechanical stability, stiffening effect
Cycloaliphatic	Thermal and mechanical stability, stiffening effect
Alkyl	Moisture and chemical resistance
Alkyl, long chain	Moisture resistance, plasticization, and toughening effect
Ether linkages	Chemical resistance, flexibilizing effect, moisture sensitivity
Amide	Toughening and adhesion
Hydroxyl	Adhesion, moisture sensitivity
Di- and multi-functional vinyl	Curing, three-dimensional network

Based on the above considerations, we have developed flexibilized diallyl phthalate and DAP/Epoxy IPN adhesive systems.

Flexibilized Diallyl Phthalate Adhesives

The excellent properties of polymerized DAP are well known.[3] These include high heat and moisture resistance, and exceptional dimensional stability. Neat DAP monomer tends to crack during polymerization due to shrinkage and produces a rigid product. To avoid cracking, mineral or fiber filling is commonly used in DAP molding compounds. In an adhesive formulation, mineral filling can be used. However, the use of a plasticizing monomer appears to offer an attractive alternative.

Copolymerization of diallyl phthalate with alkyl methacrylates has been treated earlier by the author.[4] The copolymerization of DAP and an alkyl methacrylate by a free radical mechanism at low polymerization temperatures is less practical due to unfavorable reactivity ratios. We considered the rates and reactivity ratios dependence on temperature to get an insight into the method necessary to prepare such a copolymer. Copolymerization of DAP with an alkyl methacrylate can be done at elevated polymerization temperature, e.g., 125°C.

To modify rigid polymerized DAP we evaluated butyl methacrylate (BMA), 2-ethylhexyl methacrylate (EHMA), isodecyl methacrylate (IDMA), lauryl methacrylate (LMA), stearyl methacrylate (SMA), and hydroxypropyl methacrylate (HPMA) plasticizing comonomers. BMA polymerizes quite readily with DAP but is too volatile to be useful in the open. EHMA plasticized DAP, and only slight monomer loss was encountered during curing. IDMA, LMA, and SMA were found to polymerize satisfactorily with DAP. IDMA on copolymerization with DAP, gave clear products. LMA and SMA on copolymerization with DAP gave from clear to opaque resins depending upon the amount of LMA (SMA) and the polymerization conditions.

Various ratios of DAP to comonomers, different type and amounts of initiator, and various methods of preparing the resin were investigated. The preferred mixtures were found to be DAP EHMA/HPMA (80/15/5) and DAP/LMA (80/20). HPMA copolymerized readily with DAP, and the hydroxyl groups should contribute to increased adhesion.

Flexibilized DAP can be developed into suitable adhesive forms, e.g., film form in which the complete formulated adhesive is carried on a lightweight woven scrim fabric or mat, and paste form, in which the adhesive and curing agent are maintained as separate thick pastes until ready for use.

Three methods of paste (syrup) preparation were investigated. In bottle polymerization (DAP + flexibilizing monomer + 0.2 per cent benzoyl peroxide under nitrogen blanket at 65-70°C for 24-48 hours), viscosity was difficult to control from batch to batch and occasionally gelation was encountered during syrup preparation. Another disadvantage of the bottle method is that the prepared syrup drifts up in viscosity due to catalyst residue remaining in the syrup. Use of DAP prepolymer (Dapon 35) with DAP monomer produced straw colored syrups which are suitable for adhesive formulations. The resin kettle method was very reproducible. It involved first preparing a high viscosity DAP syrup and then diluting it subsequently with EHMA/HPMA or LMA. Slightly more uniform copolymer was made with a slower and delayed addition of flexibilizing monomer to a heated and stirred DAP. To 1600g DAP monomer, 3.3g benzoyl peroxide was added and polymerization conducted under nitrogen blanket with stirring at 80±2°C. In about five hours the viscosity of DAP reached approximately 10 Pa sec. At this point the contents were quickly cooled to less than 45°C and 400g of flexibilizing monomer, e.g., LMA was sequentially added and mixed. A typical DAP/LMA (80/20) syrup had a viscosity of 1.13 Pa sec.

The following cycles were developed for DAP/EHMA/HPMA (80/15/5) and DAP/LMA (80/20) syrups:

- Slow cycle (in air and using 0.2 percent each of benzoyl peroxide and DiCup catalyst): 90°C/12 hrs. + 130°C/12 hrs.

- Medium fast cycle (in nitrogen and using 0.2 percent each of benzoyl peroxide and DiCup catalyst): 90°C/2 hrs. + 100°C/1 hr. + 110°C/1 hr. + 120°C/2 hrs. + 130°C/2 hrs. (Total cure time = 8 hrs.)

- Medium fast cycle (in air and using 0.3 percent each of benzoyl peroxide and DiCup): 125°C/12 hrs.

- Fast cycle (in air and using 1 percent DiCup): 125°C/4 hrs.

- Fast cycle (in nitrogen and using 1 percent DiCup): 125°C/1 hr.

- Super fast cycle (in air and using 1 percent DiCup): 150°C/1 hr.

- Super fast cycle (in nitrogen and using 1 percent DiCup): 150°C/30 min.

Other Allylic Resins

Closely resembling DAP are two other monomers; namely, diallyl adipate (DAA) which polymerizes very slowly like DAP, and triallyl citrate (TAC), a hydroxyl-containing trifunctional monomer which polymerizes about ten-fold faster than DAP or DAA. We have developed a DAP/DAA (80/20) resin wherein DAA provided flexibilizing action. Also developed were DAP/TAC (80/20) syrup, by polymerization at 80°C with 0.2 percent benzoyl peroxide, under constant stirring and with a nitrogen blanket. This syrup resin was cured by 0.3 percent benzoyl peroxide and 0.3 percent DiCup. DAP/DAA/TAC (80/10/10) was another promising resin we developed. The curing cycle for these resins was not optimized.

Optimization of DAP and Allylic Adhesive Systems

The flexibilized DAP and allylic adhesive optimization can consist of the following components:

- Type and level of flexibilizing monomer.
- Preparation of DAP/Alkyl Methacrylate (or other) syrup.
- Determination of catalyst type (e.g., benzoyl peroxide and dicumyl peroxide) and concentration (0.06 to 2.0 percent) and development of suitable curing cycles.
- Water "getter" fillers (e.g., aluminum and glass powders) formulation study.
- Flow control agent, adhesion promoters, and other additive formulation study.

In Figure 1 a schematic of the DAP adhesive chemistry and technology is presented.

DAP/Epoxy IPN Adhesives

IPNs have been known for many years and have been apparently "rediscovered" several times.[5-7] IPNs are mixtures of two or more distinct polymeric phases that cannot be separated by conventional physical resolution techniques. They offer another method of modifying a polymer's architectural design, like graft and block copolymerization. By varying the network composition it is possible to effect the resulting properties. Optimum properties (maxima or minima) can be obtained with binary or ternary IPNs.

Most IPNs, on the molecular scale, are heterogeneous, with phase domains rich in one or more of the polymer components. Most IPNs, on a colloidal scale, appear to be compatible (i.e., homogeneous) blends of the component networks. Thus IPNs are a unique class of molecular topology that represent different ways of embedding molecules in three-dimensional space.[5-7] Because permanent

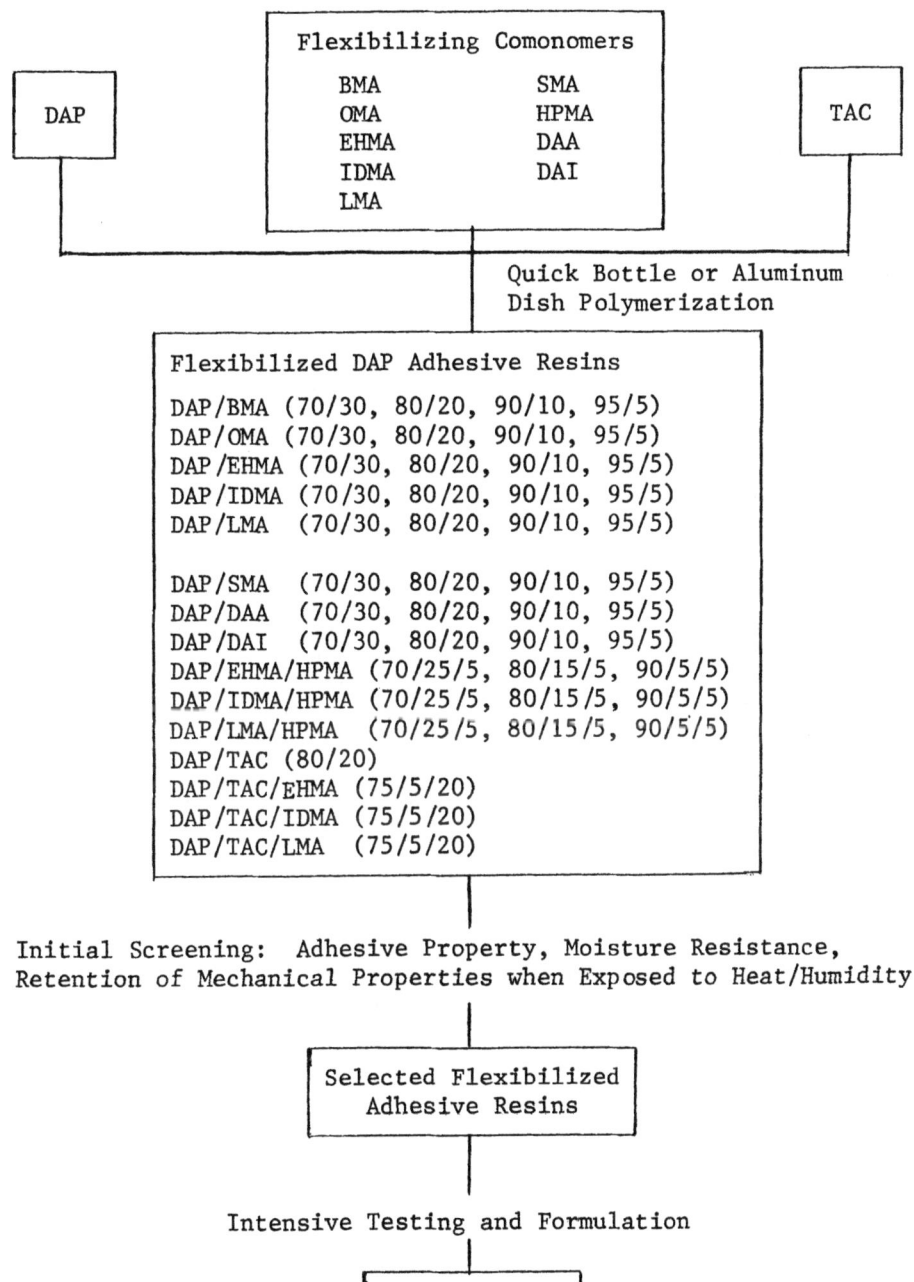

Figure 1. Schematic of DAP adhesive optimization program.

entanglements will restrict motions of segments, they would be expected to nearly mimic covalently bonded chemical crosslinks. We took advantage of these mimicking crosslinks to develop superior adhesives.

We have developed flexibilized DAP/polyepoxy type IPNs. The rate of DAP/LMA or DAP/EHMA/HPMA polymerization by a free radical initiator is about the same as the rate of an epoxy reacting with an anhydride. When both mixtures are mixed and allowed to cure, an IPN was formed. In our work we used an aromatic epoxy with hexahydrophthalic anhydride. Cycloaliphatic epoxies were less suitable. DAP/LMA (80/20) with 20, 30, and 40 percent epoxy resin plus anhydride were studied. However, no property optimization was done. In Figure 2, a schematic of DAP/Epoxy IPN optimization is presented.

CONCLUSIONS

Flexibilized DAP adhesive resins were developed that have potentially superior properties over the currently used epoxy materials. DAP/EHMA/HPMA (80/15/5) and DAP/LMA (80/20) are good compositions. DAP/epoxy IPNs were successfully explored. None of the compositions were optimized but schematics of optimization have been presented.

Since these materials represent a relative new technology of potential importance, investigators are encouraged to study these systems.

ACKNOWLEDGMENTS

The author appreciates help and support of Dr. A.G. Maadhah and Mr. P. F. Cooke (both of the Research Institute) in preparation of this chapter. Technical discussions provided by Drs. J. M. Butler and I. O. Salyer (University of Dayton) is acknowledged. Special thanks to Nihal Ahmad (Research Institute) for typing this work.

REFERENCES

1. K. L. Mittal, Pure Appl. Chem., $\underline{52}$, 1295 (1980).
2. J. C. Bolger, in "Treatise on Adhesion and Adhesives", (R. L. Patrick, Editor), Vol. 3, Chap. 1, Marcel Dekker, Inc., New York, 1973.
3. S. H. Hamid and A. M. Usmani, Polym. Plastics Technol. Eng. in press.

STRUCTURAL ADHESIVES BASED ON DIALLYL PHTHALATE

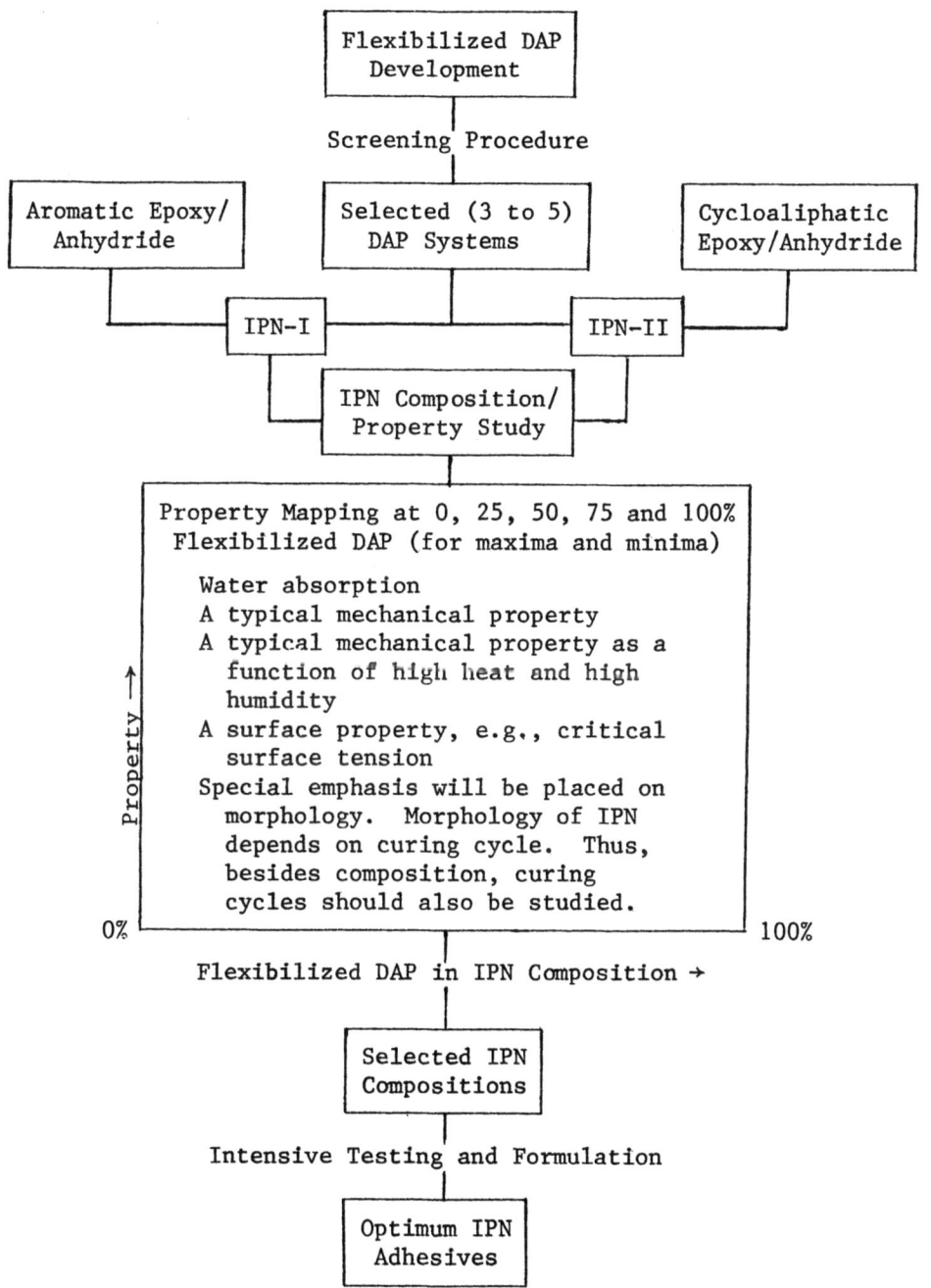

Figure 2. Schematic of IPN optimization.

4. A. M. Usmani, J. Mater. Sci. Letters, 1, 75 (1982).
5. D. Klempner, T. K. Kwei, M. Mutsuo, and H. L. Frisch, Polym. Eng. Sci., 10, 327 (1970).
6. L. H. Sperling and D. W. Friedman, J. Polym. Sci., A-2, 7, 425 (1969).
7. A. M. Usmani, J. Elastomers Plastics, 13, 170 (1981).

POSSIBILITIES OF INTEGRATING SURFACE TREATMENT OF BONDING PARTS IN THE ADHESIVE BONDING PROCESS

Klaus Ruhsland

Central Institute of Welding
Engineering of the GDR
4030 Halle (Saale) GDR

The conventional technique of adhesive bonding of metals is characterized in most cases by a time-consuming and expensive surface treatment of bonding parts before applying the adhesive and involves industrial safety and environmental problems.

The research results of ZIS Halle described here show that it is possible by means of chemical and/or physical methods to integrate the hitherto separate surface treatment, to a great extent, into the bonding process. It can be done, in principle, by the adhesive itself and/or the adhesive bonding technique.

INTRODUCTION

According to the conventional technique for the adhesive bonding of metals, the surface treatment of the bonding parts is always done before applying the adhesive in order to clean, roughen and activate the areas to be bonded. However, the methods that can be applied for this purpose, depending on the bonding parts, require much time and expenditure and involve industrial safety and environmental problems.

On the other hand, there is no doubt that the surface treatment of the bonding parts decisively determines the strength and stability of an adhesive-bonded metal joint.

Therefore, it has not been the object of the research work carried out by ZIS Halle within the last few years to call in question the importance of surface treatment, in principle, but merely to integrate it as much as possible into the process of joining with adhesives. It can be accomplished, in principle, by the adhesive itself or with the help of the adhesive bonding process.

The methods developed for this purpose, which should be useful in making a rational choice for the surface treatment of adhesive bonding of metals shall be presented in the following.

STATE OF DEVELOPMENT OF TECHNOLOGY

At the beginning of the research work at ZIS Halle it was merely known that a few groups of adhesives, due to their chemical composition and/or their setting and hardening conditions, were able to dissolve and to absorb the oil or grease films existing on the surface of the bonding parts.

Thus, for example, it has already been pointed out[1] that solvent-containing adhesives exhibit oil and grease-dissolving properties. This is especially evident when PVC plastisols are applied on oily material surfaces[2]. Here, the plasticizer present in the PVC plastisol can, in principle, also be considered as a solvent, and the grease-dissolving ability of these adhesives is intensified by the high setting temperature required.

For the latter reason, a better oil and grease compatibility can undoubtedly be found also with hot-setting epoxide resin adhesives, hot-setting thermo-plastic hot-melt adhesives, and also with cold-setting epoxide resin adhesives, if they are hardened at high temperatures[3-5]. However, as our own investigations have shown, a high setting or hardening temperature alone is not sufficient to make high-grade adhesive-bonded joints, because their water resistance is poor.

Cold-setting unmodified epoxide resin adhesives are especially sensitive to an imperfect surface treatment, particularly to impurities caused by oil or greases. The loss in stability compared to an optimum surface treatment amounts to approximately 50 to 100% (depending on the type of adhesive and oil) for this group of adhesives at a degree of greasing of $X = 24$. A degree of greasing, X, of 10 corresponds approximately to 1 µm thickness, so $X = 24$ signifies about 2.4 µm layer of grease.

It has been the object of the joint research work of ZIS Halle and VEB ASOL-Chemie Berlin to decisively improve the efficiency of a cold-setting epoxide resin adhesive on greased surfaces by chemical modification. In addition, how far the cleaning effect can be increased by physical methods has also been investigated.

MODIFIED EPOXIDE RESIN ADHESIVE

The adhesive developed (trade name: Epasol FV/ZIS 939) is a modified cold-setting two-component adhesive on the basis of epoxide resin which, because of its composition (GDR Patent WP C 09 j/123672), is able to satisfactorily wet oily surfaces, to dissolve and to absorb oil films already present and to activate the joint component surfaces[6-9].

Its efficiency on an oily Al Mg 3 alloy compared to an unmodified epoxide resin adhesive can be seen from Figure 1. Apart from the oils mentioned in Figure 1, its excellent adhesion has also been shown on the following greasing agents: ASTM test oil No. 1, engine oil, machine oil, hydraulic oil, drilling oil emulsion, silicone spray 200 (silicone separating agent), diesel fuel and Siliron (industrial soap). Its absorbing capacity for greases is small; however, it can be considerably improved by hardening at high temperature, or a "preliminary cleaning" of the bonding surfaces with wash or flushing oils, which means a coarse cleaning with oils of low viscosity.

The adhesive Epasol FV/ZIS 939 is standardized and is continuously being controlled on greased surfaces and on surfaces pretreated in an optimum way, e.g., pickling process for aluminum. The result of the quality control of the first 10 production batches is given in Table I showing that the difference in strength varies from 0 to 20%. The standard deviations are almost the same and the coefficient of variation, is in all cases, below ± 10%.

In case of hot setting of the adhesive the difference in strength between optimum surface treatment and oily surface is almost negligible or is within the measurement errors (see Figure

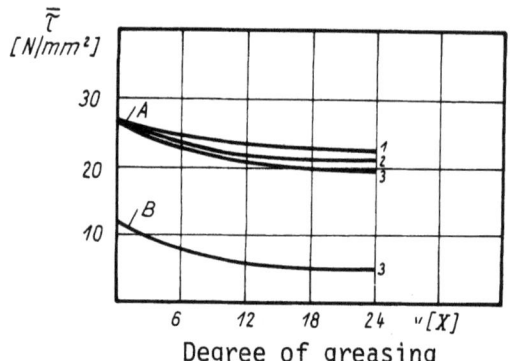

Figure 1. Tensile shear strength at 20°C as a function of greasing agent and the degree of greasing.

A = Epasol FV/ZIS 939
B = unmodified epoxide resin adhesive
Hardening conditions: 24 h at 20°C + 2 h at 90°C
v = greased
x = degree of greasing (X - 10 ≘ 1 μm grease layer thickness on the bonding parts

Greasing agent =

1 = paraffin oil;
2 = silicone oil NM 4-500
3 = gear oil GL 125.

2), the mechanical and thermal properties of the adhesive-bonded joints are distinctly improved (See Figure 3) and the already very good water resistance (see Figure 4) is further improved (see Figure 5).

The adhesive Epasol FV/ZIS 939 can be applied for adhesive bonding on oily surfaces of the following materials[8]:

aluminum and aluminum alloys, structural steels and Cr-Ni steels, grey cast iron, glass, ceramics, concrete, asbestos cement, glass-fiber reinforced polyesters, Sprelacart,

Table I. Tensile Sheer Strength as a Function of Surface Treatment and Batch Number.

Batch-No.	$\bar{\tau} \pm s (N/mm^2)$		$\Delta \bar{\tau}$ (%)
	e + g	v	
1	27,1 ± 0,6	21,9 ± 0,8	− 19,2
2	26,5 ± 0,5	21,8 ± 1,1	− 17,7
3	25,4 ± 1,1	25,7 ± 1,1	+ 1,2
4	27,0 ± 0,7	24,8 ± 1,3	− 8,1
5	23,6 ± 1,0	21,6 ± 0,9	− 8,5
6	23,7 ± 0,7	22,0 ± 0,9	− 7,2
7	27,9 ± 0,7	23,6 ± 0,6	− 15,4
8	24,0 ± 1,0	24,5 ± 1,1	+ 2,1
9	23,6 ± 0,5	23,0 ± 1,7	− 2,5
10	27,0 ± 0,8	21,8 ± 1,1	− 19,3

Adhesive = Epasol FV/ZIS 939
Hardening conditions = 24 h at 20°C + 2 h at 90°C
e + g = degreased and pickled for 30 min. at 60°C in a pickling bath containing 65% distilled water, 27.5% H_2SO_4 and 75% $Na_2Cr_2O_7 \cdot 2H_2O$.

Greasing agent = gear oil GL 125.

pressboard, phenoplasts and aminoplastics, clutch facing, wood, Ecotal sheet, unplasticized PVC, acrylonitrile-butadiene styrene, polystyrene and polyurethane foamed plastics together or combinations of any of these materials.

The maintenance sector, the building industry, shipbuilding, machine and apparatus construction and structural adhesive bonding in various branches of industry are the main fields of application of this adhesive. Apart from this, it can also be applied as a casting and laminating resin, as a coating and as an adhesion primer.

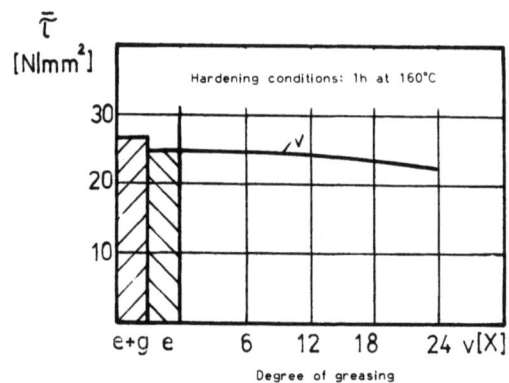

Figure 2. Tensile shear strengh at 20°C as a function of surface treatment.

Adhesive = Epasol FV/ZIS 939
e + g = degreased and pickled
e = degreased
v = greased
X = degree of greasing
Greasing agent = gear oil GL 125

VIBRATIONAL ADHESIVE BONDING

Another possibility for joining contaminated materials by adhesives without prior surface treatment is provided with the so-called vibrational adhesive bonding[10] (GDR Patent WP C 09 j/120042).

The term "vibrational adhesive bonding" signifies an adhesive bonding method according to which a "hard" filler (added to the adhesive) in combination with a mechanical relative movement of the bonding parts (e.g. by means of ultrasonics) causes cleaning and roughening of the bonding surfaces after the application of the adhesive and joining. To do this, the hardness of the filler must be greater than that of the bonding material or its surface contaminants.

INTEGRATING SURFACE TREATMENT OF BONDING PARTS

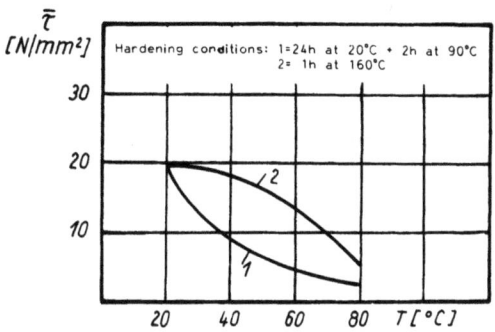

Figure 3. Tensile shear strength as a function of test temperature and hardening conditions.
Adhesive = Epasol FV/ZIS 939; Surface treatment = greased (X = 24); Greasing agent = gear oil GL 125.

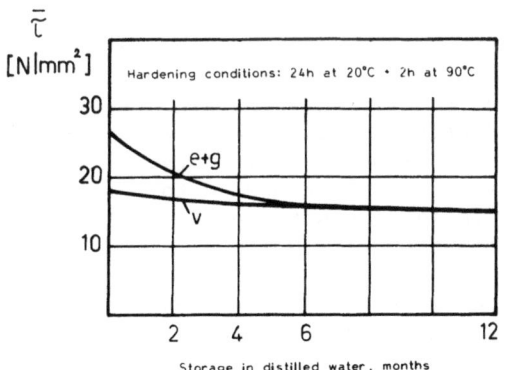

Figure 4. Tensile shear strength at 20° C after storage in distilled water as a function of surface treatment.
Adhesive = Epasol FV/ZIS 939; e + g = degreased and pickled; v = greased (X = 24); Greasing agent = gear oil GL 125.

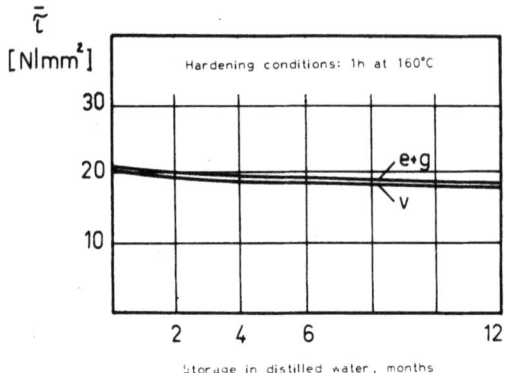

Figure 5. Tensile shear strength at 20°C after storage in distilled water as a function of surface treatment. For explanations, see Figure 4.

The investigations made by ZIS Halle have shown that already after 2 seconds of ultrasonic treatment maximum stability data for the adhesive bonds on oily surfaces are attained which exceed those of the initial adhesive using conventional surface treatment of the bonding parts by approximately 45% (curve 2) and 75% (curve 3) respectively (See Figure 6). Compared to the efficiency of the initial adhesive on an oily surface (X = 24) this means an increase in stability of about 400 or 500%.

As can be seen in Figure 7, the waster resistance of such adhesive bonds is also much improved.

Considering the present capacity of the ultrasonic welding equipment, the application of the technique of vibrational adhesive bonding is first of all possible for adhesive bonding of small and medium structural members (bonded area up to about 10 cm^2) in mass production, e.g., electrical engineering/electronics, precision mechanics/optics, precision instrument engineering, the toy industry, the jewelry trade and similar.

Making of longitudinal seams by means of continuously working ultrasonic equipment also seems possible.

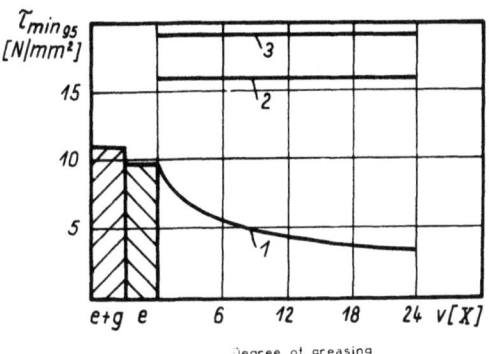

Figure 6. Tensile shear strength at 20°C as a function of surface treatment.

Adhesives: 1 = unmodified epoxide resin
2 = 1 + 40% filler
3 = 2 + 8 silane adhesion promoter

Hardening conditions: 24 h at 20°C + 2 h at 90°C
e + g = degreased and pickled
e = degreased
v = greased

Degree of greasing X = 24
Greasing agent = gear oil GL 125
Ultrasonic treatment = 2 s (for 2 and 3)
τ min 95 = τ - 2 s (mean value - 2x standard deviation

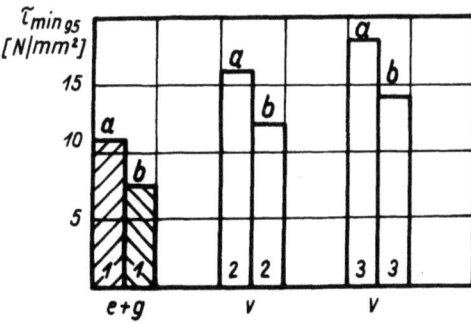

Figure 7. Tensile shear strength at 20°C before and after 60 days of storage in distilled water as a function of surface treatment. a, before storage; b, after storage. For explanations, see Figure 6.

REFERENCES

1. H. Baumann, "Glues and Contact Adhesives", Springer Verlag, Berlin, 1967.
2. G. Gierenz, "Adhesive Bonding of Thin Sheet-metal Structures with PVC Plastisols", Notice of Deutsche Forschungsgesellschaft für Blechverarbeitung und Oberflächenbehandlung E.V., 19, No. 7, 95-98 (1968).
3. "Rational Joining with Adhesives", Publication of 3 M Co., St. Paul, MN, USA.
4. K. Ruhsland and B. Winkler, "Process for Making High-grade Adhesive Bonds without Surface Treatment of Bonding Parts", GDR Patent WP 112 658; C 09 j, 5/02.
5. S. Semerdjiev and P. Panov, "On Adhesive Bonding of Metal Parts on Unclean Surfaces", Paper presented at the meeting for experts for adhesive and casting resin technology, Halle, GDR, 1969.
6. K. Ruhsland and B. Winkler, "Adhesive Bonding of Metals without Surface Treatment of Bonding Parts", Adhäsion, 21, No. 1, 6-9 (1977).
7. K. Ruhsland, "FV Adhesive Bonding Technique and its Possibilities of Application", Fertigungstechnik und Betreib, 28, No. 5, 296-298 (1978).
8. K. Ruhsland, "Results of Application-technical Testing of Epasol FV/ZIS 939", ZIS Notice, 21, No. 10, 1070-1078 (1979).
9. K. Ruhsland, "Comparison of Properties of Oily Metal Adhesive Bonds", ZIS Notice, 22, No. 10, 1172-1178 (1980).
10. K. Ruhsland, "Vibrational Adhesive Bonding", Adhäsion, 23, No. 6, 184-188 (1979).

Part III
Evaluation and Characterization

EVALUATION OF ADHESIVE TEST METHODS

G.P. Anderson,* K.L. DeVries** and G. Sharon**

*Morton Thiokol/Wasatch Division
  Brigham City, Utah 84302
**University of Utah
  Salt Lake City, Utah 84112

Three approaches to bond strength analysis are discussed in this article: those based upon linear elastic stresses, those based upon linear fracture mechanics parameters, and those based upon nonlinear analyses. The first two approaches are then discussed with respect to standard adhesion test methods. The stresses in linear lap shear test specimens are shown analytically and experimentally to be concentrated at the bond termination and to be strongly dependent upon adherend thickness for standard tests. The bond stress intensity factors and energy release rates near the bond termination are given and discussed. Adherend thickness should be at least 4 times larger than that recommended by standard test techniques. The stresses in butt tensile tests are also discussed. It is shown that test specimen alignment problems can lead to low debond forces and excessively large data scatter. Stress intensity factors are also determined for butt tensile tests.

INTRODUCTION

Most designs require that component parts be joined. Adhesive joining has several advantages over mechanical connections (e.g., bolts, rivets, screws, etc.). However, some factors have tended to limit wide usage of adhesives. These include problems associated with: (1) predicting joint strength, (2) identifying and detecting factors that affect the quality and reliability of an adhesive bond, (3) evaluating strength of an adhesive, and (4) determining how the parameters obtained from the measurements in (3) can be used to design an "optimum joint." These problems are largely mechanical in nature and have been the subject of extensive testing and analysis. Testing is important in all aspects of material science and engineering but in no area is it more important than with adhesives.

A large number of adhesive tests have been proposed and used. These might be categorized into two rather general classifications. In the first of these, which we will call "average stress criteria," there is an implied assumption that failure is controlled by the magnitude of the stress and tests are devised to measure this stress (generally an average) at which failure occurs. The other classification of tests makes use of the concepts of fracture mechanics to evaluate the quality of adhesives and to design bonded joints. Fracture mechanics puts special emphasis on the presence of stress risers in nucleating failure. It is hypothesized that failure is more dependent on the magnitude of local stresses and energy dissipation mechanisms than on an "average stress" in the bond.

Most standard tests (by ASTM or other standardizing groups) fall in the first of these categories. However, as we will show later, these standard tests can be used to obtain fracture mechanics properties. In addition, there has recently been a method adapted (ASTM D3433) based on fracture mechanics and another (ASTM D3762) that has some fracture mechanics ramifications. Standard tests might be further divided into three groups: (1) tensile tests, e.g., ASTM D897; (2) shear tests, e.g., ASTM D1002 and ASTM D3165; and (3) peel tests, e.g., ASTM D1876. Experimental results for the first two of these are usually given as the force per unit area. The stresses for such samples are very nonuniform and this average stress may or may not be closely related to the failure stress in a practical design. In fact, while the failure stress given for lap shear is usually the average shear stress, failure may often be more closely related to the cleavage stresses that generally occur near the ends of the overlap.

This paper contains numerical and analytical analyses of some features of stresses and stress intensity factors in adhesive joints. Particular emphasis is placed on lap shear joints and to a less extent on a butt tensile joint. Experimental evidence for the existence of complex stress states in such joints is also included. Peel tests were discussed in earlier papers by the authors.[1]

FACTORS AFFECTING ADHESION

The geometries for the adhesive butt joint and lap shear test configurations appear deceptively simple. Superficially, one might anticipate that the calculation of stress (or energy required to produce a unit of new surface) would be comparatively straight forward. Such is not the case. Even leaving aside for the moment intricacies of molecular structure and the singularities at imperfections or cracks, one finds that the differences in moduli between the adhesives and adherends, the physical constraints they place on each other, the effects of free edges, and the influence of minute misalignment can result in extremely complex stress states. When we add to these largely geometric effects the influence of macroscopic and microscopic material properties, we find that interpreting test results in too simplified a manner can be more misleading than enlightening. This paper treats several aspects relating to these factors.

Adhesive thickness has been shown by a number of investigators[1,2,3] to have a pronounced effect on joint "strength." The effect (even its sign) is found to be a function of the joint geometry; e.g., for a given adhesive, a butt joint's strength might be found to decrease with increasing thickness, while the strength of a lap shear joint increases to a maximum as the thickness is increased. These effects and some of the possible causes are reviewed in Reference 1. It would be too lengthy to review them here. Suffice it to note that the adhesive thickness effect may be due to any or a combination of a number of geometric factors and material properties. As a case in point, the energy required to create a unit of debonded surface is generally much larger than that required to rupture a "plane" of bonds. One reason for this lies in the demonstrable fact that in all except for perfectly elastic systems, energy is dissipated at distances well removed from the crack tip. In some cases, the thickness effect is closely related to the size of this damage zone near the crack tip. If the adhesive is not thick enough for this zone to fully develop, less energy is dissipated with an associated decrease in the adhesive fracture energy.

Alternatively, the relative magnitude of the cleavage and shear stress at or near interfaces depends not only upon the difference in moduli and Poisson's ratios for the adhesive and adherend but also on the exact joint configuration. While they have not been as extensively studied, adherend thickness, stiffness, etc., can also have a pronounced effect on bond "strength."

LAP SHEAR TEST, BACKGROUND

There has been no dearth of research in an effort to analyze the stresses in adhesive joints. As early as 1938 Volkersen[4] obtained expressions for the stresses in a lap shear joint by considering the differential displacements of the adherends and neglecting bending. This was followed in 1944 by the now classical treatment of Goland and Reissner.[5] They used standard beam theory and other strength of material concepts to obtain expressions for the joint stresses. Plantema[6] combined the results of these two earlier investigations to include shear effects in the system.

Since the stress state of the lap shear joint is so complex and doesn't lend itself to closed form solutions, it is only logical that as numerical methods become available, researchers would apply these to analyze adhesive joints.

Wooley and Carver,[7] for example, used finite element methods to calculate the joint stresses. They compared their results with the results obtained by Goland and Reissner and reported very good agreement. Adams and Peppiatt[8] used a two dimensional finite element code to analyze the stresses in a standard lap shear joint. They too reported good agreement with Goland and Reissner. These authors also investigated the effect of a spew (triangular adhesive fillet) on the calculated stresses. A nonlinear finite element analysis of the single lap joint was completed by Cooper and Sawyer[9] in 1979. They too compared their results with Goland and Reissner as well as with experimental results from photoelastic models. They concluded that while nonlinear behavior can at times have a sizable effect on joint stresses, the Goland and Reissner method is usually sufficiently accurate for prediction of midsurface stresses.

The double lap shear joint has the advantage of greatly reducing the bending stresses. This joint has also received analytical attention. Amijima, Yoshida, and Fujii[10] used finite element methods to calculate the stresses. They were able to claim some agreement between predictions of joint stresses obtained in their numerical analysis and experimentally observed strength.

Sen and Jones[11,12] also used finite element techniques for the analysis of double lap joints. They considered a viscoelastic adhesive. They report that the magnitude of the maximum adhesive stress decreases as the ratio of the adhesive modulus to adherend modulus decreases from one. They also investigated the effect of adhesive thickness and length of overlap on the stress distribution along the joint. They report that decreasing the thickness results in a more uniform stress distribution.

The <u>measurement</u> of the exact stress state in an adhesive joint is also a difficult task. There have, however, been several intensive efforts in this respect. Sharp and Muha[13] used a laser technique to measure the adhesive shear strain. The technique is only suitable for optically clear linearly elastic adhesives. They reported stresses in good agreement with Goland and Reissner.

Several investigators have suggested modifications to the standard ASTM lap shear test geometries. Leaving an adhesive spew and rounding adherend corners as analyzed by Adams and Peppiatt[8] (noted previously) can eliminate the mathematical singularity at the end of the overlap thereby reducing the maximum stress, increasing the joint strength, and reducing test variability. Renton[14] suggested thickening the adherends, over those recommended in ASTM D3165, thereby reducing bending and hence cleavage stresses in the lap joint. He describes a technique of determining adhesive strain by measuring changes in a magnetic field resulting from the displacement of the steel parts of the lap joint and suggests optimum specimen designs for various adhesive-adherend combinations. In the most thorough study of this type with which the authors are familiar, Guess, Allred and Gerstle[15] present finite element and experimental evidence that failure in standard lap shear specimens is often controlled by the tensile strength of the adhesive rather than its shear strength. As will be discussed in more detail later, they report an apparent increase in shear strength with increasing adherend thickness. River[16] also describes methods, and test results, for reducing the stress concentration at the end of the overlap and obtaining relatively accurate measurement of the shear properties of an adhesive.

Because of the singularity and other problems associated with stress criteria of failure, several groups have recently applied the principle of fracture mechanics to adhesive joints.[1,3,17,18] Of particular interest to the lap shear joint is the attempt by DeVries, et al.[19] to apply a Griffith type energy balance. They were able to claim reasonable agreement with experimental data and noted an apparent dependency of the adhesive fracture energy on mode of stress at the crack tip.

LAP SHEAR TEST, STRESS ANALYSES AND TEST DATA

For practical reasons, the designer usually attempts to load adhesives in shear. As a consequence, the most commonly used adhesive test is the lap shear test. Specifications published by ASTM and other sources describe a variety of different tests but the most commonly used lap shear bond strength tests are ASTM D1002 and ASTM D3165. The adherend thickness recommended in these standards for most metal sheets is 0.16 cm (0.0625 in.) with 1.3 cm (0.5 in.) overlap. Generally the quantity reported as the "lap shear strength," or "tensile shear strength," is the breaking load per unit bond area. However, failure initiation in materials is usually a localized phenomenon that is more dependent on maximum stresses (or energy) at a point reaching some critical value than on the average induced values. As mentioned previously and reinforced by plane strain finite element stress analyses in this study, the standard lap shear specimen shows a variation in shear stress within the adhesive layer and in fact can exhibit a singular behavior at the bond termination.

The authors have recently completed linear elastic stress analyses of lap shear joints using a plane strain finite element computer program. These analyses have particularly concentrated on the stresses near the bond termination. Lap shear specimens with ASTM D3165 geometry, adherend thicknesses of 0.16 cm (0.0625 in.) to 1.3 cm (0.50 in.) and adhesive thicknesses of 0.025 cm (0.01 in.) and 0.16 cm (0.06 in.) were analyzed. The specimens were 16.5 cm (6.5 in.) long with 1.3 cm (0.5 in.) overlap. The plane strain approximation is justified since Adams and Peppiatt[20] have shown that the tensile and shear stresses in the direction of the applied load are not significantly influenced by transverse stresses caused by Poisson's ratio strains in the adherends.

The finite element code used quadrilateral elements which were internally divided into four linear displacement triangular elements. A grid network consisting of 2,016 quadrilateral elements was used in each analysis with elements concentrated near the adhesive-adherend interface termination as shown in Figure 1. The smallest element width was 0.000025 cm (0.00001 in.) for stress and stress intensity factor analyses and 0.048 cm (0.019 in.) for energy release rate analyses. Axial displacements were applied and transverse displacements set equal to zero at the specimen ends to simu-

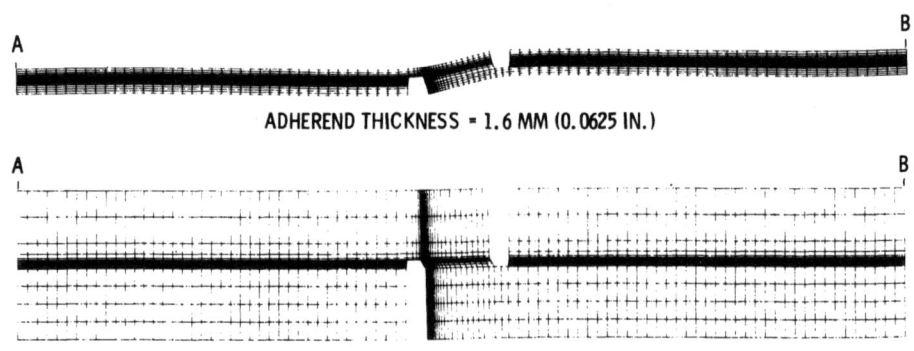

Figure 1. Lap shear deformed grids, displacements magnified ten times.

late clamped ends. The following material properties were used:

	Modulus, MPa (ksi)	Poisson's Ratio
Adherend (Steel)	207,000 (30,000)	0.30
Adhesive (Epoxy)	2,760 (400)	0.34

The adhesive ends were terminated at right angles to the adherends. Often a fillet is formed at bond edges by adhesive squeeze-out. However, the right angle geometry is obtained when the lap shear specimens are cut and notched from bonded plates as described in ASTM D3165. The effect of adhesive squeeze-out is discussed in References 8 and 21.

The computed deformed shapes for lap shear specimens with adherend thicknesses of 0.16 cm (0.062 in.) and 1.3 cm (0.50 in.) are presented in Figure 1. An axial elongation of 0.016 cm (0.006 in.) was applied and the resulting displacements multiplied by a factor of 10 to obtain the deformed grids. The amplified displacements provide a visual indication of the bending which takes place near the bond termination for thin adherend specimens. One would thus expect relatively large bond stresses (cleavage) at these points induced by adherend bending in thin adherend specimens. In addition, for linear elastic stress analyses, both the normal and shear stresses become singular as the bond termination is approached.[22,23]

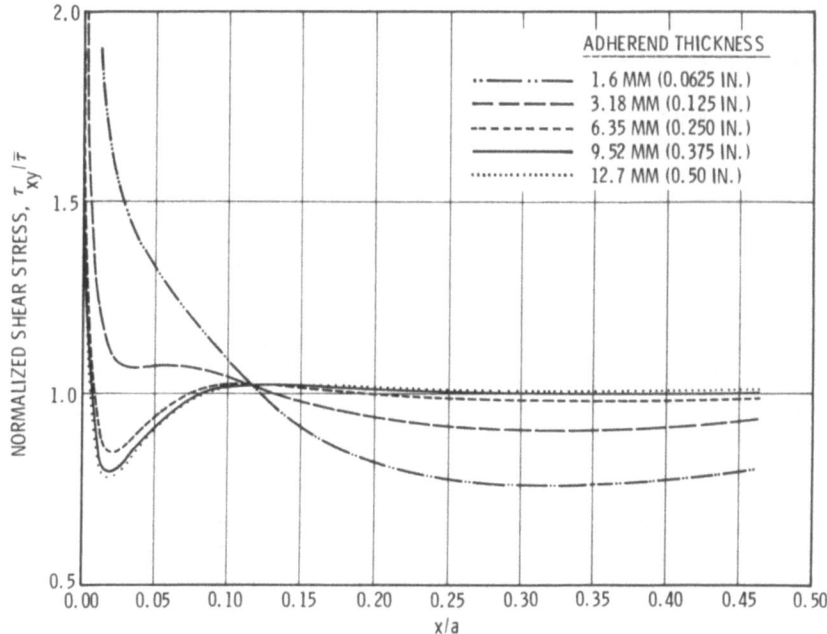

Figure 2. Bondline shear stresses in lap shear specimens.

The adhesive stresses adjacent to the nonterminating adherends are plotted in Figures 2 and 3 as a function of distance from the bond termination. The distances were normalized by dividing by the bond length, $\bar{a}$, and the stresses by dividing by the average applied shear stress, $\bar{\tau} = P/A$. Adhesive thickness was 0.025 cm (0.01 in.).

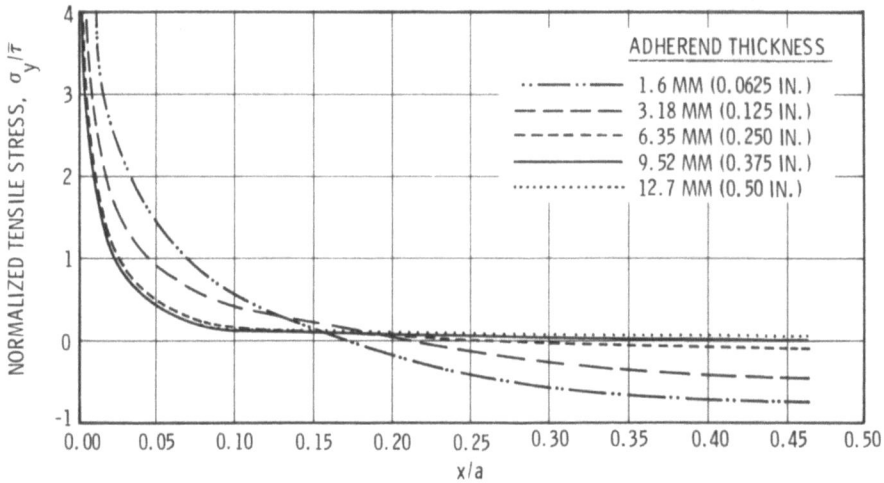

Figure 3. Bondline tensile stresses in lap shear specimens.

As expected, the stresses change radically as the bond termination is approached. The shear and tensile stresses are nearly constant and equal to the average values over 80% of the bond area for adherend thicknesses greater than 0.64 cm (0.25 in.). For thin adherends, the stresses vary continuously along the bondline.

The character of the stress singularity was examined by plotting the tensile and shear stresses in log-log form (Figure 4). The tensile and shear stresses exhibit a $1/r^n$ singularity behavior (log linear) for all elements within 0.0076 cm (0.003 in.) of the adhesive edge except for the final element. This element typically varies somewhat from the singularity pattern[1] due to the coarseness of the grid network relative to the stress changes in this area. As the distance from the bond termination increases beyond 0.076 cm (0.003 in.), the stress behavior deviates from the singular (log linear) form as expected.

Stress intensity factors and the singularity exponents are obtained from the straight line portion of the log-log plots. The "singularity exponent," n, in the equation, stress = K/r^n, is the slope of the straight line portion of the Figure 4 curves. This value was approximately 0.34 for tensile as well as shear stresses and does not depend upon adherend thickness. The stress intensity factors (intercepts of the Figure 4 curves), however, were found to depend upon adherend thickness as shown in Table I.

Table I. Stress Intensity Factors for Lap Shear Tests

Adherend Thickness cm (in.)	Shear			Tensile			
	$K_{II}/\bar{\tau}$ $cm^{0.34}$ ($in.^{0.34}$)		$K_{II}/K_{II\ 0.5}$	$K_I/\bar{\tau}$ $cm^{0.34}$ ($in.^{0.34}$)		$K_I/K_{I\ 0.5}$	K_I/K_{II}
0.16 (0.0625)	0.448	(0.326)	2.59	0.914	(0.666)	1.77	2.04
0.32 (0.125)	0.272	(0.198)	1.57	0.660	(0.481)	1.28	2.43
0.64 (0.250)	0.189	(0.138)	1.10	0.560	(0.408)	1.09	2.96
0.95 (0.375)	0.178	(0.130)	1.03	0.526	(0.383)	1.02	2.95
1.27 (0.500)	0.173	(0.126)	1.00	0.516	(0.376)	1.00	2.98

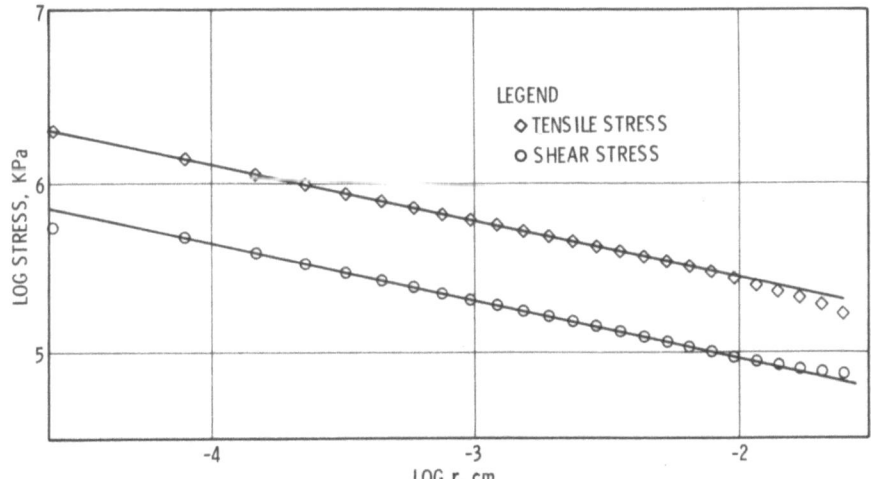

Figure 4. Log stress plot, lap shear test.

Both the Mode I stress intensity factor, K_I, and the Mode II stress intensity factor, K_{II}, per unit average shear stress, τ, increase with decreasing adherend thickness, see Figure 5. Thus one would expect higher debond forces in thick adherend specimens. The decrease in K_{II} with increasing adherend thickness is greater than that for K_I. For all adherend thicknesses, the Mode I stress intensity factor is greater than that for Mode II. As shown in the last column of Table I, the tensile component is nearly three times the shear component for large adherend thicknesses. This tendency was not expected but was consistently observed for widely varying finite element grid networks and for either plane strain or plane stress conditions.

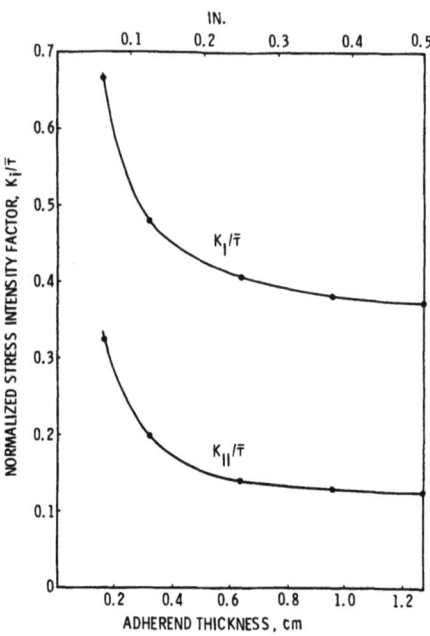

Figure 5. Lap shear stress intensity factors.

The nature of the singularity near the termination of a two material interface has been investigated by several authors.[22,23] For linearly elastic materials, the stresses sufficiently near the bond termination are shown to be proportional to r^{-n}, where r is the distance from the bond termination and the exponent n is a function of boundary conditions, material properties, and bond termination geometry. If one material is rigid and a wedge of a second material with included angle β is bonded to the rigid material, the exponent, n (from Reference 23), can be determined from the plot in Figure 6. For a 0.34 Poisson's ratio and a 90 degree bond termination angle, n = 0.32. Thus the stresses near the bond termination are singular and proportional to $1/r^{0.32}$. This is in essential agreement with the Reference 22 work and the 0.34 exponent obtained numerically during the present analyses.

For some combinations of Poisson's ratio and bond termination angles, the characteristic equation in the analytical solution has imaginary as well as real roots.[1] The imaginary roots lead to stresses which oscillate as the bond termination is approached.

An energy release rate, G, was determined for each of five different adherend thicknesses using debond depths of 0.048 cm (0.019 in.) and 0.10 cm (0.038 in.) (average x/a = 0.057). As shown in Figure 7, the square root of energy release rate is essentially constant for adherend thicknesses greater than 0.64 cm

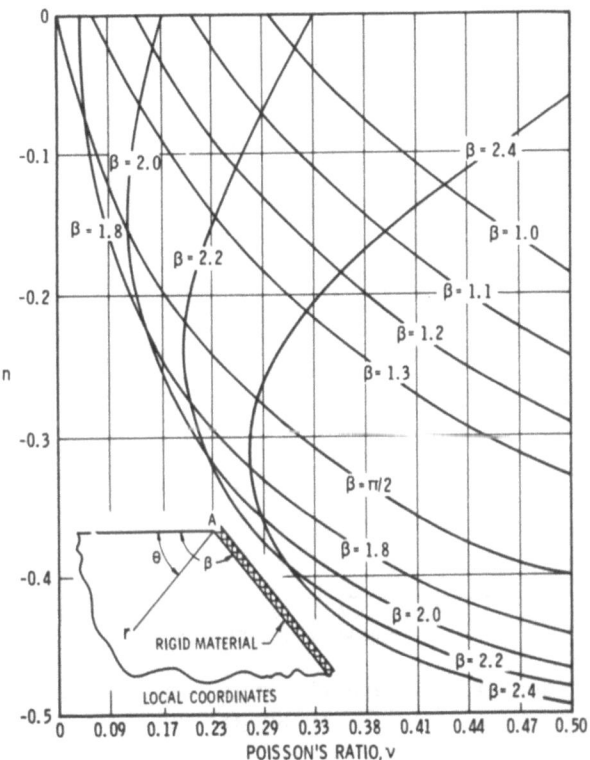

Figure 6. Real roots of the characteristic equation for a free-fixed wedge[23]

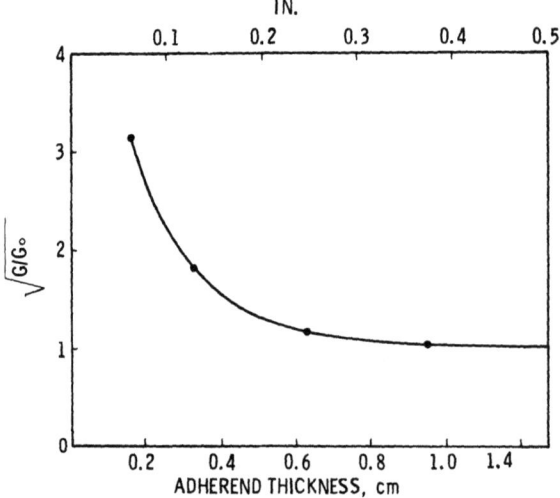

Figure 7. Effect of adherend thickness on energy release rate, G for lap shear tests.

(0.25 in.) and approximately 1/3 this value at the ASTM recommended thickness, 0.16 cm (0.0625 in.). Thus the change in energy release rate with adherend thickness for a small initial debond has the same basic characteristics as observed for stress intensity factors, i.e., nearly a constant value for adherend thicknesses greater than 0.64 cm (0.25 in.) and increasing by a factor of approximately 3 as adherend thickness is reduced to 0.16 cm (0.0625 in.).

An analysis for a 0.025 cm (0.01 in.) thick adhesive (1.3 cm, 0.5 in. adherend) showed stresses of basically the same form as for thicker adhesives. However, the stress singularity is shifted closer to the bond termination.

Testing was completed by the authors using single lap shear specimens. Steel adherends with thicknesses ranging from 0.16 cm (0.0625 in.) to 1.3 cm (0.5 in.) were used with a 1.3 cm (0.5 in.) overlap. The adhesive was a fairly rigid two part epoxy (Hysol EA 934). All testing was completed at 72°F and 0.13 cm/min (0.05 in./min) displacement rate. The resulting data are presented in Table II and Figure 8. These data show a fairly constant breaking force for specimens with adherend thickness greater than 0.64 cm (0.25 in.). At the ASTM D1002 recommended thickness of 0.16 cm (0.0625 in.), the breaking force drops by 63 percent.

Similar data were obtained by Guess, et al.[15] These authors also used steel lap shear adherends with 1.3 cm (0.5 in.) overlap. Their adhesives were EC-2214-R structural paste adhesive and FM-123-5 structural tape adhesive. Again, the breaking force was essentially independent of adherend thickness for thicknesses greater than 0.64 cm (0.25 in.). For these adherends, the EC-2214-R adhesive strength dropped by 64% and the corresponding drop for the FM-123-5 film adhesive was 23%. The amount of decrease in bond strength for thin adherend specimens is essentially the same as the increases in stress intensity factor and square root of energy release rate as would be expected if these parameters controlled debond initiation load.

The stress analyses discussed here assumed that the adhesive as well as the adherends were linear elastic materials. For some structural adhesives, particularly the elastomer based adhesives and the rubber modified epoxies, the linearity assumption is invalid and a nonlinear stress analysis is preferred. Such analyses are not difficult to complete with various nonlinear finite element computer programs which are presently available. However, the material properties required for input to such programs are not commonly available. Thus most analyses are still completed using the linearity assumption. This is a major weakness in current analytical approaches and may provide misleading data for many bond systems.

Table II. Lap Shear Test Data

Adherend Thickness, cm (in.)	Average Stress at Debond, N (psi)	Reduction (%)
0.16 (0.0625)	10,500 (1,530)	63
0.32 (0.125)	16,800 (2,440)	40
0.64 (0.25)	24,800 (3,600)	12
0.95 (0.375)	29,000 (4,200)	0
1.27 (0.50)	27,600 (4,010)	0

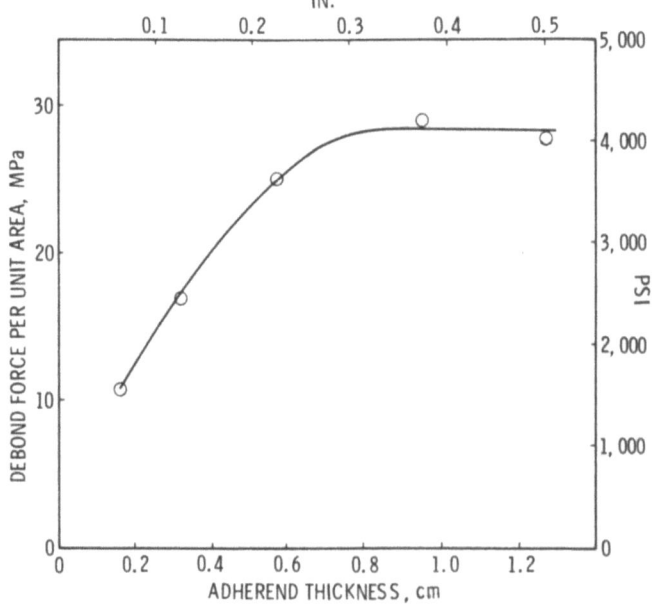

Figure 8. Debond force for lap shear tests.

In much of the literature on linear elastic analyses of lap shear specimens, the researchers have discussed the stress concentration factor at the bond terminus. When linear elastic analyses are used and the bond terminus not rounded, the stress concentration factor is infinity because of the stress singularity at this point.

Even in the idealized case where the adherends are rigid and restricted to remain parallel, the transverse stress has a singular behavior very near the bond termination. The transverse stress becomes negligibly small (as expected) and the shear stress remains relatively constant (rigid and parallel adherends) for distances greater than 1/10 the adhesive thickness from the bond termination. Whether stress intensity factors for the initial bond geometry actually control debond initiation is open to question since inherent flaws such as bondline voids could exist in the high stress area that are larger than the entire singularity region. Thus the most appropriate criteria for debond initiation may be obtained through the use of an energy release rate at the depth of the largest inherent flaw. A yet to be satisfactorily resolved question, is, by what means can this size be best determined. Microscopic examination, extrapolation from finite flaws, etc. need further exploration. Such methods as well as the use of stress failure criteria are briefly discussed in Reference 1. If one ignores inherent flaws, the problem of bond termination singularities could be avoided entirely by using properly contoured bond termination fillets and rounded adherend edges.

TENSILE BOND TESTS

Tensile bond tests are also commonly used for evaluating bond strength. The tensile strength suggested by ASTM procedure D897 is the breaking load per unit bond area. However, the bond stresses are not uniform[1,24] and, in fact, become singular at the bond edges when analyzed using linear elastic analyses[22] as discussed for lap shear tests.

If the adhesive thickness is not constant or if the grips used to attach the specimen to a tensile test machine are not perfectly aligned and rigid, the initial load is applied only at one edge of the specimen. This has two adverse effects on the resulting bond strength data: 1) The true failure load is higher than indicated and 2) since the magnitude of nonalignment is not normally the same on each sample, the data variability is much greater than for properly aligned samples.

The tensile test specimen consisting of two steel adherends bonded together with a 0.16 cm (0.060 in.) thick epoxy ($\nu = 0.34$) was stress analyzed using a finite element computer program. The maximum principal stress distributions are shown in Figure 9. The ratio of peak-to-average stress away from the edge singularity and ratio of the edge stress intensity factor (using Reference 1 techniques) to average applied stress for symmetric and edge loads are:

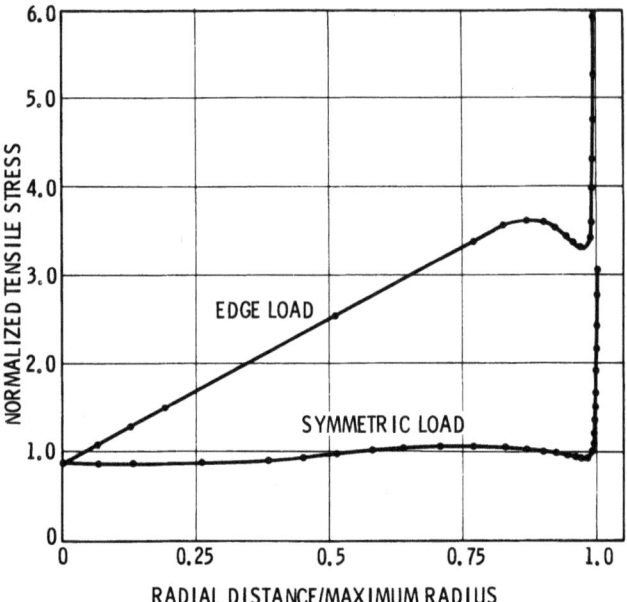

Figure 9. Stresses in tensile tests.

	Symmetric Load	Edge Load
Stress Ratio (away from edge singularity), σ/σ	1.1	3.59
Stress Intensity Factor Ratio, K_I/σ, $cm^{0.30}$ ($in.^{0.30}$)	0.291 (0.220)	0.964 (0.729)

Thus whether stresses or stress intensity factors are considered, the load carrying capability of the bond is reduced by a factor of 3.3 by edge loads. Even with symmetric loading the peak stress is 10% higher (to 100% higher for incompressible materials)[1] than the average value generally reported for such tests. The application of a symmetric load is experimentally not an easy task and in actual structural joints is usually impractical. Very small sample, grip, or test machine misalignments can result in substantially larger stresses than the average. Slight differences in adhesive thickness across the joint can have similar effects. Experiments were conducted in which "standard techniques" were used and compared with tests where extraordinary care was taken to obtain symmetric loading. In addition, a special test fixture grip was used to reduce the effects of specimen misalignment in tension or shear. This fixture and additional analyses will be more fully described in a subsequent paper. However, the fixture basically makes use of a rubber plate to reduce the effects of both shear and bending misalignment.

Test results using this fixture and standard grips (similar to those recommended in ASTM D897) are shown below:

	Load at Failure, N (lb)	Coefficient of Variation (%)
Standard Grips	14,700 (3,310)	36
Modified Grips	25,900 (5,820)	6

These data show a very large decrease in variability as well as an 80 percent gain in apparent load carrying capability when most of the effects of edge loading are removed.

CONCLUSIONS

The authors have tried to summarize and discuss some of the efforts in their laboratories as well as others to reduce uncertainties associated with adhesive bond testing. The numerical study reported here places particular emphasis on the regions very close to the bond terminations. It is in this region that failure is generally thought to originate. Numerical stress analyses demonstrated that at regions near the bond terminus, the normal and shear stress distributions exhibited the trend toward singular behavior reported by others. They have, for instance, the same singularity exponent as determined analytically by Dill, et al.[23]

It appears fairly certain that for many "standard" lap shear specimens failure initiation is not as much a function of the reported "average" shear stress as it is the induced cleavage stresses. Experimentally, higher strengths and less variability result from using thicker specimens thereby reducing bending in the adherends and hence these cleavage stresses.

For butt joints, the stress state is again more complex than it superficially appears. Unless there is a perfect match of moduli between the adhesive and adherends, shear stress will generally accompany the tensile stresses in the adhesive. In addition, the tensile stress along the bondline is not constant. It varies with distance from the bond center and exhibits a singular behavior as the bond terminus is approached. Even more critically important, however, are any misalignments of grips or testing machine, slight variation in sample thickness, etc. Where extreme care is taken to control these factors, it is demonstrated that typically much higher strength with less variation is obtained.

The authors are of the opinion that adhesive fracture mechanics somewhat along the lines described in References 1, 25, 26, and 27 has the best potential of resolving many of the problems suggested in this paper. Reference 27 is of particular interest to the current authors since the study of this paper introduced them to the application of fracture mechanics to adhesive bonds. Tests based on fracture mechanics concepts provide a measure of adhesive quality in terms of basic material parameters (adhesive fracture energy or equivalently critical stress intensity factor, moduli of adhesive and adherends, and induced or inherent flaw size). Perhaps of even more importance, once these parameters are obtained, adhesive fracture mechanics points the way in which these parameters might be systematically used to design practical joints, select optimal joint configuration, and predict adhesive bond performance.

ACKNOWLEDGEMENT

Portions of this effort were supported by NSF - Polymer Program DMR-79 25390.

REFERENCES

1. G. P. Anderson, S. J. Bennett and K. L. DeVries, "Analysis and Testing of Adhesive Bonds," Academic Press, New York, 1977.
2. J. J. Bikerman, J. Soc. Chem. Ind. $\underline{60T}$, 23, 1941.
3. S. Mostovoy and E. J. Ripling, J. Appl. Polym. Sci., $\underline{15}$, 661 (1971).
4. O. Volkersen, Luftfahrtforschung, $\underline{15}$, 41 (1938).
5. M. Goland and E. Reissner, J. Appl. Mechanics, $\underline{11}$, 17 (1944).
6. F. J. Plantema, "De Schuifspanning in eme Lijmnaad," Report M1181, Nat. Luchtvaartlaboratorium, Amsterdam (1949).
7. G. R. Wooley and D. R. Carver, J. Aircraft, $\underline{8}$, No. 10, 817 (1971).
8. R. D. Adams and N. A. Peppiatt, J. Strain Analysis, $\underline{9}$, No. 3, 185 (1974).
9. P. S. Cooper and J. W. Sawyer, "A Critical Examination of Stresses in an Elastic Single Lap Joint," NASA Technical Report 1507, NASA Scientific and Technical Information Branch (1979).
10. S. Amijima, A. Yoshida and T. Fujii, "Proceedings of the Second International Conference on Composite Materials," Toronto, Canada, Metallurgical Society of AIME, 1185 (1978).
11. J. K. Sen and R. M. Jones, AIAA J., $\underline{18}$, No. 10, 1237 (1980).

12. J. K. Sen and R. M. Jones, AIAA J., 18, No. 11, 1376 (1980).
13. W. N. Sharpe and T. J. Muha, in Proceedings of the Army Symposium on Solid Mechanics, Army Materials and Mechanics Research Center, Watertown, Mass. AMMRC-MS-74-8, 23 (1974).
14. W. J. Renton, Experimental Mechanics, 16, No. 11, 409 (1976).
15. T. R. Guess, R. E. Allred and F. P. Gerstle, J. Testing and Evaluation, 5, No. 2, 84 (1977).
16. B. H. River, Adhesives Age, 24, No. 12, 30 (1981).
17. A. N. Gent, "Strength of Adhesive Bonds - Plastic and Viscoelastic Effects," Proceedings - 46th National Colloid Sym., Amherst, Mass. (1972).
18. W. D. Bascom and R. L. Cottington, J. Adhesion, 1, 333 (1976).
19. K. L. DeVries, M. L. Williams and M. D. Chang, Society for Experimental Stress Analysis preprints, 21 Bridge Square, Westport, CT, Cleveland Meeting (1972).
20. R. D. Adams and N. A. Peppiatt, J. Strain Analysis, 8, 134 (1973).
21. A. D. Crocombe and R. D. Adams, J. Adhesion, 13, 141 (1981).
22. V. L. Hein and F. Erdogan, Int. J. Fracture Mechanics, 7, 317 (1971).
23. E. H. Dill, A. L. Deak and W. F. Schmidt, in "Handbook for the Engineering Structural Analysis of Solid Propellants," Ch. 5, Chemical Propulsion Information Agency Publication 214, Applied Physics Lab, Silver Spring, MD, May 1971.
24. H. Nakayama, K. Tozawa, A. Hirano, and O. Ohkubo, "Effect of Adhesive Layer Thickness on Strength Characteristics of Adhesive - Bonded Soft Joint Specimen Under Static Tensile and Push-Pull Fatigue Load Conditions," Proceedings 22nd Japan Cong. on Mater. Res. - Metallic Materials, Society of Materials Science, Kyoto, Japan (1979).
25. W. D. Bascom, R. L. Cottington and C. O. Timmons, J. Appl. Polymer Sci. Applied Polymer Symposium 32, 165 (1977).
26. R. J. Chang and A. N. Gent, J. Polymer Sci.: Polymer Phys. Ed., 19, 1619 (1981).
27. M. L. Williams, Bulletin Seismological Soc. America 49, No. 2, 199 (1959).

A CRITICAL APPRAISAL OF DENTAL ADHESION TESTING

W.J. O'Brien and S.T. Rasmussen

Surface Science Laboratory, Dental Research Institute
The University of Michigan, Ann Arbor, MI 48109 and
School of Dentistry, Case Western Reserve University
Cleveland, OH 44106

The various methods used to investigate dental adhesion are reviewed. Concepts from fracture mechanics are used to interpret adhesive and cohesive failure of adhesives used in dentistry and to analyze bond strength testing techniques. The practical strength of an adhesive interface is a complex relationship between the stress distribution, specific adhesive fracture energy and the size and shape of interfacial defects. Bond strength tests are not designed to investigate these separate aspects related to adhesive failure. The problem of defects in dental adhesive failure are unique in that the number, size and shape of defects which result in failure are mainly technique related. However, the investigation of dental adhesive failure can generally be characterized as ignoring the relationship that specific adhesive fracture energy and defects have on the practical strength of adhesive systems. Based on theoretical considerations we propose that dental adhesion testing be reevaluated from a theoretical, experimental and most significantly from a practical viewpoint.

INTRODUCTION

The importance of adhesion investigation is widely recognized. However, adhesion is a complicated problem which cannot be thoroughly investigated by any simple technique. Dental researchers interested in adhesion have generally looked at: 1) wettability, 2) surface preparation, 3) fracture path, 4) fracture energy and 5) bond strength.

Wettability is considered important for a fluid material to adhere
to a solid. Additionally, it also measures a fluid's ability to
displace other liquids and gases and to spread over the surface so
as to produce an interface with a minimum of voids.[1,2,3,4] While
wettability is necessary, it is not sufficient for long-term success
in the oral environment. For example, good wetting may be obtained
between tooth structure and a fluid resin, but the restorative resin
may separate during usage due to tensile forces resulting from polymerization shrinkage or the humid oral environment.[1,5]

Surface preparation is also important to obtaining good adhesion
and has received extensive attention in dentistry. Figure 1 shows
a ceramic-metal crown with adhesive failure. The enameling is
achieved through firing a ceramic-powder paste to the metal. The
paste consists primarily of a feldspar powder with water which forms
a thick slurry capable of holding its shape when applied to the
metal. Surface preparation of the metal casting has been found to
improve the bonding of dental porcelain to the metal substrate.
Surface roughness, preoxidation out-gassing and metal conditioners
have all been extensively investigated for their roles in improving
bonding of porcelain to metal.[2,6,7,8]

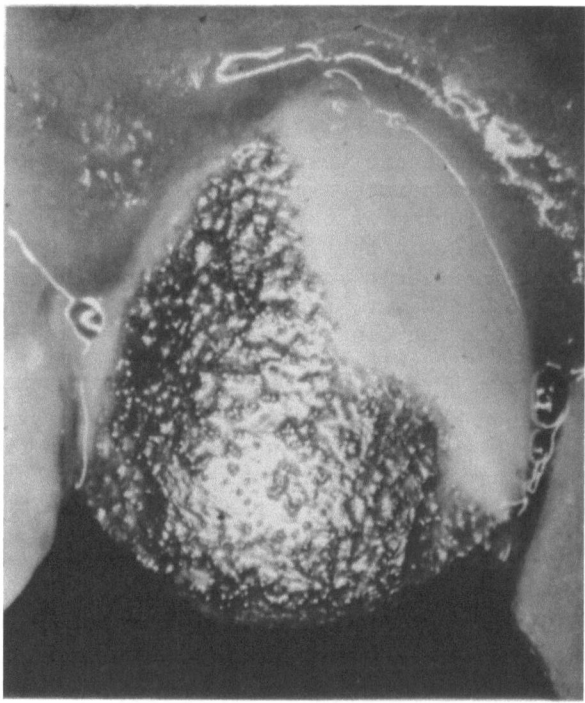

Figure 1. Ceramic-metal crown with adhesive failure. According
to the classification system of Figure 6 this is a Type I failure.

Similarly, the adhesion of resins and composites to enamel and dentin have been improved by surface pretreatment with various chemicals and by maintenance of a dry, clean surface prior to applying the resin.[1,3,9,10,11,12,13] Acid etching improves the mechanical retention of resins. Acid is applied to tooth enamel so as to produce a porous surface to which resin is applied as shown in Figure 2. The surface roughness produced by etching also improves the resins ability to cover the enamel surface.[3] Scanning electron microscopy (SEM) has proved useful in the investigation of surface morphologies of tooth surface following acid etching. Coupling agents or liners have also been reported to improve bonding of resins (composites) to tooth surfaces.[12,13,14,15]

The fracture path approach to dental adhesive failure has received attention in recent years. In metal-porcelain systems the various types of failure have been specified.[16] The basic premise is that it does little good to report bond strength if cohesive failure can be verified by analysis of the fracture path, since the values obtained reflect the strength of the material in which cohesive failure occurred. It has also been noted that specification of the actual zone in which failure occurs is not a simple matter.[17] Diffusion during firing of the porcelain to metal, results in a variation in composition at the interface which cannot be clearly defined into different zones. Fracture-path studies can help identify the weak areas so that research can be directed at them for possible improvement. The fracture-path approach to adhesion research for resin-tooth structure systems has also received attention.[10,18,19] The fracture of resin-enamel systems is not considered to be pure interfacial since deeply penetrating resin tags (Figure 2) are broken off.[19]

The application of fracture mechanics to dental adhesion has been limited. The area under stress-vs.-strain curves has been used to compare various porcelain specimens.[20] However, this approach is limited in scopy, since the fractured area is not known. A method called the "blister technique"., has been applied to barnacle and several dental cements.[21,22] This approach is particularly attractive, since once the specific adhesive-fracture energy is known, the failure stress for any situation can, in principle, be predicted.[22] Widespread application of this technique appears unlikely due to two major reasons. First, it is not clear that cohesive-failure in this test implies that the adhesive bond is stronger than the fractured material. Secondly, it would be difficult to apply the blister technique to adhesion between resins and enamel or dentin due to the small size of the specimens. Adhesion to enamel and dentin has been found to be dependent upon their surface morphology, and in order to control this variable, the specimens must be small.[23,24,25]

A work of fracture technique, which gives the interfacial work of fracture (W_i), has been applied to dental adhesion.[26] This method shows some promise in that it can be applied to small specimens and cohesive failure appears to indicate that the adhesive bond is stronger than the material in which the failure occurred.

Wettability and surface preparation studies are generally followed up by bond-strength measurements. The bond strength test configurations are generally designed such that the principal stress was claimed to be either tensile[7,9,10,13,15,23,24,27-30] or shear.[2,7,8,14,30] However, some tests were: 1) transverse,[2,6] 2) torsion,[2] or 3) combinations of shear and tensile.[31] Of the various means of adhesion investigation, bond-strength tests are probably the most popular. Unfortunately with this emphasis a multiplicity of techniques have been developed. Consequently, there has been considerable interest in standardizing a bond strength test.

In this paper we would like to present a thesis that current bond strength tests are, in general, inadequate for the purpose of evaluation and selection of adhesive materials. Consequently, we propose that dental adhesion testing be reevaluated from a theoretical, experimental and most significantly from a practical viewpoint.

ADHESIVE FAILURE

From theoretical considerations, most materials are expected to have ideal strengths approaching E/10 (E=elastic modulus) which amounts to approximately 10^6 psi for ordinary window glass and 2.8 x 10^6 psi for iron.[32] Unfortunately, these immense ideal strengths are reached only for freshly drawn glass fibers and single crystal whiskers of defect free ceramics and of relatively dislocation free metals. Except for extremely ductile materials, most useful materials are notch sensitive under certain condition. This notch sensitivity arises because of stress concentration about the sharp tips of the notch or cracks. This stress can easily exceed the forces required to break the atomic bonds. Griffith[33] showed that for brittle materials enough energy must also be supplied to equal the surface energy of the newly formed surfaces, otherwise materials with sharp cracks would fail spontaneously under low loads. His approach has been further modified to apply to semibrittle materials. The following equation gives the critical tensile stress (σ_{cr}) for failure of an isotropic-uniform plate with an elliptical crack (length 2c, cf. Fig. 3a) subjected to plain stress:

$$\sigma_{cr} = \left(\frac{2E\gamma_e}{\pi c}\right)^{1/2} \quad (1)$$

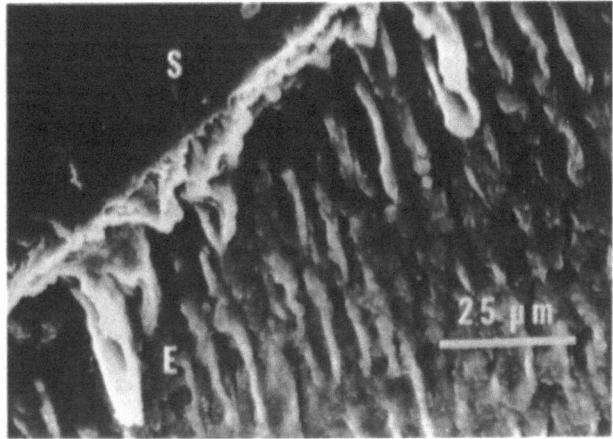

Figure 2. Scanning electron micrograph of an enamel-resin (sealant) system. The specimen was prepared so as to show the resin tags of the sealant (S) which penetrated the porous enamel (E). The enamel-sealant interface runs diagonally across the upper left hand corner.

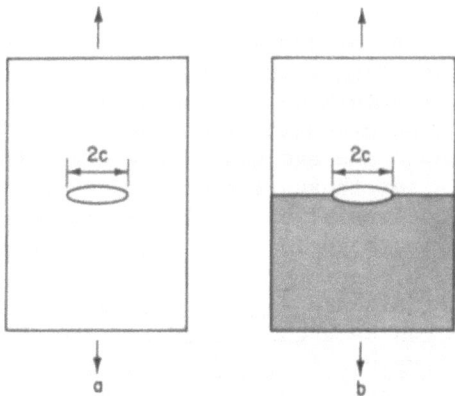

Figure 3. Uniform plates with an elliptical crack of length 2c. The plates' dimensions are large in comparison to the crack. a) Isotropic plate b) A two component plate with a crack at the interface between the two materials.

where E and γ_e are the materials elastic modulus and effective surface energy (also called specific surface energy) respectively.[34,35] The term γ_e includes contributions from surface energy, plastic flow, pullout for composites or any other means of absorbing energy during initiation of crack growth. Equation (1) along with others developed for different geometries and loading conditions have proved very successful in explaining the fracture toughness of many materials.[32]

In a similar fashion adhesive failure is linked to stress concentration about interfacial defects (Figure 3b). The formula expressing the critical pressure for adhesive failure in the "blister" technique[21,22,35] of adhesion testing is:

$$P_{cf} = \sqrt{\frac{32}{3(1-\mu^2)} \left(\frac{h}{c}\right)^3 \left(\frac{E\gamma_a}{c}\right)} \quad (2)$$

where γ_a is the specific adhesive fracture energy and the other parameters are defined in Figure 4. In this case the plate (Figure 4) is considered rigid in comparison to the adhesive. A more complicated formula arises when the plate properties must be considered. Inspection of Equation (2), shows that the failure pressure is very dependent upon the defect size. Consequently, the strength of adhesive bonds can also be seen to be highly dependent upon defect size. The larger the interfacial defect, the weaker the adhesive interface will be. Other techniques have been developed for investigation of the specific adhesive fracture energies of adhesives to various materials.[35] All of these techniques take into account both the specific adhesive fracture energy and the size of the defect which results in failure.

The investigation of dental adhesive failure can generally be characterized as ignoring this close interrelationship between the specific adhesive fracture energy and defect size. This is particularly true for "bond strength" tests where results are interpreted as measuring, in some average sense, the strength of the actual bonds at the interface. Yet these tests are most critically sensitive to the wide range of defect sizes resulting from preparation and/or application of adhering material.

When two materials are joined adhesively, the interfacial resistance to fracture can be greater than that of the weaker or more brittle material involved if the joint is free of defects.[35,36] This is because of changes in stress distribution at the adhesive interface resulting from a mismatch of elastic modulii at the interface. Additionally one of the adhering phases will likely be intrinsically more resistant to brittle fracture making the interface potentially less brittle than the more brittle phase.

ADHESIVE FAILURE IN DENTISTRY

Many researchers feel that a sound adhesive joint will fail cohesively rather than adhesively.[36] Unfortunately, this state of adhesive perfection has not progressed to all adhesive joints prepared in dentistry. At present, failure of adhesive joints involving resins(composites) and enamel or dentin can be considered to fail adhesively.[19] Of course the environmental factors are severe for both preparation and long term service. It has been noted that even properly made engineering joints involving epoxy can fail adhesively after long time exposure to water.[36] While porcelain to metal restorations can be prepared in the laboratory with homogeneous well prepared surfaces, these restorations are however, subjected to design limitations of the natural shape of a tooth and the remaining tooth structure.[37] Consequently the bulk of the porcelain cannot be freely increased to reduce stress concentrations. The porcelain must be thick enough to give a natural appearance but not so thick that the supporting tooth structure is severly weakened.

The investigation of specific adhesive fracture energy (γ_a) and defects must be considered as highly crucial to the improvement and evaluation of resin adhesives in dentistry. Interfacial defects are practical problems arising from adverse environmental conditions and techniques of manipulations as well as properties of the materials involved. Consequently, tests should be made on specimens prepared under simulated oral conditions so that the results will properly reflect the effect of defects on the practical success of the adhesive joint. It is likely that a comparison of γ_a obtained under optimum laboratory conditions with those simulating practical conditions could lead to improvement of surface preparation techniques.

Adhesion to enamel and dentin is thought to be dependent upon their surface morphology.[23,24] It is also likely that it may be dependent upon the chemical variations[38] which occur on the surface and through the depth of enamel. Certain regions of enamel or dentin may result in poor adhesion and thus be an area from which failure is initiated. Testing techniques which could be applied to small areas would be highly desirable. Interfacial work of fracture (W_i) technique[26] is attractive from this point of view, since it can be applied to small specimens. The fractured surface area in Figure 5b can be around 1mm^2 in size. W_i is only an approximation to the specific adhesive fracture energy (γ_a) for initiation of crack growth since its value reflects contributions from both initiation of crack growth and for crack propagation. However, in absence of a γ_a technique applicable to small specimens, the W_i method appears to be an important adjunct to dental adhesion research.

Many of the materials involved in restorative dentistry (enamel, dentin, porcelain, composites) can be considered as brittle materials

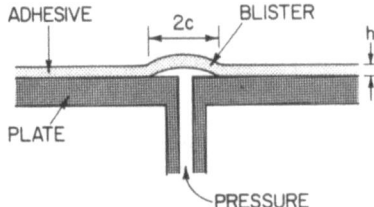

Figure 4. Schematic of the blister technique showing the adhesive with elastic modulus E and Poisson's ratio μ and the adherend (plate). Pressure is increased until failure occurs.

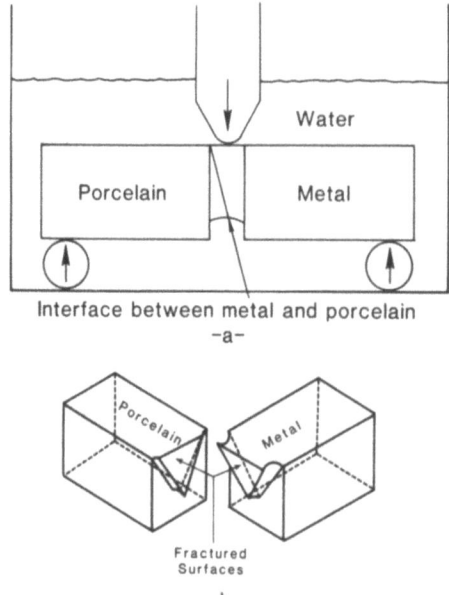

Figure 5. Schematic of porcelain-metal specimens. Specimens were notched at the interface between the two materials.

because of their low strengths[39,40] or low work of fractures.[25,26] Consequently, these materials as well as adhesive joints involving them are likely to have strengths which are highly defect sensitive.

For our purposes, defects can be classed loosely as those which are unavoidably part of the materials involved (Type I defect) and those which arise during preparation (Type II defect). For porcelain, Type I defects are those which are related to the grain size of the quartz, alumina and etc. For dental composites, a Type I defect may be bubbles of gases trapped during manufacture or the particles themselves may act as sources of stress concentration. Type I defects determine a material's maximum strength. Defects arising from preparation, Type II defects, will most likely lower the practical strength of the materials.

It should be evident that defects play an important role in any investigation of the strength of materials or adhesive joints since the researcher determines the number and size of defects by his preparation technique. The problem of defects in dental materials are unique in that the number and size of Type II defects are mainly technique related. There is likely to be a wide variability in an adhesives lifetime under mechanical stress due to differences in manipulation. In dentistry a material with average material properties may give the best results because its manipulative characteristics give either smaller or fewer large Type II defects. It is our contention that researchers of dental materials must address themselves to this problem where applicable since it is the largest defect or number of defects which will limit the cohesive or adhesive strength of an adhesive system.

Specific adhesive fracture energy (γ_a) and interfacial work of fracture (W_i) measurements are advantageous since they are, in principle, independent of defects. γ_a measurements begin with a predetermined defect size[21,22,35] and give values independent of the defect size (i.e. γ_a will be the same regardless of defect size). Interfacial work of fracture (W_i) measurements, while in principle are not defect size dependent, do give in practice values which are related to the crossectional area of the defects intersected by the fractured surface.[26] The effect is proportional to the defect area as compared to the total fractured area. If the defect area is a large percentage of the total fractured area it could be measured and be eliminated from the measured W_i. γ_a or W_i measurements could be coupled with statistical studies of defect distributions to give the effect of defects on the strength of an adhesive joint or a materials strength. The defect investigation should be conducted on specimens which are most representative of materials prepared under practical conditions. In fact old restorations would be best from this point of view.

Porcelain-metal restorations are likely to fail cohesively.[16,39] For adhesive joints, where cohesive failure is likely, the fracture path approach is mandatory since it does little good to report the fracture energy or strength of the fractured material as being a measure of the quality of adhesion. The measurements reflect characteristics of the material in which the fracture occurred and not the adhesive joint. In fact it is important to determine the fracture path for all specimens tested for adhesive failure. Figure 6 shows the various types of failure which can occur in metal-porcelain systems. While it is thought[19] that adhesive failure will occur for resin-enamel system cohesive failure is also possible and will become more so as adhesives are improved. Figure 7 shows some of the single phase type failures which might occur for an enamel-composite system.

An important aspect of the fracture-path approach for cohesive failure is that under certain conditions cohesive failure indicates that the adhesive bond is "stronger" than the material in which failure occurred. If the stress distribution, based upon a defect free specimen, is such that the tensile stresses at the adhesive interface are equal to or greater than those in the region of failure the adhesive bond can be considered sufficient. However, specimens are not defect free and failure is initiated at defects. One must decide how the role of defects fit into interpretations of cohesive failure.

The aim of many adhesive investigations is improvement of the adhesive system. From this viewpoint it is not enough to determine if failure is adhesive or cohesive, one should also determine the source of failure. Several potential situations involving cohesive failure will illustrate the importance of this statement. First consider an enamel-composite system in which defects resulting from mixing lead to cohesive failure. Here a change in the technique of mixing the composite components together might reduce the size of the defects in the composite leading to improved strength. Conceivably, the strength could improve enough that adhesive failure could occur. That is, the adhesive system is improved by changing the failure mode from cohesive to adhesive. Second consider a metal-porcelain system in which failure is initiated at a large interfacial defect but cohesive failure occurs (cf. Figure 8). The adhesive interface is the weak link of the system and efforts should be aimed at reducing the size of interfacial defects. Consequently, a change in the methods of surface preparation and/or a change in the means of applying the porcelain powder to the metal could lead to a substantial improvement of a system in which cohesive failure occurs.

A CRITICAL APPRAISAL OF DENTAL ADHESION TESTING

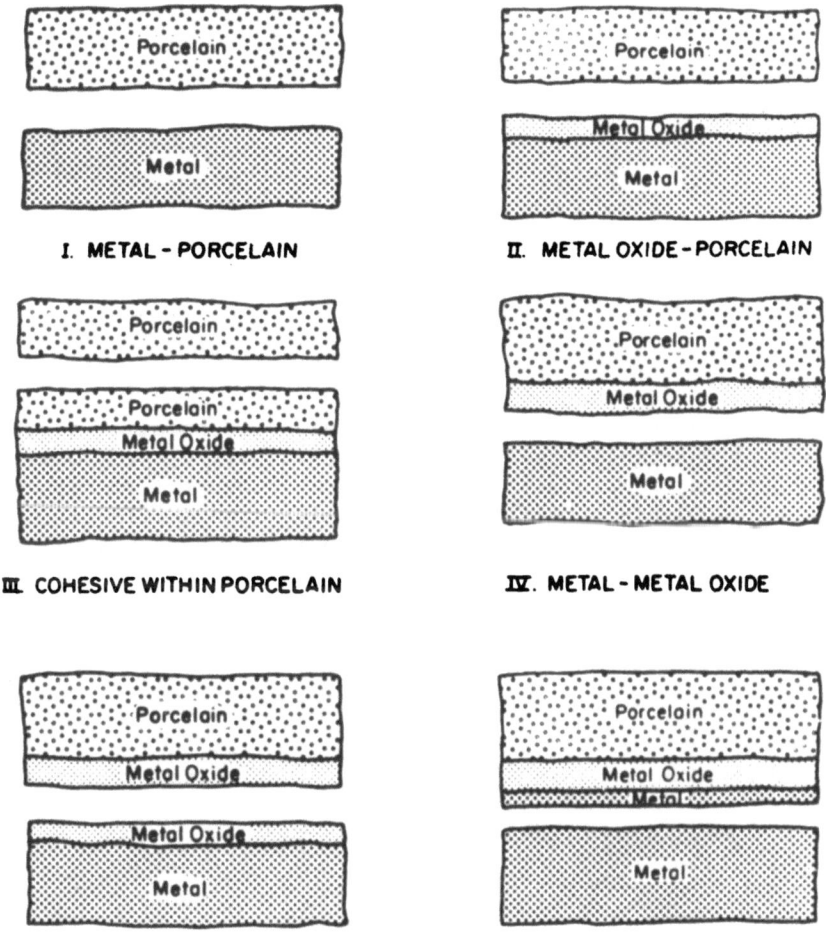

Figure 6. Porcelain-metal failures. Classification of failure is based upon the region in which the interface formed.

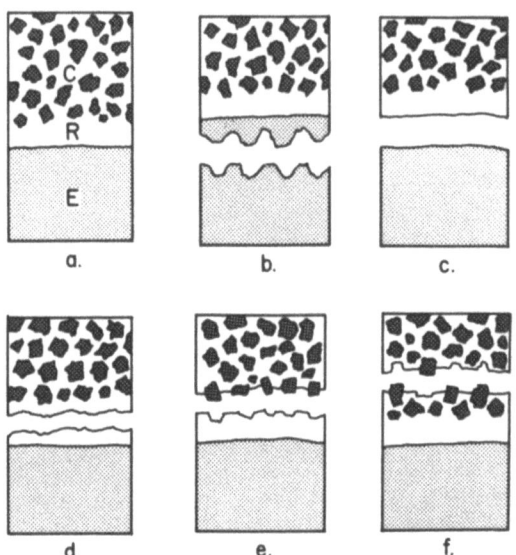

Figure 7. Enamel-composite failures: a) solid specimen showing enamel (E), particle free resin (R), resin-particle composite (C); b) cohesive within enamel; c) resin-enamel interfacial failure; d) cohesive within resin; e) interfacial at resin film-filler interface; and f) cohesive within composite. The cohesive failure of the enamel (b) is drawn to characterize the surface found for failure of enamel.[25]

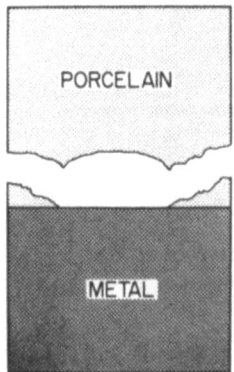

Figure 8 A composite plate, similar to Figure 3b, showing how an interfacial defect could initiate cohesive failure.

BOND STRENGTH TESTS

The investigation of dental adhesive failure can generally be characterized as ignoring the close interrelationship between the specific adhesive fracture energy and the defect size. This is particularly true for "bond strength" tests where results are interpreted as measuring, in some average sense, the strength of the actual bonds at the interface. Yet these tests are most critically sensitive to the wide range of defect sizes resulting from preparation and/or application of adhering material.

Another major problem with bond strength investigations is that a stress analysis was not made of the test geometry taking into account the elastic properties of the materials involved. Figures 9 and 10 depict various geometries used in testing porcelain to metal or resin to tooth structure bonds respectively. As is evident, the test geometries are, in general, not simple and the combination of different materials makes interpretation of the failure stress most difficult. That is, one cannot say intuitively what the failure stress is. Researchers claim a particular type of stress (tension shear etc.) resulted in failure and that the failure stress was given by the ratio of the breaking load to the fractured area (references 2,6,7,8,10,13,15,23,24,27,28,29,30,31). Recently,[17,41] some porcelain-metal bond-strength test configurations have been subjected to finite element stress analysis which verified these intuitive criticisms of bond strength testing in dental adhesive research.

Other problems with bond-strength tests can be related to specimen preparation or testing. The measured tensile strengths of brittle materials can be highly dependent upon specimen preparation and alignment of testing apparatus. This is particularly true for many dental restorative materials because in addition to many of them being brittle or semibrittle, the necessity of preparing specimens similar in size to actual restorations magnifies problems due to preparation and alignment of testing apparatus. These problems arise for brittle and semibrittle materials because of their sensitivity to stress concentration about sharp defects. Similar conditions exist for bond-strength tests.[21,35]

Most bond strength testing techniques have been designed for testing in air at room temperature rather than environments simulating oral conditions. It is a well-known fact that the environment can have a large effect on the strength of materials.[36,40]

Bond strength measurements of dental cements have been found to be highly dependent upon film thickness. Additionally it is difficult to control this variable.[23,29,30] Film thickness may also be a problem for composites placed directly on the enamel or dentin surface since there will most likely be a particle free region next

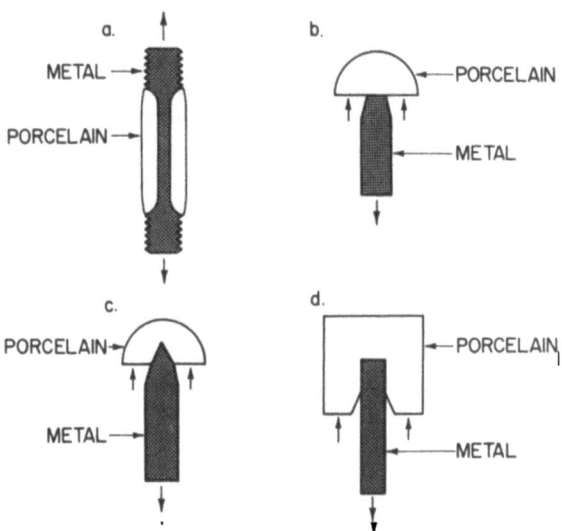

Figure 9. Schematic drawings of various test geometries that have been used to test the porcelain-metal interface.

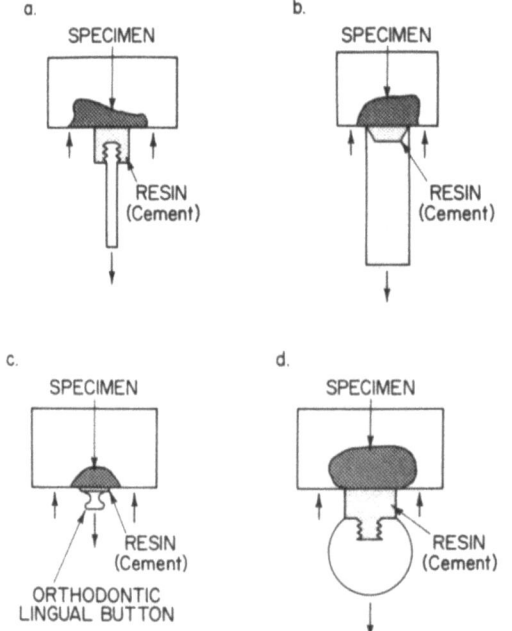

Figure 10. Schematic drawings of various test geometries that have been used to test the resin-enamel/dentin interface.

to the surface. This particle free region could be considered a film of resin (cf. Figure 7a).

The use of bond strength measurements for ranking/comparing of adhesives would be appropriate if the actual stresses were known and the specimens were prepared and tested under simulated oral conditions. The values obtained would reflect contributions from both specific adhesive fracture energy and defects. However, if one desires to improve an adhesive, independent information about both specific adhesive fracture energy and defect size would be preferable.

SUMMARY

Concepts from fracture mechanics were used to interpret adhesive and cohesive failure of adhesives used in dentistry. It was concluded that current bond strength tests are inadequate for evaluating and improving adhesion to the extent of emphasis currently placed on them. Until recently,[17,41] bond strength tests have not been subjected to stress analysis and consequently the meaning of their measurements are questionable. The practical strength of an adhesive interface is a complex relationship between the stress distribution (determined by the load, geometry and the elastic properties of the materials involved), specific adhesive fracture energy and the size and shape of interfacial defects. Bond strength tests are not designed to investigate these separate aspects related to adhesive failure. That is they are affected by all of them and consequently it is difficult to determine the major source of failure.

The problem of defects in dental adhesive failure are unique in that the number, size and shape of defects which result in failure are mainly technique related. However, the investigation of dental adhesive failure can generally be characterized as ignoring the relationship that specific adhesive fracture energy and defects have on the practical strength of adhesive systems. Consequently, independent knowledge of specific adhesive fracture energy and defect characteristics could lead to substantial improvement in the practical strength of dental adhesive systems.

ACKNOWLEDGEMENT

Dr. O'Brien wishes to acknowledge the support of Research Grant DE02731 from the National Institute of Dental Research, National Institutes of Health, Bethesda, MD 20014

ADDENDUM

Currently, the poorest adhesive system in restorative dentistry is that of the resin-dentin system. However, some recent research indicates potential improvement in the bonding between resins and dentin. The references are: 1) R.L. Bowen, E.N. Cobb, and J.E. Rapson, J. Dent. Res., 61, 1070 (1982); and 2) N. Nakabayashi, K. Kojima, and E. Masuhara, J. Biomed. Mater. Res. 16, 265 (1982).

We have recently received an update on the blister method. It was used successfully to measure the specific adhesive fracture energy of a gingival cell line to agar. The diameters of the blisters (cf. Figure 4) for the gingival cell line tested ranged from 0.4 to 1.6 mm. Considering the small diameters of the blisters used in the gingival cell investigation, it may also be possible to apply this technique to testing the adhesion of resins to enamel or dentin. The blister technique deserves further consideration for its potential use in dental adhesive investigations. The reference is: R.K. Fletcher, S.M. Breitling, J.K. Adamich, and M.L. Williams Jr., in Vitro, 13, 857, (1977).

REFERENCES

1. G.M. Brauer, in "Scientific Aspects of Dental Materials". T.A. von Fraunhofer, Editor, Butterworths, Boston, (1975).
2. R.C. Vickery and L.A. Badinelli, J. Dent. Res. 47, 683 (1968).
3. H.D. Moskowitz, G.T. Ward, and E.D. Woolridge, Editors, "Dental Adhesive Materials," Proceedings of the Symposium held November 8-9, 1973, Department of Health, Education, and Welfare, Washington, D.C., (1974).
4. W.T. O'Brien and G. Ryge, J. Prosthet. Dent. 15, 1094 (1965).
5. R.L. Bowen, J. Amer. Dental Assoc., 74, 439 (1967).
6. M.H. Lavine and F. Custer, J. Dent. Res., 45, 32 (1966).
7. J.W. McLean and I.R. Sced, Trans. J. Brit. Ceramic Soc., 72, 229 (1973).
8. J.W. McLean and I.R. Sced, Trans. J. Brit. Ceramic Soc., 72, 235 (1973).
9. G.M. Brauer and D.J. Termini, J. Dent. Res. 51, 151 (1972).
10. D.H. Retief, J. Dent. Res. 52, 333 (1973).
11. N.A. Wickwire and D. Rentz, Am. J. Orthodont. 64, 499 (1973).
12. R.L. Bowen, J. Dent. Res. 44, 903 (1965).
13. J.B. Moser, D.B. Dowling, E.H. Greener, and G.W. Marshall J. Dent. Res. 55, 411 (1976).
14. R.L. Patrick, C.M. Kaplan, and E.R. Beaver, J. Dent. Res. 47, 12 (1968).
15. H.L. Lee, A.L. Cupples, R.J. Schubert, and M.L. Swartz J. Dent. Res. 50, 125 (1971).

16. W.J. O'Brien, in "An Outline of Dental Materials and Their Selection", W.J. O'Brien and G. Ryge, Editors, pp. 180-194, Saunders, Philadelphia, 1978.
17. K.J. Anusavice, R.D. Ringle, and C.W. Fairhurst, J. Prosthet. Dent. 42, 417 (1979).
18. J.A. Rensch, Am. J. Orthodont., 63, 156 (1973).
19. D.H. Retief, J. Oral Rehab. 1, 265 (1974).
20. F.J. Knap and G. Ryge, J. Dent. Res. 45, 1047 (1966).
21. M.L. Williams, K.L. De Vries, and R.R. Despain, J. Dent. Res. 52, 517 (1973).
22. P.R. Despain, K.L. De Vries, R.D. Luntz, and M.L. Williams, J. Dent. Res. 52, 674 (1973).
23. R.L. Bowen, J. Dent. Res. 44, 690 (1965).
24. R.W. Phillips, M.L. Swartz, and B. Rhodes, J. Amer. Dental Assoc. 81, 1353 (1970).
25. S.T. Rasmussen, A.H. Heuer, R.E. Patchin, and D.B. Scott, J. Dent. Res., 55, 154 (1976).
26. S.T. Rasmussen, J. Dent. Res. 57, 11 (1978).
27. M.L. Swartz and R.W. Phillips, J. Amer. Dental Assoc. 50, 172 (1955).
28. H.L. Lee, M.L. Swartz, and G. Culp, J. Dent. Res. 48, 211 (1969).
29. G.T. Eden, R.G. Craig, and F.A. Peyton, J. Dent. Res. 49, 428 (1970).
30. E. Misrahi and D.C. Smith, Br. Dent. J. 127, 371 (1969).
31. J.S. Shell and J.P. Nielsen, J. Dent. Res. 41, 1424 (1962).
32. A.H. Cottrell, "The Mechanical Properties of Matter," pp. 223-225, 336, John Wiley and Sons, New York, 1964.
33. A.A. Griffith, Phil. Trans. Roy. Soc. (London) A221, 163 (1920).
34. H.G. Tattersall and G. Tappin, J. Materials Sci. 1, 296 (1966).
35. G.P. Anderson, S.J. Bennett, and K.L. De Vries, "Analysis and Testing of Adhesive Bonds," Academic Press, New York, 1977.
36. "Handbook of Adhesive Bonding", Charles V. Cagle, Editor, Mc Graw Hill Book Co., New York, 1973.
37. R.G. Craig and J.W. Farah, Oral Sci. Rev. 10, 45 (1977).
38. J.A. Weatherell, C. Robinson, and Hallsworth, J. Dent. Res. 53, 180 (1974).
39. W.J. O'Brien, in "Dental Porcelain: The State of the Art-1977," H.N. Yamada, Editor, pp. 137-141, University of Southern California School of Dentistry, Los Angeles, 1977.
40. C.A. Sherrill and W.J. O'Brien, J. Dent. Res. 53, 683 (1974).
41. K.J. Anusavice, P.H. De Hoff, and C.W. Fairhurst, J. Dent. Res., Special Issue A 57, Abstract No. 869 (1978).

ADHESIVE JOINT CHARACTERIZATION BY ULTRASONIC SURFACE AND INTERFACE WAVES

S.I. Rokhlin

Department of Materials Engineering
Ben-Gurion University of the Negev
P.O.B. 653, Beer Sheva 84105, Israel

Methods of measurement of the velocity and attenuation of ultrasonic waves have been used for a long time for investigating structural and mechanical properties of materials. The main purpose of this article is to show that a properly updated ultrasonic method can be successfully used for investigating surface and interface films, in particular, of adhesively bonded structures. Theoretical analysis of the propagation of elastic waves in a multi-layered medium can be used for relating the velocity of surface and interface waves with the elastic properties of the film and substrates. If the elastic properties of the substrates are known, then the viscoelastic properties of films can be determined from measurements of the velocity and attenuation of these waves. The velocity of surface and interface waves is a function of not only the film properties, but also of adhesion of the film to the substrates. This makes the suggested technique potentially promising for the evaluation of adhesion properties of thin surface and interface films.

INTRODUCTION

Ultrasonic methods have been used extensively for investigating the properties of materials and for process control. Various aspects of this problem were reviewed in the survey by Papadakis.[1] Ultrasonic waves used for testing properties of materials consist of small-amplitude mechanical vibrations and, depending on the vibratory mode being utilized, may produce both longitudinal and shear stresses in the solid body. Information about the structural properties of a substance can be obtained by measuring both the velocity and the attenuation of the ultrasonic wave. The phase velocity of the wave is controlled by the elastic constants of the body. In an isotropic solid medium, which has only two independent elastic moduli, there exist two elastic waves: the longitudinal with velocity V_ℓ and shear, with velocity V_t, which are obtained from the expressions:

$$V_t = \sqrt{\mu/\rho} \quad ; \quad V_\ell = \sqrt{(\lambda + 2\mu)/\rho} \tag{1}$$

where ρ is the density of the material, μ is the shear modulus and λ is the Lame parameter.

Ultrasonic velocities can be expressed in terms of other elastic constants of the solid body on the basis of the coupling equations:

$$\lambda = K - \frac{2}{3}\mu \quad ; \quad \lambda = \frac{\mu(E-2\mu)}{3\mu-E} \quad ; \quad \lambda = \frac{E\nu}{(1-2\nu)(1+\nu)}$$

$$E = \frac{9\mu K}{3K+\mu} \quad ; \quad \mu = \frac{E}{2(1+\nu)} \tag{2}$$

where K is the bulk modulus, E is Young's modulus and ν is Poisson's ratio. For liquids $\mu = 0$ and the velocity of the longitudinal wave is controlled only by the bulk modulus, i.e., $V_\ell = \sqrt{K/\rho}$. For rubber-like materials $K \gg \mu$, and $E \simeq 3\mu$.

Measurement of V_ℓ and V_t is one of the most exact methods for measuring elastic constants. Three kinds of bulk elastic waves may propagate in an anisotropic solid body: a quasi-longitudinal, and two quasi-transverse waves, differing in polarization and velocity. To determine the matrix of elastic constants one must measure the phase velocity in several different directions relative to the crystallographic axes.

The attenuation of an ultrasonic wave is associated with absorption of elastic waves (anelastic effect) and the scattering of elastic waves by structural inhomogeneities. Scattering may be decisive in polycrystalline, composite and ceramic materials. In the course of scattering, elastic energy is lost by the ultrasonic beam in the form of a stochastically scattered field, which is gradually absorbed. The latter is associated with conversion of elastic into thermal energy, as a result of various anelastic effects termed internal friction.

The velocity and attenuation of elastic waves are affected by the same microstructural factors of the material as the strength. Using this fact, Vary[2] is currently developing an application of the ultrasonic method for evaluation of the mechanical properties of materials.

The ultrasonic method is extensively used for evaluation of adhesively bonded structures. A detailed survey of the latest achievements in this field was given recently by Segal and Rose.[3] The majority of techniques developed until recently (these shall be referred to as standard techniques) are an upgrading of methods used for evaluation of bulk specimens. At the same time, testing of adhesively bonded structures and coatings has a number of distinguishing features:

1) It is necessary to estimate the properties of a very thin layer of material, the thickness of which is much smaller than the thickness of the adherends or substrates. The standard method which employs a normally incident ultrasonic beam cannot provide accurate information, since the length of the acoustic path is very small. The spectral method yields a lower accuracy in measuring the wave velocity and attenuation than conventional ultrasonic methods.

2) The mechanical properties of the coating-substrate or adhesive-adherends system is controlled not only by the "bulk" properties of the film, but also to a significant degree, by the properties of the film-adherend interface. The majority of standard methods[3] cannot, in principle, solve this problem.

The purpose of this paper is to discuss the possibility of application of ultrasonic surface and interface waves for characterizing thin coatings and interface films. It appears that these new techniques have great potential possibilities. These waves, termed guided waves, propagate along the surface or interface, and hence the length of the acoustic path becomes the same as in bulk measurements. For specially selected testing conditions these waves produce shearing strains in the bond-line region, and thus provide information about the adhesion properties of the adhesive-adherend interface.

Surface acoustic waves have only recently come into use for evaluation of thin films. As of now the principal results using this method were obtained for metal films. There are only few results on evaluating the properties of polymer coatings. Conversely, the applications of interface waves were primarily developed for evaluation of adhesively bonded structures. Below we consider certain methodological aspects of using these methods and describe analytical and experimental results, pertaining primarily to the use of interface waves.

SURFACE ACOUSTIC WAVES (SAW)

As shown by Rayleigh, an elastic surface wave may propagate along the surface of a solid body. The elastic energy of surface waves is localized in a subsurface layer with thickness of the order of wavelength λ. Figure 1 shows[4] the energy flux distribution of a surface acoustic wave with depth X, normalized to the wavelength. At a depth of the order of λ there is virtually no transport of elastic energy. Figure 2 depicts schematically the deformations in the subsurface layer induced by a surface wave. It is seen that both longitudinal and shear elastic displacements occur in the wave. The magnitude of surface displacement u in the figure is intentionally magnified, while actually it is in the order of tens of Å and is several orders of magnitude smaller than the wavelength λ ($u \ll \lambda$).

Velocity V_r of the surface wave in an isotropic solid body is always lower than the velocity of the shear wave and is found from numerical solution of the equation:[5,6]

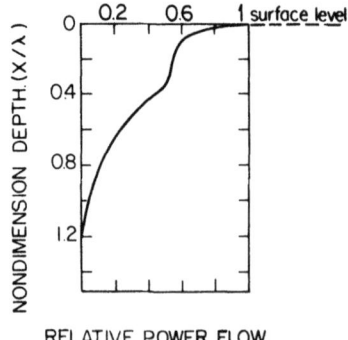

Figure 1. Relative energy flux of surface acoustic waves as function of depth (Data from Farnell[4]).

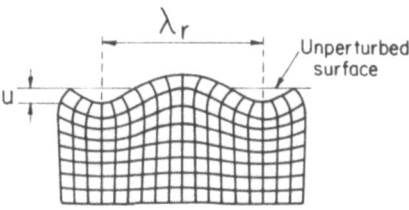

Figure 2. Deformation in the subsurface layer on propagation of a Rayleigh surface wave.

$$\Delta_r = W^2 - 4\alpha^2\beta\gamma = 0 \tag{3}$$

where

$$W = 2\alpha^2 - 1 \quad ; \quad \beta = (\alpha^2 - 1)^{1/2} \quad ; \quad \gamma = (\alpha^2 - \xi^2)^{1/2} \quad ;$$

$\alpha = V_t/V_r$ is the nondimensional wave number of the surface wave, $\xi = V_t/V_\ell$ is a quantity which is a function only of the Poisson ratio. With an accuracy better than 0.5% the velocity of a surface wave can be calculated from the approximate equation:[5,6]

$$V_r = V_t \frac{0.87 + 1.12\nu}{1+\nu} \tag{4}$$

It is seen from Equations (3) and (4) that the velocity of a surface wave is a function of both elastic constants of the isotropic material.

The presence of a thin film on the substrate surface changes the surface stiffness and also loads the surface. Figure 3 shows schematically that a thin film, whose thickness is much less than a wavelength, is subjected to a compression-tension strain in the course of propagation of a surface wave. The strain arising in the film is similar to strains in a thin plate upon propagation of a longitudinal plate mode. In this case the velocity of the surface wave depends not only on the substrate properties, but on the stiffness $E_o/(1-\nu_o^2)$ and the density ρ_o of the film (E_o is Young's modulus, ν_o is the film Poisson ratio). Achenbach and Keshawa[7], and Tiersten[8] used an approximate theory of thin plates for finding the velocity of a surface wave in a system consisting of a substrate

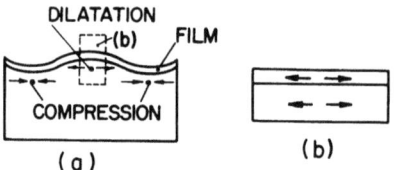

Figure 3. a) Deformation of a surface of a solid body with coating; b) Longitudinal strain of a surface film on propagation of a Rayleigh wave.

and a single coating. The corresponding equation for the surface wave velocity is best written in the form

$$\Delta_r + (\beta+\gamma)\bar{h}\theta - a_1(\bar{h}\theta)^2 + \frac{\alpha^2 E_o \bar{h}}{(1-\nu_o^2)}(a_1\bar{h}\theta-\beta) = 0 \qquad (5)$$

where

$$\theta = \rho_o/\rho; \quad \bar{h} = K_t h; \quad \alpha = K/K_t; \quad K = \omega/V_s; \quad a_1 = \alpha^2-\beta\gamma$$

V_s is the velocity of the surface wave in the coated structure, h is the coating thickness, μ is the shear modulus and ρ is the density of the substrate material. This equation can be extended[9] to the case of a multilayer coating.

Measurements of the surface wave velocity can yield information about the coating properties and about the condition of the bond between the coating and the substrate. Two methods of measurement are possible:

1) direct measurement of the velocity and attenuation of the surface wave;

2) the Rayleigh angle method.

DIRECT MEASUREMENT OF THE VELOCITY AND ATTENUATION OF SURFACE ACOUSTIC WAVES. APPLICATION TO EVALUATION OF COATING PROPERTIES

In spite of the wide application of thin films in surface acoustic wave (SAW) electronic devices, there is very little known about the use of surface waves for thin film property measurements. The measurement of SAW attenuation[10-13] was used for the study of properties of thin films as a function of magnetic field or temperature. Measurements of the velocity and attenuation of surface acoustic waves during deposition of thin films were also described.[14,15]

A number of different methods[6] can be used for excitation of surface acoustic waves. For relatively low ultrasonic frequencies (to 10 MHz) one can use wedge transducers, whereas at high frequencies interdigital transducers are used. As an illustration Figure 4 shows a specimen with two pairs of interdigital transducers. The test film is placed between one pair of transducers, whereas the second pair of transducers without a film between them is used as the reference channel. The velocity of the surface wave for the film region can be determined from the difference in phase velocities of surface waves in both channels. The film properties can then be evaluated from Equation (5).

Frequently surface films are multilayered or have nonuniform elastic properties through the thickness. (Such an inhomogeneous film can be treated as a multilayered film.) The pertinent extension of Equation (5) to the case of a multilayered surface film can be obtained by the matrix method.[9] Using the matrix approach, we obtained a series expansion of the exact solution with respect to the thickness of the successive layers as small parameters. We

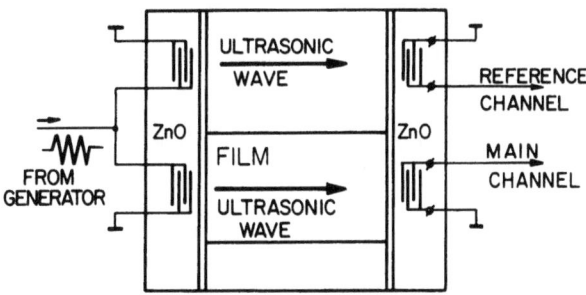

Figure 4. Schematic drawing of the experimental specimen with two pairs of interdigital transducers.

compared the analytic solution thus calculated for a single surface layer with the Tiersten equation. We found that Tiersten's equation corresponds to our expansion only up to the linear term. It is reasonable to expect that this approximation should yield the same accuracy as in the multilayered case, provided that the total thickness Σh_i of the multilayered system is not greater than the allowed film thickness h in the Tiersten equation. By carrying out the appropriate calculations for the characteristic equation of surface waves, we obtain:

$$\Delta_r + (\beta+\gamma)\Sigma \overline{H}_i \theta_i - (\alpha^2 - \beta\gamma)[\Sigma \overline{H}_i \theta_i]^2 + \alpha^2 \Sigma \frac{E_i \overline{H}_i}{\mu(1-\nu_i^2)}$$

$$\cdot [(\alpha^2 - \beta\gamma)\Sigma \overline{H}_i \theta_i - \beta] = 0 \qquad (6)$$

where

$$\Delta_r = W^2 - 4\alpha^2\beta\gamma; \quad W = \alpha^2+\beta^2; \quad \beta^2 = \alpha^2-1; \quad \gamma^2 = \alpha^2-\xi^2;$$

$$\xi = K_\ell/K_t; \quad \theta_i = \rho_i/\rho; \quad \overline{H}_i = K_t h_i; \quad \alpha = K/K_t; \quad K = \omega/V_s$$

V_s is the velocity of the surface wave in the multilayered structure, E_i and ν_i are the Young modulus and Poisson ratio, respectively for the ith layer, μ is the shear modulus and ρ the density of the substrate material and ρ_i the density of the ith layer. K_ℓ is the wave number for the longitudinal waves in the substrate.

Equation (6) is commutative with respect to the order of the successive layers. Physically this is due to the fact that as $\Sigma h_i \ll \lambda_s$ (where λ_s is the wavelength of the surface wave), the longitudinal strain in the layers is close to being homogeneous.

As an illustration of the application of this method, let us consider the results of Rokhlin, Ronen and Dariel,[9] who employed surface acoustic waves for analyzing the elastic properties of an Au-Al thin-film diffusion couple. A thin film of the Au_2Al intermetallic compound appears in the course of the diffusion reaction at the interface between the Au-Al films. When diffusion terminates, the Au-Al sandwich becomes an Au_2Al film. This structural modification induces a change in the surface wave velocity, which can be measured during the experiment.

An example of the results of the conductivity and ultrasonic SAW measurements as a function of time of diffusion is shown in Figure 5. The results indicate that the conductivity and, consequently, the thickness of the growing phase (Au_2Al), depend linearly on $\sqrt{t}$ after a certain incubation period. After this incubation time, the phase shift of the SAW also becomes a linear function of $\sqrt{t}$.

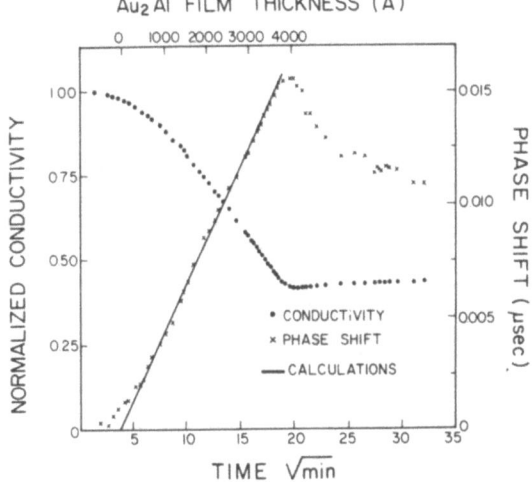

Figure 5. SAW phase shift and electrical conductivity changes as a function of the annealing time in the Au-Al diffusion couple. When the growing phase Au_2Al reaches the surface of the substrate the previous monotonic behavior of the SAW velocity changes suddenly. This phenomenon is explained by a break of the adhesion bonds on the film substrate interface (from Rokhlin, Ronen and Dariel[9]).

When the growing phase Au_2Al reaches the surface of the underlying Cr layer, the previous monotonic behavior of the experimental curves changes suddenly. The curve of the phase shift of the SAW passes through a maximum and then drops by approximately 30% of its maximum value. The electrical conductivity curve passes through a minimum and then increases slowly until stability is attained. This effect is in agreement with the results of Weaver and Parkinson[16] who measured the electrical conductivity and adhesion to a substrate of an Au-Al couple. The shape of the phase shift curve, following its maximum is similar to Weaver and Parkinson's curve representing their adhesion measurements. A suggested mechanism for these changes is related to the diffusion-induced strains present at the various interfaces.

Glass and Cr flow at significantly higher stresses than Al and Au. As the annealing treatment goes on and the compound layer reaches the Cr surface, the elastic stresses at this interface are unable to relax by plastic flow. At this stage, we may expect a break of the adhesion bonds at either the Au_2Al-Cr or the Cr-glass interface. The simultaneous increase of the conductivity is attributed to the relaxation of stresses associated with the loss of adhesion.

As a result of the reduced adhesion to the substrate the effect of the presence of Au_2Al on the velocity of the surface wave decreases. This results in an increasing velocity of the surface wave since the velocity of the SAW at a substrate with an overlying Au_2Al layer is lower than the velocity of the SAW at a free substrate.

Using the velocity data and Equation (6) the effective elastic modulus $E/(1-\nu^2)$ of the Au_2Al intermetallic compound was determined.

LEAKY SURFACE WAVES

It was noted above that, for the isotropic case, the velocity of the surface wave is lower than the velocity of bulk waves in the half space, and hence the wave energy is localized near the surface. The situation changes when the elastic half space interfaces with a liquid. The velocity of the surface wave is higher than the velocity of the acoustic wave in a liquid (for example, V_r in steel is about 3 km/sec, whereas the velocity of sound V_o in water is about 1.5 km/sec). As shown in Figure 6 this results in radiation (leakage) of elastic energy of the surface wave to the liquid. For this reason such a wave is termed a leaky surface wave; the radiation angle of this wave is obtained from Snell's law.

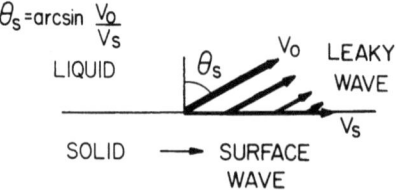

Figure 6. Leakage of elastic energy of the surface wave to the liquid (leaky surface wave). The radiation angle is obtained from Snell's law. V_o - is the ultrasonic velocity in the liquid. V_s - is the velocity of the surface acoustic wave.

An elastic surface wave also may become transformed into a leaky wave if the substrate is coated by a very stiff coating, and also in a free anisotropic half space (when there is no liquid half space). In the anisotropic case the surface wave velocity is a function of the propagation angle[17] and over a certain range of angles, its velocity may exceed the velocity of the slow shear wave. This will then cause leakage of energy from the surface to the substrate.

RAYLEIGH ANGLE METHOD

When the ultrasonic wave is incident from the liquid onto a liquid-solid interface longitudinal and transverse refracted waves are produced in the solid, as shown in Figure 7. Upon increasing the incidence angle θ the refraction angles β_ℓ and β_t increase, so that at $\theta = \theta_\ell = \arcsin V_o/V_\ell$ we have $\beta_\ell = 90°$. This angle is termed the longitudinal critical angle or the first critical angle. At $\theta = \theta_t = \arcsin V_o/V_t$, which is termed the second critical angle, we have $\beta_t = 90°$ and complete internal reflection occurs. At $\theta > \theta_t$ the absolute value of the reflection coefficient should, according to the classical theory, be equal to unity. For illustration Figure 8 shows by a solid line the energy reflection coefficient[18] for the case of a water-steel interface. The calculations were performed for the reflection of a plane incident wave and absence of attenuation in the solid. The points on this figure represent Neubauer's [18] experimental results. Near the angle $\theta_r = \arcsin V_o/V_r$ (V_r is the velocity of the surface wave) a steep drop is observed in the experimental reflection coefficient. The above effect is due to differences between the experimental and theoretical conditions: 1) presence of attenuation in the specimen; 2) incidence of a bonded ultrasonic beam, which is due to the finite dimensions of the radiator.

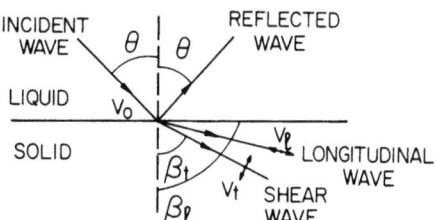

Figure 7. Reflection and refraction of an acoustic plane wave at the liquid-solid interface.

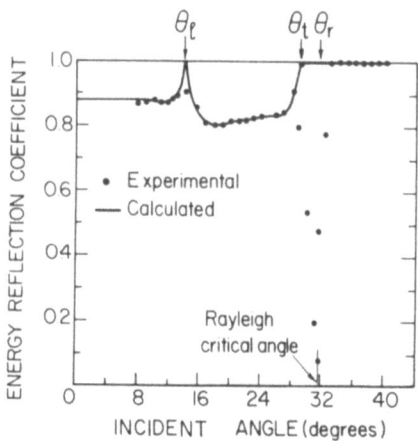

Figure 8. The energy reflection coefficient for incident plane waves in water on a water-steel interface (solid line). The points represent the experimental data (from Neubauer[18]).

Bertoni and Tamir[19] gave a theoretical explanation of the experimentally observed effects: 1) lateral beam displacement Δ of the reflected ultrasonic beam; 2) nonhomogeneous intensity distribution in the reflected beam; 3) change in the amplitude and phase of the reflected beam near the Rayleigh angle. They showed that if the width of the incident beam is more than four times greater than the value of the lateral displacement Δ, the distortion in the profile of the reflected beam vanishes and the value of Δ becomes equal to the displacement width predicted by Schoch.[20] The reflected field is produced by interference of two wave contributions:[19,21] 1) reflected beam according to geometric acoustics, and 2) contribution associated with the leaky surface wave, which is excited by the incident beam and in the course of propagation radiates energy to the liquid (Figures 9a,b).

Becker and Richardson[22] showed that the velocity and attenuation of a surface wave can be determined with sufficient accuracy from the measured amplitude and phase of the reflected beam.

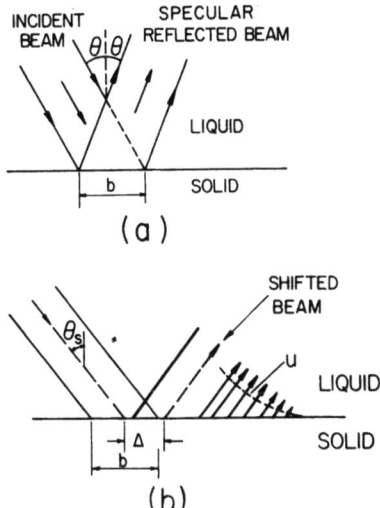

Figure 9. a) Specular reflection of the bounded ultrasonic beam from the liquid-solid interface. b) Nonspecular reflection of the bounded ultrasonic beam at the Rayleigh angle.

The measuring principle is indicated in Figure 9. A bounded ultrasonic beam with width b much greater than its displacement Δ is incident at an angle θ onto the test specimen. Angle θ can be changed with a high accuracy. The receiving transducer receives the reflected wave at the same angle. This transducer is shifted by an amount Δ to compensate for the effect of the lateral displacement of the ultrasonic beam. The value of angle θ_r which gives an amplitude minimum and phase reversal in the reflected signal is measured. From angle θ_r is it possible to calculate the velocity of the leaky surface wave $V_r = V_o/\sin\theta_r$, whereas the attenuation can be found from the amplitude of the reflected signal.[22]

All the above remains valid for the case of a coated substrate.

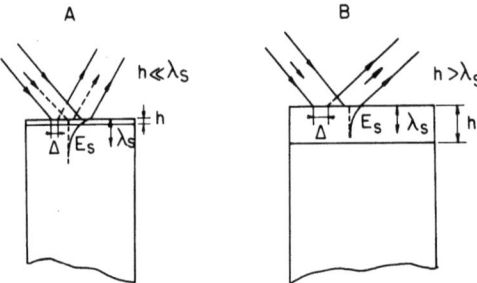

Figure 10. a) Beam reflection at angle θ_s in the case of a thin coating ($\lambda_s \gg h$). b) Beam reflection at angle θ_s in the case of a thick coating ($\lambda_s < h$).

From the measured angle θ_s which gives a minimum of the reflection coefficient the velocity of the surface wave $V_s = V_o/\sin\theta_s$ is calculated. For a thin coating with thickness $h \ll \lambda_s$ parameter $E_o/(1-\nu_o^2)$ can be determined using Equation (5). In the case of a thin multilayered coating one can use Equation (6). Figure 10a represents schematically the reflection of the beam near angle θ_s. The curve E_s illustrates the penetration of elastic energy into the solid body (corresponding to the depth of penetration of energy in the surface wave, Figure 1). Since the film thickness is small, the principal part of the elastic energy is localized in the substrate.

The situation is different for a thick coating, $h \gg \lambda_s$, Figure 10b. In this case almost all elastic energy is concentrated in the coating and the effect of the substrate is insignificant. The value of $V_s = V_o/\sin\theta_s$ in this case approaches the velocity of Rayleigh waves in the coating material.

The Rayleigh angle method was employed by Jonath[23] for determining the elastic properties of an adhesive film on a substrate. The details of the experimental technique and the experimental apparatus have been given by Knollman and Hartog.[24] It was pointed out that for materials such as adhesives, the amplitude minimum was impossible to measure at frequencies above 10 MHz. Thus, the phase reversal phenomenon was used for determining the Rayleigh angle. By measuring layers of different thicknesses,[23,24] the reduction of the shear modulus was found close to the adhesive-adherend interface. More recently Knollman and Hartog showed[25] that adherend surface treatments do not affect the gradient in the adhesive interfacial zone, whereas cure time (and/or temperature of cure) may play an important role. Samples cured more than twice display[25] a near constant shear modulus throughout the interfacial region. In determining the surface wave velocity the effect of the substrate was neglected, i.e., it was implied that $\lambda_s \ll h$.

The Rayleigh angle method was used by Adler and McCathern[26] for measuring elastic properties of coatings with thickness of the order of or less than the wavelength. A comprehensive theoretical analysis was carried out for this case by Chiment, Nayfeh and Buter.[27]

A new technique based on the effect of nonspecular reflection at the Rayleigh angle was developed by Weglein[28,29] using the acoustic microscope.

ELASTIC INTERFACE WAVES

The term elastic interface waves is used to denote waves propagating along the interface between two media and localizing energy in a band with a width of the order of several wavelengths.

An interface wave of a special type, termed the Stoneley wave, may propagate along the interface between two elastic half spaces in rigid contact with one another (Figure 11a). The velocity of the Stoneley wave should be lower than the velocities of the shear waves in the half spaces. This condition imposes rigid limitations on the possible pair of half spaces (on the ratio of their elastic moduli and densities[6,30]). In the general case the Stoneley wave is a leaky wave and its energy is radiated in the form of a shear wave into the half space with lower shear wave velocity (slow half space), analogously to what is shown in Figure 6 for surface leaky waves.

Lee and Corbly[31] observed experimentally the gradual transformation of a Rayleigh wave into a Stoneley wave due to the strong pressing together of two solids. Similar effects were measured by Claus and Palmer[32] by an optical method with direct recording of the Stoneley wave. Rigid contact at the interface was obtained by moderate compression of optically polished surfaces. Recently these results were theoretically analyzed by Kumar and Murty[33] within the framework of a phenomological model[34] of a loosely bonded interface.

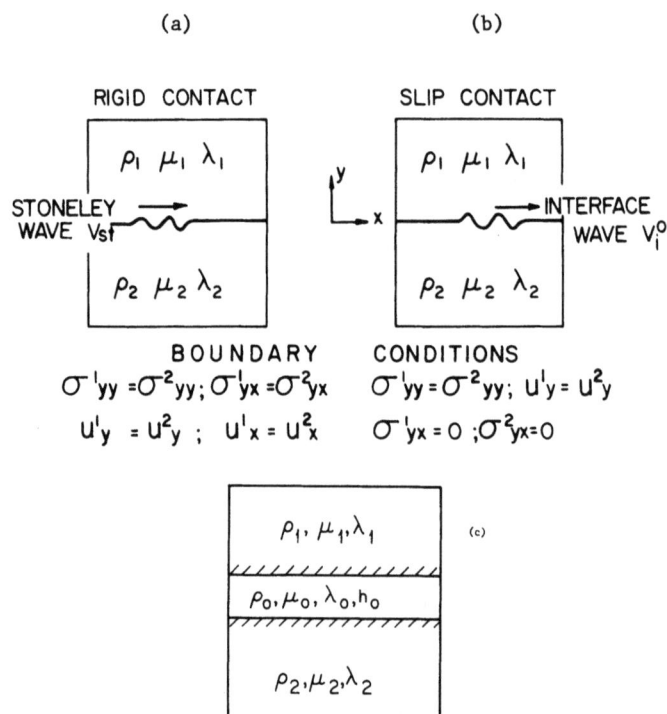

Figure 11. Different boundary situations at the solid-solid interface. a) Rigid contact (continuous stresses and displacements). b) Slip contact (shear stresses vanishing at the interface). c) Viscoelastic layer at the interface between two solids.

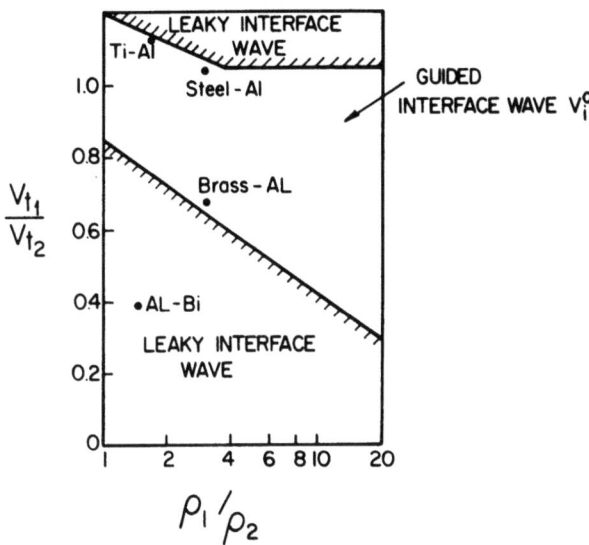

Figure 12. Existence conditions for the interface wave on a solid-solid interface with slip boundary conditions. (Based on data from Lardat, Minot, Tournois[36].)

The second limiting case of an interface wave is that of slip contact at the interface (Figure 11b) (frictionless contact). This type of boundary condition is characterized by absence of shear stresses at the interface, and corresponds to the presence of an infinitesimally thin layer of an ideal liquid at the interface. The case of a liquid layer at the interface of solids was discussed by Staecker and Wang.[35] It was found that the conditions for existence of this kind of interface waves are much less rigorous than the conditions for the existence of Stoneley waves. Figure 12 shows, in coordinates V_{t_1}/V_{t_2} vs. ρ_1/ρ_2, the domain of existence of guided interface waves, obtained on the basis of results of Lardat, Minot and Tournois.[36] The term guided points to localization of the elastic energy near the interface. Outside of this domain the interface wave is a leaky wave.

A more general case of the existence of interface waves was recently studied by Rokhlin, Hefets and Rosen.[37] They analyzed the case of propagation of interface waves in a system of two half spaces separated by a viscoelastic layer (Figure 11c). Shear modulus

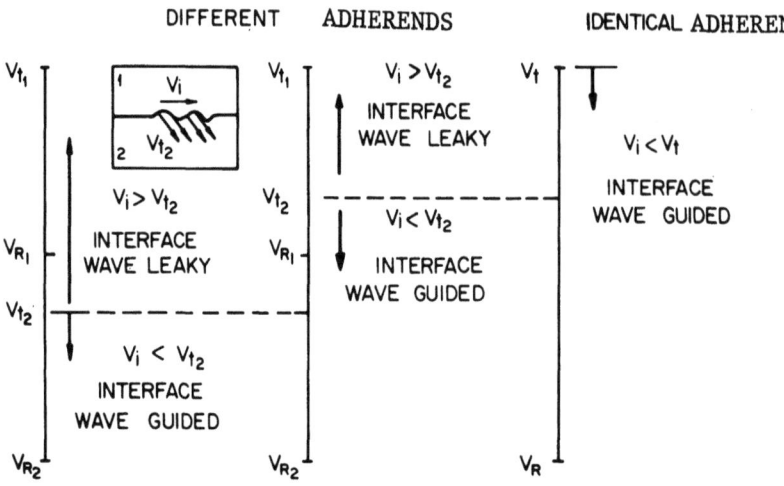

Figure 13. The conditions for the existence of an interface wave for the case of two half spaces separated by a viscoelastic layer.

μ_0 of the layer was assumed to be smaller than the shear modulus of the half spaces. Such a model satisfactorily approximates an adhesively bonded specimen, provided that the adhesive thickness is much smaller than that of the adherends. The conditions for the existence of an interface wave for this case are clarified in Figure 13. The interface wave exists as a guided wave, when its velocity is lower than the shear velocities in the half spaces. If the shear modulus of the adhesive is lower than the shear modulus of the half spaces, then in the case of identical half spaces there always exists a guided interface wave.

The deformation of a thin solid interface film by the interface wave is illustrated schematically in Figure 14. Figure 14a shows the deformation produced by the surface wave in the subsurface layer. It shows tension and compression regions separated from one another by a half wavelength. Figure 14b shows the deformation of an interface layer, the thickness of which is much smaller than the wavelength. Due to antisymmetry of motion, the interface layer is subjected to shear strain (Figure 14c).

A possible method for obtaining interface waves is shown schematically in Figure 15. The surface wave, excited in the lower substrate, is transformed in the interface region into an interface wave.

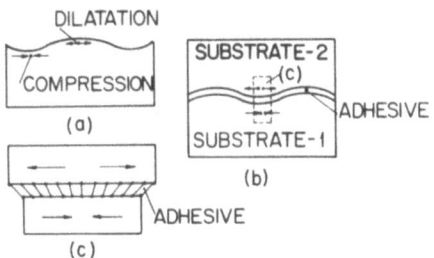

Figure 14. a) Deformation of the surface of a solid body upon propagation of a Rayleigh wave. b) Deformation of an interface film upon propagation of an interface wave. The thickness of the interface film is much smaller than the interface wavelength. c) Shear strain of the interface film. The figure corresponds, on a magnified scale, to element C singled out in Figure 14b.

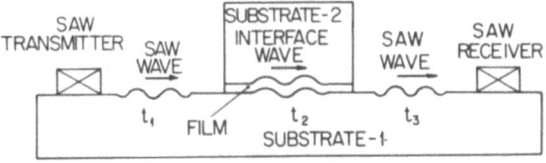

Figure 15. Illustration of the interface wave method.

Due to the closeness of velocities of the surface and interface waves, only a small part of the energy is transformed into bulk waves. The interface wave leaving the interface zone is retransformed into a surface wave, and is sensed by a receiver.

One can estimate the viscoelastic properties of the interface layer by measuring the velocity and attenuation of the interface wave. We showed[37] that the complex shear modulus of an interface wave μ_o is related by a simple expression to the interface wave velocity V_i.

$$\frac{\mu_o}{\mu} = \frac{\bar{h}\Delta_r}{\beta - \bar{h}\frac{\rho_o}{\rho}a_1 - 2\bar{h}\alpha^2 a_2} \qquad (7)$$

where $\alpha = V_t/V_i$ is the normalized wave number for the interface wave; $a_1 = \alpha^2 - \beta\gamma$; $a_2 = 2\beta\gamma - W$; $W = \alpha^2 + \beta^2$; $\beta = (\alpha^2-1)^{1/2}$; $\gamma = (\alpha^2 - V_t^2/V_\ell^2)^{1/2}$; $\Delta_r = (W^2 - 4\alpha^2\beta\gamma)$; Δ_r is the characteristic function for the Rayleigh wave, V_t and V_ℓ are the shear and longitudinal wave velocities in the substrate, ρ_o is the density of the interface film material, ρ and μ are the density and the shear modulus, respectively, of the substrate, f is the frequency and $\bar{h} = 2\pi fh/V_t$ is the nondimensional film thickness, 2h is the thickness of the film.

Equation (7) is valid at $h_o \ll \lambda_t^o$, where λ_t^o is the length of shear waves in the material of the interface film. When this condition is not satisfied, shear modulus μ_o, determined from Equation (7) can be refined on the basis of the more exact equations.[37] To determine the complex shear modulus μ_o, one should substitute into Equation (7) the complex wave number α, which is calculated on the basis of measured velocity and loss factor of interface waves.

Numerical analysis shows that the sensitivity of this method is high in the range of film thicknesses which are small as compared with the length of interface waves. Hence at film thicknesses from 100 to 1 μm the measurements should be performed at relatively low ultrasonic frequencies from 0.5 to 5.0 MHz.

It is possible to single out the principal features of the method, which indicate that the use of the interface waves for the evaluation of thin interface layers (in particular adhesively bonded structures) is promising.

1) The interface wave produces shear stresses at the interface; these stresses are most sensitive to variations of adhesion quality.

2) The interface wave propagates along the interface and hence is sensitive to small changes in the properties of the adhesive and

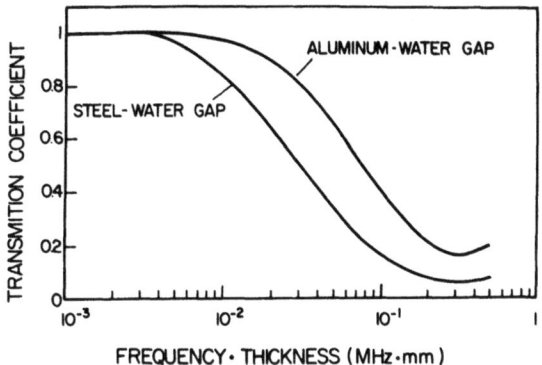

Figure 16. The effect on the transmission of normally incident longitudinal waves of a thin water gap in metal.

of the bond between the adhesive and the adherends.

3) The interface wave can be used for evaluation of very thin layers, when $2h/\lambda_t < 0.01$, where $2h$ is the thickness of the interface layer and λ_t is the length of the shear wave in the substrate. Conversely, bulk longitudinal waves, usually employed for evaluation of adhesively bonded structures, are insensitive to the existence at the interface of a thin liquid layer which exhibits no shear resistance. This is illustrated by Figure 16, which shows the effect of a thin water gap in a metal on the transmission of normally incident longitudinal waves. When the gap widths is too small, the transmission is complete, i.e., the bulk longitudinal wave is insensitive to the presence of a thin film of a nonviscous fluid on the interface. At the same time such an interface exhibits no shear resistance. As was previously noted, the existence of such a break in material continuity is reliably detected by interface waves.

4) As shown below, interface waves can be used for evaluation of multilayered interface films, which simulate the adhesion

properties at the adhesive-adherend interface.

5) In the case when the thickness of the adhesive layer is much greater than the wavelength, a leaky interface wave can be excited at the adhesive-adherend interface. This wave can, in principle, be used for evaluating the properties of the adhesive-adherend interface, provided that its attenuation, associated with the leakage of energy, is not too high.

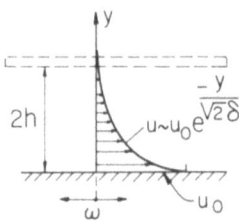

Figure 17. Hydrodynamic boundary layer in viscous liquid near the oscillating plate.

APPLICATION OF INTERFACE WAVES FOR INVESTIGATING THE CURING AND VISCOELASTIC PROPERTIES OF ADHESIVES

In the course of polymerization of adhesive films the latter undergo a transition from the liquid to the solid state. Let us first consider qualitatively the behavior of an interface wave in the course of polymerization of the interface film.

Let us assume that two elastic substrates are separated by a thin layer of a liquid adhesive. The thickness $2h$ of this layer will be taken to be much smaller than the wavelength λ of the interface wave. The fluid in the interface gap can be regarded as ideal, provided that the gap thickness is greater than the thickness of the hydrodynamic boundary layer,[37] $\delta = (\nu_o/\omega)^{1/2}$ or $\delta = |\text{Im}(\mu_o)|^{1/2}/[\omega(\rho_o)^{1/2}]$, where $\nu_o = |\text{Im}(\mu_o)|/(\rho_o\omega)$ is the kinematic viscosity of the fluid, $\mu_o = \mu_o' - i\mu_o''$ is the shear modulus, ω is the angular frequency, and ρ_o is the density of the adhesive film. So we have frictionless or slip contact at the interface between the two solids (Figure 11b). In the course of interface-film polymerization the viscosity of the adhesive rises, which increases the thickness of the hydrodynamic boundary layer. When the boundary-layer thickness is comparable to the thickness of the film, the shear stresses start being transmitted through the film from one substrate to the other. This effect is illustrated in Figure 17, which shows how the viscous fluid is set into oscillation with frequency ω due to vibrations of the lower plate. Displacement u falls across the liquid layer in an exponential manner. As viscosity rises (increasing δ) or the film

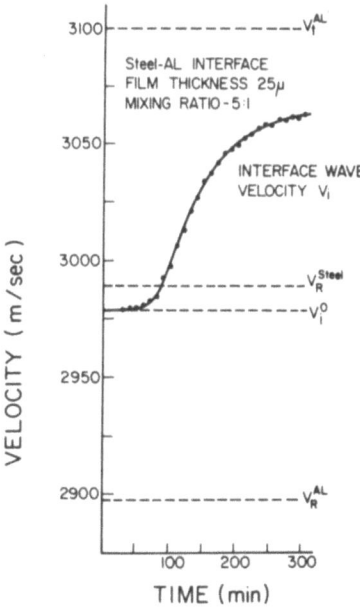

Figure 18. Changes in the phase velocity V_i of the interface wave in the course of curing of the epoxy film at the steel-aluminum interface.

thickness 2h falls, the upper plate starts "feeling" the vibrations of the lower plate. This results in a rise in the phase velocity of the interface wave due to the rising resistance of the interface to shear strain.

This means that solidification of the interface film is accompanied by an increase in the phase velocity of the interface wave which increases from the velocity V_i^o to some velocity V_i, controlled by the interface-film parameters.

These considerations are illustrated by the experimental results shown in Figure 18, which shows the variation in the phase velocity V_i of the interface wave in the course of polymerization of a thin (25 µ) epoxy film at the interface between a steel and an aluminum slab. In its initial state the epoxy film viscosity is low and the velocity of the interface wave is $V_i \simeq V_i^o$, where V_i^o is the velocity of the interface wave at the interface between these half spaces at frictionless contact. This velocity lies between the velocities of Rayleigh waves for the half spaces ($V_R^{Al} < V_i^o < V_R^{steel}$). The velocity of the interface wave rises in the course of the film's polymerization and becomes greater than V_R^{steel}. However, its value still does not attain the shear rate of the "slower" of the half spaces V_t^{Al} and, hence, in this case, the interface wave is

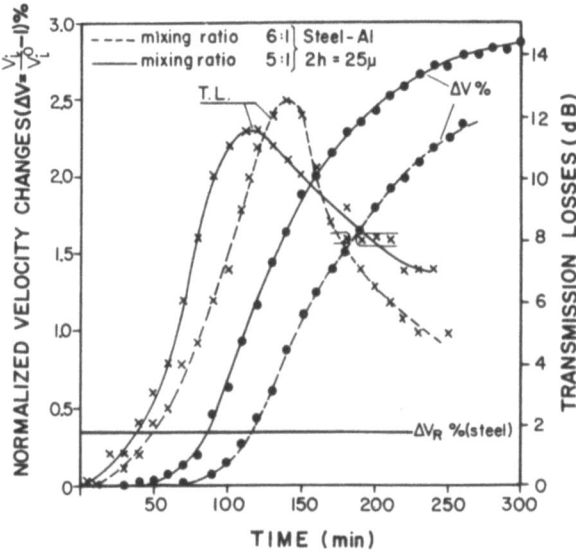

Figure 19. The change in the relative velocity ΔV (•) and the transmission losses (x) of the interface wave as a function of time of curing. Steel-aluminum interface; two mixing ratios of epoxy resins.

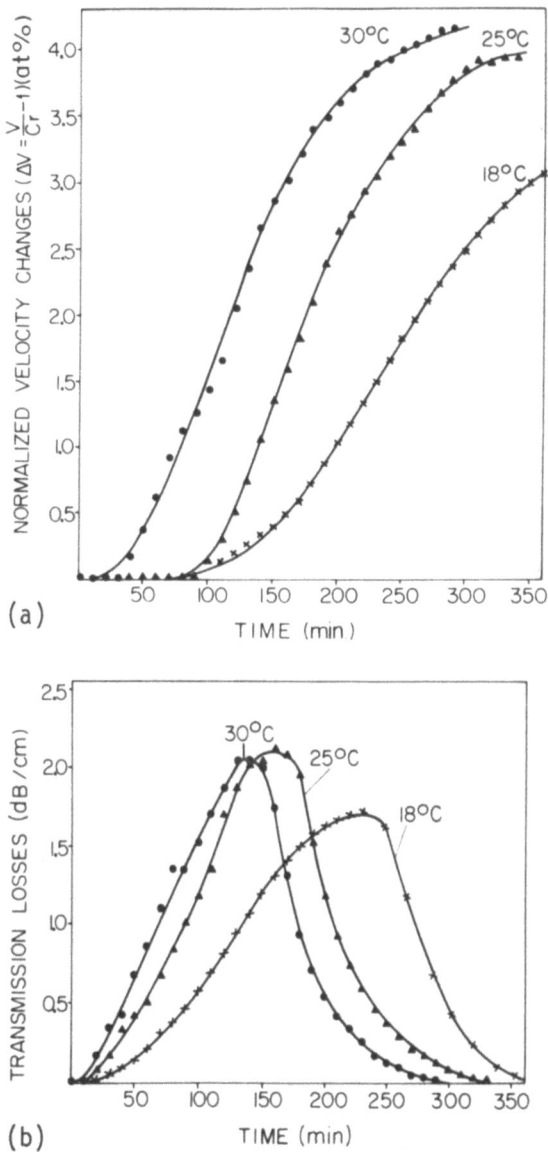

Figure 20. a) The change in the relative velocity, and b) the transmission losses of the interface wave as a function of time of curing for various temperatures. Both adherends are steel. V_r is the Rayleigh wave velocity. (From Rokhlin, Hefets, and Rosen[37].)

always a guided wave. If the shear modulus μ_o of the film is increased, then the velocity of the interface wave may intersect the line of $V_i = V_t^{Al}$ and the interface wave becomes a leaky wave.

The measured velocity and attenuation of an interface wave can be used for estimating changes in the complex shear modulus in the course of curing of a thin interface film.

Figures 19 and 20 show the measured relative changes in phase velocity and transmission losses of the interface wave as a function of the polymerization time of the interface layer at a frequency of 1.2 MHz. The adhesive used was EPOFIX, produced by Scientific Instruments, Denmark; this adhesive polymerizes in several hours after addition of a solidification agent. The static viscosity of the starting mixture was about 550 cp at 25°C.

The data in Figure 19 were obtained at 25°C for two different mixing ratios of the adhesive components. The film thickness was 25 µ, the lower adherend was steel, the upper - aluminum. The data in Figure 20 were obtained for several temperatures at a mixing ratio of 5:1; the film thickness was 12 µ. Both the upper and lower adherends were steel.

The experimental data can be smoothed and translated[37] to the complex wave number of the interface wave, and then to the complex shear modulus of the film. Thus, the calculated complex shear modulus of the film normalized to the steel shear modulus is plotted against the time of polymerization (Figure 21).

As a first approximation the shear modulus was calculated from Equation (7). These values were used as the initial guesses for the numerical solution[37] of the exact equations. The moduli obtained from the exact equation and from Equation (7) are identical to within the resolution of the plot. The dashed horizontal line in Figure 21 represents the normalized dynamic film shear modulus after complete polymerization. It was obtained by direct measurement of the bulk shear velocity, and is the asymptote of the curves of $Re(\mu_o/\mu)$ at the measured temperatures.

The velocity and attenuation of the interface wave depend on the elastic and dissipative properties of the film and half spaces. In particular, the maximum on the transmission loss curve and the point of inflection on the velocity curve of the interface wave characterize the entire system (film and substrate) as a whole. Hence the maximum on the transmission loss curve is shifted relative to the maximum on the curve of the imaginary part of the shear modulus in the adhesive material. The viscoelastic properties of the interface film are determined by calculations from experimental data. The measurements are performed at ultrasonic frequencies in the 1 MHz range, which requires consideration of the correspondence between

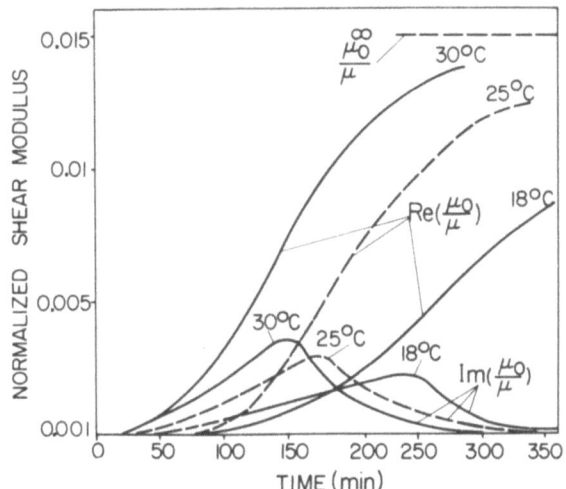

Figure 21. The normalized complex shear modulus μ_0/μ calculated from smoothed measured values of the interface wave velocity and the transmission losses (Figure 20). The dashed horizontal line represents direct measurements of the normalized film shear modulus μ_0^∞/μ after full polymerization. The steel shear modulus was taken as $\mu = 0.809 \cdot 10^{12}$ dyn/cm^2. (From Rokhlin, Hefets and Rosen[37].)

the static and dynamic viscoelastic quantities in the course of curing. In particular, one must consider the kind of correspondence between the change in the interface wave velocity and increasing strength in the course of curing of the given adhesive.

Such a comparison was performed for the FM-73 structural adhesive.[38] The measurements were performed for a standard cure cycle at a frequency of 0.5 MHz. The velocity of the ultrasonic interface wave was measured during the entire cure cycle as a function of time. The bond strength was measured over different time intervals from the start of the cure cycle. In order to immediately stop the cross-linking reaction, the shear strength of the specimen was measured when the latter was rapidly cooled to room temperature.

Figure 22 compares the shear strength data obtained upon rapid cooling with results obtained by ultrasonic measurements. The open rectangles in this graph were obtained by parallel translation of

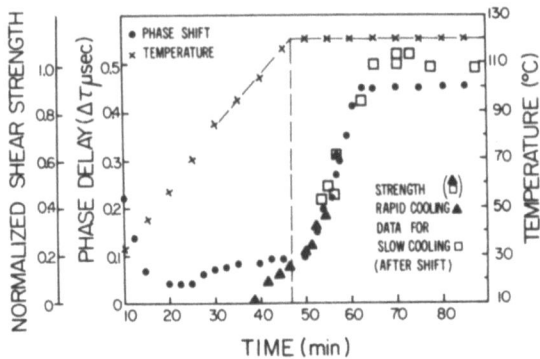

Figure 22. Comparison between the change in the interface wave velocity and the shear strength of the bond in the course of curing FM-73 structural adhesive.

the corresponding point for the shear strength, obtained for slow cooling. It is seen that a rise in the velocity of the interface wave (rise in the shear modulus of the adhesive) corresponds precisely to the time interval of the bond-strength growth.

The effectiveness of the ultrasonic method lies in the fact that it is a nondestructive technique, which shows the growth of the adhesive strength as a function of time for a single cure cycle. Obtaining analogous data by mechanical testing requires a large number of experiments with different specimens, corresponding to different times in the cure cycle. Mechanical testing is particularly time consuming when one wishes to determine the effect of different parameters, such as temperature, pressure and polymerization time on bond strength. The use of the ultrasonic interface wave technique as a supplement to mechanical testing can significantly reduce the volume and time of the experimental program.

These experimental results were obtained under nonisothermal conditions. Temperature changes modify the elastic moduli of the adhesive film and substrates and, by virtue of thermal expansion,

CHARACTERIZATION BY ULTRASONIC WAVES

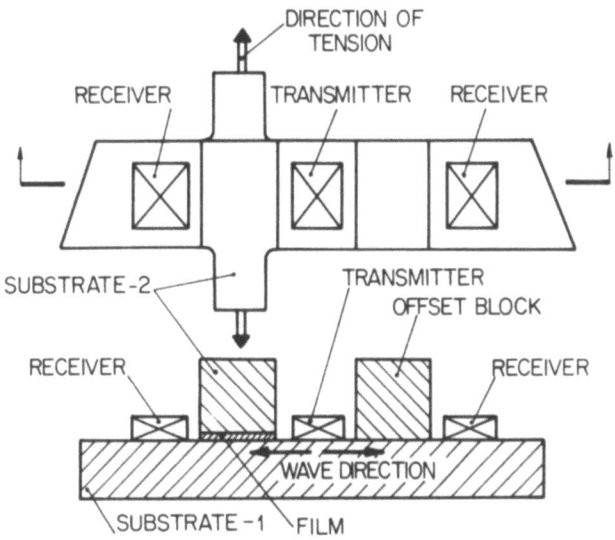

Figure 23. Schematic of the specimen for differential ultrasonic measurement and strength testing.

change the density. This results in a change in the velocity of the interface wave. In addition, the thermal expansion of the substrate modifies the acoustical path length.

To measure the properties of the film under nonisothermal conditions, it is necessary to eliminate the effect of temperature variations in the substrates and transducers. For this purpose[39] a differential measuring arrangement, in the form of an acoustic bridge, was developed (Figure 23). The middle transducer (transmitter) radiates an ultrasonic surface wave symmetrically into two arms of the bridge. In the left arm, the surface wave incident on the boundary between the free surface of the lower substrate and the interface region, is transformed into an interface wave. After leaving the interface region, the interface wave is transformed into a surface wave, and is sensed by a receiver. A surface wave propagates along the entire right arm. An offset block which is identical to substrate 2 is placed without bonding on substrate 1, in order to equalize the thermal masses of both arms (Figure 23). Due to surface roughness there is no acoustic contact between the substrate and the offset block and it does not affect the propagation of the surface wave.

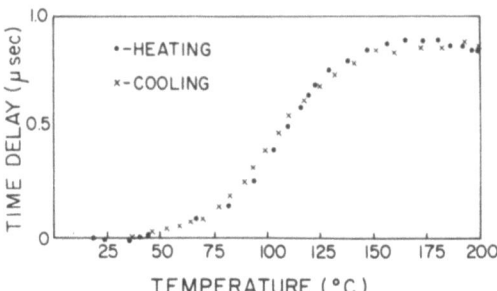

Figure 24. Example of the measurement of thermomechanical properties of an FM-73 adhesive interface film. Phase-delay changes of the interface wave due to the temperature cycling of an adhesive-bonded specimen.

If the dimensions of both arms of the bridge and of the transducers are the same, then the use of the differential arrangement eliminates the effect of thermal changes of the acoustical path. It is clear that the difference in the delay time between the two channels will depend slightly on temperature changes in the substrates, since the equations expressing the velocities of the surface and the interface waves as a function of the properties of the substrates are not entirely identical. Thus, there arises the question of the accuracy of temperature compensation in the bridge. The desired estimate cannot be obtained experimentally, since the film properties are also a function of the temperature (particularly in the case of polymer interface films). The estimate was obtained analytically in Reference 39, where the experimental arrangement was also described in detail.

The method is illustrated by the determination of the thermoelastic properties of the FM-73 structural adhesive as shown in Figure 24. The relative phase change of the interface wave at 0.5 MHz during the temperature cycle of the cemented aluminum specimen is shown. The bond line thickness was 90 μm. An increase in the delay time corresponds to a reduction in the shear modulus of the adhesive. These changes correspond to the ordinary thermomechanic behavior of the polymer, the mechanical properties of which are strongly temperature dependent. At high temperature the polymer exhibits a low-modulus, rubber-type behavior, whereas at low temperatures it behaves as a high-modulus glass-type material.

THE CONCEPT OF THE EFFECTIVE SHEAR MODULUS - STRENGTH PREDICTION

An adhesive bond may fail in two ways: 1) Failure inside the adhesive, which is termed cohesive failure. The cohesive strength of a given adhesive correlates with its elastic modulus. 2) Failure along or close to the adhesive-adherend interface, which is termed interfacial or adhesion failure.

The cohesive strength of a given type of adhesive is governed by its chemical structure and by the conditions of the curing process (in particular, by the percentage of crosslinking), and by the presence of various kinds of microdefects (e.g., microcracks or voids). These quantities also determine the shear modulus and the ultrasonic loss factor of the adhesive. By virtue of this fact, for a given type of adhesive, a change in the elastic modulus correlates with a change in its strength.

There is no rigorous physical theory of interfacial failure. From the phenomenological point of view,[40,41] interfacial failure can be identified with cohesive failure of a weak boundary layer (WBL). The WBL can be represented by any low-strength region at the adhesive-adherend interface, and if the thickness of the WBL tends to zero, then the interface proper can be regarded as the WBL.

It was shown experimentally[23-25,42] that the viscoelastic properties of the adhesive are not uniform over its thickness, and that a reduction in the elastic modulus may occur at the adhesive-adherend interface. It is hence convenient to consider a multilayered model of a bond line, consisting of an adhesive and WBL (Figure 25). It will be assumed that WBL characterizes the adhesion properties between the adhesive and the adherend.

According to this model, the shear modulus of the adhesive, calculated on the basis of the measured velocity and attenuation of interface waves is the effective shear modulus μ_{eff}. It characterizes the effective elastic properties of the multilayered adhesive-WBL system. The concept of the effective dynamic shear modulus of the bond line was introduced by Rokhlin, Hefets and Rosen.[43]

The qualitative description of the properties of the interface wave in the presence of a WBL can be given on the basis of the matrix method.[44,45] We obtained a series expansion of the exact solution with respect to the thickness of interface films and compared this solution with the solution for the case of a single film.[37]

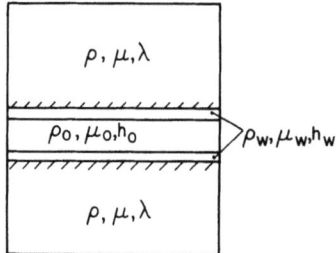

Figure 25. A multilayered model of a bond line, consisting of an adhesive and weak boundary layers.

According to this comparison, we obtained Equation (7) for a single interface film when only linear terms are retained in the matrix expansion. A similar approximation for the multilayered system (substrate/WBL/film/WBL/substrate) makes it possible to write the characteristic equation for the velocity of the interface wave in the form

$$\mu_{eff}/\mu = \bar{h}\Delta_r/(\beta - \bar{h}\frac{\rho_{eff}}{\rho}a_1 - 2\bar{h}\alpha^2 a_2) \tag{8}$$

where

$$\rho_{eff} = \rho_o(1 + \frac{\rho_w h_w}{\rho_o h_o})/(1 + \frac{h_w}{h_o}) \quad ; \quad h = h_o + h_w$$

and

$$\mu_{eff} = \mu_o(1 + \frac{h_w}{h_o})/(1 + \frac{h_w \mu_o}{h_o \mu_w}) \tag{9}$$

$2h_o$ - is the thickness of the adhesive layer. h_w, ρ_w, μ_w - are the thicknesses, density and shear modulus of the weak boundary layer. The other terms have the same meanings as in Equation (7).

When contact between an adhesive and substrates is not ideal the interface wave velocity decreases. Hence, the effective shear modulus μ_{eff} found from Equation (8) on the basis of the experimentally measured velocity will be smaller than the actual modulus of the film.

For example, in the case of absence of shear bonding between the adhesive and at least one of the substrates (slip contact) the velocity of the interface wave will be close to the Rayleigh wave velocity and the calculated μ_{eff} will be equal to zero. It is thus seen that the interface-wave velocity will characterize not only the elastic properties of the adhesive and its cohesive strength, but also the bond between the adhesive and the substrates, i.e., the interfacial (adhesion) strength. This means that the ratio μ_{eff}/μ_o can be used as a criterion of the bond strength. If the effective shear modulus μ_{eff} is measured for a given bond, and the shear modulus μ_o of the adhesive is measured on a reference specimen, then Equation (9) can be used for estimating[45] the properties of WBL.

For the study of the correlation between the strength of the bond and the effective shear modulus,[43] the metal-adhesive interface was weakened by modification of the surface of the substrate by a thin lubricant film. The velocity and attenuation of the interface wave were measured in the course of polymerization of the adhesive. After completion of the polymerization process the specimen was tested in order to determine the shear strength of the bond. Figure 26 shows a normalized plot of the shear strength of the bond vs. the effective shear modulus measured by the interface-wave technique. The general transmission loss factor η

$$\eta = 20 \lg \left(\frac{A_m A_m}{A_o A_F}\right)$$

shown in Figure 27 serves as another parameter correlating with the shear strength. Here A_m is the amplitude of the transmitted signal, read in all the experiments at the time corresponding to the loss maximum in reference bonds, A_o is the amplitude of the signal at the initial stage of the polymerization, and A_F is the amplitude of the signal at the final stage of polymerization. Quantity A_F makes allowance for scattering of the interface by micro-inhomogeneities of the adhesive layer and of adhesive-adherend interfaces. It was shown by Claus and Kline[46] and Claus and Palmer[47] that the attenuation of interface waves is significantly affected by the quality of the surface topography of the adherends.[48] The correlation between η and strength shows that information on the final adhesion strength

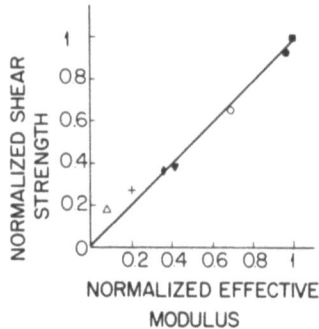

Figure 26. Relation between the normalized effective shear modulus and the normalized shear strength (from Rokhlin, Hefets, and Rosen[43]).

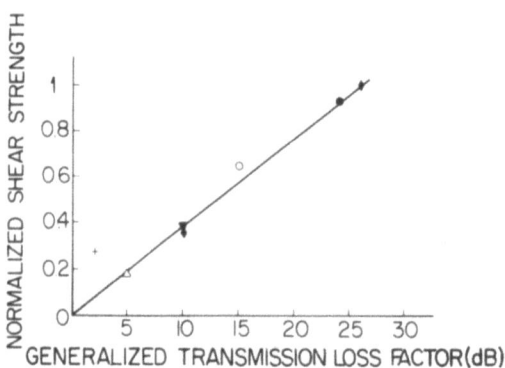

Figure 27. Relation between the generalized transmission loss factor and the normalized shear strength (from Rokhlin, Hefets, and Rosen[43].)

of the bond is present at the intermediate stages of the cure (relation between the strength and A_m).

Let us clarify the physical basis of the above correlation between the parameters of the interface wave and the bond strength. At each of the points along the wave path the velocity and the attenuation of the interface wave are characterized by the adhesion properties of the metal-adhesive interface and the elastic properties of the adhesive proper. The measurements were performed along an interface segment of length L. Hence the velocity and attenuation are integral characteristics of the effective elastic properties of the interface not only over the thickness of the adhesive but also along the total acoustic path. The velocity represents results obtained by averaging the velocities at two different lengths: 1) at segments with good adhesion, the velocity which is equal to the velocity of the interface wave in the reference specimen; 2) at segments with inferior adhesion, the velocity which is lower, and lies between the velocity at the reference specimen and the velocity of the interface for slip contact. Hence the total velocity decreases as compared with that at the reference specimen.

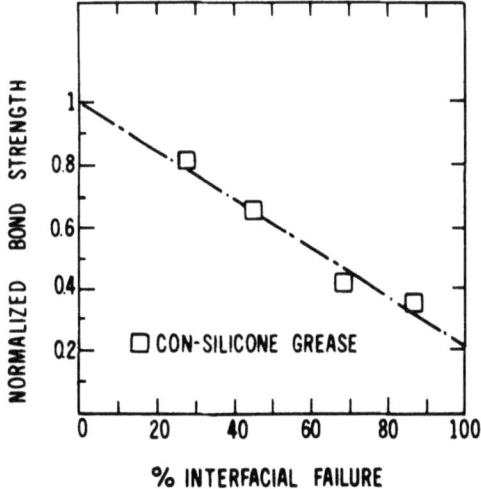

Figure 28. The normalized shear strength of the bond as a function of the percentage of interfacial (adhesion) failures (data from Smith and Smith[49].)

The mechanical shear-lap joint test is performed perpendicular to the direction of propagation of the interface wave. Since the interface regions with inferior adhesion have a lower shear stiffness the effective width of the lap joint decreases. Hence the critical value of the failure stress at good parts of the interface is attained at lower test forces as compared with the reference specimen. The transmission of shear forces from one adherend to another through the adhesive film is controlled by the effective shear modulus. It is therefore natural to expect a correlation between the interface wave velocity and the bond strength.

The above is confirmed by the results of Smith and Smith.[49] Figure 28 shows the relation obtained by them between the shear strength of the bond and the percentage of the overall surface of a specimen with an interfacial (adhesion) mode of fracture. As in our experiments, the adhesion strength was modified by introducing specified amounts of impurity to the adherend surfaces prior to bonding. It appears to us that the correlations shown in Figures 28 and 26 are well supported and complement one another.

CONCLUSION

The surface wave method has only recently been introduced for investigating the properties of thin films. The surface wave produces longitudinal strains in the film and its velocity is related to Young's modulus of the film. Although most of the results until now were obtained for metal films, it is clear that this method should be useful for investigating photoresist materials and the curing of polymeric coatings. The Rayleigh-angle method appears promising for investigating solid polymer coatings.

Unlike surface waves, interface waves were immediately utilized for the study of adhesively bonded structures. The interface wave produces shear strains in the interface layer and it is hence sensitive to the state of the adhesion contact between the film and the substrates. Interface waves can play an important role in <u>in situ</u> study of the curing process and of viscoelastic properties of adhesives and for strength prediction of adhesive bonds.

The difference in the form of elastic deformations produced by surface and interface waves result in a number of basic differences in their applications:

1) Surface waves can be used for measuring Young's modulus, whereas interface waves can be utilized for obtaining the shear modulus.

2) The sensitivity of the surface wave method to film proper-

ties decreases with reduction in film thickness. In the interface wave method the sensitivity can be high particularly for small thicknesses of interface films.

3) The effective stiffness for a multilayered surface film is obtained as a successive combination of the stiffnesses of the individual films, whereas the effective modulus in a multilayered interface film is equivalent to a parallel combination.

4) Interface waves are more sensitive to the state of adhesion at the film-substrate interface, since the weak coupling "shunts" the elasticity of the system as a whole. At the same it was shown that reduction of the adhesion between the film and the substrate can also be observed by the surface wave method.

REFERENCES

1. E.P. Papadakis in "Physical Acoustics," W.P. Mason and R.N. Thurston, Editors, Vol. 12, p. 277, Academic Press, New York, 1976.
2. A. Vary in "Mechanics of Nondestructive Testing," W.W. Stinchcomb, Editor, p. 123, Plenum Press, New York, 1980.
3. E. Segal and J.L. Rose in "Research Techniques in Nondestructive Testing," R.S. Sharpe, Editor, Vol. 4, p. 275, Academic Press, London, 1980.
4. G.W. Farnell in "Surface Wave Filters," H. Matthews, Editor, p. 20, J. Wiley & Sons, New York, 1977.
5. I.A. Viktorov, "Rayleigh and Lamb Waves," Plenum Press, New York, 1967.
6. B.A. Auld, "Acoustic Fields and Waves in Solids," Vol. II, J. Wiley & Sons, New York, 1973.
7. J.D. Achenbach and S.P. Keshava, J. Appl. Mech., 34, 397 (1967).
8. H.F. Tiersten, J. Appl. Phys., 40, 770 (1969).
9. S.I. Rokhlin, Z. Ronen, and M.P. Dariel, Thin Solid Films, 89, 109 (1982).
10. M. Toguchi and F. Akao, Ultrasonics Symposium Proceedings, IEEE Cat. 79CM1482-9SU, 439 (1979).
11. M. Levy, H. Salvo, Jr., D.A. Robinson, M. Maki and M. Tachiki, Ultrasonics Symposium Proceedings, IEEE Cat. 76CH1120-SSU, 633 (1976).
12. D.R. Snider, H.P. Fredricksen and S.C. Schneider, J. Appl. Phys., 52, 3215 (1981).
13. I. Feng, M. Tachiki, C. Krischer and M. Levy, J. Appl. Phys., 53, 177 (1982).
14. E. Harnik and E. Sadar, J. Appl. Phys., 52, 3705 (1981).
15. G. Gorodetsky and I. Lachterman, Rev. Sci. Instrum., 52, 1386 (1981).
16. C. Weaver and D.T. Parkinson, Philos. Mag., 22, 377 (1970).

17. G.W. Farnell in "Physical Acoustics," W.P. Mason and R.N. Thurston, Editors, Vol. 6, p. 109, Academic Press, New York, 1970.
18. W.G.N. Neubauer in "Physical Acoustics," W.P. Mason and R.N. Thurston, Editors, Vol. 10, p. 61, Academic Press, New York, 1973.
19. H.L. Bertoni and T. Tamir, J. Appl. Phys., 2, 157 (1973).
20. A. Schoch, Acustica, 2, 17 (1952).
21. L. Adler, Revue du Cethedec NS80-2, 116 (1980).
22. F.L. Becker, R.L. Richardson in "Research Techniques in Nondestructive Testing," R.S. Sharpe, Editor, Vol. 1, p. 91, Academic Press, London, 1970.
23. A.D. Jonath in "Adhesion and Adsorption of Polymers," L.H. Lee, Editor, Vol. A, p. 175, Plenum Press, New York, 1980.
24. G.C. Knollman and J.J. Hartog, J. Appl. Phys., 53, 1516 (1982).
25. G.C. Knollman and J.J. Hartog, J. Appl. Phys., 53, 5514 (1982).
26. L. Adler and D.A. McCathern, J. Appl. Phys., 49, 2576 (1978).
27. D.E. Chiment, A.F. Nayfeh, D.L. Buter, J. Appl. Phys., 53, 170 (1982).
28. R.D. Weglein, Appl. Phys. Lett., 34, 179 (1979).
29. R.D. Weglein, Appl. Phys. Lett., 35, 215 (1979).
30. R.N. Thurston, J. Acoust. Soc. Am., 64, 1 (1978).
31. D.A. Lee and D.M. Corbly, IEEE Trans. Son. Ultrason., SU-24, 206 (1977).
32. R.O. Claus and C.H. Palmer, Appl. Phys. Lett., 31, 547 (1977).
33. V. Kumar and G.S. Murty, IEEE Trans. Son. Ultrason., SU-29, 138 (1982).
34. G.S. Murty, Phys. Earth and Planet. Interiors, 11, 65 (1975).
35. P.W. Staecker and W.C. Wang, J. Acoust. Soc. Am., 53, 65 (1973).
36. C. Lardat, J.P. Minot and P. Tournois, IEEE Trans. Son. Ultrason., SU-22, 16 (1975).
37. S.I. Rokhlin, M. Hefets and M. Rosen, J. Appl. Phys., 51, 3579 (1980).
38. S.I. Rokhlin, J. Composite Mater., in press.
39. S.I. Rokhlin, J. Acoust. Soc. Am., in press.
40. R.J. Good in "Adhesion Measurement of Thin Films, Thick Films, and Bulk Coatings," K.L. Mittal, Editor, p. 18, American Society for Testing Materials, Philadelphia, 1978.
41. J.J. Bikerman in "Adhesion Measurement of Thin Films, Thick Films, and Bulk Coatings," K.L. Mittal, Editor, p. 30, American Society for Testing Materials, Philadelphia, 1978.
42. Y.S. Lipatov in "Adhesion and Adsorption of Polymers," L.H. Lee, Editor, Vol. B, p. 601, Plenum Press, New York, 1980.
43. S.I. Rokhlin, M. Hefets and M. Rosen, J. Appl. Phys., 52, 2847 (1981).
44. S.I. Rokhlin, (1981), unpublished.
45. S.I. Rokhlin and M. Rosen, Thin Solid Films, 89, 143 (1982).
46. R.O. Claus and R.A. Kline, J. Appl. Phys., 50, 8066 (1979).
47. R.O. Claus and C.H. Palmer, IEEE Trans. Son. Ultrason., SU-27, 97 (1980).

48. R.O. Claus and R.T. Rogers in "Physicochemical Aspects of Polymer Surfaces," K.L. Mittal, Editor, Vol. 2, p. 1101, Plenum Press, New York, 1983.
49. T. Smith and P. Smith in "Adhesion and Adsorption of Polymers," L.H. Lee, Editor, Vol. A, p. 123, Plenum Press, New York, 1980.

NONDESTRUCTIVE EVALUATION OF SOME BONDED JOINTS

T. C. Ward, Margaret Sheridan, and D. L. Kotzev[*]

Chemistry Department and Polymer Materials and Interfaces Laboratory, Virginia Polytechnic Institute and State University, Blacksburg, Virginia 24061

Free films and lap-shear bonded metal joints were compared with respect to their linear dynamic viscoelastic behavior. Thermosetting and thermoplastic adhesives were investigated. Damping phenomena and relative shear moduli indicated a broader mechanical relaxation occurring at higher temperatures in the bonded assemblies. Triblock copolymer model adhesives were used to illustrate the importance of sample preparation in the evaluation of bonded materials.

INTRODUCTION

In terms of a total systems analysis approach to adhesive bonding[1] there exists, among others, one clear role with which polymer scientists might identify, namely, that of evaluating proper material response functions and their connection to joint failure events. This paper was motivated, to some extent, by a desire to explore a particular response function, the linear viscoelastic quantity tan δ, which is widely used by polymer researchers to <u>characterize</u> primary and secondary relaxations in bulk polymers, most notably the glass transition, T_g.

---

[*] Permanent Address: Scientific-Industrial Center for Special Polymers, Zheljo Voivoda 4A, 1156, Sofia, Bulgaria

It is important to recognize that the molecular relaxations associated with maxima in the tan δ function for glass containing composites are highly time (or frequency) dependent in any geometry of testing. These maxima indicate changes in the response of the entire assembly undergoing testing; hence, they may originate from a number of sources, but <u>not</u> from a transition in the viscoelastic response of the polymer. Only the kinetic events associated with the relaxations are modified by a new time-temperature set of variables. At more rapid strain rates, polymeric materials respond as if the temperature had been lowered. Thus, it is well accepted that quantities such as high speed impact strength of bulk polymers are quite sensitive to test rate and temperature. In particular, for amorphous glasses, such as epoxies, functions based on the reduced temperature, $T-T_g$, best fit the evidence concerning the variation of impact properties on heating of the polymer once test rates are specified. Hence, any mechanically perceived alteration of the glassy relaxation is of paramount importance in bonded joint performance.

One of the fundamental questions in adhesion science is whether or not the results from a bulk property test will reliably represent the response behavior of the same polymer in a thin bond line (typical of the type encountered in a structural adhesive application) and of the entire bonded structure. Over the past few years investigations have been underway in our labs which suggest that some caution is appropriate when transferring test results obtained on free polymer samples into situations where the polymer may be (geometrically) under more severe constraints. Other workers have cited data to indicate that adhesive joints often do not perform as predicted from properties of the bulk polymer adhesive.[2] However, some researchers[3] have reported good correlation between <u>in situ</u> and bulk properties, on FM-73 in the cited case. In our present paper, observations on different bonded and free polymer adhesives are presented to help clarify this situation, particularly in the case of rapid strain rates (110 Hz) being applied at low amplitude strains.

Overall, two major objectives were defined for this investigation. First, nondestructive evaluation using very small strain deformations of both the free and the bonded adhesives was of interest. Although fracture and failure must be accurately modeled in real systems by nonlinear mechanics, it is generally appreciated that linear viscoelastic behavior is indicative not only of the kinetics of molecular motions in a Boltzmann superposition sense but may also be related to many of the experimentally observed ultimate properties.

Furthermore, most of the design of polymer molecular
architectures is based on knowledge gained from examining
unsupported polymers. In other words, our ideas about
structure/property correlations in polymeric adhesives
originate many times from research on samples which, when
tested, differ physically from the configurations encountered
in actual applications.

A second objective of this work was to focus on bond
thickness and bond thermal and stress history vis-a-vis the
nondestructive type of test. There are many citations in the
literature which deal with adhesive bond line thickness[4,5], of
course, but usually with regard to ultimate properties; G_{1C}'s
and time's-to-failure are good examples. Work on epoxy bonded
systems at the Naval Research Labs[6,7] over the years has
emphasized that, depending on testing rate and temperature,
mode I fracture energies may be optimized by selection of bond
thicknesses in the range of 0.02-0.07 cm. The viscoelastic
natures of these adhesives was often cited in the models which
were formulated by these workers to explain the observed
performance of the bonds. Indeed, various master curves were
developed relating qualities such as tan δ to test frequency
and temperature. Shifts of a few degrees in the T_g for CTBN
modified epoxies were found to vary the reduced frequency
(time) scale's of the unbonded adhesive by as much as a
decade.[7] From the point of view of the impact strength of an
adhesively bonded system, such shifts would be of major
significance. The occurence of such shifts, or changes of the
breadth and shape of the tan δ vs temperature curves
themselves by virtue of the bonding process was also an item
to be explored in our work. Particularly when long term
environmental effects on adhesively bonded joints are of
interest, it is important to know how the composite joint
behaves viscoelastically relative to the bulk adhesive in
order that accelerated testing procedures might be developed;
the shifts in relaxations would play a role in the analysis.

There is conflicting evidence on the influence of a rigid
substrate or filler on the small strain viscoelastic
properties of a thin bond line. Recent modeling has predicted
a shift in the relaxation maxima of tan δ curves which was
attributed to changes in boundary layer properties of the
polymer molecules.[8] Such modeling involves the application of
a three-phase system, effectively a composite, for predicting
conditions for the shift and resolution of relaxation maxima
of tan δ curves. The three phases are the boundary-layer
polymer, the bulk polymer, and the filler. Calculation of the
glass transition temperature of filled polymers follows from
two assumptions. First, the boundary layer, having properties

differing from the bulk polymer due to the action of the filler surface, and the bulk polymer have different glass transition temperatures. Second, the filler, whose concentration determines the concentration ratio of boundary layer polymer to bulk polymer, also affects the shape of tan δ curves because of its high modulus.[8] Experimental work by Lipatov et. al.[9] showed the addition of glass beads into epoxy resin resulted in a shift in the glass transition temperature to higher temperatures and a decrease in the maxima values of tan δ. Similar results were observed for epoxy resins filled with quartz powder. For poly(butyl methacrylate) filled with glass beads there was a considerable decrease and some broadening of the relaxation maxima. However, there was no shift in the glass transition temperature for this latter system. Their results were explained in terms of a decreased molecular mobility of the boundary-layer polymer due to the effect of the filler.

The idea of an "interphase" of polymer with modified properties has frequently been referred to in reports on filler or substrate modification of polymer response near T_g's.[10] However, there is comment in the literature which seriously questions the ability of a high modulus surface to influence macromolecules at any significant distance from the interface.[11] Indeed, it is clear that because of the high possibility of residual stress fields due to thermal or curing operations in the preparation of filled or bonded polymer systems it may be quite difficult to identify mechanistically the origin of any anamolous behavior. Nevertheless, mathematical modeling of the interphase to allow for increases or decreases in the T_g of a filled composite has now appeared.[12] Theocaris and Spathis predict that strong bonding between a filler and its matrix material results in a higher composite T_g, weak bonding leading in the opposite direction.[12]

In summary of the above review, it is apparent that one may not easily anticipate the outcome of a linear viscoelastic test scheme on a bonded joint assembly, even if the polymeric adhesive has well documented properties. There is one salient feature of such comparisons, on the other hand, which must be kept in mind. As was clearly pointed out recently concerning the mechanical response of a nonhomogeneous viscoelastic glassy composite undergoing heating, the relaxations (or dispersions) of the entire assembly may mistakenly be solely identified with macromolecular motions; whereas, in fact, an interjection of instrumental factors into the observed behavior may have occurred.[13,14] These papers on the torsion braid experimental method emphasize that a fiber glass braid/ amorphous polymer composite exhibits quite different tan δ functions of temperature at approximately constant frequency than does the neat resin itself.[13,14] In order to avoid this kind of misinterpretation it must be emphasized that in the present research we were investigating from precisely the opposite point of view: namely, while concerned about molecular dispersions in the resin, the response of the joint as a whole was of major concern. Clearly, this is an important question from a design criterion, one seldom addressed in the past.

EXPERIMENTAL

Since adhesive joints are most commonly designed, tested and used in lap shear mode, we chose a test specimen which in some cases was formed from two degreased stainless-steel plates adhering at the overlap. The thickness of adhesive layer (h) was usually 0.02 cm, but could be in the range of 0.01 to 0.06 cm. In the initial phase of the study the bond thickness was not controlled, but simply allowed to vary about the 0.02 cm value. Details of the bonded assembly appear in Figure 1. Tensile forces were applied to the non-lapped ends of the joint by a Rheovibron.

In the later work, only on the triblock styrene/isoprene/styrene adhesive described below, a second design for the joint which is shown in Figure 2 was developed. Shims were used to control bond thickness at exactly 0.3, 0.5 or 0.7 mm. Each end of the joint then was fitted into new clamps on the Rheovibron having slots offset as indicated in the figure. The entire assembly was rigidly screwed together. Titanium 6,4 alloy (phosphate fluoride etch) was used for the bonded substrates in this case.

The Rheovibron viscoelastometer, introduced by Takayanagi,[15] has found broad application for dynamic tensile testing of polymeric films and fibers. Recently it has been used by Murayama[16,17] for measurement of the dynamic shear and dynamic compression mechanical properties of materials. For this purpose special shear and compression grips were developed. Two other reports of use of the Rheovibron in a mode where a viscoelastic material was sheared between parallel plates have been cited; however, in both cases the temperatures were such that only rubbery and liquid-like materials were examined.[18,19] In this present paper, the tensile motion of the Rheovibron instrument clamps was transformed into strains in the adhesive that closely approximate those of simple shear, with a calculated magnitude of approximately 10^{-3}. A frequency of 110 Hz and a heating rate of 1°/min were selected for all testing except on the

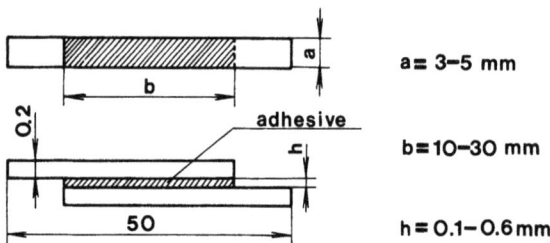

Figure 1. Lap shear bond assembly for mounting in Rheovibron. All dimensions in mm's.

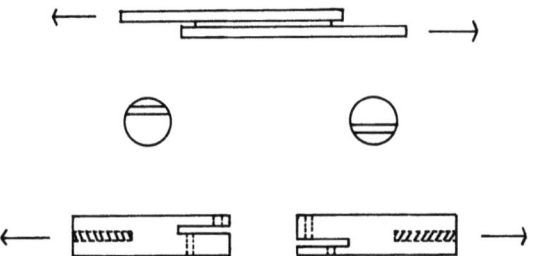

Figure 2 Modified lap shear assembly with controlled bond thickness for mounting in Rheovibron. Dimensions are similar to those in Figure 1. Middle and bottom views are of new clamps for Rheovibron.

triblock copolymers where the rate was 1/2°/min. Inertial effects and bending moments in the joint were not observed during any tests. In the materials examined in this study the Massa correction[20] was not applied.

After the adhesive material was subjected to sinusoidal shear strains, calculations of modulus proceeded according to theory.[21] Taking into account the construction of the Rheovibron the dynamic complex shear modulus, G^* in dynes/cm^2, would be[22]

$$G^* = \frac{10^9 \, h}{ADab}$$

where A is the amplitude factor, D is the corrected value of the dynamic force, h is the thickness of adhesive layer, a is the width, and b the length of the adhesive layer. After recording the tan δ value, the storage and loss moduli were calculated in the usual fashion. In some cases, however, due to a discrepancy between the sample geometry and the instrument's working range, the calculated values of G^* were observed to be too low. But, decreasing the values of "a" and "b" led to poor result reproducibility, while the magnitude of h was limited by the co-axiality of the clamps and by the actual thickness of the adhesive line for practical comparisons. We found that good qualitative results were obtained by plotting the ratio of the observed value of the shear modulus to its initial value at room temperature, G/G_{in}, as a function of increasing temperature.

Two crosslinking adhesive systems were investigated. First a Shell Epon 828 resin was cured with a stoichiometric amount of bis(p-aminocyclohexyl) methane (PACM-20, DuPont) for 2.5 hr. at 150°C. Also a recently synthesized crosslinking allyl 2-cyanoacrylate was examined.[23] Use of this material as opposed to the alkyl ester is known to produce improved thermal resistance, created by thermally induced crosslinking of the anionically cured 2-cyanoacrylate monomer.[24]

Several thermoplastic adhesives were melted between the apppropriate substrates in a compression mold and cooled to form the adhesive joint. These were: (1) poly(ethyl 2-cyanoacrylate), (2) styrene/isoprene/styrene (S/I/S) triblock copolymers of varying composition kindly provided by the Phillips Petroleum Co., and (3) a novel polysulfone/polyester block copolymer having the following structure[25], with a 50/50 (wt.) % composition and an overall molecular weight of 25,000 g/mol. The joint preparation conditions corresponding to the numbers above were (1) room temperature, slight pressure;

$$\left[-O-\underset{\underset{O}{\|}}{\overset{\overset{O}{\|}}{\underset{}{\bigcirc}-S-\bigcirc}}-O-\bigcirc-\underset{\underset{CH_3}{|}}{\overset{\overset{CH_3}{|}}{C}}-\bigcirc- \right]_x$$

$$\left[-\underset{\underset{}{}}{\overset{\overset{O}{\|}}{C}}-\bigcirc-\underset{\underset{CH_3}{|}}{\overset{\overset{CH_3}{|}}{C}}-\bigcirc-O-\overset{\overset{O}{\|}}{C}- \right]_y$$

(2) various pressure and temperature cycles which are discussed below, (3) 220°C, 2000 psi, slow cooling.

The S/I/S samples were selected to have narrow molecular weight distributions, $M_w/M_n = 65000/63000$, and $M_w/M_n = 86000/83000$ containing 50% and 40% by weight of styrene, respectively. These samples were found to be highly microphase separated in structure and to have a lamellae type morphology.

For polymer tested in the bulk (free) form in the usual Rheovibron tensile geometry, typical sample dimensions of 0.2 to 0.6 mm thickness, 3-5 cm length and 0.5 cm width were chosen. Thus, the polymer part of the composite joint was comparable in size to the free film. Calculations of tan δ and storage modulus E' proceeded in the usual way for free films.

RESULTS

Figures 3 and 4 show the results obtained for the common epoxy resin tested as a film in a tensile geometry and then as an adhesive bond between metal substrates in the shear mode. The tan δ maximum appeared at 187°C in the bulk neat material and at 5°C higher than this in the bonded adhesive system. However, there was a substantial broadening of the latter peak, reminiscent of the changes produced by adding mineral fillers to thermosets. The maximum damping in the bonded joint is lower by almost a factor of ten. Less dramatic was the larger drop in the free film's storage modulus before a plateau appeared in the relaxation.

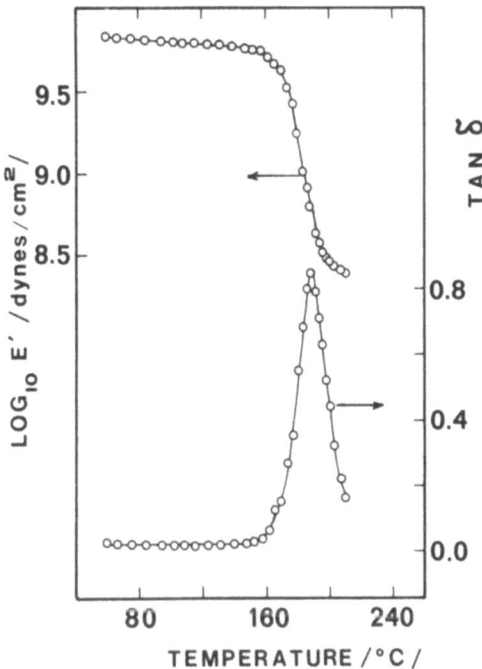

Figure 3. Dynamic mechanical storage modulus and damping for epoxy free film (at 110 Hz).

Comparisons of the polysulfone/polyester block copolymer mechanical spectra appear in Figs. 5 and 6, with the free film results in the former. The single glass transition indicated at above 200°C was intermediate between those recorded for either homopolymer also tested at 110 Hz and heated at 1°/min. As observed in the epoxy systems, the maximum in tan δ appeared 5°C higher when the lap shear geometry was employed for testing. Furthermore, the difference in the magnitude of the mechanical damping in the two samples was even larger than for the epoxy. Again, the temperature range of the relaxation was broadened as detected in the joint, while the storage modulus drop was approximately equivalent to that for the unconstrained film.

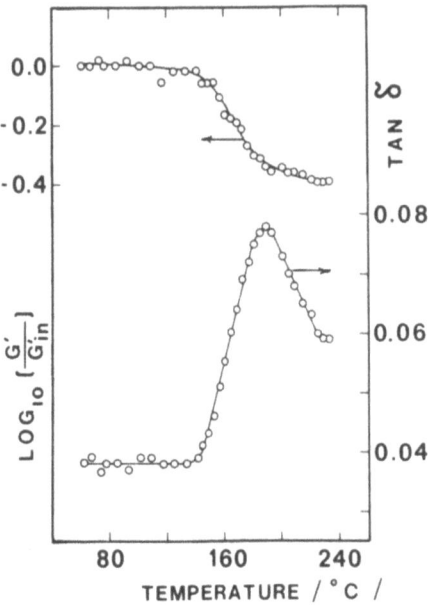

Figure 4. Dynamic mechanical relative shear modulus and damping for epoxy bonded joint assembly (at 110Hz).

Dynamic mechanical testing of poly(ethyl 2-cyanoacrylate) produced results shown in Figures 7 and 8 in the unbonded and bonded configurations, respectively. The sharp relaxation with a maximum at about 110°C for the free film was modified in the bonded joint to become a quite broad temperature response at this test frequency, with a more complex shape. The magnitude of the higher temperature damping maximum in the joint is approaching the absolute values described above for the free films. In addition, the first peak found on heating is about 20°C higher than in the unrestricted film. In previous work we have noted that the poly(ethyl 2-cyanoacrylate) was found to have very poor thermal stability, showing a total loss of tensile lap shear strength on aging for 24 hours at 150°C.[24]

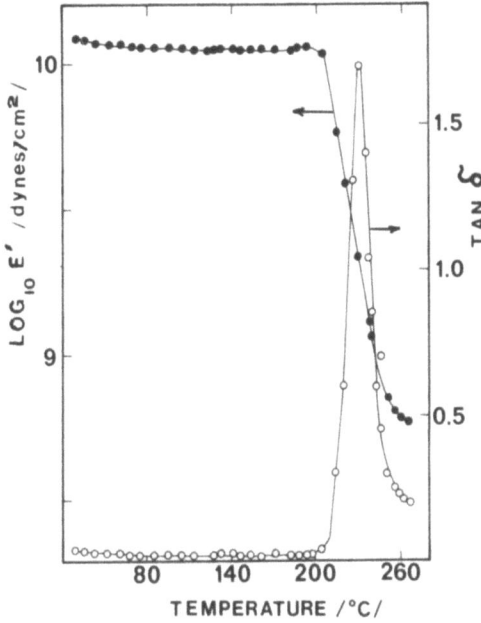

Figure 5. Dynamic mechanical storage modulus and damping for polyester/polysulfone block copolymer film (at 110Hz)

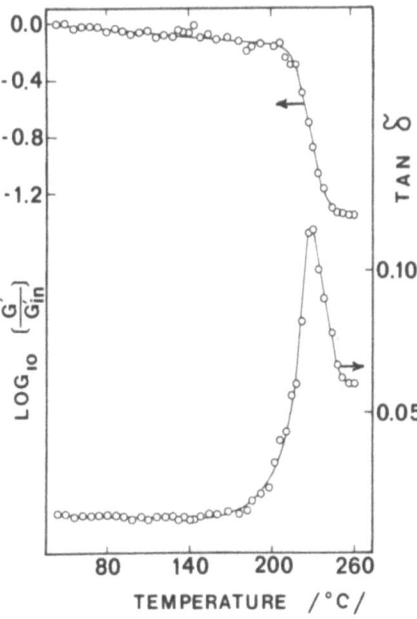

Figure 6. Dynamic mechanical relative shear modulus and damping for polyester/polysulfone block copolymer adhesively bonded joint (at 110 Hz).

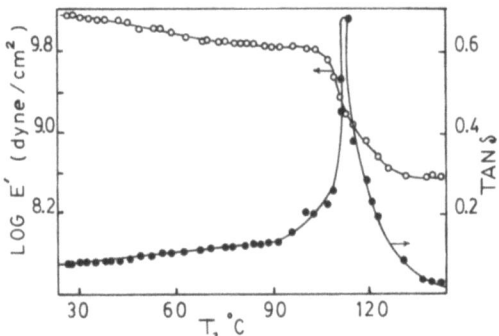

Figure 7. Dynamic mechanical storage modulus and damping for poly(ethyl 2-cyanoacrylate) film (at 110 Hz).

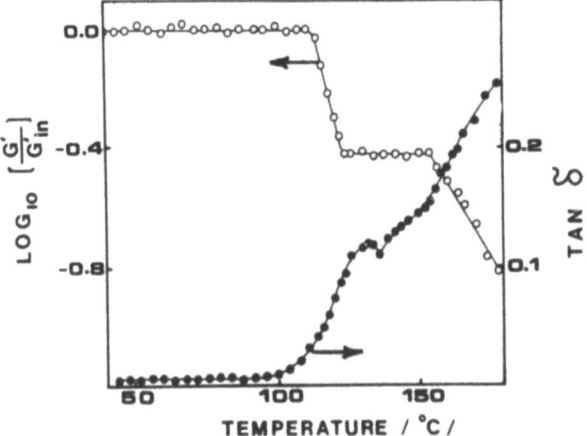

Figure 8. Dynamic mechanical relative shear modulus and damping of poly(ethyl 2-cyanoacrylate) bonded adhesive joint (at 110 Hz).

In contrast to the alkyl 2-cyanoacrylate adhesive, the allyl version had quite different mechanical response as is presented in Figure 9. The maximum in tan δ for the bonded system is about 8°C higher in temperature than that seen for the alkyl version, but then decreases gradually with continued heating. More striking is the recovery of the inital shear storage modulus to its original value on heating. Previously, we determined that a thermally induced crosslinking was responsible for this behavior.[24] As the vitrification due to network formation proceeded, the glassy storage modulus was approached, in spite of the increase in temperature at 1°C/min (which apparently was slower than the progress of the chemical reaction). The fully-cured allyl ester adhesive was found to have a T_g greater than 200°C.[24]

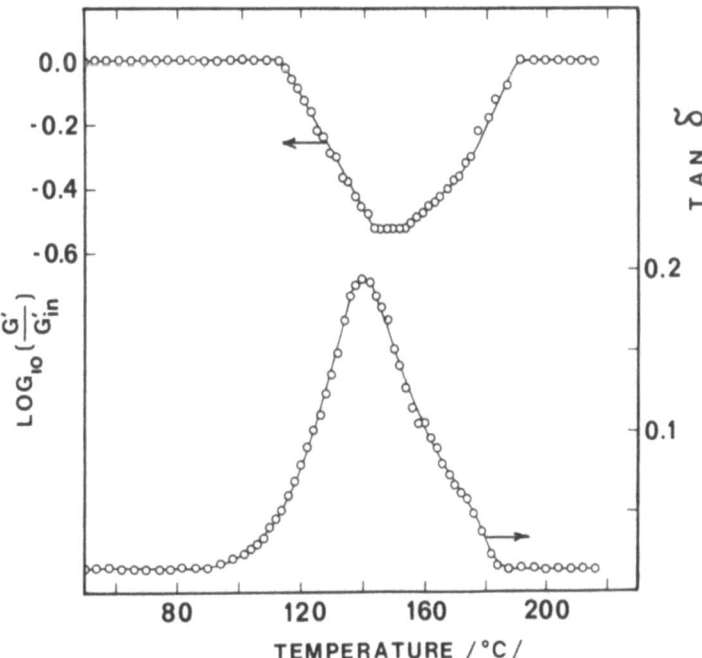

Figure 9. Dynamic mechanical relative shear modulus and damping of a poly(allyl 2-cyanoacrylate) bonded adhesive joint (at 110 Hz).

The final experimental results were obtained on the triblock (styrene/isoprene/styrene) thermoplastics, and represent a second phase of the research in which bond thickness was tightly controlled rather than varying around a nominal value of 0.2 mm. The design of Figure 2 was employed for all of this work.

Preparation of the adhesively bonded joints with the S/I/S polymers was extremely important because of the sensitivity of the polymer morphology to thermal history. For example, we were able to extrude films of similar triblock materials which showed different free film storage moduli (almost a factor of two) in the extruded and transverse directions.[25] Also, using DSC we sometimes noted unexpected thermal response of these materials at about 20°C below the glass transition temperature which was attributed to a "stress release phenomenon." It was demonstrated that differences in the compression molding operation could be associated with either the presence or the absence of such stress release events.[25] Thus, the S/I/S materials served as model adhesives with which to evaluate the trends in the results for the other polymers described above.

Results of experiments on S/I/S free films and on S/I/S bonded joints with the thermoplastic glue line at 0.3, 0.5 and 0.7 mm appear together in Figure 10. All preparations were made with the 50% styrene content polymer. The preparation temperature was 150°C and at least 5000 psi was applied in the molding. Pressure was maintained during cooling. The Tan δ's are enlarged in Figure 10 compared to the usual presentation in order to emphasize sample differences. An increase in the rate of the molecular relaxations seems to appear in the bulk free film (B) on heating, with a very rapid increase in damping in the vicinity of 100°C. For this unbonded film the Tan δ passed out of the range of the instrument (for these sample dimensions) at about 108°C. In contrast, for all three bond thicknesses investigated the damping of the joint assembly was lower than that of the free film at equivalent temperatures, while the onset of very high damping shifted upward, approximately 20, degrees depending on bond thickness. It is apparent that in the very thin bond-line joints there are indications of several temperature ranges where small maxima or details in the damping appear. These results were independent of heating rate and reproducible as long as the same joint preparation conditions were observed. A major question of concern was whether the small peaks were real or artifacts.

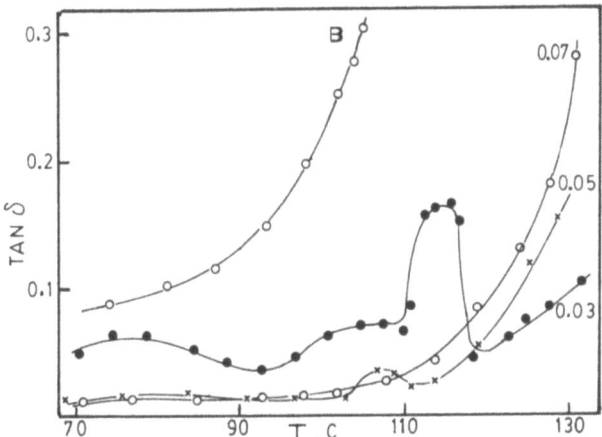

Figure 10. Dynamic mechanical damping of 50% styrene S/I/S bonded joint and free film at 110 Hz. Bond thicknesses shown on curves are in cm for joints. Free film = B.

When the experiments represented by Figure 10 were repeated with the 40 wt. % styrene S/I/S block copolymer, similar observations were again recorded. These results appear in Figure 11. Preparation conditions were the same as in Figure 10. As expected, the magnitude of the damping was higher overall due to the increased isoprene fraction in the copolymer. However, the major features of the previous curves were retained. These are (1) higher temperatures required for appearance of damping maxima in joints as opposed to free films (2) unusual behavior of the Tan δ prior to very high damping phenomena, which seems to depend on bond thickness in the joints as opposed to the free film (B).

Further experiments were conducted on the 50% styrene content S/I/S copolymer. Modifications of the molding procedure for the S/I/S adhesives were made in an attempt to

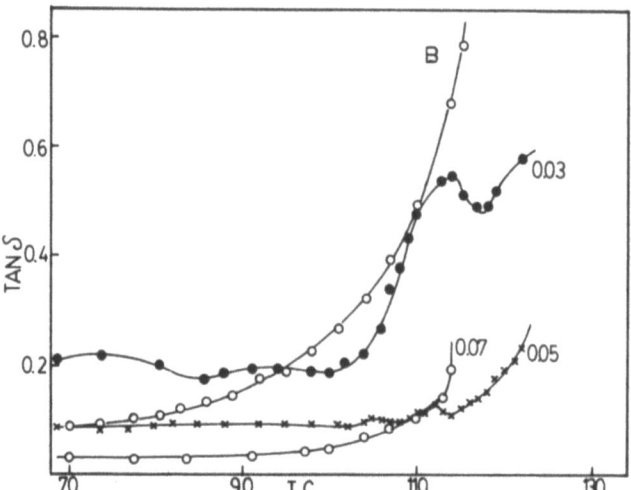

Figure 11. Dynamic mechanical damping of 40% styrene S/I/S bonded joint and free film at 110 Hz. Bond thicknesses shown on curves are in cm for joints. Free film = B.

understand the behavior of the fabricated joints as a whole. The major changes were to provide for slow cooling of the free films with and without any application of external pressure and to vary sample thickness. As a consequence of this new manufacturing process, the mechanical behavior was also altered. It was reasoned that if pressure-induced frozen stress changed mechanical response in bonded composites, a very thin free film prepared under conditions identical to those in the joints might also exhibit similar response. Confirmation of this hypothesis appears in Figure 12, where data on two free films are shown. The curve labeled P,0.1 represents results on a 0.1 mm thick, cooled-with-pressure sample. In the other curve (B), a 0.6 mm thick, cooled-with-pressure, material response is indicated. These

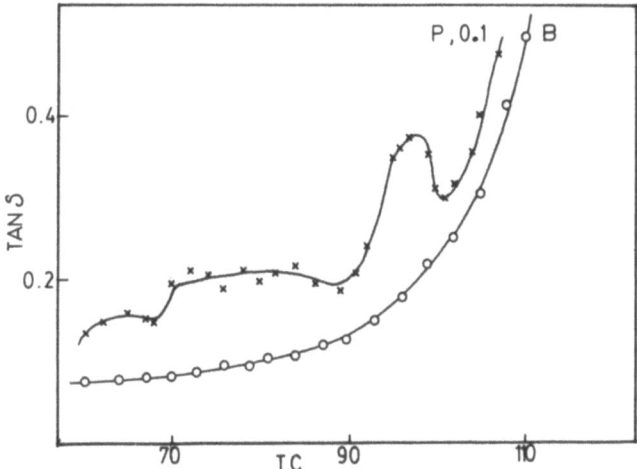

Figure 12. Dynamic mechanical damping of 50% styrene S/I/S free films, cooled with > 5000 psi pressure. B = 0.6 mm, P,0.1 = 0.1 mm thick.

latter data are replicates of those in Figure 10. Influence of the sample thickness is striking in this figure.

A final set of experiments on the 50% styrene S/I/S materials is summarized in Figure 13. Both curves shown depict the outcome of tests on free films. Now, holding the sample dimensions in each run almost the same with a thickness of 0.1 mm, preparation of the specimens differed only in that cooling was accomplished after molding with either > 5000 psi pressure applied (P) or with no pressure (NP). The difference in Tan δ was dramatic again and parallels that found when a thick specimen (B) was used with pressurized cooling, Figure 12, instead of the thin one. Neither of the comparisons made in Figures 12 or 13 causes a shift of the tan δ curves to higher temperatures as was found when bonded joints were investigated.

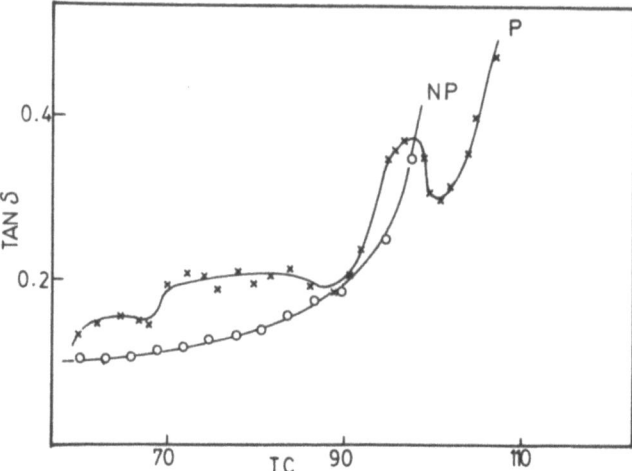

Figure 13. Dynamic mechanical damping of 50% styrene S/I/S free films of 0.1 mm thickness. P = cooled with > 5000 psi pressure, NP = cooled without pressure.

DISCUSSION AND CONCLUSIONS

In both the thermosetting and thermoplastic adhesives there were always different Tan δ (at 110 Hz) functions of temperature when free and bonded films were compared at equal heating rates. Generally, the viscoelastic response of the composite joint occurred over a wider temperature range than that of the free film, again heating rates being equal. Substantially different magnitudes of the monitored relaxation events were observed in each case. Non-equilibrium effects in the testing procedure do not seem to be responsible for any of these observations, non-equilibrium states in the joints may be.

At this time no firm conclusions may be put forth concerning the origins of the variations in dynamic mechanical properties with respect to free film vs. bonded lap shear joints; however, some facts are clear. In the thermoset materials and the polyester-polysulfone single phase block copolymer the differences in performance of bonded and free films are not as large and may reflect normal shifting due to the change in the monitored viscoelastic function in each case (tensile elongation vs. shear). Since simple shear is a constant volume process, and all other types of deformations require change in volume (Poisson's ratio ≠ 0.5), a correct constitutive equation description of the polymer's response would predict different relaxation behavior at constant temperature, or different temperature profiles in isochronous experiments. In the investigations of the present paper this effect should be quite small, and difficult to calculate.

A second view of the difference in mechanical behavior of the free and bonded films would attribute the physical variations to frozen stresses (originating from shrinkage, non-equilibrium morphology and crosslinking). Certainly the results on the S/I/S block copolymers suggest that thermal history must not be ignored in evaluating damping in a polymerically bonded structure placed in oscillatory shear fields. As the bond thickness was decreased the magnitude of the changes in viscoelastic response usually increased, depending on the polymer. Further experimentation at other frequencies and under certain isothermal conditions where stress release may be measured would be informative in clarifying the mechanism in this case.

With regard to long range polymer modulation (e.g. limitation of conformation isomers, damping of normal modes) by a highly rigid surface, the present experiments lead us to make no statement. The substrate surface area to volume of polymer ratio is so small that one finds it difficult to believe that any variation of dynamic mechanical relaxations of the overall bonded structure would manifest themselves in such simple experiments. Even in much more sensitive experiments on thin films, no mention is made of instrumentally induced limitation of molecular motion.[27]

Finally, the series of investigations on the S/I/S block copolymers reveals some interesting conclusions vis-a-vis dynamic mechanical testing of compression molded heterophase polymers. It appears that either free films or bonded joints may be manufactured which will show a fine structure in the Tan δ vs. temperature experiment, prior to the actual maximum damping of the film or assembly. Cooling thin films with applied pressures greater than 5000 psi was required to produce these details. Using thick films, or thick bond lines, or cooling without pressure would completely remove these smaller damping peaks. Confirming our DSC work, there was always a "bump" in these curves in the region of 70°C. The presence of the substrates, furthermore, when a thin bond line was used, shifted the large Tan δ damping to substantially higher temperatures, free films never showed this behavior.

Because of their nice model adhesive character, the S/I/S copolymers' behavior helped clarify the response observed in the other adhesives of this study. The generally broadened peaks of mechanical energy damping, occurring at higher temperatures, for these other samples may be manifestations of similar unrelaxed stress fields in the bonded joints. The caveat which was suggested on completion of this work reinforces the importance of sample preparation technique in systems where glass formation occurs during joint manufacture (such as is the case for structural adhesives).

ACKNOWLEDGEMENTS

The support of the Office of Naval Research and the Army Research Office is gratefully recognized. Also the synthesis of polymers by Dr. James E. McGrath and his research group significantly aided this project.

REFERENCES

1. W. B. Jones, Jr., Organic Coatings and Applied Polymer Science Proceedings, 47, 247 (1982).
2. D. W. Dwight, E. Sancaktar and H. F. Brinson, in "Adhesion and Adsorption of Polymers", L. H. Lee, Editor, Part A, 141, Plenum Press, New York, 1980.
3. G. Dolev and O. Ishai, J. Adhesion, 12, 283 (1981).
4. L.R.F. Rose, J. Adhesion, 14, 93 (1982).
5. R. W. Hylands and E. H. Sidwell, J. Adhesion, 11, 203 (1980).
6. W. D. Bascom, R. L. Cottington and C. O. Timmons, J. Appl. Polym. Sci., Appl. Polym. Symp., 32, 165 (1977).
7. D. L. Hunston, S. S. Wang, A. J. Kinloch, Organic Coatings and Applied Polymer Science Proceedings, 47, 408 (1982); also see J. T. Bitner, J. L. Rushford, W. S. Rose, D. L. Hunston and C. K. Riew, J. Adhesion, 13, 3 (1981).
8. V. F. Babich and Y. S. Lipatov, J. Appl. Polym. Sci., 27, 53 (1982).
9. Y. S. Lipatov, V. F. Rosovizky and V. V. Shifrin, J. Appl. Polym. Sci., 27, 455 (1982).

10. G. J. Howard and R. A. Shanks, J. Macromol. Sci.-Phys., B19(2), 167 (1981).
11. P. Peyser, Polym - Plast. Technol. Engr., 10(2), 117 (1978).
12. P. S. Theocaris and G. D. Spathis, J. Appl. Polym. Sci., 27, 3019 (1982).
13. D. J. Plazek, J. Polym. Sci., 20, 1533 (1982).
14. D. J. Plazek and G-F. Gu, J. Polym. Sci., 20, 1551 (1982).
15. M. Takayangi, in "Proceedings of the Fourth International Congress on Rheology," C. Klason and J. Kubat, Editors, Part I, p. 161, Swedish Soc. Rheo. Publisher, Gothenburg, 1965.
16. T. Murayama, J. Appl. Polym. Sci., 19, 3221 (1975).
17. T. Murayama, in "Proceedings of the Seventh International Congress on Rheology," E. H. Lee and A. L. Copley, Editors, p. 402, Interscience, New York, 1977.
18. B. H. Shah and R. Darby, Polym. Eng. Sci., 16, 46 (1976).
19. P. F. Erhardt, J. J. O'Malley and R. G. Crystal, ACS Polym. Preprint, 10(2), 812 (1969).
20. D. J. Massa, J. Appl. Phys., 44, 2595 (1973); also A. R. Ramos, F. S. Bates, R. E. Cohen, J. Polym. Sci., Polym. Phys. Ed., 16, 753 (1978).
21. S. Timoshenko, "Strength of Materials," Part I, Van Nostrand, Princeton, 1955.
22. Instruction Manual No. 68, Rheovibron Model DDV-II-C, Toyo Bladwin Co., Ltd. Tokyo (1973).
23. D. L. Kotzev, C. Konstantinov, P. Novabov, V. Kabaivanov, Bulg. Pat. No. 23321 (1977).
24. D. L. Kotzev, T. C. Ward, D. W. Dwight, J. Appl. Polym. Sci., 26, 1941 (1981).
25. J. E. McGrath, Private Communication.
26. A. Wood, M. S. Thesis, Virginia Polytechnic Institute and State University, Blacksburg, Va. (1982).
27. J. D. Ferry, "Viscoelastic Properties of Polymers," 3rd Ed. Chapter 5, Wiley, New York, 1980.

ULTRASONIC ASSESSMENT OF CURE RATE EFFECTS IN BONDED HONEYCOMB

STRUCTURES

R. A. Pike and R. S. Williams

United Technologies Research Center
Silver Lane
East Hartford, Connecticut 06108

The effect of variations in cure temperature on fillet formation in adhesively bonded honeycomb structures has been defined in terms of fillet length, degree of cure and Tg of the cured neat adhesive. The results are correlated with predicted behavior based on ultrasonic NDE analysis prior to testing. The ultrasonic test method uses advanced pattern recognition analysis of attenuation and velocity data extracted from a multi-frequency Lamb wave propagated along the face sheet of the honeycomb. Laboratory applications of this test method are described.

INTRODUCTION

The increasing awareness that adhesive performance is controlled not only by the condition of the adherend surface but also the condition or "state" of the adhesive and the process parameters used during fabrication is expected to result in improved reliability as well as bond performance. Improved reliability will also result from the implementation of a well defined NDE analysis directly relatable to the intrinsic properties of the cured adhesive. The increasing application of honeycomb adhesively bonded systems as structural components wherein the adhesively bonded surface is considerably less than in a normal metal-to-metal system underscores the necessity of obtaining a complete definition of the factors which affect bond performance and quality.

Recognition of the fact that assurance of bond quality and integrity involves many facets, a multidisciplinary approach that combines NDE, adhesive technology and fracture mechanics has been employed at United Technologies Research Center to form the basis for a comprehensive understanding of adhesive performance. A description of the NDE method employed to define the bond strength of metal-to-metal bonds involving the ultrasonic measurements and mathematical method, RESID (Recursive Structural Identification) has been reported.[1] An adaptation of this NDE method for bonded honeycomb structures is described in the following sections of this paper.

Heat-up or cure rate during fabrication has been identified as one of the major factors which control bond performance. This effect has been previously alluded to by a number of workers in terms of stress levels,[2] rheological cure behavior[3] and tensile lap shear strength.[4] In-house investigations have involved the correlation of adhesive chemistry, strength and damping characteristics with heat-up rate.[5]

The objective of the study described here was to correlate ultrasonic wave response, NDE, with the pertinent physical and mechanical properties of adhesively bonded honeycomb structures using cure temperature as the primary fabrication variable. The effect of heat-up rate on bonded honeycomb performance and NDE response is also being investigated and will be reported when complete.

EXPERIMENTAL

The reticulatable (unsupported) and scrim supported EA-9649 adhesive film used in the study was kindly supplied by the Hysol Division of Dexter Corporation, Pittsburg, California. The adhesive, designated as a 178°C (350°F) curing system, has been previously described.[6]

Honeycomb specimens were fabricated from 5.0mm thick, 10.16 cm^2 solid and perforated 2024 aluminum plates using 8mm hexagonal aluminum honeycomb. All surfaces were acetone rinsed, vapor degreased, etched (FPL) and primed prior to bonding. Supported film bonded the solid face sheet; unsupported film, the perforated plate.

Thermomechanical analysis (TMA) was carried out using a DuPont 1090 Thermal analyzer at a heating rate of 10°C/min.

Dielectric analysis of neat adhesive was performed as previously described [6] using a Tetrahedron, Inc. Audrey Model 203 dissipation factor bridge with a Model 1501 automatic press. Temperature was automatically recorded by using a thermocouple inserted directly into the two-ply adhesive lay-up. A heat-up rate of 2.5°C/min was used with the 7.6cm^2 sample.

Bonding was carried out at 50psi pressure at four different cure temperatures, namely 140°, 160°, 180° and 200°C using a 2.5°C/min heat-up rate. The time to reach the specified temperature thus varied from 46, 54, 62 and 70 minutes, respectively.

Adhesive flow or fillet length was determined by averaging ten fillet lengths measured in millimeters under 10X power magnification.

The ultrasonic plate wave technique developed for honeycomb structures is shown in Figure 1. A longitudinal wave from the transmitter is propagated through the Lucite into the skin of the honeycomb at the Rayleigh critical angle. By the process of mode conversion, a surface or plate wave is generated that propagates along the face sheet of the honeycomb. At the other end of the fixture, this process is reversed and the wave detected by the receiver. Since the wave travels over a known distance, the attenuation per unit distance and the velocity is easily measured.

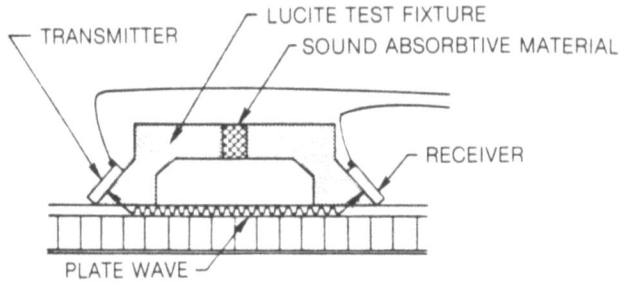
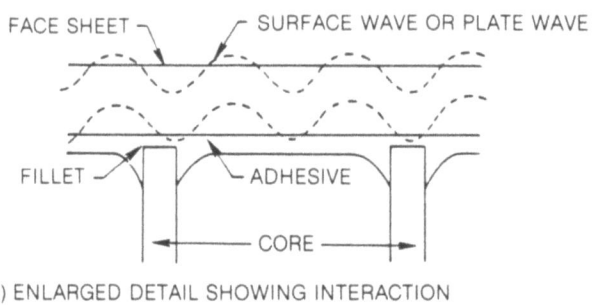

Figure 1. Plate wave transducer.

The level of interaction with the sheet adhesive, the fillet, and/or the honeycomb is determined by the wavelength (or frequency). At frequencies greater than 15-20 MHz, a surface wave is generated that will be affected little by the bond. At lower frequencies (2-5 MHz), a plate wave will be generated that will interact strongly with the adhesive adhering to the lower surface of the face sheet. At still lower frequencies (500 kHz-2 MHz), the plate wave will interact with the fillet and the honeycomb. A RESID algorithm, as discussed below, is used to correlate the ultrasonic data obtained with the actual measured properties of the adhesive or bonded structure.

RESULTS AND DISCUSSION

Three methods were used to assess the effect of varying cure temperature on both supported and unsupported EA-9649 adhesive film in terms of flow, gel time and final Tg. These were dielectric and TMA analysis of the neat adhesive and fillet length measurements of the reticulated unsupported adhesive film in a bonded honeycomb structure.

The dissipation factor profile curves obtained on the two types of adhesive film are illustrated in Figures 2 and 3. The pertinent data are listed in Table I.

Table I. Cure Temperature Effects on EA-9649 Adhesive-Dielectric Analysis [a].

	Cure Temp, °C	ISA[b] (%flow)	Time to gel, minutes	T_g, °C[c]
Supported Adhesive	200	97	72	246
	180	70	81.5	220
	160	89	86	203
	140	71	101	139
Unsupported Adhesive	200	45	70	239
	180	25	81	205
	160	35	90	185
	145	42	102.5	142

a. Analysis run at 1000 Hz, 50psi pressure, 2 ply lay-up.
b. ISA = Increase in surface area.
c. Measured by TMA on the cured sample.

For the unsupported adhesive an increase in Tg point and decrease in time to gel with increasing cure temperature was observed as expected. The flow, as measured by the increase in surface area during cure, was found to be minimum at the specification 180°C level. Higher flow was experienced above and below this temperature. As previously shown[6,7] changes in the major melting point and gel times were influenced by the changes in the condition of the adhesive as well as fabrication parameters. The supported adhesive in general gave a similar cure response with varying cure temperature. Minimum flow was again found to occur at 180°C. The low flow associated with the 140°C cure is probably due to the effect of the support scrim coupled with the higher viscosity of the adhesive. Accompanying these changes in resin behavior was the higher final dissipation factor obtained by samples cured at temperatures other than the specification 180°C. This, as has been previously discussed for aging[6] and moisture[7] effects on EA-9649 adhesive, is indicative of a different final molecular

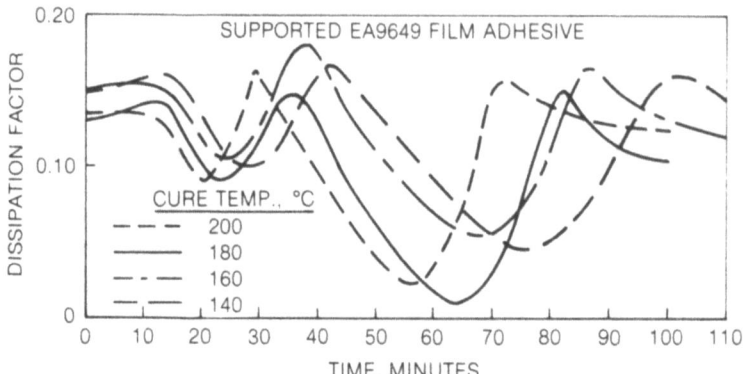

Figure 2. Dielectric analysis - effect of cure temperature, Supported EA-9649 film adhesive.

structure in either a fully or partially cured state. DSC analysis of the cured adhesive showed the presence of unreacted catalyst at the 140°C and 160°C temperatures. This was apparent from the slight endotherm (melting) which occurred at approximately 220°C. The melting point of dicyandiamide is 216°C at the heat-up rate used for the analysis. Thus, at

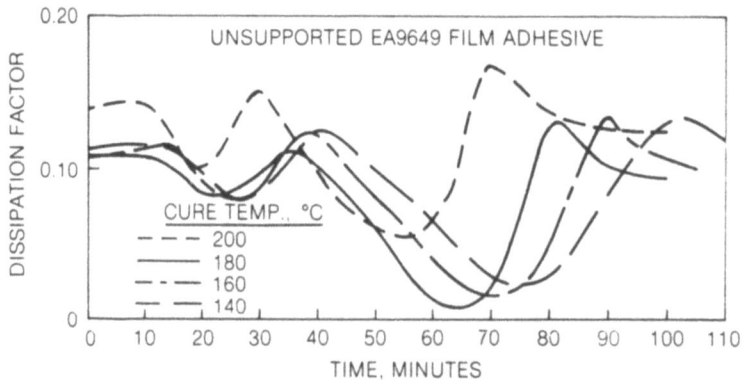

Figure 3. Dielectric analysis - effect of cure temperature, Unsupported EA-9649 film adhesive.

temperatures below 180°C the adhesive appears to be only partially cured. The marked effect of cure temperature on flow related to resin viscosity, cure kinetics and temperature rise indicates the necessity of careful monitoring of process parameters to achieve reproducibility in bonded structures.

In order to correlate the flow response obtained by dielectric analysis with fillet formation in a bonded honeycomb structure, measurements of fillet length were made on the reticulated unsupported adhesive using specimens fabricated at the four curing temperatures with primed and unprimed honeycomb. The results are shown graphically in Figure 4. The degree of flow obtained on the primed honeycomb with perforated plate on top was in exact agreement with the dielectric analysis results, i.e., a minimum flow was obtained at the 180°C cure temperature with longer fillets being formed at cure temperatures above and below the specified cure temperature. The flow on the unprimed honeycomb showed a slight minimum at 160°C indicative of the different response of the spreading characteristics of the adhesive to a metal or resin surface. Reversing the honeycomb construction so that the perforated plate was on the bottom (adhesive flows up) gave a minimum flow condition at 180°C on the unprimed surface. Thus, varying cure temperatures produce trends in adhesive flow and final cured state properties which should affect not only the mechanical properties of a bonded joint but the response of the adhesive to ultrasonic wave stimulation.

The RESID algorithm approximates a complex nonlinear relationship with a network of simple binary quadratic functions. The interconnections of these elements proceed in a selective and intelligent manner so that,

(1) Dissimilar features are optimally combined.
(2) Higher order feature combinations are created as they are needed.
(3) Inappropriate combinations of features are automatically excluded.
(4) Overfitting the data set is minimized.

The RESID, Recursive Structure Identification, algorithm is based on the polynomial theory of complex systems introduced nearly a decade ago by the Russian academician, A.G. Ivakhnenko [8]. The theory has since been implemented by a number of investigators.

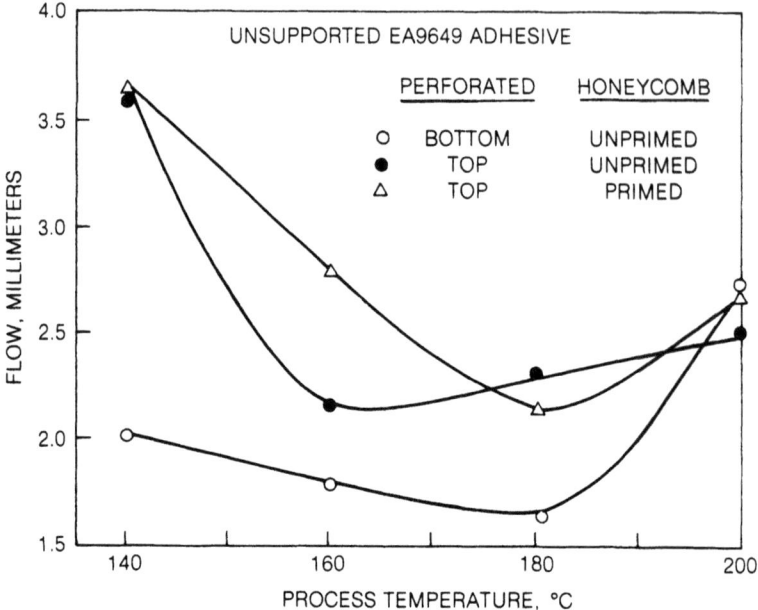

Figure 4. Adhesive fillet length (in terms of flow) vs processing temperature.

The motivation for RESID is threefold. First, the relationship between features and the quantity to be determined, or item to be classified, is unknown and often nonlinear. Second, the feature set may be quite large and may contain different or even qualitative types of features. Third, it is desirable to train the classifier using as small a data set as possible.

In general, the nonlinear relationships between the material properties and the ultrasonic features can be represented by a multinomial power series which in this case is impractical to use.

The approach used for this investigation was to approximate a complex function with an interconnection of simple functions. A suitable elemental function is the binary quadratic function,

$$X = b_0 + b_1\omega_1 + b_2\omega_2 + b_3\omega_1\omega_2 + b_4\omega_1^2 + b_5\omega_2^2$$

which contains six coefficients and is shown schematically in Figure 5. With this strategy, it is possible to achieve fourth power relationships with as few as eighteen coefficients.

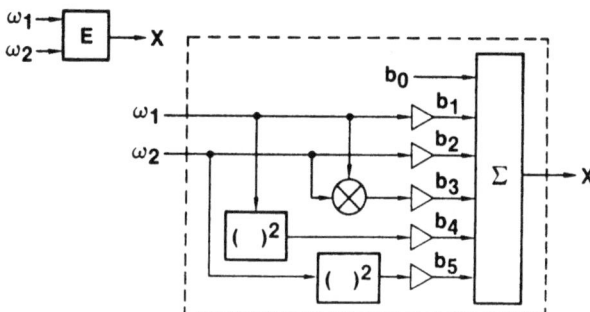

Figure 5. Network for predicting material property.

There are several approaches to obtaining the element coefficients and interconnections. One strategy is to fix the network structure, and depend on many of the coefficients becoming small enough to ignore. The coefficients can be computed using either a global random search or a deterministic least squares fit. A second strategy is to grow the network layer by layer until adequate mixing of the variables has occurred. RESID uses the second method.

For the first layer, RESID examines all pairwise combinations of features. Those elements having minimum square error fit to the output variable are retained. The six coefficients in each element are computed using deterministic matrix operations. Subsequent layers are added by examining all pairwise combinations of previous layer outputs. Those elements having errors less than either of the errors of its parents are retained. When this condition is not met, the layer generation terminates.

A key step in the RESID algorithm is the partitioning of the data set into three mutually exclusive subsets: training subset, selection subset, and evaluation subset. The training subset is used to determine the coefficients in each of the network elements. The selection subset is used to eliminate those elements that have low correlation with the known output. The evaluation subset is used to give an independent assessment of overall network performance. The training and selection subsets are applied alternately as each level of the network is grown. The network stops growing when the algorithm starts to fit the noise type variations in the data.

Single feature ultrasonic data (in this case plate wave attenuation at 1 MHz) versus the fabrication temperatures and measured Tg of the adhesive are shown in Figure 6. Although there is a slight general correlation, i.e., increasing Tg with increasing cure temperature, the data exhibit considerable scatter. This is expected since no single ultrasonic feature is uniquely a function of the adhesive intrinsic properties such as Tg.[1] Additional ultrasonic features must be used to provide multi-dimensional data for processing by the RESID algorithm. Some of these data, ultrasonic plate wave attenuation at several frequencies and power levels are shown in Figure 7. It should be noted that maximum attenuation for two of these features occurs at the 180°C cure temperature which corresponds to the minimum fillet length results.

Application of pattern recognition methods (RESID) to the multiple ultrasonic data greatly improves the correlation. Using the RESID network, as described above, the scatter can be reduced such that the maximum prediction error is 10 percent as shown in Figure 8. The horizontal axis is the predicted cure temperature and resulting Tg using the RESID network with five ultrasonic attenuation and velocity features, the vertical axis is the actual cure temperature and measured Tg. It should be noted that the graph contains both training as well as evaluation specimens. The former were used to determine the parameters required to finalize the RESID network.

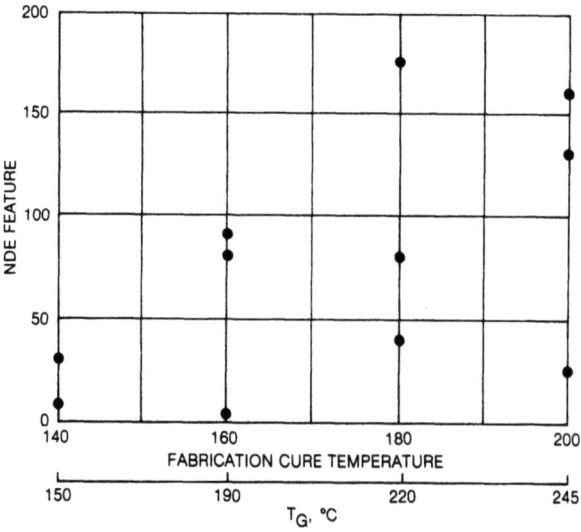

Figure 6. Honeycomb cure property vs NDE feature.

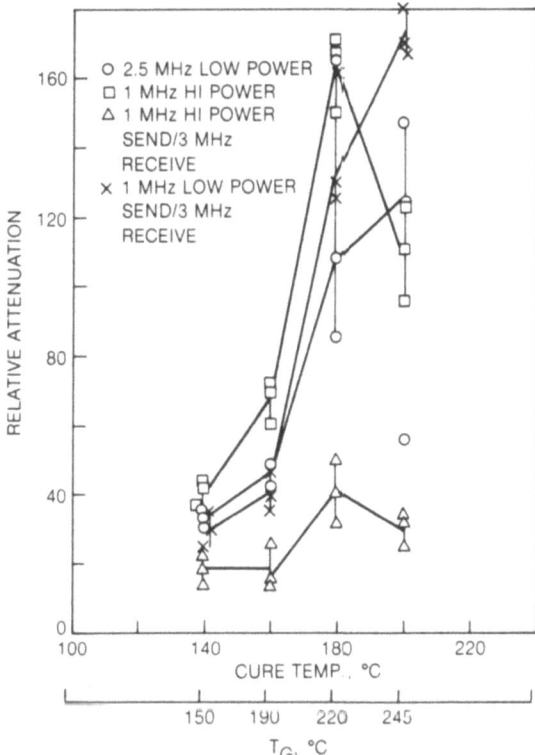

Figure 7. Cure temperature vs attenuation.

This approach has also been successfully used to predict a number of other metal-to-metal adhesive bond properties, including bondline thickness, porosity, glass transition temperature, degree of cure, and bond strength.[1] For each property, a different RESID network is derived, although the ultrasonic features for each network may be selected from the same set of data. For example, network I to predict property I may use five features from a possible sixteen; network II to predict property II may use six entirely different features or it could share some of the features used for network I. In all cases a network can, by design, accommodate variations in the adhesive system; however dramatic changes in the adhesive/adherend system will require the use of a different set of networks.

Correlation between ultrasonic wave response and flatwise tensile data for the bonded honeycomb specimens is in progress. Preliminary results are very encouraging and will be reported when complete.

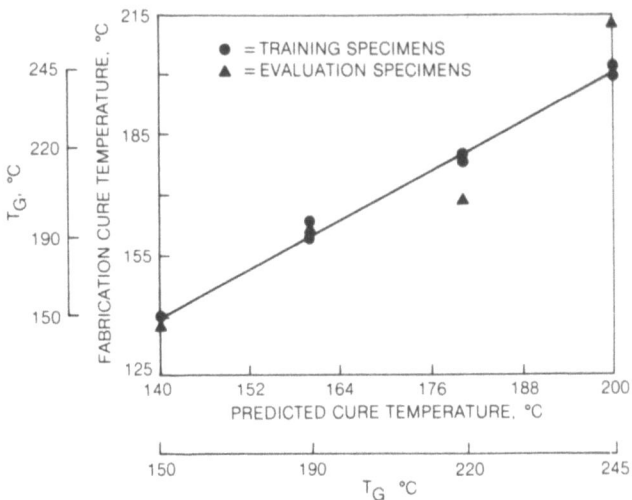

Figure 8. Honeycomb cure property prediction.

CONCLUSIONS

A test method using experimental ultrasonic measurements and recursive structure identification (RESID) to analyze the data has been developed that is potentially capable of predicting adhesive bond properties in honeycomb bonded structures. Cure temperatures should be strictly monitored to ensure optimum adhesive performance.

ACKNOWLEDGEMENTS

The authors wish to acknowledge the technical contributions of Dr. P. E. Zwicke in the development of the RESID algorithms and Mrs. J. P. Pinto for the determination of adhesive properties.

REFERENCES

1. R.S. Williams and P.E. Zwicke, Materials Evaluation 40,312 (1982).
2. C.L. Mahoney, paper (No.790151) presented at the Soc. of Automotive Engineers, Congress & Exposition, Detroit, February 26,1979.
3. J. Thuen and R. Hinrichs, in "Proc. 25th National SAMPE Symposium & Exposition", held May 6, 1980, p. 126.
4. E. Sancaktar, H. Jozavi and R.M. Klein, paper presented at the 5th Annual Adhesion Society Meeting, Mobile, AL, February 22, 1982.
5. R.A. Pike, F.P. Lamm and J.P. Pinto, to be published.
6. R.A. Pike, F.P. Lamm and J.P. Pinto, J. Adhesion 12,143 (1981).
7. R.A. Pike, F.P. Lamm and J.P. Pinto, J. Adhesion 13,229 (1982).
8. A.G. Ivakhnenko, IEEE Trans. on Systems, Man and Cybernetics, SMC-1, No. 4, October 1971, p. 364.

THE THREE-POINT BEND TEST FOR ADHESIVE JOINTS

N. T. McDevitt and W. L. Baun

Mechanics and Surface Interactions Branch
Air Force Wright Aeronautical Laboratories
Wright-Patterson Air Force Base, Ohio, 45433

Unless a large amount of degradation has occurred in the interfacial region of a metal-to-metal adhesive bond, the T-peel and wedge opening test usually provide only mode I information on the bulk adhesive. The single-lap shear test shows primarily mixed-mode failure. However, the thickness of the adherend must be carefully considered in order not to have a large amount of mode I failure. When these three tests failed consistently to find defects that were incorporated into the interfacial region of test specimens, the three-point bend test geometry was evaluated. This initial study, utilizing the short beam shear test geometry, has shown that several pieces of information may be obtained from the experimental data. The initial portion of the test curve appears quite linear and the slope of the curve seems directly related to the bulk properties of the adhesive. The portion of the curve where a definitive failure occurs appears directly related to interfacial failure of the bonded joint.

INTRODUCTION

Historically the effect of stress and environment on the overall strength of metal-to-metal adhesive bonds has been determined primarily by three mechanical tests; (1) peel, (2) wedge opening, or (3) lap shear. However, attempts to correlate data from accelerated laboratory experiments to actual service life, using these mechanical tests, have met only with marginal success.[1-4] Actual in-service failures of an adhesive bonded structure have been reported to occur in the region of the adhesive-oxide interface.[5] A recent study using surface instrumentation techniques shows failure sites also occurring at the oxide-metal interface. Therefore, this interphase region of adhesive-oxide-metal should be of concern in laboratory studies of adhesive bond line failure. Since the joint geometry of the peel and wedge opening tests provide only crack opening data, primarily toughness properties of the adhesive are obtained and very little information is generated about the subtle changes that may occur in the interphase region. The lap shear test is capable of generating data from the interphase region but care must be taken in the proper choice of thickness of the test adherend in order to eliminate most of the peel forces.

When dealing with high strength aluminum alloys the aspect of corrosion has to be considered as a potential problem. In most corrosion processes a thin film is involved at some stage. The controlling factor in the growth of this film is generally the type of environment and length of time of exposure of the test specimen. In order to study the subtle effects of surface preparation and corrosion films on the strength of the adhesive-oxide and/or oxide-metal interfaces our interest was directed to a test geometry that would be simple to fabricate and would load the bond line with as much pure shear force as possible.

We chose the three-point bend test for our study and the purpose of this paper is to illustrate the general aspects of this test method on metal-to-metal adhesive bonds. This study, to our knowledge, is the first use of this test method on metal sandwich structures.

EXPERIMENTAL

Aluminum Alloy Specimens

Rectangular specimens were cut from a sheet of 7075-T6 bare aluminum alloy. Each specimen was 10 x 15 x 0.35 centimeters. All of the specimens were degreased with an acetone wipe, then alkaline etched with 0.1N NaOH for three minutes at room temperature. Specimens were deoxidized with a solution of 5:1 HNO_3-HF for two minutes at room temperature, then desmutted with 50% HNO_3 for 30 sec.

Duplicate panels were then anodized in a 1.0M H_3PO_4 bath according to the conditions described in Table I. Each pair was then bonded with FM 123-2 adhesive (in tape form), cured at 250°F and 25 psi, without the aid of a primer. Specimens 3.8 x 1.3 x 0.7 centimeters were cut from the bonded panels. Duplicate specimens of each surface preparation were subjected to the environmental conditions described in Table II.

Table I. Anodization Conditions for 1.0M H_3PO_4

Surface Preparation	Voltage	Time in Bath	Average Oxide Thickness
1	10	0.5 min	225Å
2	10	2.0	1000
3	10	10.0	3200
4	10	16.0	3600
5	40	2.0	1200
6	40	5.0	3000

Table II. Type of Curing and Exposure for Test Specimens.

Test	
A	specimens cured then tested within 24 hours
B	specimens cured then post cured at 220°F for 8 hours
C	specimens cured then stored in desiccator under ambient conditions for 90 days
D	specimens from Test C subjected to dry heat, 100°C, for 63 hours
E	specimens from Test C subjected to SO_2 environment at R.T. for 2160 hours
F	specimens from Test C subjected to SO_2 environment at 100°C for 72 hours
G	specimens from Test C stressed to 1000 lb for two minutes then subjected to 100% R.H. and 100°C for 8 hours

Mechanical Test

The data were obtained for each specimen using an Instron test machine. The three-point bend test procedure aligns the specimen so the resultant of the applied load is perpendicular to the adhesive bond line, see Figure 1. The support noses were adjusted to a span four times the average specimen thickness. The test was carried out by loading each specimen at a crosshead speed of 0.02 inch per minute until interfacial failure occurred. All tests were stopped after 6 minutes. All tests were performed at room temperature.

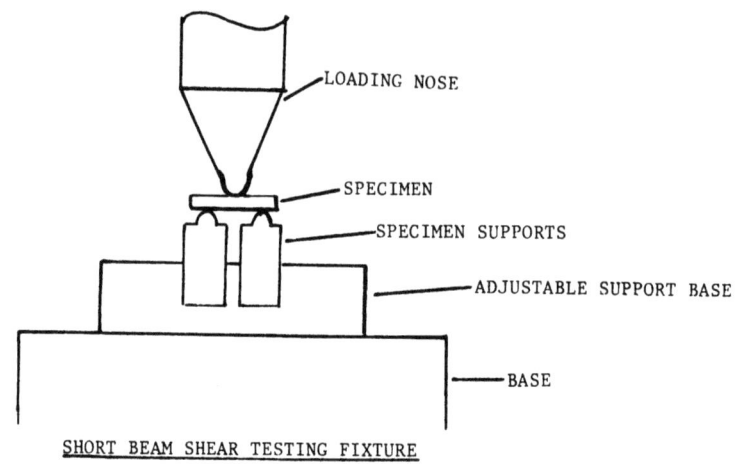

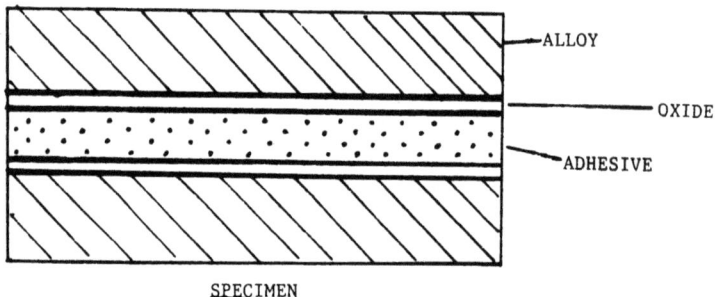

Figure 1. Illustration of short beam shear test geometry

ANALYSIS OF DATA

The three-point bend test, usually called the short beam shear test, is one of the most widely used test methods for evaluation of the shearing strength of composite materials. Several assumptions are made in applying beam theory to this test, one of which is that the shear stress S is distributed across the transverse face of the specimen with a maximum at the center according to the following Equation (1):

$$S = 0.75 \, (L/wt) \tag{1}$$

where L is the load applied at the center of the beam, w is the width of the beam, and t is the thickness. The fracture load (L) is obtained from the test record, and the value of S calculated from the equation is assumed to represent the shearing strength of the composite.

This same test has been used to study the sandwich structure of metal-to-metal adhesive bonds. This is apparently the first attempt to apply this geometry to metal adhesive joints. No attempt has been made to mathematically analyze our specimen from beam theory. In the present study, the test was used only to differentiate between the interfacial properties of various surface preparations that had been subjected to changing temperatures and/or environments.

Figure 2 represents the test record of a completely nonbonded specimen. This load-displacement curve was obtained from a specimen that was prepared by physically laying a film of FM 123-2 adhesive between two adherends of 7075-T6 aluminum. The zero displacement point is obtained by dropping a tangent along the linear portion of the curve. The displacement of the specimen is plotted as a function of time. A load value, M, is obtained from the linear portion of the curve where it intercepts with the one minute mark of the abscissa. After two minutes the stiffness slope changes and the load values represent slight metal deformation mixed with the energy required to overcome the friction of the adhesive tape opposing the slide of the two opposite faces of the adherends. The load value, M_n, from this curve represents our baseline for a completely nonbonded structure.

Figure 3 represents the type of curve obtained from a specimen where the adhesive has been cured and has not been subjected to any environmental tests. This is a baseline curve for a strongly bonded specimen. The load value, M_b, (or slope) is obtained from the linear portion of the curve after one minute. After two minutes the slope changes and the load take up is more gradual. In this region of the bonded specimen the load values represent a slight metal deformation mixed with the energy required to overcome the adhesion of the adhesive-oxide-metal interfaces. The energy required to overcome these interfacial forces is distinguished by a definitive break in the test record (L). The load value at this point is taken as the yield strength of the bonded interface (YSI). After the break in the curve a definite fall off in load value is recorded, a second curve will show another load value being recorded. This second portion of the record represents the load value of the deformation of the adherends plus the friction generated by the type and degree of failure of the interface. The type of failure is distinguished by cohesive and/or adhesive. The larger the degree of adhesive failure in the interfacial region the closer the second value will parallel the curve obtained from the nonbonded specimen.

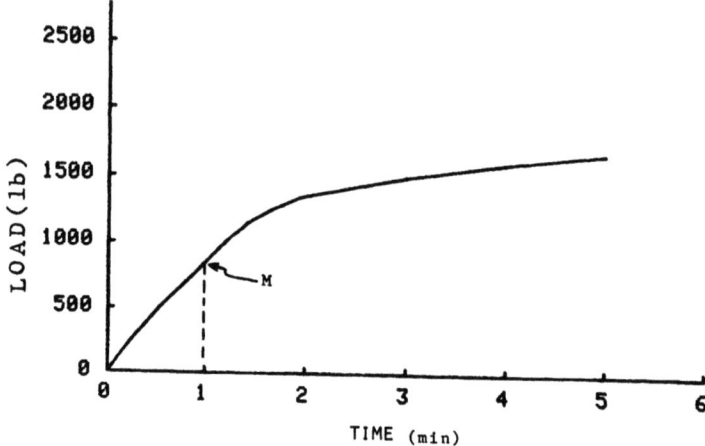

Figure 2. Load-displacement curve for non-bonded specimen.

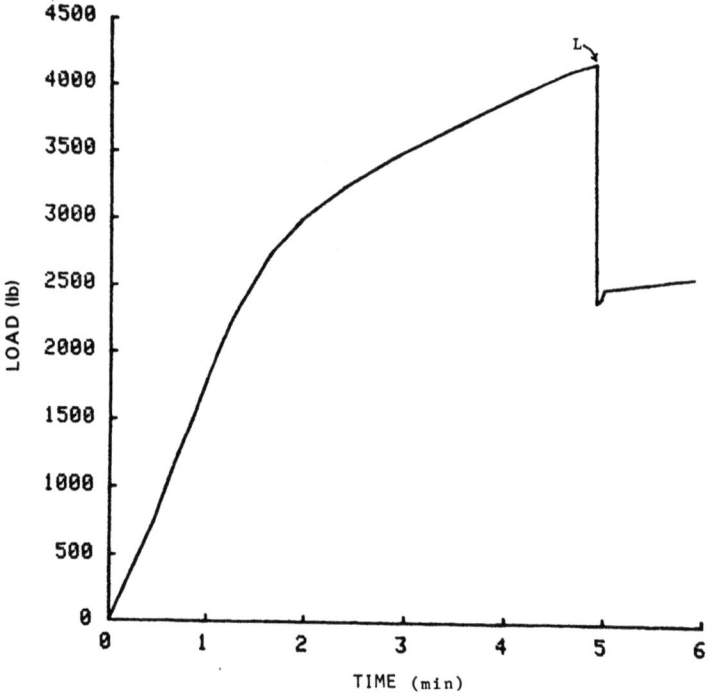

Figure 3. Load-displacement curve for bonded and cured specimen.

We can evaluate bond line stiffness (BLS) from the linear portion of these curves by using Equation (1) and relating the data from bonded specimens to the nonbonded specimens, S_b/S_n, as in Equation (2):

$$BLS = \frac{S_b}{S_n} = \frac{M_b}{M_n} \qquad (2)$$

RESULTS AND DISCUSSION

The following data represent the preliminary step in the development of a testing method that will detect subtle changes in the chemistry of surfaces and interfaces when they are exposed to harsh environments. Of particular interest was the determination of changes occurring at the adhesive-oxide-metal interfaces. Figures 2 and 3 show typical test curves of a nonbonded and bonded test specimen. The bonded specimen was post cured and was not exposed to any environmental tests. It represents a good bonded joint. Figure 4 shows a typical curve from a bonded specimen that has been subjected to environmental tests. This test record is characteristic of a degraded bond line. The YSI break in the curve is still present but at a very low value. The last type of curve recorded in this study is shown in Figure 5. This represents a bonded specimen that has totally failed interfacially due to the environment test. Although the test record is similar to the nonbonded specimen curve (Figure 2), Figure 5 does show a larger area under the curve as would be expected.

Two types of information can be obtained from the test records generated by this study. One type deals with the definitive break in the curve and is denoted the yield strength of the interface (YSI). The second utilizes the initial linear portion of the curve. When the load value, obtained after one minute of displacement, is referenced to the nonbonded specimen, a bondline-stiffness (BLS) ratio value for the cured specimen is obtained.

The fact that the stress generated by the three-point bend test is driven to the interfacial region can be easily observed by a visual inspection. The determination of failure at a specific interface, adhesive-oxide [6] or oxide-metal, can only be determined by surface instrumentation. Figure 6 shows the surfaces of specimens that were completely opened after Instron testing. These three surfaces are responsible for the test records shown in Figures 3, 4, and 5. Figure 6a is the surface of the failed specimen considered to be a good bond. This good bond still shows a reasonable amount of adhesive failure (oxide metal surface exposed). Since a lap shear and a wedge opening test on a similarly prepared bond line showed only cohesive failure, apparently more stress is generated in the interfacial region in the short beam shear test. Figure 6b shows the surface of the low strength bond line with considerably more adhesive failure. Figure 6c shows the surface of the totally failed interface. A harsh sulfur dioxide atmosphere has generated corrosion products that are visible on these surfaces.

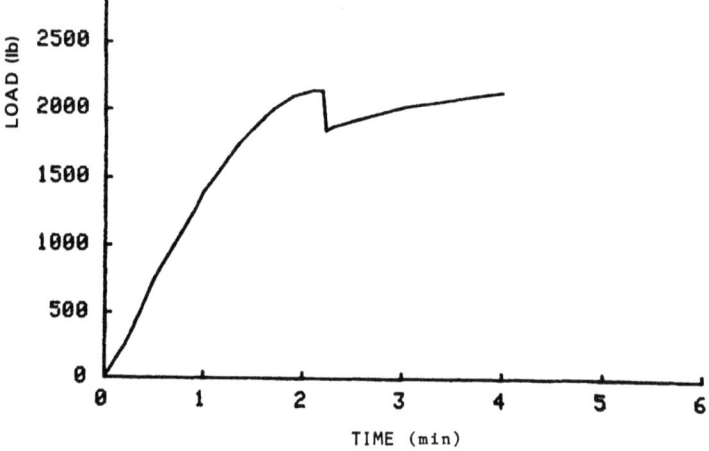

Figure 4. Load-displacement curve for test specimen with low strength interface.

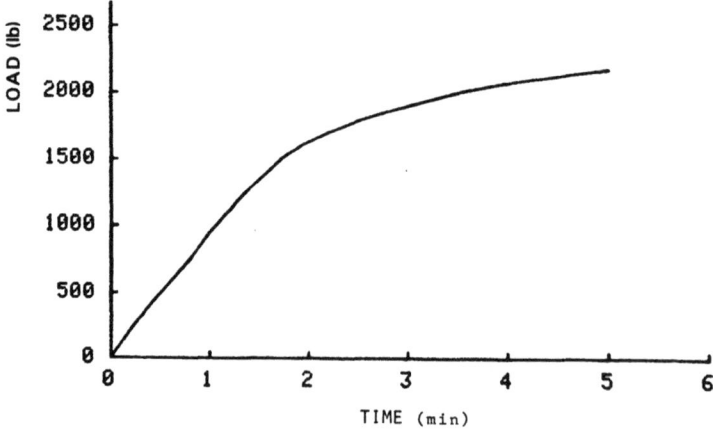

Figure 5. Load-displacement curve for test specimen with interfacial failure.

THE THREE-POINT BEND TEST FOR ADHESIVE JOINTS

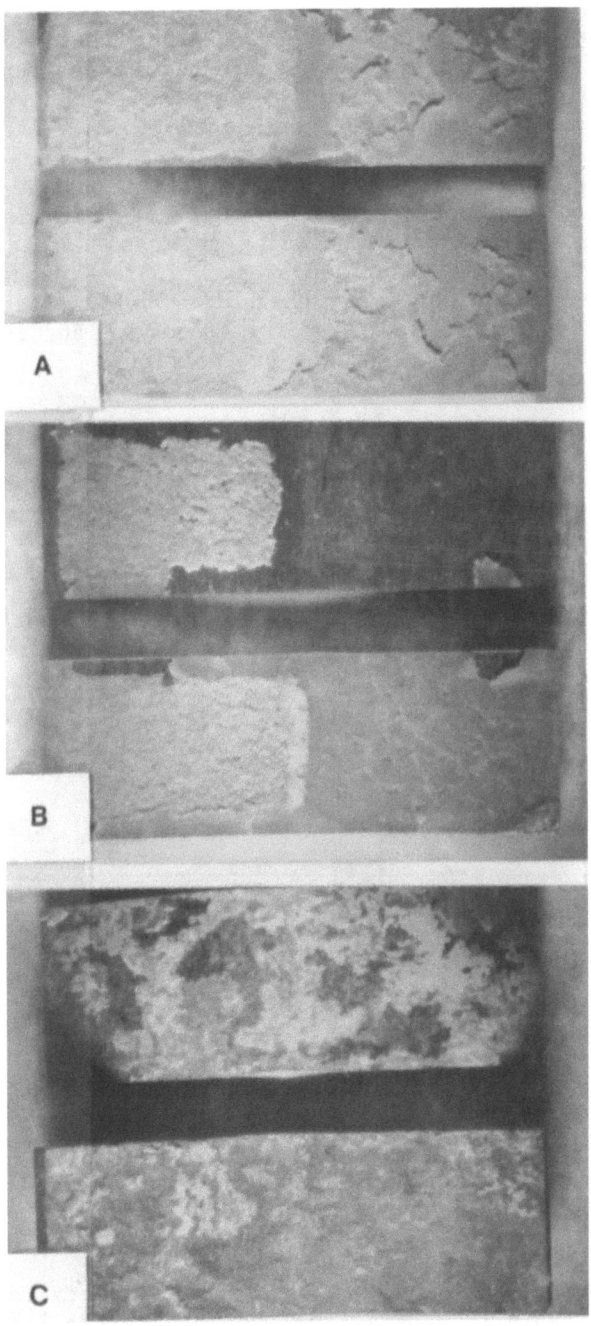

Figure 6. Low magnification photographs of test specimens (a) specimen from Figure 3, (b) specimen from Figure 4, (c) specimen from Figure 5.

The test data were generated from two groups of specimens. Table III reports* the data from the group that was cured under various conditions but were not subjected to environmental tests. Table IV reports the data from the specimens that were subject to various temperatures and atmospheric conditions.

The data spread in the YSI values for Groups A and C is very small; therefore, the average value for these data is taken as the control value. These data including the data spread are shown in Figure 7 plotted against surface preparations. The data population is small for these groups, but there is a definite indication that the 10 volt, 10 minute and 10 volt, 16 minute anodizations should provide the more reliable surface preparation.

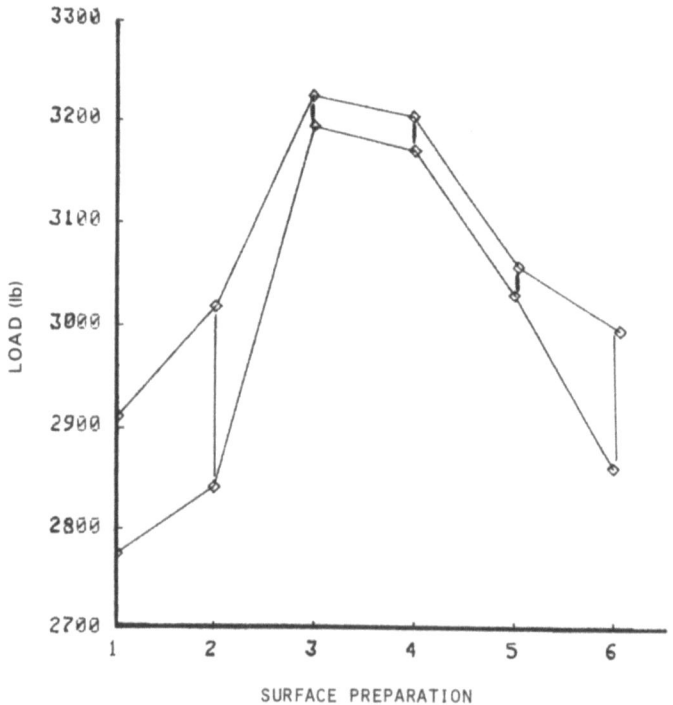

Figure 7. Spread of load values for control specimens.

Data from the specimens subjected to the environmental tests are shown in Table IV. Data from the more severe tests (E, F, and G) were averaged and plotted against surface preparation (Figure 8). The data spread is larger for the specimens exposed to environmental test, but this would be expected. The data population again is small; however, a definite trend toward lower

Table III. Yield Strength of the Interface (YSI) and Bondline Stiffness (BLS) Data from Control Specimens.

Surface Prep	A		B		C	
	BLS	YSI (lbs)	BLS	YSI (lbs)	BLS	YSI (lbs)
1	1.53	2825	2.06	3050	1.84	2758
2	1.59	3012	2.15	3552	1.80	2793
3	1.55	3200	2.12	4120	1.86	3222
4	1.58	3175	2.12	4202	1.92	3202
5	1.53	3025	2.01	3725	1.84	2900
6	1.59	2875	1.98	3650	1.85	2940

Table IV. Yield Strength of the Interface (YSI) and Bondline Stiffness (BLS) Data from Specimens Exposed to Environmental Tests.

Surface Prep	D		E		F		G	
	BLS	YSI(lb)	BLS	YSI(Lb)	BLS	YSI(lb)	BLS	YSI(lb)
1	1.53	3100	1.76	2520	----	NIF*	1.62	2455
2	1.62	2960	1.82	2760	1.76	2475	1.70	2590
3	1.60	2780	1.81	2945	1.70	2820	1.73	3045
4	1.65	3210	1.82	3145	1.70	2900	1.53	2500
5	1.59	3000	1.88	2910	1.70	2515	1.56	2640
6	----	NIF*	1.82	2400	1.69	2690	1.65	2847

*No Interface Failure (See Figure 5)

values is consistent and follows the curve for the control specimens, that is the 10V, 10 minute and 10V, 16 minute surfaces give overall better results.

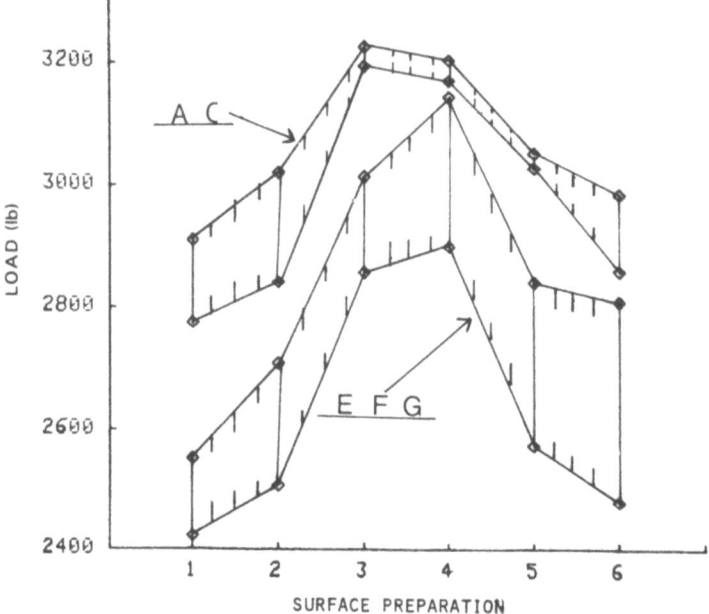

Figure 8. Comparison of the spread of load values for control and test specimens.

Before any data were obtained from the bonded specimens it was presumed that the ratio of the slopes,

$$\text{BLS} = \frac{M_b}{M_n}$$

would also give some indication as to changes in the surface preparation. Examination of the BLS values in Tables III and IV does not show any trend with respect to changes in the Surface. Instead the BLS values remain reasonably constant for each individual group of specimens. Therefore, the bond line stiffness value represents a figure of merit for the bulk adhesive and its response to each particular environment.

The fact that the YSI values give information on the interfacial region of the bond line and the BLS values relate primarily to bulk adhesive properties can be shown from the following analogy. Data were obtained previously from a wedge opening test study using an SO_2 environment. These specimens were exposed to this environment for 168 hours at room temperature.

Zero crack growth was recorded and no interfacial failures observed. Since we know the wedge opening test generates data primarily from the bulk adhesive, a zero crack growth would indicate the SO_2 environment is not very detrimental to the adhesive strength properties. At the same time we know the SO_2 environment is quite harsh for a phosphoric acid anodized aluminum alloy surface as can be seen in Figure 9[6]. More than likely the SO_2 environment has some effect on the oxide-metal interface of these specimens but the stress generated at this interface by the wedge opening test was not sufficient to cause failure.

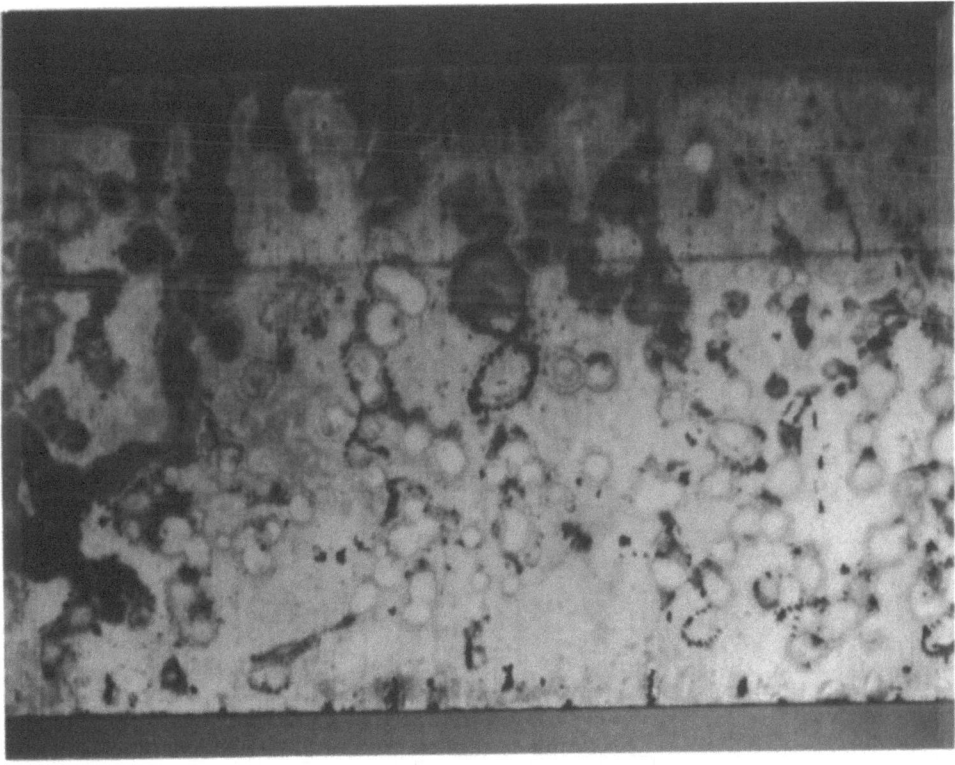

Figure 9. Phosphoric acid anodized surface of 7075-T6 bare aluminum alloy subjected to SO_2 environment.

In the present study a number of specimens were subjected to this same environment (Test E and F) and then evaluated using the short beam shear test. Since the SO_2 environment is essentially selective, having very little effect on the bulk adhesive properties but detrimental to the oxide-metal interface, we may take the BLS data as indicative of the bulk adhesive properties and the YSI data as indicative of the interfacial properties. If we use the average value of the BLS data from Test C as our control we can see there is a 1.7% change in the average BLS value of Test E. On the other hand the average value for YSI (Test E) changes by 6.4% when compared to the same value for Test C. Comparing F (the more harsh conditions) to C we can see a change of 7.5% in BLS while the YSI value changes 9.7%. It is apparent from these data that the SO_2 environment is more detrimental to the oxide-metal interface than to the bulk adhesive as was indicated by the previous study.[7] However, it is also obvious that as the effect of the environment becomes greater on the bond line (Test F) the two pieces of data obtained from this test cannot remain independent from each other.

CONCLUSION

The data obtained from this preliminary study indicate the short beam shear test can be a valuable mechanical test for adhesively bonded structures. This test appears capable of following subtle changes that may occur in the interphase between the adhesive-oxide and oxide-metal regions of a bonded structure. Analysis of detailed stress-strain data appears to give information on each component of the adhesive bond, rather that just on the resultant bonded structure.

REFERENCES

1. N. L. Rogers, National SAMPE Technical Conference, Vol. 2, p. 571 (1970).
2. N. L. Rogers, in "Proc. Symposium on Durability of Adhesive Bonded Structures," p. 57, Picatinny Arsenal, Dover, NJ, October 1976.
3. V. S. Srinath, in "Proc. Conference on Structural Adhesives and Bonding", p. 24, arranged by Technology Conferences Associates, El Segundo, CA, March 1979.
4. W. D. Bascom, Adhesives Age, p. 28, April 1979.
5. W. L. Baun, Air Force Materials Laboratory, TR-79-4126, Wright-Patterson AFB, OH, September 1979.
6. W. L. Baun, these proceedings.
7. N. T. McDevitt, W. L. Baun, and J. S. Solomon, Air Force Materials Laboratory, TR-77-184, Wright-Patterson AFB, OH, October 1977.

ACCELERATED AGING PROCEDURES FOR GLUE-WOOD BONDS AND THEIR USE FOR IN-PLANT QUALITY CONTROL

R.B. Jathar

Elmendorf Board Corporation
RFD #2, Box 303
Claremont, N.H. 03743

Bond degradation in a particulate material like waferboards or oriented strand boards bonded with synthetic adhesives like phenol formaldehyde or urea formaldehyde can be explained as a failure or the breaking of the three dimensional mechanical branching of the cured adhesive due to the stresses generated by shrinking and swelling of wood as a result of moisture changes, as well as due to the chemical degradation of the adhesive which is hydrolytic in nature. In a service environment, the principal factors causing glue-wood degradation are (1) heat, (2) moisture, (3) stresses (generated during shrinkage and swelling), (4) microorganisms, and (5) chemicals.

Various bond degradation procedures -- more commonly known as "accelerated aging procedures" have been developed over the years to assess the extent of degradation of the adhesive joint caused by factors that influence degradation in service. This paper discusses the approaches used in devising the various accelerated aging test procedures, their practicality in routine quality control and their limitations. The experimental part deals with the comparison of four different aging methods and has suggested one of them for use in in-plant quality control.

INTRODUCTION

Ever since the introduction of exterior particleboard, there has been a lack of precise information about several factors with respect to its use. Areas in which the information was needed were the strength properties, dimensional stability and durability (the ability of the boards to retain their strength properties at an acceptable level after extended periods of exposure to the weather). To seek this information it was necessary to establish reasonable levels of initial strength as well as acceptable levels of these properties after exposure.

Because of the relatively recent introduction of exterior quality composite boards into structural applications, there was not time enough to gather information resulting from long-term exposure tests for those boards put under severe conditions. Thus, products have sometimes been used on the basis of an intuitive feeling regarding their long-term durability in particular use situations.

The studies involving the problem of predicting the long-term durability have followed two broad approaches in the glued wood products industry. The first approach is to identify the factors causing degradation and attempt to determine the effect of the range of intensities of each of these factors occurring in service environments. The principal factors causing glue wood bond degradation in products using Urea Formaldehyde, Melamine Formaldehyde or Phenol Formaldehyde are: a) Heat; b) Moisture; c) Stresses generated in the glue-wood bonds during shrinkage and swelling (which are a result of moisture changes in wood); d) Micro-organisms; and e) Chemicals. Several studies (Gillespie[10], Heebink[15], Bryan and Schniewind[5], Northcott, Kreibich and Currier[18], Freeman and Kreibich[9], and Troughton[23]) have established the above factors to be the causes of degradation. An adhesive bonded wood joint in a product like plywood or in a reconstituted panel like Oriented Strand Board or Waferboard, when in service may degrade at a rate controlled primarily by only one of the above factors or more likely due to a combination or a synergistic interaction of these factors; Gillespie[10].

In light of the above, the logic behind the design of laboratory aging procedures (provided that all the materials in use are affected by the degrading factors in broadly the same way and the interactions between these factors are not markedly different in natural and artifical exposure) is that it should be possible to devise a laboratory test incorporating all the salient features of and producing the effects of several years of exposure to weather in a small fraction of this time. This is usually attempted by increasing the amount and frequency of exposure to these degrading influences. For example, increasing exposure to extreme temperatures or ultra-violet radiation or cyclic humidity conditions, without increasing their intensities beyond those occurring in natural exposure. This has also been called the performance evaluation approach, Carrol[7].

The second and purely empirical approach has been to select severe laboratory conditions which produce changes in properties which can be correlated reliably with those which occur in practice. Any one, or more than one, degrading factors chosen in such a laboratory test may, if necessary, exceed in its intensity beyond what is found in natural exposure, to bring about a change which is equivalent to and correlatable to the effect of prolonged natural exposure. Thus, the laboratory conditions chosen to accelerate the bond degradation may be totally unrealistic in nature. The present ASTM D-1037 accelerated aging test is a typical example of this approach (termed as the performance simulation approach). There is no theoretical basis for subjecting the particleboard specimens to steaming at 95°C in order to determine the probability that the bond will withstand long-term exposure to weather. Other examples of such test procedures are the German standard DIN, which is a two hour boil test, and the standard for exterior plywood PS 1-74, consisting of a four hour boil, dry and four hour boil procedures. Test procedures like these prompted remarks from manufacturers like, "We don't boil our houses," which is clearly the result of a feeling that such a test for plywood is too severe to be realistic.[8,9]

There are other problems associated with the design of a standard. One of them is the conflict between the need for a quick test to detect mismanufacture and the need for more prolonged tests with exclusion ability, i.e. one which can distinguish and exclude adhesive types known to be unsuitable in service. The second problem is the necessity of setting up glue line performance requirements or the bonding quality control tests to detect mismanufacture by simulating actual use conditions. But this procedure becomes impractical from the point of a routine quality control aspect. Thus, bonding quality requirements can be set up by using a performance simulation test procedure which can be performed within a very short period of time.

Thus, it is clear that there are several requirements that are to be satisfied by accelerated test procedures and a clear distinction between the purposes that an accelerated aging test is designed to serve is necessary. The purposes that a standard procedure can serve are:
1) A simple, low-cost procedure that provides rapid feedback information and detects mismanufacture of the sort that is known to impair the performance of the product.

2) A test procedure that distinguishes and excludes products made with adhesives known to be unsuitable in applications for which the product will be used.

3) A performance simulation test procedure that can be used to evaluate the effect of known or unknown manufacturing variables on the long-term durability of the product made with a known adhesive.

4) Similar to 3), but for products made with an unknown or new adhesive.

It is very clear that all the test procedures which would be used in a routine, in-plant quality control would fall in the first category. The objective of the experimental part of this study was, in fact, to design and select a test procedure which will satisfy the requirements of such a test for quality control.

Before going into the experimental details, a review of the various accelerated aging (A-A) treatments studied by several researchers is presented in the following section. This review also explains the reasons why I selected the four accelerated aging procedures for my experiments and comparisons.

Review of Accelerated Aging Procedures

Lehman[16] collected data from outdoor exposure tests and attempted to correlate these data with results of standard and other laboratory exposure tests. Standard exterior-grade particle-boards from four manufacturers were used for this study. Standard tests for strength properties were made in exposed and unexposed specimens and on specimens exposed to laboratory aging. The exposed specimens were tested after one-half, one, and two years. The laboratory accelerated aging procedures consisted of:

1) Vacuum Pressure Soak and Dry (VPSD) Test; five cycles.
   a) Dry at 220°F for 22 hours.
   b) Soak in water at 70°F with 25-30-inch vacuum for 30 minutes.
   c) Submerge in water at 70°F with 75 psi for 60 minutes.
2) ASTM accelerated aging; six cycles (AA).
3) West Coast Adhesive Manufacturers Association (WCAMA) Test; consisting of six cycles.
   a) Submerge in water at 70°F with 27-inch vacuum for 30 minutes.
   b) Boil in water at 210°-212°F for 3 hours.
   c) Dry at 220°F for 20 hours.

All the specimens were reconditioned at 70°F and 65% relative humidity before testing for strength properties.

It was found that the WCAMA Test was the most severe, followed in decreasing order of severity by ASTM AA and VPSD. Regression analysis was done on the data and the correlation coefficients in all cases were found to be very low except in case of VPSD vs. two year exposure. The study concluded that VPSD is the best predictor of these properties after exposure. Thus, this study proposed a new test which serves purpose Number 3, listed earlier, but is still not very much different from the standard ASTM procedure in terms of its capability to be a quick quality control test procedure which can satisfy Purpose 1.

Gressel[11] studied the behavior of phenolic bonded particleboards subjected to two-hour boil, four-hour boil, six-hour boil, and 1, 2, and 3 cycles of the French CTB Test.[26,27] The boiled specimens were tested wet, as well as after reconditioning at $20°C$ and 65% relative humidity. Matched specimens were also exposed for seven months, one and two years.

The CTB procedure is obviously not suitable as a routine quality control procedure, due to the time required to execute it (3 weeks). The boil tests showed that the strength properties did not deteriorate further after a certain period of boiling which, in case of phenolic bonded particleboards, was found to be about two hours. He also concluded that the particleboards bonded with a boil-proof adhesive which reach a plateau level of their strength properties after a certain boiling period can, in all probability, be classified to be "as weather resistant as a phenolic bonded particleboard." The reduction in strength properties after two hour boil and wet tests were found to be consistent with the reduction obtained approximately after five years of exposure. The two hour boil test was also found to be a very suitable procedure for quality control, providing rapid feedback of reliable information about the durability of the boards and was proposed for adoption as a standard test. It is now the German standard DIN 68761.

Beech[6] subjected exterior-quality particleboards from five manufacturers to outdoor exposure for two years, as well as the following laboratory exposures.

1. Weatherdrum: Panels of chipboards fixed on a ten-sided rotating drum (3 rph) were exposed to ultraviolet light and rows of water jets, which operated intermittently for 16 out of each 24 hours. The total exposure was about 3,000 hours.
2. Cyclic Test: Exposure to three days of water soak at $20°C$, one day freeze at $-12°C$ and three days drying at $70°C$. After three cycles, recondition at $25°C$ and 65% relative humidity.
3. Two hour boil and wet test, and also test after reconditioning.
4. Hot water soak ($70°C$ for five hours), both wet and dry testing.

The cyclic test results correlated very well with two years exposure. None of the other tests were as successful. Although the boiling and hot water tests were not very effective in prediction, internal bonds (IB) on wet samples gave highly significant correlations with data obtained from outdoor exposure. This suggested that two hour boil could be a simple and rapid test for predicting the suitability of exterior particleboards.

Lehman[17] designed and carried out a study to provide data on the relationships between long-term durability and accelerated aging tests for a variety of structural panels including waferboards. Outdoor exposure consisted of 13 inch by 22 inch specimens mounted vertically facing south. The accelerated aging treatments used were:
  1. ASTM D-1037 accelerated aging, 1-24 cycles.
  2. VPSD, 1-24 cycles.
  3. Spray-dry ASTM D-2829 (SD), 1-24 cycles. This consisted of four hours spraying with water at 70°F (21°C), four hours drying under heat lamps at 150°F (66°C), repeat spray dry and eight hours rest period.
  4. Two hour boil (2HB), 1 cycle.

After completion of the last wet cycle, prior to testing in all accelerated aging tests with the exception of wet boil test, all specimen were dried for three days at 120°F (49°C) to a moisture content (M.C.) of three or four percent and then tested.

All the specimens from the first and second cycle of spray-dry (SD) and AA test were found to be almost equal to six or 12 months of outdoor exposure, in respect to Modulus of Elasticity (MOE). At the end of the 24 cycles, the tests ranked in the order of severity were AA, VPSD, 2HB (1 cycle), and SD, with MOE retentions of 71, 80, 81 and 82 percent, respectively. In case of Modulus of Rupture (MOR), all the accelerated tests were found to be more severe than one year exposure, even in the first cycle. The ranking in severity after 24 cycles was found to be the same as in the case of MOE, except that 2HB was almost equivalent to AA. Internal Bonds (IB) also showed the same ranking. No correlations were found between natural and accelerated weathering tests or between the various accelerated aging tests. Lehman concluded after the study that in most cases, one or two cycles of accelerated aging were more severe than six or 12 months of natural exposure, which is the maximum period for which the boards are exposed unprotected during building construction (houses, etc.). Thus, considering the minimal losses of strength properties in this period, little concern should be placed on the durability of these panels during the construction period, provided minimal exposure to liquid water occurs after construction. Hence, he suggested that a larger emphasis be placed on a test of assurance of bond quality, which based on this study could be 2HB or VPSD, or a combination such as vacuum pressure soak -- 30 minutes under 70°F (21°C) water and 28 in. (711 mm) Hg vacuum, 60 minutes at 60 psi (0.41 mPa), two hours boil and wet test or dry at 120°F (49°C), and test.

He proposed that, for durability testing, some cyclic test with conditions closely approaching the natural exposure be adopted.

Based on the conclusions of the above discussed studies, it was decided that 2HB be used as one of the accelerated aging tests for waferboards for the purpose of this study.

An extensive amount of work has been done on the behavior of the plywood glue-line durability, its behavior in short-term accelerated aging tests, as well as in actual exposures. Northcott, Kreibich and Currier[19] compiled a review of 94 such studies dealing mostly with plywood and some dealing with laminated timber beams. This work was prompted by the Steering Committee for Accelerated Testing of Adhesives (SCATA) which was formed specifically with an objective of developing a test procedure, or several of them, by which the service life of any glued product in any specified service environment can be accurately predicted from short (a few weeks or months) laboratory tests.

In an effort to attain this goal, Northcott, Kreibich and Currier[18] studied the efficiency of eleven bond-degrade accelerating (BDA) systems, in order to select the most potential estimator of bond service life. BDA systems were methods (such as repeated cycles of wetting and drying, or continuous soaking in water) for accelerating the degradation of the bonds. The criterion for acceptability of a BDA system was that it must rank glues in the order of their durabilities in exterior service. Hence, the study used a range of highly durable to non-durable glues. The material used to evaluate the BDA systems was 3/8 inch thick plywood shear specimens. The criterion chosen for comparison of BDA systems was the number of tretment days required to reduce the breaking load of plywood shear specimen to 100 pounds. The system which required the least time would be most desirable. The study found that BDA #6, consisting of cycles of 10-minutes boil in water, followed by immersion in ice-cold water for 3.75 minutes, followed by 60 minutes of drying in a forced-air oven at 225°F, was the most promising of the eleven systems studied. BDA #6 ranked the glues closely to the predicted order of exterior durability and attained strength reduction in the least amount of time.

Based on the above study, a multiple boil-dry cycle test was developed and was adopted as a standard test (ASTM D3434). This test, however, is intended for comparing potential long-term durability of exterior adhesives rather than for use as a quality control test.

It was on the above basis that one cycle of BDA was included as one of the accelerated aging treatments for this waferboard study. From an exploratory test using commercially produced 7/16 inch thick waferboards, it was found that one cycle of BDA and then immediate measurement of MOR reduced the MOR by about 50%, which is more than the strength reduction obtained after the ASTM D-1037 standard treatment.

Shen and Wrangham[20] observed that the German standard (DIN) two-hour boil test for evaluating the internal bond retentions was not applicable to particleboards made with thick flakes, e.g., waferboards, because of the uneven swelling produced by boiling. The same conclusion was reached from studies at the Ottawa Forest Products Laboratory, done to examine the applicability of the German Standard to waferboards. Non-uniform swelling resulted in non-uniform loading of the IB specimens giving lower test values. Based on this study, the Canadian Standard applies the two-hour boil test to a bending specimen. Shen and Wrangham, however, proposed a torsion shear technique as suggested by Gertjejansen and Haygreen.[13] A good correlation between the failing torque and internal bond of the particleboard specimen was found.

The Canadian Standard for waferboards CSA 0188(75), consisting of two-hour boil, cool, and static bending test, when wet took about three hours and apparently the waferboard manufacturers were not entirely satisfied with this procedure. Hence, upon the request of the waferboard manufacturers, the Eastern Forest Products Laboratory undertook a study to develop an even quicker, yet reliable, accelerated aging test that would fulfill the requirements of an in-plant quality control test.

In an attempt to achieve the above, Shen[21] adopted the torsion shear method in combination with two-hour test treatment as a standard laboratory procedure for evaluating the bonding of exterior grade phenolic waferboard. The experimental work confirmed the existence of a correlation between torsion shear and Modulus of Rupture only when both are wet-tested. In experiments done to assess the influence of boiling period on torsion shear, it was found that the first 20 minutes of boiling of the one-inch square torsion shear specimen swelled and weakened the specimen to a great extent and further boiling did not further deteriorate the strength properties considerably. Thus, the study concluded that the two-hour boiling can be shortened to 20 minutes and the entire test completed in less than 40 minutes.

In a similar study by Gaudert and Szabo[14], to assess the effect of boiling period on other properties of waferboard, a striking result was observed. Data obtained from specimens taken from 10 commercial waferboard panels of three thicknesses showed that a plateau condition was reached in bending strength in the first five minutes of boiling. This pattern was found to be quite consistent within each panel, although in individual panels the bending strength retention varied from a low of 39% to a high of 58%, probably as a result of density variations between the boards.

So, based on the above observations, it was decided to use a 20-minute boil, cool and wet test procedure in the present study.

EXPERIMENTAL DESIGN

The number of observations per treatment to be used in the present study, or in other words the sample size, was determined by using the method outlined by Cook and Larntz.[28] For the purpose of sample size determination it was assumed that the mean of two treatments will be considered unequal if a difference of more than 75% of the standard deviation per observation is detected for any property.

The sample size indicated by this method was 40 observations per treatment. The four treatments compared were ASTM aging (AA), BDA, 2 hour boil (2HB) and 20 minute boil (20MB); the controls served as a reference. Considering the easy availability and the small expense involved, it was decided to use a sample size double the one determined statistically.

As the objective of the experiment was to study the behavior of waferboards under the four accelerated aging treatments, it was decided that a cross section of commercially produced waferboards be taken for the study.

Waferboards manufactured by two different manufacturers were collected for this study. The waferboards were taken from different press loads manufactured on different days and times of the day, and the test specimen cut in such a way that they came from different locations in the press.

The following four physical properties were evaluated after subjecting the samples to the four accelerated aging treatments. The physical properties of the control (untreated) samples were also measured.

1. MOR (Modulus of Rupture) was determined by the static bending test procedure as described in ASTM D-1037.
2. MOE (Modulus of Elasticity) was determined from the load-deflection curves obtained in static bending.
3. IB (Internal Bond), or the tensile strength perpendicular to the surface was determined from a 2" x 2" specimen per ASTM D-1037.

It was observed in the case of waferboards that the bending specimens fail visibly due to shear and that the shear spreads across the entire length of the specimen. This is especially true in the cases of those which have been subjected to accelerated aging treatments. For this reason, the test specimens for internal bond were taken from separate samples instead of cutting them from the bending specimen, as is done conventionally.

4. The Internal Torsion Shear Strength of 1-inch square samples was also determined. The testing apparatus was fitted to the Instron testing machine, as directed by Gertjejansen and Haygreen.[13]

For treatments involving boiling, a special apparatus was fabricated. Perforated steam pipes were arranged in a tray in such a way that the pipes circulated across the length and width of the tray. This ensured that there was no drop in temperature after the specimens were immersed in boiling water and also maintained the water at boiling point.

As some difficulty was encountered in testing the wet internal bond specimens, the 2-inch square specimens were face-sanded and the faces allowed to dry out to facilitate their gluing to the testing blocks. All the wet IB's were tested at moisture contents above the fiber saturation point.

The manufacturing data for the boards from the two manufacturers are given below.

Manufacturer A
> Board Thickness: 7/16 inch.
> Avg. Density: 42 pounds per cubic foot (pcf).
> Resin Content: 2.7% (solid) based on oven dry (O.D.) weight of wafers.
> Wax: 1.15%, based on O.D. weight of wafers.
> Wafer Dimensions: 0.03 in. thickness (0.75 mm.), 1-1½ inch square.

Manufacturer B
> Board Thickness: 7/16 inch.
> Avg. Density: 42.5 pcf.
> Resin Content: 2.5% to 3%.
> Wax: 2%.
> Wafer Dimensions: 0.025 in. thickness (0.63 mm.), 1-1½ inch square.

STATISTICAL ANALYSIS

Analysis of Variance (ANOVA)

As the specimens for this study were taken from different boards (15 from Manufacturer A and 3 from Manufacturer B) and also from different positions within a board, a pilot analysis was done to evaluate the board effect, the position effect, and the possible interactions of treatment by board, treatment by position, board by position and treatment by board by position. These analyses showed the board effect present in the case of IB's only. However, for the overall analysis it was decided to neglect the board effect and analysis based on a split plot design, as laid out in Figure 1, was adopted.

Manufacturer	Treatment	1	2	3	Whole Plots 4	....	30
A	AA						
	BDA						
	2HB						
	20 MB						
B	AA						
	BDA						
	2HB						
	20 MB						

Figure 1. Split Plot Design for Analysis of Variance.

In the analysis of variance, the main plot analysis is that of the two manufacturers. The sub plot analysis contains the sum of squares for the four accelerated aging treatments, the sum of squares for the manufacturer by treatment interactions and the sub plot error.

The ANOVA was performed using the program IVAN (Weisberg and Koehler[25]) using the model:

$$Y(IJK) = M(I) + E1(IJ) + T(K) + MT(IK) + E2(IJK)$$

In the above model M equals Manufacturer, T equals accelerated aging treatment, and MT equals the Manufacturer by treatment interaction. El and E2 are the errors associated with the whole plot and sub-plots respectively. The whole plot error sum of squares is used to test the hypothesis that the boards from the two manufacturers had equal strength properties averaged over treatments. The sub-plot error sum of squares was used to test the hypothesis of equality of the treatment effects and also to test the significance of the manufacturer by treatment interaction.

In the above analysis, a significant manufacturer by treatment interaction was seen to be present in the case of all properties except MOE. As the presence of a significant interaction complicates interpretation of the main effects, a separate analysis of variance was also done for each of the manufacturers to evaluate the treatment effects more accurately. The IVAN model used for this analysis was: $Y(IJ) = T(J) + E(IJ)$.

Regression Analysis

Linear regression analysis was done to find out if a simple linear relationship exists between the values after the ASTM accelerated aging and the three short accelerated aging methods investigated in this study. The regression analysis was done separately for the samples from the two manufacturers, as well as on both the manufacturers together. The following pairs of treatments were analyzed to find out if a linear relation exists.

1. ASTM Acc Aging (Y)   vs.   BDA (X)
2.   "   vs.   2HB (X)
3.   "   vs.   20MB (X)

The analysis was done separately for each of the four properties evaluated in the study, viz.: MOR, MOE, IB, and IB by the torsion shear method.

RESULTS AND DISCUSSION

Table I lists the means and coefficients of variation for the thickness, density and percent thickness swelling for the specimens from the two manufacturers.

It was observed that the specimens from Manufacturer B exhibited greater variability in the original thickness and density, but a much lower variability in the percent thickness swelling after the four accelerated aging treatments.

Table I. Means and Coefficients of Variation (C.O.V.) for Thickness, Density, and Percentage Thickness Swelling For Specimens from Manufacturers A and B.

Treatment		Thickness (in.) Mean	C.O.V.	Density (lbs./in^3) Mean	C.O.V.	% Thickness Swell Mean	C.O.V.	
Control	A	0.423	4.25	43.5	4.18	----	----	
	B	0.439	11.40	43.0	5.99	----	----	
Accelerated Aging	A	0.425	4.50	43.2	4.80	34.9	49.7	(AA)
	B	0.451	11.5	42.5	5.4	26.7	21.6	
Bond Degrade Accelerated Aging	A	0.422	4.7	43.4	3.7	33.4	20.0	(BDA)
	B	0.433	10.2	42.7	5.8	29.4	11.0	
Two Hour Boil	A	0.422	4.7	42.9	4.5	50.5	20.3	(2HB)
	B	0.446	11.2	42.9	5.2	45.0	11.4	
20 Minute Boil	A	0.422	4.7	43.2	3.5	37.5	30.5	(20MB)
	B	0.445	11.2	42.1	10.0	34.8	16.4	

The thickness changes shown in Table I illustrate that the two hour boil (2HB) produced the maximum thickness swelling (averaging over both Manufacturers) followed in decreasing order of magnitude: 20 MB, BDA and AA. For all treatments, the average thickness swelling of the specimens from Manufacturer B was observed to be smaller than that for Manufacturer A. Higher percentage of wax used by Manufacturer B probably explains this phenomenon.

Table II contains the means, coefficients of variation and the percent strength retentions after the four treatments for the Manufacturers A and B.

The results of the analysis of variance on the values of the four strength properties obtained after the four accelerated aging treatments are discussed in the following sections.

Modulus of Rupture: (Table III)

It was observed that the specimens from the two manufacturers did not differ significantly with respect to their response averaged over the four accelerated aging treatments. Comparison of the four treatment means averaged over manufacturers by linear contrasts showed that the ASTM accelerated aging treatment (AA) was significantly different and least severe compared to the average of BDA, 2HB and 20 MB. The comparison between BDA and the average of 2HB and 20 MB did not show any significant difference. The third comparison was between 2HB and 20 MB and the difference was found to be insignificant.

Comparison between the four treatment means (averaged over the two manufacturers) by Bonferroni method (Bonferroni Significant Difference or BSD) indicated that only AA was significantly different (least severe) from all the other treatments. There was no significant difference between BDA, 2HB and 20 MB.

The analysis of variance done separately for each of the manufacturers showed that for both the manufacturers AA was the least severe treatment and significantly different from the others. BDA, 2HB and 20 MB did not differ significantly.

The same conclusions were reached from the data which were transformed into logarithms. Transformation of the data was necessary, in the case of MOR, because a pattern was observed in the plot of the treatment means and the standard deviation within a treatment.

Regression analysis showed that there was no statistically significant relationship between the results of the four treatments.

Table II. Table of Strength Properties of Controls and Accelerated Aged Specimens

Treatment		MOR (psi) Mean	(C.O.V.)	%Retention	I.B. (psi) Mean	(C.O.V.)	%Retention	I.B. by T.S. (psi) Mean	(C.O.V.)	%Retention	MOE (psi) Mean	(C.O.V.)	%Retention
Control	A	3467	(19.6)	--	80.14	(26.9)	--	94.62	(25.0)	--	643,400	(15.6)	--
	B	3231	(28.4)	--	66.31	(27.9)	--	85.56	(19.3)	--	624,200	(25.0)	--
AA	A	2216	(27.3)	63.9	17.44	(66.5)	21.8	17.45	(63.2)	18.4	484,600	(21.8)	75.3
	B	2652	(26.8)	82.1	28.82	(34.0)	43.5	27.93	(43.0)	32.6	460,300	(34.9)	73.7
BDA	A	1870	(18.4)	53.9	27.30	(51.3)	34.0	28.33	(55.1)	29.9	397,500	(17.1)	61.8
	B	1620	(22.0)	50.1	22.08	(32.2)	33.3	22.15	(37.1)	25.9	326,000	(24.9)	52.2
2HB	A	1683	(21.6)	48.5	15.26	(64.4)	19.0	16.18	(49.6)	17.1	425,000	(15.7)	66.1
	B	1756	(28.1)	54.3	20.46	(31.3)	30.9	19.15	(43.2)	22.4	346,700	(22.4)	55.5
20MB	A	1742	(16.8)	50.2	23.54	(53.9)	29.4	20.67	(41.3)	21.8	407,600	(14.6)	63.4
	B	1600	(19.8)	49.5	24.94	(29.3)	37.6	21.24	(33.6)	24.8	317,100	(19.0)	50.8

Table III. Modulus of Rupture.

i.) Analysis of Variance

Source	d.f.	S.S.	M.S.	F	
Manufacturer	1	54753	54753	.1891	
Error-1	58	.16790E+08	.28949E+06		
Treatment	3	.24618E+08	.82061E+07	39.97	*
Manu. X Tr.	3	.41638E+07	.13879E+07	6.76	*
Error-2	174	.35723E+08	.20531E+06		
Total	239	.81350E+08			

*significant at 5%

ii.) Comparison Between Treatment Means for MOR using Bonferroni Significant Difference (BSD).

a) Averaged Over Manufacturers A and B

20 MB	BDA	2HB	AA
1676	1734	1757	2458

b) Means for Manufacturer A

20 MB	BDA	2HB	AA
1764	1825	1703	2245

c) Means for Manufacturer B

20 MB	BDA	2HB	AA
1587	1615	1811	2672

(Underlined means are not significantly different at 5% level).

iii.) R^2s of Regression for Modulus of Rupture

Manufacturer	AA - BDA	AA - 2HB	AA - 20 MB
A	0.0001	0.0366	0.0279
B	0.0408	0.0158	0.0705
Overall	0.0000	0.0005	0.0160

In the analysis of variance: d.f.: Degrees of Freedom; s.s.: Sum of squares of deviations; M.S.: Mean Square; F: Variance Ratio.

Internal Bond: (Table IV)

It was found from the analysis that the two manufacturers did not differ with respect to their average response over the four treatments, i.e., the mean of IBs averaged over the four treatments for Manufacturer A was not significantly different from that of B.

Using the BSD criterion, it was found that 2HB differs significantly, i.e., it is the most severe test in the case of IB. The BDA treatment was found to be least severe. However, there was not significant difference between AA, BDA and 20 MB.

From the analysis of variance done individually for A and B, it was found that 2HB was the most severe treatment on IB in the case of both manufacturers. However, for Manufacturer A, 2HB did not differ significantly from AA and 20 MB and for Manufacturer B, 2HB did not differ significantly from BDA and 20 MB. No correlation was found between any of the pairs of treatments in the case of IB's.

IB Computed by Torsion Shear Method: (Table V)

Results obtained from the combined analysis (Manufacturers A and B combined) followed exactly the same order as in the case of internal bond strengths measured by the standard method.

Results from the analysis done separately on the two manufacturers also followed the same pattern as in the case of conventional IB's. Regression analysis also gave the same results as in the case of conventional IB's.

Modulus of Elasticity: (Table VI)

In the case of MOE, the specimens from the two manufacturers were found to be significantly different. The average MOE for A averaged over the four treatments was significantly higher than the mean for B. There was no interaction between manufacturers and treatments. The mean MOE after each of the treatments was consistently higher in the case of A.

The comparison between the four treatment means averaged over the two manufacturers, using the BSD criterion, indicated that AA was significantly different (least severe) from BDA, 2HB and 20 MB. BDA was the most severe of the four treatments, but not significantly different from 2HB or 20 MB.

Because no treatment by manufacturer interaction was present, separate analyses were not performed for MOE.

Table IV. Internal Bond.

i.) Analysis of Variance

Source	d.f.	S.S.	M.S.	F	
Manufacturer	1	203.18	203.18	.8497	
Error-1	56	13391	239.12		
Treatments	3	1682.8	560.93	9.498	*
Manu. X Tr.	3	1544.0	514.67	8.715	*
Error-2	168	9921.7	59.057		
Total	231	26742			

*significant at 5%

ii.) Comparison Between Treatment Means for IB using BSD.

a) Averaged over Manufacturers A and B

```
        2HB    20 MB    AA    BDA
        18.11  23.22    23.85 25.23
               ─────────────
```

b) Means for Manufacturer A

```
        2HB    AA     20 MB   BDA
        16.25  19.15  22.89   28.01
               ─────────────
```

c) Means for Manufacturer B

```
        2HB    BDA    20 MB   AA
        19.97  22.44  23.55   23.19
               ─────────────────────
```

(Underlined means not significantly different at 5%)

iii.) R^2s of Regression for IB

Manufacturer	AA - BDA	AA - 2HB	AA - 20 MB
A	0.0640	0.3913	0.2107
B	0.0455	0.0242	0.0130
Overall	0.0067	0.2743	0.1129

Table V. Internal Bond by Torsion Shear.

i.) Analysis of Variance.

Source	d.f.	S.S.	M.S.	F
Manufacturer	1	.78478E+01	.78478E+01	.3848E+03
Error-1	44	8973.0	203.93	
Treatments	3	1466.7	488.91	7.362 *
Manu. X Tr.	3	1580.8	526.94	7.935 *
Error-2	132	8765.8	66.408	
Total	183	20786		

* significant at 5%

ii.) Comparison Between Treatment Means for IB by TS using BSD.

a) Averaged over Manufacturers A and B

2HB	20 MB	AA	BDA
18.8	22.0	22.8	26.73

b) Means for Manufacturer A

2HB	AA	20 MB	BDA
18.32	18.71	22.51	30.87

c) Means for Manufacturer B

2HB	20 MB	BDA	AA
19.28	21.48	22.59	26.89

(Underlined means are not significantly different at 5%)

iii.) R^2s of Regression for I.B. by Torsion Shear.

Manufacturer	AA - BDA	AA - 2HB	AA - 20 MB
A	0.4849	0.1838	0.0104
B	0.0252	0.0006	0.2604
Overall	0.0754	0.0537	0.2091

Table VI. Modulus of Elasticity

i.) Analysis of Variance

Source	d.f.	S.S.	M.S.	F	
Manufacturer	1	.27881E+12	.27881E+12	21.08	*
Error-1	42	.55545E+12	.13225E+11		
Treatments	3	.42987E+12	.14329E+12	14.51	*
Manu. X Tr.	3	.37502E+11	.12501E+11	1.266	
Error-2	126	.12439E+13	.98720E+10		
Total	175	.25455E+13			

* indicates significance at 5%

ii.) Comparison Between Treatment Means for MOE using BSD.

Averaged Over Manufacturers A and B

	BDA	20 MB	2HB	AA
	<u>350800	354100	376200</u>	472300

(Underlined means not significantly different at 5%)

iii.) R^2s of Regression for MOE.

Manufacturer	AA - BDA	AA - 2HB	AA - 20 MB
A	0.0073	0.0414	0.0387
B	0.0003	0.1443	0.0258
Overall	0.0011	0.0386	0.0123

No correlation was found between any of the pairs of treatments in the case of MOE's.

CONCLUSION

The study has shown that for the properties MOR and MOE, the three alternative accelerated aging tests, viz., 2HB, 20 MB and BDA are significantly more degradative in effect than the present standard ASTM D-1037-72a accelerated aging treatment. This behavior was found to be quite consistent irrespective of the manufacturer from which the specimens were taken, the boards from which the specimens were cut and the position of the specimens in the press.

The three alternative treatments, viz., 2HB, 20 MB and BDA were almost equally degradative in effect.

It can be concluded that all of the alternative treatments investigated in this study could be used as substitutes for ASTM accelerated aging test with respect to the bond degrading effect on MOE and MOR. If values of MOE and MOR obtained after ASTM aging indicate mismanufacture in the test specimens, the alternative aging treatments investigated in this study will also yield values of MOR and MOE which will lead to the same conclusion.

In the case of internal bond strength (measured conventionally and by the torsion shear method), the 2HB (two hour boil, cool and test wet) treatment was found to be significantly more degradative in effect when the values for the manufacturers were averaged together. The IB values, however, exhibited a variation depending on the boards from which the specimens came and also to some extent the position of the specimens in the press. This highly variable behavior of waferboards with respect to internal bond strength is not unusual and has been observed by others. For this reason, Shen and Wrangham[20] suggested that the two hour boil test should not be applied to IB specimens.

Thus, if a test procedure were to be selected based on the findings of this study, any one of the three alternative treatments will satisfy the requirements of a quick in-plant quality control test providing rapid feedback regarding bond quality. Also, these treatments are inexpensive to perform as compared to ASTM aging treatment. These tests could be coupled with static bending providing a measure of MOE and MOR. The study also shows that 20 MB can accomplish the quality control objectives as well as the 2HB treatments and faster. Thus, 2HB could be eliminated in any choice to be made between the three alternative tests.

Between 20 MB and BDA, from a practical standpoint, BDA has an advantage over 20 MB, because the one hour drying in the BDA treatment removes the difficulty encountered in wet testing of IB specimens (if necessary). Also, the BDA treatment (10 minutes boil, 4 minutes in ice cold water and 1 hour drying at 212°F) does not take much more time than the 20 MB (20 minutes boil, 30 minutes cooling with water and wet test).

ACKNOWLEDGMENTS

I wish to express my sincere appreciation to Dr. John G. Haygreen, Department of Forest Products, University of Minnesota, for the guidance provided for the above study.

I would also like to gratefully acknowledge the encouragement and support provided by Paul J. Heenan, President, Elmendorf Board Corporation, Claremont, New Hampshire, to participate in the ACS symposium in Kansas City.

REFERENCES

1. American Society for Testing and Materials, "Standard Methods of Evaluating the Properties of Wood Base, Fiber and Particle Panel Materials," ASTM Designation D 1027-72a, 1972.
2. Deutsche Normen, "Flachpresplaten fur das Bauwesen," DIN 68763, Beuth Vertrieb GmbH, Berlin 30, 1971.
3. Anonymous, "Mat Formed Wood Particleboard," Commercial Standard CS-236-66, U.S. Department of Commerce, 1966.
4. Anonymous, Forest Products J., 16(6), 19 (1966).
5. E. L. Bryan and A. P. Schniewind, Forest Products J., 15(4), 143 (1965).
6. J. C. Beech, "An Accelerated Test for Predicting the Performance of Chipboards for Exterior Use", Research Paper IS, Princess Risborough Laboratory, Buckinghamshire, U. K., 1973.
7. M. N. Carrol, Forest Products J., 28(5), 23 (1980).
8. M. N. Carrol, "We Still Don't Boil Houses: Part II. Test Procedures for Particleboard used in General Building Construction", Paper presented at Washington State University Particleboard Symposium, 1980.
9. H. G. Freeman and R. E. Kreibich, Forest Products J., 18(7), 39 (1968).
10. R. H. Gillespie, Forest Products J., 15(9), 369 (1965).
11. P. Gresel, "Utersuchungen an Freiberwitterten Spanplatten. Holz als Roh-und-Werkstoff", 27(10), 142 (1968).
12. P. Gresel, "Utersuchungen an Freiberwitterten Spanplatten. Holz als Roh-und-Werkstoff", 27(10), 366 (1969).

13. R. Gertjejansen and J. G. Haygreen, Forest Products J., 21(11) 59(1971).
14. P. Gaudert and T. Szabo, "Technical Report", Forintek Canada Corp., Ottawa, 1980.
15. B. G. Heebink, Forest Products J., 17(1), 59 (1967).
16. W. F. Lehman, "Durability of Exterior Particleboard", Washington State University Particleboard Symposium, No. 2, 275-305, 1968.
17. W. F.Lehman, "Durability of Composition Board Products", Proceedings of Particleboard Symposium, No.11, 351-368, 1977.
18. P. L.Northcott, R. E. Kreibich and R. A. Currier, Forest Products J., 18(5), 58 (1968).
19. P. L. Northcott, R. E. Kreibich and R. A. Currier, "Supplemental Bibliography and Literature Review Report No. 1", Task Group on Accelerated Treatment Design, SCATA, Unpublished Report, 1964.
20. K. C. Shen and D. Wrangham, Forest Products, J., 21(5), 230 (1971).
21. K. C. Shen, "A Proposed Rapid Accelerated Aging Test for Exterior Waferboards", Eastern Forest Products Lab, Ontario, Information Report OP-X-160E, 1977.
22. G. W. Snedocor and W. G. Cochran, "Statistical Methods, Sixth Ed.," Iowa State University Press, Ames, Iowa, 1967.
23. G. E. Troughton, J. Wood Sci., 1(3), 172 (1967).
24. G. E. Troughton and S. Z. Chow, J. Wood Sci., 4, 29 (1968).
25. S. Weisberg and K. J. Koehler, "IVAN Users Manual", University of Minnesota, School of Statistics, 1976.
26. Centre Technique du Bois, "Marque de Qualitie des Panneau de Particles", CTB-H, Agglos 454, C.T.B., Paris, 1975.
27. Association Francaise de Normalization, "Epreuve de Vieillissment Accelere Par La Method Dit V313", NF 51263, AFNOR, Paris, 1972.
28. R. D. Cook and K. Larntz, "Sample Size Determination for 'Fixed Effects' ANOVA Models", Technical Reprot #212, University of Minnesota, 1973.

PEEL-STRENGTH AND ENERGY DISSIPATION

T. Igarashi

Department of Technology
Gunma University
Kiryu, Gunma 376, Japan

The effect of energy dissipation on peel-strength is discussed on the basis of mechanics. In the case of T-peeling of specimens consisting of two flexible tapes bonded by elastic adhesive, steady peel-strength is simply given as $2f = \Gamma$, where f is the peeling force per unit width of the tape and Γ is twice the surface energy of the fracture surface. However, in the cases where the energy is dissipated in the adhesive layer or in the adherends, another term w' representing the energy dissipation is to be added to the peeling force, i.e. $2f = \Gamma + 2w'$. The energy w' associated with the deformation of the adhesive layer is given by the work done at the tip of the crack per unit length of the adherend during a cycle of the elongation to fracture and subsequent recovery of the adhesive. This energy w' can be estimated from an area u' enclosed by the stress-strain hysteresis curve obtained from extention test of the adhesive and the thickness h of the adhesive layer, as $2w' = hu'$. The values of u' obtained from experimental studies on the dependence of peeling force on the thickness of the adhesive layer agree with these obtained from extention test of the adhesives. On the other hand, the energy w' associated with the energy dissipated in the adherends is given by the area enclosed by moment-curvature hysteresis curve obtained from a bending test of the adherends. Especially in the case of elastic-plastic adherends such as mild steel, w' is approximately given as $w' = 3EI\theta_e\theta_p$, where E is the Young's modulus, I is moment of inertia of the cross section, θ_e is the bending curvature at yielding, and θ_p is the curvature of the adherends during the peeling. Experiments show that the difference between the measured peeling force for the specimens consisting of the mild steel adherends and for the specimens consisting of flexible canvas adherends, which corresponds to the energy dissipation in the mild steel adherends, agree with the values calculated using the relation $w' = 3EI\theta_e\theta_p$.

LIST OF SYMBOLS

- E Young's modulus of adherend plate
- f peeling force per unit width
- F free energy of a system
- F_e free energy of elastic deformation of a system
- F_s surface free energy of a system
- h thickness of adhesive layer
- I moment of inertia of cross section of adherend plate
- M bending moment of adherend plate
- M_e bending moment of adherend plate at which yielding of the plate just takes place
- M_p bending moment of adherend plate during steady peeling
- R radius of curvature of adherend
- t thickness of adherend plate
- u elastic strain energy per unit volume of adhesive to be recovered during the recovery of the deformation
- u' energy dissipated per unit volume of adhesive during deformation and the subsequent recovery
- U' energy dissipated in a system during deformation and the subsequent recovery
- w' energy dissipated per unit width and unit length of adhesive layer or adherend plate during deformation and subsequent recovery i.e. U' for peeling specimen with unit width
- x distance measured along direction of propagation of peeling
- y distance measured from neutral line of bending plate
- y_e y at elastic-plastic boundary
- Γ twice surface energy per unit area of fracture surface
- ε normal strain, i.e., elongation
- $\dot{\varepsilon}$ $d\varepsilon/dt$
- ε_b fracture elongation
- ε_e yield elongation
- ε_p maximum elongation corresponding to maximun curvature of adherend during steady peeling
- θ curvature of adherend
- θ_e curvature of adherend plate at which yielding of the plate just takes place
- θ_p maximum curvature of adherend plate during steady peeling
- θ_r maximum residual curvature of adherend plate after peeling

Θ   peel angle
σ   stress
σ_b fracture stress
σ_e yield stress
σ_p maximum stress corresponding to maximum curvature of adherend during steady peeling

INTRODUCTION

The original study of peeling of flexible tapes on the basis of mechanical point of view was carried out by Hata.[1] For the peeling of a flexible, inextensible tape bonded to a rigid surface, he proposed a very simple and elegant relation,

$$f = \Gamma/(1 - \cos\Theta) \qquad (1)$$

where f is the peeling force per unit width of the tape and Θ is the peel angle. Γ is named as the work of adhesion, which is equivalent to twice the surface energy of fracture surface (Figure 1).

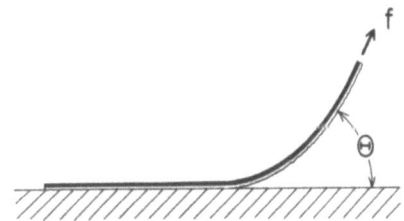

Figure 1. Peeling of flexible tape from a rigid surface.

This important relation is the basis of later studies of peeling.
Further studies on peeling were made by Bikerman,[2] Spies,[3] Kaelble,[4] and Yurenca.[5] In their treatments, the deformation behavior of the adherends was assumed to obey an elastic equation for bending plates. Gardon[6] pointed out that steady peel strength depends on the energy of fracture, and derived a relation between the glue line thickness, the diameter of the tear tip, and the tear energy. Orowan[7] has mentioned that the strength of adhesive joints is derived from the viscoelastic deformation of adhesives. He also pointed out the importance of energy dissipation. Gent and Petrich[8] have discussed the role of the energy dissipation during the deformation of viscoelastic materials.

In the present paper, we will consider the effect of energy dissipation in both of the following cases: case one when the energy is dissipated in the adhesive layer[9], and in the other it is dissipated in the adherends.[10]

BASIC EQUATION OF PEELING

In the case of T-peeling of specimens consisting of two flexible tapes, such as canvas, bonded by an elastic adhesive (Figure 2), the steady peeling force, i.e., force required to continue the peeling, is derived from the following thermodynamic condition

$$\delta F = \delta F_e + \delta F_s = 0 \qquad (2)$$

where F is the free energy of the system, F_e is the free energy of the elastic deformation and F_s is the surface free energy.

As the work done by the external force f is stored as the potential energy i.e. F_e, the following expression is obtained for the specimen with unit width

$$-F_e = 2\int_0^{(\varepsilon_b)} f dx \text{(application)} \qquad (3)$$

where the integration with respect to the displacement x of the point of the application is taken to the point at which the elongation of the adhesive layer reaches the breaking point. The variation of the surface free energy accompanied by the progression of the crack dx is given by

$$\delta F_s = \Gamma \delta x \text{(crack)} \qquad (4)$$

where Γ is twice the surface free energy per unit area. From equation (3)

$$-\delta F_e = -2\delta \int f dx \text{(application)} \qquad (5)$$

$$= -2 \int \frac{\partial f}{\partial x \text{(crack)}} dx \text{(application)} \delta x \text{(crack)} \qquad (6)$$

Because the adherends are assumed to be inextensible,

$$dx \text{(application)} = dx \text{(crack)} \qquad (7)$$

Equation (6) is integrated with respect to df to the peeling force $f(\varepsilon_b)$,

$$-\delta F_e = -2f(\varepsilon_b)\delta x \text{(crack)} \qquad (8)$$

From Equations (2),(4) and (8), omitting the parentheses for F and x, one can obtain

$$2f = \Gamma \qquad (9)$$

For peeling accompanied by energy dissipation one may adopt the following expression instead of Equation (2)

PEEL-STRENGTH AND ENERGY DISSIPATION   423

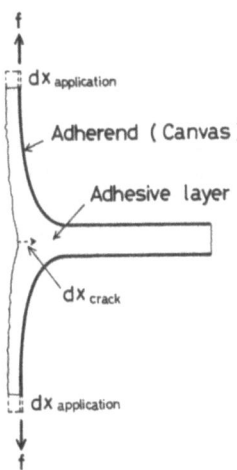

Figure 2. T-peeling specimen with flexible, inextensible adherends and rubbery adhesive layer.

$$\delta F = \delta F_e + \delta F_s + \delta U' = 0 \qquad (10)$$

where U' is dissipation energy. If the energy is dissipated in the adhesive layer, U' becomes for the specimen with unit width

$$\delta U' = hu\delta x(crack) \qquad (11)$$

where h is the thickness of the adhesive layer and u' is the energy dissipated per unit volume of the adhesive during the deformation to fracture and subsequent recovery, i.e. the area enclosed by stress-strain hysteresis loop obtained from extention test of the adhesive (Figure 3).

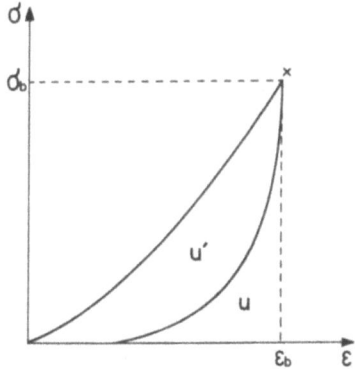

Figure 3. Stress-strain curve of adhesives. Materials consisting of adhesive layer are prepared for simple extension test. Specimen is first ex-

tended to the elongation just before its breaking point and is then returned immediately. In actual peeling process, energy corresponding to u' is dissipated in the adhesive layer and that to u is stored as elastic strain energy of residual adhesive layer.

With the same consideration as mentioned above, Equations (10) and (11) lead to the following relation instead of Equation (9)

$$2f = \Gamma + hu' \qquad (12)$$

On the other hand, let us consider peeling of specimens consisting of inelastic adherends bonded with elastic adhesive. In these cases energy may be dissipated not in the adhesive layer but in the adherends during the propagation of peeling. With propagation of peeling, any part of the adherend is subjected to one cycle of deformation of bending and the subsequent recovery(Figure 4). The

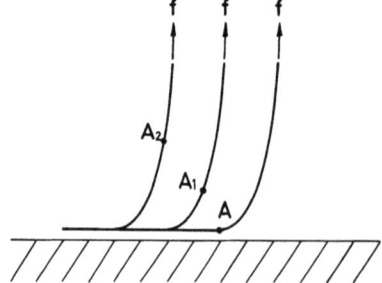

Figure 4. Propagation of peeling. Let us fix our eyes on any point A fixed on the adherend. With the start (A), is bent (A_1), and then it is placed from A to A_1 and A_2. Accordingly the adherend at this point, which is straight at the start(A), is bent (A_1), and then it is straightened again (A_2).

energy dissipation may be associated with this deformation. Then same treatment as above, peeling force becomes

$$2f = \Gamma + 2w' \qquad (13)$$

where w' is the energy dissipation per unit width and unit length of the adherend during the bending cycle.

DETERMINATION OF DISSIPATION FUNCTION FOR BENDING

The energy dissipation function w' can be obtained from the relation between the bending moment M and bending curvature θ of the adherend as

$$w' = \oint M d\theta \qquad (14)$$

where the integration should be carried out for a cycle of the moment- curvature hysteresis loop(Figure 5). Let us estimate this

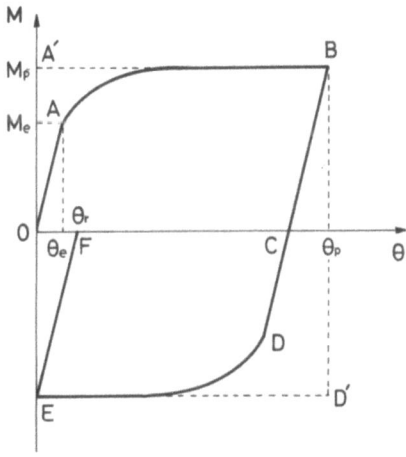

Figure 5. Bending moment and curvature of adherend. Corresponding to elastic deformation of adherend, bending moment of adherend plate increases linearly with increase in its curvature(OA). After beginning of yielding, increase in the bending moment levels off to M_p corresponding to steady peeling (AB). With propagation of peeling, curvature of adherend decreases. Curve BC corresponds to "spring back" of the adherend plate.
Curve CE corresponds to "stretching" process and shape of the curve resembles bending process. θ_r is the residual curvature of adherend of the specimen after it has been removed from the cramp.

value for mild steel adherend assuming it behaves as an elastic-plastic body. In Figure 6 a schematic stress-strain curve for mild steel is shown with a solid line. To reduce the difficulties in mathematical analysis it may be convenient to approximate the stress-strain curve for mild steel to the broken line representing elastic-plastic property.

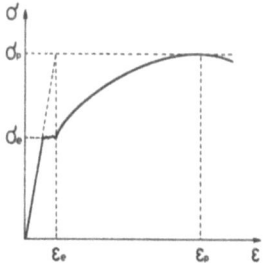

Figure 6. Stress-strain curve of mild steel. Actual stress-strain curve is approximated to broken line in calculation. ε_e means effective strain and σ_e is the corresponding stress. ε_p and σ_p are strain and stress respectively corresponding to maximum curvature of adherend during steady peeling.

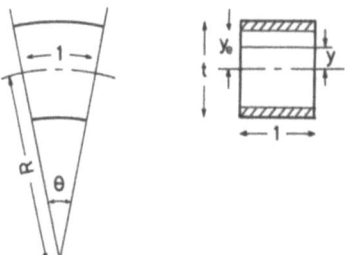

Figure 7. Side view and cross section of the adherend plate. As unit length of the adherend is shown in this side view, θ means the curvature and R is the radius of curvature of the adherend. The specimen has unit thickness. Yield areas are shown by hatch in the cross section. y_e is the coordinate of elastic-plastic boundary.

The bending moment of a plate of the adherend is generally given by

$$M = \int_{-t/2}^{t/2} \sigma(y)\,dy \tag{15}$$

where t is thickness of the adherend plate, y is the distance measured from the neutral line, and $\sigma(y)$ is the stress directed along the length of the plate (Figure 7).

Within the linear proportional part of the stress-strain curve in Figure 6, the relation

$$\sigma(y) = \varepsilon(y)E \tag{16}$$

is valid, where $\varepsilon(y)$ is the elongation and E is Young's modulus of the adherend. From Equations (15),(16) and the following geometrical relation from Figure 7:

$$\varepsilon(y) = y\theta, \tag{17}$$

we obtain the relation,

$$M = EI\theta \tag{18}$$

where the moment of inertia of cross section I of the adherend is given by

$$I = t^3/12 \tag{19}$$

Equation (18) corresponds to the linear part OA of the M-θ curve in Figure 5.

After yielding of the adherend has taken place, the integration of M should be separated into two terms; one is for the elastic part and the other is for the plastic part, i.e., the yielded part. In the latter, the σ-ε relation is assumed to be

$$\sigma(x) = \text{const.} = \varepsilon_e E \tag{20}$$

where ε_e is yield elongation. Then,

$$M = E\theta \int_{-y_e}^{y_e} y^2\,dy + 2\varepsilon_e E \int_{y_e}^{t/2} y\,dy \tag{21}$$

$$= \frac{2}{3} E\theta y_e^3 + \varepsilon_e E \left(\frac{t^2}{4} - y_e^2 \right) \tag{22}$$

where y_e is y at the elastic-plastic boundary. From Equation (17), y_e is expressed as

$$y_e = \varepsilon_e/\theta \tag{23}$$

On the other hand, by representing the curvature at the initiation of yielding by θ_e, also from Equation (23), and substituting $t/2$ for y_e one obtains

$$t/2 = \varepsilon_e/\theta_e \tag{24}$$

Hence, from Equations (19),(23) and (24), M becomes

$$M = \frac{1}{2} EI\theta_e \left\{ 3 - \left(\frac{\theta_e}{\theta}\right)^2 \right\}, \quad (\theta \geq \theta_e) \tag{25}$$

Equation (25) gives the curve AB in Figure 5, where θ_p is the maximum curvature of the adherend during the peeling.

Here, let us summarize the properties of the mild steel plate used as adherends in our experiments. The mechanical properties are shown in Table I. The relation between stress and strain is

Table I. Mechanical Properties of Mild Steel.

Tensile Strength	$35 kg/mm^2$
Yield Strength	20 "
Yield Elongation	0.08–0.1 %
Young's Modulus	$21,000 kg/mm^2$

just like that shown in Figure 6. At an elongation of 0.15 - 0.2%, about $30 kg/mm^2$ of stress is observed, and then the stress gradually increases with increase in strain. ε_e in Figure 6 is thought to be the effective yield strain and is estimated to be 1.5×10^{-3} (0.15%). The value of curvature at the initiation of yielding θ_e is estimated to be $0.1 (cm^{-1})$ for an adherend with thickness 0.3mm from Equation (24). The value of EI is estimated to be 4.7kgcm from Equation (19) and the value of Young's modulus shown in Table I.

Equation (25) indicates that value of the moment will increase with increase in θ after the initiation of yielding and will level off for $\theta > 2\theta_e$. However, the value of the maximum curvature θ_p is estimated to be about 2 (cm^{-1}). Hence the value of the moment after initiation of yielding can be regarded as constant, which is given from Equation (25) and is denoted with M_p:

$$M_p = \frac{3}{2} EI\theta_e \tag{26}$$

Further, taking account of the experimental fact that

$$\theta_p/\theta_e = 20 \tag{27}$$

one can finally approximate the M-θ hysteresis loop to a rectilinear curve O→A'→B→D'→E in Figure 5. Thus the dissipation function w' of Equation (14) is simply given as

$$w' = 2 M_p \theta_p \quad (28)$$

$$= 3EI\theta_e \theta_p \quad (29)$$

EXPERIMENTAL AND RESULTS

Specimens for the peel tests are prepared from hot-pressed Vinylon-Nylon canvas (adherend No.1-No.4 in Table II) and mild steel plates (adherend No.5 in Table II) bonded with a rubbery adhesive layer (Table III). The canvas is treated with "Resorcinol-formaline-latex" prior to press. On the other hand, the surface of mild steel plates are treated with "Chemlock"(Hughson) to promote affinity for rubbery adhesive layer. The rubber vulcanizates used as the adhesive layer are tested by simple extension to fracture. The specimens are prepared with 2mm width and cut with an ASTM C dumbell die. The average values of fracture elongation and stress (ε_b and σ_b) are first measured. Then, other specimens are elongated to these average values, and if they did not fracture they are immediately unloaded at the same speed to obtain stress-strain curve and to evaluate energy dissipation. Results of the tensile tests

Table II. Preparation and Properties of Specimens.

Specimen No.	Adherend	Adhesive		Steady Peeling Force (kg/cm)
		No.	h(mm)	
1	Canvas	G1	1 ↓ 3	dep. on thick.#
2	Canvas	G2	0.5 ↓ 3	dep. on thick.#
3	Canvas	G3	0.5 ↓ 3	indep. of thick.# 4.0
4	Canvas	G4	1	indep. of thick. 3.6 → 5.3
5	Mild steel	G4	1	10.9 → 13.4

Cross-head speed:100mm/min for No.1-No.3, 50mm/min for No.4&No.5
#: for the dependence of peeling force on thickness, see Figure 8.

Table III. Composition of Rubber Adhesives.

	PHR for			
	G1	G2	G3	G4
BR	—	40	40	—
SBR	100	60	60	100
HAF Carbon	45	—	—	—
FEF Carbon	—	1	—	—
MT Carbon	—	—	20	—
Diatomaceous Earth	30	20	—	—
CaCO$_3$ (surface-treated)	—	40	—	—
Chalk	—	50	—	—
Process Oil	30	15	—	—
ZnO	5	5	5	5
Stearic Acid	1.5	1.5	1.5	1.5
S	2	2	2	2
TMTD	0.1	0.1	3	0.2
CZ	1	1	—	1

150°C, 15min Press Cure.

Table IV. Results of Tensile and Peeling Tests.

Specimen No.	Tensile			Peel	
	σ_b(kg/cm)	ε_b	u'(kg/cm^2)	Γ(kg/cm)	u'(kg/cm^2)
1	155	7	413	9.0	220
2	25.6	7.3	29	8.5	27
3	20	1.0	0	4.0	0

Strain rate($\dot{\varepsilon}$)=50min^{-1}.

for these rubber vulcanizates are shown in Table IV. Results of the peeling tests with the specimens No.1– No.3 are shown in Figure 8 and Table IV. The dependence of peeling forces on the thickness of adhesive layers lead to the values of u' using Equation (12). The values of Γ are obtained using extrapolation (h→0) in Figure 8.

In the cases of the peeling tests with specimen No. 4 and 5, difference in the peeling forces resulting from the difference of their adherends become the point of consideration (Table V). To calculate the values of 2w' using Equation (29), the maximum values of curvature of adherends during peeling θ_p are observed photographically.

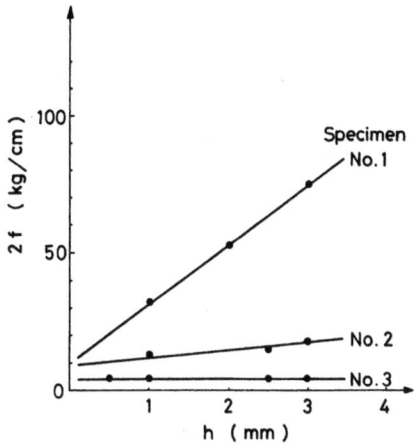

Figure 8. Dependence of peeling force on thickness of the adhesive layer. Adhesive layer with considerable energy dissipation:No.1, with medium energy dissipation:No.2, without energy dissipation: No. 3. All adherends of the specimens in this figure are made of canvas.

Table V. Effect of Difference in Adherends.

v [#] (mm/min)	θ_p (obs.)	2f(kg/cm) (obs.)		2w'	
		Specimen No.		Calculated using Eq.(29)	obs. [##]
		4 (Canvas)	5 (Steel)		
1000	2.1	5.3	13.4	5.9	8.1
50	1.8	5.3	13.2	5.1	7.9
5	2.2	3.6	10.9	6.1	7.3

[#] v:cross-head separation speed.
[##] Observed values of 2w' are obtained from the difference between observed 2f for (Steel) and that for (Canvas).

DISCUSSION

The relation between peeling force and thickness of the adhesive is found experimentally to be linear as expected from Equation (12) in the case of specimens with canvas adherends. In particular with the adhesive with no energy dissipation, the strength is independent of the thickness of the adhesive as expected from Equation (12) by assuming u'=0. These results signify that in the case of fracture of materials with energy dissipation, external force is required to do the additional work associated with the deformation to break the materials even in the steady peeling.

Agreement between the values of u' obtained from stress-strain curve of simple extension tests and those obtained from the dependence of peeling force on the thickness using Equation (12) is fairly good (Table IV).

In the same manner, energy dissipated in the adherends will also make a contribution to the peel strength. Here the difference in the peeling force between specimen No.4 and No.5 is thought to result from the energy dissipated in the elastic-plastic adherends of mild steel (Table V). Agreement of 2w' calculated using Equation (29) with those obtained from experiment is rather satisfactory. The "tough" materials will become "tough" composites when they are bonded with the same adhesive.

Thus the significant role of energy dissipation in fracture strength is apparent from this study.

Finally it should be noted that the problems discussed here are exclusively concerning the propagation of peeling. The force required to initiate the peeling is generally not equal to the steady peeling force [11,12].

REFERENCES

1. T.Hata, Kobunshi Kagaku (J.Polym.Chem.Japan) $\underline{4}$, 67 (1947)
2. J.J.Bikerman, J.Appl.Phys. $\underline{28}$, 1484 (1957)
3. G.J.Spies, Aircraft Eng. $\underline{25}$, 64 (1953)
4. D.H.Kaelble, Trans. Soc. Rheol. $\underline{3}$, 161 (1959); ibid. $\underline{4}$, 45 (1960)
5. S.Yurenca, J.Appl.Polym.Sci. $\underline{6}$, 136 (1962)
6. J. L. Gardon, in "Treatise on Adhesion and Adhesives", Vol. 1, R. L. Patrick, Ed., Marcel Dekker, Inc., New York, 1967.
7. E.Orowan, J.Franklin Inst., $\underline{290}$, 493 (1970)
8. A.N.Gent and R.P.Petrich, Proc. Royal Soc. (London) Ser.A $\underline{310}$, 433 (1969)
9. T.Igarashi, J.Polym.Sci.Polym. Phys.Ed. $\underline{13}$, 2129 (1975)
10. T.Igarashi, Nihon Setchaku Kyokai Shi (J.Adhesion Society Japan) $\underline{9}$, 1 (1973)
11. T.Igarashi, J.Polym.Sci.Polym.Phys.Ed. $\underline{16}$, 407 (1978)
12. T.Igarashi, in "Adhesion and Adsorption of Polymers", L.-H.Lee, Editor, p.421, Plenum Press, New York, 1979.

THE ESTIMATION OF ADHESION IN FILLED POLYMER SYSTEMS

Yu. S. Lipatov, T. T. Todosiychuk, P. K. Tsarev and
L. M. Sergeeva

Institute of Macromolecular Chemistry
The Ukrainian SSR Academy of Sciences
252160 Kiev, USSR

Some new methods of estimating the adhesion strength between filler particles and polymer matrix are proposed. All the methods are based on the determination of stresses at which the continuity of samples is disturbed.

INTRODUCTION

The problem of adhesion strength between the filler particles and the polymer matrix is one of the most important considerations for development of composite materials. The thermodynamic approach to its solution might be more rigorous but its application is limited[1-3]. Though the thermodynamic estimation describes the adhesion characteristics rather well, it does not provide the real value of adhesion strength which depends on numerous factors, in addition to thermodynamic considerations. Adhesion strength is not an equilibrium characteristic but a kinetic one depending, e.g., on the rate of peeling. When adhesive joints are destroyed, their strength is usually determined under nonequilibrium conditions causing the discrepancy between the thermodynamic work of adhesion and the adhesion strength determined experimentally[4,5]. Nevertheless, adhesion strength between filler particles and matrix is a very important characteristic for filled polymers and its determination is of great importance. There are available some results relative to quantitative measurements of adhesion characteristic in composite materials during deformation[6,7]. For coatings and films, a method of adhesion estimation under quasi-equilibrium conditions was developed[8,9]. However, this method cannot be applied to composite materials containing fillers.

The current concepts regarding mechanism of polymer reinforcement allow us to assume that the usual viscoelastic deformation of a polymer matrix, accompanied by polymer separation from the filler surface, takes place during deformation of the composite. The failure of adhesion bonds in these cases may have a character of interfacial or cohesive break in a weak layer near the interface. Practically, both types of adhesion failures have been observed. However, it is very difficult to determine precisely the type of adhesion failure, especially in polymers filled with dispersed fillers[10-12]. But the adhesion strength in such systems may be experimentally determined without knowing the exact mechanism of failure.

We have made some attempts to perform a quantitative estimation of adhesion strength in filled polymers using both model and industrial samples. Our methods of adhesion strength determination are based on the following hypothesis: if a sample with high filler content undergoes stretching from zero stress to destruction of materials then such stress where the continuity of sample structure is disturbed corresponds to a quantitative measure of adhesion strength[13,14]. Thus, our task is to find the stress at which continuity disturbance ensues while undergoing deformation. This stress is supposed to correspond to break of the majority of adhesion bonds in the system.

ESTIMATION OF ADHESION IN FILLED POLYMER SYSTEMS

The value of adhesion strength (A) may be calculated if the stress-strain dependence is determined $\sigma = f(\varepsilon)$ and the stress value corresponding to transition from the linear dependence to curvilinear one is known:

$$A = \sigma/2\phi_v^{2/3}$$

where the volume fraction of filler (ϕ_v) is expressed as

$$\phi_v = \frac{\rho_p}{\rho_p + \rho_f(1-\phi_w)/\phi_w}$$

Here ϕ_w is the weight fraction of filler (f), ρ are densities of filler (f) and polymer (p).

The various methods discussed below allowed us to determine the starting point of continuity disturbance in filled systems during their deformation.

1. Compensation Method

The system consists of calibrated, by tension, elastic element joined with the composite sample investigated (Fig. 1,2). This system is elongated at constant rate. The point of force applica-

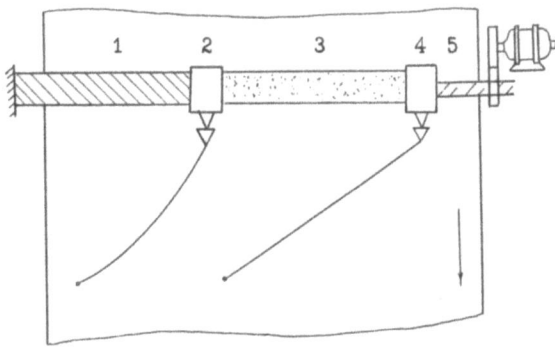

Figure 1. The schematic of the device for the determination of the stress causing the local break of composite material continuity. 1 - elastic element, 2 - mechanism recording the deformation of elastic material, 3 - experimental sample, 4 - the mechanism for recording the total deformation of the system, 5 - the stretching device.

tion 4 records the total deformation whereas point 2 records the difference curve, which allows one to determine the deformation of each element.

If the modulus is not changed during deformation, both points 2 and 4 record straight lines (OB and OA, Figure 2). At a definite tension the break of adhesion bonds proceeds resulting in modulus change. Here the curve recorded by point 2 deviates from linearity (AC, Figure 2). From this curve the tension corresponding to adhesion bonds breakage may be calculated

The corresponding stress at the deviation point characterizes the strength of adhesion between the dispersed filler and the polymer binder.

The samples of rubber (trademark SKMS-30) with various carbon black (PM-100) contents were investigated by this method. The corresponding stress-strain curves are given in Figure 3 and calculated values of adhesion strength are summarized in Table I.

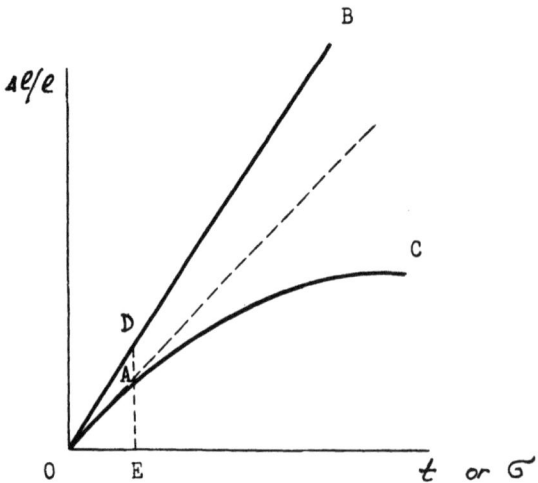

Figure 2. The change of deformation with stress or time for elastic element-composite material element (ODB) and for elastic element alone (OAC). AE-deformation of elastic element, AD-deformation of composite material.

ESTIMATION OF ADHESION IN FILLED POLYMER SYSTEMS

Table I. Adhesion Strength in the Systems with Various Amounts of Carbon Black in Rubber.

Sample No.	Filler Amount wt. %	Adhesion 10^5 N/m^2
1	5	6.38
2	20	4.37
3	30	4.10
4	35	3.83
5	40	3.72

Table I shows that measured adhesion falls off with higher filler loading, and this phenomenon may be explained by flocculation of filler at high concentration. It would seem then that low filler loading would give the best indication of true adhesion.

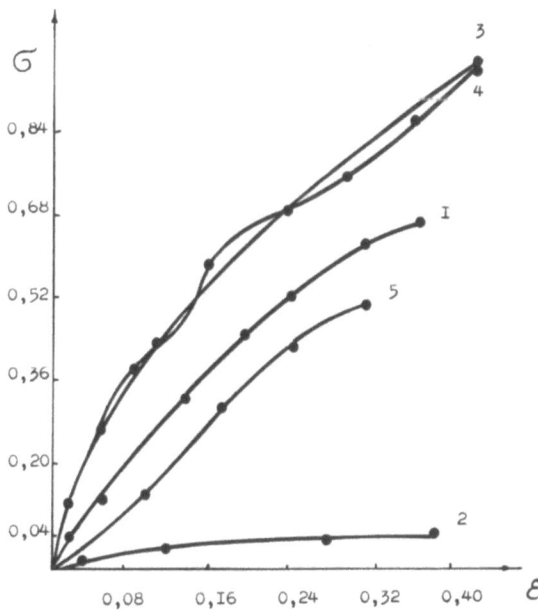

Figure 3. The stress-strain curves for samples presented in Table I.

It is obvious from the results in Table I that the increase of filler amount causes the decrease of adhesion strength by a factor of 1.5. This fact may be explained in terms of the flocculation of carbon black particles with its increasing amount. As a result the real surface of contact between the filler particles and the polymer diminishes and adhesion strength drops.

2. Estimation of Adhesion from the Change in Volume of the Sample while Undergoing Deformation

We have investigated the deformation of polybutadiene rubber filled with aluminum powder. When the sample was stretched, the failure of the adhesion contact between the polymer binder and filler particles was noted. The microphotographs of filled sample slit suffering continually growing deformation illustrate the apparent increase in filler particle size due to formation of some vacuoles around.

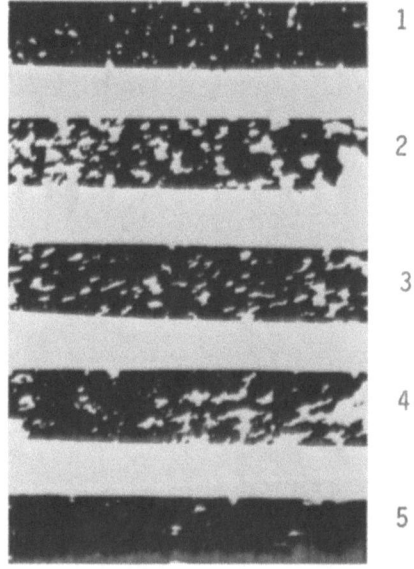

Figure 4. Microphotograph of the slit of deformed sample at various loadings (in arbitrary units): 1-0, 2-8, 3-12, 4-16, 5-20.

The initial sample 1, where the filler particles are closely incorporated into the polymer, transforms after some deformation in a state where some filler particles are separated from the polymer. The number of particles separated is increased if the stress is growing. So the filler becomes surrounded by voids stretched and directed along the stress direction. At high stresses, the voids collapse and change into slots under the action of lateral compression. These slots are noted as thin lines on the photographs. The formation of voids results in an increase in volume of the sample while it is stretching.

Let us define the initial volume as $V_o = h^2 \ell$ where h and ℓ are lateral and longitudinal dimensions of the sample. Then, the volume of the deformed sample V_1 may be calculated by the following expression:

$$V_1 = V_o + \Delta V - (h - \Delta h)^2 (\ell + \Delta \ell)$$

where ΔV, Δh and $\Delta \ell$ are the increases in volume and geometrical dimensions of the sample. The relative change in the volume is expressed by

$$1 + \Delta V/V = (1 - \Delta h/h)^2 (1 - \Delta \ell/\ell)$$

After taking logs:

$$\ln(1 + \Delta V/V) = 2 \ln(1 - \Delta h/h) + \ln(1 + \Delta \ell/\ell)$$

Here $\ln(1 + \Delta V/V)$, $\ln(1 - \Delta h/h)$ and $\ln(1 + \Delta \ell/\ell)$ are the relative deformation of the volume and sizes in logarithmic form. After substituting the Poisson's ratio

$$\mu = \ln(1 - \Delta h/h)/\ln(1 + \Delta \ell/\ell)$$

in this formula and having done some transformations, we obtain the next expression:

$$\Delta V/V = (1 + \Delta \ell/\ell)^{1-2\mu} - 1$$

The change in sample volume while undergoing deformation may now be calculated from the changes in samples size. The calculated and experimental curves of volume changes for the stretched filled sample of polybutadiene rubber, depending on the value of deformation, are given in Figure 5.

As is evident, the calculated and experimental curves of volume changes differ when deformation of the sample is increased. The deviation of experimental curve from the calculated one is the result of additional increase in sample volume due to appearance of discontinuity. At subsequent elongation both experimental and calculated curves coincide. This is a result of slot collapse.

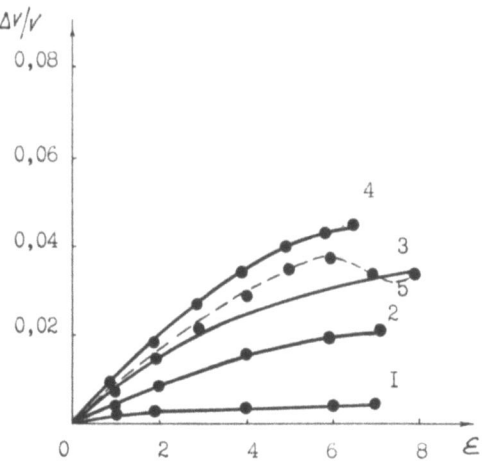

Figure 5. The dependence of volume change on deformation for filled polybutadiene rubber; 1,2,3,4 - calculated and 5 - experimental curves.

The experimental dependence $\Delta V/V = f(\Delta \ell /\ell)$ allows us to determine the deformation and stress corresponding to the critical point, where the experimental and calculated curves deviate. This stress will characterize the adhesion strength between the polymer matrix and filler particles. A special device was constructed to determine the volume changes of the sample during its deformation (Figure 6). It consists of a barrel filled with inert liquid and is contained in a thermostat. The cover fits snugly to the barrel and has a stretching device and calibrated capillary. When the sample is stretched, the liquid in the capillary ascends and the volume increase is fixed. After calculating the stress and deformation corresponding to the sharp increase in volume we can determine the adhesion strength from curves of $\Delta V/V = f(\Delta \ell /\ell)$. For the polybutadiene rubber-carbon black (45.5% by weight) its value is equal to 1.23×10^5 N/m^2 and for polybutadiene rubber-aluminum powder it is (43.2%) - 3.97×10^5 N/m^2.

3. Acoustic Methods

These methods are principally based on the usage of ultrasound waves for detection of discontinuities in the deformed filled composition[15,16]. The polymer volume changes because of the growing porosity in the material due to formation of voids between the polymer matrix and filler particles. As a result, the volume concentration is changed after the deformation of the filled system. The changing concentration of filler is accom-

Figure 6. The photograph of device for measuring the sample volume during its deformation.

panied by attenuation of ultrasound waves and is characterized by two parameters: α_1 - the parameter involving internal friction, heat conductivity, elastic hysteresis, plastic flow and relaxation phenomena and α_2 - the parameter relative to Rayleigh scattering in the filled system. The total attenuation is expressed as

$$\alpha_{total} = \alpha_1 f + \alpha_2 f^4$$

where f is the frequency. The value α_2 is expressed as:

$$\alpha_2 = (2\pi^2/C^4) V (\Delta E/E)^2$$

where C is the sound velocity, V is the volume of a scattering particle, ΔE is the difference between the filler and matrix modulus, E. The ratio $\Delta E/E$ is calculated by full solid angle integration. This value and the volume of a particle allows us to calculate the concentration of the filler in the system.

At composite sample deformation, as it was mentioned above, volume fraction of filler changes, leading to changing acoustic characteristics. The characteristics of transitional process are the most reliable and simplest; they are obtained when the investigated sample is periodically excited by acoustic vibrations. In our experiments, the vibration periodicity was 50 periods/sec and the frequency 275 KHz. The picture of transition process changes in dependence on the degree of sample deformation. The energy of free vibration may characterize the transition process. The value of free vibration amplitude (b) is expressed, as a rule, by the exponential function $b = \exp(-pt)$, where p is an indicator of attenuation. ν - the rate of amplitude decrease is expressed as: $\nu = -p \exp(-pt) = pb$.

The energy of the signal is

$$\varepsilon = \int_0^\infty \nu^2(t)dt = \int_0^\infty p^2 b^2 dt$$

being identical to kinetic energy. p may be calculated from the value of amplitude of forced vibration (a):

$$a = 1 - b = 1 - \exp(-pt)$$

Having carried out a series of expansion of $\exp(-pt)$ we obtain

$$a = pt - p^2 t^2/2$$

Taking into account the first and the second terms (as the process is rather quick) we obtain $p = a/t$. This value is used in the above mentioned equation, which yields:

$$\varepsilon = \int_0^\infty (a^2 b^2)/t^2 dt, \quad t = n \cdot \Delta t$$

where n is the number of indications, Δt is the time between measurements. The following equation may be used for graphical integration:

$$\varepsilon = \sum_{k=1}^{n} (a^2 b^2)/n \cdot \Delta t)^2 = 1/(\Delta t)^2 \sum_{k=1}^{n} \frac{a_k^2 b_k^2}{n^2}$$

For some filled systems we have determined the energy for various tensions. As a result, the dependence of free vibration energy on tensile stress was obtained (Figure 7).

It should be noted that the free vibration energy does not change or changes slightly if the tension is low. At a certain value of the stress, the energy decreases due to the decrease in the filler volume concentration; and subsequently the separation of the polymer matrix from filler particles proceeds leading to

ESTIMATION OF ADHESION IN FILLED POLYMER SYSTEMS

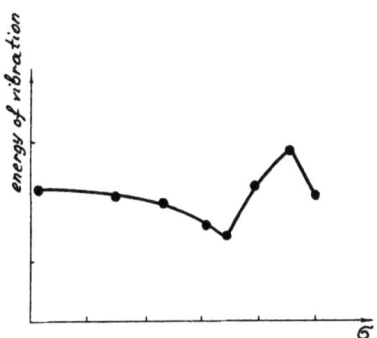

Figure 7. The dependence of the energy of free vibrations on the tension due to stretching.

void formation, which increases the free vibration energy. Having reached the minimum, the filler concentration begins to grow due to collapse of voids and the energy is diminished again.

The critical tension, corresponding to minimum of free vibration energy, corresponds also to the beginning of destruction of adhesion bonds, i.e. this value characterizes the adhesion strength.

The deformation and tension corresponding to the beginning of void formation are determined from data resulting from the dependence of energy on stress.

Adhesion strength may be calculated from this tension. In order to calculate the adhesion strength the stress value determined is related to the unit area occupied by the filler in 1 cm^2 of sample cross-section. Volume concentration is calculated and the number of filler particles in a unit volume is also estimated as:

$$n_v = \phi_v / (4/3)\pi r^3$$

where ϕ_v is the volume concentration and 2r is the particle diameter. For regular geometrical arrangement of particles in the rib of cube of 1 cm^3:

$$n_1 = n^{1/3}$$

and an area of 1 cm^2, we shall have

$$n_2 = n_1^2 = \left(\frac{\phi_v}{4/3\pi r^3}\right)^{2/3}$$

The total area of filler surface is

$$S_f = \pi r^2 n_2 = (3/4\pi r^3)^{1/3} \pi r^2 \cdot \phi_v^{2/3} = 2.0 \phi_v^{2/3}$$

Having found the tension at which discontinuity appears and knowing the geometrical contact area between the filler and the polymer we should be able to estimate the adhesion strength.

The experimental results given in Figures 2, 5 and 7 have been used to calculate the adhesion strength in the system polybutadiene rubber - Al powder using the three values of stresses found in three types of experiments and calculated contact surface. The results are given in Table II.

Table II. Calculation of Adhesion Strength from Data obtained by Various Methods.

Method No.*	Vol. Fraction of Filler	Stress 10 N	Contact Area $10^4 m^2$	Adhesion Strength $10^5 N/m^2$
1	0,432	1,23	0,34	3,60
2	0,432	1,30	0,35	3,72
3	0,432	0,76	0,19	3,98

* 1 - compensation method, 2 - volumetric method, 3 - acoustic method.

Table II shows that there is a satisfactory agreement between the adhesion strength (between the filler particles and the polymer matrix) values obtained by different methods. Thus the proposed methods might serve as the basis for discussion of practical application to evaluate the effectiveness of filler treatment.

4. Method of Transfer Functions

One of the non-destructive methods of estimating the adhesion strength is based on the usage of transfer functions of ultrasound vibrations (275-400 KHz) modulated by rectangular pulses (50 Hz) and transmitting through the sample[17,18]. The rectangular ultrasound signal passing through a sample changes its parameters depending on the medium characteristics.

The growth of amplitude in time may be represented by a transitional function which may be found from approximation of this curve and expressed as an exponential polynom. The transitional function is the ratio of two polynoms of P operator which represents the transfer function. It provides the integral characteristics of filled polymeric systems and consists of transfer functions of some characteristics. The transfer function of adhesion interaction in filled systems cannot be separated from the total transfer function. For its performance an auxiliary sample identical to the one under investigation is used where the adhesion bonds are weaker due to filler surface treatment by antiadhesive. Such sample is compared with the sample under investigation and the transfer functions are obtained.

The transfer function of the investigated sample characterizes the properties of the system and adhesion bonds present. To obtain the transfer function which characterizes the adhesion properties $F(P)$ the transfer function of investigated sample $F''(P)$ should be divided by function of auxiliary sample $F'(P)$. The transfer function of adhesion properties then is

$$F(P) = \frac{F''(P)}{F'(P)} \quad \text{or} \quad F(P) = \frac{A_n P^n + A_{n-1} P^{n-1} + \ldots + A_1 P + A_0}{B_m P^m + B_{m-1} P^{m-1} + \ldots + B_1 P + B_0}$$

where A_n, B_m are coefficients of the polynom, and P is the operator.

Having synthesized this function we can, using electromechanical analogies, transform it into a mechanical model, and calculate the elastic, viscous and inertial properties, characterizing adhesion interaction in the filled system. Using this method, the adhesion characteristics were determined for polyurethanes filled with kaoline

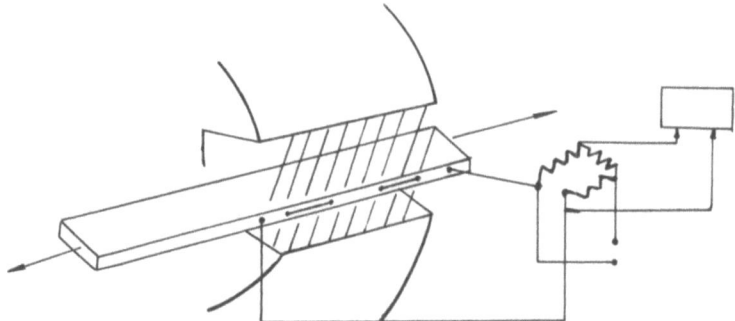

Figure 8. The schematics for fixing the discontinuity in polymers reinforced with carbon fibers during the deformation in a magnetic field.

5. Estimation of Adhesion Strength in Systems Reinforced with Fibers

In order to determine the adhesion strength in the polymeric systems reinforced with fibrous carbon, the method of conductor vibration may be used. A conductor with current is subjected to a magnetic field to detect the discontinuities during deformation.

Here, the property of fibrous carbon to be a conductor of electric current is used. The schematic set up is shown in Figure 8. For experimental purposes, filaments of carbon fiber were pressed into the epoxy resin. The filaments were disposed transverse to the deformation direction.

The filament terminals from the sample were connected to an alternating current source of frequency 20 KHz. The second diagonal of the bridge was connected with the recorder after the rectifier. The sample is put into a magnetic field (up to 8000 Oe and stretched. During this process at a certain stress the polymer matrix is separated from the filler surface. Under the influence of the alternating current, the filament starts vibrating in the magnetic field. The opposing electromotive force is induced in the filaments. This force changes the balance of the bridge arm, where the sample filaments are included in. As a result of fixed discontinuity, the electric tension appears in the second bridge diagonal. This stress is determined both by the tension at elongation and adhesion bonds strength. This force converted to unit area determines the adhesion strength.

The adhesion strength in the system mentioned was determined using this method.

ESTIMATION OF ADHESION IN FILLED POLYMER SYSTEMS

Figure 9. The schematic of sample polymer reinforced with fibrous carbon disposed at an angle of 45°.

The method of shear stresses is simpler for calculation of carbon filament adhesion to polymer binder. If the fabric is formed from carbon filaments, the plate with 2-3 layers of fabric should be cut so that the sheet filaments are disposed at an angle of 45° (see Figure 9). The prepared sample is stretched and the force needed for break is measured. Figure 9 shows that the sheet filaments release from the entanglements, being shifted relative to each other along the binder. When the area of destruction is known (it is equal to the sample square width) the shearing force needed for adhesion failure can be calculated.

The shearing force is related to the breaking force by the Poisson ratio:

$$E/2G = 1 + \mu$$

where E is modulus of elasticity and G is shear modulus. μ is the Poisson ratio equal to 0.34 in our experiments. So the adhesion strength σ_{ad} is expressed in terms of shearing force as

$$\sigma_{ad} = 2.68 \sigma_{shear}$$

Using this method the value of adhesion strength was determined for the composite made of epoxy resin reinforced by carbon fibers. It was found that the adhesion strength is equal to $42-37,52 \cdot 10^5$ N/m².

CONCLUSIONS

The proposed methods of adhesion strength determination in polymers filled with disperse fillers provide rather accurate idea of adhesion interaction and are widely used in our experimental work. At the same time their usage is based on one important parameter; i.e. the stress at the failure of continuity of filled materials caused by polymer separation from the filler surface is taken as a measure of adhesion strength. As a rule, the character of adhesion failure (pure interfacial type, or cohesive or mixed type) is unknown. According to the well-argued viewpoint developed by Bikerman[19] the break of adhesive joints is seldom interfacial in character, rather as a rule it proceeds in a weak layer near the interface. Nevertheless, this argument cannot change the importance of the values determined by the above mentioned methods. The correctness of these methods is proved by the following theoretical arguments. Earlier[20] a new correlation between adhesion and cohesion strength was developed, i.e.,

$$W_{ad} = W_c$$

where W_{ad} is the thermodynamic work of adhesion and W_c the cohesion work for the component with lesser strength, the binder being such a component. The correlation proves that adhesion as a thermodynamic value is determined only by the cohesive strength of the binder when there are no specific interactions at the interface. However, W_c represents the cohesive strength of the binder in the boundary layer. The numerous data obtained by us and other authors[21,22] testify to the considerable changes in properties of polymer boundary layers in contact with a solid surface. They are caused by adsorption interaction with the surface of filler. These changes display the processes occurring at the interface when adhesion bond is formed. In particular, some physical and technological[24] factors cause the formation of "weak" (using Bikerman's terminology) layers at the interface, and this is where the adhesion bond failure takes place.

From what is mentioned above one can easily conclude that the real conditions of adhesion failure are very complicated. Even in the case when the adhesion bond break has a cohesive character, the principal equation relating the works of adhesion and cohesion is still valid. This gives a theoretical basis to adhesion estimation based on the determination of the stress corresponding to the appearance of the discontinuity in the sample under investigation. Hence the combination of strict theoretical approach to adhesion in combination with concepts of physics and mechanics of polymer fracture makes it possible to resolve one of the most important problems in the processing of composite materials, i.e., the problem of estimating the adhesion interaction without knowing the precise locus of adhesion failure.

REFERENCES

1. A. A. Berlin and V. E. Basin, "Principles of Polymer Adhesion (Russ.)", Chimija, Moscow, 1974.
2. B. V. Derjagin, N. A. Krotova and V. P. Smilga, "Adhesion of Solids (Russ.)", Nauka, Moscow, 1973.
3. S. S. Voyutsky and B. V. Derjagin, Kolloid Zh., 27, 626 (1965).
4. K. L. Mittal, Polymer. Eng. Sci., 17, 467 (1977).
5. K. L. Mittal, in "Adhesion Science and Technology", L. H. Lee, Editor, Vol. A, pp. 129-168, Plenum Press, New York (1975).
6. G. P. Sendeckyj, Editor, "Mechanics of Composite Materials", Academic Press, New York (1974).
7. L. E. Nielsen, "Mechanical Properties of Polymers and Composites", Marcel Dekker, Inc., New York (1974).
8. A. D. Zimon, "Adhesion of Films and Coatings (Russ.)", Chimija, Moscow, 1977.
9. P. J. Zubov, et al., Lakokracochnye materialy i ikh primenenie, No. 3, p. 28 (1964).
10. K. L. Mittal, Pure Appl. Chem., 52, 1295 (1980).
11. W. L. Baun, in "Adhesion Aspects of Polymeric Coatings", K. L. Mittal, Editor, pp. 131-146, Plenum Press, New York (1983).
12. W. L. Baun, these proceedings, pp. 3-17.
13. Yu. S. Lipatov and L. M. Sergeeva, "Adsorption of Polymers", John Wiley, New York (1974).
14. Yu. S. Lipatov and P. K. Zarev, Mekhanika polymerow, No. 2, p. 195 (1972).
15. P. K. Zarev and Yu.S. Lipatov, Vysokomolek. soed., A 12, 282 (1970).
16. P. K. Zarev and Yu. S. Lipatov, Vysokomolek. soed., A 17, 717 (1979).
17. P. K. Zarev and Yu. S. Lipatov, Vysokomolek. soed., A 21, 514 (1979).
18. A. M. Yaglom and I. M. Yaglom, "The Probability and Information (Russ.)", Nauka, Moscow, 1973.
19. J. J. Bikerman, Uspekhi chimii, 41, 1431 (1972).
20. Yu. S. Lipatov and A. E. Feinerman, J. Adhesion, 6, 165 (1973).
21. Yu. S. Lipatov, "Interphase Phenomena in Polymers (Russ.)", Naukova Dumka, Kiev, 1980.
22. Yu. S. Lipatov, in "Adhesion and Adsorption of Polymers", L. H. Lee, Editor, Vol. B, p. 601, Plenum Press, New York (1980).
23. Yu. S.Lipatov, Pure Appl. Chem., 43, 273 (1975).
24. J. J. Bikerman, Ind. Eng. Chem., 5, No. 9, 40 (1967).

Part IV
Durability or Stability Aspects

ADHESION AND DURABILITY OF METAL/POLYMER BONDS

J. D. Venables

Martin Marietta Laboratories
1450 South Rolling Road
Baltimore, Maryland 21227

This paper reviews the results of a comprehensive investigation made at the author's laboratories to determine those factors responsible for promoting the integrity and long-term durability of metal/polymer bonds used in the fabrication of aircraft and aerospace structures. Using a multidisciplinary approach and a variety of surface analytical techniques, such as extended resolution scanning electron microscopy (XSEM), X-ray photoelectron spectroscopy (XPS), ellipsometry, and a new technique we call surface behavior diagrams (SBD), we have evolved several important concepts. First, we have determined that the initial integrity of metal/polymer bonds used for structural applications depends critically upon the morphology of the surface oxide on the metal. In the case of the metals studied, Al and Ti, we have observed that certain etching or anodization pretreatment processes produce oxide films on the metal surfaces which, because of their porosity and microscopic roughness, mechanically interlock with the polymer forming much stronger bonds than if the surface were smooth. Second, we have shown that the long-term durability of metal/polymer bonds is determined to a great extent by the environmental stability (or lack of stability) of the same oxide which is responsible for promoting good initial bond strength. For Al, the effect of moisture intrusion at the bond line is to cause the oxide to convert to an hydroxide with an accompanying drastic change in morphology and bond strength. For Ti, the evidence suggests that the oxides formed on it are much more stable than those on Al but under certain circumstances the oxide can undergo a polymorphic transformation which may lead to bond degradation. Third, a major finding of these investigations is that significant improvements in durability of adhesive bonds to Al can be achieved using an extremely simple treatment in which monolayer films of certain organic acids are applied to the adherend oxide to protect it against the effects of moisture.

INTRODUCTION

Although adhesive bonding has been widely used in the aerospace industry for many years, the reduction in weight and cost savings possible with properly designed adhesive joining systems have recently added impetus to perform sophisticated research in the field. More widespread use of bonding will necessarily involve its use in primary structures, those responsible for keeping the major structures intact. The problem of developing confidence in the reproducibility of bonding operations and long-term reliability of the bonds must be addressed before such use can occur.

Major efforts, such as the Air Force sponsored Primary Adhesively Bonded Structure Technology (PABST) program,[1] have focused attention on pretreatment of the metal parts to be joined as the most important step in guaranteeing both initial bond strength and long-term durability. Using new and highly sophisticated surface analysis tools, workers at Martin Marietta Laboratories have been investigating what the pretreatments do to metal substrates, particularly Al and Ti, since they are the most widely used metals in the industry. These efforts have lead to a new understanding of how strong adhesive bonds are formed with metals, what causes their degradation over time, and how their performance can be improved by treatments which stabilize the oxides present on the metal against the effects of moisture.

INITIAL BOND INTEGRITY

The initial integrity of metal/polymer bonds used for structural applications is critically dependent upon the morphology of the surface oxide on the metal. In the case of Al and Ti, the metals studied, we have observed that certain etching or anodization pretreatment processes produce oxide films on the metal surfaces which are extremely rough and porous on a microscopic scale.

For example using XSEM (SEM done on a STEM at ~ 30 Å resolution) we have observed[2] that the Forest Products Laboratories (FPL)[3] process and phosphoric acid anodization (PAA)[4] process used for preparing Al produce surface oxide structures of the type shown in Figures 1 and 2. The microscopic interlocking roughness exhibited by the structures is apparently a crucial factor determining adhesion at the epoxy-oxide interface in bonded aircraft structures. To demonstrate this, we have intentionally added 500 ppm fluorine to the FPL etch bath and observed that the surface oxide morphology is drastically modified, so that it is much less interlocking in nature. This surface, which exhibits undulations, but does not interlock with polymeric coatings, may be readily separated from an overlying adhesive or primer coating to allow examination of the polymer side of the interface. When this is done, it is observed that the polymer retains a perfect replication of the original oxide features. This situation contrasts in two respects with the behavior exhibited when attempts are made to separate a normally prepared FPL surface from an overlying adhesive. First, separation at the oxide/polymer interface is much less easy to achieve if the metal is properly prepared, but can be done if the Al is first bent very sharply. Second, we have observed that the separation of good bonds, when forced to occur at the oxide/polymer interface, is accompanied by an extreme amount of deformation of the polymer which is generally badly torn and ripped. One consequence of this different mode of separation is that the bond strength, as measured by a climbing drum peel (CDP) test, for example, may be as much as a factor of three different, with the interlocking surface yielding the highest strength levels.

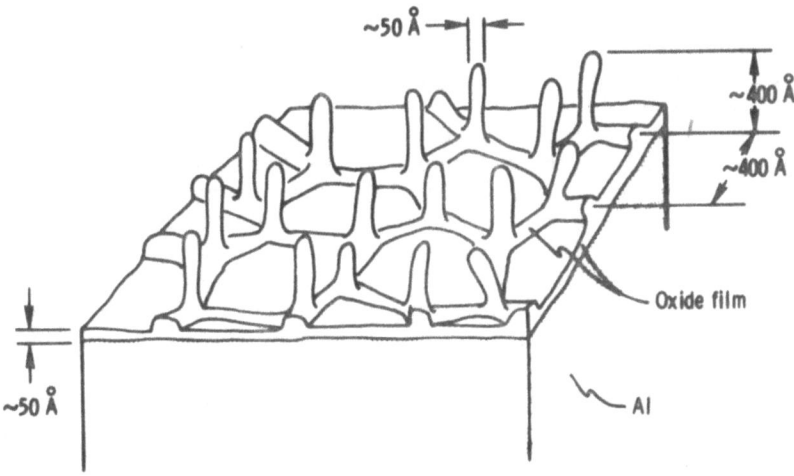

Figure 1. Oxide morphology produced on Al surface by the FPL process, as observed by stereo XSEM (Ref. 3).

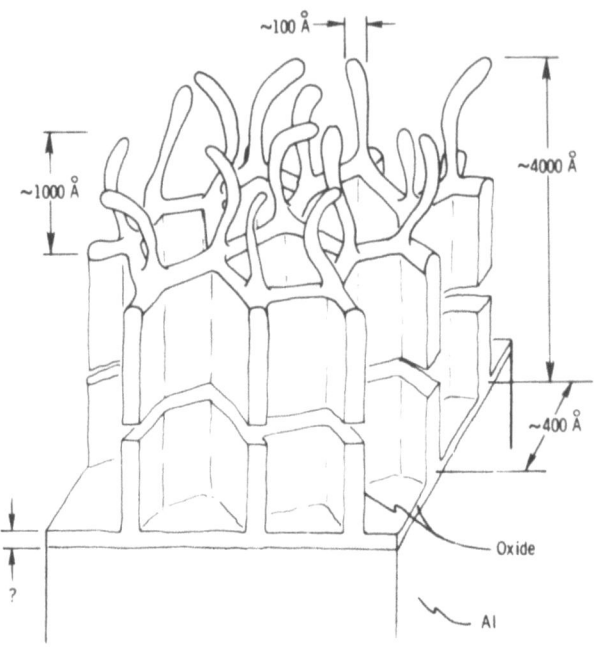

Figure 2. Oxide morphology produced on Al surface by the PAA process (Ref. 4).

Another consequence of not having an interlocking morphology is that the peel strength can be lowered further simply by placing a drop of water in the crack developed during the test. We interpret this to mean that in the absence of mechanical interlocking, when the bond strength is determined principally by chemical forces across the interface (e.g., van der Waals or dispersion forces), the presence of water can disrupt these bonds readily thereby reducing the interfacial strength. A similar effect is observed for mica[5] which is bonded across the layer planes by Van der Waals forces. When cleaved in the presence of water, the cleavage energy ($\phi = 250$ ergs/cm^2) is lower than when done in a dry environment ($\phi = 1200$ ergs/cm^2). For adhesive bonds, however, water has no significant short-term effect when interlocking is present. (The long-term effect of moisture on properly prepared bonds will be discussed in the next section).

It might be argued that fluorine picked up from the contaminated bath and deposited on the oxide surface might be responsible for the degradation in bond strength observed in the above experiment. However, examination by Auger/XPS of surfaces prepared in fluorine contaminated baths indicated that the surface concentration of fluorine was very low (less than 3% surface coverage). Prior work,[6] in which fluorine was added to rinse water (in which case, much higher concentrations of F can be adsorbed on the surface) has demonstrated that such low concentrations of F, per se, do not significantly degrade bond strength so long as the oxide morphology is not altered.

Although considerable success can be achieved in bonding polymers to Al using either the PAA or FPL pretreatment process, the situation is not as straightforward for Ti. Prior attempts to develop a pretreatment for Ti that would be as successful as those for Al have yielded a multitude of processes, but not until recently have any shown promise. Because of the increasing interest in using Ti for advanced aircraft structures we felt it was important to know more about the types of surfaces generated by these processes with the hope that this information would provide guidelines for future improvements. Accordingly, using the techniques we employed for studying Al adherends, Ditchek et al.[7] characterized surfaces of Ti prepared according to a variety of different pretreatment processes. The results were then compared with mechanical properties measurements performed by other investigators[8] in a coordinated U.S. Navy (NAVAIR) program as discussed below.

Table I lists the various surface pretreatments investigated and provides some comments on the types of surfaces generated on Ti-6Al-4V alloys. The appearance of these surfaces varied considerably, but it was possible to classify them into three groups according to similarities in their surface morphologies. Group I surfaces, which include those resulting from the PF and MPF treatments (notation defined in Table I), display little macro- or micro-roughness.* Group II surfaces, which derive from the DA, LP, TU and DP treatments, all exhibit a large degree of macro-roughness and a small degree of micro-roughness on the LP and TU surfaces. Group III surfaces, which include those generated by chromic acid anodization[9] (CAA) at 5 or 10 volts are characterized by having no macro-roughness, but a high degree of micro-roughness associated with a porous oxide.

The Group III surface morphologies are of particular interest because they bear a marked resemblance to those produced on Al by the PAA or FPL process. For example, the CAA surface exhibits a porous oxide with protruding whiskers similar to the FPL structure and is approximately the same thickness if the anodizing

---

* A macro-rough surface is defined as an uneven surface with characteristic bumps or jagged features about 1.0 μm or greater. Micro-rough surfaces have fine structure with dimensions 0.1 μm or less.

TABLE I. Morphological Characteristics and Chemical Contaminants Associated with Various Ti Pretreatment Processes

Process	Process Code	Oxide Thickness (Å)	Group[a] Number	Comments
1. Phosphate fluoride	PF	200	I	F contamination
2. Modified phosphate fluoride	MPF	80	I	F contamination
3. Dapcotreat	DA	60	II	No apparent fine structure; Cr on surface
4. Dry hone PASA JELL 107	DP	100–200	II	Deformed surface with embedded Al_2O_3; fluorine contamination
5. Liquid hone PASA JELL 107	LP	200	II	Embedded alumina; fluorine contaminant; Cr on surface
6. Turco 5578	TU	175	II	Fe containing particles on surface
7. Chromic acid anodize	CAA+F	5V:400 10V:800	III	Porous oxide with protruding whiskers; fluorine contamination
8. Alkaline peroxide	AP	450–1350 depending on chemistry, temperature, time	III	Porous oxide

[a] For a definition of group numbers, see text.

potential is 5 volts and the anodization time is 20 minutes. When the anodizing potential is raised to 10 volts (and the time remains the same), the surface morphology becomes somewhat intermediate between that of FPL and PAA oxides in both appearance and thickness. Because of these similarities, it would therefore be expected that the CAA oxides would interlock with polymer coatings providing interfacial bond strengths comparable to those associated with Al prepared by the FPL and PAA processes.

Evidence that this is the case is provided by wedge tests performed by Brown[8] whose results are shown in Figure 3. For these tests the Ti surfaces were prepared in exactly the same manner as those samples examined by XSEM. A standard wedge test configuration was employed using the BR127/FM300 primer/adhesive system to bond the Ti test strips together. The test conditions are indicated in Figure 3.

The test results clearly demonstrate a correlation between surface roughness and bond strength. Thus, the Group I surfaces which exhibit no macro- or micro-roughness, exhibit very poor behavior in the wedge test with all of them failing adhesively at the primer/metal oxide interface. On the other hand, the Group III

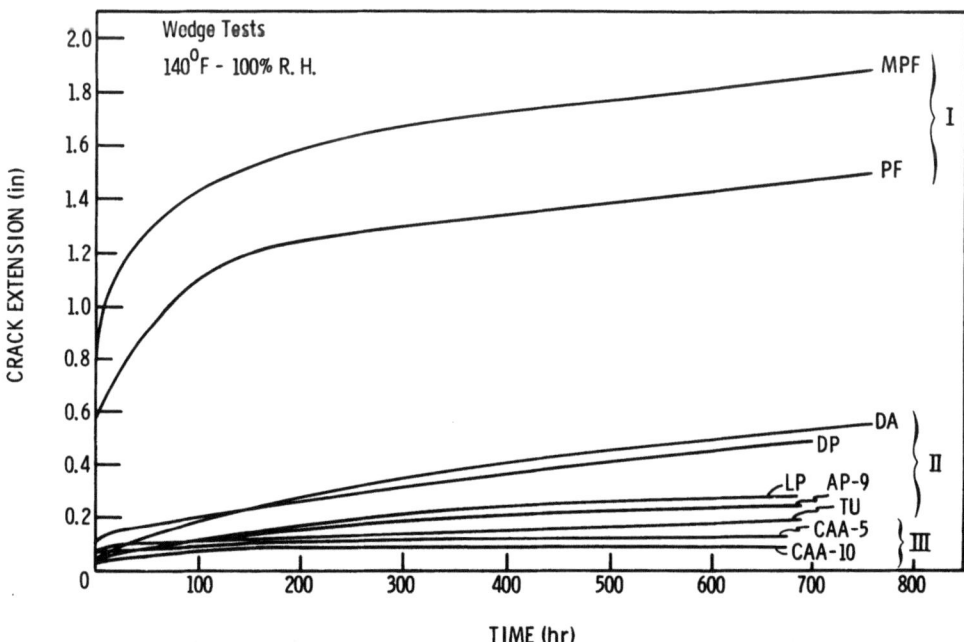

Figure 3. Wedge test crack length as a function of exposure time @ 100% R.H., 60°C for Ti-6Al-4V adherends prepared by eight different processes described in Table I. Definitions of Groups I, II and III can be found in the text.

surfaces, which exhibit no macro-roughness but a high degree of micro-roughness in the form of porous oxides exhibited almost no crack growth except for a small amount of cohesive failure in the adhesive during early stages of the test. The wedge test values corresponding to the Group II category lie intermediate between those of Group I and III consistent with the fact that Group II oxides exhibit more macroroughness than Group I but less microroughness than Group III. Evidently, the presence of a microrough porous oxide which can interlock with the polymer is just as important a factor determining the strength of polymer bonds to Ti as is the case for Al.

LONG-TERM DURABILITY

The long-term durability of metal/polymer bonds is determined to a great extent by the environmental stability (or lack of stability) of the same oxide that is responsible for promoting good initial bond strength. For Al, moisture intrusion at the bond line causes the oxide to convert to an hydroxide with an accompanying drastic change in morphology,[10] Figure 4. The resulting hydrated material adheres poorly to the Al beneath it and, therefore, once it forms, the overall bond strength may be severely degraded. The proposed failure model is shown schematically in Figure 5.

Figure 4. Morphology of aluminum hydroxide (pseudo-boehmite) produced on Al surface during exposure to moisture. The change in morphology which accompanies the conversion of the original oxide to this hydrated form results in a drastic reduction in adhesive bond strength (Ref. 10).

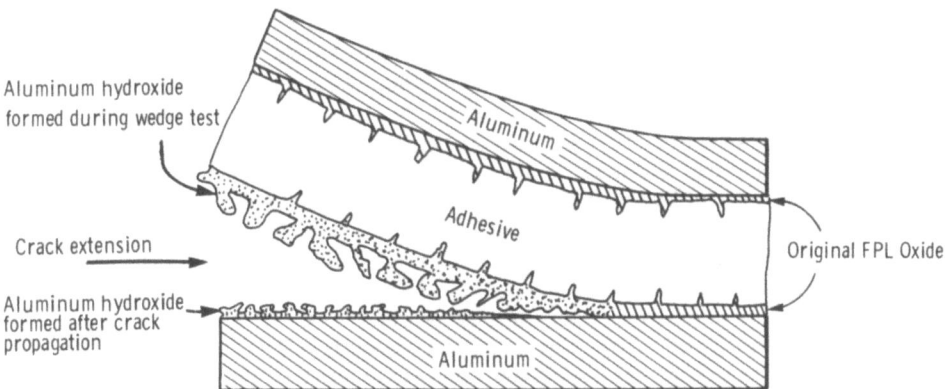

Figure 5. Schematic drawing of the mechanism deduced for crack propagation during wedge testing. In the humid environment, the original FPL oxide is converted to a hydroxide which adheres poorly to the aluminum substrate. The crack propagation rate is much faster here than in a dry atmosphere, when the crack propagates directly through the adhesive (Ref. 10).

For Ti, the evidence suggests that its oxides are much more stable than those of Al. However, preliminary evidence[11] suggests that under some circumstances the oxide, which is originally amorphous, undergoes a polymorphic transformation to anatase which, because of volume changes and accompanying morphology changes, may lead to bond degradation just as the oxide-to-hydroxide conversion process does for bonds to Al. This transformation, which occurs at temperatures above 200°C in a dry environment and above 85°C in the presence of moisture, is being studied further because it could be an important degradation mechanism in future applications, for example, in attempts to increase service temperatures through the use of polyimide adhesives.

The proposed degradation model for Al/polymer bonds, Figure 5, is supported by evidence showing a correlation between incubation times for the oxide-to-hydroxide conversion process and wedge test results; the longer the incubation time of oxides prepared (and treated) in various ways, the better the wedge test results. This appears to be the first time that such a direct correlation has been observed between a bond durability test and a measurable physical parameter of the metal adherend.[10]

A new technique, the surface behavior diagram, was developed by Davis et al.[12] during this work. This type of diagram is analogous to phase diagrams for bulk phases but is intended to represent effects that are specific for surfaces, e.g., reactions between a surface, an adsorbate, and the environment. In the present case, the technique was used to show that the greater stability of PAA surface oxides, relative to FPL oxides, is due to the presence of adsorbed phosphate (from the electrolyte) which inhibits the oxide/hydroxide conversion process until the phosphate itself becomes extensively hydrated and is lost by dissolution. Thus, as shown in Figure 6, XPS data were used to determine the composition of the PAA surface, and therefore its position on the Al_2O_3-$AlPO_4$-H_2O behavior diagram as a function of exposure time to hot water. The fact that the surface reaction follows the path shown in the figure, rather than the alternative path of proceeding first to Al_2O_3 and then to the hydroxides, suggests that the effectiveness of the inhibitor is eventually lost not because it goes directly into solution but because it first hydrates and then desorbs. Evidently, choosing an inhibitor that is more resistant to hydration and subsequent desorption is an approach that would provide greater stabilization of the surface against the effects of moisture. This example suggests that SBD's may be generally useful for surface science studies particularly in the fields of adhesive bond durability, metal corrosion and corrosion inhibition.

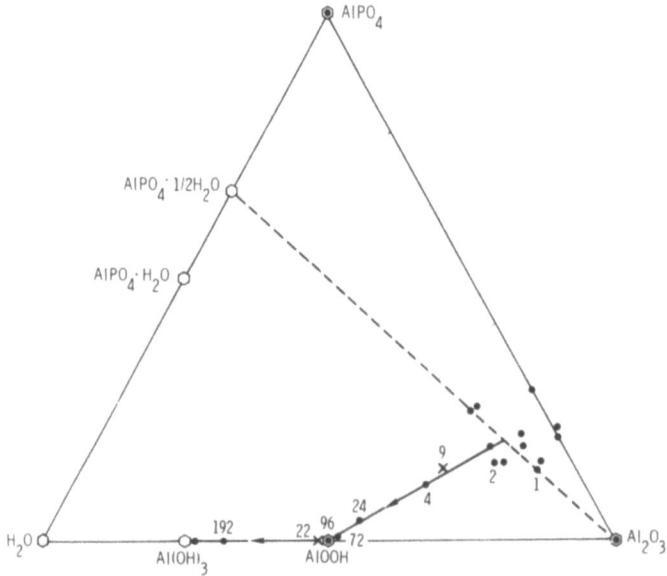

Figure 6. The ternary behavior diagram of fresh and hydrated PAA Al oxide surfaces. The cluster of unnumbered points are data taken on freshly prepared surfaces. The numbers by some points denote the exposure time in hours to 100% relative humidity at 50°C (solid points) or at 60°C (crosses) (Ref. 12).

INHIBITORS

A major finding of these investigations is that significant improvements in durability of adhesive bonds to Al can be achieved using an extremely simple treatment whereby monolayer films of certain organic acid molecules are used to protect adherend oxides from the effects of moisture.[13] Specifically, we have shown that an adsorbed monolayer of an amino phosphonic acid, nitrilotris methylene phosphonic acid (NTMP), can improve the stability of FPL-treated Al so that its performance in wedge tests is almost comparable to Al treated by the PAA process (Figure 7). Moreover,

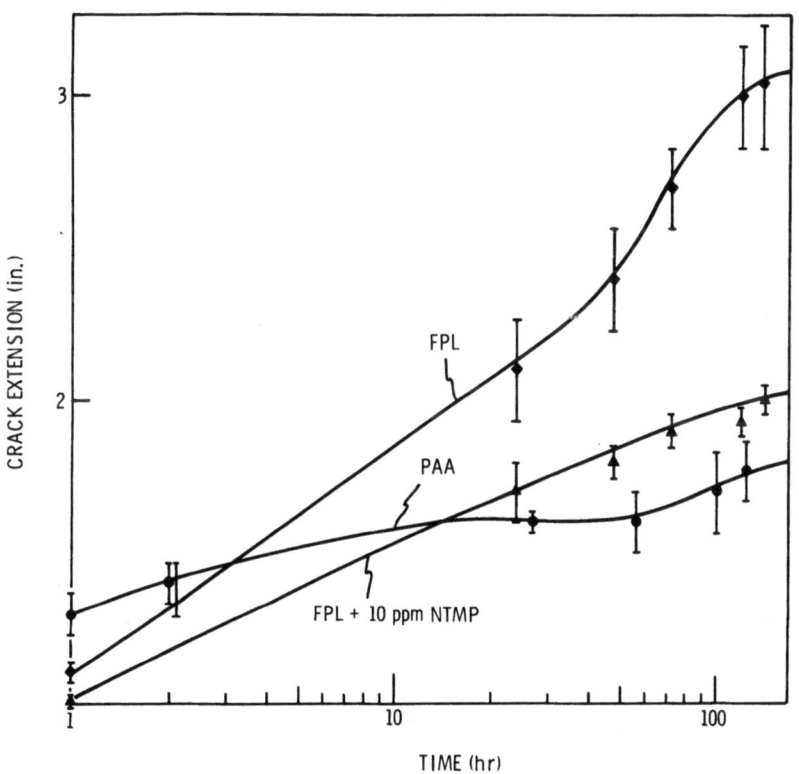

Figure 7. Wedge test crack length as a function of exposure time in a humidity chamber @ 100% R.H., 60°C. The data demonstrates that the performance of FPL-treated adherends under these test conditions is improved significantly when the adherend is treated before bonding in an inhibitor solution containing just 10 ppm of nitrilotris (methylene) phosphonic acid (NTMP). In fact, after the treatment, the performance approaches that of PAA-treated adherends. (Ref. 13).

those oxides formed by the PAA process, which are normally stabilized to some extent against moisture by adsorbed phosphate, can be stabilized even more effectively by the amino phosphonic acid treatment.[14] When this is done, the wedge test performance of PAA adherends is improved by almost a factor of two under the conditions of the tests employed (Figure 8). The attractiveness of the inhibitor treatment is enhanced by the fact that since the inhibitor is used in monolayer form, very large quantities of Al can be treated with only small quantities of inhibitor solution. We also note that another potential approach for improving durability is suggested by the following observation we have made. In an attempt to perform some specialized hydration experiments on bulk aluminum oxide, we were surprised to find that crystalline Al_2O_3, whether in the form of powders or single crystals, is very much more stable than FPL or PAA oxides, which we determined, by our electron diffraction studies, to be amorphous. In fact, we observed that whereas the amorphous oxides hydrate in a matter of minutes or hours when immersed in 80°C water, no evidence of hydration could be found by XSEM on crystalline material that had been exposed for over a week in water heated to temperatures up to the boiling point. Evidently, the degree of crystallinity of Al_2O_3 has a profound influence on its stability against the effects of moisture, which clearly suggests another avenue of investigation for improving the performance of metal/polymer bonded structures.

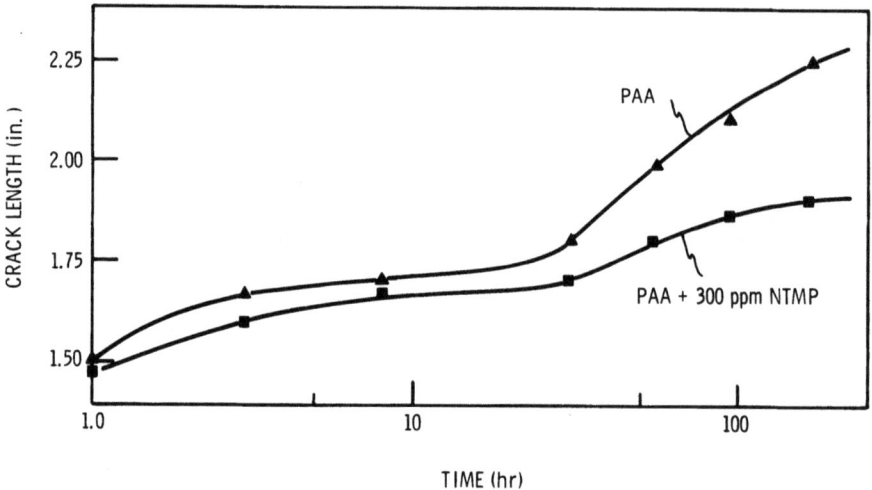

Figure 8. Wedge test crack length as a function of exposure time in a humidity chamber @ 100% R.H., 60°C showing the improvement in performance when PAA adherends are treated with NTMP inhibitor. (Ref. 14).

DISCUSSION AND CONCLUSIONS

In summary, our work indicates that the initial integrity of polymeric bonds to metals is significantly enhanced if the adherends are prepared in such a manner that the surface oxide is rough and porous on a microscopic scale. For Al, the PAA and FPL pretreatment processes provide porous oxide structures which interlock with the overlying polymer to form strong adhesive bonds. For Ti, the CAA process provides similar oxide morphologies and good bond strengths. In humid environments, we find that the long-term durability of metal/polymer bonds is determined to a great extent by the stability (or lack of stability) of the same oxide which is responsible for promoting good initial bond strength. For Al, the degradation mechanism involves an oxide-to-hydroxide conversion process which leads to drastic surface morphology changes and an accompanying loss of bond strength. For Ti, the evidence suggests that the oxides formed on it are much more stable than those on Al but that morphological changes do occur albeit at more elevated temperature. Since the environmental conditions may be more severe for applications which require the use of Ti, however, durability problems could be just as troublesome as they are for Al although much more work is required to define those circumstances which would threaten bond integrity.

Considerable improvement in the durability of bonds to Al can be achieved using a simple treatment procedure in which monolayer films of certain organic acids are applied to the adherend oxide to protect it against the effects of moisture. These experiments demonstrate that organic compounds can be just as effective, and potentially perhaps much more effective, than inorganic inhibitors (e.g., chromates) which are currently used as additions to corrosion inhibiting primers. We are currently using surface behavior diagrams, a new technique developed during this work for examining the interaction between surfaces, adsorbates and environment, to study organic inhibitors further.

Although the emphasis of this work has been on metal/polymer bonds, the results appear to have implications that go beyond this technology area. Specifically, the work on durability has clear relevance to the general field of corrosion and corrosion inhibition. For instance the reaction between water and a surface oxide to form a hydroxide cannot be considered strictly a corrosion process because the metal itself is not attacked in the initial stages of the reaction. Nevertheless, we suggest that this reaction is an important precursor step that leads to eventual corrosion of the metal. In support of this we have observed that when an Al sample is exposed to water, the oxide film initially passivates and protects the metal from attack during an "incubation" period. Following this period, intense hydration activity occurs, during which the original oxide is converted to a hydroxide. We observe evidence of true corrosion only after hydroxylation when the evolution of gas suggests that the Al metal itself is reacting with the water and forming hydroxide and hydrogen gas. Evidently, the protection provided by the oxide layer is disrupted once the oxide hydrates, which suggests that procedures designed to reduce the hydroxylation rate would also be effective in corrosion protection of Al at least in certain pH ranges. If so, we suggest that the concepts and techniques developed during these investigations could be of considerable benefit if properly applied to the general field of metal protection against environmentally induced degradation.

ACKNOWLEDGEMENTS

We wish to thank the following agencies for supporting the indicated aspects of this work: The surface behavior diagram work was performed under contract F49520-78-C-0097 to AFOSR; the Ti work under contract N00019-80-C-0508 to NASC; the inhibitor work under contract N00014-80-C0718 to ONR/AROD.

REFERENCES

1. E.W. Thrall, in "Proc. 10th National SAMPE Tech. Symp.," Kiamesha Lake, NY, Oct. 17-19, p. 73 (1978).
2. J.D. Venables, D.K. McNamara, J.M. Chen, T.S. Sun, and R.L. Hopping, Appl. Surface Science $\underline{3}$, 88 (1979).
3. H.W. Eichner and W.E. Schowalter, Forest Products Laboratory Report No. 1813 (1950).
4. G.S. Kabayaski and D.J. Donnelly, Boeing Co. Report No. DG-41517 (February 1974).
5. B.V. Deryagin and M.S. Mesik, Soviet Phys.-Solid State (Eng. Transl.) $\underline{1}$, 1393 (1960).
6. J.M. Chen, T.S. Sun, J.D. Venables and R.L. Hopping, in "Proc. 22nd Nat. SAMPE Symp.", p. 25 (1977).
7. B.M. Ditchek, K.R. Breen, T.S. Sun, J.D. Venables and S.R. Brown, in "Proc. 12th Nat. SAMPE Tech. Conf.", Oct. 7-9, 1980, Seattle, WA, Vol. 12, pp. 882-885 (1980).
8. S.R. Brown, in "Proc. 27th Natl. SAMPE Symp.," San Diego, CA, May 4-6, p. 363 (1982).
9. Y. Moji and J.A. Marceau, "Method of Anodizing Titanium to Promote Adhesion," U.S. Patent No. 3,959,091, May 25, 1976.
10. J.D. Venables, D.K. McNamara, J.M. Chen, B.M. Ditchek, T.I. Morgenthaler, T.S. Sun, and R.L. Hopping, in "Proc. 12th Nat. SAMPE Tech. Conf." Oct. 7-9, 1980, Seattle, WA, Vol. 12, pp. 909-923 (1980).

11. M. Natan and J.D. Venables, paper presented at the 2nd International SAMPE Conf., Stresa, Italy (1982).
12. G.D. Davis, T.S. Sun, J.S. Ahearn and J.D. Venables, J. Mat. Sci. $\underline{17}$, 1807 (1982).
13. J.S. Ahearn, G.D. Davis, T.S. Sun and J.D. Venables, in "Adhesion Aspects of Polymeric Coatings," K.L. Mittal, Editor, pp. 281-299, Plenum Press, New York (1983).
14. D.A. Hardwick, J.S. Ahearn and J.D. Venables, unpublished work.

FAILURE MECHANISMS IN THE BOUNDARY LAYER ZONE OF METAL/POLYMER SYSTEMS

W. Brockmann, O.-D. Hennemann and H. Kollek

Fraunhofer-Institut für angewandte Materialforschung
D-2820 Bremen 77, W.-Germany

A systematic optimization of the long term stability of polymer/metal systems in the boundary zone requires the exact knowledge of the mechanisms producing the so-called adhesional failure particularly under the influence of humidity. Strong and water-stable adhesion, for example, between organic primers and aluminium alloys is obtainable if the metal surface is coated with a porous oxide layer, which can be produced by special etching and anodizing processes. The relatively small molecules of the primers can penetrate practically the whole oxide structure. But despite this, it is easy to demonstrate that very small differences in the parameters of the surface treatment, leading to small differences in the surface morphology, can change the water stability of the boundary zone considerably. The transition zone between the oxides and primers in such systems was investigated by using the wet peel test, transmission-electron microscopy of ultra-thin cross sections from the boundary zone before and after delamination and ESCA on the fracture surfaces. It was found that in bonds showing adhesional failure in the wet peel test very near to the oxide layer there is a zone whose properties appear to be different from that of the bulk polymer. After an adhesional failure of these bonds, in many cases, relatively high amounts of polymer remain on top of the oxides, and on the polymer surface no metal oxides are detectable, if an anodized oxide is present. These observations lead to the conclusion that at least partly an adhesional failure in reality is a cohesional failure in weak boundaries of the polymer, whose properties are influenced by the state of the oxide surface.

INTRODUCTION

One of the problems in metal-polymer systems like adhesively-bonded metal joints or metals coated with organic coatings is the stability of the adhesion in humid environments. In many cases bonded joints showing high initial adhesional strength after exposure to humid environments fail in the form of so-called "adhesional failures". At first glance it appears as a clear separation between the metal surface and the polymer. The sensitivity of metal/polymer systems against humidity penetrating the polymer and weakening or destroying adhesion can be influenced by special surface treatments of the metals prior to the resin application.

Particularly the aircraft industry uses for the different aluminium alloys some empirically developed chemical surface treatments to obtain water-stable adhesion in adhesively bonded joints and coated surfaces. Common are etching processes in chromic sulphuric acid solutions at elevated temperatures with different etching times (in Europe 30 min (CSA); in USA 10 - 15 min (FPL)). In some production lines of the European aircraft industry the CSA etching followed by an anodization process in chromic acid (CAA) as a surface treatment has been used for nearly 30 years. In the United States 10 years ago an anodizing process in phosphoric acid (PAA) was developed and is widely used today.[1,2]

These different surface treatmens produce on the aluminium surfaces oxide layers of very different morphologies as can be seen in the transmission electron micrographs in Figure 1. The TEM-micrographs were taken from ultra-thin cross sections of the transition zone between metal and polymer in bonded aluminium joints.

With all these surface treatments it is possible to produce highly water-stable adhesion between the polymer and the metal. But experience shows that in some cases bonded joints of aluminium alloys in aircraft delaminate as adhesional failures after service times of 4 or 5 years if the surface pretreatment was only an etching process like FPL or CSA. This is not observed with CAA- or PAA-treated metals bonded by adhesives. Both anodizing processes are practically equal in achieving bonded joints of high long term stability under humid environmental conditions.[3]

The good adhesional properties of the oxide layers produced by the anodizing processes are attributed to their high porosity.[4]

FAILURE MECHANISMS IN THE BOUNDARY LAYER ZONE

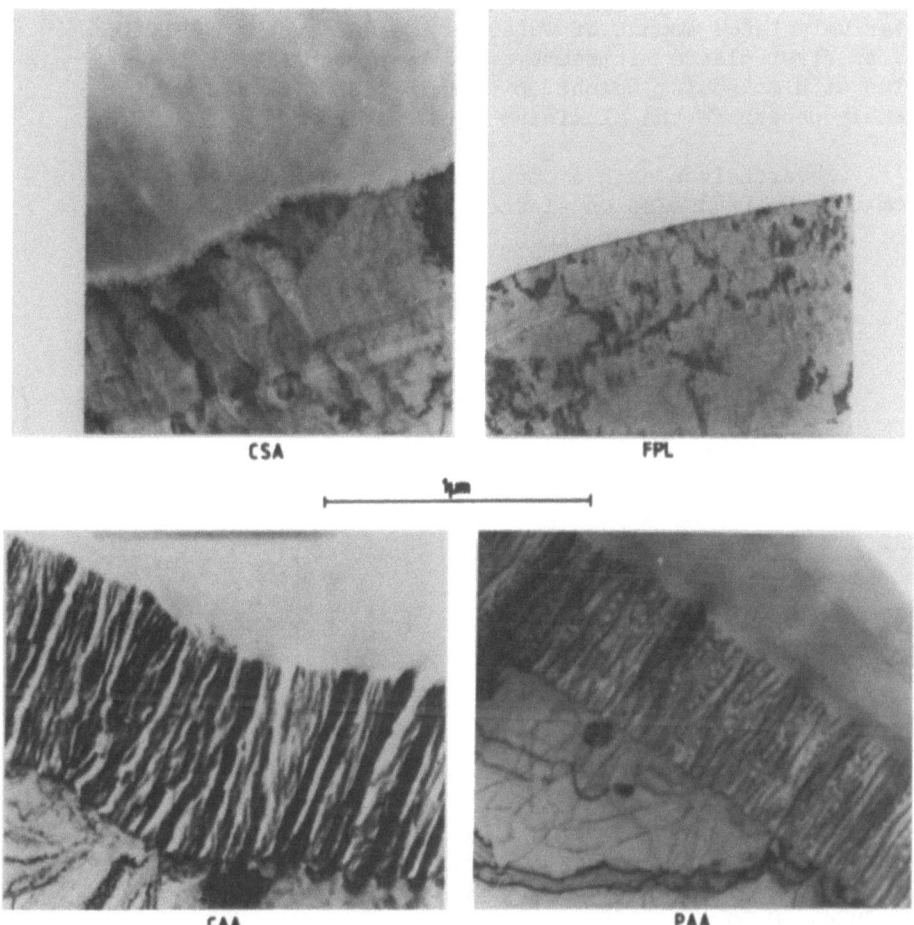

Figure 1. TEM of cross sections from the transition zone between metal and polymer in bonded aluminium joints after different surface treatments.

Compared with oxides produced by CSA or FPL they have greater stability against hydration,[2] and high chemical reactivity.[5] Especially the hypothesis of a micromechanical interlocking between polymer and oxides as an important mechanism, besides chemical reactions, has attracted some attention.

THE RESIN-OXIDE TRANSITION ZONE

A micromechanical interlocking can only become effective if the molecules of the primer or the adhesive during and after application and before curing penetrate the oxide layers produced by the surface treatment. Even after drying of the treated surfaces a relatively large amount of water is present in the oxide layer.[6] So, at first glance, it seems not very probable that organic molecules with molecular weight up to more than a thousand can completely penetrate the aluminium oxide layer.

Nevertheless such a total penetration seems to be possible from the two transmission electron micrographs, shown in Figure 2.

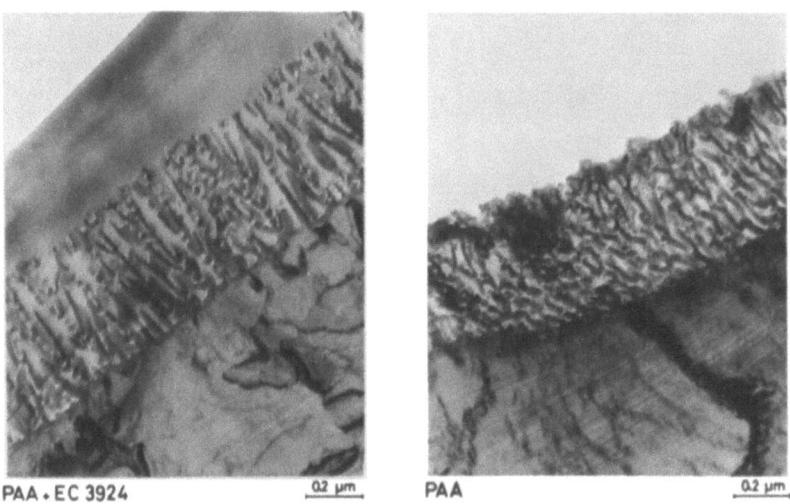

Figure 2. TEM of cross sections from PAA-oxides prepared from a metal surface without primer, right; and with cured primer (EC 3924, 3M Co.), left.

If an ultra-thin cross section is taken from close to the surface of a PAA-treated aluminium, deformations in the oxides structures can occur as can be seen in the right micrograph. If the oxide layer is coated by a primer before the cutting process and this primer is cured as usual, a deformation of the oxide by the cutting process is not observed, as can be seen in the left TEM-picture. From the differences in the pictures it may be concluded that the primer strengthens the oxide against deformation in this preparation process. The same strengthening effect by a primer is also observed on surfaces pretreated by etching or by the chromic-acid-anodizing (CAA)-process.

The conclusion that large portions of the oxides on the aluminium are penetrated by molecules of the primer is confirmed by results of the electron energy loss spectroscopy (EELS) on ultra-thin cross sections. One result of such investigations is shown in Figure 3.

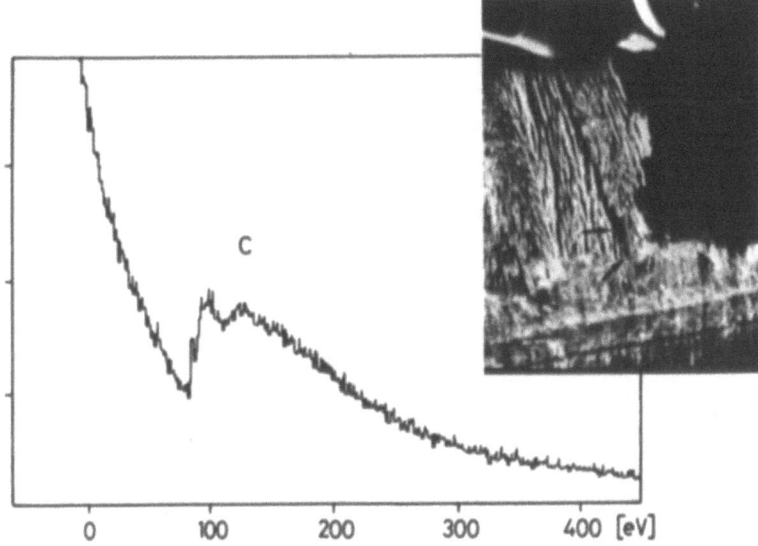

Figure 3. Polymer identification in CAA-oxide by EELS by analysis of carbon.

The points of identification in the oxide structure are marked and are shown in the TEM Figure 3. Today only those analytical techniques with lateral resolution in the range of less than 100 Å allow identification of organic material in different ranges of the oxide structure. If the cross section is prepared on an anodized metal without primer coating, carbon cannot be detected by EELS.

Assuming the presence of a partly or completely cured resin in the pores of the oxide layer, it is hard to understand how invading water should cause an adhesional failure in this zone. Especially considering the rapid failures induced by water this is not easy to visualize. Such fast aging effects are known from the wedge-test and the wet peel test.[2,7] In the last test within seconds a cohesional crack can be turned into adhesional failure only by adding water.

For the examination of strength and durability of metal bonds, the wedge-test has been used as a short-time test method for some years.[2] Today this test method is accompanied in some cases by the wet peel test. In this test using a normal roller-peel-specimen the pulling-off procedure is started in a normal climate of 20 °C and 60 % rel. humidity. After peeling-off half the length of the bond line, the peel-process is stopped and in the tip of the cohesive crack in the bond-line a drop of water (20 °C) containing 0.5 % detergent, for complete wetting, is applied. 10 sec later the peeling process is continued. If the adhesion between primers or adhesives and oxides on the metal surface is sensitive against humidity, the peel-strength decreases after the water application and the fracture surfaces show totally or partly adhesional failure, as can be seen in Figure 4 on peeled specimens, whose metal parts were anodized under different conditions before primer application.

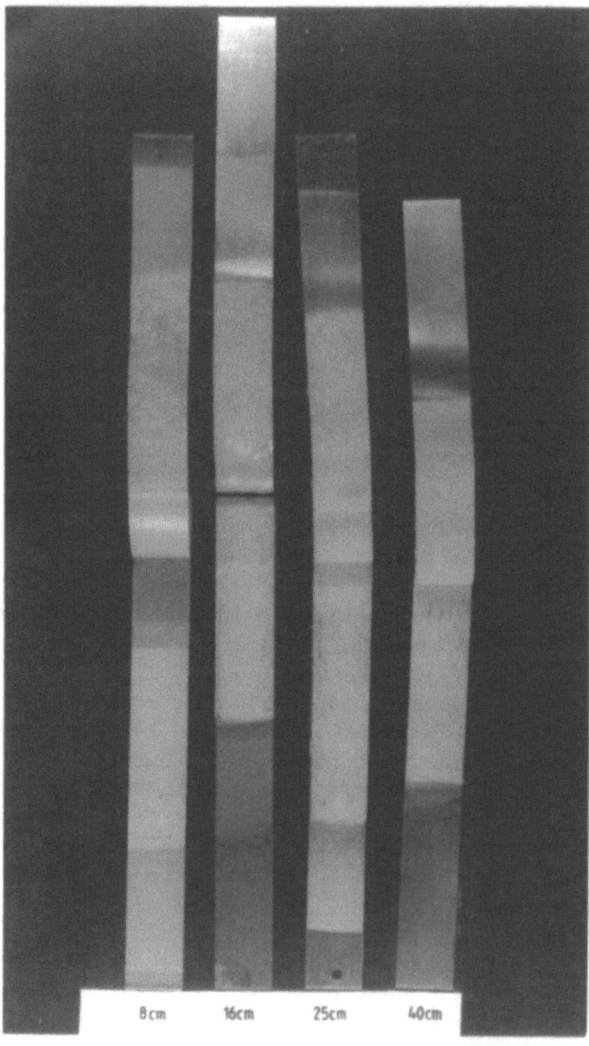

Figure 4. Fracture surfaces of peeled specimens, aluminium CAA-anodized with different anode-cathode distances.

The peel strength measured on these specimens in dry and wet states is plotted in Figure 5.

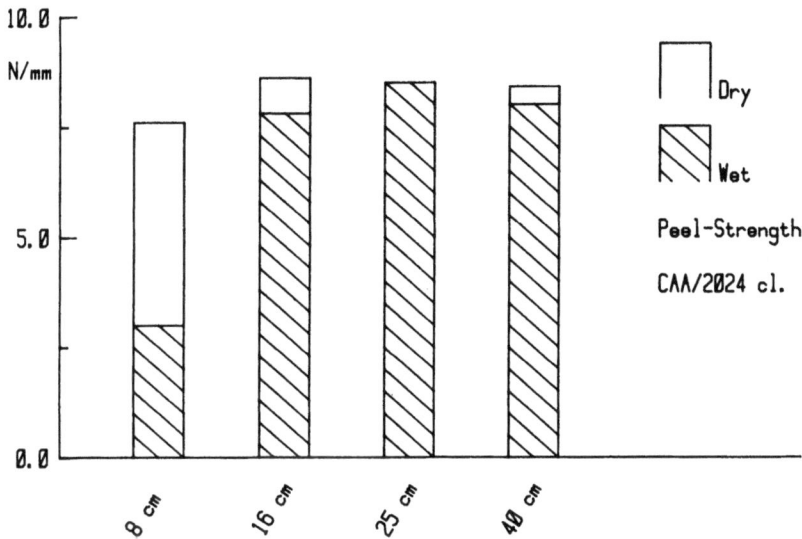

Figure 5. Peel strength of metal bonds after different CAA-anodizing processes.

In the case of sufficient adhesion stability, the peel strength remains practically constant in the wet state and no adhesional failure is detected on the fracture surfaces. This test has proved to be more sensitive than the wedge-test in our and other investigations[7] and seems to be one of the best short-time tests to determine the long term stability of metal-polymer bonds against humidity.

The roller-peel specimens of the tests mentioned here were produced using the Alloy 2024 T3 clad, corrosion inhibiting aircraft-primer EC 3924 containing different polymer components like epoxy-resins, epoxy-novolak-resins, special curing agents and strontium-chromate as pigment, and the nitrile modified epoxide-adhesive AF 126. Prior to bonding, the metal surfaces were degreased, etched in chromic sulphuric acid and anodized in chromic acid at 40 °C with a stepwise increasing of voltage up to 50 Volts.[3] For the anodizing process, different anode-cathode distances were used to reach different current densities, which are shown in Table 1. The results show that only under relatively low current densities adhesion of sufficient stability is obtainable.

Table I. Dependence of Current Density on Anode-Cathode Distance during Chromic Acid Anodizing.

Anode-cathode distance cm	Current density A/dm²
8	2
16	1,6
25	1,2
40	0,4

FAILURE MECHANISMS

Because scanning electron micrographs of the surfaces after different anodizing processes (Table 1) showed practically no differences in the oxide topography, ultra-thin cross-sections from the boundary zone of undestroyed bonded joints were prepared and examined in a transmission electron microscope. Some results of these investigations are shown in Figure 6.

Recognizable in the micrographs are, at first, differences in the oxide morphology depending on the current density during the anodizing process. Second and probably more important is the presence of bright zones in the primer area near the oxide in the case of oxides produced by anodizing with 2 A/dm² and 1.6 A/dm².

Only after anodizing with 0.4 A/dm² such a bright zone is not recognizable. A comparison between these micrographs and the fracture surfaces in Figure 4 lead to the conclusion that a correlation between the appearance of such bright zones and water-unstable adhesion may exist. A first assumption that these bright zones in the polymers are due to insufficient wetting during the primer application does not seem very probable, because if this was the case then adhesional failure must occur in the peel-test in the dry state, too, and not only in the wet state. But the initial peel-strength of all the bonds is practically the same, and in the dry state cohesional failures occured in all cases.

A second assumption is that the bright areas in the microtomed samples are due to fractures at the polymer/oxide interface or extremely thin spots both, for example, caused as artifacts by the cutting process which can only be carried out in contact with water. If this is true, we have to conclude that the resistance of the polymer near the oxides against mechanical deformation in contact with water must depend on the state of the oxides on which it was cured because using the same cutting parameters artifacts occur on some oxides and not on others.

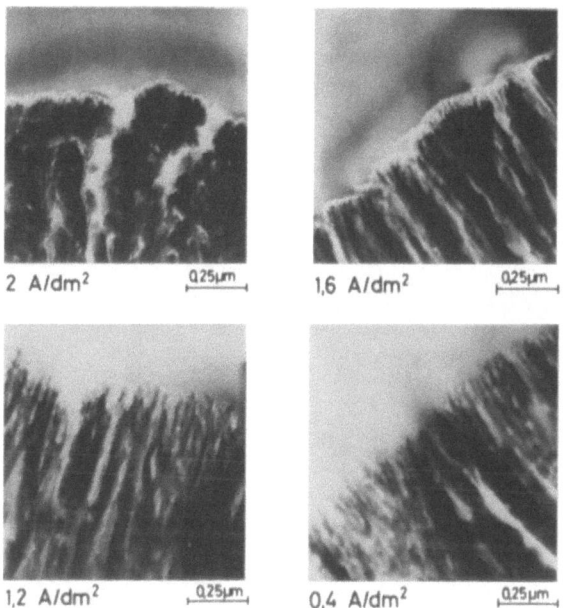

Figure 6. TE-micrographs of cross sections from oxide/primer transition zones. CAA-oxide produced with different current densities; primer EC 3924.

At last it may be that the bright zones are caused by material of much lower molecular weight than the surrounding materials, oxides and cured polymer. Both, differences in resistance against the cutting process or differences in molecular weight of the polymer zone near the oxides may be induced by selective chemical reactions of the organic molecules with the oxide or selective adsorption mechanisms of reactive portions of the primers invading the oxides, whereas portions of lower reactivity or larger size remain at the surface and cannot crosslink as tight as the other polymer in the curing process.[8]

The question now is whether the so-called "adhesional

failures" in the wet peel tests could be correlated directly to the appearance of the bright zones.

To demonstrate this, ultra-thin cross sections from the boundary area of a metal/polymer bond after an "adhesional failure" induced by wet peeling was prepared. The TEM of this cross section are shown in Figure 7.

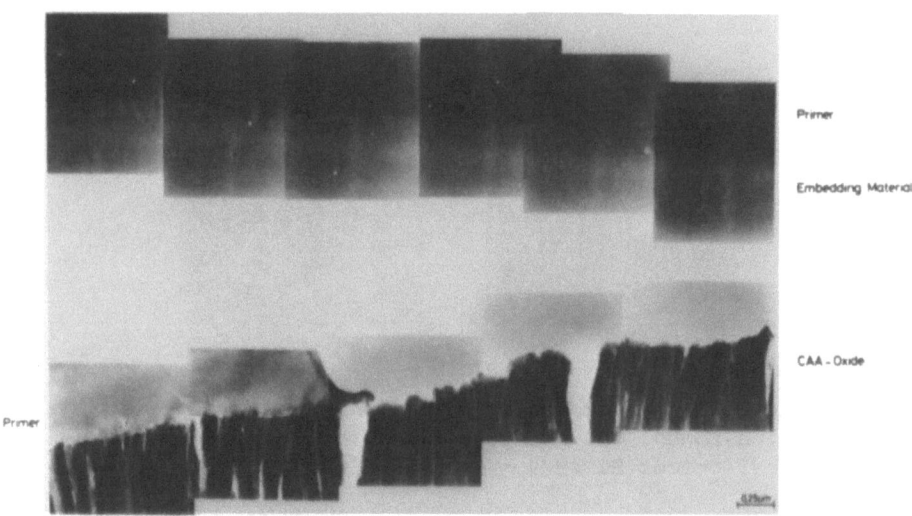

Figure 7. TEM of cross sections from the polymer and the metal sides of an adhesively bonded aluminium joint after "adhesional failure". Surface treatment CAA. Upper Figures: Transition zone between the primer and the embedding material used for microtoming. Lower Figures: Transition zone between CAA-oxides and the primer. This part was microtomed without embedding material.

In the lower part of the Figure the typical CAA-oxide-structure is recognizable on which remains, particularly on the left side, a primer in which again a bright zone in the boundary layer can be seen. At the right, the crack changes from a cohesional failure to an adhesional mode, but on the top of the oxides some material remained, which is perhaps of different nature than the polymer bulk itself. On the other hand, in the upper part of the

Figure there is the area of the primer surface which had to be
embedded before the cutting process. Between the primer and the
embedding material practically no dark areas caused by the presence of oxide particles are to be seen. The picture leads to the
conclusion, that the "adhesional failure" in reality occurred as a
a cohesional failure in the polymer near the top of the oxides.
This interpretation of the transmission electron micrographs is
confirmed by results from ESCA, by investigating the fracture
surfaces after such an adhesional failure. Some of the results
are shown in Table II.

Table II. ESCA investigations of an "Adhesional Failure" in an
Adhesively-Bonded Aluminium Joint.

Element	Metal Surface %	Adhesive Surface %
O	21,3	15,3
N	1,1	1,3
C	65,4	78,8
Al	6,1	0,5
S	3,3	2,2
Na	2,9	1,6

Surface Treatment: CAA
Adhesive/Primer: AF 126/EC 3924

These results show that in the case of adhesional failure
on the metal side, relatively high amounts of carbon are detectable; and on the primer or adhesive side there are very low
amounts of aluminium or aluminium-oxides.

The carbon containing material on the metal surface may be
partly residual detergent on the oxide surface as a result of the
wet peel test procedure. But the high carbon/sulphur and carbon/
sodium ratio makes it probable that at least portions of layer are
remains of the primer.

The situation is very different, if, for example, only the
chromic sulphuric acid etching as a surface treatment was used.
After an adhesional failure, not only the resin remains on the
metal side but oxides particles are also seen on the primer surface, Figure 8.

From the TEM's in Figure 8 it is not possible to decide
precisely, whether the boundary failure occurs only in case an
influenced polymer zone is present. But it cannot be excluded that
between the thin oxide whiskers of the CSA-layer, polymer with
properties different from that in the bulk is existing.

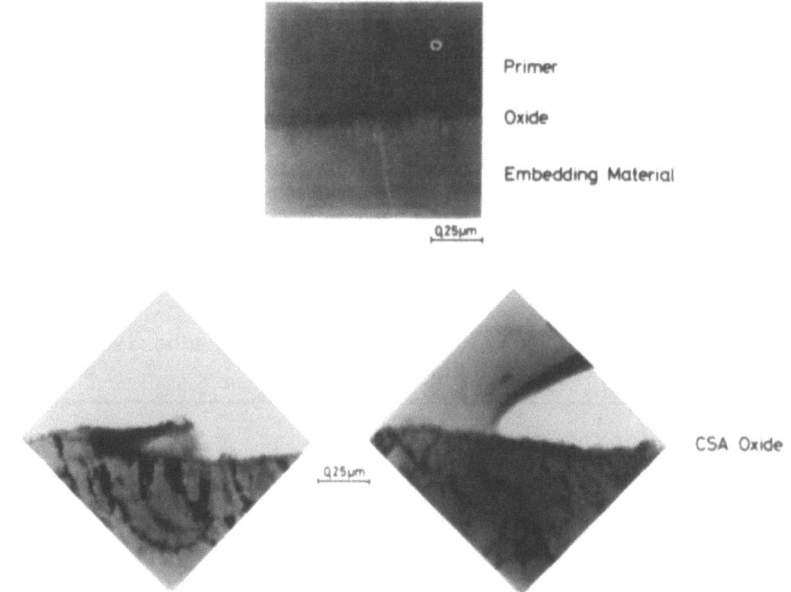

Figure 8. TEM of cross sections from the polymer and the metal sides of an adhesively bonded aluminium joint after "adhesional failure". Surface treatment CSA.

These statements are true only for the system aluminium with corrosion inhibiting primers; however some other results show that on steel surfaces, too, in the case of adhesional failure on the metal, polymer remain whose chemical and mechanical properties seem to be influenced by the state of the metal surface.[9]

The presence of weak boundary layers in polymers as an important reason for the water instability of the adhesion in such systems is contradictory to the conclusions of Venables in the foregoing chapter of this volume[10]. From his results he concluded that an adhesional failure in aluminium/polymer bonds is mostly due to hydrolysis of the oxides followed by crack in the hydroxylated zones. But in comparing his results and the pictures shown here, the different aging conditions in the two investigations have to be taken into account.

It is true that a hydration process in the oxides can also produce delaminations in an aluminium-polymer system or bond. But the stability of certain anodic oxide layers against hydration is relatively high. This is measurable by storing, for example, unprotected anodized aluminium parts in a humid environment of 40 °C and 95 % relative humidity without condensation of water. Under these conditions it needs hundreds of hours until hydration starts. To initiate hydration in CAA-oxides coated with a 0.2 mm thick primer layer (EC 3924) cured under normal conditions very severe environmental conditions are needed. Hydration of the oxide layer under the polymer as shown in Figure 9 was only possible, if the specimen was stored for over 7 days at 133 °C in water vapour with a pressure of 2 bars.

Figure 9. TEM of cross sections after an aging process from CAA-oxides coated with 0.2 mm thick primer EC 3924. Aging conditions: 7 days at 133 °C in H_2O vapour.

For example, in adhesively bonded aluminium joints with anodized surfaces and epoxide adhesives with corrosion inhibiting primers aged over a year in humid climate of 40 °C and 95 % rel. humidity we could not detect partly hydrated oxides as shown in Figure 9. Further we investigated recently an adhesively bonded aluminium joint from the fuselage of British civil aircraft Trident. The surfaces were shotblasted, chromic-sulphuric etched and chromic acid anodized, and the adhesive was Redux 775, a modified phenolic resin. The fuselage was tested in a pressure tank under water over 13 years and 139 500 simulated flight load cycles. After this test the fuselage was stored without protection for over 4 years on a British airfield. Despite this long aging process no hydration in the CAA-oxide layer under the phenolic resin was detected, as shown in Figure 10. The specimen investigated was microtomed in an area which is 5 mm away from the unprotected edge of a bond between the stringer and the outer cover of the fuselage.

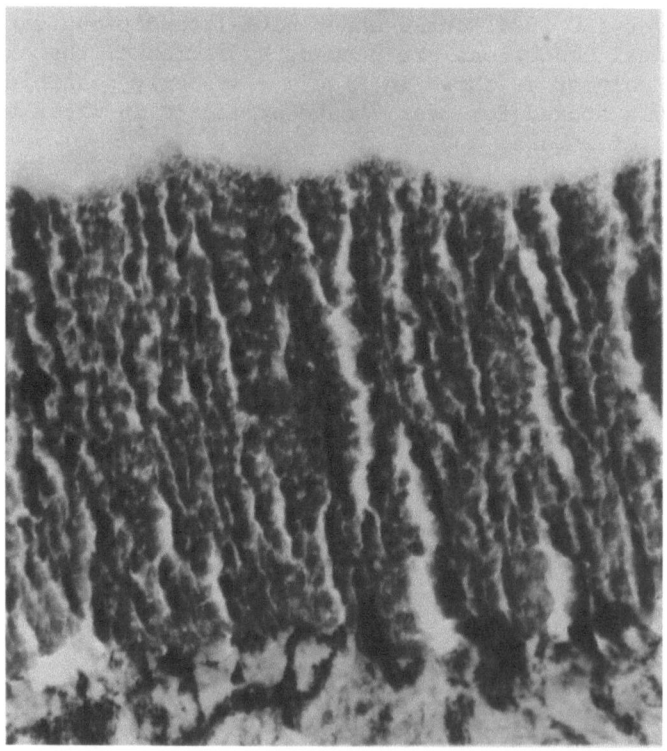

Figure 10. TEM of a cross section from a long time aged adhesively bonded aluminium joint.
Surface treatment: CAA
Adhesive: Redux
Aging time: 13 years under load in H_2O
  +4 years in natural climate.

CONCLUSIONS

Insufficient water stability in the boundary layer of adhesively bonded aluminium joints is easily detectable by the wet peel test. Particularly, in the case of chromic acid anodized surfaces, a water-sensitive zone in the polymer on the top of the oxides can exist. The polymer in this zone is of higher water sensitivity than the bulk polymer and its properties seem to be influenced by the state of the oxides. High stability in the boundary zone can only be observed if such a weak boundary layer does not exist. The failure mechanism in the weak polymer zones described here can only be induced in the presence of water by high plastic deformations of the polymer. Under these conditions microcracks can accelerate the migration velocity of water[11,12]. In the dry state this weak boundary layer is not of importance for the behaviour of the bonds. If wet peel specimens after a peeling process in the presence of water are redried, the crack again changes from the boundary layer back into the adhesive as a cohesional failure. This partial reversibility of the strength in the boundary layer in the presence or absence of water is an indication too, that the weakening mechanism in the boundary layer is not due to hydration processes of the oxides, which are not reversible under all aging processes.

All of the statements given in this report are valid only for aging processes of polymer/metal systems without primary corrosion attack of the metal parts. If primary corrosion occurs, for example, under condensing humidity at unprotected edges of the joints and aging times of more than several days, "bond line corrosion failures" are observed in some cases which are induced by other failure mechanisms, in which the chemical stability of the oxides plays an important role.

REFERENCES

1. P.F.A.Bijlmer and R.J. Schliekelmann, SAMPE-Quarterly 3, 13 (1973).
2. J.C. McMillan, "Surface Preparation, The Key to Bondment Durability", in: Bonded Joints and Preparation for Bonding, AGARD-Lect. Ser. 102, 1979.
3. O.-D. Hennemann and W. Brockmann, J. Adhesion 12, 297 (1981).
4. A. Hartman, Report of Nationaal Luchtvaart-Laboratorium Amsterdam 1961.
5. W. Brockmann, Adhesives Age 20, 30 (1977).
6. W. Wernick and R. Pinner,"The Surface Treatment and Finishing of Aluminium and its Alloys,"Fourth Edition, R. Draper, London 1972.
7. K.K. Knock and M.C. Locke, in "Adhesion Aspects of Polymeric Coatings", K.L. Mittal, Editor, p. 301, Plenum Press, New York 1983.
8. J.A. Marceau, "An SEM Analysis of Adhesive Primer Oriented Bond Failures on Anodized Aluminium", SAMPE Quarterly 10, 1, (1978)
9. D.M. Hercules, J. Electron Spec. Rel. Phen. 5, 811 (1974).

10. J.D. Venables, these proceedings, pp. 453-467
11. C.L. Leung and D.H. Kaelble in "Resins for Aerospace", C.A. May, Editor, ACS Symp. Series 132, p. 419, Washington 1980
12. R.J. Morgan and E.T. Mones in "Resins for Aerospace", C.A. May, Editor, ACS Symp. Series 132, p. 232, Washington 1980.

JOINT DURABILITY STUDIES WITH ABRADED, ETCHED, COATED AND

ANODIZED ALUMINUM ADHERENDS

J. Dean Minford

Product Engineering Division
ALCOA Laboratories
Alcoa Center, PA   15069

The effect of surface pretreatment on the aluminum oxide surface is critical and intimately related to the bond permanence of aluminum adhesively bonded joints. It is difficult, however, to distinguish the benefit of different surface treatments in the absence of some weathering test condition since the joints will fail cohesively for many different surface-treating conditions when tested without such exposure. Aluminum alloy 6061-T6 was pretreated by various procedures including sanding, grit blasting, hot acid etching, heat-cured epoxy paint priming, Alodining, and sulfuric acid anodizing. Both room-temperature-curing two-part and one-part heat-curing epoxy adhesives were used to fabricate the test specimens. The weathering evaluation procedures included both accelerated laboratory-type and natural atmospheric exposures. The list of weathering conditions included: immersion in room-temperature water, exposure to continuously condensing humidity at 52°C (125 F), and hot water soak/freeze/thaw cycling; as well as, exposure in an industrial-type atmosphere and a seacoast atmosphere.

INTRODUCTION

While many theories have been proposed to explain the various aspects of the adhesion phenomena, no universal theory of adhesion exists.[1,2] It is during the adsorption phase of the bonding process that the forces of adhesion become effective and whether highly durable joints are formed at this time is significantly influenced by the characteristics of the adherend surface.[3-8, 10-20]

Two types of debonding are generally recognized for weathered metallic adherend joints, as studied by Gledhill and Kinloch.[9] The first of these is promoted by the presence of water in the environment and can be considered a preferential desorption of the organic polymer adhesive by the highly polar water molecules at the interface. In this process, the aluminum oxide interface generally is converted to a cohesively weak boundary layer of aluminum hydroxide resulting in a joint failure appearance described as being interfacial in nature to the adherend. Another effect of the water may be to weaken the cohesive strength of the cured adhesive itself; in which case, the breaking load on the joint can be significantly below the original value, yet the nature of the appearance of the failed joint may be largely cohesive in nature. The second type of debonding is where corrosive water is present, and there can be gross corrosion of the aluminum as the failure mechanism. These failure mechanisms would indicate that to resist joint failure, more stable oxides and stronger interfacial forces need to be forged from better surface preparations.

It is the purpose of this paper to make suitable comparisons regarding the long-term durability tests between a variety of surface pretreatment conditions on aluminum including abrasion, etching, coating and anodizing techniques. The direct comparisons between surface pretreatments can only be made because the experimental variables of alloy, adhesives, curing conditions and weathering exposures were maintained as constants in experiments conducted over a twenty-year period of investigation.

MATERIALS AND METHODS

Tensile-shear 6061-T6 aluminum lap specimens in 3.18 mm (0.125") thickness were prepared according to ASTM D1002-72. This adherend has been used for more than 20 years at Alcoa because of its wide use in aluminum structures both as rolled sheet or extrusions. It also affords comparable joint strengths in the sheet thickness used here to those obtained with the prescribed thickness of Alclad 2024-T3 aluminum recommended in ASTM D1002-72.

For comparison purposes between surface pretreatments, the same two-part, room-temperature-curing epoxy (National Starch 82-0688/84-9023) and one-part, nitrile-modified, heat-curing epoxy (3M EC-2086) were used to bond each differently prepared adherend surface. These same adhesives have been used for more than 15 years in Alcoa investigations that have been carried out to compare a variety of different bonding parameters.

Similarly, the same long-time exposure weathering conditions consisting of immersion in room-temperature deionized water, soaking in continuously condensing humidity at 52°C (125°F), repeated cycling for consecutive 24-hour periods in 74°C (165°F) water, -34°C (-30°F) freezing air and 77°C (170°F) hot air oven, or an industrial or seacoast-type natural atmosphere have been continued in use.

Surface Preparations

The following surface preparations were investigated:

A. Vapor degreasing in refluxing perchloroethylene vapor.

B. Belt sand abrasion with Aloxite 120 grit cloth followed by vapor degreasing in perchloroethylene.

C. Blasting with silica grits to uniform appearance followed by vapor degreasing.

D. Alcoa 3 (chromic-sulfuric acid etching) - The surfaces were degreased in perchloroethylene vapor and treated by immersion in a solution of 35 grams of chromic acid and 100 ml of concentrated sulfuric acid in a liter of distilled water followed by a tap water and deionized water rinse and 121°C (250°F) forced air dry. The solution temperature was 82°C (180°F) and immersion time was 5 minutes.

E. Alodining - As prescribed by Amchem, the supplier of Alodine 1200.

F. Alumilite 215 (sulfuric acid anodizing) - The surface was pretreated by cleaning for 5 minutes in Diversey 914 at 66°C (150°F) and etched 8 minutes in Alcoa 1 (5 percent Pennwalt AE16 caustic at 49°C (120°F) followed by desmutting three minutes in 8-10 percent Sanfax MT324 at room temperature. This thick architectural-type coating was applied by anodizing in 16 percent sulfuric acid at 21°C (70°F) at 12 amps/ft^2 for 60 minutes to produce a coating about 0.95 mil thick. A tap water rinse was followed by a 10-15 sec. hot deionized water rinse (66-82°C, 150-180°F) and air dry. Sealing was in pH 6 deionized boiling water for 30 minutes.

G. Vapor degreasing in refluxing perchloroethylene. Spray application of Midland Chemical V-805 epoxy-phenolic coating and heat curing as recommended by manufacturer.

RESULTS AND DISCUSSION

Initial Joint Strengths

The average initial joint strengths obtained for both adhesives and ten different conditions of surface pretreatment are shown in Table I and have been graphed in Figure 1. A joint code number which refers to a fixed set of bonding parameters including alloy, surface pretreatment, adhesive, and curing condition has also been assigned in Table I. This code number has been used in subsequent tables along with only metric units of temperature and joint strength to conserve space in reporting the durability test results. It can be observed that a wide range of joint strengths result from the various surface pretreatments. For the room-temperature-curing epoxy, the range was from 13.90 MPa to 20.19 MPa, whereas the heat-cured epoxy joints were in a significantly wider and higher joint strength range from 17.02 MPa to 44.65 MPa.

Table I. Joint Strength with Different Surface Pretreatments.

Joint Code No.	Adhesive	Pretreatment	Curing Time	Temp,°C	Average Joint Strength, MPa
1C	82-0688/84-9023	Vapor Degreased	7 days	25	14.68
2C	EC-2086	Vapor Degreased	15 min.	204	29.68
1A	82-0688/84-9023	Belt Sanded	7 days	25	13.99
2A	EC-2086	Belt Sanded	15 min.	204	33.55
3A	82-0688/84-9023	Silica Grit Blast	7 days	25	13.90
1E	82-0688/84-9023	CrO_4/H_2SO_4 Etch	7 days	25	19.50
2E	EC-2086	CrO_4/H_2SO_4 Etch	15 min.	204	36.72
1CO	82-0688/84-9023	Alodine 1200	7 days	25	7.79
2CO	EC-2086	Alodine 1200	15 min.	204	17.02
3CO	82-0688/84-9023	Epoxy-Phenolic	7 days	25	20.19
4CO	EC-2086	Epoxy-Phenolic	15 min.	204	36.52
1AN	82-0688/04-9023	Alumilite 215[1]	7 days	25	17.43
2AN	EC-2086	Alumilite 215[1]	15 min.	204	22.53

1 - Alcoa Anodize Trademark

The magnitude of the joint strengths obtained with the various surface pretreatments merits some discussion. Any increase in joint strength due to abrading the adherend is strongly influenced by the nature of the adhesive. For example, belt sanding or silica grit blasting obviously produced physical roughening of the surface. No increase in joint strength resulted, however, as compared with the smoother, vapor degreased surface when using the more brittle, room temperature-curing epoxy adhesive. A significant increase in joint strength did occur, however, when the tougher, more crack-resistant, heat-cured epoxy adhesive was used. Chemical deoxidizing of the adherend produced significant joint strength gains using both adhesives because of a significant enhancement in the chemical wettability of the surface. The conversion coating presents a unique problem greatly influenced by a combination of coating thickness and adhesive curing conditions. Increasing thickness increases the possibility of finding lower joint load-to-failure values and weak boundary layer conditions present in the coating cross section. This is most evident in the very low joint strength of the room temperature-curing epoxy adhesive. A further dramatic change in physical cohesive strength of the conversion coating can occur where the adhesive is cured at elevated temperatures, as shown by the EC-2086 heat-cured joints. Heat-cured paints used as primers are capable of developing joint strengths equivalent to those of the best deoxidized surface pretreating. The anodized aluminum surface again presents a thickness-sensitivity situation, as shown for the conversion coatings. The decreasing joint strength accompanying increasing anodizing times for aluminum has been shown by Minford[6] to be an example of variable thickness anodic oxide layers breaking through planes of weaker structure.

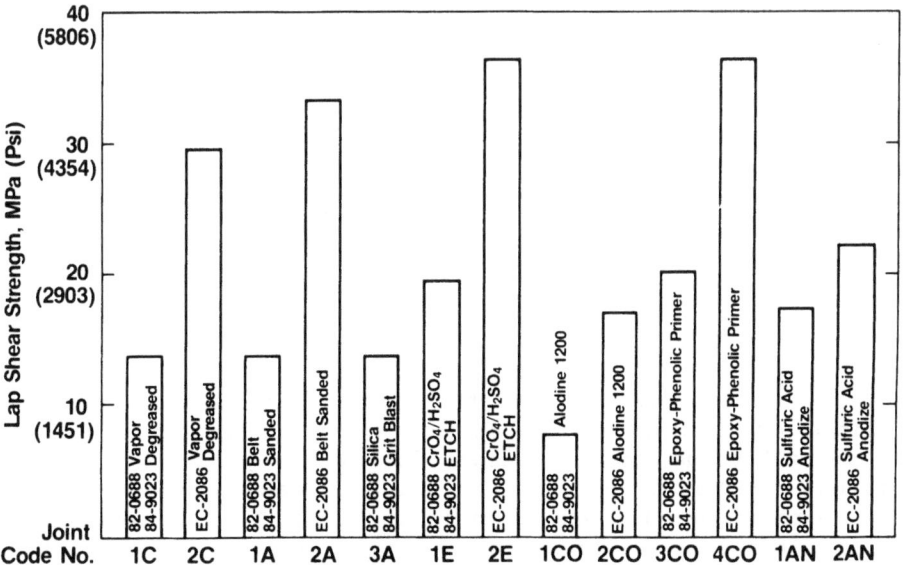

Figure 1. Effect of adhesive and surface pretreatment on joint strength.

Bond Durability in Room Temperature Water

The relative joint durability performances after 24 months of continuous soaking in deionized water are shown in Table II and have been plotted in Figures 2 and 3. No joints lost as much as 50 percent of their initial strength although the vapor degreased-only surface pretreated joints degraded 47 percent. Six of nine surface pretreatment condition joints showed less than 20 percent strength loss in the two-year period. The highest overall joint strength retention was 19.29 MPa for the two-part epoxy and 30.32 MPa for the heat cured one-part epoxy. In each case, the adherend pretreatment was a heat-cured epoxy phenolic paint over a vapor degreased-only surface.

Table II. Room Temperature Water Immersion Tests.

Joint Code No.	Average Joint Strength After Indicated Exposure Time, MPa				% Strength Retention
	3 Mo.	6 Mo.	12 Mo.	24 Mo.	
1C	8.96	7.34	8.47	7.79	53
2C	21.15	13.78	10.82	8.06	27
1A	12.73	12.88	12.35	*	88
2A	25.98	23.63	21.36	14.95	45
3A	14.88	*	12.88	14.68	100
1E	20.46	*	17.43	18.40	94
2E	30.80	30.32	28.73	24.32	66
1CO	7.85	6.75	5.99	4.69	60
2CO	17.02	14.95	14.95	11.51	67
3CO	20.88	20.19	20.88	19.29	96
4CO	34.93	32.38	31.49	30.32	83
1AN	17.57	17.02	15.16	14.95	86
2AN	17.02	14.59	16.54	16.74	74

*No Test Result

In spite of the fact that the highest strength retention was obtained with heat-cured EC-2086 joints, one should recognize from the overall comparatively steeper decline in joint strength for the heat-cured EC-2086 joints in Figure 3, as compared with the room temperature-cured epoxy joints in Figure 2, that performance for this adhesive is more dependent on good surface preparation. Thorough surface deoxidizing, priming with a paint, anodizing, or conversion coating are required for good EC-2086 joint durability even in relatively mild room temperature water-soaking conditions.

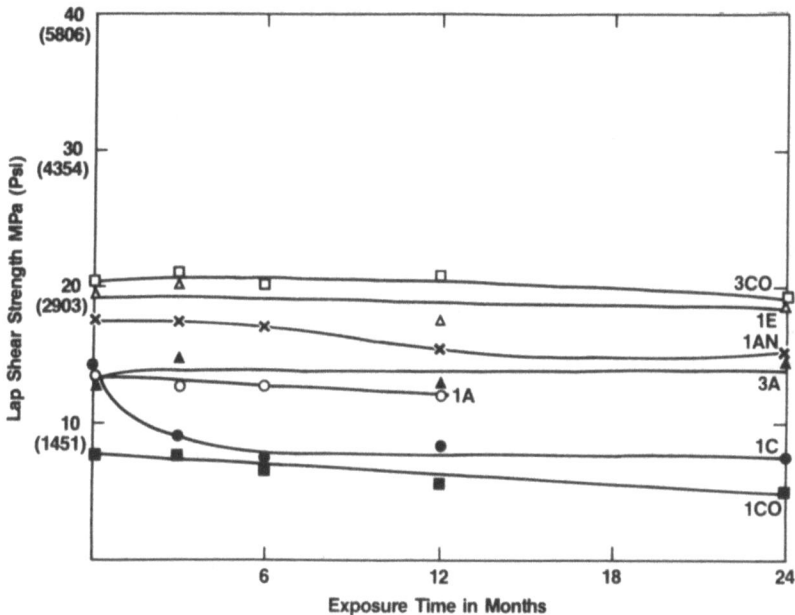

Figure 2. Durability of room-temperature-curing epoxy joints in room temperature water-soaking exposure (for code no., see Table I).

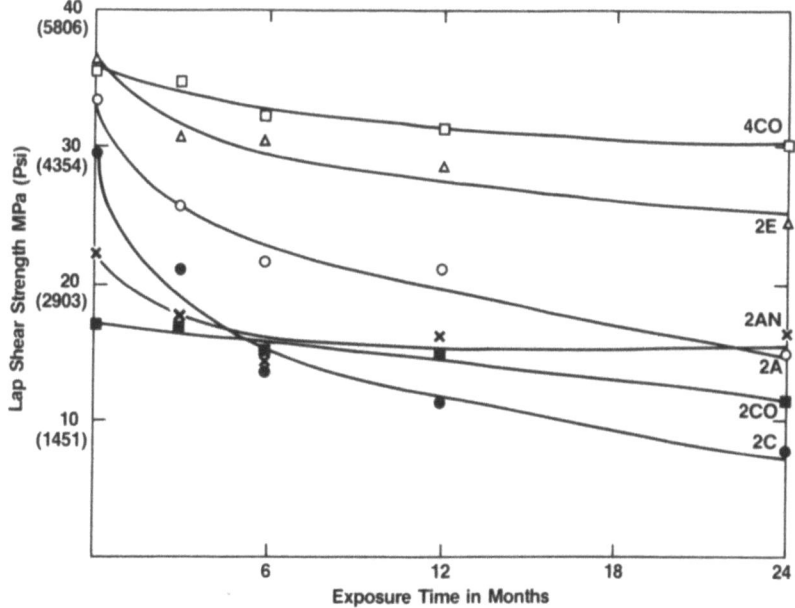

Figure 3. Durability of heat-cured epoxy joints in room-temperature water-soaking exposure (for code no., see Table I).

492   J.D. MINFORD

Bond Durability in Condensing Humidity at 52°C (125°F)

The results are summarized in Table III and Figure 4. Although a limited number of surface pretreatment joints were exposed to this condition, but some valid comparisons can still be made.

Table III. Exposure to 100% Relative Humidity at 52°C (125°F).

Joint Code No.	Average Joint Strength After Indicated Exposure Time, MPa				% Strength Retention
	3 Mo.	6 Mo.	12 Mo.	24 Mo.	
1A	10.32	10.80	11.07	*	79
2E	14.68	8.75	4.55	6.10	17
3CO	19.77	18.60	17.91	17.43	86
4CO	28.73	25.70	22.25	19.50	53
1AN	14.33	12.40	13.09	15.16	87
2AN	18.33	16.54	18.12	17.91	80

*No Test Result

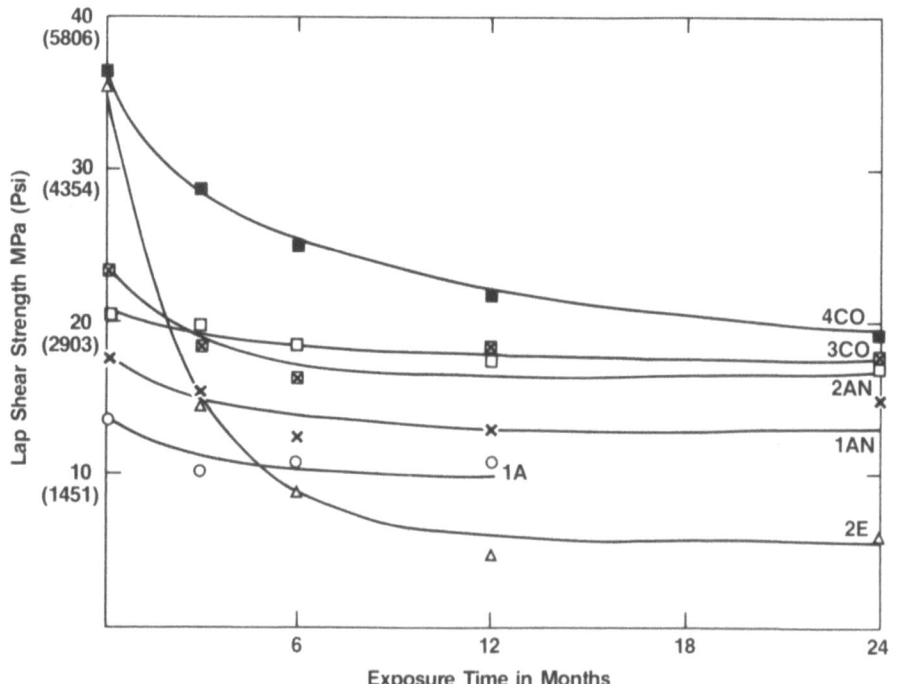

Figure 4. Durability of heat-cured epoxy joints in condensing humidity at 52°C (125°F) exposure (for code no., see Table I).

Strength retention of 80 percent or higher was shown by at least one example of an abrasion, etching, coating or anodizing procedure with the room temperature-curing epoxy. Only the sulfuric acid anodizing treatment achieved this percentage retention using the heat-curing epoxy. Once again the epoxy phenolic paint-primed joints remained strongest overall at 17.43 MPa for the two-part epoxy joints and at 19.50 MPa for the heat-cured epoxy joints.

It should be noted that thorough deoxidizing of the adherend was not an effective treatment for promoting a high level of joint durability with heat-cured EC-2086 adhesive in this environment. Since this laboratory weathering condition is closely related to accelerated tropical-exposure conditions, one would conclude that the adherend surface would need to be primed, anodized, or conversion coated for long-time tropical service.

Bond Durability in Soak/Freeze/Thaw Cycling

The data summarized in Table IV and plotted in Figures 5 and 6 show no surface pretreatment could overcome a tendency for significant joint strength loss for the heat-cured epoxy joints. As expected, the Alcoa 3 (similar to FPL etch) pretreatment afforded the highest strength retention in this aggressive water weathering exposure for the two-part room temperature-cured epoxy joints.

Table IV. Soak/Freeze/Thaw Cycle Testing

Joint Code No.	Average Joint Strength After Indicated Exposure time, MPa				% Strength Retention
	3 Mo.	6 Mo.	12 Mo.	24 Mo.	
1C	0.83	0	0	0	0
2C	13.09	11.51	7.30	3.45	12
2A	12.4	11.23	9.65	0	0
3A	8.82	8.75	6.44	0	0
1E	17.71	13.71	13.57	14.47	74
2E	11.71	10.54	9.16	0	0
1CO	5.72	5.55	5.27	0.83	11
2CO	17.08	12.94	16.54	0	0
1AN	13.57	12.88	10.54	8.41	48
2AN	15.64	13.99	15.16	6.41	28

An important fact that needs to be recognized when testing a room temperature-curing epoxy adhesive is that the water temperature of the test environment or the air temperature in part of a cyclic weathering condition can have strong influence. When a room temperature-cured epoxy adhesive is exposed to a significantly higher air- or water-temperature condition in test, there is a tendency for additional cross-linking to take place, and this occurs before any appreciable degrading action due to other factors that may be present. In the cyclic weathering environment used in this investigation, the adhesive, already cured at room temperature, experiences both 74°C (165°F) water and 77°C (170°F) air during each three-day cycle. Without special surface preparation the joint is still degraded to the point of failure within three months; but with special surface preparations, the room-temperature-curing joint performances, as shown in Figure 5, are significantly superior to that predicted from the EC-2086 data in Figure 6. One should also note from Figure 5 that a thoroughly deoxidized surface preparation is equal in performance to an anodizing preparation using the room temperature-curing epoxy. In contrast, this surface preparation had been inadequate even in hot humidity when using the heat-cured nitrile modified epoxy adhesive.

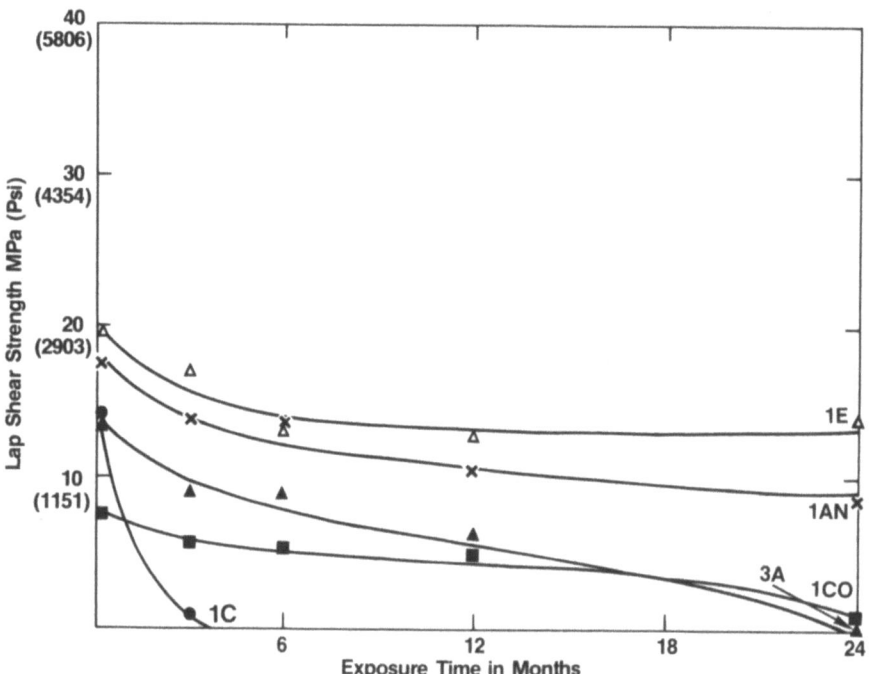

Figure 5. Durability of room-temperature-curing epoxy joints in soak/freeze/thaw cycling exposure (for code no., see Table I).

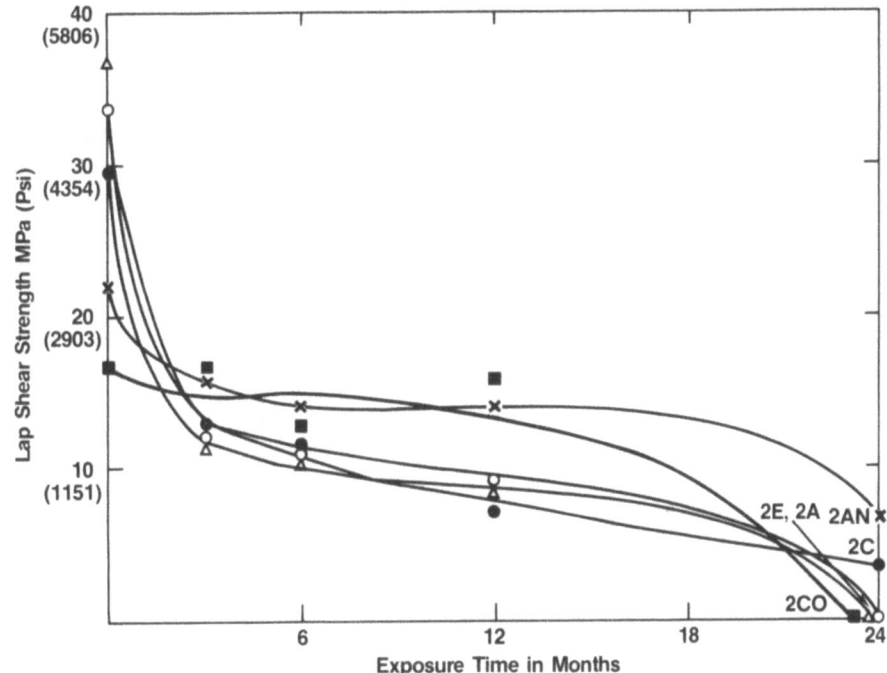

Figure 6. Durability of heat-cured epoxy joints in soak/freeze/thaw cycling exposure (for code no., see Table I).

Bond Durability in Industrial Atmosphere

The data are summarized in Table V and plotted as strength-retention curves over an eight-year exposure time in Figures 7 and 8. It is clear that the industrial atmospheric exposure does not deteriorate adhesive-bonded aluminum joints to the same degree as shown in earlier accelerated laboratory-exposure tests. Added confidence in the acceptable durability of such joints comes from the fact that this exposure condition represents the real service condition for the majority of structures. Since no edge or top coating protection was provided for these joints, there is the possibility of some acceleration still occurring in this weathering condition.

From the room temperature-curing epoxy data in Figure 7, it is clear that all forms of surface pretreatment attempted lead to durability enhancement as compared to the vapor degreased-only surface condition. An identical summary statement can be made after studying the results in Figure 8 obtained with the one-part heat-cured epoxy.

Table V. Results of Industrial Atmospheric Weathering.

Joint Code No.	Average Joint Strength After Indicated Exposure Time, MPa				% Strength Retention
	1 Year	2 Years	4 Years	8 Years	
1C	12.20	10.54	9.09	6.25	43
2C	22.05	21.36	19.29	3.10	10
2A	29.63	28.46	25.98	23.43	70
3A	13.78	14.26	13.99	13.99	100
1E	18.60	20.67	20.23	18.71	100
2E	37.21	36.52	35.83	32.59	59
1CO	8.75	8.27	8.96	7.79	100
2CO	18.60	17.91	18.60	19.09	100
3CO	*	19.50	20.77	19.70	98
4CO	*	33.97	30.56	33.35	91
1AN	16.05	14.68	12.88	10.92	63
2AN	19.29	19.09	19.50	18.09	80

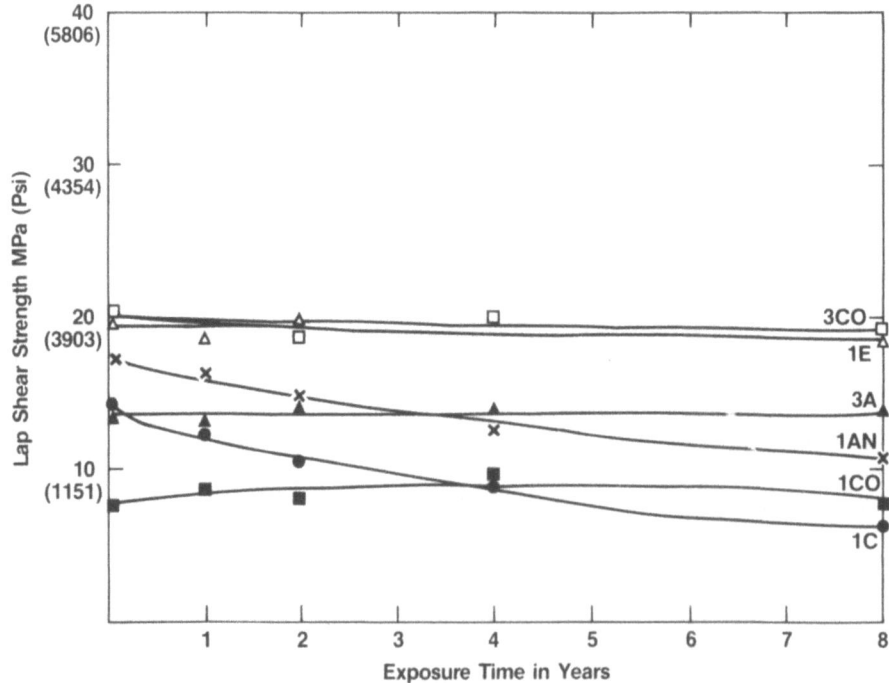

Figure 7. Durability of room-temperature-curing epoxy joints as a function of industrial atmospheric exposure (for code no., see Table I).

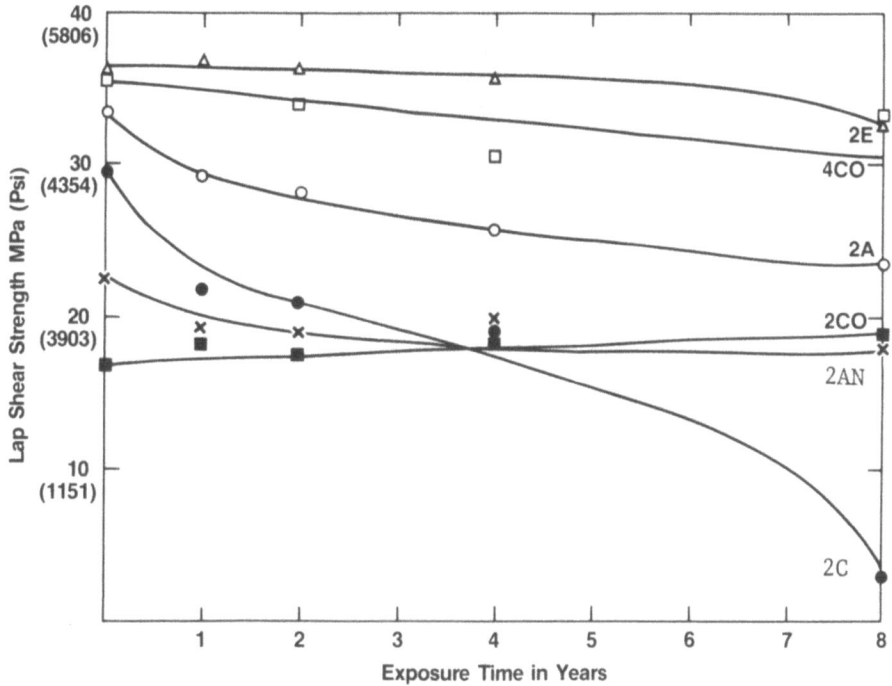

Figure 8. Durability of Heat-Cured Epoxy Joints in Industrial Atmospheric Exposure.

Bond Durability in Seacoast Atmosphere

The data are shown in Table VI and plotted in Figures 9 and 10. As indicated in the Introduction section of this paper, we must consider that bonded metallic joints can fail due to debonding by water penetration to the interface, or gross pitting corrosion of the adherend can destroy the adhesion by converting the metallic surface to corrosion product. In most cases, the cured adhesive may be left unaffected because the factors creating the condition for metallic corrosion are not necessarily aggressive to the cured polymer. As with the industrial atmospheric exposure just discussed, we must consider that exposure to a seacoast atmosphere is the real world for joints whose application brings them into contact with corrosive saltwater conditions.

The results shown for the room temperature-curing epoxy joints in Figure 9 fall into predictable and nonpredictable categories based on prior experience. The experience to which we refer is whether the surface, as treated, is inherently resistant to corrosive chloride ion attack in the sheet surface condition.

Table VI. Results of Seacoast Atmospheric Exposure.

Joint Code No.	Average Joint Strength After Indicated Exposure Time, MPa				% Strength Retention
	1 Year	2 Years	4 Years	8 Years	
1C	0	(All failed 71 days)			0
2C	0	(All failed 270 days)			0
2A	11.99	(All failed 760 days)			0
3A	16.74	12.40	10.82	9.30	67
1E	17.43	0 (All failed 760 days)			0
2E	34.66	23.22	0	(All failed 1440 days)	0
1CO	7.43	4.58	2.76	(All failed 1475 days)	0
2CO	16.74	14.68	16.26	17.78	100
3CO	*	11.09	5.17	(All failed 1497 days)	0
4CO	*	31.90	30.04		71
1AN	15.64	14.81	14.29	13.94	80
2AN	19.29	14.67	16.26	17.77	79

Of the various treatment conditions, one would expect both the vapor degreased and the deoxidized 6061-T6 surface to develop pitting readily, and these two kinds of joints did degrade first. We can explain the significantly longer survival time of the deoxidized surface joints as due to the much higher order of initial surface wetting[5] and adhesive spreading on the deoxidized surface. It was unexpected, and still not completely understood, to find the silica grit blasted joints surviving eight years exposure with no significant joint strength degradation. We might relate this result to the probable deposition of a silicate barrier film on the surface which resists undercutting pitting on the adherend. We have many years of experience demonstrating the effectiveness of anodizing films on aluminum as barriers to pitting corrosion, and this effect is clearly evident in Figure 9. The Alodine 1200 chromate-conversion coating has been an effective barrier to undercutting corrosion of heat-cured paint coatings in products like aluminum siding. This protection was only partially effective through four years of exposure (see Figure 9). We relate this to the fact that a change in the durability properties of the conversion coating results from heat curing after an adhesive or coating is put in place; but this change is not realized where a room temperature-curing adhesive is used. Finally, we believe the failure of the paint-primed, room temperature-curing joints was due to the use of only vapor degreasing as a pretreatment under the paint primer and less-than-optimum wetting of the primer by the room temperature-curing adhesive.

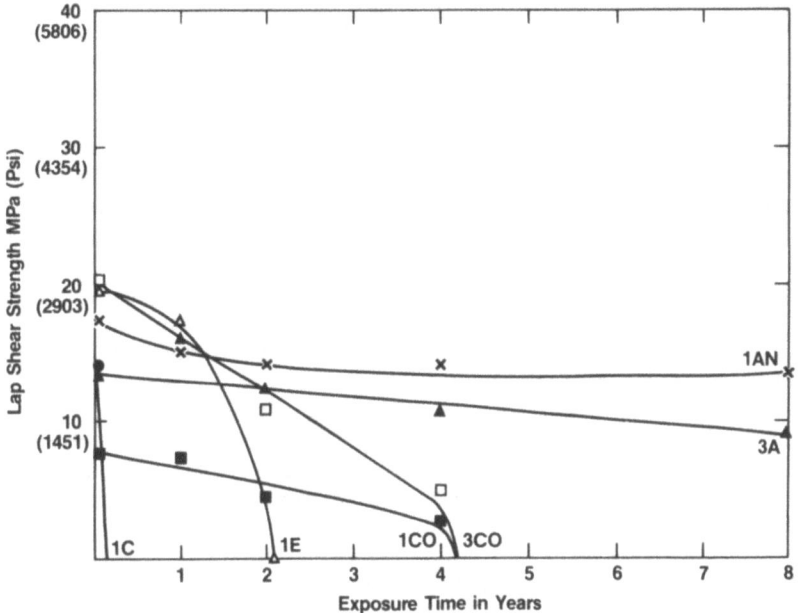

Figure 9. Durability of room temperature-curing epoxy joints in seacoast exposure (for code no., see Table I).

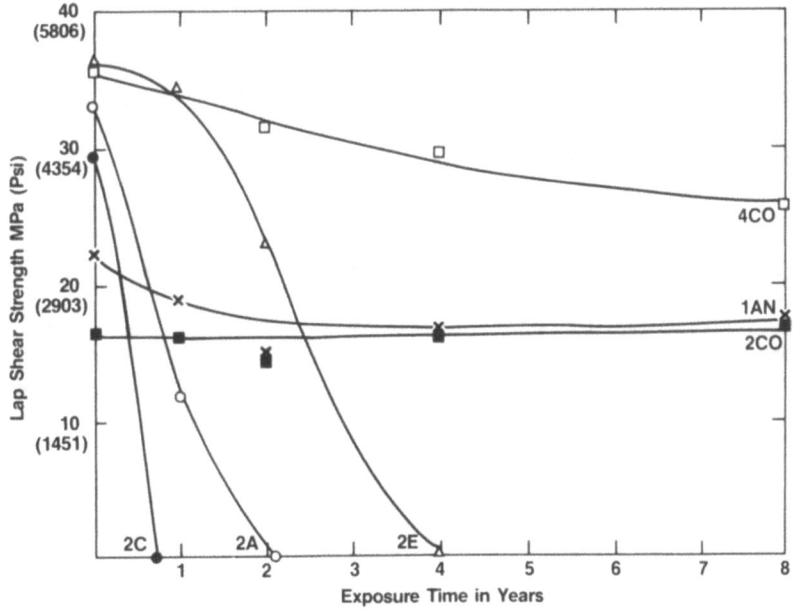

Figure 10. Durability of heat-cured epoxy joints in seacoast exposure (for code no., see Table I).

Some of the same comments used to explain the results of Figure 9 also apply to Figure 10. The poor durability performances of the vapor-degreased, belt-sanded, and deoxidized surface joints is expected because the surfaces are all sensitive to pitting corrosion. One needs to note, however, that the times to failure are all significantly extended over the room temperature-curing epoxy results because of the better opportunity of the adhesive to wet and spread over these surfaces due to heat curing. The anodized surface joints are virtually unaffected in eight years, as anticipated. The effect of heat curing of an adhesive over the Alodine 1200 conversion coating is striking, with the result that the durability is equal to that produced by anodizing. Finally, we also note a significant effect produced by heat curing the adhesive over the heat-cured paint primer, which originally had been applied over vapor degreased-only adherends. In a recent technical paper, Minford[10] has reported that this long-term durability of EC-2086 adhesive over epoxy-phenolic paint- primed aluminum has continued over a ten-year exposure period.

CONCLUSIONS

1. A wide variety of joint strength responses to various weathering conditions was achieved through surface abrading, etching, coating or anodizing procedures.

2. All of the pretreatments employed for this investigation showed acceptable joint strength retention in continuous water soaking exposures for two years when using a room temperature-curing epoxy. Only the acid etching, conversion or paint coating and anodizing pretreatments were acceptable for a heat-cured epoxy.

3. Only etching and anodizing pretreatments were satisfactory for room temperature-curing epoxy joints in soak/freeze/thaw cycling. None worked for the one part heat-cured epoxy.

4. All four kinds of pretreatment were effective in eight year industrial atmospheric exposure. Anodizing and conversion or paint coating procedures were most satisfactory in seacoast weathering.

REFERENCES

1. B. O. Bateup, Int. J. Adhesion Adhesives, 233 (July 1981).
2. R. J. Good, J. Adhesion, $\underline{8}$, 1(1976).
3. J. D. Minford, Adhesives Age, $\underline{17}$ (7), 24 (1974).
4. J. D. Minford, in "Treatise on Adhesion and Adhesives", R. L. Patrick, Editor, Vol. 3, Chap. 2, Marcel Dekker, New York, 1973.
5. J. D. Minford, in "Treatise on Adhesion and Adhesives", R. L. Patrick, Editor, Vol. 5, Chap. 3, Marcel Dekker, New York, 1981.
6. J. D. Minford, SAMPE Quarterly, $\underline{7}$, 18 (1978).
7. J. D. Minford, Paper presented at the ASM/SAMPE Conference on Specialized Cleaning, Finishing, and Coating Processes, Los Angeles, February 1980.
8. J. D. Minford, Adhesives Age, $\underline{23}$ (10), 36 (1980).
9. R. A. Gledhill and A. J. Kinloch, J. Adhesion, $\underline{6}$, 315 (1974).
10. J. D. Minford, Paper presetned at the ASM Surtech and Surface Coating Exposition-82, Dearborn, May 1982.
11. H. W. Eickner, Forest Products Laboratory, Madison, WI, WADC Technical Report 54-447, Part I, 1955.
12. H. W. Eickner, Forest Products Laboratory, Madison, WI, WADC Technical Report 54-447, Part II, 1957.
13. H. W. Eickner, Forest Products Laboratory, Madison, WI, WADC Technical Report 54-447, Part III, 1958.
14. H. W. Eickner, Forest Products Laboratory, Madison, WI, WADC Technical Report 59-567, Part I, 1960.
15. H. W. Eickner, Forest Products Laboratory, Madison, WI, WADC Technical Report 59-564, Part I, 1962.
16. R. F. Wegman, W. M. Bodnar, E. S. Duda and M. J. Bodnar, Adhesives Age, 22 (October 1967).
17. G. F. Carter, ASTM STP 401, 28 (August 1966).
18. R. F. Wegman, Picatinny Arsenal Technical Report 4169, Dover, NJ, June 1971.
19. J. A. Marceau and W. Scardino, Air Force Materials Laboratory Report, AFML-TR-75-3, Wright-Patterson Air Force Base, Ohio, February 1975.
20. J. C. McMillan, AGARD Lecture Series No. 102, Wright-Patterson Air Force Base, Dayton, Ohio, October 1979.

COMPARATIVE STUDY OF ALUMINUM JOINT STRENGTH AND DURABILITY WITH VARYING THICKNESS, BOEHMITE-TYPE OXIDE SURFACES

J. D. Minford

Product Engineering Division
Alcoa Laboratories
Alcoa Center, PA 15069

The presence of boehmite oxide on aluminum surfaces has often been associated with weather-durable bonding. In this work, a considerable variance was shown in aluminum joint durability involving such oxide surfaces with varying thicknesses as generated in the absence or presence of electrolyte in boiling water; i.e., deionized or tap water. The changes in oxide were generated starting from a deoxidized surface base by varying the exposure times in boiling water.

Joint strengths tended to decrease with increasing oxide thickness; i.e., longer exposure times, and could be as much as 50 percent lower in strength than the deoxidized surface control joints. Variable durability patterns resulted from varying weathering conditions; also, when stress was simultaneously imposed.

Bonding to deionized water-produced oxide surfaces showed enhanced durability using unstressed joints in strictly water-soaking conditions or corrosive salt fog cycle exposures. Durability generally increased as the oxide buildup in deionized water increased and was inversely related to initial joint strength. In addition to generally lower durability in these exposures, the tap water-produced oxide bonds had a different pattern, peaking in durability after ten minutes' exposure.

With simultaneously imposed stressing conditions, the tap water-produced oxide surface joints assumed the greater durability. These joints peaked in durability with the five-minute generation oxide condition. Performing overall with lower durability potential, the deionized water-produced oxide joints had their best response at ten-minute oxide generation level.

Very favorable durability responses were found for these predominantly boehmitic oxide-type compared to similarly fabricated and tested joints reported by Alcoa using acid deoxidized, and chromic or phosphoric-acid-anodized 6061-T6 adherends.

This procedure could have some practical application for low-cost general manufacturing operations where the simple boiling water bath treatment would be replacing more expensive aerospace-type surface preparations, like modified FPL and chromic or phosphoric acid anodizing procedures.

INTRODUCTION

Wegman et al.[1,2] studied how the oxides on 2024-T3 aluminum affect the bond strength of joints by deoxidizing the adherend in hot sulfuric acid-sodium dichromate solution and establishing that a new oxide developed after immersion in hot deionized or tap waters. Low joint strengths and adhesive-type failures resulted when bonding to the former. High joint strengths and cohesive-type failures were found with the latter treating conditions. They concluded that a thick, hydrated, cohesively weak surface oxide formed in deionized water from which strong bonds could not be obtained. The slight acidity present that promotes multivalent compounds in tap water was said to prevent generation of this hydrated layer. No durability test data were offered in these investigations. More recently other investigators, among them Wu and Bowen[3,4], have attached specific significance to generating the boehmitic form of oxide for most durable bonding with high initial strengths. It is the intent of this investigation to study some aspects of both considerations by studying the bond durability of joints fabricated with surfaces generated from either hot deionized or tap water immersion processing. The water temperature was deliberately maintained at 100°C (212°F) in order that any oxide generated would essentially be the boehmitic type. Additional variances from the Wegman work included substitution of 6061-T6 for 2024-T3 alloy and deoxidizing in hot caustic (Alcoa 1 Process) instead of sulfuric acid-sodium dichromate solution. Finally, because Russell and Garnis[5] had demonstrated significant differences in joint strength after varying water rinse conditions, a decision was made to study the boehmite generation effects using a range of water-exposure times.

Harrington and Nelson[6] first reported the film substance formed on aluminum by boiling water treatment was boehmite. They further found that temperatures above 80°C (176°F) were necessary. Bryan[7] and Altenpohl[8] found that with alloys of aluminum, the reaction was initially very rapid, with fairly quick slowing of the reaction afterwards, until virtual cessation after four to eight hours. The boehmite films formed were intermediate in thickness between air-formed films and anodic films. In the immersion times used for this study (one- to 60-minute range) the oxide film increase would be virtually linear according to Hart[9], who also found the rate seemed to be the same whether oxygenated or deoxygenated water was used.

METHODS AND MATERIALS

Individual lap joints were fabricated and tested according to ASTM D1002-72. Aluminum alloy 6061-T6 was vapor degreased and deoxidized by etching in a 66°C (150°F) 5 percent sodium hydroxide solution for two minutes. The resulting smut was removed by immersion in room temperature 40 percent nitric acid, rinsing in cold tap water, and immediately drying in an air jet; then immersing in boiling New Kensington, PA tap water or deionized water for either 1, 5, 10, or 60 minutes. Immediately after hot air drying, joints were made with EC-2086 nitrile epoxy adhesive cured in the bondline for 15 minutes at 204°C (400°F).

In order to make comparisons among such a wide variety of adherend pretreatments and different weathering exposure times, it was pertinent to use an Alcoa-developed technique which has been described.[11] In brief, exposure specimens are periodically tested to 50 percent of their estimated initial joint strengths based on actual initial test specimen values as shown in Table I. Joints which continue to pass this criterion continue to be exposed until they are deliberately stressed to failure at the end of a sufficiently long exposure time. The Alcoa Durability Factors shown in the tables of this paper are based on calculations which consider how much of the initial strength was retained for how long an exposure time and can be written as:

$$\text{Durability Factor} = \frac{\text{\% Retained Joint Strength} \times \text{Days in Exposure}}{1000}$$

In the evaluation of performance under simultaneous stressing and weathering exposure conditions the above formula was altered to read:

$$\text{Durability Factor} = \frac{\text{Imposed Stress in psi} \times \text{Joint Failure Time in Days}}{1000}$$

RESULTS AND DISCUSSION

Initial Joint Strengths

The average initial joint strengths obtained from 11 different adherend pretreatment conditions are shown in Table I and graphed in Figure 1. The deoxidized surface permitted highest joint strength development as expected. We assume increasing exposure time in water correlates with increasing oxide residual on the surface. We should expect the overall joints to decrease in strength as the oxide layer thickens. However, in Figure 1 we note that the relationship between exposure time and joint strength is quite different in deionized as compared to tap water. In deionized water, the average joint strength with increasing soaking time follows an exponential decay-type curve; whereas there does not appear to be any predictable relationship between joint strength level and soaking time through ten minutes for tap water-treated joints. After an exposure time of 60 minutes, however, joint strength is significantly lower and similar for joints derived from either water treatment. Also, we do not observe, as Wegman did, any striking difference between the types of bondline failures noted for joints pretreated in the two different waters; for example, joints treated 1, 5, and 10 minutes in each water averaged 80 to 90 percent cohesive-appearing failure.

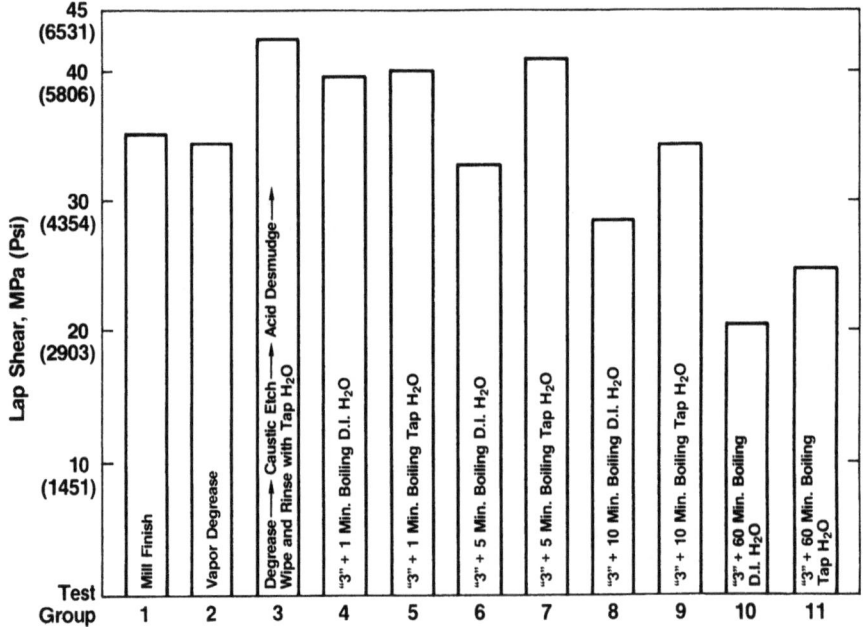

Figure 1. Initial Joint Strengths Obtained with Varying Surface Oxide Conditions.

Table I. Initial Joint Strength Comparisons.

Test Group	Adherend* Pretreatment	Boiling H₂O Treatment Type H₂O	Boiling H₂O Treatment Time (min.)	Average Shear Strength MPa	Average Shear Strength psi	Average Type Failure Adhesive (%)	Average Type Failure Cohesive (%)	% of Mill Finish Joint Strength
1	None	None	0	35.28	5120	50	50	100
2	A	None	0	34.59	5020	60	40	98
3	A-D	None	0	42.58	6180		100	121
4	A-D	Deionized	1	39.48	5730		100	112
5	A-D	Tap	1	33.97	4930	15	85	96
6	A-D	Deionized	5	32.73	4750		100	93
7	A-D	Tap	5	40.86	5930	5	95	116
8	A-D	Deionized	10	28.18	4090	40	60	80
9	A-D	Tap	10	34.45	5000	10	90	98
10	A-D	Deionized	60	20.46	2970	100		58
11	A-D	Tap	60	24.63	3575	100		70

*A = Degrease; B = Caustic Etch; C = Acid Desmudge; D = Wipe and Rinse under Tap Water

A treatment of 60 minutes in either water resulted in a significantly lower joint strength accompanied by 100 percent failure in the adherend oxide, which could be expected if a significantly thicker oxide layer was present. Joints produced by deionized water-treated adherends averaged 30.2 MPa (4385 psi), only 10 percent below the tap water-treated joints at 33.5 MPa (4859 psi). We believe the fact that Wegman used a two-part, room temperature-curing epoxy as compared to our use of a 177°C (350°F) curing epoxy may be a major factor in accounting for the different observations made in the two separate investigations.

Durability Results in Room Temperature Water

The results of soaking the various joints for up to two years in room temperature deionized water are shown in Table II, and the durability factors are plotted in Figure 2. It is clear that the durability of joints with a deoxidized or built-up boehmite oxide layer surface are superior to mill finish or vapor-degreased adherend joints. Increasing the overall time to develop the oxide surface generally favored increased durability, with the 60-minute thicker-oxide joints showing the highest durability factors.

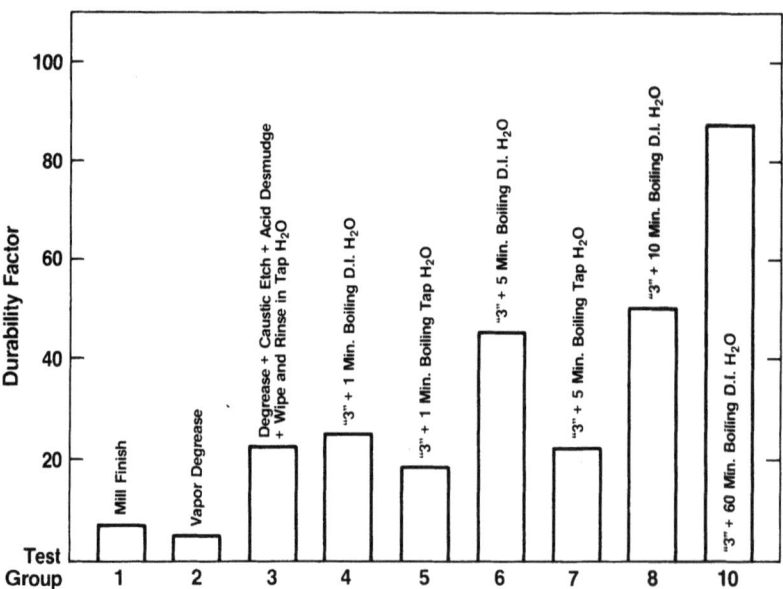

Figure 2. Joint Durability Results in Room Temperature Water Exposure.

Table II. Durability Results in Room Temperature Water.

Test Group	Surface Condition	% Average Retained Joint Strength	Average Exposure Time (days)	Average Type Failure Adhesive (%)	Average Type Failure Cohesive (%)	Durability Factor
1	Mill Finish	35	182	97	3	6.37
2	Vapor Degrease	40	126	95	5	5.04
3	Caustic Etched	42	532	50	50	22.34
4	Deionized H_2O (1 min)	40	630	85	15	25.20
5	Tap H_2O (1 min)	35	518	95	5	18.13
6	Deionized H_2O (5 min)	63	714	50	50	44.98
7	Tap H_2O (5 min)	40	560	85	15	22.40
8	Deionized H_2O (10 min)	69	714	40	60	49.27
10	Deionized H_2O (60 min)	123	714	N.A.	N.A.	87.82

This is particularly interesting in view of the fact that these joints showed the lowest initial joint strengths with highest apparent adhesive-type failure. We must assume, however, that the lower stress capable of fracturing the 60-minute oxide, as compared to the thinner oxide joints, was responsible for the low strength and the apparent adhesive-type failure. The increasing percentage of apparent adhesive failure after approximately two years of water soaking observed for the thinner-oxide joints indicated either these oxides were now fracturing at lower stress levels near the interface[20] or microscopic fragments of water-weakened adhesive were being broken out of the surface at a lower stress level than initially had been necessary to fail the bulk adhesive cohesively.[22] In contrast, the long-timed-soaked, thicker-oxide adhesive joints required an even higher stress to fail the joints than initially present, so neither oxide nor adhesive near the interfacial area seemed to be weakening.

Durability Results in 100 Percent Relative Humidity at 52°C (125°F)

The results of soaking in elevated temperature condensing humidity for a year are summarized in Table III and the durability factors plotted in Figure 3. The deoxidized adherend joints did not perform differently from the mill finish or vapor-degreased joints in this accelerated tropical-type exposure. We know, from earlier Alcoa investigations, that deoxidizing with chromic-sulfuric acid instead of caustic has significantly increased durability using this same adhesive and adherend. All the boehmite oxide oriented joints showed superior durability with thicker layered oxide generally increasing the durability performance. Comparisons can be made between the performance of joints made from adherends treated for 1, 5, 10, and 60 minutes in both deionized and tap waters, with the result that the former averaged 36 percent higher durability factors.

Durability Test Results in Salt Fog Cycle

It should be mentioned that aluminum adherend joints can fail by two different mechanisms, and the testing for durability in a salt fog cycle environment relates to that fact. While the progression of water from the environment into the interfacial joint area is the mechanism promoting joint failure in most weathering conditions, there is also the possibility of joint failure due to gross corrosion of the adherend. Salt water has a special tendency to produce a pitting type of corrosion on aluminum, and the effect of a progression of such pitting across the adherend interface promotes catastrophic joint failures.

Table III. Durability Results in 100 Percent Relative Humidity at 52°C (125°F).

Test Group	Surface Condition	% Average Retained Joint Strength	Average Exposure Time (days)	Durability Factor
1	Mill Finish	22	140	3.08
2	Vapor Degrease	37	84	3.11
3	Caustic Etch	38	84	3.19
4	Deionized H_2O (1 min)	45	294	13.23
5	Tap H_2O (1 min)	31	196	6.08
6	Deionized H_2O (5 min)	54	266	14.36
7	Tap H_2O (5 min)	43	224	9.63
8	Deionized H_2O (10 min)	58	336	19.49
9	Tap H_2O (10 min)	73	364	26.57
10	Deionized H_2O (60 min)	91	364	33.12
11	Tap H_2O (60 min)	54	308	16.63

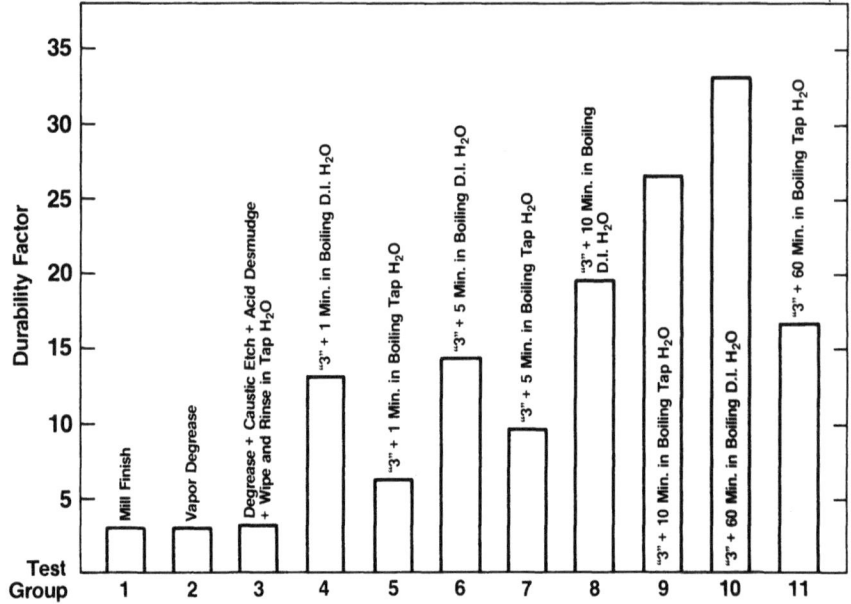

Figure 3. Joint Durability Results in 100 Percent Relative Humidity at 52°C (125°F).

Several ways to protect against this type of failure have been to anodize or apply a surface conversion coating to the aluminum or use special primers. The dramatic effect on improving joint durability resulting from even a one-minute exposure to boiling deionized water is clear from the data reviewed in Table IV and graphed in Figure 4. It must be remembered, however, that developing a predominantly bohemitic surface in boiling water is merely another type of conversion coating. Also, one means known to the aluminum industry to increase the corrosion resistance of aluminum surfaces has been to build up boehmite oxide surfaces by exposing aluminum to water in an autoclave, so it is not totally unexpected to find high resistance to undercutting corrosion in these joints.

Deoxidizing of the aluminum adherend produces a highly significant increase in resistance of the joint to undercutting corrosion, which is the major mechanism for joint failure in the salt fog cycle or a marine atmospheric exposure.[10-12] This is shown by the comparative durability factor average of 0.25 for mill finish or vapor-degreased surface joints as compared to 7.18 for the deoxidized surface joints. The most significant increase in joint resistance to corrosive salt water appeared when the

Table IV. Durability Results in Salt Fog Cycle.

Test Group	Surface Condition	% Average Retained Joint Strength	Average Exposure Time (days)	Durability Factor
1	Mill Finish	17	28	0.48
2	Vapor Degrease	0.5	70	0.04
3	Caustic Etched	27	266	7.18
4	Deionized H_2O (1 min)	78	364	28.39
5	Tap H_2O (1 min)	28	168	4.70
6	Deionized H_2O (5 min)	77	364	28.03
7	Tap H_2O (5 min)	45	266	11.97
8	Deionized H_2O (10 min)	89	364	32.40
9	Tap H_2O (10 min)	85	364	30.94
10	Deionized H_2O (60 min)	103	364	37.49
11	Tap H_2O (60 min)	66	364	24.02

deoxidized surface was exposed to boiling deionized water for only one minute. The retention of 78 percent of initial joint strength after one year cycling under conditions where untreated adherend joints fail within 30 days is impressive. Further increases in the boehmite oxide thickness induced by increasing the soaking times did not significantly change the durability potential. It might be mentioned, however, that the retention of 100 percent of initial joint strength after one year of saltwater cycling by the joints fabricated with adherends pretreated with a 60-minute soak in deionized water represents perfect protection against the corrosive weathering conditions.

The relative effects of treating in deionized or tap water can also be considered. It is clear from Table IV and Figure 4 that the durability potential of the boehmite oxide form created in tap water does not develop as readily as in deionized water. Ten minutes in tap water is required to achieve similar joint durability to one-minute treatment in deionized water. If we compare the average durability of 5-, 10-, and 60-minute coating joints, we see that the average durability factor for those treated in deionized water is 46 percent higher.

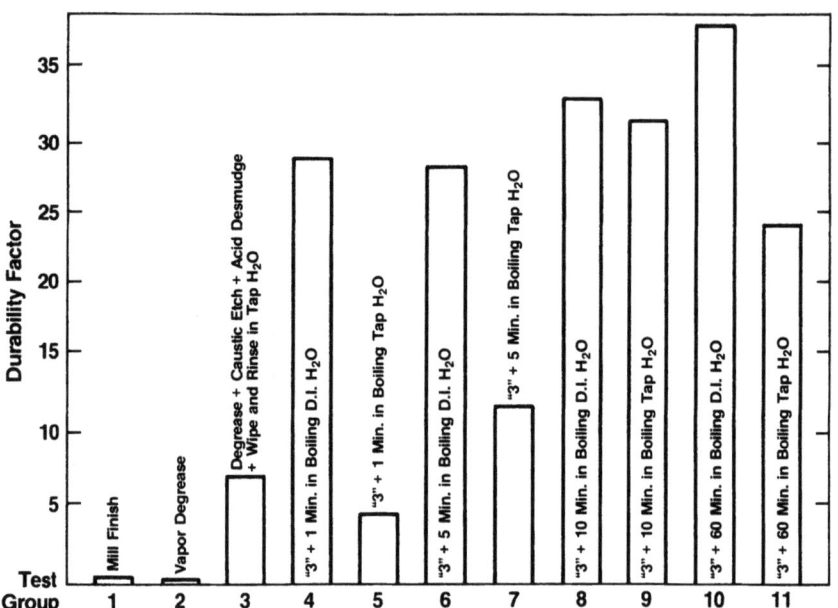

Figure 4. Joint Durability Results in Salt Fog Cycle Exposure.

Durability Data from Simultaneous Stress and Humidity Exposure

The significant added effect of simultaneously imposed stress on weathering exposure with epoxy bonded aluminum joints has been discussed in many technical articles.[10,11,13-15] The benefit of better surface pretreatment in extending the survival time of highly stressed joints under hot humidity soaking conditions has often been less than predicted from separate water durability testing. In Table V and Figures 5 and 6 the data for joints with adherend pretreatments in boiling tap or deionized water have been summarized. By plotting the data obtained at two different percentage levels of initial joint strength for each set of different adherend pretreatments, we can show the relative effectiveness of the treatments as compared to mill finished, vapor degreased, and deoxidized adherend joints.

The most significant observation is that either deoxidizing the adherends or establishing boehmite oxide surfaces enables the survival time at the same 35 percent of stress level to be increased over mill finish or vapor-degreased adherend joints by approximately 1700 times. The survival time potential for all the boehmite oxide-type surfaces was, however, equal to or better than that obtained with deoxidized-only adherends. Finally, the comparison of the plots in Figures 5 and 6 clearly show that the tap water-induced boehmite oxide surface joints show increased joint durability over those induced in deionized water. This is a reversal from all the previous test results in room temperature and hot humidity deionized water soaking and corrosive saltwater testing using nonstressed joints. Comparing the actual durability factors from Table V, we find that, at the 35 percent stressing level, the average durability factor for the tap water-treated adherend joints was 46 percent higher than for the corresponding deionized water-treated adherend joints.

This result would support the conclusion that the oxide formed in deionized water permitted joints with more water and pitting corrosion resistance, which argues for better wetting and spreading of the adhesive.[14] However, the oxide deposited in tap water had the higher resistance to simultaneous stress and water soaking conditions.

Comparisons with Other Aluminum Surface Pretreatments

Because Alcoa has employed the same aluminum alloy and elevated-temperature-curing EC-2086 epoxy adhesive in all its publications since 1972, it is possible to draw a direct comparison of the results of this investigation as regards comparative joint durabilities afforded by different aluminum surface pretreatments. These results are shown in Table VI.

Table V. Durability Data from Simultaneous Stress and Humidity Soak Exposure.

Test Group	Surface Condition	% Stress	Average Joint Failure Time (days)	Stress/Humidity Durability Factors
1	Mill Finish	35, 15	0.005, 1.12	0.009, 0.859
1	Mill Finish	7.8, 4.9	118, 123	47.20, 30.75
2	Vapor Degreased	35, 15	0.009, 0.519	0.015, 0.391
2	Vapor Degreased	8, 5	104, 120	41.6, 30.00
3	Caustic Etch	35, 15	8.5, 53.0	18.39, 49.13
4	Deionized H$_2$O (1 min)	35, 15	9.5, 45.3	19.05, 38.94
5	Tap H$_2$O (1 min)	35, 15	20.0, 115	34.50, 67.85
6	Deionized H$_2$O (5 min)	35, 15	18.5, 62.5	30.75, 44.5
7	Tap H$_2$O (5 min)	35, 15	17.0, 124	35.28, 110.24
8	Deionized H$_2$O (10 min)	35, 15	18.5, 122	26.47, 74.79
9	Tap H$_2$O (10 min)	35, 15	23.0, 135	40.25, 101.25
10	Deionized H$_2$O (60 min)	35, 15	33.5, 135	34.81, 61.43
11	Tap H$_2$O (60 min)	35, 15	42.0, 147	52.54, 78.79

Table VI. Comparative Durability Performance with Varying Aluminum Surface Pretreatments

Surface Pretreatment	Initial Strength (psi)	Exposure Conditions*							
		A		B		C		D	
		Exposure Time (days)	Strength Retention (psi)	Exposure Time (days)	Strength Retention (psi)	Exposure Time (days)	Strength Retention (psi)	Retention Stress (lbs.)	Survival Time (min.)
Chromic-Surfuric Deoxidize[1]	5330	730 (49.64)[3]	3624	365 (6.06)	885	N.A.	N.A.	800	100,000
Chromic-Surfuric Deoxidize[2]	5430			364 (10.92)	1629				
Chromic Acid Anodize	5513	730 (42.34)	3198	365 (22.20)	3363	365 (29.93)	4521	900	100,000
Phosphoric Acid Anodize (Boeing)	6480	730 (47.45)	4212	365 (20.08)	3551	365 (28.94)	5119	1050	100,000
5 Min. D.I. H_2O	4750	714 (44.98)	2993	266 (14.36)	2565	266 (11.97)	2138	710	100,000
60 Min. D.I. H_2O	2970	714 (87.82)	3653	364 (33.12)	2703	364 (37.49)	3059	700	100,000

*Exposure Condition:
A = Room Temperature Deionized Water.
B = 100% Relative Humidity at 52°C.
C = Salt Fog Cycle.
D = Stress/Humidity.

1 = Vapor degrease plus 5 min. at 180°F in chromic-sulfuric.
2 = Ridoline 53 alkaline cleaner plus 5 min. at 180°F in chromic-sulfuric.
3 = All figures in parentheses are calculated durability factors based on the data shown in this table.

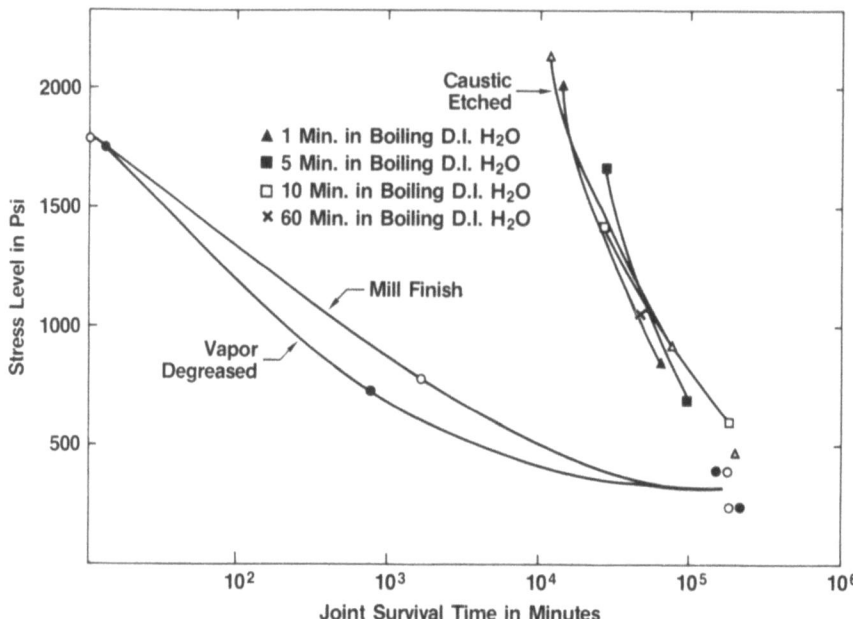

Figure 5. Stress-Endurance Curves for Joints Made to Boehmitic Oxide Surfaces Generated in Boiling Deionized Water.

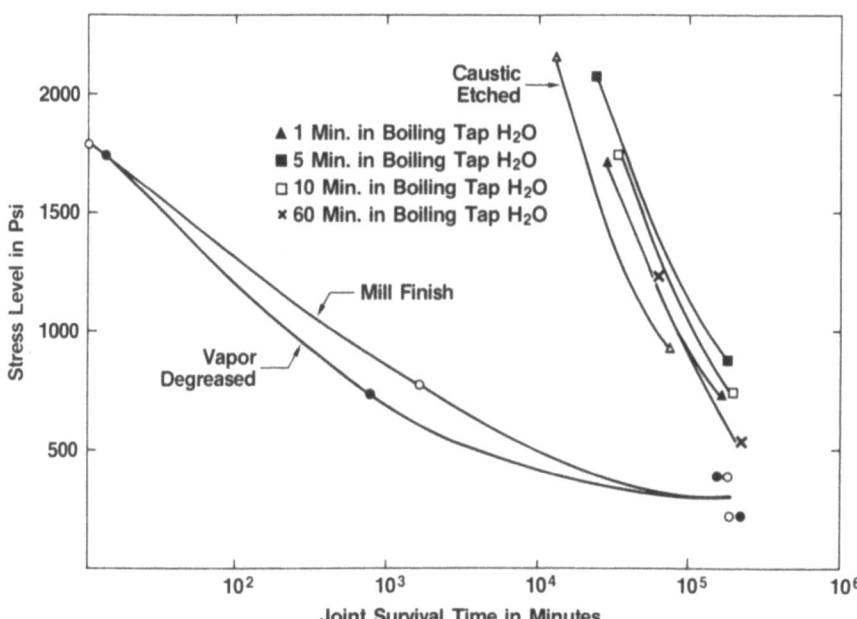

Figure 6. Stress-Endurance Curves for Joints Made to Boehmitic Oxide Surfaces Generated in Boiling Tap Water.

Comparative data involving the deoxidized and anodized surface joints have been taken largely from an Alcoa publication of 1978.[15] The joints with boehmite oxide generated after 60 minutes in deionized water were selected for comparison because of their generally superior performance in most exposure conditions. A performance comparison is also offered with joints prepared after only a five-minute treatment time in deionized water as a more economical and fast procedure for general manufacturing.

The boehmitic oxide surface joints perform very favorably overall as compared with the higher state-of-the-art surface pretreatments whether water soaking, corrosive water cycling, or stress and hot humidity exposure environments are considered. We do not propose that these relatively simple-to-prepare, built-up oxide joints are the equal in all respects of the more expensive and difficult-to-control aerospace procedures, but their performances are creditable enough for strong consideration of acceptance.

DISCUSSION

When considering how adhesively bonded aluminum joints fail, consideration must be given to the initial surface condition of the adherends and how the adhesive interacts with this environment. A considerable amount of work in recent years has been published or announced about the micromorphology of aluminum surface oxides and the stability of those oxides to attack by moisture.[16,17] Additional investigations have focused on how the polymeric adhesive can interact with this morphology.[18-21] The emphasis in these papers has primarily been focused on the so-called FPL etch or PAA (phosphoric acid anodize) aluminum surface pretreatment procedures. The particular oxide surface generated in the present investigation has not been so specifically studied. The FPL etched surface condition has been described by Bijlmer[22] as a "microetch-type pitted surface" and is almost uniquely produced in only a few ratio combinations of the total matrix of many possible different combinations of chromic and sulfuric acid that could be considered. The caustic-type deoxidizing used as the pretreatment for generating boehmite films in this present investigation is a much more vigorous, gas-evolving, aggressive attack of the aluminum surface, leaving much larger, low-power-magnification-viewable craters in the treated surface and generating considerable smut that is readily removed in a dilute acid rinse. The oxide generated by the boiling waters is built up over this pock-marked landscape. At the time these experiments were conducted many years ago, there was no predisposition to analyze surfaces as is commonplace today. Accordingly, we must confess to having left that part of this study uninvestigated.

Because of the remarkable durability results we have obtained (see Table VI), we appreciate that such a study should be conducted and compared with the observations already made on other prepared surfaces as noted above.

If the theory of aluminum bond failure due to generation of a weak boundary layer of aluminum hydroxide is considered,[23] we must assume from the hot humidity data in Table VI that most of our boehmite converted surfaces are less sensitive to that conversion to hydroxide than found even with FPL-type etched surfaces. Or we may consider the weak boundary layer in the polymer theory that the surface morphology influences the orientation of the polymer or prepolymer molecules during the adsorption phase as a major factor in establishing resistance to joint failure under adverse weathering conditions.[22] The comparatively good durability found for these boehmitic oxide joints under all the weathering and stress/weathering conditions would indicate we are dealing with an oxide surface that has not significantly altered the normal cross-linking of the polymer in the vicinity of oxide as proposed in this weak boundary adhesive layer theory.

CONCLUSIONS

1. Significant improvement in joint durability under stressed and nonstressed weathering conditions can result from pretreating 6061-T6 aluminum to produce varying boehmitic oxide surface films as compared to mill finish, vapor degreased, or alkaline-deoxidized surface adherend joints.

2. Further distinctions in joint durability result from using boiling deionized water or New Kensington tap water as well as varying exposure times in either.

3. Thicker boehmitic oxide conditions generally produce higher durability responses in room temperature water or hot humidity exposure conditions with unstressed joints.

4. Oxide generated in the presence of deionized water produced the higher durability performances when using unstressed joints in either water soaking or corrosive saltwater cycling exposure conditions.

5. With simultaneous stressing and hot humidity exposure testing conditions present, the oxide produced in boiling tap water showed the higher joint durability potential.

REFERENCES

1. R. F. Wegman, W. M. Bodnar, M. J. Bodnar and M. J. Barbarisi, Technical Report 3495, Feltman Research Laboratories, Picatinny Arsenal, Dover, NJ, May (1967).
2. R. F. Wegmen, Adhesives Age, 10(1), 20, January (1967).
3. B. B. Bowen, R. E. Herfert and K. C. Wu, AFML Contract No. F33615-74-C-5-27, Northrop Corp., Hawthorne, CA, March (1975).
4. K. C. Wu and B. B. Bowen, Interim AFML Report IR-854-5 (I), Contract No. F33615-75-C-5083, Northrop Corp., Hawthorne, CA, June-November (1975).
5. W. J. Russell and E. A. Garnis, SAMPE Quart. 7(3), 5, April (1976).
6. R. A. Harrington and H. R. Nelson, Amer. Inst. Min. Met. Eng., Tech. Publ. 1158 (1940).
7. J. M. Bryan, J. Soc. Chem. Ind., 69, 169 (1950).
8. D. Altenpohl, Aluminium, 29, 361 (1953).
9. R. K. Hart, Trans. Faraday Soc., 53, 1020 (1957).
10. J. D. Minford, Metals Eng. Quart., November (1972).
11. J. D. Minford, in "Treatise on Adhesion and Adhesives", R. L. Patrick, Editor, Vol. 3, Chap. 2, Marcel Dekker, New York, 1973.
12. J. D. Minford, Adhesives Age, 17(7), 24 (1974).
13. J. D. Minford, J. Appl. Polym. Symp., 32, 91 (1977).
14. J. D. Minford, in "Treatise on Adhesion and Adhesives", R. L. Patrick, Editor, Vol. 5, Chap. 3, Marcel Dekker, New York, 1981.
15. J. D. Minford, SAMPE Quart., 18-27, July (1978).
16. J. D. Venables et al., Appl. Surf. Sci., 3, 88 (1979).
17. J. S. Ahearn, G. D. Davis, T. S. Sun and J. D. Venables, in "Adhesion Aspects of Polymeric Coatings", K. L. Mittal, Editor, pp. 281-299, Plenum Press, April, 1983.
18. A. V. Pocius, in "Adhesion Aspects of Polymeric Coatings", K. L. Mittal, Editor, pp. 173-192, Plenum Press, April, 1983.
19. K. K. Knock and M. C. Locke, in "Adhesion Aspects of Polymeric Coatings", K. L. Mittal, Editor, pp. 301-318, Plenum Press, April, 1983.
20. D. E. Packham, in "Adhesion Aspects of Polymeric Coatings", K. L. Mittal, Editor, pp. 19-44, Plenum Press, April, 1983.
21. W. Brockmann, in "Adhesion Aspects of Polymeric Coatings", K. L. Mittal, Editor, pp. 265-280, Plenum Press, April, 1983.
22. P. F. A. Bijlmer, J. Adhesion, 5, 319 (1973).
23. J. D. Venables et al., in "Proc. 12th Natl. SAMPE Tech. Conf.", Seattle, WA, p. 909 (1980).

EPOXY ADHESION TO COPPER, PART II: ELECTROCHEMICAL PRETREATMENT

Jae M. Park and James P. Bell

Department of Chemical Engineering and
Institute of Materials Science
University of Connecticut
Storrs, Connecticut 06268

Electrochemical methods of pretreating copper surfaces for adhesive bonding were studied. The objective of the electrochemical pretreatment was to control the oxide layer under the copper-coupling agent complex film, since this layer was found to be the locus of failure of copper/epoxy bonds aged in boiling water. Both cathodic and anodic pretreatments resulted in improved durability of the joint after exposure. Experiments were also performed to understand the mechanism of weakening of copper/epoxy bonds in boiling water.

INTRODUCTION

We previously reported that the durability of copper/epoxy bonds was enhanced by pretreating the copper surface with substituted benzotriazole (BTA) coupling agents.[1] In the previous studies several BTA coupling agents capable of bonding with both copper and epoxy adhesive were prepared and the shear strength of treated adhesive joints was evaluated by a torsional test method. It was found that joints treated with 5-amino-benzotriazole (5-ABTA) were 500% stronger than a similarly prepared control after a long term immersion test in boiling water. Our previous work also showed that the copper-BTA (CuBTA) complex layer remained on the epoxy adhesive surface after the joint was broken. This result clearly indicated that the locus of failure of the treated copper/epoxy joint was the oxide layer under the complex film (Figure 1). The existence of an oxide layer under the complex

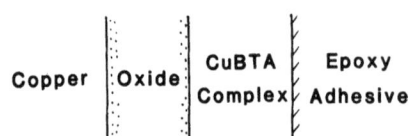

Fig. 1. BTA treated copper/epoxy bond.

film was previously reported by Ogle and Poling.[2] Mansfeld et al.[3] explained that an unreacted oxide layer remains under the complex film; the oxide layer is the substrate for the CuBTA complex formation and provides the copper ions for the reaction with BTA. In other words, the oxide layer is partially replaced with CuBTA complex. Chadwick and Hashemi also found that the CuBTA layer formed on a cathodically polarized copper surface (oxide free copper) was too thin to be detected by multiple reflection infrared spectroscopy.[4]

The objective of pretreating a copper surface electrochemically is to control the oxide layer under the complex film, which seems to be weaker than any other part of the bond. According to the thermodynamic phase diagram of copper in water,[5] the isothermal state of copper oxidation in water changes with applied potential and the solution pH. Copper tends to remain as cuprous oxide at higher pH (alkaline) and as cuprous ions at lower pH (acidic) when no potential is applied. If a cathodic potential is applied, the copper surface is reduced and remains as elemental copper.

Our cathodic pretreatment of copper in an aqueous medium containing coupling agent consists of two steps, i.e., 1) reduction of the surface oxide by a cathodic potential and 2) complex formation after the potential is removed. When an oxide-covered copper surface is dipped in an aqueous BTA solution the complex formation starts at the top layer of the oxide surface, but the entire oxide layer is not completely changed into CuBTA complex.[3] By applying a cathodic potential, the oxide layer is removed almost completely. It is, therefore, possible to have the complex layer grow very near to the elemental copper surface.

The complex formation reaction is a controlled corrosion reaction. During the complex formation step two reactions are expected. First, the elemental copper is oxidized into its ion forms. Secondly, some of the oxide is transformed to CuBTA. The resulting layer is possibly a mixture of oxide and the complex, whose composition depends on the relative rates of the two

reactions. During the complex formation step, it is possible to
apply a small anodic potential in order to increase the dissolution
rate of copper. It was found that a thick CuBTA complex layer
(several thousands of angstroms thick) could be formed by applying
an anodic potential to copper in an aqueous solution of BTA
containing sodium chloride.[2] It was also found that the complex
film grown anodically protected the copper from corrosion more
effectively than the naturally grown film.

In order to understand the importance of corrosion effects
(resulting in interfacial instability) of copper on the durability
of copper/epoxy bonds, an electrochemical aging experiment was
also performed. The glued joints were placed in boiling water in
an electrochemical cell. Various potentials were applied between
the joints and a counter-electrode. The change of joint strength
with the applied potential was examined.

EXPERIMENTAL

The torsional joint geometry designed by Lin and Bell[5] was
used for the present work. This type of joint was used successfully in the previous work[6] and proved to yield more useful information on the bond strength than other current methods.[7] The two
parts of the joint were polished with 0.5 micron alumina slurry on
a rotating micro-polishing cloth. The polished joints were washed
with a copious amount of distilled water and dried with reagent
grade methanol. It was important to keep the gap between two parts
of the joint constant (0.254 mm). In order to keep the gap constant
and to prevent the edge from becoming rounded by repeated polishing,
the joints were remachined on a lathe after they were used three
times. A weight of 44.0 ± 0.1 mg of the adhesive was applied on
the rim-part of the joint (Figure 3) and the flat-part was placed
over it. The joints were tested in shear on an Instron^R tensile
test machine equipped with a torsion device as previously
described.[5,7]

The adhesive used for the present work was a mixture of 4
parts of dicyandiamide and 100 parts Epon 828^R (Shell Chemical Co.).
Reagent grade dicyandiamide (Aldrich Chemical Co.) was used without purification. Dicyandiamide was mixed with liquid Epon 828
by ball-mill mixing followed by mechanical mixing.[6] The adhesive
was cured at 175°C for 2 hours. Some of the experiments were
performed with P108^R curing agent (Shell Chemical Co.). Mixing
solid P108 curing agent with Epon 828 was done by mechanical mixing.

The electrochemical cell used for the pretreatment of the
polished joints is shown in Figure 2. Four holes were made in
the cylinder wall of a 350 ml Pyrex^R Buchner Funnel to fit the

joints. The funnel contained a F grade glass filter. The two stopcocks were prepared to bubble gas and to drain liquid. A rubber seal was used between the hole and the joint to prevent leaking of the electrolyte. The platinum anode was placed at the center of the Plexiglas® cover. Another hole was prepared in the cover for the reference electrode (saturated calomel electrode, not shown in Figure 2). For cathodic polarization, a Model 6204 Power Supply by Harrison Laboratories was used. For anodic polarization, the mini-potentiostat designed by Greene et al.[8] was used. A Simpson 260 multimeter and a Keithley 606 Digital Electrometer were used to measure the potential and the current, respectively. 0.3% aqueous solution of potassium chloride or sodium sulfate was used as the electrolyte. In order to minimize the contact time of the wet joints with the air, the joints were washed with a copious amount of distilled water while the potential was applied. The liquid drain was opened while fresh distilled water was poured into the cell. The washed joints were dried with reagent grade methanol. It should be noted that the precipitation of Cu-ABTA complex was observed after several minutes during the anodic polarization. The anodized surface was covered with the precipitate when it was removed from the solution. The treated joints were placed in an ultrasonic cleaner, cleaned for 5 minutes to remove the precipitate and dried.

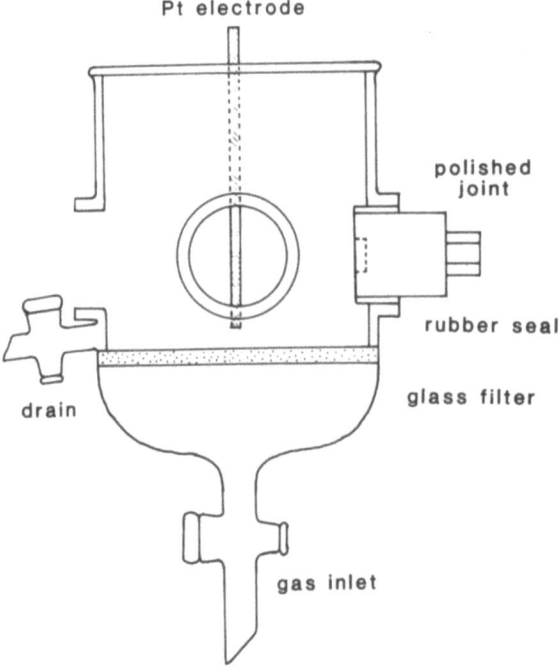

Fig. 2   Electrochemical cell for joint pretreatment.

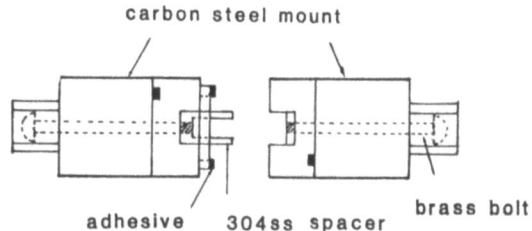

Fig. 3 Demountable joint for aging test.

For the electrochemical aging test, the joint assembly was modified as shown in Figure 4. The geometry of the modified joints was the same as previously, except that the two parts in contact with the adhesive were mounted on steel supports and could be removed to facilitate exposure to boiling water, etc. The joint was reassembled before strength testing.

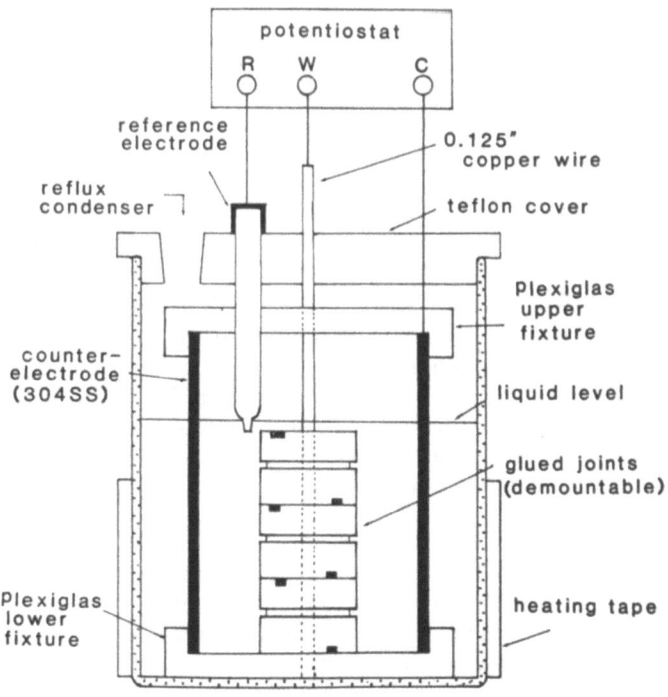

Fig. 4 Polarization cell for aging test.

The glued joints were stacked at the center of an electrochemical cell, shown in Figure 4. The cylindrical cell was heated by a flexible heating tape. The cover of the cell was made of Teflon^R. Four holes were prepared for the working electrode (joints), reference electrode, counter-electrode, and refluxing condenser respectively. A 2 1/2" dia. x 3" long piece of stainless steel (304) tubing was used as the counter-electrode. The potentiostat designed by Greene et al[8] was used as the power supply. By using the electrochemical cell and the power supply, the potential between the joints and the counter-electrode could be kept constant within a variation range of less than 2.0%. A 1.0% aqueous sodium sulfate solution was used as the electrolyte. The joints were aged in the electrolyte at 100°C for 24 hours while constant potential was applied. After aging, the joints were cooled to room temperature and washed with distilled water. The joints were tissue dried and reassembled for testing.

RESULTS AND DISCUSSION

It was pointed out in the Introduction section that the oxidation state of copper in water changes with applied potential and solution pH. The solution pH is, therefore, important during the complex formation period after the cathodic reduction. Results summarized in Table I, show that an acidic solution was less effective for strong adhesion. The reason seemed to be the difference in thicknesses among the CuBTA layers. Ogle and Poling defined a limiting pH for BTA solutions, associated with the thickness of CuBTA film formed on the copper surface.[2] Thin films with a limiting thickness of 60 Å were formed in solutions with a pH higher than the limiting pH. Thick films, about 4000 Å in thickness, were formed in solutions with a pH lower than the limiting pH. They suggested that the limiting pH was between 3.5 and 4.0.

It was also found that a large portion of a thick film could be easily wiped off and that thin films showed better corrosion protection than thick films. The experiments on the effects of solution pH on the joint strength were not repeated with 5-ABTA. The amino group of 5-ABTA was expected to form $-NH_3^+$ in acidic solution, which would not react with the epoxy adhesive.

Different potentials were applied between the platinum anode and the polished joints during the polarization process. No significant change in the joint strength was observed among the joints pretreated with different potentials (Table II). The results are consistent with the fact that the cathodic process, $Cu^+ + e^- = Cu$, is rapid. When the potential is increased, the current also increased. The time required to reduce the oxide layer is shortened when a higher current flows between the electrodes. Assuming the diffusion effect around the electrode is negligible, the time required for the reduction of the oxide was calculated according the Faraday's law. It was also assumed that cuprous oxide has a simple cubic structure with a lattice spacing of 2.0 Å. The result showed that only 2.0 seconds are required for the reduction of the polished joint surface when the current is 10 mA. When a potential of 1.0 volt was applied, the measured current was 2.0 mA. A reduction time of 5 minutes seemed quite enough for a complete reduction even when a potential of only 1.0 volt was applied.

The concentration of BTA may be important if the CuBTA complex-forming reaction is controlled by the diffusion of BTA into the complex layer. There was no statistically significant difference between joints treated in 0.02M and 1.25×10^{-5}M solutions (Table III). It seems reasonable to conclude that the effect of BTA concentration was not appreciable.

It was also pointed out in the Introduction that the substrate for the complex formation reaction is the oxide layer, which provides the copper ions for the reaction. If it is true that the substrate for the reaction is the oxide, the thickness and structure of the complex film should be affected by the amount of oxygen dissolved in the solution. Table IV shows the importance of oxygen during the pretreatment. The joints treated in a nitrogen-bubbled solution were weaker than those treated in an air-bubbled solution. This strengthens the evidence that BTA compounds react with copper ions, not with the elemental copper. The requirement of oxygen reduction to meet the charge neutrality of the oxidation reaction of copper necessitates oxygen in the complex forming reaction.

Table I. Effects of Solution pH on Joint Strength.

	pH 3.1*	pH 5.9	pH 11.0**
Breakage Force (lbs)***	264.0	288.8	282.8
Standard Deviation	14.6	19.3	13.2

Electrolyte: 0.3% aqueous KCl solution prepared with double-distilled water, containing 1×10^{-4}M BTA
Polarization: Cathodic 5 volts for 10 min. and 10 min. without potential (with air-bubbling)
* pH was adjusted with conc. H_3PO_4
** pH was adjusted with 10% NaOH solution
*** After 24 hours immersion in boiling water

Table II. Effect of Different Cathodic Potentials for Reduction on Joint Strength.

Electrolyte: 0.02M aq. BTA solution containing 0.3% KCl (neutral)

Polarization: Various cathodic potentials for 5 min. and 10 min. without potential (with air-bubbling)

Joint No.	Breakage Force (lbs)*		
	1 volt	5 volt	10 volt
1	279	289	284
2	286	288	302
3	298	300	282
4	296	304	293
Avg.	289.8	295.3	290.3
S.D.	8.9	8.0	9.2

*After 24 hours immersion in boiling water.

Table III. Effect of BTA Concentrations on Joint Strength.

Electrolyte: 0.3% aq. KCl solution containing different amounts of BTA

Polarization: Cathodic 5 volts for 10 min. and 10 min. without potential (with air-bubbling)

Joint No.	Breakage Force (lbs)*	
	0.02M	1.25×10^{-5} M
1	287	306
2	295	273
3	292	322
4	266	287
Avg.	285.0	297.0
S.D.	13.1	21.0

*After 24 hours immersion in boiling water.

Most of the experiments reported thus far were performed with BTA. Table V shows a comparison of the results of joint tests performed with BTA and 5-ABTA. The results show that there is no significant difference in joint strengths between BTA and 5-ABTA treatments when the joints are treated electrochemically. The reason seemed to be the susceptibility of 5-ABTA to air oxidation. When an aromatic primary amine is exposed to the air, it acquires a dark red color. The color is developed as a consequence of atmospheric oxidation. When an aromatic amine is allowed to react with an oxidizing agent stronger than air, a variety of reactions, some of them very complex, occur[9]. The same phenomenon was observed with 5-ABTA. When an aqueous 5-ABTA solution was exposed to the air, the solution changed to red in a few hours. Because the experiments were performed with air-bubbling during the pretreatments, an extensive oxidation of 5-ABTA could be expected, although the change in color could not be observed because of the low concentration of 5-ABTA. When all of the amino groups of 5-ABTA were oxidized, the effect of 5-ABTA treatment would be no better than BTA treatment.

A set of torsional joint tests were conducted to examine the effect of complex formation time on the joint strength. After cathodic reduction, the joints were allowed to remain in the solution for different lengths of time for complex formation. It was expected that a thicker complex film would be formed on the copper surface at longer reaction time. The results in Table VI, showed that the joints reacted with 5-ABTA for 15 minutes were strongest; longer immersion time did not yield stronger bonding. The rate of oxide film growth on a copper surface in water was studied by Kruger.[10] It was found that very rapid oxidation took place initially in unstirred water in equilibrium with 1 atm. oxygen, followed by a leveling off at 65-95 Å. This oxidation behavior was similar to that observed by Rhodin[11] for a copper surface oxidized at room temperature in gaseous oxygen. The only difference was the limiting thickness, which was greater for the oxidation in water. Another important feature of Kruger's study was that the relative phase retardation changed rapidly after 90 minutes. The change in this optical parameter was associated with the formation of a new component in the oxide film. Kruger concluded that the change was due to the formation of cupric oxide. A loosely held cupric oxide layer, which could be wiped off easily, was observed after 18 hours immersion. The rate of oxide formation in an aqueous solution of 5-ABTA containing sodium sulfate should not be comparable to Kruger's results. However, the formation of cupric ions explain the lower joint strength at longer immersion time. It is possible that a loosely held Cu(II)-ABTA complex was formed on the copper surface in the same manner as the cupric oxide formation in pure water.

Table IV. Effects of Oxygen Content of Solution on Joint Strength.

Electrolyte: 0.3% KCl aq. solution prepared with double-distilled water, containing 0.02M BTA

Polarization: Cathodic 5 volts for 10 min. and 10 min. without potential (with air- or nitrogen-bubbling)

Joint No.	Breakage Force (lbs)*	
	air-bubbling	N_2-bubbling
1	287	260
2	295	245
3	292	240
4	266	250
Avg.	285.0	248.8
S.D.	11.3	7.4

*After 24 hours immersion in boiling water.

Table V. Effect of 5-ABTA Treatment on Joint Strength.

Electrolyte: 0.3% aq. KCl solution containing 1×10^{-4}M BTA (pH 5.9) or 5-ABTA (pH 6.1)

Polarization: Cathodic 5 volts for 10 min. and 10 min. without potential (with air-bubbling)

Joint No.	Breakage Force (lbs)*	
	BTA	5-ABTA
1	276	291
2	264	310
3	310	277
4	305	284
Avg.	288.8	290.5
S.D.	19.3	12.3

*After 24 hours immersion in boiling water.

Tables VII, VIII, and IX and Figure 5 are summaries of long term test data from electrochemically treated joints. An anodic potential of -0.1 volt vs saturated calomel electrode (SCE) was selected because the corrosion potential of BTA treated copper was found to be between -0.2 and -0.3 volt vs SCE.[12] A potential of -0.1 volt seems high enough to enhance the dissolution rate of copper. However, the rate of copper dissolution, which is directly related to the rate of complex formation, may change the properties (i.e. crystallite size, etc.) of the resulting complex film. Little difference in dry strength was observed among the treatment methods (dipping, cathodic, and anodic pretreatments). After 24 hours immersion, both the cathodic and the anodic treatments showed about 12% increase in strength over the dipping treatment. In immersion up to 100 hours, the cathodically treated joints were perhaps slightly stronger than those treated by dipping. The long immersion tests (840 hours immersion in boiling water), showed that all of the 5-ABTA treated joints were much stronger than the untreated control. However, after extended water immersion, any differences between electrochemical treatments and the dipping treatment were indistinguishable.

The importance of the corrosion effects on the durability of a copper-epoxy bond was also studied by an electrochemical aging experiment. The glued joints were placed in boiling water in the

Table VI. Effects of Reaction Time after Cathodic Polarization on Joint Strength.

Electrolyte: 1×10^{-4} M aq. 5-ABTA solution containing 0.3% Na_2SO_4

Polarization: Cathodic 5 volts for 10 min. and for various lengths of time without potential (without air-bubbling)

	Breakage Force (lbs)*			
Joint No.	5 min.	15 min.	30 min.	60 min.
1	238	278	273	278
2	258	263	253	272
3	235	284	247	251
4	269	271	258	226
Avg.	250.0	274.0	257.8	256.8
S.D.	14.1	7.8	9.6	10.4

*After 24 hours immersion in boiling water.

Table VII. Effects of Cathodic Polarization on Joint Strength.

Electrolyte: 1×10^{-4} M aq. 5-ABTA solution containing 0.3% Na_2SO_4 (neutral)

Polarization: Cathodic 10 volts for 10 min. and 10 min. without potential (with air-bubbling)

Adhesive: 4 parts dicy + 100 parts Epon 828

Curing: 2 hours at 175°C

Joint No.	Breakage Force (lbs)			
	Dry	24	48	96*
1	354	294	283	246
2	366	291	256	279
3	339	288	275	255
4	348	279	261	267
Avg.	351.8	288.0	268.8	261.8
S.D.	9.8	5.6	10.8	12.4

*Hours in boiling water.

electrochemical cell of Figure 4. Various potentials were applied between the counter-electrode and the glued joints. The changes in joint strength as a function of applied potential were examined.

The results of the joint tests are summarized in Figure 6. The joint strength showed a maximum at 0.0 volt vs SCE. This potential seemed to be within the potential region where cuprous oxide is stable. The low strength at -1.0 volt indicates that the oxide layer or the complex layer was almost completely reduced to the elemental copper. The decreased strength at +0.1 volt indicated that the potential was high enough to oxidize cuprous oxide into cupric oxide. A potential higher than 0.1 volt was not tried because an extensive dissolution of copper was observed within an hour.

It was pointed out earlier that an oxide layer exists under the complex layer. Because the joints were polished and pretreated at room temperature, the oxide layer is primarily cuprous oxide, as found by Rhodin[11] and Kruger.[10] When a cathodic potential is applied, the cuprous oxide layer tends to be reduced to the elemental copper. The reduction of cuprous oxide into the elemental copper means that a phase transformation of copper occurs at one part of the bond, which eventually leads to a change in

EPOXY ADHESION TO COPPER

specific volume. The volume change will result in the weakening of joint strength. When a positive potential which is high enough to oxidize cuprous oxide to cupric oxide is applied, we can expect another phase transformation leading to the weakening of joint strength.

The role of oxygen and water in the weakening of copper/epoxy bonds was confirmed by the following experiments. One set of joints was aged in boiling water purged with nitrogen for 24 hours. The joints boiled in the nitrogen-purged water were much stronger than those boiled in the air-saturated water (Table X). Some joints were heated in an oven at 100°C for 24 hours. A large difference in strength was observed between the joints aged in boiling water and the joints aged in the air (Table XI). The joints aged in hot air were stronger than the unaged control, showing that the loss of strength in boiling water is not a thermal aging effect. The increased strength of the aged sample can be

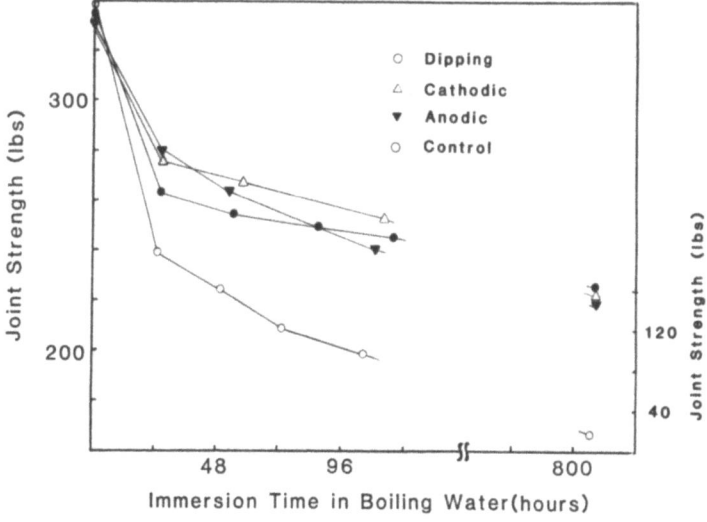

Fig. 5 Effects of electrochemical treatments on joint strength.

explained in two ways: First, the frozen-in stress built up during the cooling period may have been relieved while the joints were aged in the oven. Second, the curing reaction of the adhesive may not have been completed in 2 hours at 175°C, and the reaction may have continued as the postcure of the adhesive. The results of two experiments, immersion in nitrogen-purged water and aging in 100°C air, showed that both oxygen and water are required for the weakening of a copper/epoxy bond. The results also indicated the importance of substrate corrosion on the joint durability. When a copper/epoxy joint is exposed to air saturated water, oxidation of copper, $Cu^+ = Cu^{++} + e^-$ and $Cu = Cu^{++} + 2e^-$, seems to occur as an anodic reaction. The reduction of oxygen, $O_2 + 4H^+ + 2e^- = 2H_2O$, seems to be required as the cathodic reaction to meet the requirement of electrical neutrality.

CONCLUSIONS

Electrochemical methods of pretreating copper surfaces for adhesive bonding have been studied. The durability of a

Table VIII. Effects of Anodizing Copper Surfaces on the Joint Strength.

Electrolyte:	1×10^{-4}M aq. 5-ABTA solution containing 0.3% Na_2SO_4 (neutral)
Polarization:	Cathodic 10 volts for 10 min. and anodic -0.1 volt vs. SCE for 15 min. (without air-bubbling)
Adhesive:	4 parts dicy + 100 parts Epon 828
Curing:	2 hours at 175°C

	Breakage Force (lbs)			
Joint No.	Dry	24	48	96*
1	347	296	281	273
2	352	282	252	264
3	338	294	274	241
4	355	300	259	258
Avg.	348.0	293.0	266.5	259.0
S.D.	6.4	6.7	11.5	11.7

*Hours in boiling water

Table IX. Long-term Immersion Test.

Pretreatment

5-ABTA treated: Dipped in a 2×10^{-4} M aq. 5-ABTA solution for 10 min. at room temp.

Cathodically Polarized: Cathodic 10 volts in a 1×10^{-4} M aq. solution containing 0.3% Na_2SO_4 and another 10 min. without potential (with air-bubbling)

Anodized: Cathodic 10 volts for 10 min. and anodic -0.1 volt vs. SCE for 15 min. (without air-bubbling) in a 1×10^{-4} M aq. 5-ABTA solution at room temperature.

Adhesive: 4 parts dicy + 100 parts Epon 828

Curing: 2 hours at 175°C

Joint No.	Breakage Force (lbs)*			
	Control	Dipped	Cathodic	Anodic
1	18	109	158	122
2	35	171	144	151
3	43	181	102	98
Avg.	32.0	153.7	134.7	124.3
S.D.	10.4	31.8	23.8	22.5

*After 840 hours immersion in boiling water

copper/epoxy bond in boiling water was enhanced by pretreating the copper surface electrochemically in a BTA coupling agent solution. After 24 hours immersion in boiling water, both the cathodic and the anodic treatments showed about 12% increase in strength over the dipping treatment. Longer immersion tests showed that all of the 5-ABTA treated joints were much stronger than the untreated control. However, after extended water immersion, differences between the electrochemical treatments and a dipping treatment were indistinguishable. The presence of oxygen in the solution during the complex formation period was essential.

The results of the electrochemical aging tests and the required presence of water and oxygen for the weakening of copper/epoxy bonds suggested that the chemical stability of interface is an important factor for the durability of the adhesive bond.

Table X. Effect of Oxygen Dissolved in Boiling Water on Joint Strength.

Pretreatment: No

Adhesive: 100 parts Epon 828 + 4 parts P108 (based on dicyandiamide)

Curing: 2 hours at 175°C

Joint No.	Breakage Force (lbs)		
	Dry	Wet*	Wet**
1	360	257	295
2	352	255	314
3	360	266	301
4	367	263	321
5	–	256	–
6	–	248	–
Avg.	359.8	257.5	307.8
S.D.	5.3	5.8	10.3

*Immersed in boiling water for 24 hours
**Immersed in boiling water with nitrogen purging for 24 hrs.

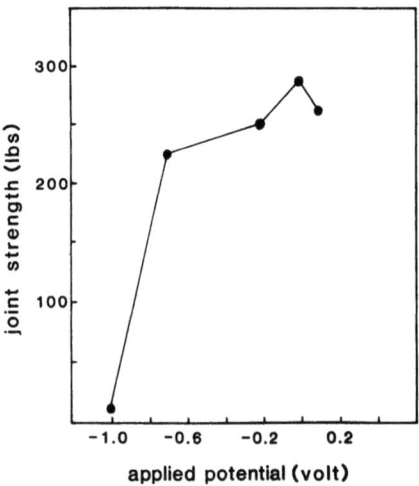

Fig. 6. Changes in joint strength as a function of applied potential.

Table XI. Effect of Aging in Hot Air on Joint Strength.

Pretreatment: No

Adhesive: 100 parts Epon 828 + 4 parts P108 (based on dicyandiamide)

Curing: 2 hours at 175°C

Joint No.	Breakage Force (lbs)		
	Unaged	Aged*	Wet**
1	360	435	257
2	352	406	255
3	360	399	266
4	367	388	263
5	-	379	256
6	-	-	248
Avg.	359.8	401.4	257.5
S.D.	5.3	19.2	5.8

*Heated in a 100°C oven for 24 hours
**Immersed in boiling water for 24 hours

ACKNOWLEDGEMENT

Financial support by Shell Development Co. for the present work is gratefully acknowledged by the authors.

REFERENCES

1. J. M. Park and J. P. Bell, in "Adhesion Aspects of Polymeric Coatings", K. L. Mittal, Editor, pp. 205-224, Plenum Press, New York, April 1983.
2. I. C. Ogle and G. W. Poling, Canad. Metall. Quart., 14(1), 37 (1975).
3. F. Mansfeld, T. Smith and E. P. Parry, Corrosion Sci., 27, 289 (1971).
4. D. Chadwick and T. Hashemi, Corrosion Sci., 18, 39 (1979).
5. C. J. Lin and J. P. Bell, J. Polym. Sci., 16, 1721 (1972).
6. J. M. Park, Master's Thesis, University of Connecticut, 1980.
7. W. T. McCarvill and J. P. Bell, J. Adhesion, 6, 185 (1974).

8. N. D. Greene, G. A. Moebus and N. H. Baldwin, Corrosion, $\underline{29(6)}$, 234 (1973).
9. R. Q. Brewster and W. E. McEwen, "Organic Chemistry", 3rd Ed., p. 238, Prentice-Hall, Englewood, NJ, 1961.
10. J. Kruger, J. Electrochem. Soc., $\underline{108}$(6), 503 (1961).
11. T. N. Rhodin, J. Am. Chem. Soc., $\underline{73}$, 3143 (1951).
12. M. Osawa and W. Suetaka, Corrosion Sci., $\underline{19}$, 709 (1979).

HYDROTHERMAL STABILITY OF TITANIUM/EPOXY ADHESIVE JOINTS

F. J. Boerio and R. G. Dillingham

Department of Materials Science and Metallurgical Engineering
University of Cincinnati
Cincinnati, Ohio 45221

γ-Aminopropyltriethoxysilane (γ-APS) was shown to be an extremely effective primer for improving the hydrothermal stability of titanium/epoxy adhesive joints when applied to the adherends by adsorption from dilute aqueous solutions prior to adhesive bonding. The breaking strength of lap joints prepared from unprimed adherends decreased slowly during immersion in water at 60°C and was only about 960 psi after 60 days. The breaking strength of joints prepared from adherends primed with γ-APS at pH 10.4 and 8.0 decreased very little during similar hydrothermal aging and was about 1750 psi after 60 days. Joints prepared from adherends primed with γ-APS at pH 5.5 were somewhat less durable and had a breaking strength of about 1180 psi after 60 days in water at 60°C. The performance of γ-APS as a primer was very different for titanium and iron adherends even though the molecular structures of the primer films formed on the two substrates were very similar. As a result, the performance of γ-APS primer films could not be related to the overall molecular structure of the films but was related to acid/base interactions between γ-APS and the oxidized surfaces of titanium and iron.

INTRODUCTION

It is well known that proper pretreatment of metal surfaces is essential for the successful use of adhesive bonding for joining metals. Pretreatments for titanium adherends include a phosphate-fluoride conversion coating, a modified phosphate-fluoride process, an alkaline peroxide etch, and chromic acid-ammonium fluoride anodization[1]. The most useful pretreatments, the alkaline peroxide etch and the chromic acid anodization, produce a relatively thick, porous oxide[2].

The use of organofunctional silanes as primers or coupling agents for improving the wet strength of glass fiber reinforced plastics is well known[3]. More recently it has been shown that certain organosilanes are extremely useful as primers for improving the hydrothermal stability of iron/epoxy[4] and aluminum/epoxy[5,6] adhesive joints. Comparatively little information is available regarding the use of silanes as primers for adhesive bonding of titanium. However, Schrader and Cardamone[7] have considered the use of γ-aminopropyltriethoxysilane (γ-APS) as a primer for improving the hydrothermal stability of lap joints prepared from titanium-6Al, 4V adherends and an anhydride cured epoxy adhesive. The dry strength of lap joints prepared from adherends pretreated with 1% aqueous solutions of γ-APS prior to adhesive bonding was always about 25% greater than that of joints prepared from unprimed adherends. After 24 hours immersion in boiling water, the shear strength of joints prepared from primed adherends was about 50% higher than that of joints prepared from unprimed adherends. The improved hydrothermal stability was attributed to possible covalent bonding between the silane and the adhesive and the oxide[7].

The objectives of this research were to determine the molecular structure of films formed by γ-APS adsorbed onto titanium-6Al, 4V from dilute aqueous solutions and to determine the relationship between such structure and the performance of γ-APS as a primer for titanium/epoxy adhesive joints. The molecular structure of primer films formed by γ-APS adsorbed onto titanium was determined using reflection-absorption infrared spectroscopy (RAIR) and ellipsometry. RAIR is a technique for obtaining infrared spectra of thin films on metallic substrates by reflecting radiation polarized parallel to the plane of incidence from the substrate at a large, nearly grazing angle of incidence. Ellipsometry is a technique for determining the thickness and refractive index of a thin film on a metallic surface by reflecting plane polarized visible light from the surface and measuring the relative amplitude reduction and phase shift for components polarized parallel to and perpendicular to the plane of incidence. RAIR[8,9] and ellipsometry[10] have been thoroughly described in the literature and no further discussion will be given here.

EXPERIMENTAL

Sample mirrors for infrared spectroscopy were prepared by mechanically polishing titanium-6Al, 4V coupons. After polishing, the mirrors were rinsed several times in distilled water and then dried in a stream of nitrogen. γ-APS films were formed on freshly polished mirrors by immersing such mirrors into 1% aqueous solutions of γ-APS for approximately 30 minutes, withdrawing the mirrors, and blowing off the excess solution using a strong stream of nitrogen.

Infrared spectra were obtained using an external reflection accessory (Harrick Scientific Co., Ossining, NY) and a Perkin-Elmer 180 infrared spectrophotometer. The reflection accessory was configured to provide one reflection at an angle of incidence equal to 80°. A silver bromide wire grid polarizer was placed in front of the entrance slit to the monochromator and oriented to transmit only radiation polarized parallel to the plane of incidence at the sample mirrors.

In some cases the sample mirrors were examined with an ellipsometer (Rudolph Research Model 436) immediately before and after adsorption of the γ-APS films. The results obtained indicated that the adsorbed films were typically about 100A in thickness.

Titanium/epoxy lap joints were prepared according to ASTM standard D1002. Titanium adherends (4" x 1" x 0.063") were mechanically polished, rinsed, and dried in nitrogen. Pairs of adherends were then bonded together using an adhesive consisting of an epoxy resin (Epon 828, Shell Chemical Co.) and a tertiary amine curing agent (Ancamine K-61B, Pacific Anchor Chemical Co.). The adhesive was cured for four days at room temperature and one hour at 100°C. Prior to adhesive bonding, some of the adherends were pretreated by immersion in 1% aqueous solutions of γ-APS for 30 minutes and then blown dry with nitrogen. All of the lap joints were immersed in a water bath at 60°C. At appropriate intervals, joints were withdrawn from the bath and tested to determine their shear strength. In some cases, the fracture surfaces were examined using a scanning electron microscope (Cambridge 600) or an optical microscope.

RESULTS AND DISCUSSION

The infrared spectrum shown in Figure 1A was obtained from films formed by γ-APS adsorbed onto titanium mirrors as described above. The spectrum is characterized by a strong band near 1070 cm^{-1} and by weaker bands near 1570, 1470, and 1300 cm^{-1}. Infrared spectra of γ-APS monomer are dominated by intense bands near 1105, 1080, and 960 cm^{-1} that are characteristic of SiOC bonds.[11] The

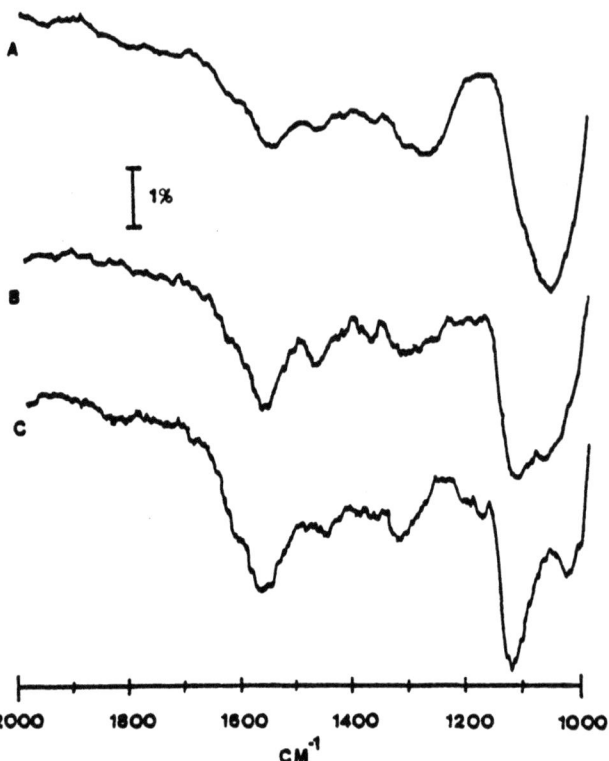

Figure 1. Infrared spectra of thin films formed by γ-APS adsorbed onto titanium mirrors from 1% aqueous solutions at pH 10.4: (A) - as formed, (B) - after five hours exposure to the laboratory atmosphere, and (C) - after three days exposure. Reproduced by permission from reference 8.

absence of these bands from the spectrum shown in Figure 1A indicates that γ-APS was adsorbed as highly hydrolyzed oligomers. However, the band observed near 1070 cm^{-1} in Figure 1A is assigned to an SiOSi stretching vibration, indicating that the adsorbed silane polymerized on the surface to form a siloxane polymer.

The bands near 1570, 1470, and 1300 cm^{-1} have been the subject of considerable debate[11-13]. However, we have recently shown that these bands are related to the formation of amine bicarbonate species by absorbed carbon dioxide[14]. The bands observed near 1570 and 1470 cm^{-1} are, therefore, assigned to the deformation modes of protonated amino (NH_3^+) groups while the band near 1300 cm^{-1} is assigned to the symmetric stretching mode of bicarbonate ions.

The as-formed films were stable in a dry atmosphere but several changes were noted during exposure to the laboratory atmosphere.

(see Figures 1B and 1C). The band originally observed near 1070 cm^{-1} gradually split into components near 1130 and 1040 cm^{-1}, indicating additional polymerization of the adsorbed silane. The band initially observed near 1470 cm^{-1} decreased in intensity and the band near 1300 cm^{-1} increased in frequency to about 1330 cm^{-1}, indicating that the bicarbonate species was unstable in the presence of atmospheric moisture.

The spectra shown in Figure 2 were obtained from films formed by γ-APS adsorbed onto titanium mirrors before and after heat treatment in an oven at 100°C for 20 minutes. The bands originally observed near 1570, 1470, and 1330 cm^{-1} (Figure 2A) are not observed after heat treating (Figure 2B), indicating that the absorbed carbon dioxide is easily removed from the as-formed films. Moreover, the band originally observed near 1130 cm^{-1} has increased somewhat in frequency to about 1145 cm^{-1} and the band near 1040 cm^{-1} has become more prominent, indicating additional polymerization within the film.

The spectra shown in Figure 3 were obtained from films formed by γ-APS adsorbed onto titanium mirrors from 1% aqueous solutions that were acidified by addition of HCl. These spectra are dominated by an intense band near 1120 cm^{-1} that is assigned to an SiOSi stretching mode, again indicating formation of siloxane polymers.

Figure 2. Infrared spectra of thin films formed by γ-APS adsorbed onto titanium mirrors from 1% aqueous solutions at pH 10.4: (A) - before and (B) - after heat treating at 100°C for 20 minutes.

Figure 3. Infrared spectra of thin films formed by γ-APS adsorbed onto titanium mirrors from 1% aqueous solutions at (A) - pH 10.4, (B) - pH 8.5, and (C) - pH 5.5.

The weaker bands near 1600 and 1500 cm^{-1} are assigned to the deformation modes of amino groups that are protonated as amine hydrochlorides ($NH_3^+Cl^-$).

The stability of the air-formed oxide on titanium mirrors during immersion in aqueous solutions of γ-APS has also been considered. Several mirrors were polished as described above and then heated in air at about 300°C for one hour. Results obtained from ellipsometry indicated that the oxide on the mirrors was about 150A in thickness. RAIR spectra from such mirrors were characterized by an absorption band near 850 cm^{-1} that was assigned to the rutile form of TiO_2[15]. The intensity of the oxide band near 850 cm^{-1} was then determined as a function of immersion time in 1% aqueous solutions of γ-APS at pH 10.4 (see Figure 4). The results obtained indicated that the air-formed oxide on the titanium mirrors was stable during immersion in such solutions. Similar results were obtained at pH 5.5.

These results indicate that there are only minor differences in structure between films formed by γ-APS adsorbed onto titanium mirrors from aqueous solutions at pH 10.4 and those formed by adsorption from aqueous solutions at lower pH values. The films are

HYDROTHERMAL STABILITY

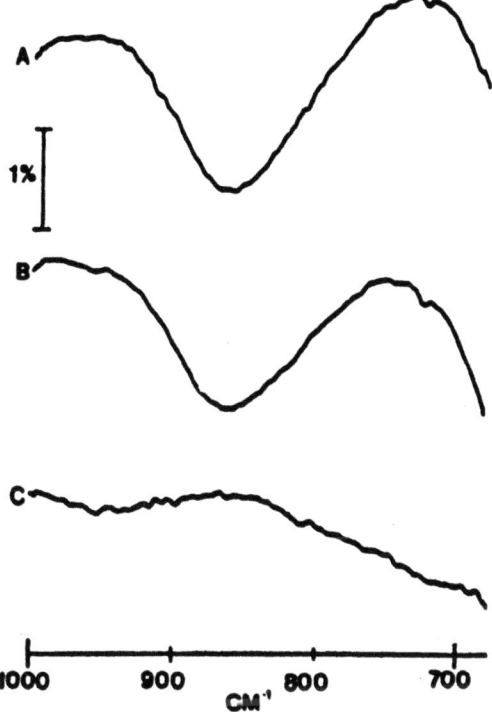

Figure 4. Effect of immersion in 1% aqueous solution of γ-APS at pH 10.4 on infrared spectra of oxide on titanium mirrors: (A) - before immersion, (B) - after immersion for thirty minutes, and (C) - difference.

always composed of siloxane polymers. However, at pH 10.4 the amino groups form bicarbonate salts with adsorbed carbon dioxide. At pH values less than 10.4, the amino groups are protonated as amine hydrochlorides.

As indicated above, the use of γ-APS as a primer for improving the hydrothermal stability of titanium/epoxy lap joints has also been considered. The results are shown in Figure 5. The average shear strength of joints prepared from unprimed adherends decreased slowly with immersion time and was only about 960 psi after 60 days. However, γ-APS was an extremely effective primer. The breaking strength of joints prepared from adherends primed with γ-APS at pH 10.4 and 8.0 hardly decreased during water immersion and was still about 1750 psi after 60 days. γ-APS was considerably less effective as a primer when applied to the adherends at pH 5.5. The breaking strength of such joints was about 1180 psi after 60 days immersion in water at 60°C.

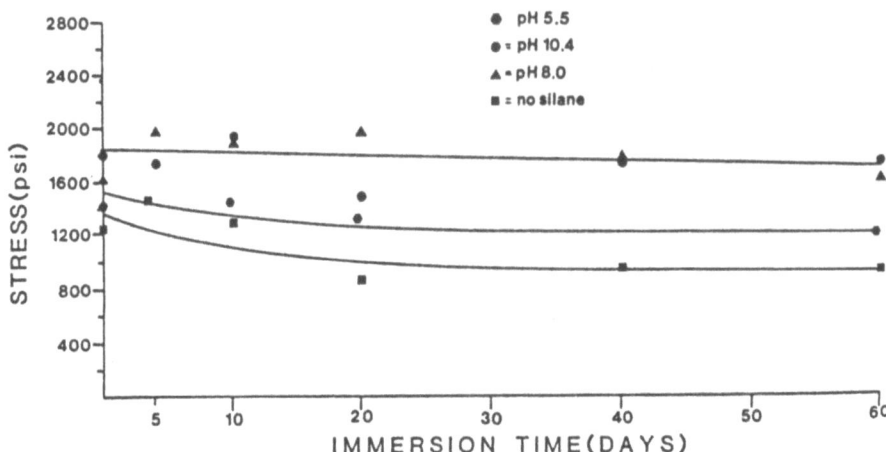

Figure 5. Breaking strength versus immersion time in water at 60°C for titanium/epoxy lap joints.

It is very interesting to compare the results described above with those previously obtained using iron substrates[4,11,12]. Although some evidence for etching of the oxidized surface of iron in aqueous solutions of γ-APS at pH 10.4 has been obtained[16], the infrared spectra of thin films formed by γ-APS adsorbed onto iron mirrors are very similar to the spectra of films formed on titanium, indicating that the structures of the films are also similar. However, the hydrothermal stability of iron/epoxy lap joints was very different from that of titanium/epoxy lap joints[5]. The breaking strength of iron/epoxy lap joints always decreased rapidly during the first ten days of immersion and then continued to decrease at a much slower rate for times as long as 60 days (see Figure 6). γ-APS was an extremely effective primer for iron/epoxy lap joints, especially when applied from acidified solutions[4]. Iron/epoxy lap joints prepared from unprimed adherends retained about 25% of their initial strength after 60 days in water at 60°C. Joints prepared from adherends primed with γ-APS at pH 10.4 retained about 50% of their strength after 60 days but joints primed with γ-APS at pH 8.0 retained about 70% of their strength after such immersion (see Figure 6).

It is evident that the properties of γ-APS primer films are very different on titanium and iron substrates even though the molecular structures of the films are very similar. The performance of γ-APS primer films on iron improves substantially when the pH at which the primer is applied is lowered from 10.4 to 8.0 even though there are only minor differences in structure between films formed at pH 10.4 and those formed at 8.0. Moreover, the performance of γ-APS primer films on titanium is essentially the same

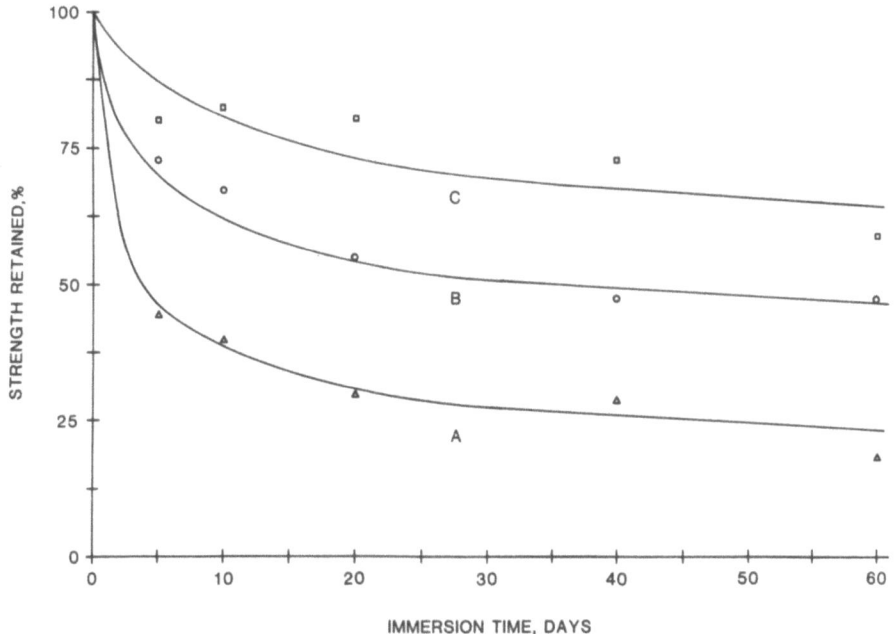

Figure 6. Breaking strength versus immersion time in water at 60°C for iron/epoxy lap joints: (A) - no primer, (B) - γ-APS primer at pH 10.4, and (C) - γ-APS primer at pH 8.0. Adapted by permission from reference 4.

when the primer is applied at pH 10.4 and 8.0 even though some minor structural changes are observed between films formed at pH 10.4 and those formed at pH 8.0. As a result, it must be concluded that the performance of γ-APS primers for titanium/epoxy and iron/epoxy adhesive joints is not governed by the overall molecular structure of the primer films. However, it is considered that the performance of the primers can be largely explained in terms of acid/base interactions.

Bolger[17] has suggested that an organic compound having ionizable functional groups will be stably adsorbed onto the oxidized surface of a metal during immersion in water as long as the pH of the water is between the isoelectric point of the oxide (IEPS) and the pk_A of the ionizable functional group. Otherwise, the organic compound will be displaced from the surface by water.

As noted earlier, γ-APS is hydrolyzed in aqueous solutions to form silanol species. Both the silanol and amino groups in hydrolyzed γ-APS are ionizable and have pk_A values of about 3.0 and 10.0, respectively[18]. IEPS for the oxidized surfaces of iron and titanium are approximately 10.0 and 6.0 respectively[18], and the pH of the water used for hydrothermal aging of the lap joints was about 6.2. At any pH value less than about 9.0, γ-APS should be adsorbed onto iron (IEPS 10.0) through the silanol groups ($pk_A \sim 3.0$). During exposure to water at pH 6.2, joints primed with γ-APS at pH values less than about 9.0 should be stable and that is exactly what has been observed (see Figure 6). At pH 10.4, γ-APS is probably adsorbed onto the oxidized surface of iron through either the amino groups ($pk_A \sim 10.0$) or the silanol groups ($pk_A \sim 3.0$). During immersion in water at pH 6.2, the silanol groups should be stably adsorbed but the amino groups should be displaced by water. As a result, γ-APS should be less effective when applied to iron at pH 10.4 and that is also observed (see Figure 6).

IEPS for the oxidized surface of titanium is about 6.0. At either pH 10.4 or 8.0, γ-APS should be adsorbed onto the oxidized surface of titanium through the amino groups ($pk_A \sim 10.0$). During immersion in water at pH 6.2, the amino groups are just stably adsorbed. γ-APS should be equally effective as a primer for titanium/epoxy lap joints when applied at pH 10.4 or 8.0 and that is observed (see Figure 5). γ-APS should be adsorbed onto the oxidized surface of titanium from aqueous solutions at pH 5.5 through both the amino and silanol groups. During immersion in water at pH 6.2, the amino groups should just be stably adsorbed but the silanol groups should just be displaced by water. As a result, γ-APS should be somewhat less effective as a primer for titanium/epoxy adhesive joints when applied to the substrate at pH 5.5 and this is also observed (see Figure 5).

Some support for the mechanism described above was obtained from visual inspection of the fracture surfaces of titanium/epoxy lap joints that were tested after immersion in water at 60°C for long periods of time. Unprimed joints always failed very near the interface, leaving a small patch of adhesive near one end of the joint and an apparently bare area near the other end (see Figure 7A). Such behavior is characteristic of lap joints[19] and is related to the large peeling stresses near the ends of a lap joint. The fracture surfaces of the primed joints were similar in that failiure occurred near the interface and that a small patch of adhesive was left near one end of the joint. However, some adhesive was always observed in the "bare" area near the other end of the joint (see Figure 7B) and it was evident that failure of the primed joints was mostly within the adhesive even after long immersion times.

When scanning electron microscopy (SEM) and optical microscopy were used to examine the fracture surfaces of titanium/epoxy lap joints that were tested after 60 days immersion in water at 60°C, an even better indication of the effects of γ-APS primers on the locus of failure was obtained. Virtually no adhesive was observed at the "bare" end of joints prepared from unprimed adherends (see Figure 8), indicating that the adhesive was in fact displaced

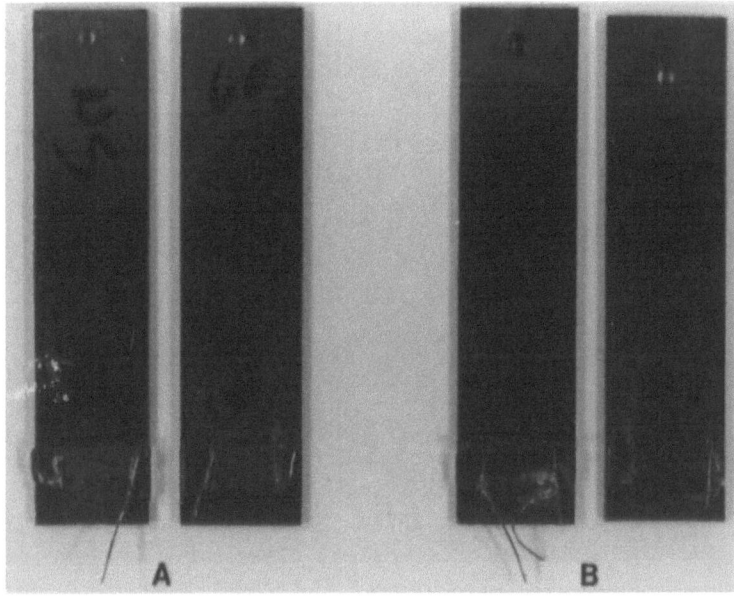

Figure 7. Fracture surfaces of titanium/epoxy lap joints that were tested after 60 days in water at 60°C: (A) - no silane primer; (B) - γ-APS primer at pH 10.4.

from the surface by water. However, some adhesive was always evident in the "bare" area of joints prepared from primed adherends (see Figure 9), indicating that the adhesive was not readily displaced from the surface in such cases.

CONCLUSION

γ-aminopropyltriethoxysilane (γ-APS) is an extremely effective primer for improving the hydrothermal stability of titanium/epoxy adhesive joints when applied to the adherends by adsorption from dilute aqueous solutions prior to adhesive bonding. However, γ-APS primers are most effective when applied to titanium adherends from aqueous solutions at pH values near or slightly below 10.4. γ-APS is also an effective primer for iron/epoxy adhesive joints but is most effective when applied to iron adherends from acidified solutions. The performance of γ-APS as a primer for improving the wet strength of titanium/epoxy and iron/epoxy adhesive joints is not related to the overall molecular structure of the primer films but is explained in terms of acid/base interactions between γ-APS and the oxidized surfaces of titanium and iron.

Figure 8. Optical micrograph of "metal" fracture surface of titanium/epoxy lap joint prepared from unprimed adherends and tested after 60 days in water at 60°C; 500X.

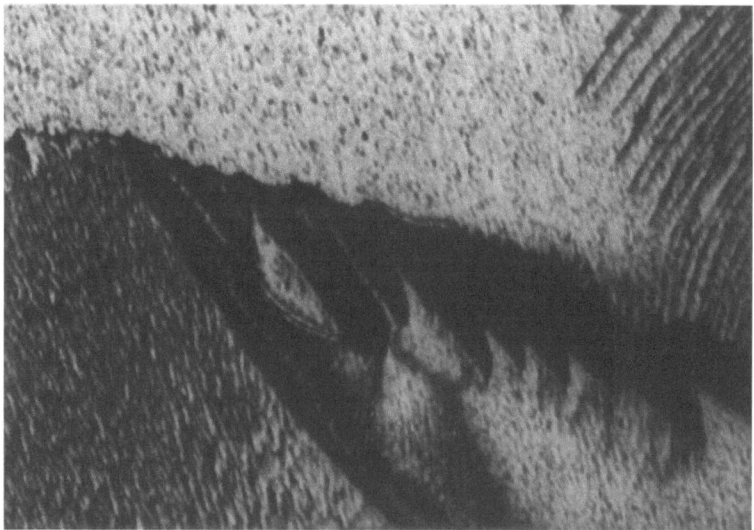

Figure 9. Optical micrograph of "metal" fracture surface of titanium/epoxy lap joint prepared from adherends primed with γ-APS at pH 10.4 and tested after 60 days in water at 60°C; 500X

ACKNOWLEDGMENTS

This research was supported in part by a grant from the Office of Naval Research. The assistance of Union Carbide Corp. in providing γ-aminopropyltriethoxysilane and of Timet, Inc. in providing titanium samples is gratefully acknowledged.

REFERENCES

1. M. C. Locke, K. M. Harriman, and D. B. Arnold, in "Proc. of National SAMPE Symp. and Exhibition," pp. 1-12, 1980.
2. B. M. Ditchek, K. R. Breen, T. S. Sun, and J. D. Venables in "Proc. of National SAMPE Symp. and Exhibition," pp. 13-24, 1980.
3. P. W. Erickson and E. P. Plueddemann, in "Composite Materials," L. J. Broutman and R. H. Krock, Editors, Vol. 6, Ch. 1, Academic Press, New York, 1974.
4. F. J. Boerio and J. W. Williams, Appl. Surf. Sci. 7, 19 (1981).
5. F. J. Boerio and C. A. Gosselin, in "Proc. 36th Ann. Tech. Conf., SPI Reinforced Plastics/Composites Inst.," Sec. 2G, 1981.
6. R. L. Patrick, J. A. Brown, N. M. Cameron, and W. G. Gehman, Appl. Polymer Symp. 16, 87 (1981).
7. M. E. Schrader and J. A. Cardamone, J. Adhesion 9, 305 (1978).
8. F. J. Boerio, C. A. Gosselin, R. G. Dillingham, and H. W. Liu, J. Adhesion 13, 159 (1981).
9. F. J. Boerio and C. A. Gosselin, in "Instrumental and Physical Characterization of Macromolecules," C. D. Craver, Editor, American Chemical Society, Washington, DC, in press, 1982.
10. F. L. McCrackin, E. Passaglia, R. R. Stromberg, and H. L. Steinberg, J. Res. Natl. Bur. Stds. 67A, 363 (1963).
11. F. J. Boerio, L. H. Schoenlein, and J. E. Greivenkamp, J. Appl. Polymer Sci. 22, 203 (1981).
12. F. J. Boerio, L. Armogan, and S. Y. Cheng, J. Colloid Interface Sci. 73, 416 (1980).
13. H. Ishida, S. Naviroj, S. K. Tripathy, J. J. Fitzgerald, and J. L. Koenig, J. Polymer Sci. 20, 701 (1982).
14. F. J. Boerio, J. W. Williams, and J. M. Burkstrand, J. Colloid Interface Sci. 91, 485 (1983).
15. S. Thibault, Thin Solid Films 35, L33 (1976).
16. F. J. Boerio and J. W. Williams, to be published.
17. J. C. Bolger in "Adhesion Aspects of Polymeric Coatings," K. L. Mittal, Editor, pp. 3-18, Plenum Press, New York, 1983.
18. E. P. Plueddemann in "Composite Materials," L. J. Broutman and R. H. Krock, Editors, Vol. 6, Ch. 6, Academic Press, New York, 1974.
19. W. Russell, SAMPE Quarterly, p. 8, July, 1978.

A STUDY ON ELASTOMER/METAL BONDS APPLICABLE IN

UNDERWATER SONAR SYSTEMS

>Robert Y. Ting
>
>Naval Research Laboratory, USRD
>P. O. Box 8337
>Orlando, FL 32856

Recent emphasis on the long-term performance of sonar systems requires that investigations be carried out to improve the durability of elastomer-to-metal adhesive bonds in water. A screening test of some commercially available adhesive systems was performed. ASTM-D429 test methods using both conical button specimens and peel-strips were employed. The 90-degree peel test was modified such that the effects of stress corrosion, water salinity and gas content could be examined. Most commercial systems were found to give adequate dry bond strength. Preliminary environmental aging studies also showed that thermal aging did not produce any measurable effect in accelerating the degradation of the elastomer/metal bond. However, chemical attacks from salt water under stress caused great reductions in the bond strength, especially when the oxygen supply was abundant. The weakened bonds were found to show increasingly large portions of fracture failure at the adhesive/metal interface rather than rupture failure of the bulk rubber.

INTRODUCTION

In addition to applications for joining the structural components in hydrofoil boats, surface effect vehicles and other advanced ships, adhesive-bonding technology is also applied in many underwater systems such as vibration damping mounts, acoustic windows for sonar transducers and other acoustic conditioning devices. Elastomer-to-metal bonding usually appears in one of two forms. Either the cured rubber components are bonded to a metal substrate or the uncured rubber is molded in place onto metal surfaces which have been prepared in advance with primers and adhesives. The latter technique is extremely popular, especially when the area of coverage is reasonably small. Essentially all the rubber face plates of wet-end sonar transducers are fabricated in this fashion. For large rubber structures such as sonar dome windows which contain steel reinforcements, adhesives are often applied in-between plies during the lay-up process before autoclave cure. In this case, the adhesive is used to promote adhesion between the cured rubber surfaces. In the discussion here, however, the emphasis will be on the elastomer/metal bond prepared by direct vulcanization of uncured rubber to a metal substrate.

The life-time requirement of many sonar devices may be as long as ten to fifteen years. Since the device is constantly exposed to rigorous environmental stresses due to chemical attack of sea water, temperature extremes, pressure cycling plus mechanical stresses, it is at first glance very unlikely that an elastomer-to-metal bond will survive over such a long period of time. Many factors may contribute to this pessimistic observation. First of all, the bonds are normally prepared in manufacturers' production warehouse or in shipyard environments where a "clean room" condition is probably difficult to maintain, if not impossible. Subtle changes in bonding procedure or sheer negligence of the workers in matters such as surface treatment, adhesive mixing and application, and vulcanization temperature control could easily lead to an unsatisfactory bond. Furthermore, primers, adhesives and rubber compounds are all typically complex chemical mixtures. Their compositions are often kept by the manufacturer as secretive as possible under proprietary claims. At the same time, since these materials usually have narrow profit margins, the formulator will frequently change suppliers or replace ingredients in order to maintain a low cost basis. Without extensive studies of the long-term effect of such changes, these formulation variations, even minor at times, often lead to compromises in performance. Degraded adhesive bonds cause serious water-leakage, corrosion, loss of mechanical integrity and ultimately the failure of the complete sonar system. The result is ever-increasing cost of maintenance, repair and replacement for these systems, which also adversely affects fleet readiness. This makes the reliability consideration an utmost important matter.

In order to improve the reliability of elastomer-to-metal bonds in sonar devices, it is believed that steps toward stringent quality control must be taken. It would be necessary to specify the chemical composition of adhesives and elastomers, and to develop required bonding and molding procedures. The materials will eventually have to be developed to replace proprietary commercial products. Quality control methods need also be investigated, both for chemical analysis and for performance testing. In this paper, the first effort made

toward achieving this long-range goal will be described. The
concerns are testing methodology, adhesive screening and preliminary
environmental aging studies.

TEST METHODS

For quality control, the specification of adhesive chemical
composition is necessary, but unfortunately not sufficient. The
performance of adhesives depends not only on their chemical contents
but also on how the ingredients are mixed together and how the
adhesive bond is fabricated. Performance evaluation is therefore
essential. Although there are many mechanical tests in general
use at the present time, there is a need to determine their relevance to field performance, and to standardize the test procedure.
For these reasons, two test methods have been investigated.

The first was the ASTM-D429-73 Method C, using a 45-degree
conical button specimen. This method, in addition to being a
standard technique, was also cited by ASTM as statistically more
meaningful and more reproducible than other methods[1]. These ASTM
testing methods have also recently been reviewed by Cutts[2]. Figure 1
shows the schematic of the conical specimen. Conical buttons were
prepared from 2.5-cm metal bar stock. They were first vapor
degreased, grit blasted using 120 grit silicon carbide, and then
degreased again prior to coating with adhesive. The manufacturer's
recommendations were followed for each adhesive system used in order
to achieve the suggested nominal coating thickness. After sufficient
drying time, the thickness was checked with a magnetic type film
thickness gauge. An electrically heated transfer mold was used to
fabricate the test specimens. An injection molding technique was
applied for introducing the rubber into the mold for final vulcanization. The fully cured specimens had a controlled distance of
11.5 $\pm$ 1 mm between the tips of the conical end pieces. The metal
cones were pulled away from each other in an INSTRON machine at a
crosshead speed of 5 cm/min. Stress concentration at the tip of
the cones caused failure to initiate at the tip, then progressed
along the conical face until rupture. The required failure load
was recorded.

The second method employed in this investigation was a 90-degree
peel test in either air or water. The test was designed in a fashion
similar to that given in ASTM-D429-73 Method B, but with some modifications. The method was essentially developed by the Navy[3], and
has been applied extensively in the industrial community[4,5]. It
basically involves hanging a weight on the end of a rubber peel
strip that is bonded to either a steel or an aluminum base, (see
Fig. 2). The specimen may be exposed to hot air for accelerated
life testing, or be immersed in water to study the environmental
effect on·peeling. The rate of peel was measured visually with an

optical microscope for a given weight. Alternatively, the weight
could be increasingly added to initiate peeling and the rate was
monitored for a specific time period. Besides being a simple method,
this technique has a main advantage in that it is essentially a
stress corrosion test. The chemical action of water accelerates the
debonding process while the bond is under sustained high stresses
from the load. This test method allows one to simulate one of the
more severe service conditions for many underwater rubber-to-metal
joints. Effects of parameters such as temperature, stress, salinity
and gas content of water may be easily studied with varying combinations of elastomer, adhesive and metal.

ADHESIVE SCREENING

There are many commercially available rubber bonding agents.
These materials are normally offered as single-source solutions
for handling a broad range of adhesion problems. The idea is, of
course, to supply the fewest different adhesive formulas for
industrial applications covering a wide range of substrates and
elastomer formulations. It is therefore not surprising to find
that many adhesives are described as "general purpose cover coat
or one-coat adhesive". However, this approach is not acceptable
when the optimum long-term performance is sought for Navy's underwater systems. For instance, an adhesive may be adequate for
bonding many household rubber items without showing any measurable
difference because the performance requirement is very minimal.
But for applications in underwater sonar systems, the performance
requirement is severe and a different combination of elastomer/adhesive/substrate could be very critical in terms of the initial
strength or the lifetime of the bond. Table I is an example to
illustrate this point. Chemlok 304 has been recommended for bonding
rubbers to a steel surface. Four different kinds of rubber were
tested using the static wet peel test in distilled water. For a
2.54-cm wide strip of butadiene rubber, the peel rate was 0.125
cm/day under a 20-kg load. The natural rubber specimen showed a
greater peel rate of 0.25 cm/day at a reduced load of 16 kg. For
neoprene and nitrile rubbers, the load was further reduced in order
to obtain a meaningful measurement. The peel rate was found to be
even greater at 0.5 cm/day in these cases.

It is clear, therefore, that an initial screening test is
necessary. The rubber was chosen to be a Navy neoprene formulation,
and the following commercial adhesives were examined:

            1. Chemlok 205/236A
            2. Chemlok 205/220
            3. Chemlok 205/234B
            4. Chemlok 252
            5. Thixon P6-1/OSN-2
            6. Thixon D21243/508

7. Thixon OSN-2

The Chemlok materials are the products of Hughson Chemicals, Lord Corporation, and the Thixon adhesives were manufactured by the Dayton Coatings and Chemicals Division, Whittaker Corporation. Single designations mean one-coat adhesives, whereas double designations are combinations of primer/adhesive systems with the first number being the primer. These adhesives were tested by using conical button samples, which were prepared from AISI 4130 steel bar stock.

Table II shows the result of this series of tests. Each of the failure load data represents the average of eight measurements. The fracture surface was visually inspected. The percentages of bulk rubber tear (R), rubber/adhesive (R/A) and adhesive/metal (A/M) failures were estimated, following the classification method developed by Peterson[6]. With the only exception of Chemlok 205/234B, all adhesive systems led to bulk rubber rupture. This indicates that the adhesive bond is stronger than the bulk rubber, and the bond strength should be recognized as greater than that indicated in Table II.

Static 90° peel test was also carried out in distilled water for a few selected adhesive systems. Not all the adhesives were examined because the peel rate was generally very low, and a complete investigation would be very time-consuming. Figure 3 shows the test result for the same neoprene formulation with Chemlok 205/234B, which was the least effective system as determined by the conical button tests. The plot shows the observed crack length as a function of time under a static load of 17 kg. The slope therefore gives the peel rate. The initial peel rate was always very large, which was perhaps an artifact of the test. In five days the peel rate reached a steady state value, which in this case was 0.05 cm/day. For Chemlok 205/236A, a better adhesive according to Table II, the crack showed practically no growth for five days. At that point, the rubber pulled apart at the holder of the dead weight, a masked unprimed portion of the specimen. This forced the test to be terminated, but gave an indication that the bond strength was greater than the strength of the bulk neoprene, which was in agreement with the observation given in Table II.

ENVIRONMENTAL AGING

Since the applications for underwater acoustic systems generally have very long life-time requirements, the question of aging needs to be addressed with accelerated life testing techniques. It is common to assume that elevated temperature translates to longer time. But such an assumption is valid only if the kinetics of the aging process is known and the interrelationship between time and temperature can be established. The long term durability of an

Table I. Result of Wet 90° Peel-Test with Chemlok 304.

Rubber	Load (kg)	Peel Rate (cm/day)
Butadiene	20	0.125
Natural	16	0.25
Neoprene	4.5	0.5
Nitrile	4.5	0.5

Table II. Neoprene-Steel Bond Strength as Determined by Conical Button Samples at Room Temperatures.

Adhesive	Failure Load (kg)	Fracture Surface Estimation (%)		
		R	R/A	A/M
Chemlok 252	353 ± 22	100	0	0
Chemlok 205/220	339 ± 13	100	0	0
Thixon P6-1/OSN-2	336 ± 11	100	0	0
Thixon D21243/508	329 ± 7	99	0	1
Chemlok 205/236A	321 ± 13	99	0	1
Thixon OSN-2	314 ± 14	100	0	0
Chemlok 205/234B	310 ± 16	86	7	7

A STUDY ON ELASTOMER/METAL BONDS

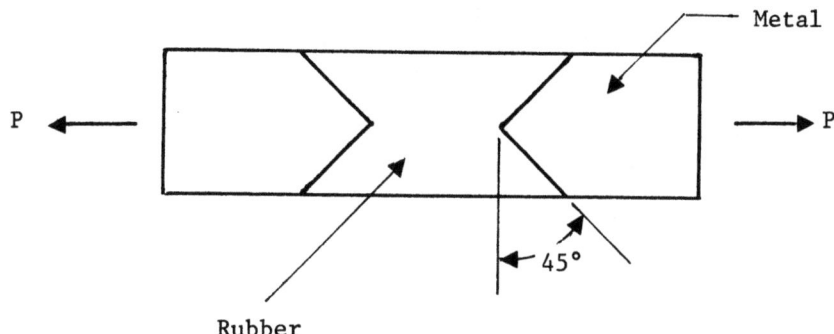

Figure 1. Conical button specimen for rubber/metal adhesion test.

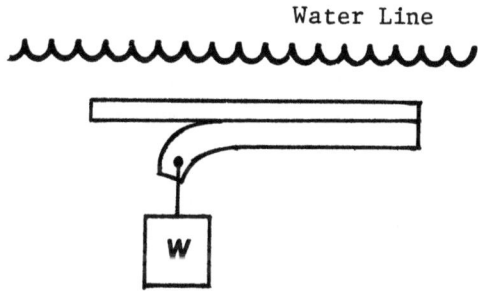

Figure 2. The schematic for static wet 90°-peel test.

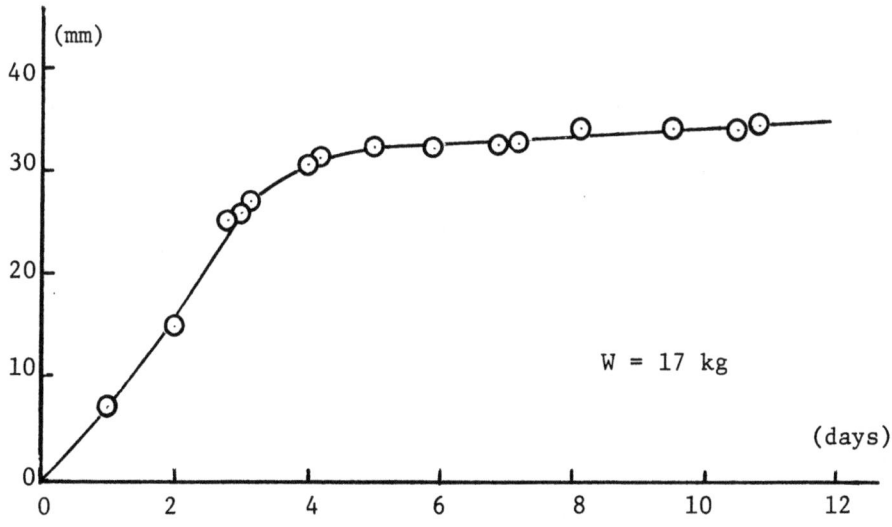

Figure 3. Crack length as a function of time in wet peel test for Chemlok 205/234B.

adhesive bond in sea water probably involves at least two distinct processes: (1) bond line stress corrosion and (2) oxidative or other chemical degradation. Obviously, these processes are themselves very complex and may even be closely related to each other. As a first step, simple attempts were made to obtain some understanding of the aging effect on adhesive bonds, and preliminary results are presented here.

The result of the screening test shows that most of the existing adhesive systems give an adequate dry bond strength in that the bond strength is greater than the tear strength of the rubber itself. The conical button specimens may then be used to examine the effects of thermal aging and chemical degradation. Specimens were exposed to either distilled water or 3.5% salt water at elevated temperatures, then tested at room temperature for their bond strength. By doing so, any effect of high temperature or chemical oxidation due to salt water attack may be isolated, free from any stress-loading complication. Up to this point, specimens aged at 50°C for 200 hours have been tested, and the results are shown in Table III. The effect of 50°C aging in distilled water is almost unnoticeable except for Chemlok 205/234B and 205/236A, in which case the failure loads were reduced by about 13% and 11%, respectively. The deterioration of the Chemlok bonds may be attributable mainly to the effect of water rather than heat, because in a separate static 90°-peel test for the same bond placed in a 75°C dry air oven under a load of 10 kg no crack growth was found in 14 days. The slight increase in failure load in some cases may be related to the post curing effect on rubber due to 50°C aging. The immersion in 3.5% salt water at 50°C had a more pronounced degradation effect in that the failure loads for Thixon D21243/508, Chemlok 205/236A and 205/234B were reduced. Coupled with this reduction, it was observed that the fracture surface showed increasing adhesive/metal interface failure as a result of the salt water immersion. No observable degradation effect was found for the Chemlok 252, 205/220 or Thixon P6-1/OSN-2 systems. As of the writing of this report, the aging time was still too short to manifest any such effect yet.

The oxidative degradation effect of water was further examined by using the wet 90°-peel test. The water tank was saturated with either nitrogen or air, and the crack length measured as a function of time. Table IV shows the result of a Chemlok 205/234B bonded neoprene/steel specimen. Under the same loading, crack growth was enhanced by the saturated air in salt water. As the load was increased from 7 to 9 kg, the crack growth was further accelerated, as another indication of the stress corrosion phenomenon in this type of adhesive bonds.

CONCLUDING REMARKS

Adhesive bond evaluation techniques using conical button

Table III. Effect of 200-hr. Wet-aging at 50°C on the Bond Strength Failure Load of Conical Button Specimens.

Adhesive	Control* (kg)	Distilled water (kg)	3.5% Salt water (kg)
Chemlok 252	353	356	351
Chemlok 205/220	339	347	350
Thixon P6-1/OSN-2	336	346	350
Thixon D21243/508	329	328	295
Chemlok 205/236A	321	285	307
Thixon OSN-2	314	327	310
Chemlok 205/234B	310	271	302

*Room temperature, dry

specimens and peel strips were found to be useful in determining the strength of elastomer-to-metal bonds under various conditions. The static wet peel test has the main advantage of incorporating the stress corrosion effect with other variables such as water salinity and gas content in the same test. Many commercially available adhesive systems were found to provide adequate dry bond strength. But chemical degradation by salt water under stress loading was shown definitely to reduce bond strength. This degradation effect seemed to be enhanced when the oxygen content in water was high. These failures mainly occurred at the adhesive/metal interface.

Table IV: Effect of Gas Content in 3.5% Salt Water on Neoprene-Steel Bond.

Static Load (kg)	Crack Length in 4 Weeks (cm)	
	Nitrogen	Air
7	0	0.2
9	0.15	0.8

REFERENCES

1. ASTM Standard D429-73, Part 37, p. 82 (1981).
2. E. Cutts, in "Development in Adhesives-2," Ed. A. J. Kinloch, Appl. Sci. Publ. London, pp. 367-404 (1981).
3. Mare Island Naval Shipyard, Rubber Laboratory, Rept. MI-93-15, Feb. 1967, Mare Island, California.
4. General Dynamic Corporation, Electric Boat Division, Rept. C443-77-082, 1977, Groton, Connecticut.
5. Bolt, Beranek and Newman Inc., Tech. Memo NL-055, May 1982, New London, Connecticut.
6. C. H. Peterson, Adhesives Age, $\underline{7}$, 30 (1964).

Part V
Stress Analysis and Performance Aspects

LIFE PREDICTION METHODOLOGY FOR ADHESIVELY BONDED JOINTS

J. Romanko[1], K. M. Liechti[1] and W. G. Knauss[2]

[1] General Dynamics
Fort Worth Division
Fort Worth, TX 76101

[2] California Institute
of Technology
Pasadena, CA 91125

A comprehensive integrated methodology for adhesive bonded joint life predictions is outlined. The guiding assumption is that the useful life of bonded joints is determined by failure in the adhesive interlayer, and this is the basis for a systematic analysis of information and the techniques required to provide valid predictions. Emphasis is placed on time-dependent fracture mechanics procedures, including detailed through-the-adhesive thickness viscoelastic finite element calculations of the stress-strain distributions, and on analytical methods involving the constitutive relations of the adhesive interlayer. Use is made of instrumented bonded joint data obtained from structural overtest methods under a number of U.S. Air Force-sponsored programs. A logical program rationale is outlined which has sufficient generality to apply to any adhesive bonded joint structure including metal and composite adherends, various loading/environmental conditions, which apply to high-performance aircraft. The five main modules of the integrated methodology are: (Definition of) Loads/Environments; (Adhesive) Material Properties; Stress Analysis; Fatigue Mechanisms and Failure Criteria; and Structural Failure Analysis (Life Predictions). The necessary contents of each module, including input and output data, are specified and a systems approach is taken interconnecting the modules into a logical sequence suitable for making service life predictions for bonded structures.

INTRODUCTION

The use of polymeric adhesives to structurally join metals and to join or repair composite aircraft structural components offers significant cost and weight savings. Initial costs are reduced through design simplicity and reduced part count as proven recently in at least two advanced technology programs.[1,2] Operational costs are potentially reduced through longer service lives. Maintenance costs are reduced through demonstrated damage tolerance and fatigue endurance far superior to that of riveted joints. Fuel savings can result from reduced weight. To properly evaluate the cost savings through improved service life of the bonded structures, and to be able to more accurately estimate the total cost of ownership prior to commitment to hardware, several programs have been sequentially initiated as described below. They are highly representative of fundamental understanding of bonded joint fatigue endurance and damage tolerance. The backbone program of one such series of closely coordinated programs conducted sequentially is the "Integrated Methodology" program discussed herein.

INTEGRATED METHOD

Figure 1 shows the five major elements or modules (I through V) of the predictive method and their inter-relationships as applied to adhesively bonded structures. This integrated methodology parallels the systems approach for solid rocket service life prediction first proposed by Kelley and Trout[3]. Extension of Kelley's ideas from solid rocket propellants to adhesively bonded structures is possible in view of the time/temperature (and moisture) dependence of the constituent material (polymer) properties and similar failure modes (crack propagation) in both cases. The various feeder sub-programs, which must be accomplished for full implementation of the ① Integrated Methodology (IM) program, are labeled ② through ⑧ in Figure 1 and are being carried out under AFWAL/MLBC-sponsored contracts F33615-: 79-C-5117 ② (Residual Stress Analysis; Texas A&M Univ.); 80-C-5093 ③ (Basis for Accelerated Testing; TRI, Austin, TX); 80-C-5167 ④ (Viscoelastic Stress Analysis; General Dynamics, Fort Worth Division, TX); 81-C-5114 ⑤ (Time Dependent Fracture; United Technologies, Chemical Systems Div., CA); 82-R-5067 ⑥ Fracture Analysis Boundary Value Problems; ⑦ Fracture Data Base; and ⑧ Constitutive Relations with Damage.

LIFE PREDICTION METHODOLOGY

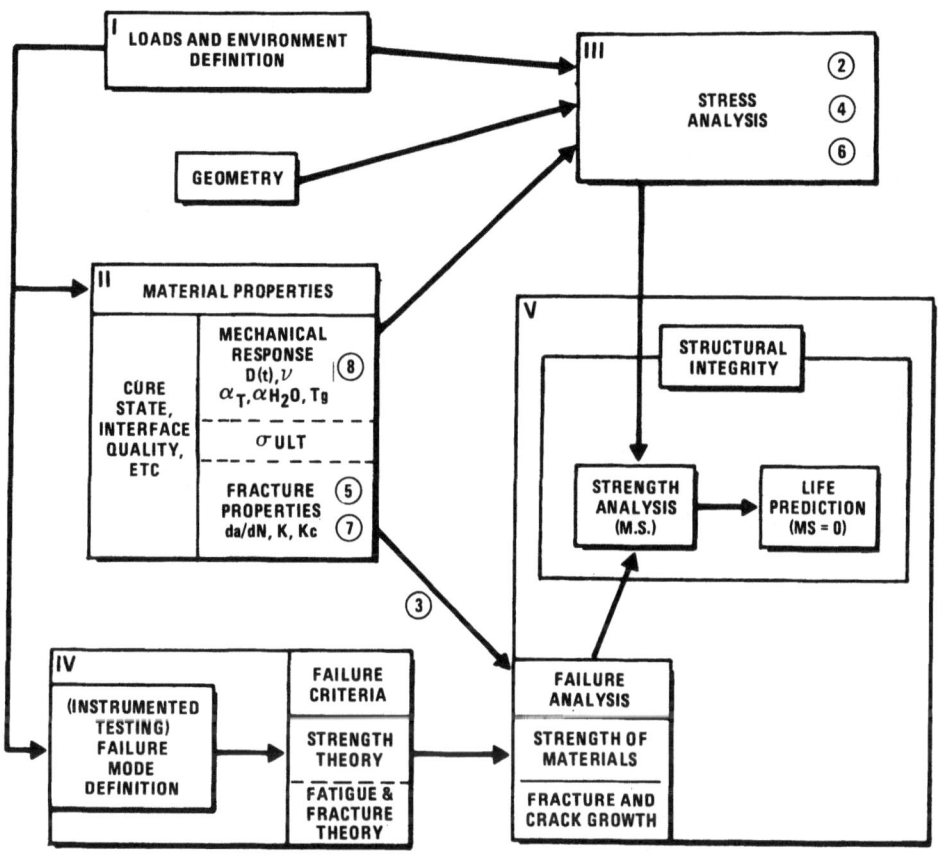

Figure 1. Modules and Flow Path of Predictive Method: ① "Integrated Methodology for Adhesive Bonded Joint Life Predictions".

The format of the IM program highlights setting up the framework and its constituent modules which would be used to make the life prediction(s). The IM program would deal with the individual modules to some extent, with the specific details of the modules to be developed by the feeder sub-programs. In any case, priorities were established on which of the elements of each module were necessary with an indication of the gaps in the availability thereof for the life prediction.

The contents of the five main modules (viz., analytical methods and/or questions that the module needs to address) and their inputs and outputs have been established, and are duly recorded in great detail in Reference 4. Included therein is an outline of the experimental test plan for developing the required data, in addition to an overall integration test plan for effecting the life predictions of a bonded joint of a real structural aircraft component.

In view of the time and space limitations of this paper, we will endeavor to highlight some of the more important modules, supplemented by pertinent comments concerning the remaining modules which are of basic importance to the methodology. First, we will make general observations regarding certain of the modules which set them apart from the others.

We note that physical laws govern the contents of a number of the modules, with only engineering approximations possible in describing the phenomena involved. Also, basically there are no modifications possible for these except for economic trade-offs in these approximations. Modules in this category commanding top priority include "Stress Analysis" and "Failure Mechanisms and Criteria".

Modules which allow descriptions not directly governed by physical laws allow system choices which affect the final result. Two such modules include "Loads and Environments" and "Material Properties" which are of top priority in our scheme. Geometry, if it can be considered a module at all, commands secondary emphasis at this time.

Material properties represent the only element sensitive to aging time although there is naturally a time element inherent in the environmental loading rate, duration and frequency. Since storage time is a primary variable in the service life prediction scheme, the overall rationale should apply as well to the zero-time design point which considers all material properties in the unaged condition. Another important breakout of material properties, when considering the impact of aging, is the separation of response properties from the fracture properties; response properties being a principle input to the stress analysis and fracture properties applying primarily to the strength analysis.

The response and fracture characteristics of bonded structures are known to be dependent on loading rate and on history in such a complicated fashion that analytical procedures must be supplemented with adequate laboratory testing to provide the necessary design confidence. The guiding assumption in this integrated methodology scheme is that the useful life of a bonded structure is determined by the initiation and relatively slow propagation of cracks emanating from corner singularities and reaching a critical size at which time catastrophic failure occurs.

The emphasis in this integrated methodology scheme is on the "deterministic" approach, with full recognition that a "probabilistic" treatment will eventually have to be incorporated in the methodology to realistically predict mixes of missions and mission profiles and to account for experimental error in the data used. However, since a deterministic approach provides the basis for a statistical representation and in view of the time and budget resources of this program, the deterministic approach was the first priority.

We proceed with a description of the contents of a number of the five main modules and the corresponding input and output information required. Of special interest is the central position of the Stress Analysis module (III) and the dominance of the Failure Mechanisms and Criteria module (IV) in focussing all related activities. Accordingly, we will highlight these two modules in the latter stages of this discussion.

The experimental data required to validate the concepts in the modules is indicated herein and the overall integration of the method for the analysis of a test joint or an "aircraft", with associated Management Plan is also presented.

Order of priority of some items, insofar as being accomplished in the Integrated Methodology program is indicated by the designation $P(n)$ after the time in the Tables, with $n=1, 2, 3,...$, in order of decreasing priority. Of course, the aforementioned feeder subprograms will eventually develop the necessary inputs for full implementation of the methodology for a general case.

Loads/Environments (Module I)

Often the types of loads under consideration are directly related to failure modes (as far as the way engineers do it); one tends to forget that often loads may produce different, possibly longer term failures. It is important to delineate the kinds of "loads" so that the types of histories are accounted for in material characterization (stress and failure response).

Table I summarizes the contents of this important module, emphasizing the time, temperature, moisture, stress level and sequence of events in defining the impressed loads, whatever their origins may be. The magnitude and frequency of the periodic mechanical load and steady and monotonic non-periodic mechanical loads are of concern in the Integrated Methodology program.

Table I Loads/Environments, Contents*.

A. MECHANICAL LOADS
    LOAD HISTORIES (DEDUCED MISSION PROFILES)
        MISSION PROFILE
        MANEUVER SEQUENCE
        FLIGHT ENVELOPE
        (a) PERIODIC (FATIGUE) – MAGNITUDE
                                           FREQUENCY
        (b) NON-PERIODIC   – STEADY
                                           MONOTONIC

B. THERMAL ENVIRONMENT
    (a) EFFECT ON $\underline{A}$ (THROUGH STRESS ANALYSIS) UNDER CONSIDERATION OF DIFFUSION TIMES
    (b) EFFECT ON MATERIAL BEHAVIOR

C. MOISTURE ENVIRONMENT
    (a) EFFECT ON $\underline{A}$ (THROUGH STRESS ANALYSIS) UNDER CONSIDERATION OF DIFFUSION
    (b) EFFECT ON MATERIAL BEHAVIOR

D. SEQUENCE OF MECHANICAL LOADS, THERMAL MOISTURE ENVIRONMENT

*Details are recorded in Reference 5.

Table II summarizes the input and output information relating to this module. Traditionally, the Loads Group has developed an expertise in these types of problems for rate independent structures, with some modification effected by "composite" needs. For adhesives, we envisage the requirements indicated for failure analysis purposes (output from the module and raw loads data input). It must be noted that the listing of input and output items is not necessarily complete at this time, although we have tried to be as inclusive as possible.

Table II Loads/Environments, Inputs and Outputs*.

INPUT	OUTPUT
MECHANICAL	*** TIME DEPENDENCE OF LOADS**
AIRPLANE RECORDS ON ACCELERATION WITH EXPLICIT TIME	(A) STEADY
(A) GROUND REST	(B) CYCLIC WITH WHAT FREQUENCY
(B) TAKE-OFF	
(C) LANDING	(C) "BLOCKS" FOR STATISTICAL ANALYSIS
(D) IN-FLIGHT MANEUVERS	
DETERMINISTIC	
STATISTICAL (SPECTRA)	
THERMAL	***(TIME DEPENDENT)**
GROUND STORAGE TIME SCALE	TEMPERATURE DISTRIBUTION
FLIGHT: STAGNATION-INDUCED TIME SCALE	TEMPERATURE GRADIENTS
ENGINE – DUCT INDUCED TIME SCALE	(INCLUDING SKIN TEMPERATURE)
ATMOSPHERIC TIME SCALE	RADIATION OF SUN
MOISTURE (AND OTHER SOLVENTS)	*** TIME DEPENDENT**
DIFFUSION AS A FUNCTION OF TEMPERATURE	H_2O (AND/OR SOLVENT)
GROUND STORAGE STRESS DEPENDENCE	CONCENTRATION AND GRADIENTS
FLIGHT DRY-OUT	POWER SPECTRAL RESPONSE

*Details are recorded in Reference 5.

Material Properties (Module II)

The local stress distribution developed in a bonded joint will be greatly influenced by the material properties which define their deformation characteristics. The "response" properties of the adhesive, such as the creep compliance and Poisson's ratio will dictate, by and large, these deformational characteristics of the bonded joint and in general will exhibit time, temperature, moisture and stress level dependence.

After cyclic loading of the environmentally conditioned (temperature and humidity) bonded joints, the "fracture" or "failure" properties of the adhesive will control the residual strength capabilities of the structure. Typical fracture properties of the adhesive (joint) include the crack growth rate with cyclic frequency, viz., $da/dn = a'$, or the equivalent $da/dt = \dot{a}$ representation. Failure or fracture "properties" are dependent on the current knowledge of failure analysis (properties are determined in tests prescribed by failure analysis).

A comprehensive review of these time-dependent response and fracture properties of adhesives required for this program is summarized in Reference 6. The physical and mechanical behaviors considered included constitutive relations under monotonic and cyclic loading conditions, dilatations during cure, cooling and moisture infusion, aging, mechanisms of damage accumulation, fracture toughness and ultimate strength spectra, among others.

Table III and Table IV summarize the response and fracture properties contents of the Materials Property module, and their sensitive dependence on time, temperature, moisture, and stress level, among other parameters[6].

Physical, chemical and structural aging effects are distinguished and these effects are being covered in the program ③ on "Basis for Accelerated Testing," currently in progress.

The importance of the dependence on response and fracture properties of the state of the adhesive, whether it is neat-neat (without scrim), or neat (with scrim) is acknowledged.

The fracture properties of the adhesive layer describe the relationship between a fracture parameter and the crack growth rate in the adhesive layer. Thus, since the fracture properties are known, the history of the crack growth can be obtained by integration of the crack growth rate. Failure can then be defined by the achievement of some critical length crack. Since the geometry changes as a crack grows in a joint, the relationship between the crack length and the chosen fracture parameter must also be determined in order to evaluate the integral. The variation of fracture parameter with crack length is determined by stress analysis which is the subject of module III. The choice of fracture parameter and a discussion of failure mechanisms and criteria comprise module IV.

Stress Analysis (Module III)

A thorough, through-the-thickness stress analysis is vital to the success of the Integrated Methodology program. It allows point-

Table III Materials Properties,* Contents (Response).

RESPONSE
$D(t)$, or $E(t)$ or $J(t)$
$k(t)$ or $\nu(t)$ (WHAT MINIMUM ACCURACY REQUIRED?)
TRANSITION TEMPERATURE (GLASS OR MELT TEMP FOR CRYSTALLINE COMPONENTS IN AN ADHESIVE)
SHIFT FACTORS ϕ DUE TO TEMPERATURE ϕ_T
MOISTURE ϕ_p
STRESS LEVEL ϕ_σ
COEFF OF EXPANSION DUE TO TEMPERATURE α_T
MOISTURE $\alpha_{p(x,t)}$
STRESS LEVEL   ($\equiv k(t)$)
CONSTITUTIVE FORMULATION; NON-LINEAR VE $<$ VOLUME / OCTAHEDRAL SHEAR
AGING: PHYSICAL (STRUIK)
CHEMICAL (DARK EDGE)
STRUCTURAL (MICROCRACKING)
NEAT vs NEAT-NEAT BEHAVIOR*
*NEAT TAPE IS WITH MATTE DACRON SCRIM. NEAT-NEAT TAPE IS WITHOUT SCRIM.

*Details are recorded in Reference 6.

by-point description of the displacements and stresses (both peel and shear) on a "micro level" in attempting to account for material displacements (creep) and damage mechanisms occurring in the adhesive interlayer on a realistic basis. Averaging of stresses through the thickness of the adhesive interlayer leads to overly-conservative results.

Table V summarizes the requisite content of the Stress Analysis module, including the validation of the stress analysis by holographic interferometry. This validation (described in Reference 7, Appendix 1) is an important part of the methodology since it will allow first-hand confirmation of the finite element procedures adopted in this program.

Table IV Materials Properties,* Contents (Fracture).

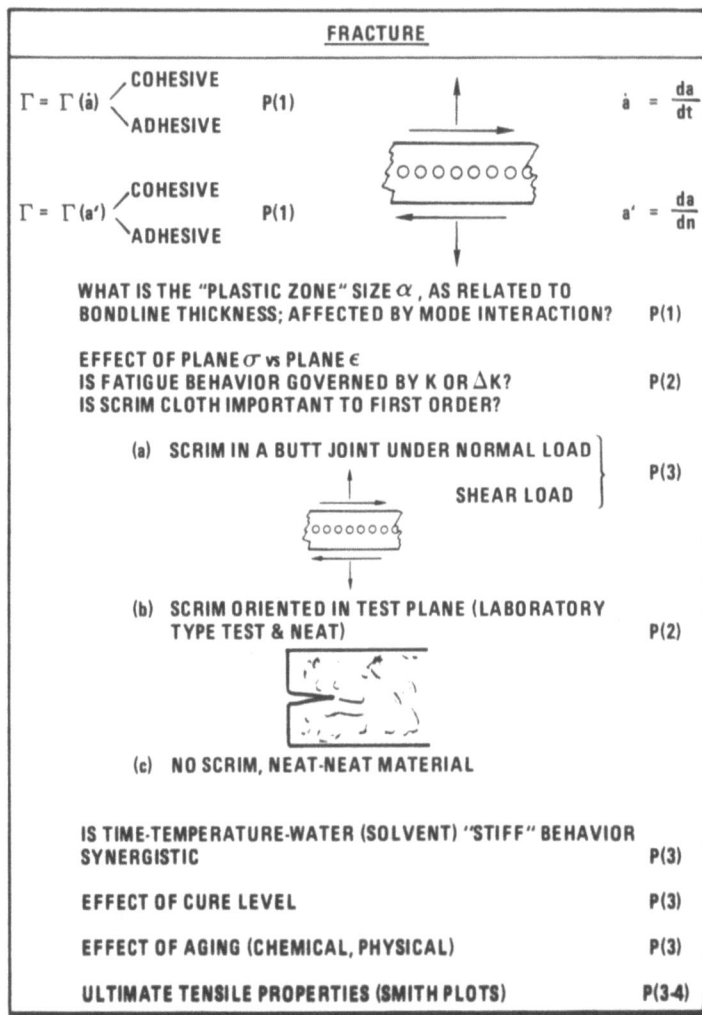

*Details are recorded in Reference 6.

Table VI details the input and output information required for and expected from the Stress Analysis module. Note the selection of the appropriate cost-effective, finite element stress analysis program for the calculation at hand. In some cases, the more extensive VISTA program with generalized boundary conditions being developed under program ④ the Viscoelastic Stress Analysis program is prescribed. In other cases, the original MARC program or its recent PRONY series modification may be more cost-effective,

Table V Stress Analysis, Contents.

HIERARCHY OF REQUIREMENTS AND CAPABILITIES:	WHICH PROGRAM:	
A. MATERIAL REPRESENTATION		
LINEARLY ELASTIC	MARC	VISTA
ELASTIC – PLASTIC (WITH AND WITHOUT STRAIN HARDENING)	M	
LIN VE	M	V
NON-LINEAR VE		V
VISCOELASTIC NON-LINEARITY (GEOMETRY UPDATING)	M	V
DILATATIONAL BEHAVIOR OF MATERIAL DUE TO TEMPERATURE AND OR WATER (SOLVENTS)	M	V
B. GEOMETRY REPRESENTATION:		
CRACK TIP ELEMENT DEAL WITH INTERACTION	M	V
INTERFACE CRACK TIP ELEMENT		V
END TERMINATION (CORNER SINGULARITY)	M	V
VOIDS AND POROSITY		

C. DISTRIBUTION OF
    a) TEMPERATURE
    b) WATER
		V

–VALIDATION– (HOLOGRAPHIC INTERFEROMETRY)*		

*Details are described in Appendix 1 of Reference 7.

and, of course, the much more cost-effective MacNeal Schwendler Corp. NASTRAN program is recommended for gross analyses.

At the present time, the MARC finite element program is the most flexible and powerful program for performing the stress analysis of adhesive bonded joints. It has many significant capabilities that are directly applicable to the Integrated Methodology program including: plasticity, viscoelasticity, large deflections, and J-integral calculation. The addition of the generalized Prony series expansion for viscoelastic material behavior makes MARC one of the few programs readily available for performing viscoelastic analysis. Finally, MARC is well supported, documented, and relatively easy to use considering that it is a very sophisticated finite element program.

Table VI Stress Analysis, Inputs and Outputs.

INPUT	MARC	VISTA	OUTPUT

A. GEOMETRY INCLUDING FRACTURE PATH
  (a) INTERFACE
  (b) COHESIVE
	MARC	VISTA	ENERGY RELEASE RATE OR J INTEGRAL
			  − (THICKNESS AVERAGED)
			  − DETAILS

		V	STRESS INTENSITY FACTOR
B. MATERIAL RESPONSE CHARACTERISTICS		V	CORNER SINGULARITY STRENGTH
CONSTITUTIVE BEHAVIOR ACCOUNTING FOR TEMP AND WATER	M	V	STRESS DISTRIBUTION MAPS FOR STRESS COMPONENTS PLUS

C. BOUNDARY CONDITION ON
  (a) LOADS
  (b) TEMPERATURE
  (c) WATER
			(a) PRINCIPAL STRESSES AND THEIR
			(b) ORIENTATIONS
			(c) OCTAHEDRAL
			(d) DILATATIONAL ($\sigma_1 \pm \sigma_2 \pm \sigma_3$)

D. EXTREME EXAMPLE: TIME MARCHING PROBLEM	M	V	NODAL DISPLACEMENTS
		V	ENERGY RELEASE FOR MINIMUM CRACK P(2)
			DETERMINATION OF CRACK PATH TO BE ACCOMPLISHED BY TRIAL AND ERROR FOR MAXIMUM ENERGY RELEASE RATE AS A FUNCTION OF INCREMENTAL CRACK EXTENSION
NASTRAN WILL BE USED FOR GROSS ANALYSIS			

Currently a more sophisticated version of MARC, called the MARC Prony program has become available. It permits a more accurate representation of the viscoelastic behavior of adhesives. The more general program VISTA is being developed with more generalized boundary condition capabilities.

An example of MARC finite element modeling of the adhesive interlayer of a bonded joint, e.g., a structural lap joint, SLJ, is shown in Figure 9 of Reference 8. Details of the mesh refinement at the free edges of the adhesive where stress concentrations are possible are shown in the enlarged views therein. As cracks grow during the tension-tension fatigue cycling, the corresponding finite element models can be modified so that the J-integral calculations can be performed. This requires local refinement of the mesh similar to that done to the model joint (MJ) in an early phase of the Integrated Methodology program.[5] An $r^{-\frac{1}{2}}$ strain singularity is created by using a ring of isoparametric quadrilaterals with collapsed sides at the crack tip. The J-integral is calculated by determining the change in stiffness due to a differential change in crack length and multiplying this by the displacement vector created by the applied load. This gives the change in energy due to crack advancing. When divided by the length, a value for the J-integral can be determined.

Failure Mechanisms and Criteria (Module IV)

The contents and inputs and outputs of this important module are shown in Tables VII and VIII, respectively.

There are a number of reasons that crack growth in an adhesive bond could differ from that in metals. First, we recognize that adhesives are polymers and therefore potentially viscoelastic and time dependent. Second, the adhesive layer is heavily constrained by the metal adherends, giving rise to the possibility of geometric and material nonlinear behavior. Finally, fatigue crack growth must be understood under conditions of varying degrees of mode interaction.

The fracture parameter chosen to characterize fatigue failure in adhesively bonded joints in this program is the J-integral, as described in the previous section. Not only can it be compared directly with the linear elastic fracture mechanics (LEFM) approach

Table VII Failure Mechanisms and Criteria, Contents.

CRITERIA OF FAILURE — BRITTLE, "DUCTILE"; VISCOELASTIC			
	(A)	CRACK GROWTH RATE EXCESSIVE	P(1)
	(B)	AN INSTABILITY DEVELOPS (DISCONTINUOUS GROWTH RATE OF CRACK COALESCENCE)	P(2)
	(C)	CRACK SIZE EXCEEDS ALLOWABLE VALUE	P(3)
VIA	(A)	THICKNESS AVERAGED ENERGY RELEASE	
	(B)	LOCAL ENERGY RELEASE OR STRESS INTENSITY INCLUDING LOCAL CRACK TIP DISPLACEMENT CRITERIA FOR GENERATION OF SMALL BUT FINITE CRACK	P(1)
	(C)	MINIMAL ENERGY PATH FOR ADHESIVE AND COHESIVE FRACTURE	P(2-3)
VECTORIAL DISPLACEMENT CRITERION:		$K_I^2 + K_{II}^2 = \Gamma$ ALONG INTERFACE, OR ALONG SCRIM PLANE	

Table VIII Failure Mechanisms and Criteria, Inputs and Outputs.

INPUT	OUTPUT
ENERGY RELEASE RATE G	(A) CURRENT RATE OF CRACK GROWTH
STRESS INTENSITY FACTORS K_I, K_{II} AND ΔK_I, ΔK_{II}	(B) CRACK SIZE AS A FUNCTION OF TIME
FRACTURE PROPERTIES, $\Gamma(\dot{a})$, $\Gamma(a')$	(C) INTRODUCTION OF MULTIPLE CRACK LEADING TO INSTABILITY
FRACTURE PATH (COULD BE CALCUALTED AT GREAT COST) $\dot{a} = \frac{da}{dt}$ $a' = \frac{da}{dn}$	

commonly used for metals, but it is also defined for cases of non-linear elastic, nonlinear elastoplastic and, more recently, nonlinear viscoelastic behavior[9]. The J-integral has also been found to be equivalent to the vectorial crack opening displacement, a fracture criterion which was successfully used to characterize interfacial crack propagation in an adhesive joint subjected to loading normal to or tangential to the bondline[10]. A data base for the predictive methodology is being developed using three specimen geometries; namely, the thick adherend model lap joint (MJ), the cracked lap shear joint (CLS) and the structural lap joint (SLJ) based on the plan of Table IX. Preliminary results of measurements conducted in Programs ① and ⑤ are recorded in Reference 7.

Slow fatigue crack growth portion of the crack growth history is common to all three geometries. It is characterized by a crack growth law which relates the rate of crack growth per cycle (da/dN) to the change in J integral (ΔJ) during a cycle. Catastrophic failure following slow fatigue crack growth has been observed in the MJ and SLJ. The criterion for this final mode of failure is that the shear strength of the adhesive has simply been exceeded due to the shorter overlap produced by an increase in crack length. Prediction of the crack growth history is thus obtained by a cycle-by-cycle

LIFE PREDICTION METHODOLOGY

Table IX Experimental Data Requirements.

R = 0.1

I MODEL JOINT (MJ)	ΔK & $\bar{K}$ TESTS FOR FAILURE JOINT CRACK FRONT LOCATION a(N)	"K" vs a' FROM COMPUTER DETERMINE $\bar{K}$ vs CRACK LENGTH (K_I & K_{II})	THESE TESTS USED TO FOLLOW INVESTIGATION OF ΔK & $\bar{K}$ EFFECT SINCE THIS INTERPRETATION AND APPLICATION OF I & II TO III DEPENDS ON THAT
II CRACKED LAP SHEAR (CLS)	(1) CHOOSE h_1/h_2 SO THAT THE SAME K_I & K_{II} AS FOR MJ RESULTS: DOES IT LEAD TO SAME CRACK GROWTH LAW? (2) IF CHOICE IS MADE WITH MORE DIFFICULTY THAN THAT, ASK WHETHER A DIFFERENT K_I-K_{II} COMBINATION PRODUCES THE "SAME" FRACTURE RATE BY $K_I^2 + K_{II}^2 = \Gamma'(\dot{a})$		$K_I^2 + K_{II}^2 = \Gamma'(a')$ CHECKED "IF NOT, WHAT WORKS?"
III STRUCTURAL LAP JOINT (SLJ)	NEED STRESS ANALYSIS TO DETERMINE LOADS SUCH THAT K_I & K_{II} ($K_I^2 + K_{II}^2$) ARE IN THE SAME BALL PARK AS FOR CLS & MJ. THEN PROCEED TO PREDICT NUMBER OF CYCLES TO FAILURE FOR GIVEN LOAD LEVELS USE DIFFERENT LOAD HISTORY		$\dot{a} = \dfrac{da}{dt}$ $a' = \dfrac{da}{dn}$ $\bar{K}$ = AVERAGE K PER CYCLE

integration of the crack growth law for the fatigue crack growth portion subjected to the critical crack length for catastrophic failure.

The procedure outlined above combines the finite element stress analysis, which calculates the J integral as a function of crack length, a, and the fracture property measurements, viz., da/dN versus ΔJ, for a specific joint geometry. This yields a prediction of the crack growth history, viz., crack length, a, versus number of cycles, N, as shown schematically in Figure 2. The joint failure point, N_f, is thus determined where the slope of the a versus N curve approaches infinity.

Structural Failure Analysis (Module V)

Concerning the contents of this module, if a criterion exists, it contains the onset of failure; so problems of structural failure analysis really become problems of reducing structural loads to

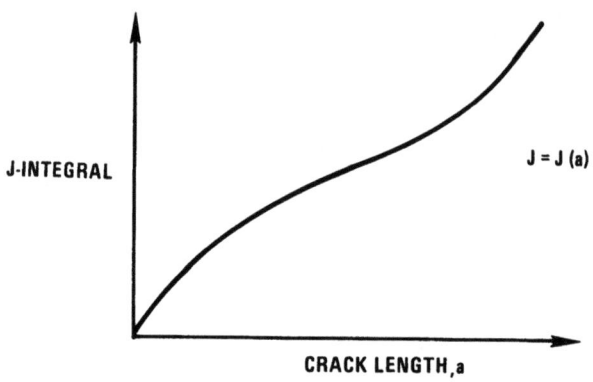

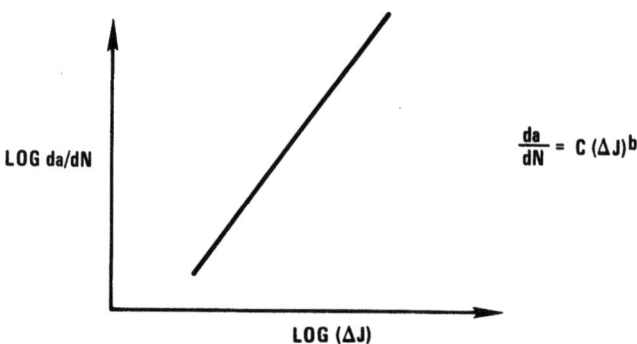

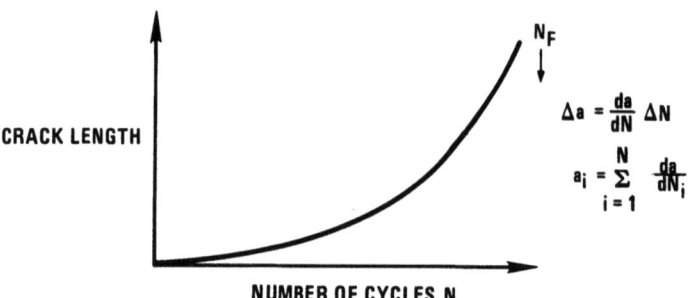

Figure 2. Crack Growth Prediction Scheme.

local stresses causing bond failure. Once the failure has occurred, one must determine whether the overall structure will fail. Of course, the extent of bond crack growth must be considered, including the question of whether the crack stops.

The inputs and outputs of this module are outlined in Table X.

Finally, the overall integration scheme for service life prediction of an aircraft bonded structure, including the management plan, is shown in Figure 3.

PRELIMINARY RESULTS

The crack growth history prediction scheme outlined above and shown schematically in Figure 2 has been computer programmed, with the code validated in Reference 11 using results obtained from the fatigue testing of center cracked panels of neat FM-73M adhesive in the Fatigue Behavior program.[12] Further, preliminary predictions of lifetimes for the SLJ geometry, and comparisons with the corresponding measured lifetimes, are recorded in Reference 7.

Table X   Structural Failure Analysis Module, Inputs and Outputs.

INPUT		OUTPUT
(1.) FAILURE CRITERION	MARGIN OF SAFETY:	BOND OR JOINT
	LET SERVICE LOAD BE L	
		(A) LIFE
(2.) OVERALL STRUCTURE (Airplane) AND PROBLEM AREA LOCATION	L GIVES RISE TO $\dot{a}_L$	
	LOOK FOR OVERLOAD L_0	(B) YES OR NO ON FLY
	WHICH PRODUCES THE	(C) MARGIN OF SAFETY
REDUCTION TO STRESSES AT BONDED JOINT	CRITICAL GROWTH RATE $\dot{a}_0$	
	AT THE EXISTING GROWTH	
(3.) LOADS	GEOMETRY	
	$MS = \dfrac{L_0}{L} - 1$	

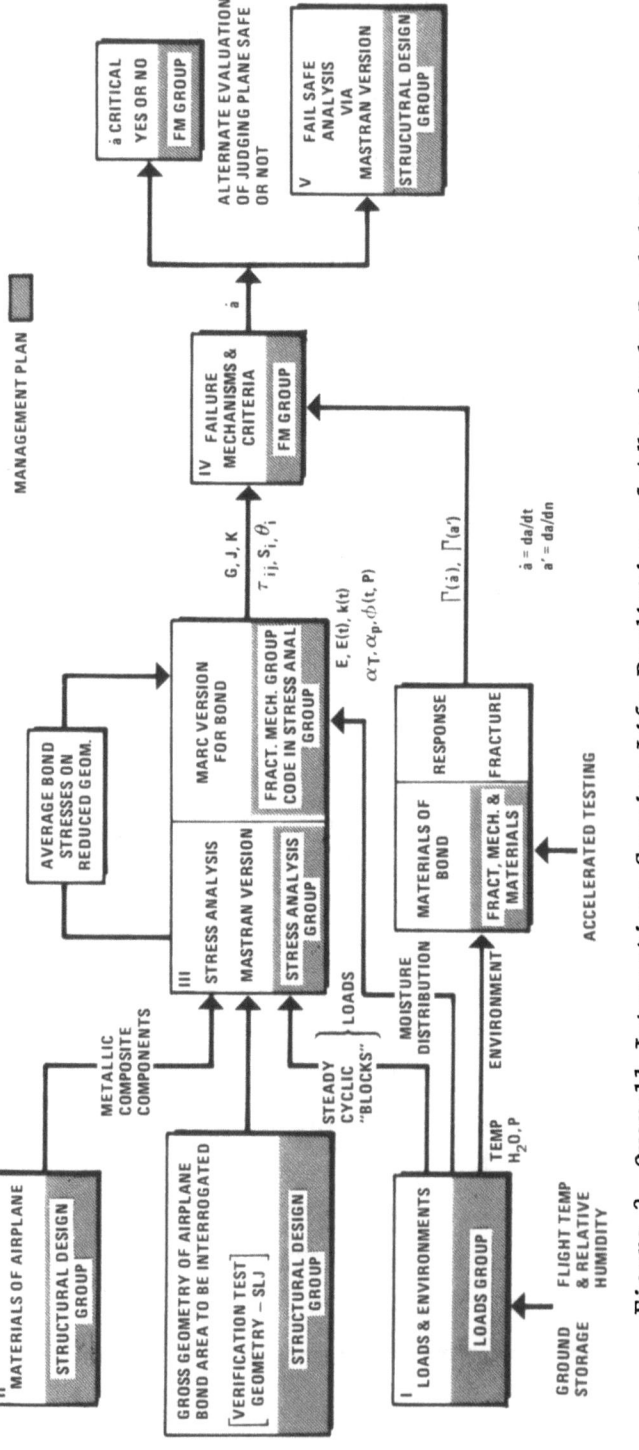

Figure 3 Overall Integration: Service Life Prediction of Adhesively Bonded Joints.

The eventual aim of this program is to predict the lifetime of a circumferentially bonded splice taken from station 703 of the PABST fuselage (Figure 24 of Reference 13). This real structure is similar in geometry to the SLJ geometry used in generating the above-mentioned baseline data.

ACKNOWLEDGEMENTS

This work was sponsored by Air Force Systems Command, Air Force Wright Aeronautical Laboratory, Wright-Patterson Air Force Base, OH, under Contract F33615-79-C-5088, "Integrated Methodology for Adhesive Bonded Joint Life Prediction." The authors especially wish to acknowledge the interest and guidance of Dr. Wm. B. Jones, Jr., AFWAL/MLBC project engineer on this program. Mr. L. R. Collins of the Stress Analysis Group of Materials and Structures Technology, General Dynamics Fort Worth Division conducted the MARC analyses. Also many scientists of the Materials and Structures Technology group contributed to the efforts in this program, including: Messrs. M. E. Tohlen, F. C. Nordquist, C. P. Fisher, and B. O. McCauley.

REFERENCES

1. Contract F33615-75-C-5178, "Laminated Wing Structures", sponsored by USAF AFML/FIBC, Wright-Patterson AFB, OH 45433, and conducted by General Dynamics, Fort Worth Division, Forth Worth, TX, June 1975 - June 1980.
2. Contract F33615-76-C-3138, "Advanced Technology Wing Structure", sponsored by USAF AFML/FIBC Wright-Patterson AFB, OH 45433, and conducted by Vought Corporation, Dallas, TX, November 1976 - March 1980.
3. F. N. Kelley and J. L. Trout, Paper presented at the AIAA/SAE 8th Joint Propulsion Specialist Conference, Paper No. 72-1085, New Orleans, LA, Nov. 29-Dec. 1, 1980.
4. Quarterly Progress Report No. 5, General Dynamics, Fort Worth Division Tech. Report, FZM-6952, December 1980.
5. Quarterly Progress Report No. 2, General Dynamics, Fort Worth Division Tech. Report, FZM-6892, March 1980.
6. "Materials Properties Handbook, FM-73", General Dynamics, Fort Worth Division Tech. Report, FZM-7055, July 1980.
7. John Romanko, K. M. Liechti and W. G. Knauss, "Integrated Methodology for Adhesive Bonded Joint Life Predictions", AFWAL-TR-82-4139, (Final Report for Period July 1979 to July 1982), Wright-Patterson AFB, OH 45433, November 1982.
8. Quarterly Progress Report No. 9, General Dynamics, Fort Worth Division Tech. Report, FZM-7010, December 1981

9. R. A. Schapery, in "Proc. Japan - U.S. Conference on Composite Materials", Tokyo, January 1981.
10. K. M. Liechti and W. G. Knauss, 1981 Advances in Aerospace Structures and Materials, AD-01, p. 51 (1981).
11. Quarterly Progress Report No. 8, General Dynamics, Fort Worth Division Tech. Report, FZM-6990, September 1981.
12. J. Romanko and W. G. Knauss, in "Developments in Adhesives - 2", A. J. Kinloch, Editor, p. 173, Applied Science Publishers, London, England, 1980.
13. Douglas Aircraft Co., Tech. Report AFFDL-TR-79-3129, Wright-Patterson AFB, OH 45433, November 1979.

STRESS ANALYSIS OF ADHESIVELY BONDED JOINTS

Ronald A. Kline*

Engineering Mechanics Department
General Motors Research Laboratories
Warren, Michigan 43090

In this study, the effect of adhesive bondline thickness on the stress distribution in an adhesively bonded joint is examined. A linear variation in bondline stresses through the adhesive thickness is used to model this effect. In this way, an important influence on bond behavior, heretofore neglected or incompletely modeled, is studied. Also presented is a parametric study of the effect of various joint parameters on bondline stress distribution when thickness effects are included.

INTRODUCTION

Adhesive bonding has long been recognized as an attractive alternative to conventional fastening techniques due to a greater uniformity in load distribution as well as reduced weight and processing ease. This is particularly true for the composite materials which are being used with increasing frequency in the aerospace and automotive industries as these materials do not lend themselves easily to conventional fastening techniques. Thus, there is a need for a firm understanding of the nature of the stress states which are found in adhesively bonded joints. Much research has been directed towards this end.

---

*New Address:
School of Aerospace, Mechanical and Nuclear Engineering
University of Oklahoma
Norman, Oklahoma 73019

The basic features of the adhesive bond problem were examined in the classic analysis of Goland and Reissner[1]. In their work, the adherends were assumed to deform as thin plates coupled by an elastic adhesive layer. Particular attention was paid to the fact that, in addition to the resulting shear stress (τ_{xz}), an appreciable transverse tensile stress or peel stress (σ_{zz}) is developed in an adhesively bonded joint. Also, for single lap shear joints, the eccentricity of load application produces a significant bending moment at the joint edges. Later work has extended this research to examine the effects of plasticity in the adherends[2], plasticity in the adhesive layer[3-5], viscoelastic adhesives[6-7], and anisotropic adherends[8-9].

One of the major drawbacks to these works is that the adhesive layer is assumed to be very thin in comparison to the adherend layer and bondline thickness effects are often neglected. The resulting stress distribution from these models represent the average of the stress state in the adhesive joint. However, thickness effects are quite important, particularly if one is concerned with failure mechanisms in composite joints. As is illustrated in Figure 1, failure in a single lap joint with composite adherends initiates at opposite edges of the joint (sites 1 and 3 rather than at opposite edges of the joint (sites 1 and 3 rather than 2 and 4). Thickness independent bond theories are insensitive to this effect. Ojalvo and Eidinoff[11] recognized this shortcoming and developed a model incorporating this effect. They modified the Goland-Reissner analysis to include a linearly varying shear stress distribution through the adhesive layer. Although exhibiting the desired thickness dependent behavior, this model has several serious deficiencies associated with it. While allowing shear stress to vary linearly through the thickness, the peel stress was modeled as being invariant in the adhesive layer. Poisson's ratio was assumed to be zero in this analysis. This represents a significant mathematical simplification but is somewhat unrealistic. In addition, moment and horizontal force resultants in the adhesive layer were neglected. As a result, their solution failed to satisfy equilibrium relations for the adhesive layer. In a later work, Carpenter[12] attempted to modify the Ojalvo-Eidinoff analysis to reflect these concerns. However, Carpenter's modifications fail to satisfactorily resolve these problems. In this work, a new model for determining the stress distribution in an adhesively bonded joint, incorporating bondline thickness effects, is presented.

THEORETICAL MODEL

The basic geometry of the single lap shear joint is shown in Figure 2. Sign conventions are indicated in this figure. The

STRESS ANALYSIS OF ADHESIVELY BONDED JOINTS

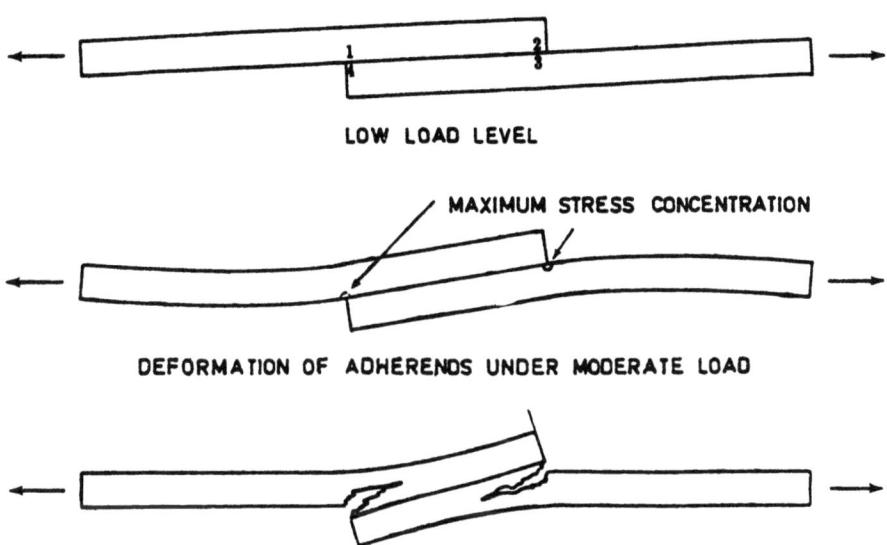

Figure 1. Composite Joint Behavior.

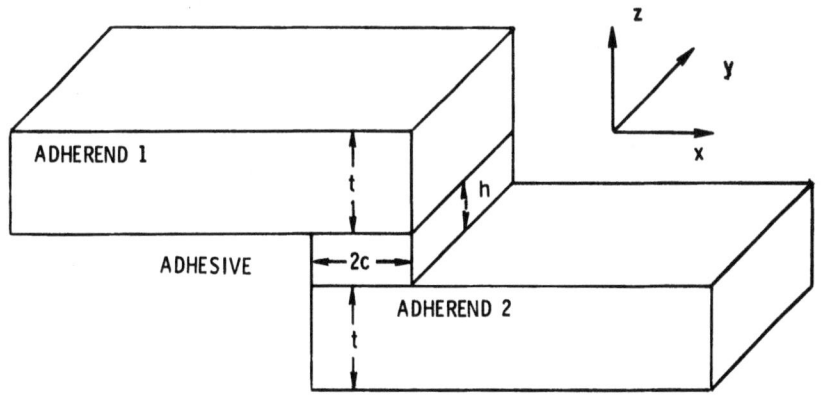

Figure 2. Adhesive Joint Geometry.

origin of the co-ordinate system is chosen so that $z = \pm \frac{h}{2}$ corresponds to the adhesive-adherend interfaces and $x = 0, 2c$ corresponds to the joint edges.

In this analysis of adhesively bonded joints, the following assumptions are employed:

1. The adherends and adhesive are linearly elastic. For simplicity the adherends are assumed to be identical. (With the proper choice of material constants, the theory can be easily modified to include differing adherends.) The constitutive equations for the adhesive layer, unlike most previous investigations in the adhesive layer[1,3-5,8,9,11, 5,8,9,11,12]. This allows for a non-zero Poisson's ratio which is more realistic for structural adhesives than the simplified model used in these studies.

2. The adherends deform as thin plates in cylindrical bending. A plane strain condition in the xz plane is assumed. The effects of transverse shear forces and deformations are included in this analysis.

3. The displacements in the adhesive are assumed to vary linearly through the adhesive thickness. Most earlier models assume that these displacements are constant through the adhesive layer[1-10,13,14].

The starting point for the development of this model is the consideration of static equilibrium of a differential element of each of the joint constituents. The force and moment resultants (N_i, Q_i, M_i) acting on each are illustrated in Figure 3 along with the shear $(\tau_{xz})_i$ and peel $(\sigma_{zz})_i$ forces which develop at the adhesive-adherend interfaces. In this analysis the subscripts 1, 2, and a will refer to the upper adherend (or upper adherend-adhesive interface), the lower adherend (or lower adhesive-adherend interface) and the adhesive layer, respectively. Static equilibrium (neglecting width effects) of the upper adherend requires that:

$$N_1 + \frac{\partial N_1}{\partial x} dx - (\tau_{xz})_1 dx - N_1 = 0 \quad \text{Horizontal Force Equilibrium}$$

$$N_1' = (\tau_{xz})_1 = \tau_1 \qquad (1)$$

$$Q_1 + \frac{dQ_1}{dx} dx - (\sigma_{zz})_1 dx - Q_1 = 0 \quad \text{Vertical Force Equilibrium}$$

$$Q_1' = (\sigma_{zz})_1 = \sigma_1 \qquad (2)$$

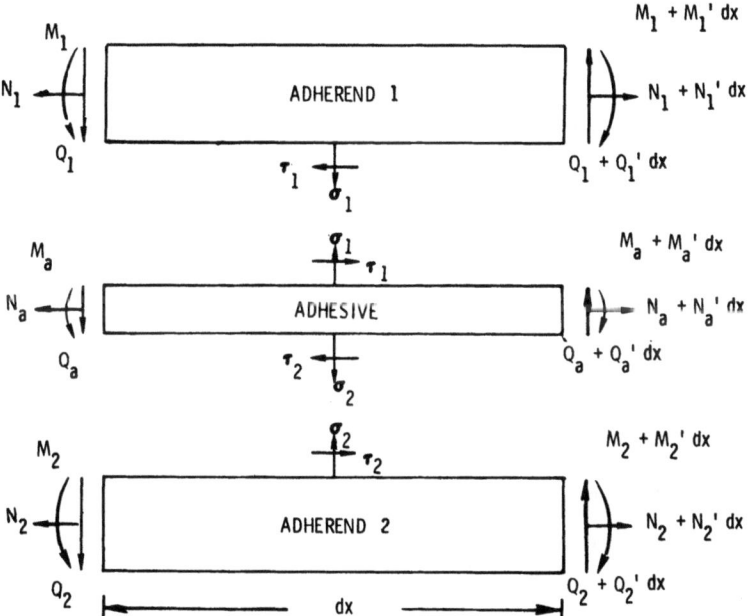

Figure 3. Balance of Forces for Adhesive Joint Element.

$$-M_1 + \frac{\partial M_1}{\partial x} dx + \left(Q_1 + \frac{\partial Q_1}{\partial x} dx \frac{dx}{2} + Q_1 \frac{dx}{2}\right)$$

$$- (\tau_{xz})_1 \, dx \left(\frac{t}{2}\right) + M_1 = 0 \qquad \text{Moment Equilibrium}$$

$$M_1' = Q_1 - \tau_1 t/2 \tag{3}$$

where quadratic terms in dx have been neglected and ´ is used to denote the partial derivative $\frac{\partial}{\partial x}$.

In a similar fashion, static equilibrium conditions for the adhesive and lower adherend require:

Adherend 2

$$N_2' = -\tau_2 \tag{4}$$

$$Q_2' = -\sigma_2 \tag{5}$$

$$M_2' = Q_2 - \tau_2 \frac{t}{2} \tag{6}$$

Adhesive

$$N_a' = \tau_2 - \tau_1 \tag{7}$$

$$Q_a' = \sigma_2 = \sigma_1 \tag{8}$$

$$M_a' = Q_a - \frac{(\tau_1 + \tau_2) h}{2} \tag{9}$$

From assumption 2, the following relations between the force and moment resultants and the displacements for identical thin orthotropic adherends (again neglecting width effects) are known to be[10]

$$u_{01}' = \frac{1 - \nu_{12} \nu_{21}}{E_1 t} N_1 \tag{10}$$

$$u_{02}' = \frac{1 - \nu_{12} \nu_{21}}{E_1 t} N_2 \tag{11}$$

$$w_{01}' = \frac{6}{5 t G_{31}} Q_1 - \beta_{01} \tag{12}$$

$$w_{02}' = \frac{6}{5 t G_{31}} Q_2 - \beta_{02} \tag{13}$$

STRESS ANALYSIS OF ADHESIVELY BONDED JOINTS

$$\beta'_{01} = \frac{12(1-\nu_{12}\nu_{21})}{t^3 E_1} M_1 \qquad (14)$$

$$\beta'_{02} = \frac{12(1-\nu_{12}\nu_{21})}{t^3 E_1} M_2 \qquad (15)$$

where the constitutive equations for the adherend are given by

$$\begin{bmatrix} \varepsilon_1 \\ \varepsilon_2 \\ \varepsilon_3 \\ \varepsilon_{23} \\ \varepsilon_{31} \\ \varepsilon_{12} \end{bmatrix} = \begin{bmatrix} \frac{1}{E_1} & \frac{-\nu_{21}}{E_2} & \frac{-\nu_{31}}{E_3} & 0 & 0 & 0 \\ \frac{-\nu_{12}}{E_1} & \frac{1}{E_2} & \frac{-\nu_{32}}{E_3} & 0 & 0 & 0 \\ \frac{-\nu_{13}}{E_1} & \frac{-\nu_{23}}{E_2} & \frac{1}{E_3} & 0 & 0 & 0 \\ 0 & 0 & 0 & \frac{1}{G_{23}} & 0 & 0 \\ 0 & 0 & 0 & 0 & \frac{1}{G_{23}} & 0 \\ 0 & 0 & 0 & 0 & 0 & \frac{1}{G_{12}} \end{bmatrix} \begin{bmatrix} \sigma_{11} \\ \sigma_{22} \\ \sigma_{33} \\ \tau_{23} \\ \tau_{31} \\ \tau_{12} \end{bmatrix}$$

u_{oi}, w_{oi} are the midplane displacement components and β_{oi} are the average midplane rotations.

For cylindrical bending we can write the displacements at the adhesive-adherend interfaces $(\bar{u}_i, \bar{w}_i, \bar{\beta}_i)$ as[16]

$$\bar{u}_1 = u_{01} + \frac{t}{2}\beta_{01} \qquad (16)$$

$$\bar{u}_2 = u_{02} - \frac{t}{2}\beta_{02} \qquad (17)$$

$$\bar{w}_1 = w_{01} \qquad (18)$$

$$\bar{w}_2 = w_{02} \qquad (19)$$

$$\bar{\beta}_1 = \beta_{01} \qquad (20)$$

$$\bar{\beta}_2 = \beta_{02} \qquad (21)$$

To complete the model, it is necessary to consider the constitutive relationship for the adhesive layer. For convenience, Lamé notation is used for the adherends. From assumption (1) we have

$$\tau_{xz} = \mu_a \left(\frac{\partial u_a}{\partial z} + \frac{\partial w_a}{\partial x}\right) \tag{22}$$

$$\sigma_{zz} = (\lambda_a + 2\mu_a)\frac{\partial w_a}{\partial z} + \lambda_a \frac{\partial u_a}{x} \tag{23}$$

Using assumption 3 and the compatibility of the displacements at the two interfaces, we can write the adhesive displacements in terms of the adherend displacements as

$$u_a = \frac{\bar{u}_1 + \bar{u}_2}{2} + \frac{z}{h}(\bar{u}_1 - \bar{u}_2) \tag{24}$$

$$w_a = \frac{\bar{w}_1 + \bar{w}_2}{2} + \frac{z}{h}(\bar{w}_1 - \bar{w}_2) \tag{25}$$

Combining Equations (22-25) yields

$$\tau_1 = \tau_{xz}|_{-h/2} = \mu_a \frac{\bar{u}_1 - \bar{u}_2}{h} + \bar{w}_1' \tag{26}$$

$$\tau_2 = \tau_{xz}|_{-h/2} = \mu_a \frac{\bar{u}_1 - \bar{u}_2}{h} + \bar{w}_2' \tag{27}$$

$$\sigma_1 = \sigma_{zz}|_{h/2} = \lambda_a \bar{u}_1' + (\lambda + 2\mu_a)\frac{\bar{w}_1 - \bar{w}_2}{h} \tag{28}$$

$$\sigma_2 = \sigma_{zz}|_{-h/2} = \lambda_a \bar{u}_2' + (\lambda_a + 2\mu_a)\frac{\bar{w}_1 - \bar{w}_2}{h} \tag{29}$$

These 32 equations can be algebraically reduced to 15 independent equations which must be solved to determine the stress distribution in the bonded joint subject to appropriate boundary conditions.

The necessary boundary conditions were determined from the applied loads at the ends of the joint; this procedure yields:

BC1	$Q_1 = -Q_0$	
BC2	$M_1 = -M_0$	
BC3	$N_1 = N_0$	at $x=c$
BC4	$Q_1 = 0$	
BC5	$M_1 = 0$	
BC6	$N_1 = 0$	at $x=c$
BC7	$Q_2 = 0$	
BC8	$M_2 = 0$	
BC9	$N_2 = 0$	at $x=c$
BC10	$Q_a = 0$	
BC11	$M_a = 0$	
BC12	$N_a = 0$	at $x=-c$

and the need to fix the displacements at one point to eliminate rigid body motions which yield:

BC13	$u_1 = 0$	at $x=-c$
BC14	$w_1 = 0$	
BC15	$\beta_1 = 0$	

The remaining boundary conditions follow from equilibrium and symmetry considerations.

This problem was solved for a variety of joint configurations using an iterative technique in conjunction with the method of shooting points[17]. Thickness dependent terms in the governing equations were multiplied by an iteration variable α. With $\alpha = 0$, the thickness independent solution (a la Goland and Reissner[1] et al.) was obtained. Using this solution as the basis for the next iteration, the parameter α was increased slightly and the next solution was obtained. This process was repeated until $\alpha = 1$, where the full problem was recovered and the final solution determined.

One principal drawback to this method is that the resulting shear stress distribution is not zero at the joint edges as required. This deficiency is common to virtually all linearized plate theory analyses of the problem, including that of Goland and Reissner[1]. Renton and Vinson[8,9] have shown that this shortcoming can be eliminated using a somewhat more elaborate model for adherend behavior with a parabolic rather than linear variation in adherend shear stress through the thickness. Their results show that, instead of reaching a maximum at the joint edges as predicted in Goland and Reissner type analyses, the actual stress distribution reaches a maximum a short distance from the edge of the joint and decreases rapidly to zero at the edge. As computing space limitations precluded incorporation of this aspect of bond behavior, this fact should be taken into account in interpreting the results presented here.

RESULTS AND DISCUSSION

Bond stress predictions based on the model described above are presented in Figures 4-6. For comparison, the values of the nondimensionalized peel $\left(\frac{\sigma_{zz}}{N_o/2c}\right)$ and shear $\left(\frac{\tau_{xz}}{N_o/2c}\right)$ stresses are shown as a function of the normalized position $\frac{x}{2c} = 7$. Stress distribution along both the upper (τ_1, σ_1) and lower (τ_2, σ_2) bondlines are presented to illustrate the variation of stress through the adhesive thickness. The symmetric nature of the problem is also reflected in the solution with $\tau_1(x) = \tau_2(2-x)$, $\sigma_1(x) = \sigma_2(2-x)$. The basic parameters used to characterize joint geometry and material composition are shown in Table I. These parameters correspond to a typical joint used in automotive applications.

Qualitatively, there is a great deal of similarity between the stress distributions predicted by the earlier models and that found in this study. Both shear and peel stress distributions are roughly parabolic with the maximum stresses being developed at the joint edges and minimum stresses being observed in the central region of the joint. Thus, the major portion of the joint experiences relatively light loads while small regions at the joint edges are more severely stressed. It is precisely in these high stress regions at the joint edges that the principal differences in the various models emerge. While peak shear stress predictions are quite similar for the models presented, larger differences in peel stress maxima were observed principally due to the more complete constitutive equation used in this study for the adhesive layer.

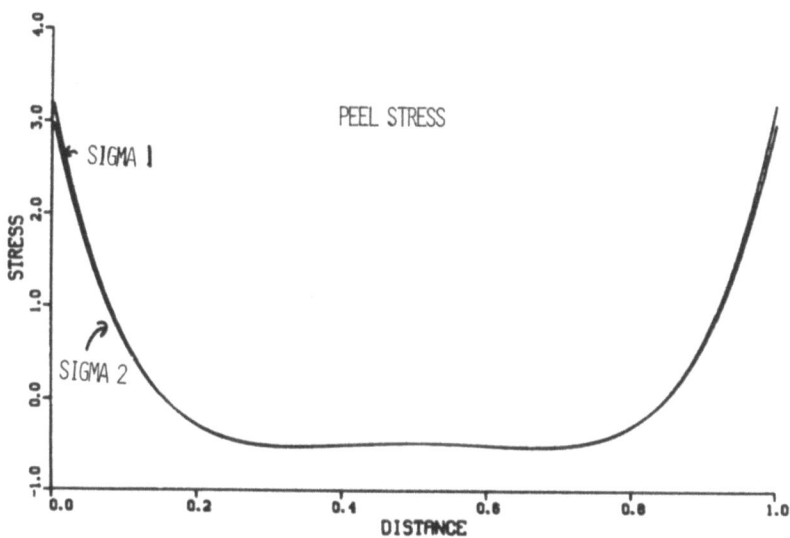
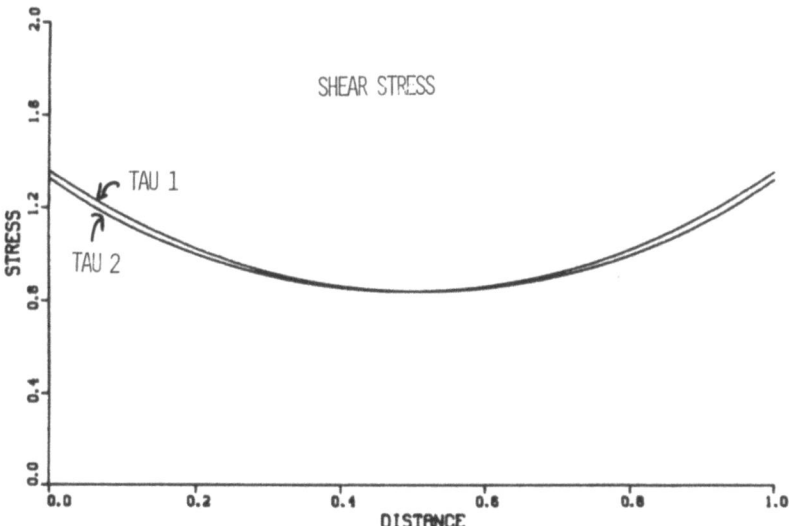

Figure 4. Stress Distribution in Adhesively Bonded Joints (Low Modulus Adhesive).

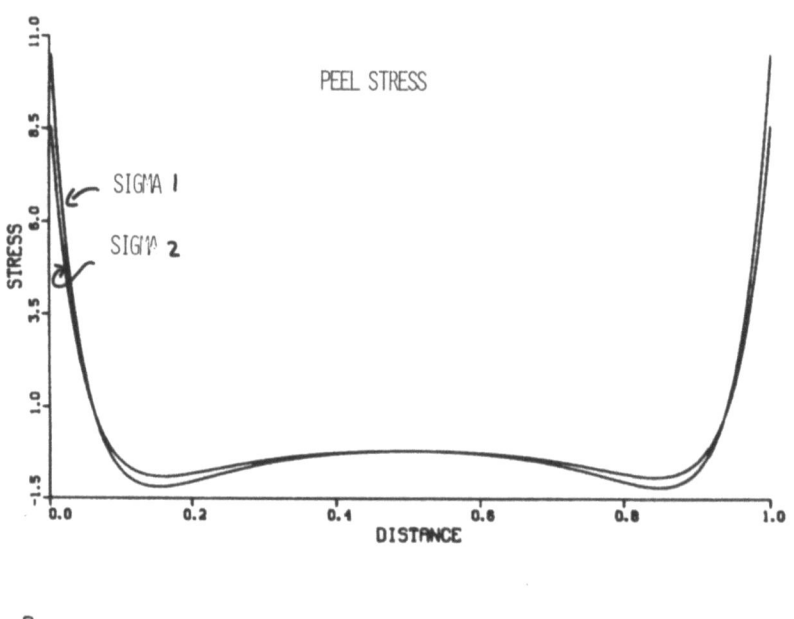

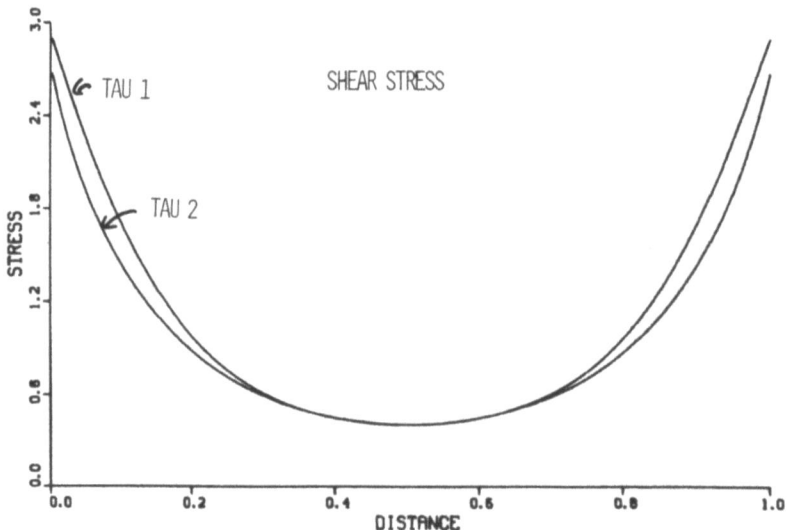

Figure 5. Stress Distribution in Adhesively Bonded Joints (High Modulus Adhesive).

STRESS ANALYSIS OF ADHESIVELY BONDED JOINTS

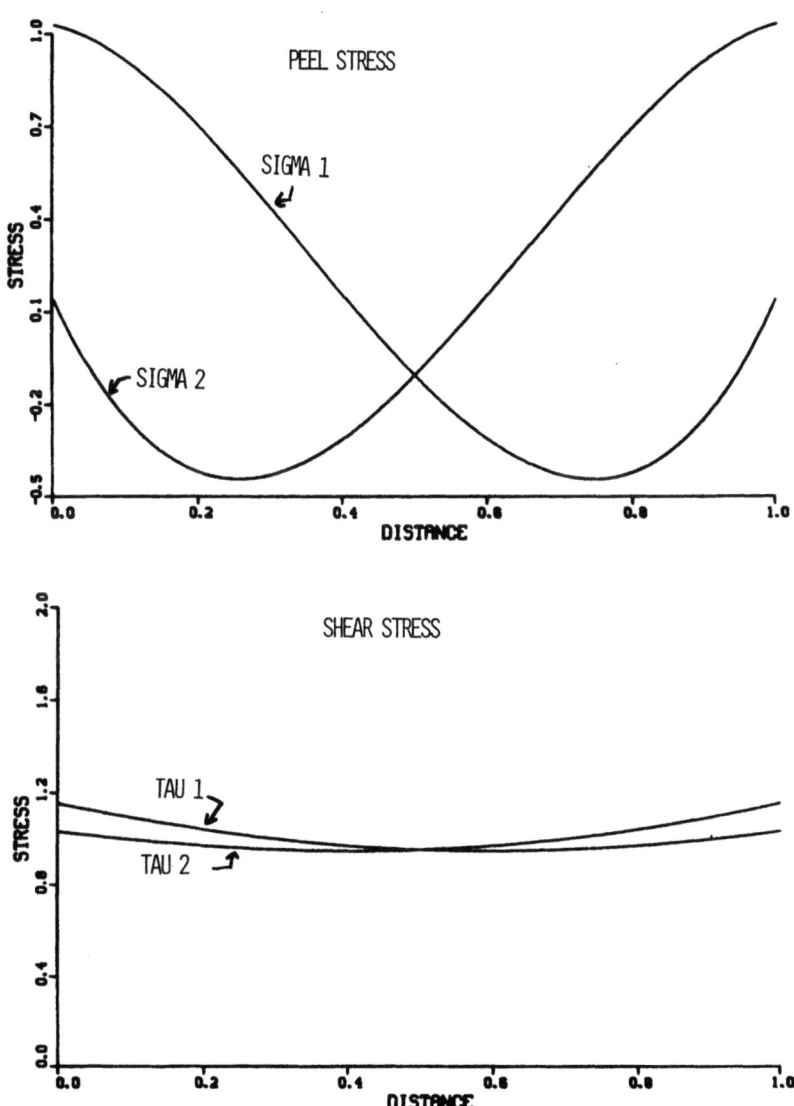

Figure 6. Stress Distribution in Adhesively Bonded Joints (Thick Adhesive Layer).

TABLE I

Parameter	Symbol	Value
Adhesive Thickness	h	.038 cm (.015")
Adherend	t	.635 cm (.250")
Overlap Length	2c	2.54 cm (1.00")
Tensile Modulus (Adherend)	E_{11}	13.24 GPa† (1.92 x 10^6 psi)
Poisson' Ratio (In-Plane, Adherend)	ν_{12}, ν_{21}	0.25†
Poission's Ratio (Out-of-Plane, Adherend)	ν_{13}, ν_{31},	0.25°
Shear Modulus (Adherend)	G_{31}	0.53 GPa* (7.6 x 10^5 psi)
Lamé Constant (Adhesive)	λ_a	.230 GPa^Δ (3.25 x 10^4 psi)
Lamé Constant (Adhesive)	μ_a	.0259 GPa^Δ (3.76 x 10^3 psi)

† R. Heimbuch and B. Sanders, "Mechanical Properties of Chopped Figer Reinforced in "Composite Materials in the Automotive Industry," pp. 111-139, American Society Mechanical Engineers, 1978.

* Measured ultrasonically using pulse-echo superposition technique.

Δ T. Wang, B. Sanders and U. Lindholm, "A Loading Rate and Environmental Effects Study of Adhesively Bonded SMC Joints," GMMD Report Number 80-044, General Motors Manufacturing Development, Warren Michigan, 1980.

° Estimated.

Stress variation through the thickness of the adhesive was found to be highly dependent on the joint geometry and properties of the materials used. Typical results are presented in Figure 4. At the edge of the joint, the difference in peel stress between σ_1 and σ_2 was approximately 9% for a relatively thin adhesive (h = 0.010"). The variation in shear stress was somewhat less, approximately 4%. Utilization of a higher modulus, higher strength adhesive tends to increase the stress variation. This is illustrated in Figure 5 for the same joint as in Figure 4 except that the higher modulus adhesive ($E_a = 1.02 \times 10^5$ psi) has been used. The peel stress variation was found to increase in this case to approximately 17% with a corresponding increase to 9% in shear stress. Extreme variations in stress distribution were observed for joints with thick adhesive layers. Of course, this is not surprising. Figure 6 illustrates this for the case of the base joint with h being increased from 0.010" to 0.250". Under these conditions the peel stresses differ by an order of magnitude at the joint edges. The variation in shear stress was significantly less. It is not known how accurate these results are for extreme cases such as this, as the assumption of a linear variation in stress through the adhesive thickness may no longer be applicable. However, it does indicate that the variation in peel stress can be considerable for certain geometries.

It is important to note that the effect of this variation in stress on composite joints is significant for predicting the failure site in a bonded joint. Thickness independent models predict that failure is equally likely to occur at any of the joint edges (sites 1,2,3,4 in Figure 1). This is in contrast to the prediction of the thickness dependent models which indicate that sites 1 and 3 should be the preferred fracture initiation sites.

Experimentally, this is precisely what is observed. Figures 7(a-d) follow the behavior of a single lap shear joint (SMC composite) during the course of a test from the beginning of the test through failure. The initial specimen damage appears (at oppposite edges as predicted using this model unlike thickness independent models) as small matrix cracks in the composite at the joint edges (Figure 7b). As the load is increased, the cracks grow in a highly non-uniform manner, pricipally via fiber-matrix debonding. This is attributable, in all likelihood, to the weakness of the fiber-matrix interface[23]. In any event, this behavior is indicative of the high stresses (particularly peel) which develop at the joint edges and the inability of composite materials to withstand out-of-plane loads. For bonds with metal adherends, a somewhat different sequence is observed[19]. The onset of fracture again occurs at adhesive-adherend interfaces at opposite edges of the joint. However, due to the (statistically) isotropic nature of the polycrystalline metal, they are not as susceptible to adherend peel stress failure as the composite material. Failure is confin-

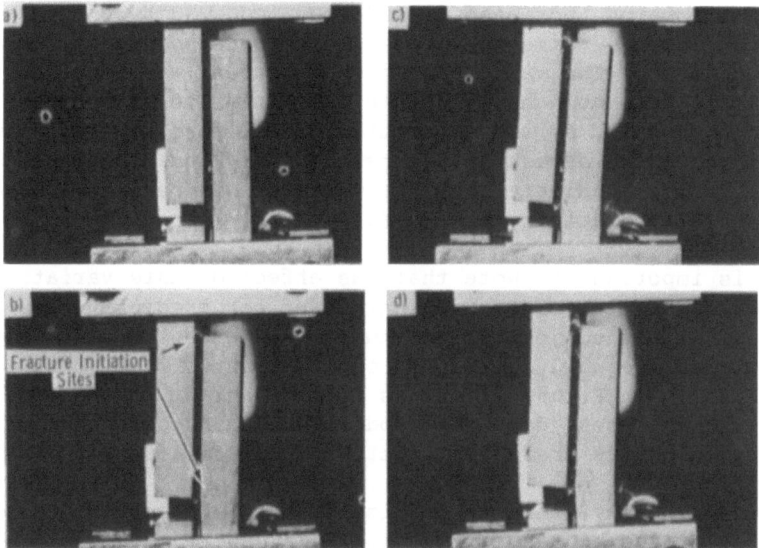

Figure 7. Failure Mechanism in Adhesively Bonded SMC Composite.

ed to the adhesive-adherend interface (adhesive failure) or in the adhesive layer itself (cohesive failure) depending on their relative strengths.

A parametric study of the effect of pertinent geometric and material factors was conducted to examine possible methods of reducing the high stresses which develop at the edges of the bonded joint. The results of this effort are illustrated in Figures 8-12 where nondimensionalized bond stress distributions for variations in adhesive modulus (E_a), adhesive thickness (h) adherend thickness (t) and joint overlap length (2c) are shown. Adhesive modulus was found to have the most pronounced effect of the four parameters investigated, the use of a lower adhesive modulus resulting in a large decrease in peak stress values at the edge of the bonded joint. Peak stress reduction can also be effected by decreasing the adhesive overlap length and increasing adherend thickness. Increasing adhesive thickness tended to reduce peak stresses provided that h << t. This is illustrated in Figures 11 and 12 for a relatively thin adherend (t $\cong$ 0.100") and a relatively thick adherend (t = 0.250"). For adhesive thicknesses comparable to or greater than the adherend thickness, increasing the adhesive thickness increased the peak stresses due to the increase in the applied bending moment $M_o = N_o$ (t + h)/2. It should also be mentioned that applied load can also affect the stress distribution observed due to the bending of the joint. Increasing load on a joint tends to reduce the peel stress observed as illustrated in Figure 13. This figure represents the stress distribution in the basic composite joint at three load levels (1 lb., 100 lb, 1000 lb). The reduction in peel stress at the joint edge was found to be approximately 13%. It should be mentioned that the peak stresses which develop at the ends of a bonded joint are only one factor in determining overall joint efficiency. The effect of changing any of these parameters on the strength of the adherend, the adhesive and the adherend-adhesive interface must be taken into account in determining whether the performance of the bonded structure will be improved. In any event, it should be stressed that it is erroneous to conclude that simply because a given joint fails in the adherend, it is well designed. Conversely, it is not necessarily true that failure of the adhesive layer in a bonded joint implies that the joint is poorly designed. In assessing the performance of a given joint, many inter-related factors (stress distribution, constituent strengths, etc.) must be considered. This is particularly true for fiber reinforced composites due to their inherent weakness when subjected to out-of-plane loads. With proper choice of bonding agent and specimen geometry, it is felt that improvements in the efficiency of adhesively bonded joints can be achieved.

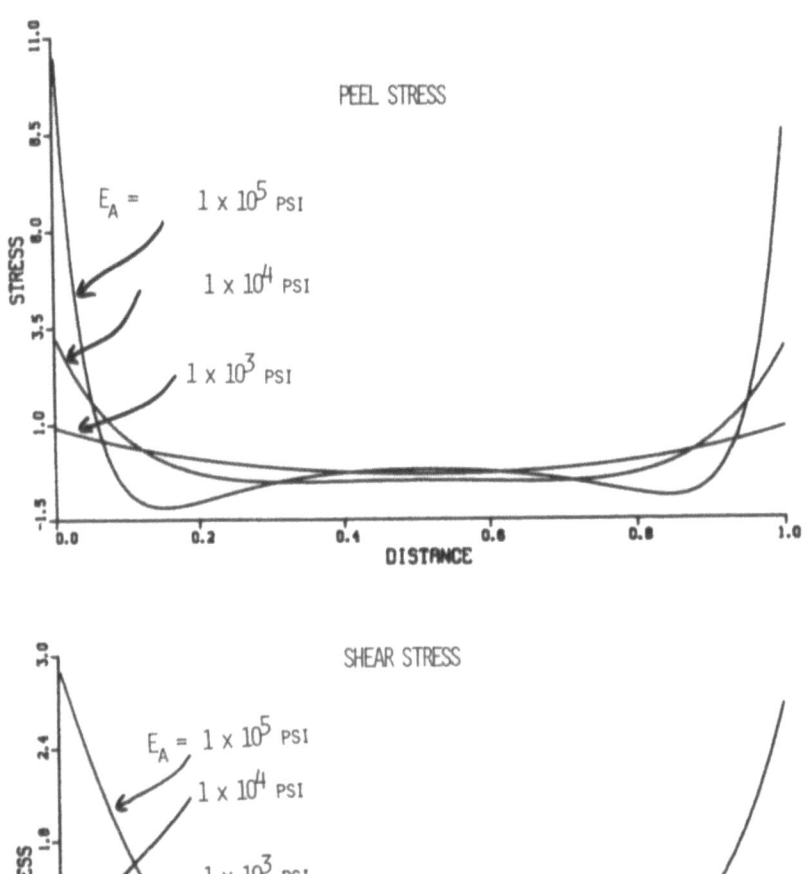

Figure 8. Effect of Adhesive Modulus on Stress Distribution.

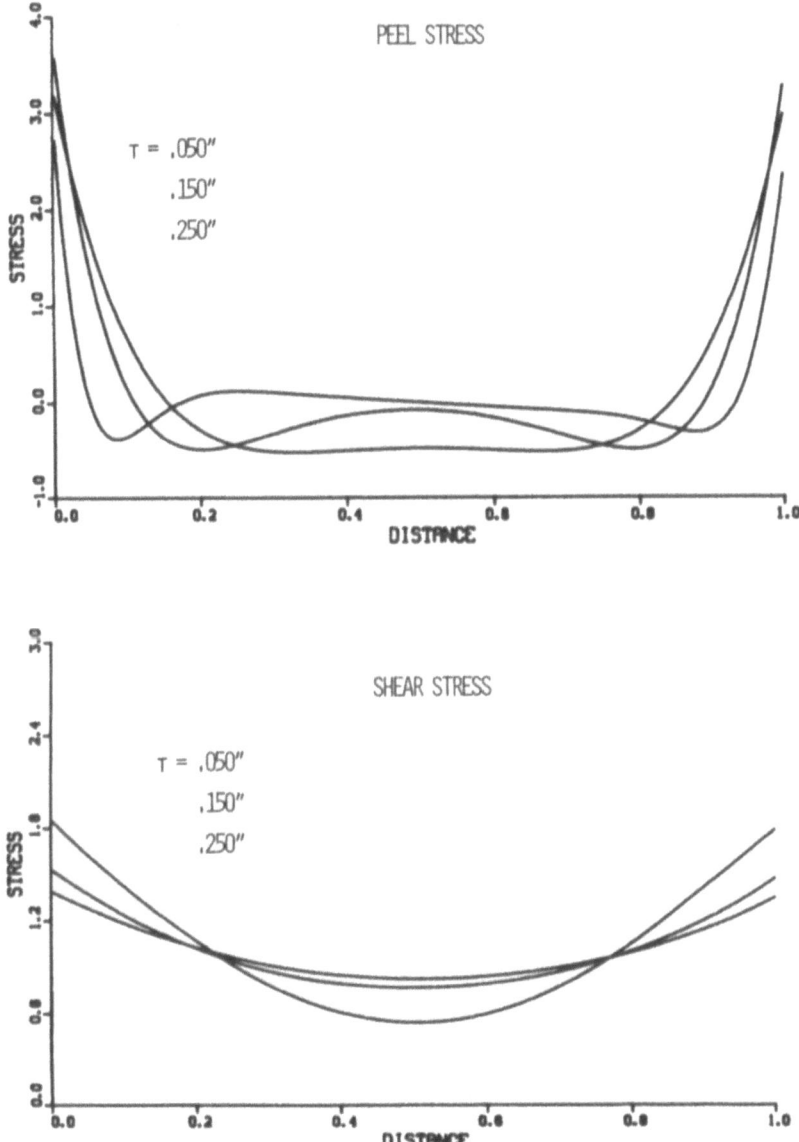

Figure 9. Effect of Adherend Thickness on Stress Distribution.

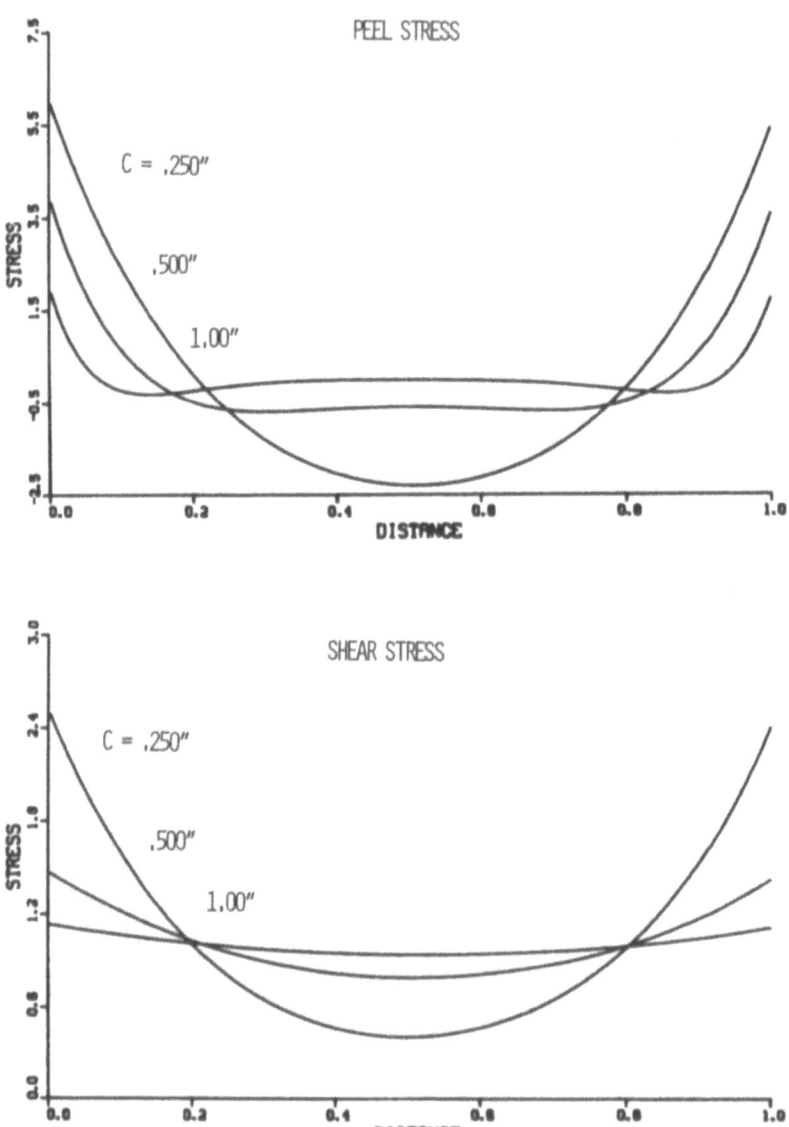

Figure 10. Effect of Adhesive Overlap on Stress Distribution.

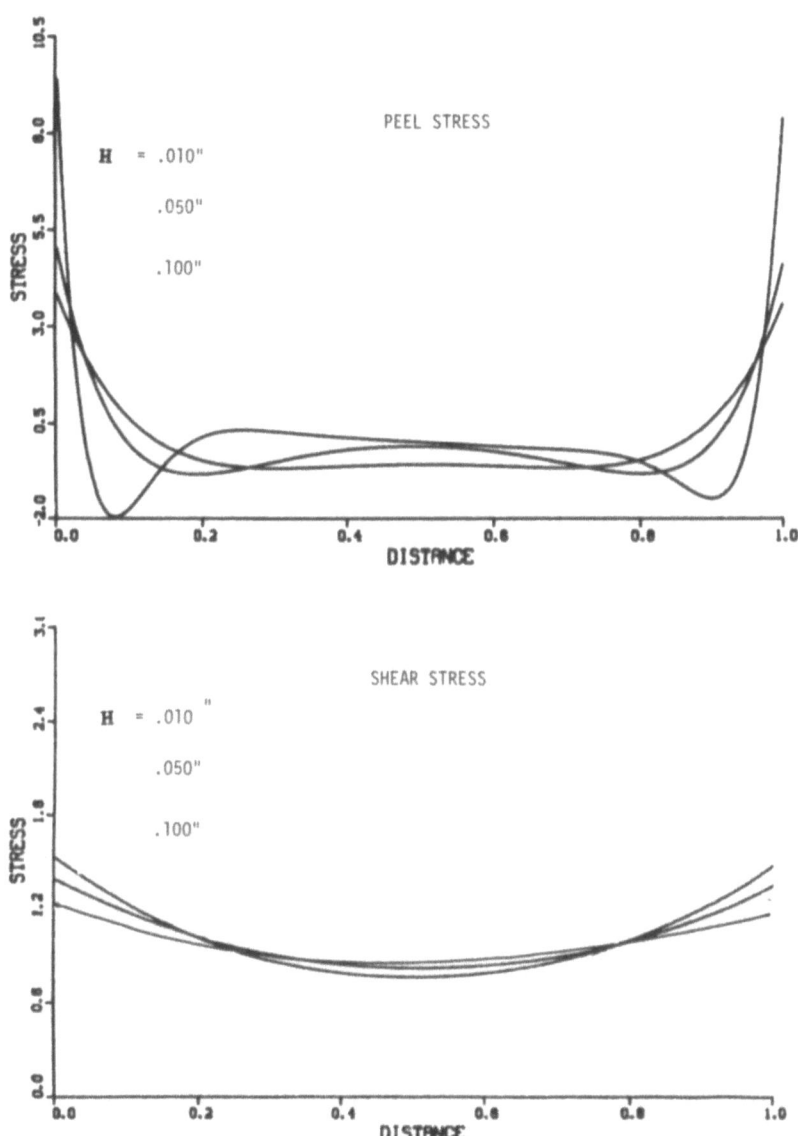

Figure 11. Effect of Adhesive Thickness on Stress Distribution (Thin Adherend t = 0.100").

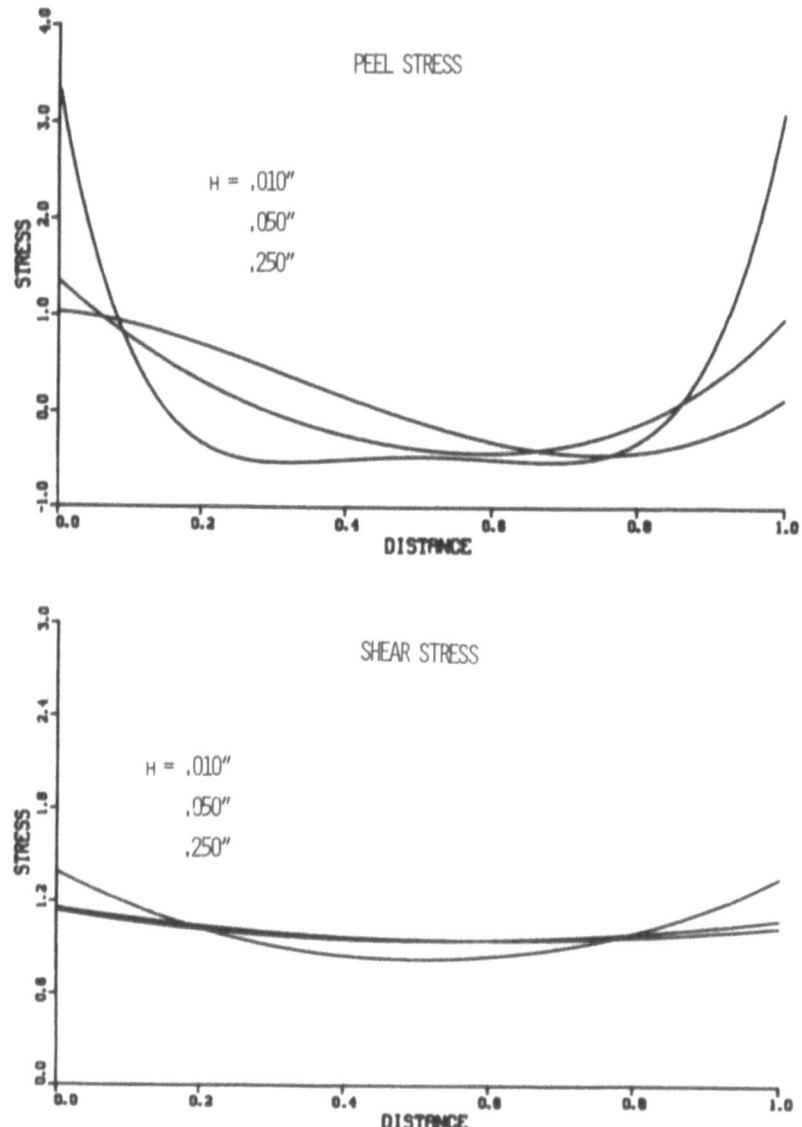

Figure 12. Effect of Adhesive Thickness on Stress Distribution (Thick Adherend t = 0.250").

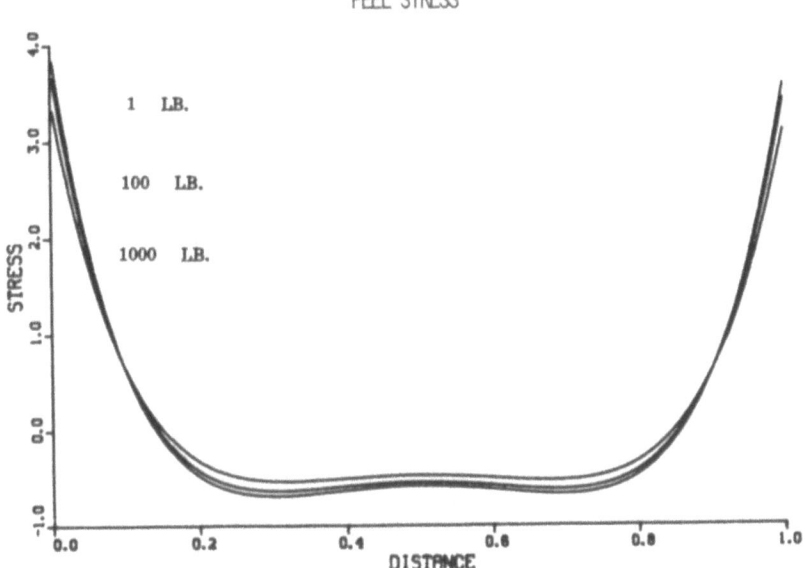

Figure 13. Effect of Applied Load on Stress Distribution.

ACKNOWLEDGMENT

The author wishes to thank his colleagues Martin R. Barone, Robert L. Frutiger and Mark F. Nelson for their many helpful technical discussions throughout the course of this work.

REFERENCES

1. M. Goland and E. Reissner, J. App. Mech., 11, A17 (March 1944).
2. M. Chang, J. Adhesion, 4, 221 (1972).
3. L. Hart-Smith, "Analysis and Design of Advanced Composite Joints", NASA 1974. Available through NTIS.
4. L. Hart-Smith, "Adhesive-Bonded Double Lap Joints", NASA CR-112235, 1973. Available through NTIS.
5. L. Hart-Smith, "Adhesive-Bonded Single Lap Joints", NASA CR-112236, 1973. Available through NTIS.

6. Y. Weitsman, J. Adhesion, 11, 279 (1981).
7. F. Delale and F. Erdogan, J. Appl. Mech., 48, 331 (1981).
8. W. Renton and J. Vinson, J. Appl. Mech., 44, 101 (1977).
9. W. Renton and J. Vinson, J. Adhesion, 7, 175 (1975).
10. F. Delale, F. Erdogan and M. Aydinoglu, J. Composite Materials, 15, 249 (1981).
11. I. Ojalvo and H. Eidinoff, AIAA J., 16, 204 (1978).
12. W. Carpenter, AIAA J., 18, 350 (1980).
13. U. Yuceoglu and D. Updike, J. Eng. Mech., 106, 37 (1980).
14. U. Yuceoglu and D. Updike, J. Eng. Mech., 107, 55 (1981).
15. W. Ewing, W. Jardetzky and F. Press, "Elastic Waves in Layered Media", McGraw-Hill Book Company, 1957.
16. S. Timoshenko and S. Woinowsky-Krieger, "Theory of Plates and Shells", 2nd ed., McGraw-Hill Book Company, 1959.
17. G. Dahlquist and A. Bjorck, "Numerical Methods", Prentice-Hall, Inc., 1974.
18. R. Kline, "The Effect of Microstructure on the Mechanical Properties of SMC", paper presented at ASTM Symposium on Producibility and Quality Assurance of Composite Materials, St. Louis, MO, 1981.

THE IMPACT STRENGTH OF ADHESIVE LAP JOINTS

J.A. Harris and R.D. Adams

Department of Mechanical Engineering
University of Bristol, University Walk
Bristol, BS8 1TR, U.K.

Structural adhesive joints may be required to withstand high loading rates. The effect of impact loading on the strength and energy absorption of single lap joints has been investigated. A rubber-toughened epoxy adhesive, in bulk form, shows significant changes in mechanical response, when tested in shear over a wide range of strain rates. Despite large changes in the response of the adhesive, the strength of lap joints is only slightly reduced by impact loading. When high strength adherend materials are used, energy absorption is relatively small; however, it is increased by using lower strength adherend materials, as significant plastic deformation can take place in the adherend prior to joint failure.

The finite element method has been used to model and predict joint strength for the various conditions from a knowledge of the bulk adhesive mechanical response under those conditions.

INTRODUCTION

With the development of modern structural adhesives, the use of adhesive bonding as a joining technique has become increasingly attractive. In particular, the joining of aluminium alloys is made possible where the use of conventional spot welding methods is not feasible on a high volume production basis. One possible area of application is in the automotive industry, where the use of aluminium alloys and plastics for car bodies represents a considerable weight saving advantage. This is the background to the problem that is being addressed here, that is, that the automobile designer must be confident that structural bonded joints will have sufficient strength under normal loading conditions and also retain that strength under high loading rate conditions, as would occur in a crash situation. The energy absorbing capability of the structure to protect the passenger is important here, and the view is taken that joints must retain sufficient strength such that the main volume of material that is between them will plastically deform and buckle. Hence, the two parameters of interest are strength retention and energy absorption under impact.

The feasibility of using adhesives in structures in which the joints may be required to be highly deformed or crumpled under an impact load has been investigated[1]. Cylinders were fabricated from sheet material by forming a tube with an overlapping seam running the length of the cylinder. The seam was either bonded or spot welded, and the tube then crushed axially under impact conditions in a drop weight rig. Both cylinders of steel and aluminium alloy which were tested showed no significant difference in energy absorbing performance between bonded and spot welded seams. The bonded seams were fractured at various points in the crumpled zone because of the continuous nature of the joint, whereas the spot welds remained intact, because of the flexibility of the joint between the welds.

The majority of test work that has been carried out on adhesives and adhesive joints has been at low loading rates or 'quasi-static' conditions. It is thus necessary to know what effect high loading rates or impact conditions have on adhesives and adhesive joint strength. The existing ASTM standard test for the impact testing of adhesive bonds (ASTM D 950-72) was not suitable. The test, which consists of shearing off a steel block bonded to another steel base block, is difficult to reproduce in the form of a quasi-static test for comparison and, because of the high stiffness of the specimen used, very much higher loading rates result than would be experienced in a realistic bonded structure, which would be very much more compliant. Thus, the single lap joint test piece was chosen to be tested under impact conditions, as it was reasonably compliant and was already commonly used as a test piece for quasi-static testing of adhesives.

THE IMPACT STRENGTH OF ADHESIVE LAP JOINTS

The problem of understanding the effects of high loading rates on bonded joints is part of a more general problem of being able to predict adhesive joint strength. The approach taken here is that if the mechanical properties of an adhesive are known and there exists a suitable analysis of the critical stresses for a particular loaded joint then, with a suitable failure criterion, joint strength can be predicted. In order to gain an understanding of the effects of impact loading on an adhesive joint, it was therefore necessary first to investigate the effects of loading rate on the adhesive as a bulk material. Then, using the finite element method, the analysis of the joint could be carried out for both quasi-static and impact loading, and compared with test results.

EXPERIMENTAL

Adhesive properties

A rubber modified epoxy material has been used, which was a diglycidyl ether of bisphenol-A, with 15 parts (by weight) per

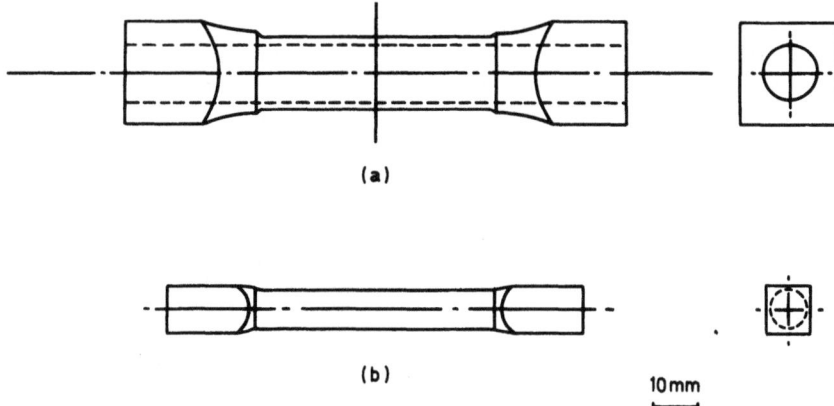

Figure 1 Bulk adhesive specimens for quasi-static torsion tests. (a) Annular; (b) Solid.

hundred of resin of carboxyl-terminated butadiene-acrylonitrile rubber and 5 parts per hundred of piperidine as the curing agent.

The rate dependent properties of the material were investigated by a series of tests, in pure shear, of bulk specimens machined from cast blocks of the adhesive. Over a range of relatively low strain rates, specimens were tested on a variable speed torsion testing machine. Two types of bulk specimens were tested: an annular specimen and a solid circular specimen, both shown in Figure 1. For the solid specimen, when a torque is applied, because the shear stress is greatest at the outside, yield of the material begins here and spreads inwards through the radius. The torque-twist curve from this specimen can be converted to a shear stress-strain curve by applying the correction due to Nadai[2]; however, this is only approximate since it is not applicable to rate sensitive materials. The use of an annular specimen avoids the use of the Nadai correction as the variation of shear stress across the annulus is small. Also, the shear strain rate varies little across the section. However, these specimens failed due to buckling of the gauge length so the ultimate shear strain could not be derived. In these tests, torque was measured by a strain gauged load cell, and the shear strain was measured by a rotary capacitative transducer. The twist was transmitted to the transducer by strings passing over pulleys attached to the specimen at each end of the gauge length. On this machine strain rates in the range 10^{-3} min^{-1} to 1 min^{-1} were achieved.

For tests at higher rates of strain, a shear impact type device was used. This together with the instrumentation used is illustrated in Figure 2. The operation of the machine was based on the running up of a flywheel to a controlled speed. A spring-loaded arm was then released to positively engage into the flywheel resulting in the application of an abrupt torque which in turn was transmitted through the specimen to the load cell attached to a rigid base. Again, an annular specimen was used as shown in Figure 3. It has castellated ends by which it is located into the cruciform ends of the load cell and loading shaft. The shear strain at the outside of the specimen was measured from strain gauges bonded to the surface. In order to record the torque and shear strain signals over the short periods of testing, a two-channel transient recorder system was used, from which the stored data could be 'replayed' on a flat bed plotter. On this instrument strain rates in the range of 10^2 min^{-1} to 10^3 min^{-1} were achieved.

Joint strengths

Single overlap shear specimens were manufactured to dimensions as specified in ASTM D1002-72 for the standard 25.4 mm wide, 12.7 mm overlap length test piece, using the rubber modified epoxy

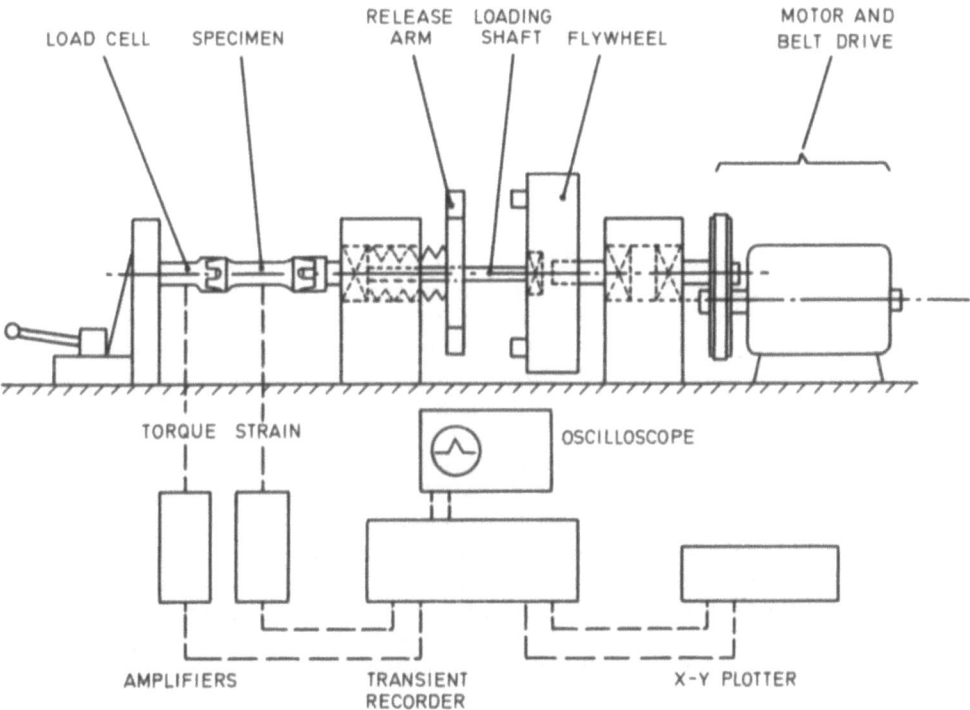

Figure 2 Impact torsion test rig.

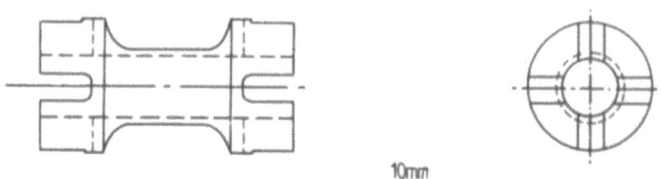

Figure 3 Bulk adhesive specimen for torsion impact test.

adhesive. A glue line thickness of 0.125 mm was maintained throughout. Initially, the adherend material used was a high strength aluminium alloy designated 2L73 which has a 0.2 per cent proof stress of 430 N mm^{-2}; but, subsequently, lower strength aluminium alloys designated BB2hh and BB2s were used with 0.2 per cent proof stresses of 220 N mm^{-2} and 112 N mm^{-2} respectively. Batches of six specimens were tested at low and high loading rates. For the low rate tests, a screw driven tensile testing machine was used with a crosshead speed of 2 mm min^{-1}. A typical time to failure of the 2L73 joints was 4 minutes, with an almost linear rise in the load to failure. For the lower strength adherends, after an initial linear rise, the loading rate rapidly decreased as the adherends yielded and considerable plastic deformation occurred in the adherends prior to failure which, for the lowest strength adherend, typically took 10 minutes to occur.

For the high loading rate tests, an existing Izod pendulum impact machine was modified to incorporate the geometry of the single lap specimen. At one end, the specimen was clamped into a piezo-electric force link which was rigidly attached to the base of the machine. At the other, a block clamped on to the specimen carried a bar which lay across the direction of the pendulum swing. The impact load was thus applied by an impactor attached to the pendulum, which had two striking heads which picked up on each side of the bar. In order to prevent dynamic oscillations being set up during a test, a layer of lead was interspersed around the impact bar which provided a cushioning effect in the initial part of the impact loading. As a result, reasonably linear loading to failure could be achieved. The energy absorbed in the test was measured from the load-end displacement response of the specimen, the end displacement being measured using a capacitative displacement probe. For large energy absorption, the loss of pendulum energy derived from the loss of height of the swing was used. Again, the two-channel transient recorder system was used to measure the dynamic load and displacement. For the 2L73 specimens, an impact velocity of about 1 ms^{-1} resulted in a time to failure of about 5 ms. For the lower strength materials, higher impact velocities were required to fracture the specimens, such that for the BB2s specimens an impact velocity of 4 ms^{-1} resulted in a time to failure of about 10 ms.

Finite element analysis.

The use of the finite element method in the analysis of the single lap joint enables more accurate modelling of the joint to be made compared with existing closed form analyses. Adams and Peppiatt[3] and Crocombe and Adams[4] have used the finite element method to model the single lap joint, including the geometry of the adhesive fillet formed at the ends of the overlap. The inclusion of the spew fillet served to significantly reduce the stress concentration in the adhesive at the ends of the overlap so that, compared to closed form analyses with a square edged adhesive layer, more realistic stress predictions resulted[4]. The direction of cracks in failed lap joints was perpendicular to the predicted maximum tensile stresses running down from the corner of the unloaded adherend at an angle of about 45^0 to the adherend surface. It was concluded[3] that failure of a lap joint is initiated by tensile failure of the adhesive within the spew fillet.

In this analysis, it was required to include non-linearities due to the yielding and plastic deformation of the adherend materials, and also due to the rotation of the overlap under load which makes the problem one of large displacements. Adams, Coppendale and Peppiatt[5] have also shown, using the finite element method for double lap joints, that, in order to predict lap joint strength, the elastic-plastic response of the adhesive must be included in the analysis.

The finite element program FELDEP, written by Dr. A.D. Crocombe, was used for the analysis. In this both large displacement effects and material non-linearities could be included. The program is based on the incremental theory of plasticity. The yield criteria used were the Von Mises criterion for the adherend and a modified Von Mises criterion for the adhesive. The modified yield criterion is of the form

$$F = \left[J_1 (S-1) + (J_1^2 (S-1)^2 + 12 J_2 S)^{\frac{1}{2}} \right] /2S - Y_T = 0$$

where J_1 and J_2 are the first and second stress invariants, Y_T the

yield stress in uniaxial tension and S the ratio of the yield
stress in uniaxial compression to uniaxial tension. A value of
1.3 for S is assumed, which is typical of a rubber-modified epoxy
type material. The adhesive properties for both quasi-static
loading rates and impact loading rates are represented by uniaxial
tensile stress strain curves, which have been measured. As has
been noted, failure in the adhesive occurs under the action of
primarily tensile stresses. For this rubber modified epoxy, a
significant amount of ductility is shown; thus, a failure criterion
based on a maximum principal strain is assumed and is made equal
to that measured in the uniaxial tensile test. Under quasi-static
conditions, this is 0.15 and, under impact conditions, 0.06.

The analysis was carried out for the high strength adherend
joints for both quasi-static loading and impact loading, and for
the lower strength materials under quasi-static loading.

RESULTS AND DISCUSSION

Adhesive properties

In Figure 4, the shear stress-strain curves for the range of
strain rates over which the tests were carried out are shown. As

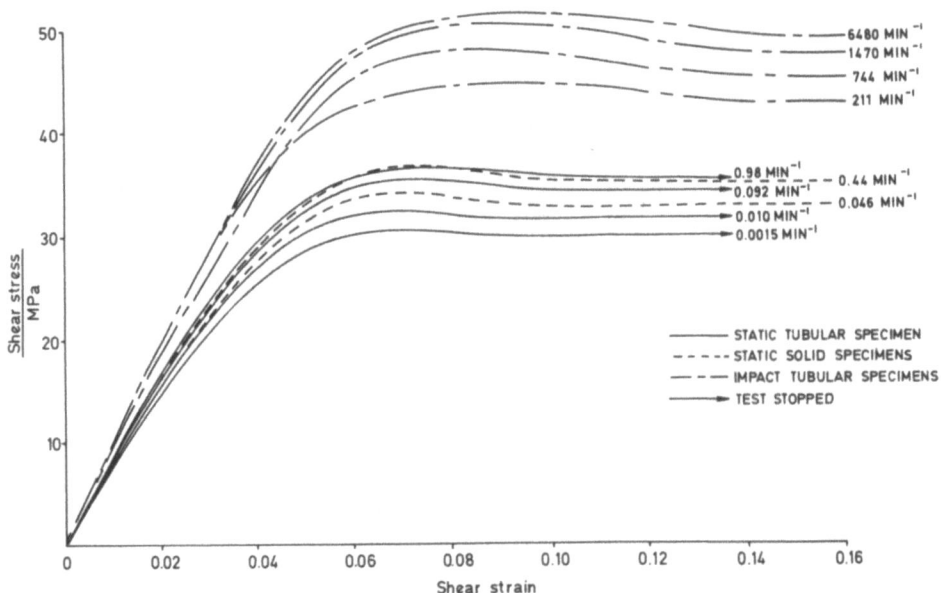

Figure 4. Shear stress-strain curves for bulk adhesive over range
of strain rates from various tests.

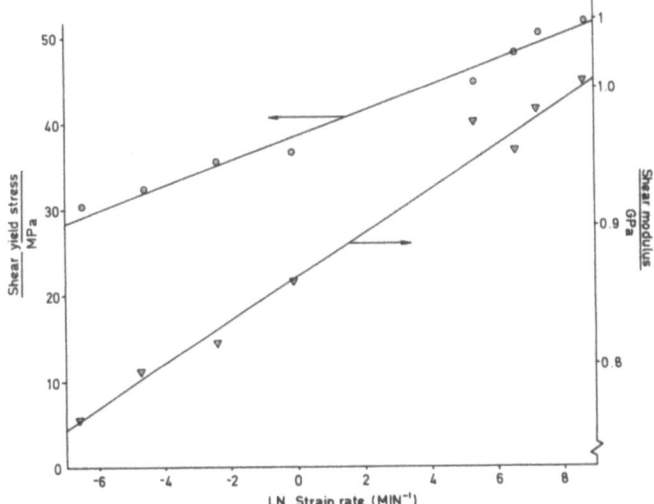

Figure 5   Variation of shear modulus and yield stress of bulk adhesive with logarithm of strain rate.

can be seen, the rubber-modified epoxy shows significant rate dependent properties over the complete range of strain rates. Both the initial shear modulus and maximum flow stress are increased by increased strain rate, indicating the material response to consist of an initial visco-elastic region and a visco-plastic region after yielding. In materials such as this the distinction between elastic and plastic deformation is not well defined, thus the definition of a yield point is not straightforward. Some polymers deforming at a constant strain rate have a stress strain curve which passes through a maximum point which often approximates to the onset of permanent plastic deformation.[6] Here, therefore, this maximum point, often defined as the "intrinsic yield point" is assumed to be the yield point for the material. In Figure 5 the values of initial shear modulus and shear yield stress are plotted against the natural logarithm of the strain rate for the annular specimens. Both show a reasonably linear relationship which, for the yield stresses, is in agreement with the flow model of Eyring.[7]

Another effect of increasing strain rate is to reduce the strain to failure of the material. The failure strain of the solid specimens at a rate of 0.04 min^{-1} is 0.55 compared with a value of 0.175 at a rate of 1470 min^{-1} in the impact test.

Thus, the effect of impact rates of loading on the mechanical properties of the adhesive material is significantly to increase the shear modulus, yield stress and failure stress, and to decrease the failure strain.

Joint strength

The results of the lap joint strengths from both the quasi-static and impact tests are shown in Figure 6 for each of the adherends. As can be seen, the effect of impact loading on each of the joints is similar in that joint strength is reduced. For the high strength 2L73 adherends, the reduction is about 11 per cent, for BB2hh adherends 6 per cent, and for BB2s adherends 4 per cent. Thus, the effect of the impact loading would appear to be less significant for lower strength adherends but in all cases the reduction is small.

The data from the impact tests have been replotted in Figure 7 as a function of the proof stress of the adherend material. Also shown is a line indicating the onset of plastic deformation in the adherend under the combined tension and bending that the adherend undergoes. This has been calculated from the results of the analysis of the single lap joint configuration by Goland and Reissner[8] which, for the adherends, is based on plate bending theory. Also shown is a line indicating the load at which the cross-section

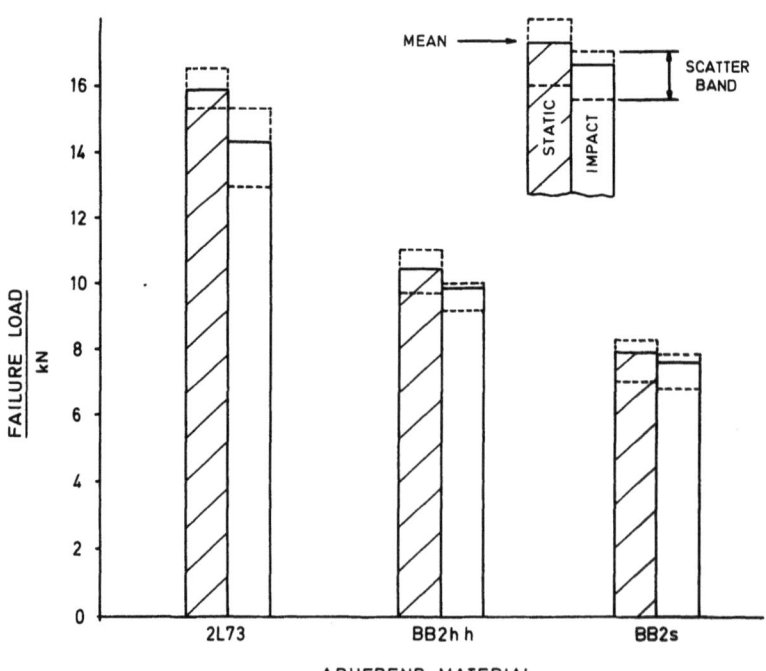

Figure 6   Comparison of single lap joint strength between static and impact loading, for various adherend materials.

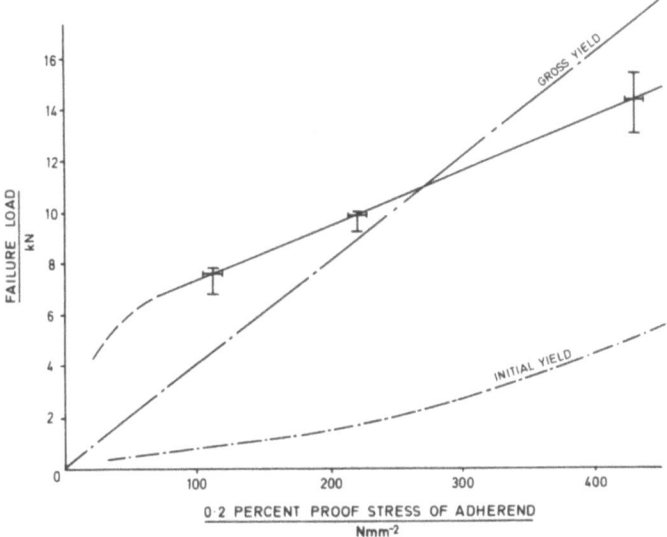

Figure 7 Variation of single lap joint strength under impact conditions with adherend proof stress.

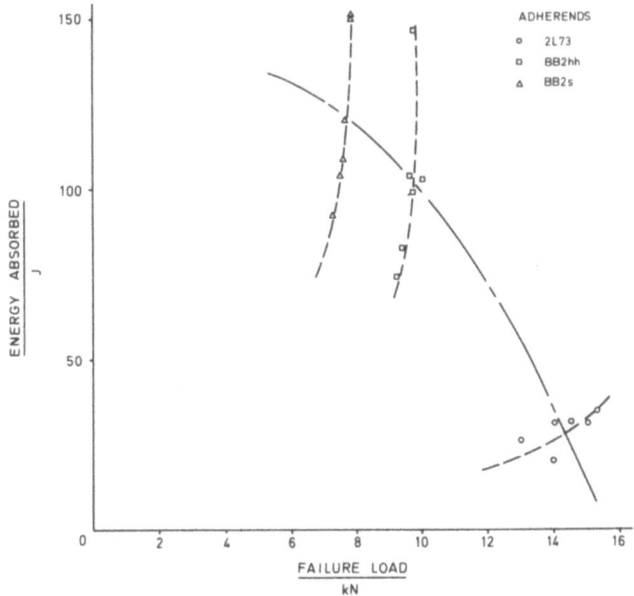

Figure 8 Variation of the energy absorbed by single lap joints under impact conditions with joint strength for various adherends.

of the adherend will reach the proof stress under uniaxial tension. Thus it can be seen that, in each case, some plastic deformation must take place due to bending in the adherends, and for the BB2hh and BB2s materials considerable plastic deformation in the adherends takes place prior to failure. This was confirmed by the significant elongation of the failed specimens that was observed.

The failure loads from the impact tests are again replotted with the energy absorbed by the joint for each of the tests in Figure 8. These results indicate that energy absorption of adhesive joints is very much dependent on the adherend material. Since the fracture energy of the adhesive itself is relatively small, significant energy absorption can only be achieved through plastic deformation of the adherend materials. It would appear that the joints can withstand large amounts of adherend plastic deformation, and thus the result of using a lower strength adherend material is to enhance the energy absorbing capability of the joint, despite the fact that joint strength is significantly reduced. The rapid increase in energy absorbed with failure load for the BB2hh and BB2s materials is again an indication of the large plastic deformation that is occurring.

From the surfaces of failed joints, it was observed that for both the quasi-static and impact tests, the locus of failure was different for the high strength 2L73 adherends as compared with that of the lower strength materials. For the high strength adherends, failure initiated with a crack in the fillet of adhesive at the edge of the overlap, running down approximately 45^0 to the adherend surface as in Figure 9 (a). For the lower strength adherends, failure was initiated at the very edge of the adhesive fillet running along the interface, as in Figure 9 (b). For convenience, these will be referred to as failure types A and B. This change in failure mode is assumed to be attributable to the plastic deformation of the adherends.

Finite element analysis

In Table I, a summary of the finite element predictions is compared with the practical results obtained. In Figure 10, the results of the analysis for the high strength adherend under quasi-static loading is illustrated by the development of the plastic zone within the adhesive layer. Initial yielding takes place at only 1.5 KN load and, before failure, the whole of the adhesive layer has yielded. Peaks in the stress and strain fields occur around points 'a' and 'b' which, as can be seen, are points where initial yielding takes place. For this particular joint, the maximum strains occur around point 'a', thus leading to the prediction of a type A failure. The same is true for the impact joint also. For these high strength adherend joints, although

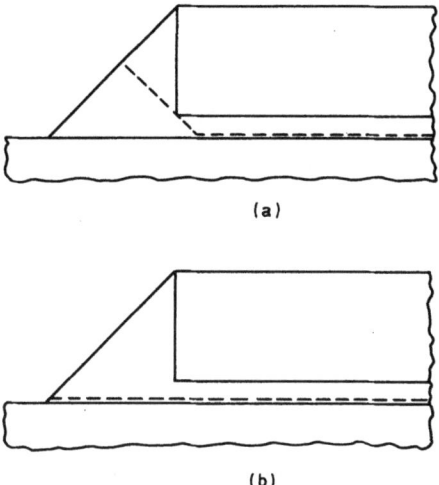

Figure 9 Locus of failure for various joints: (a) type A failure;
(b) type B failure.

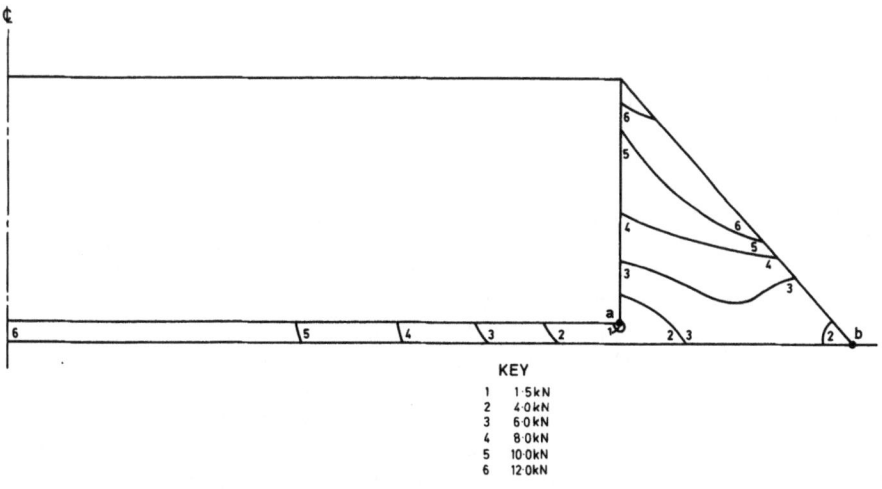

Figure 10 Finite element prediction of the growth of the plastic zone through the adhesive layer of a single lap joint with applied load.

Table I. Comparison of Finite Element Predictions and Experimental Results.

Adherend	Rate	Finite Element Predictions		Experimental Results	
		Type	Load (KN)	Type	Load (KN)
2L73	Impact	A	10.3	A	14.3
2L73	Static	A	14.7	A	15.9
BB2hh	Static	B	11.6	B	10.1
BB2s	Static	B	8.1	B	7.9

some plastic deformation of the adherend takes place (a maximum of about 2 per cent strain in the quasi-static joint at failure) this does not have a significant effect on the state of stress in the adhesive layer.

By applying the 0.15 strain to failure criterion to the quasi-static joint, a fairly accurate prediction of joint strength is made. For the impact joint, the 0.06 strain to failure does correctly predict a reduction in the joint strength because of the change in adhesive properties due to the change in loading rate. However, the prediction is somewhat low, perhaps because of error in the value for maximum strain; or because, for much lower strains in the impact case, the maximum strain to failure criterion is not as applicable.

In the analysis of the quasi-static joints with lower strength adherends, significant adherend plasticity takes place which has two effects on the adhesive stress field. Firstly, the rotation of the overlap is much enhanced, as the bending deformation in the adherends is increased. This results in a reduction in the offset of the applied load at the edge of the overlap and hence a reduction in the bending moment at the edge of the overlap, leading to a reduction in the concentration of stress around point 'a' in Figure 10. Secondly, as adherend strains become large, particularly around the region at the edge of the adhesive fillet where the adherend has tensile bending stresses superimposed on the applied tension, so then the adhesive around region 'b' in Figure 10 is constrained by the adherends such that the strains build up rapidly with applied load. As a result of this, the type B failure is predicted to occur with lower strength adherends, at a reduced load depending on the yield strength of the material.

Again using the maximum adhesive failure strain criterion, reasonable predictions of the failure loads for the lower strength adherend quasi-static joints are made as indicated in Table I.

CONCLUSIONS

The effect of a realistic rate of impact loading on a realistic bonded joint has been investigated for a rubber-toughened epoxy with adherends of various strengths.

The effect of strain rate on the adhesive properties has been demonstrated by the shear stress-strain response of the adhesive over a range of strain rate of 10^{-3} min^{-1} to 10^3 min^{-1}. Both the initial modulus and yield strain are increased linearly with the natural logarithm of the rate, and the failure strain is siginficantly reduced by higher rates of strain. Thus, the effects of impact loading rates on the strength of bonded joints is attributed to the siginifcant change in response of the adhesive material. In practice, only about a 10 per cent decrease in joint strength, due to impact loading, was measured with high strength adherends; thus, as far as joint strength is concerned, the reduction in ductility of the adhesive is compensated for to a large extent by the increase in yield and failure stress. For lower strength adherends, even smaller reductions in strength occur, for similar reasons.

The joints, both under quasi-static and impact loading rates, could sustain considerable amounts of adherend plasticity prior to failure. In fact, the lower the adherend strength, the greater the energy absorbing capability of the joint even though joint strength is reduced. Thus, it would appear that adhesive joints are suitable for the joining of structures from which high energy absorption is required; and, for a given adhesive, the energy absorbing capability is strongly dependent on the adherend strength.

The finite element method is an important tool in the analysis of bonded joints, as both geometric and material non-linearities can be included in the model. Using this method, reasonable prediction of joint strength and the effects of impact loading on it have been made. Through the prediction of the change in the position of failure initiation in the joint due to the adherend plasticity, the effect of using lower strength adherends on joints strength can be assessed. Experimental evidence is in keeping with these predictions.

From a knowledge of the adhesive properties, together with an analysis of the stress-strain state that the adhesive is under in a loaded joint, a greater understanding of the mechanism of failure in bonded joints can be achieved. By application of a suitable failure criterion the effects of various parameters on failure load have been predicted, thus, to some extent, verifying the validity of this approach to the analysis of bonded joints.

REFERENCES

1. J. A. Harris and R. D. Adams. Unpublished data, 1981.
2. A. Nadai, "Plasticity. A Mechanics of the Plastic State of Matter", pp. 128-130, McGraw-Hill, New York, 1931.
3. R. D. Adams and N. A. Peppiatt, J. Strain Anal., $\underline{9}$, 185, (1974).
4. A. D. Crocombe and R. D. Adams, J. Adhesion, $\underline{13}$, 141, (1981).
5. R. D. Adams, J. Coppendale and N. A. Peppiatt, in "Adhesion 2", K. W. Allen, Editor, pp. 105-119, Applied Science Publishers, London, 1978.
6. D. B. Bowden, in "The Physics of Glassy Polymers", R. N. Howard, Editor, pp. 279-339, Applied Science Publishers, London, 1973.
7. H. Eyring, J. Chem. Phys., $\underline{4}$, 283, (1936).
8. M. Goland and E. Reissner, J. Appl. Mech., Trans. ASME, $\underline{66}$, A17, (1944).

THE PERFORMANCE OF ADHESIVE-BONDED THIN-GAUGE SHEET METAL STRUCTURES WITH PARTICULAR REFERENCE TO BOX-SECTION BEAMS

A. Beevers and A. C. P. Kho

Oxford Polytechnic
Headington
Oxford, U.K., OX3 OBP

In an effort to reduce vehicle weight, the automotive industry is developing car body structures made from light-weight materials such as composites, plastics and aluminium alloys. Fabrication of these materials in high volume automotive applications using traditional welding techniques is not feasible and adhesive bonding is now being investigated as a potential assembly method. In order to assess performance characteristics of bonded vehicles, thin-gauge sheet-metal box-section beams have been used to simulate structural details in automotive applications such as car bodies and commercial vehicles. Beams were fabricated from flanged strips by different joining methods to form box-section structures approximately 1m. long x 60mm. square. Tests were carried out to determine torsional and flexural rigidity and ultimate flexural strengths, and in the majority of tests, bonded structures gave better characteristics than the equivalent riveted or spot-welded beams. The failures of beams under 3-point bending have been related to buckling of the side webs and further experimental tests have shown that collapse is critically dependant on flange-bend radius. Finite element techniques have been used to analyse stress distribution in the beam section and this confirms the experimental observations of beam collapse.

INTRODUCTION

The motor industry has for some time used adhesives in semi-structural applications particularly for the attachment of stiffeners to bonnets and boot lids. As well as providing a uniform load distribution, adhesive bonding provides a smooth surface finish and thus eliminates the need for expensive secondary operations prior to painting. More recently the feasibility of using adhesives in fully structural automotive body applications has been studied. Although the concept of a bonded, unitary construction car body is applicable to conventional mild steel assembly, it is potentially more relevant to the fabrication of bodies from light alloy and composites. These materials are becoming increasingly attractive for automotive applications because of their light weight but they cannot be joined by the traditional welding methods. Bonding technology is, therefore, expected to make a significant contribution to future developments in the motor industry and a considerable amount of research effort has already been committed to establish characteristics of bonded joints for such structures.

A large volume of data has been established on properties of bonded joints using relatively small test pieces [1-4] and new analytical techniques are providing a greater understanding of bonded joint performance [5,6]. However, there is little information available on the performance of structurally bonded large scale elements which might be encountered in automotive applications. It is envisaged that box sections and "top hat" stiffener configurations will feature prominently in fully-stressed body shells and these forms of structural elements are also applicable in other types of vehicle construction such as transport containers, coaches, commercial vehicle cabs, and rail carriages. The work described in this paper was carried out to determine flexural and torsional properties of box-section beams formed from thin gauge sheet metal.

The general design of the fabricated beam structure adopted for the tests was chosen to give a reasonable representation of a typical detail of a car body such as a sill or door pillar. It was expected that the test method would also provide a simulation of a variety of other sheet metal structures and the external flange arrangement enabled the beam to be made by a number of different joining methods.

BOX SECTION BEAM TESTS

Beam Manufacture

Each beam was manufactured to the same nominal dimensions using four pieces of sheet metal 910mm long by 90mm wide. To form the 'box', two pieces were shaped into a shallow channel section by bending 15mm flanges on to the edges of the strip using standard workshop bending bars. The beam was completed by joining the edges

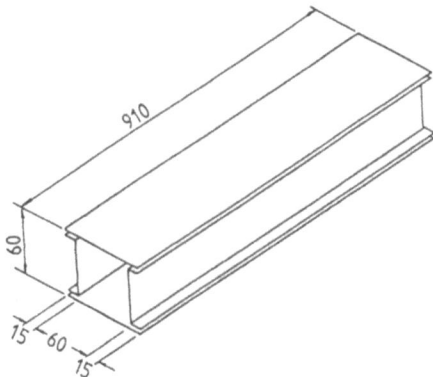

Figure 1. Configuration of box-section beams fabricated from thin sheet (dimensions in mm).

of the flat strips to the flanges so that the main section formed a 60mm square, as shown in Figure 1. The beams were made from 0.9mm thick deep drawing quality low carbon mild steel and 1.2mm thick 2.5per cent Mg. aluminium alloy.

For the adhesive bonded structures all the surface areas to be bonded were grit blasted and thoroughly cleaned and degreased with trichloroethylene. The adhesives were then applied uniformly to both surfaces. In the case of the toughened acrylic adhesive, the resin was applied to one surface and the activator to the other joint surface. The parts were then located, assembled together and held with spring clips while the adhesive cured. The single-component high-temperature-curing, epoxy adhesive was cured at 180^0C for 30 minutes, while all the other adhesives were cured at room temperature.

For the spot-welded and riveted beams the welds and rivets were placed at 25mm intervals along the centre line of the flange using standard practices as far as possible. The rivets were standard 3.1mm diameter steel-bodied blind rivets which were inserted with a pneumatic/hydraulic rivet-setting power tool.

Flexural stiffness was determined by three-point bend tests as shown in Figure 2. The beam was supported by two radiused supports

710mm apart on the bed of a compression testing machine and the load was applied at a third "centre" point on the top of the beam by a similar radiused anvil. To reduce the risk of sheet indentation and premature collapse small steel pads were placed between the beams and the support radii. During testing the bending load was applied incrementally at intervals of 1kN and the midspan deflection was measured by means of a dial test indicator.

Torsional stiffness was measured by mounting the ends of the beams into special end fixtures which rigidly clamped the four sides of the box-section to a central core. The fixtures were then located into grips on the torsion testing machine which enabled the torsional load to be measured for incremental angles of twist at 0.25 degree intervals.

Ultimate strength tests were carried out using the same machines and fixtures by continuing the loading beyond yield until failure occurred.

Beam Test Results

Although there are some inconsistencies in the results of the beam tests in Table I, it is possible to draw a number of general observations and conclusions. The most striking observation is that with only one exception the adhesive bonded structures were superior to the equivalent spot-welded and riveted beams in all test performance parameters and in some cases the stiffnesses and ultimate strength properties were more than doubled.

It is also clear that there is little correlation between the performance of a structure and the performance of a small lap shear strength test piece. In the case of bending (flexure) tests the shear force on the joint interface is relatively low and in all tests failure of the beam occurred by buckling of the inner webs.

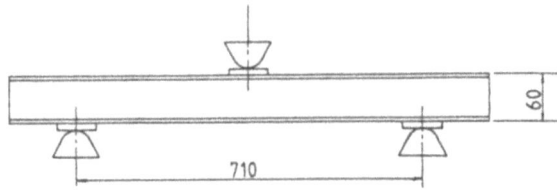

Figure 2. Arrangement for flexure tests (dimensions in mm).

Table 1. Summary of Results on Beam Tests.

BEAM MATERIAL	JOINING METHOD	FLEXURAL STIFFNESS kN/mm	TORSIONAL STIFFNESS kNm/rad	ULTIMATE FLEXURAL STRENGTH kN
0.9mm Mild Steel	Rivet	1.8	2.5	3.4
	Spot Weld	1.9	9.5	3.6
	2 Part, Cold-Cured Epoxy Adhesive	2.9	11.6	7.7
	1 Component, Hot-Cured Epoxy Adhesive	4.2	8.2	6.9
	Toughened Acrylic Adhesive	2.85	11.0	5.0
1.2mm Al. Alloy (2.5% Mg)	Rivet	1.6	2.6	7.0
	Spot Weld	1.9	5.3	6.4
	2 Part, Cold-Cured Epoxy Adhesive	2.8	6.7	8.2
	1 Component, Hot-Cured Epoxy Adhesive	3.1	11.1	8.5

If the section of the web is considered as a simple strut it will be realised that the buckling, or crippling load is a function of eccentricity of load application and rigidity of end-constraint.

This, in turn, will depend on the bend radius which is introduced during the forming of the flanges. The mild steel sheet naturally assumes a greater bend radius during forming because of its higher yield strength and thus in the ultimate flexural strength tests the mild steel beams will buckle more easily.

Extending this hypothesis it may also be reasoned that in all the adhesive bonded beams the strut end constraints are more rigid because the adhesive support extends up to, and beyond, the centre line of the web as it forms a fillet in the corner of the box-section. Compared with the more remote attachment of a spot-weld or rivet this additional support may explain greater flexural strength of bonded beams.

Theoretical Analysis of Flexural Stiffness

The deflection, δ, of a beam under three-point bending with the load at mid-span is given by the expression

$$\delta = \frac{WL^3}{48EI}$$

where W = applied load
L = length of beam between supports
E = modulus of elasticity of beam material
I = second moment of area of beam section about neutral axis

The flexural stiffness is then

$$W/\delta = \frac{48EI}{L^3} \quad (1)$$

In an idealized beam model it may be assumed that the flange/plate connections are homogeneous and continuous and the second moment of area I, of the section can be calculated on this basis. At the other extreme, if there were no joint between the flanges and plates the only shear flow from the top sheet into the webs would be by friction. The section would then approximate to a double channel configuration with no top and bottom sheets and the value of I can be similarly calculated. It might be expected that the stiffness of a fabricated beam would be between these two theoretical conditions.

Taking L = 710mm as used in experimental tests, E_{steel} = 210GN/m^2 and E_{Al} = 70GN/m^2, the values of I and flexural stiffness, W/δ, calculated from equation 1 are given in Table II. These theoretical values of stiffness are compared with experimental test results in figure 3.

It can be seen from the bar charts in figure 3, that the bonded aluminium beams give values of flexural stiffness which are close to the idealized model of the homogeneous beam. The spot-welded and riveted beams in both steel and aluminium have stiffnesses which are closer to the beam model having no flange/sheet connection. It is evident that the discrete point attachments allow some local elastic buckling of the top sheet in compression giving slight interfacial separation between each joint. This would cause a greater deflection of the welded or riveted beams compared with the bonded structures in which the joint is continuous. In the case of the riveted structures some interfacial slippage during flexure may also occur due to the clearance between the rivet body and the hole, and this may also account for the low values of torsional stiffness obtained in the experimental tests in Table 1.

Table II. Theoretical Stiffnesses of Box-Section Beams.

Theoretical beam model	Sheet thickness mm	I mm^4 x10^6	Theoretical Stiffness kN/mm	
			Aluminium	Steel
Homogeneous Continuous Connection	0.9	0.227	2.15	6.39
	1.2	0.303	2.84	8.53
No connection	0.9	0.078	0.72	2.14
	1.2	0.112	0.94	2.83

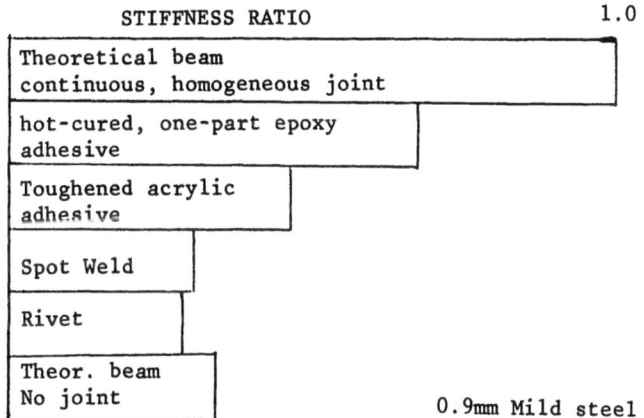

Figure 3. Comparison of beam stiffnesses with theoretical models.

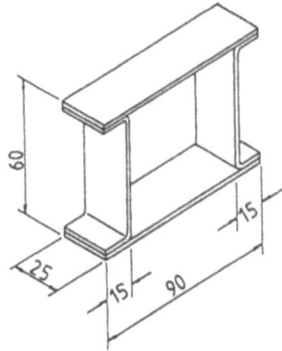

Figure 4. Beam section compression test piece (dimensions in mm).

Table III. Results of Beam Section Compression Tests.

SHEET MATERIAL	THICK-NESS mm	FLANGE BEND RADIUS mm	ULT. COMP. STRENGTH kN		
			ADHESIVE BOND	RIVET	SPOT WELD
Al.Alloy (2½%Mg)	0.9	2.5	1.03	0.51	0.51
		1.5	1.18	0.53	0.73
		1.0	1.47	0.72	0.80
	1.2	2.5	3.12	1.09	1.12
		1.5	2.87	1.19	1.49
		1.0	2.80	1.63	1.55
Mild Steel	0.9	3.0	1.89	0.54	0.59
		2.5	1.94	0.58	0.76
		2.0	2.94	1.11	1.25

In all the tests on the mild steel beams the ratio between the stiffness of the experimental beams and the idealized models are relatively lower than for aluminium beams. This may be partly attributed to the localised deflections occurring at the flange bend radii which are greater in the mild steel beams.

BEAM SECTION COMPRESSION TESTS

In order to verify the suggested explanation for the ultimate flexural strength characteristics a further series of tests was carried out on specimens which were, in effect, a short length of beam section as shown in Figure 4. Bending of the flange was carried out so that a range of flange bend radii was obtained and the strips were then joined using the same methods as in the fabrication of beams. The adhesive used was an aluminium-filled, one-component, hot-curing, epoxy resin.

The sections were subjected to compressive loading between two parallel plates and the ultimate strengths were recorded. The results of these tests are given in Table III.

In all the tests failure occurred by the inward collapse of the side plates. With the riveted and spot welded sections the smaller bend radii gave higher strengths than those with the larger radii. These observations support the inference that the flange bend radius has a significant effect on the buckling strength of the structure. With the adhesive bonded structure, the buckling strength is not so dependant on the flange bend radius. This again supports the earlier observations that the adhesive fillet provides additional support at the corner radius of the formed flange thus giving greater resistance to buckling. It should also be noted that all the adhesive bonded specimens give considerably higher strengths than the equivalent rivet and spot-welded structures.

Finite Element Analysis of Section Compression Tests

The finite element stress analysis package programme, "Pafec 75" was used to determine the distribution of stress in a box-section structure under a compressive load. A comparison of the maximum stresses sustained in the side plates of structures with different flange radii was investigated.

Since the box-section structure was axisymmetric about both the 'x' and 'y' axis, only a quarter of the structure was considered in this analysis. The eight-noded isoparametric element (code number 36210 in Pafec) was used to divide up the structure and the nodal diagram for a typical section is shown in Figure 5.

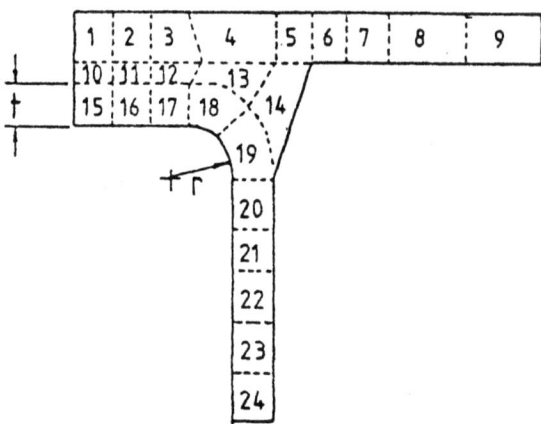

Figure 5. Nodal diagram of typical section for finite element analysis ($r/t = 1$).

Analysis was made of section models without adhesive and with adhesives of varying values of elastic moduli. The calculated stresses resulting from applied compressive loads were used to provide a measure of buckling resistance; the higher the maximum stress in any model, the lower would be its resistance to buckling. The analytical results are shown in Table IV, in which the critical dimensional features of the finite element model are expressed as a ratio between the flange bend radius and the sheet thickness, r/t. In the ideal case the bend radius would be zero, i.e. $r/t = 0$ and the tabulated values of buckling resistance are given relative to this ideal flange condition.

Table IV. Buckling Resistance of Sections from Finite Element Analysis.

r/t	Without Adhesive	With Adhesive
0	1	-
0.4	0.314	0.418
1.0	0.285	0.37
1.6	0.225	0.3

The finite element analysis of sections without adhesive shows a substantial loss of compressive strength as the value of r/t increases. In practice $r/t>1$. It is also seen that the bonded section models give greater buckling resistance, and although their compressive strengths are still dependant on the r/t ratio it is evident that the adhesive fillet has contributed to the improved performance.

In comparing these values with the results of the compressive tests in Table III, it might be suggested that the riveted and spot-welded sections are equivalent to the finite element models without adhesive. Similar trends are observed in the experimental values but the adhesive-bonded sections are proportionately better than predicted from the finite element analysis.

CONCLUSIONS

The flexural and torsional stiffness and strength properties of bonded thin-sheet box-section beams are higher than for similar beams formed by riveting or spot welding.

The improved ultimate flexural strength properties of bonded beams are partially due to the supporting effect of the adhesive fillet at the flange bend radius.

ACKNOWLEDGEMENTS

Support of this work by Permabond Adhesives Ltd. is gratefully acknowledged.

REFERENCES

1. J. D. Minford and E. M. Vader, "Adhesive Bonding Aluminium Body Sheet", Society of Automotive Engineers Paper No. 740078 (1974).
2. G. E. Nordmark, "Fatigue Performance of Aluminium Joints for Automotive Applications", Society of Automotive Engineers Paper No. 780328 (1978).
3. S. J. Thompson, "Investigation Into the Impact Properties of Adhesive Bonded Joints", Oxford Polytechnic postgraduate thesis (1979).
4. W. A. Lees, Int. J. Adhesion Adhesives, $\underline{1}$, 5 (1981).
5. G. P. Anderson, S. J. Bennett and K. L. DeVries, "Analysis and Testing of Adhesive Bonds", Academic Press, New York, 1977.
6. A. J. Kinloch, Editor, "Developments in Adhesives - 2", Applied Science Publishers, London, 1981.

CYCLIC DEBONDING OF ADHESIVELY BONDED COMPOSITES

S. Mall,* W. S. Johnson, and R. A. Everett, Jr.**

NASA Langley Research Center
Hampton, Virginia 23665

To analyze the fatigue behavior of a simple composite-to-composite bonded joint, a combined experimental and analytical study of the cracked-lap-shear specimen subjected to constant-amplitude cyclic loading was undertaken. Two bonded systems were studied: T300/5208 graphite/epoxy adherends bonded with adhesives EC 3445 and with FM-300. For each bonded system, two specimen geometries were tested: (1) a strap adherend of 16 plies bonded to a lap adherend of 8 plies, and (2) a strap adherend of 8 plies bonded to a lap adherend of 16 plies. In all specimens tested, the fatigue failure was in the form of cyclic debonding with some $0°$ fiber pull-off from the strap adherend. The debond always grew in the region of adhesive that had the highest mode I (peel) loading and that region was close to the adhesive-strap interface. Furthermore, the measured cyclic debond growth rates correlated well with total strain energy release rates G_T as well as with its components G_I (peel) and G_{II} (shear) for the mixed-mode loading in the present study.

---

*University of Maine, on leave of absence at NASA Langley Research Center.
**Structures Laboratory, U.S. Army Research and Technology Laboratories (AVRADCOM), NASA Langley Research Center.

INTRODUCTION

To increase performance and fuel economy, aerospace industries have been turning more and more to the use of advanced composites in both commercial and military aircraft. These materials offer excellent strength-to-weight and stiffness-to-weight ratios. But their efficient application requires more sophisticated joining methods than used in metallic structures. Because composites are severely weakened by fastener holes, their weight advantage may be lessened when mechanically fastened joints are used. Adhesive bonding, on the other hand, provides a desirable alternative to mechanical fastening because of the following potential advantages: (1) higher joint efficiency index (relative strength/weight of the joint region), (2) lower part count, (3) no strength degradation of basic laminate due to fastener holes, (4) less expensive and simpler fabrication techniques, (5) lower maintenance costs, and (6) potential corrosion problems avoided.

Design methods for adhesively bonded composites require criteria to predict both strength and durability. Although analytical and experimental work has been reported on the static strength of bonded composites,[1-3] very little information is available on their fatigue behavior. Several possible fatigue failure modes exist for bonded composites: cyclic debonding (i.e., progressive separation of the adhesive bond under cyclic load), interlaminar damage (delamination), adherend fatigue, or a combination of these. Therefore, life predictions require a basic understanding of the mechanics associated with each failure mode.

Many of the results obtained in cyclic debonding studies of composite-to-metal joints[4] and metal-to-metal joints[5] may be applicable to the present case of composite-to-composite joints. In the study by Roderick et al.,[4] the fracture mechanics concept of strain energy release rate was used to model the cyclic failure of bonded composite-to-metal joints. This is similar to the approach in metals where fatigue-crack-propagation rate is correlated with the strain energy release rate. The total strain energy release rate, G_T, associated with the cyclic failure of an adhesive bond can be resolved into three components G_I, G_{II}, and G_{III} associated with three debonding modes: I (opening), II (sliding), and III (tearing), respectively. However, in most practical applications, only G_I and G_{II}, due to peel and shear stresses, respectively, exist near the debond front during cyclic loading. Using metal-to-metal joints, Brussat et al.[5] developed the cracked-lap-shear specimen to study the effect of mixed-mode (G_I and G_{II}) loading on adhesive joints. Romanko and Knauss[6] have extended these fracture mechanics concepts to investigate the failure of adhesive joints under various environmental conditions involving temperature, moisture, etc. Everett[7] recently showed that the strain energy release rate G_I associated with peel stress had a significant effect on cyclic debonding.

The objective of the present study was to analyze the fatigue behavior of simple composite-to-composite bonded joints subjected to constant-amplitude cyclic loading. For this purpose, graphite/epoxy (T300/5208) cracked-lap-shear specimens were tested using EC 3445 and FM-300 adhesives. This investigation focused on the correlation of the measured cyclic debond rates with strain energy release rates calculated using a finite element analysis of the cracked-lap-shear specimen. Failure modes of this simple joint were also analyzed to investigate the relative influence of G_I and G_{II}.

NOMENCLATURE

a	length of debond, mm
$\dfrac{da}{dN}$	debond growth rate, mm/cycle
B	width of specimen, mm
c, n	curve-fit parameters
E	Young's modulus of adhesive, GPa
E_1, E_2	Young's moduli of composite, GPa
G	shear modulus of adhesive, GPa
G_{12}	shear modulus of composite, GPa
G_I	mode I strain energy release rate, J/m^2
G_{II}	mode II strain energy release rate, J/m^2
G_{III}	mode III strain energy release rate, J/m^2
G_T	total strain energy release rate (= $G_I + G_{II}$), J/m^2
N	number of cycles
P	applied load, kN
r	residual from least-square curve fit
ν	Poisson's ratio of adhesive
ν_{12}, ν_{23}	Poisson's ratios of composite

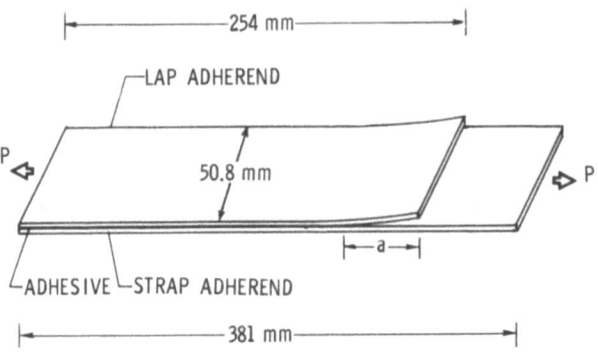

Figure 1. Cracked-lap-shear specimen.

EXPERIMENTS

Test Specimen

The cracked-lap-shear specimen, shown in Figure 1, was employed in the present study because it represents a simple structural joint subjected to in-plane loading. Both shear and peel stresses are present in the bond line of this joint. The magnitude of each component of this mixed-mode loading can be modified by changing the relative thicknesses of strap and lap adherends.[5,8] For the present study, the strap and lap adherends had 8 or 16 plies. These adherends had a quasi-isotropic layup and were typical of those currently employed in an Army program to build an all-composite helicopter airframe which is almost entirely adhesively bonded.[9]

Two bonded systems were studied—graphite/epoxy (T300/5208) adherends bonded with EC 3445 adhesive and with FM-300 adhesive. The EC 3445 adhesive is a thermosetting paste adhesive with a cure temperature of 121°C. Specimens with this adhesive were fabricated using conventional secondary bonding procedure. However, specimens with FM-300 adhesive were fabricated by a co-curing procedure whereby adherends were cured and bonded simultaneously. The FM-300 is a modified epoxy adhesive supported with a carrier cloth with a cure temperature of 177°C. These adhesives, as well as the concepts of

Table I. Adhesive Material Properties.

Adhesives	Modulus, GPa		Poisson's Ratio
	E	G	ν
EC 3445	1.81	0.65	0.4
FM-300	2.32	0.83	0.4

secondary and co-cure bonding, are also being employed in the bonded all-composite helicopter airframe mentioned previously.[9] The bonding process followed the manufacturer's recommended procedures for each adhesive. The nominal adhesive thickness was 0.10 and 0.25 mm for the EC 3445 and FM-300, respectively.

The Young's modulus of FM-300 adhesive was calculated from the shear modulus provided by the manufacturer, assuming the adhesive to be an isotropic material. Poisson's ratio was assumed to be 0.4 for both adhesives which is a typical value for adhesives. The EC 3445 adhesive is the paste version of the AF-55 adhesive film. Therefore, the Young's modulus of EC 3445 was calculated from the shear modulus of AF-55,[10] assuming the adhesive to be an isotropic material. The material properties of both adhesives are given in Table I.

The composite adherends consisted of quasi-isotropic lay-ups of $[0°/45°/-45°/90°]_S$ and $[0°/45°/-45°/90°]_{2S}$. The material properties of graphite/epoxy, presented in Table II, were obtained from reference 11. For each bonded system, two types of specimen were tested: (1) thin lap adherend of 8 plies bonded to thick strap adherend of 16 plies, and (2) thick lap adherend of 16 plies bonded to thin strap adherend of 8 plies. This arrangement provided four sets of specimens. Initially, the lengths of strap and lap adherends were 381 and 254 mm, respectively (a total of 127 mm was for grip support on both ends). However, specimens with a thick co-cured lap adherend were modified due to a pinched-off edge obtained during fabrication. This pinched-off edge, as discussed in reference 12, caused a nonuniform thickness of lap adherend and adhesive near the end of the overlap. To avoid problems due to this nonuniform thickness, the pinched-off ends were removed by machining. In the resulting modified specimen, lengths of strap and lap adherends were 305 and 203 mm, respectively.

Testing Procedure

All specimens were tested in a closed-loop hydraulic test machine at a frequency of 10 Hz. In all tests, constant amplitude

Table II. Graphite/Epoxy[a] Adherend Material Properties.

Modulus,[b] GPa			Poisson's Ratio[b]	
E_1	E_2	G_{12}	ν_{12}	ν_{23}
131.0	13.0	6.4	0.34	0.35

[a]T300/5208, fiber volume fraction is 0.63.
[b]The subscripts 1, 2, and 3 correspond to the longitudinal, transverse, and thickness directions, respectively, of a unidirectional ply.

cyclic loads were applied at a stress ratio of 0.1. The debond initiated at the lap end. Subsequent cyclic debonding was monitored throughout each test.

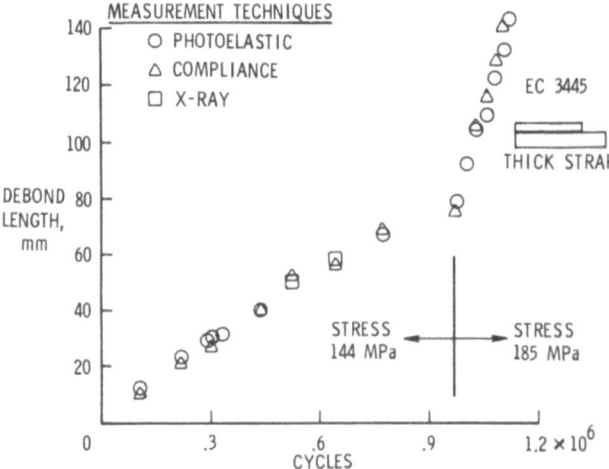

Figure 2. Comparison of different techniques to measure debond growth.

CYCLIC DEBONDING OF ADHESIVELY BONDED COMPOSITES

Three techniques to measure the cyclic debond growth were evaluated. For each technique, the location of the debond front was measured periodically to calculate debond growth rates. The first method used a sheet of photoelastic material bonded to the lap adherend of the specimen, as discussed in reference 4. Isochromatic fringes developed at the debond front as a result of the high strain gradient in that vicinity when subjected to load. These isochromatic fringes were observed through a polarizer and were used to locate the debond front. The second method involved locating the debond front with an X-ray technique using a dye penetrant, zinc iodine. The third method involved measuring the compliance of the specimen and then calculating the debond length using a crack length compliance formula. The compliance of the specimen was measured with two displacement transducers attached on opposite sides of the specimen. All three methods provided good agreement for debond growth measurement as shown in Figure 2. In the present study the photoelastic technique was selected; an automated measurement system photographed the isochromatic fringes at predetermined intervals.

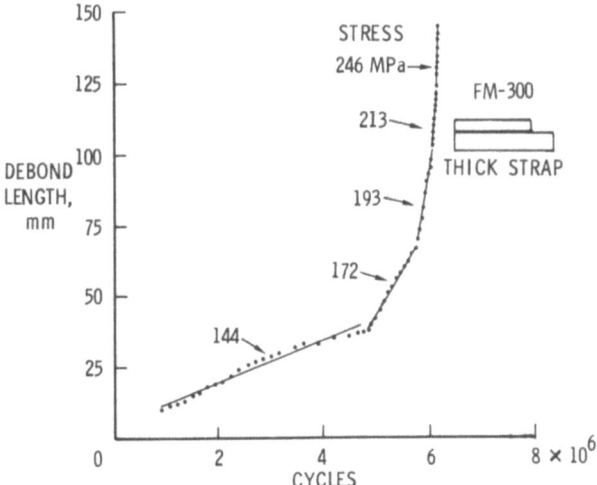

Figure 3. Typical variation of debond length with fatigue cycles at different stress levels.

The strain energy release rates (G_T, G_I, and G_{II}) are usually uniform in the cracked-lap-shear specimen for a significant debond growth region.[8] The debond data were measured over this region. This region is discussed in the subsequent section on finite-element analyses. Tests were conducted at two or more constant amplitude stress levels to obtain several values of debond growth rate (da/dN) from each specimen. At each stress level, the debond was measured as it grew over about 3 to 4 cm to ensure an accurate estimate of the debond growth rate over that region. Figure 3 shows a typical debond data for different stress levels. In all cases debonding was initially nonlinear, but became linear after the debond progressed a short distance (about 10 mm). For each stress level, the debond data were fitted with a straight line using regression analysis.

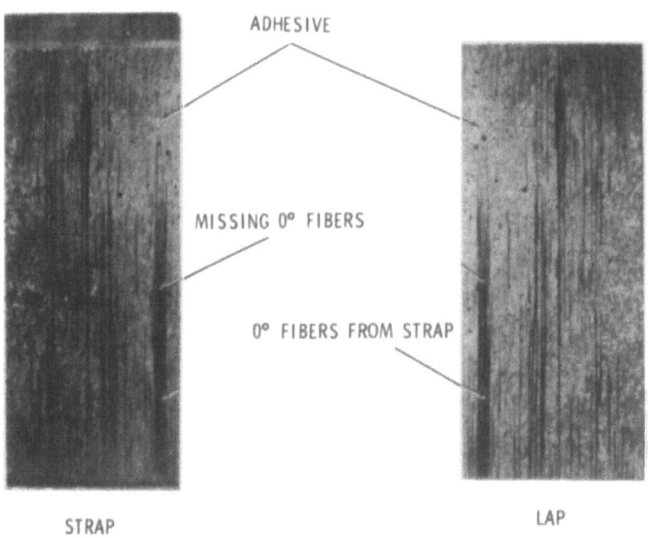

Figure 4. Debonded surfaces of cracked-lap-shear specimen (EC 3445, thick strap).

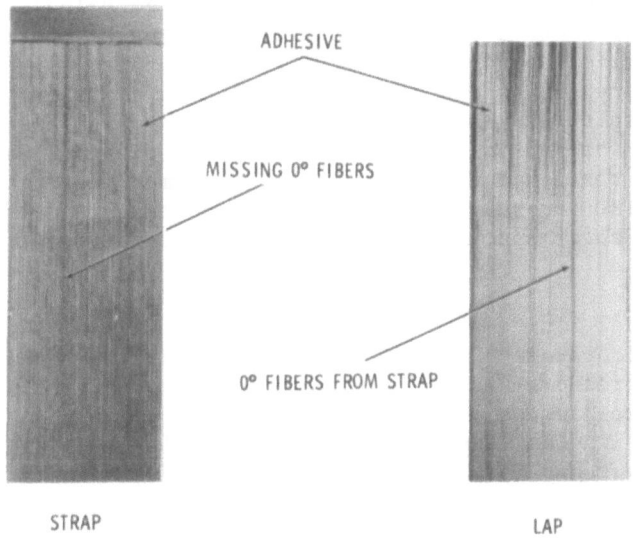

Figure 5. Debonded surfaces of cracked-lap-shear specimen (FM-300, thin strap).

Debond Surfaces

As previously mentioned, the possible failure modes in a composite bonded joint are cyclic debonding, delamination, adherend fatigue, or a combination of these. All specimens in the present study failed primarily by cyclic debonding of the adhesive. Although some adhesive remained on both the strap and lap adherends, significantly more adhesive remained on the lap adherend. Also, there was some $0°$ fiber pull-off from the strap adherend in most specimens. Typical debonded surfaces with these failure details are shown in Figures 4 and 5 for both systems. On close examination of the debonded surface, the following conclusions regarding the cyclic debond can be drawn. The debond was basically of a cohesive nature, i.e., failure was within the adhesive with some $0°$ fiber pull-off from the strap. Further, the debond was always closer to the strap than to the lap. A possible explanation for this debond characteristic is discussed in the next section.

FINITE-ELEMENT ANALYSIS

Previous studies of fatigue damage mechanisms in adhesively bonded joints have shown that the strain energy release rate, defined from fracture mechanics principles, may be useful for correlating cyclic debond growth rate.[4,5] Therefore, all four sets of specimens were analyzed with a finite element program, called GAMNAS[8] to calculate strain energy release rates. This two-dimensional analysis accounted for the geometric nonlinearity associated with the large rotations in the unsymmetric cracked-lap-shear specimen. The importance of a geometric nonlinear analysis for the cracked-lap-shear specimen has been discussed in detail in reference 8.

A typical finite-element mesh, shown in Figure 6, consisted of about 1600 isoparametric four-node elements and had about 3000 degrees of freedom. Each ply of composite was modeled as one layer in the finite-element model, except for the ply at the adhesive interface which was modeled in two or three layers. A multipoint constraint was applied to prevent rotation of the loaded end of the

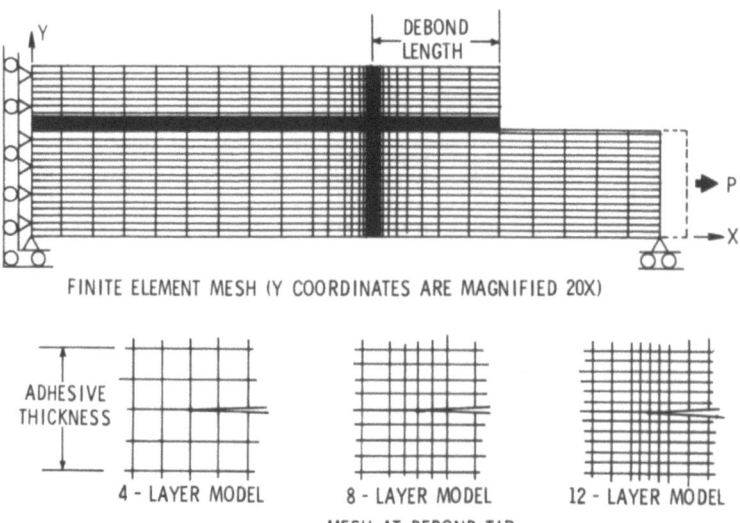

Figure 6. Finite element model.

model (i.e., all the axial displacement along the ends are equal) to simulate actual grip loading of the specimen. A double-cracked-lap-shear specimen with isotropic adherends was analyzed using a quasi-three-dimensional analysis like that in reference 13 to show that the present two-dimensional analysis should be based on the plane strain condition. The strain energy release rates G_T, G_I, and G_{II} in the analysis were computed for the maximum load in the fatigue cycle using a virtual crack closure technique.[14] The details of this procedure are given in reference 8. The calculation of strain energy rates in the present analysis depends on three parameters: (1) debond location, (2) load level, and (3) debond length. These parameters will be discussed next.

As mentioned previously, the debond grew near the interface of the adhesive and strap adherend in all tests. It was very difficult to measure the exact location of the debond, but in general the debond grew within one-fourth of the thickness of adhesive closest to interface. The effect of the location of the debond within the adhesive in the cracked-lap-shear specimen was investigated using GAMNAS.[8] Figure 7 shows the variation of calculated G_T, G_I, and G_{II} with the debond location within the thickness of the EC 3445 adhesive. These results were calculated by modeling the adhesive

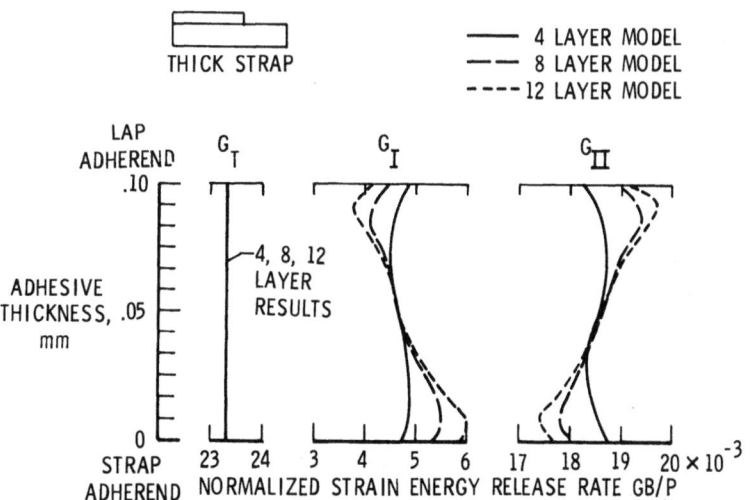

Figure 7. Variation of strain energy release rate with location of debond within EC 3445 adhesive.

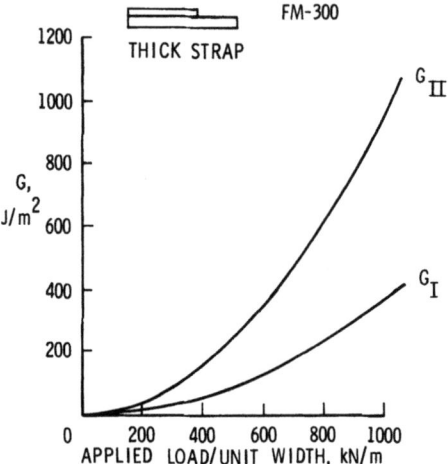

Figure 8. Variation of strain energy release rates with applied load.

with 4, 8, and 12 layers of elements, as shown in Figure 6. Figure 7 clearly shows that G_T remains constant for all locations of the debond, while G_I has its maximum value near the adhesive-strap interface and G_{II} has its maximum value near the adhesive-lap interface. The debond always initiated and grew in the region of highest G_I (near the adhesive-strap interface). This indicates that G_I has the greater influence on the debond location in adhesive joint. This is consistent with the observations that adhesives are inherently weaker under peel loading than under shear loading.[1,2,15] Additionally, these results show that an accurate evaluation of G_T can be achieved by a four layer model, while accurate evaluation of G_I and G_{II} require a more refined model. To

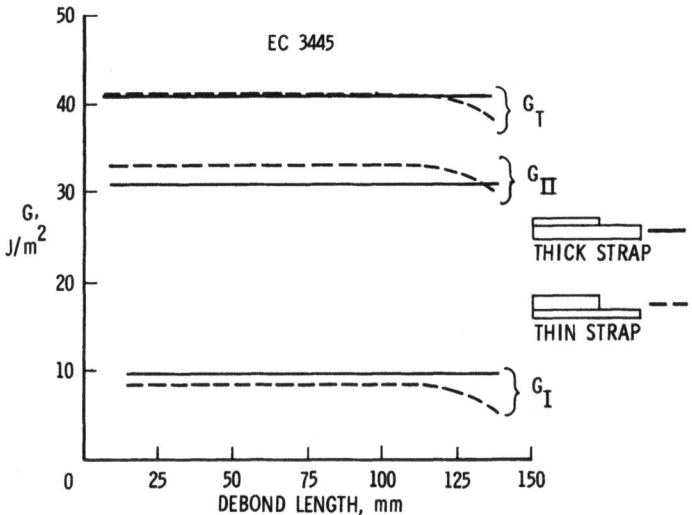

Figure 9. Variation of strain energy release rates with debond length.

analyze the experimental debond growth rates, the debond location for all subsequent calculations was selected by engineering judgement to be at one-sixth of the adhesive thickness away from the adhesive-strap interface. Also, the 12-layer model was used in these calculations.

Figure 8 shows the typical variation of strain energy release rates G_I and G_{II} with the applied load on the cracked-lap-shear specimen obtained from geometric nonlinear analyses. The G_I and G_{II} in nonlinear analysis were found to be functions of the square of applied load within one percent.

All four sets of specimens were then analyzed to determine the variation of G_T, G_I, and G_{II} with the debond length. Figure 9 shows the typical dependence of G_T, G_I, and G_{II} on the debond length for both types of specimen with EC 3445 adhesive. For specimen with thick strap, G_T, G_I, and G_{II} were constant up to 140 mm of debond length. Similar behavior was found for co-cured specimens with thick straps. For specimens with thin straps,

Table III. Strain Energy Release Rate.

Specimen Details		Strain Energy Release Rate[a] (J/m^2) for the Applied Stress of 82.0 MPa		
Adhesive	Strap Type	G_I	G_{II}	G_I/G_{II}
EC 3445 (Secondary bonding)	Thick	9.75	31.08	0.31
	Thin	8.23	33.20	0.25
FM-300 (Co-cure bonding)	Thick	11.21	29.60	0.38
	Thin	10.66	32.40	0.33

[a]Calculated with debond location at one-sixth of adhesive thickness from the strap-adhesive interface.

G_T, G_I, and G_{II} were constant up to debond lengths of 115 mm and 65 mm in secondary and co-cure bonded systems, respectively. The constant values of strain energy release rates for all four sets of specimens are provided in Table III for a specified stress level.

RESULTS AND DISCUSSION

An attempt was made to determine if one of the components of strain energy release rate (G_T, G_I, or G_{II}) dominates the cyclic debonding. The measured debond growth rates were, therefore, correlated with each of the calculated strain energy release rates G_I, G_{II}, and G_T. These correlations are shown in Figures 10 and 11 for EC 3445 and FM-300, respectively. If one component of strain energy release rate had a dominant influence, it would correlate significantly better than the others when comparing the debond data from specimens with different G_I-to-G_{II} ratios.

An equation of the form

$$\frac{da}{dN} = c(G)^n \tag{1}$$

was fitted to the data in Figures 10 and 11 by using a least-squares regression analysis. The values of c and n, as well as the sum of errors, Σr^2, are shown in the figures. For each adhesive, the values of Σr^2 are about the same for the G_I, G_{II}, and G_T. However, the Σr^2 term is lowest for G_T, indicating that G_T provided a somewhat better correlation than either G_I or G_{II}. This suggests that debond growth rate is a function of the combined

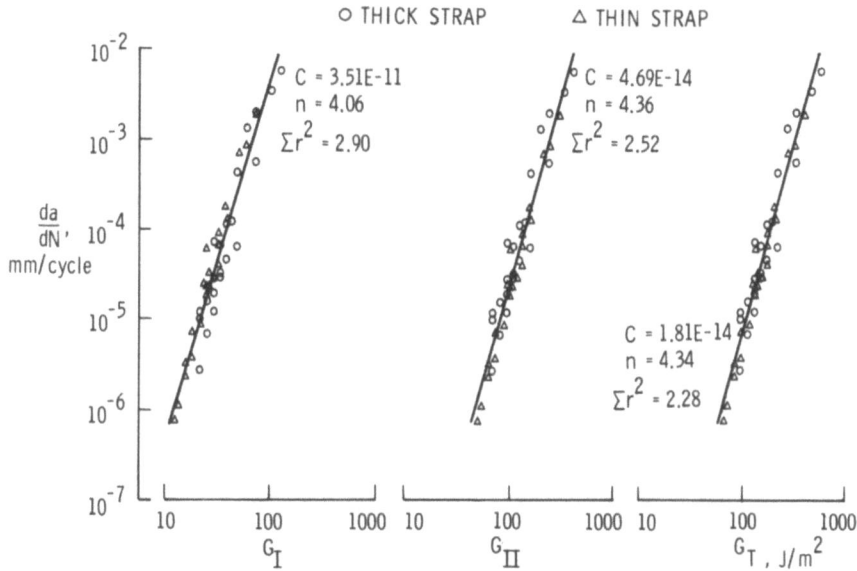

Figure 10. Relation between strain energy release rates and debond growth rate for EC 3445 adhesive.

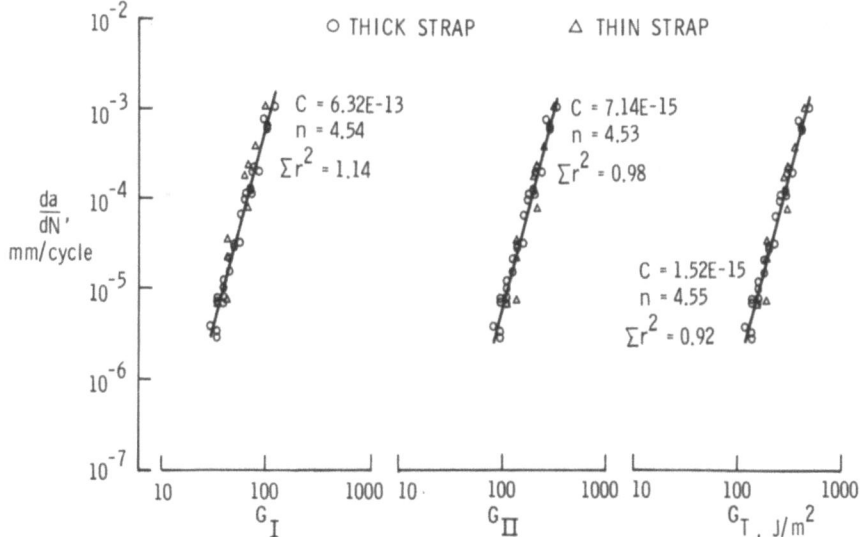

Figure 11. Relation between strain energy release rates and debond growth rate for FM-300 adhesive.

effects of G_I and G_{II}. However, the G_I-to-G_{II} ratios for the test specimens were all within the rather narrow range of 0.25 to 0.38. As a result, it is not surprising that G_I, G_{II}, and G_T all were reasonably successful in correlating the data in Figures 10 and 11. Furthermore, each figure shows that data from both specimen geometries are within an acceptable scatter band (similar to that observed in fatigue crack propagation in metals).[16] This indicates that specimen geometry did not influence the relationship between the debond growth rate and strain energy release rate. The relations between da/dN and G_T, for both adhesives, are compared in Figure 12. This figure shows that the debond rate with FM-300 is about 40 percent of that for EC 3445 for the same applied load and debond length.

Because of the log-log scale in Figure 12, the curves relating da/dN and G have slopes equal to the n term in Equation (1). The values of n found in this investigation ranged from 4 to 4.5. This is quite high compared to typical values of n for fatigue crack growth in aluminum and steel alloys that range from 1.5 to 3.[16] These steep slopes mean that a small change in applied load would cause a large change in debond growth rate. Thus, the debond propagation in adhesive joints is more sensitive to errors in design loads than are typical cracks in metallic structures. Because of these steep slopes, it may be difficult to design bonded joints for finite life. Minor design alterations or small analysis errors could cause a much shorter life than the design value. A viable alternative would involve an infinite-life approach. For this purpose, the no-growth threshold, G_{th}, (based on G_T for discussion purpose here) may be an important material property for bonded systems. If a 10^{-6} mm/cycle rate is arbitrarily assumed to be the no-growth threshold, the curves in Figure 12 show that G_{th} values for EC 3445 and FM-300 are 60.8 J/m^2 and 90.5 J/m^2, respectively. These G_{th} values are equivalent to applied stresses (based on nominal area of strap) of 100 MPa and 121 MPa for EC 3445 and FM-300, respectively.

The threshold strain energy release rate, G_{th}, appears to depend on the adhesive and not on the specimen geometry within the range of this study. As a result, a G_{th} value and an analytical method such as GAMNAS[8] could be employed during design to determine the maximum cyclic loads to obtain an infinite fatigue life for bonded structural components. An initial debond must be assumed to exist in order to calculate strain energy release rate. The size of the initial debond can be estimated from the anticipated manufacturing defects on NDI limitations. The maximum design load can then be determined to ensure that the applied G_T in the bondline is below G_{th} for the given adhesive. In the present study the values of G_{th} for EC 3445 and FM-300 were obtained under laboratory ambient conditions. However, as with other adhesive properties, it is anticipated that G_{th} will be influenced by the

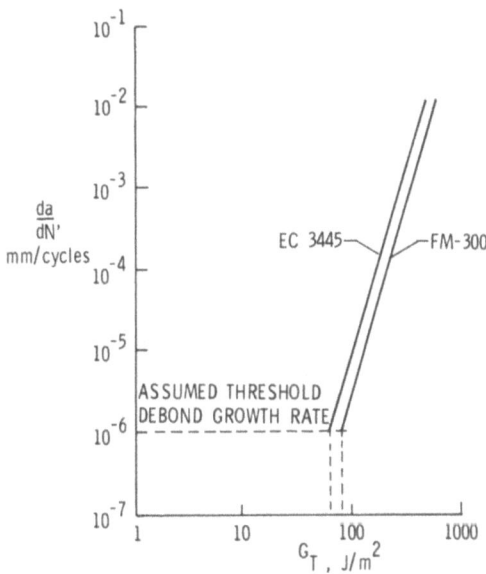

Figure 12. Comparison of debond growth rate between two adhesives.

environmental conditions. Thus, experimental data as presented in Figures 10 and 11 should be reproduced for the environment and cyclic frequency of interest.

CONCLUSIONS

A combined experimental and analytical study of the cracked-lap-shear specimen subjected to constant-amplitude cyclic loading was undertaken to analyze the fatigue behavior of simple composite-to-composite bonded joints. Two bonded systems were studied—graphite/epoxy adherends bonded with EC 3445 adhesive in a secondary bonding procedure and with FM-300 adhesive in a co-cure bonding procedure. With each bonded system, two specimen types were tested: (1) strap adherend of 16 plies bonded to lap adherend of 8 plies, and (2) strap adherend of 8 plies bonded to lap adherend of 16 plies. A finite-element analysis was conducted to calculate the strain energy release rates: G_I due to peel stress, G_{II} due to shear stress, and G_T due to both stresses. The present study led to the following conclusions:

1. Cracked-lap-shear specimens were found to provide consistent debond growth data. The cyclic debond growth rates were reproducible from one specimen to another within a scatter band comparable to that for crack growth in metals.

2. Cracked-lap-shear specimens failed predominately by cyclic debonding of adhesive, accompanied by some $0°$ fiber pull-off from the strap adherend.

3. The debond always grew in the region of the adhesive that had the highest mode I loading (peel stress or G_I). This indicates that G_I had a stronger influence than G_{II} on the debond location.

4. Debond growth rates correlated very well with G_I, G_{II}, and G_T. However, the specimen geometries tested in this study did not produce a wide enough range of mixed mode loading conditions to clearly establish the relative influences of strain energy release rates G_I, G_{II}, and G_T on debond growth rate.

5. The slopes of the debond growth rate curves are considerably higher than for crack growth rates in metals. This difference makes adhesive debond growth life much more sensitive to possible errors in design loads. Therefore, a safe-life (i.e., no debond initiation or growth) design criterion may be more acceptable. This safe-life approach would emphasize the threshold debond growth region.

ACKNOWLEDGMENTS

The first author is grateful for the support from the Army Structures Laboratory (RTL-AVRADCOM) through the In-house Laboratory Independent Research (ILIR) Program during the course of this study.

REFERENCES

1. F. L. Matthews, P. F. Kilty, and E. W. Godwin, "A Review of the Strength of Joints in Fibre-Reinforced Plastics, Part 2--Adhesively Bonded Joints," Composites, 13 (1), 29 (1982).
2. L. J. Hart-Smith, "Analysis and Design of Advanced Composite Bonded Joints," NASA CR-2218, National Aeronautics and Space Administration, Washington, DC (1974).
3. W. J. Renton and J. R. Vinson, "The Analysis and Design of Composite Material Bonded Joints Under Static and Fatigue Loadings," AFOSR-TR-72-1627, Air Force Office of Scientific Research, Washington, DC (1973).
4. G. L. Roderick, R. A. Everett, Jr., and J. H. Crews, Jr., in "Fatigue of Composite Materials," ASTM STP 569, pp. 295-306, American Society for Testing and Materials, Philadelphia, PA, 1975.
5. T. R. Brussat, S. T. Chiu, and S. Mostovoy, "Fracture Mechanics for Structural Adhesive Bonds," AFML-TR-163, Air Force Materials Laboratory, Wright-Patterson AFB, OH (1977).
6. J. Romanko and W. G. Knauss, "Fatigue Behavior of Adhesively Bonded Joints," Vol. I, AFWAL-TR-80-4037, Air Force Wright Aeronautical Laboratories, Wright-Patterson AFB, OH (1980).
7. R. A. Everett, Jr., "The Role of Peel Stresses in Cyclic Debonding," NASA TM-84504, National Aeronautics and Space Administration, Washington, DC (1982).
8. B. Dattaguru, R. A. Everett, Jr., J. D. Whitcomb, and W. S. Johnson, "Geometrically-Nonlinear Analysis of Adhesively Bonded Joints," presented at the 23rd AIAA/ASME/ASCE/AHS Structures, Structural Dynamics, and Materials Conference, New Orleans, LA, May 1982. Submitted to ASME Journal of Engineering Materials and Technology.
9. G. R. Alsmiller, Jr. and W. P. Anderson, "Advanced Composites Airframe Program - Preliminary Design," USAAVRADCOM-TR-80-D-37A, U.S. Army Aviation Research and Development Command, St. Louis, MO (1982).
10. E. J. Hughes and J. L. Rutherford, "Evaluation of Adhesives for Fuselage Bonding," Report No. KD-75-74, The Singer Company, Little Falls, NJ (1975).
11. K. N. Shivakumar and J. H. Crews, Jr., "Bolt Clampup Relaxation in a Graphite/Epoxy Laminate," NASA TM-83268, National Aeronautics and Space Administration, Washington, DC (1982).
12. L. J. Hart-Smith, "Effect of Adhesive Layer Edge Thickness on Strength of Adhesive-Bonded Joints," MDC J4675, McDonnell-Douglas Corporation, Long Beach, CA (1981).
13. I. S. Raju and J. H. Crews, Jr., "Interlaminar Stress Singularities at a Straight Edge in Composite Laminates," Computers and Structures, 14 (1-2), 21 (1981).
14. E. F. Rybicki and M. F. Kanninen, "A Finite Element Calculation of Stress Intensity Factors by a Modified Crack Closure Integral," Engineering Fracture Mechanics, 9 (4), 931 (1977).

15. F. A. Keimel, in "Kirk-Othmer: Encyclopedia of Chemical Technology," Vol. 1, 3rd ed., pp. 488-510, John Wiley & Sons, New York, NY, 1978.

16. "Damage Tolerant Design Handbook," Battelle Metals and Ceramics Information Center, Columbus, OH, 1972.

EFFECTS OF LOW CYCLE LOADING ON SHEAR STRESSED ADHESIVE BONDLINES

W. Althof

Deutsche Forschungsanstalt für Luft- und Raumfahrt (DFVLR), Institut für Strukturmechanik
D-3300 Braunschweig-Flughafen, West-Germany

Adhesively bonded thick adherend model joints were cyclically loaded on squarewave and sinusoidal modes. During the cycling loading, the real shear strains of the bondlines and, for the first time, the failure strains were quantitatively measured by use of a special shear strain extensometer. The analysis of bondline strains of a modern structural adhesive indicates dependence on time, shear stress and cycle frequency, and suggests a strengthening of the adhesive in the bondline under cyclic loads. A linear viscoelasticity can be expected only at very low bondline shear stresses. At higher stresses, residual strains build up to a critical level and exceeding this level causes failure of the bondline. Adhesive material properties obtained in the present study provide a guide to allowable shear stresses or strains to design cyclically loaded adhesively bonded structures.

INTRODUCTION

Adhesives are organic polymers and at mechanical loading the displacements of the polymers are time-dependent as a consequence of their viscoelastic properties. At a constant load, the displacements increase with time and the polymer creeps. On the other hand, the load required for a constant displacement decreases with time and the polymer relaxes. Corresponding time-dependent displacements are expected at cyclic loading of polymers and adhesives.

Low cyclic loading means periods of loading and unloading in a slow sequence. The loading and unloading can continually follow at constant or different frequencies, or after the loading and unloading the loads are held constant for a fixed time. Especially at the last cycle mode the viscoelasticity of the adhesives is of a great importance.

In the last few years the viscoelastic properties of adhesives have been intensively investigated.[1,2] Often viscoelastic properties were determined by tests on cured neat adhesive coupons. At DFVLR, however, adhesive shear properties are determined by measuring the shear strains of short bondlines between adherends of high stiffness.[3] It was proved that the stress-strain relationships of such short bondlines are representative of the adhesive sections in long bondlines.[4] This method was employed on static short time tests and also for determining viscoelastic properties for creep and relaxation.[5]

All the investigations cited established that adhesives deform linear-viscoelastically only at low stresses, and only under this condition the time-dependent stress-strain relationships can be described by simple mathematical formulae. It should be noted that the adhesive bondlines failed on exceeding a limit of shear strain.[5]

Adhesively bonded joints under sinusoidal cyclic loads were already investigated in 1963 and an accumulating of residual shear displacements was found.[6] Quantitative stress-strain-time relationships were not ascertained because methods to measure the shear displacements in bondlines of the used thin adherend joints were imperfect at that time. Investigations on cycled thick adherend lap joints were recently reported but again the bondline displacements were not measured quantitatively.[7] At DFVLR, however, this measurement is possible and new promising investigations on cyclically loaded joints were carried out dealing with the following:

Measurement of the time-dependent bondline shear strains during cycling at different times.

Analysis of the time-dependent shear strains with respect to a mathematical description.

Estimation of bondline shear stresses at which strains are negligible with respect to a long life of bondlines.

Comparison of the effects of squarewave and sinusoidal cycled bondlines.

Estimation of failure criteria of bondlines under cyclic loading.

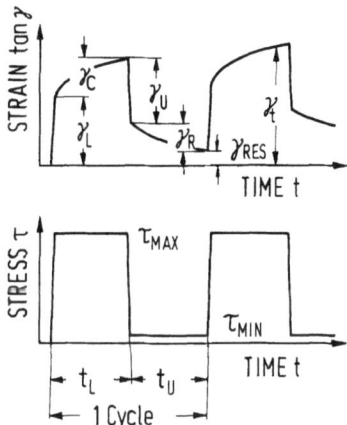

Figure 1. Squarewave stress/strain-time functions.

EXPERIMENTS

Cycle Modes, Specimens, Equipments

The tests were carried out mostly on a squarewave load-time function with parameters imitating simplified flight missions of an aircraft. The specimens were loaded and unloaded during ca. 10 seconds, the times under maximal loads (load periods) and under minimum loads (unload periods) lasted between 0.5 and 60 minutes. Figure 1 represents the stress-time and the strain-time functions. Table 1 shows the different times of load and unload periods, t_L and t_U, during squarewave cycling tests.

Table I. Times of Load and Unload Periods During Squarewave Cycling.

Load Period (minutes)	Unload Period (minutes)	Abbreviation	
60	15	60'/15'	"Boeing mission"
15	5	15'/5'	"Douglas mission"
4	2	4'/2'	
1	1	1'/1'	
0.5	0.5	0.5'/0.5'	

Thick adherend specimens bonded by a modern structural epoxy adhesive (Cyanamid "FM 73") were tested. The shear strain was measured by using a zero-gage extensometer developed at DFVLR. Figure 2 shows schematically the arrangement of the thick adherend specimen (model joint) with the clipped extensometer and the bondline shear displacement.

The selected loads caused the maximum bondline shear stresses τ_{max} = 20, 25, 30, 35, 37.5 and 40 Mpa, and minimum shear stresses $\tau_{min} = 0.01\tau_{max}$, respectively.

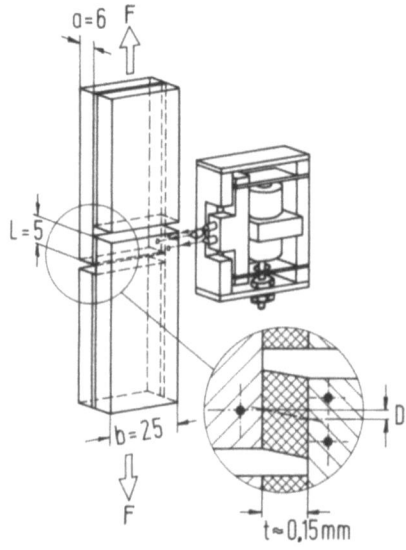

Figure 2. Thick adherend specimen (model joint) with clipped shear strain extensometer, schematically.

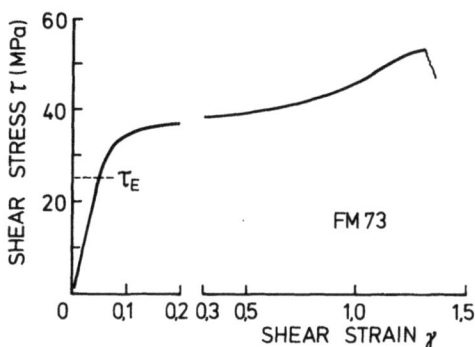

Figure 3. Basic shear stress-shear strain diagram of the tested structural adhesive "Cyanamid FM 73".

On sinusoidal cycling the frequencies f = 1 Hz and 10 Hz were used. The load amplitudes caused bondline shear stresses τ_{max} = 10, 15, 20, 25, 30, 32.5, 35 and 40 MPa, and τ_{min} = 0.1 τ_{max}, respectively.

For the adherend thickness, a = 6 mm (aluminum alloy bare 2024-T351), and the lap length, L = 5 mm, the resulting shear stress distribution in the loaded bondline is predominantly uniform. Therefore the shear stress can be calculated from

$$\tau = F/b \cdot L$$

and the shear strain is obtained as

$$\tan \gamma = D/t$$

where F = load, b = width, t = bondline thickness, D = shear displacement of the bondline which is the measured relative movement of the adherends. Figure 3 represents the basic shear stress-shear strain diagram of the adhesive tested as recorded on a static short time test.

At the cycling loading, the bondline shear strain was measured and recorded continually. On sinusoidal cycles, the failure strains were obtained by use of a transient recorder. The failure strain was defined by the maximum strain recorded at the last complete sinusoidal strain amplitude before the bondline fails.

The specimens were loaded on square wave cycles by a mandrel-driven tension tester. On sinusoidal cycles, a servo-hydraulic test equipment controlled by a microprocessor developed at DFVLR was used which also stored continually the measured strains.

RESULTS

Model Joint Squarewave Cycling

During the cycling of model joints the bondline shear strains were recorded from the start of cycling up to the failure of the bondline. Figure 4 shows a typical stress-strain relation during cycling.

The loops in Figure 5 recorded at the shear stresses τ = 35 and 25 Mpa during 60'/15'-, 15'/5'-, and 4'/2'- cycles characterize the stress- and time-dependent shear strain behavior of bondlines under low cyclic loading. The broken lines in the diagrams represent the static stress-strain curve of the adhesive tested. With the aid of these lines it is possible to discern whether the initial strain at the first loading is related to the elastic, the elastic-plastic or the plastic section of the total bondline shear strain.

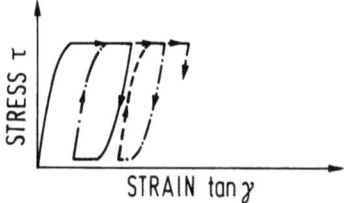

Figure 4. Shear stress-shear strain loops recorded on squarewave cycling, schematically.

EFFECTS OF LOW CYCLE LOADING

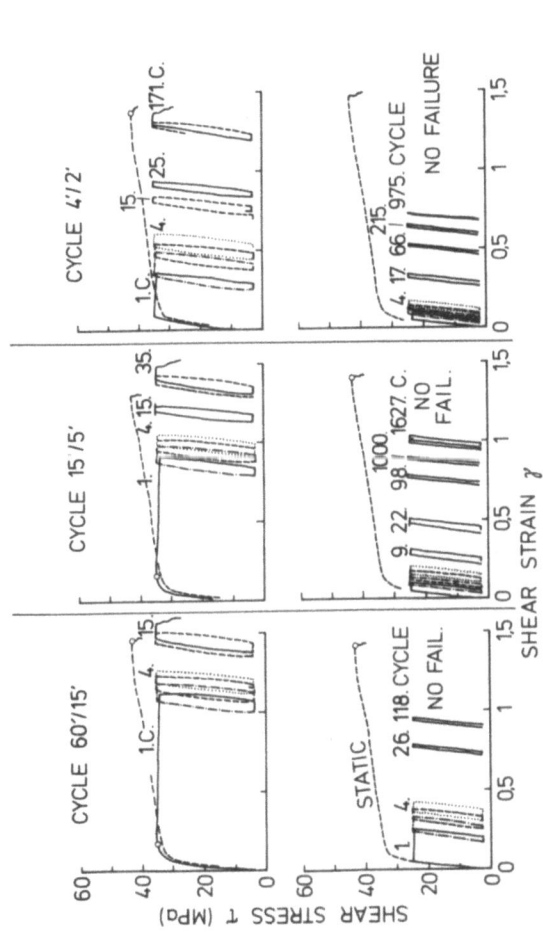

Figure 5. Shear stress–shear strain loops recorded at different shear stresses and different squarewave cycle periods.

The following characteristics can be deduced from Figure 5:

If in the first cycle the load already causes a plastic strain, the bondline survives only relatively few cycles.

Shear stresses below the quasi-elastic stress limit, τ_E, cause low strains; however, they increase continually and slowly.

The bondline creeps during the first cycles more than in the later ones.

The recovery during the unload periods is mostly lower than the creep during the load periods hence residual strains remain.

The residual strains build up to a critical strain at which the bondline fails.

The shorter the periods of loading and unloading, the more are the numbers of cycles up to failure.

Table II shows, for all the tests, the failure strains and the numbers of cycles at bondline failure. A comparison indicates:

The failure strain corresponds to the failure strain in static tests.

The number of cycles at failure increases with shorter cycle periods and lower shear stresses.

At shear stresses below τ = 20 MPa, the residual strains are low and failures are expected at very high numbers of cycles.

Table II. Test Results of FM 73-Adhesive Bondlines, Squarewave Cycled on Different Times of Cycle Periods and Different Shear Stresses.

Shear Stress max	Time of Load Period/Unload Period (Minutes)									
	60/15		15/5		4/2		1/1		.5/.5	
	N_f	γ_f	N_f	γ_f	N_f	γ_f	N_f	γ_f	N_f	γ_f
40 MPa	7	1.48	6	1.48	5	1.46	29	1.48	30	1.50
			5	1.45	25	1.60				
37.5	8	1.40	11	1.45	62	1.41	27	1.43	28	1.38
	5	1.43	13	1.52	26	1.50				
35	15	1.50	35	1.48	47	1.36	164	1.52	72	1.42
			35	1.53	56	1.40				
					171	1.45				
30	33	1.50	61	1.60	49	1.48	1765	n.f.= no failure		
	52	1.38								
25	120	n.f.	1626	n.f.	975	n.f.	1640	n.f.		
20	1270	n.f.								

N_f = numbers of cycle at failure, γ_f = shear strain at failure, On static loading the bondlines fail at γ_f = 1.50.

EFFECTS OF LOW CYCLE LOADING

Analysis of Bondline Strains

The stress-strain loops shown in Figure 5 assume that strains built up at loading are reversible, and that the time-dependent creep and recovery during the load and unload periods are of importance with respect to the life time of the bonded joints. For analysing the strain components additional specimens were tested at different shear stresses using cycle periods of 30 minutes. During cycling strains in the succesive periods were recorded.

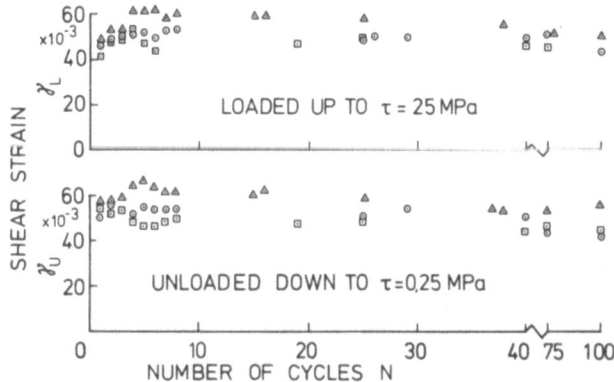

Figure 6. Shear strain after loading up to shear stress $\tau = 25$ MPa and after unloading down to $\tau = 0.25$ MPa vs. number of cycles.

In Figure 6 the measured strains at loading, γ_L, and unloading, γ_U, are compared for a stress level of $\tau = 25$ MPa. This diagram as well as other test results at lower stresses indicate that these strain components are approximately equivalent pointing to a quasi elastic strain behavior at loading up to the maximum stress and unloading down to the minimum stress.

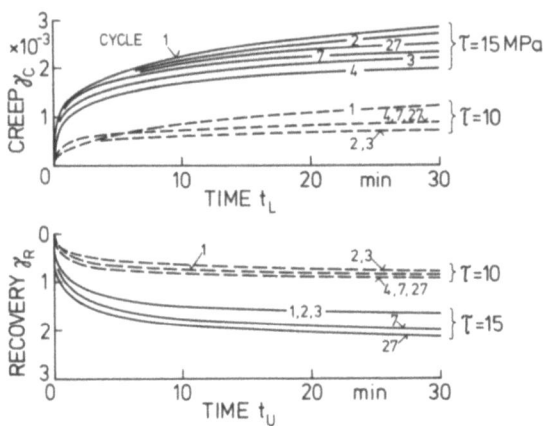

Figure 7. Creep and recovery vs. time at low shear stresses during the load and unload periods on 30'/30'- squarewave cycles.

For different stress levels the creep strain, γ_C, and the recovery strain, γ_R, at successive cycles are presented in the next figures. Figure 7 shows these strain components for the stresses τ_{max} = 10 and 15 MPa. The creep and the recovery are approximately equivalent indicating a linear-viscoelasticity at these low stresses. The time-dependence of strains at low stresses is often better described by a power function than by the use of a spring-dashpot-model:[5]

$$\gamma_{C,R} = a \cdot t^n$$

For τ = 10 and 15 MPa stresses, the formula is applicable with a = 0.00055, n = 0.19, and a = 0.0011, n = 0.23, resp.

The creep and the recovery at τ = 20 and 25 MPa are shown in Figure 8. It can be seen that already at 20 MPa the creep is higher than the recovery. The difference is more remarkable at τ = 25 MPa. Moreover, during the initial load periods the creep increases, later it decreases with increasing number of cycles. The recovery behaves similar to the creep. However, after about 100 cycles creep and recovery are approximately equivalent.

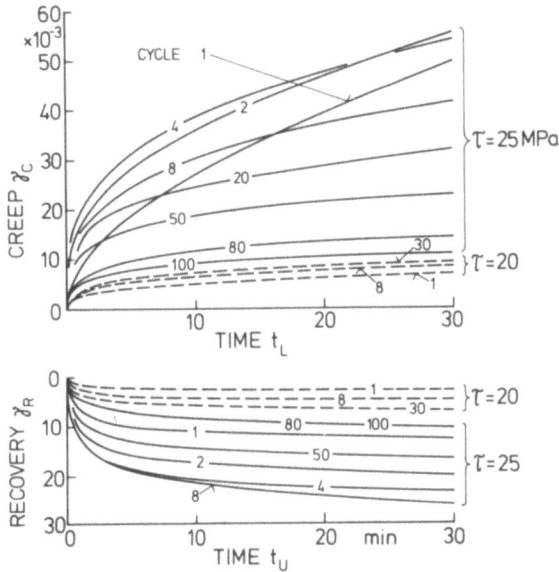

Figure 8. Creep and recovery vs. time at medium shear stresses during the load and unload periods on 30'/30' squarewave cycles.

The strain behavior described is typical of polymers which strengthen during longtime loading and then deform quasielastically. Such a strengthening can be explained by a stretchability of the initial amorphous arrangement of the molecules in the polymer caused by the longtime loading.

It is evident that for a cycle-dependent strengthening of the adhesive the time-dependence of the strains cannot be described by simple mathematical formulae. It was often recommended to ignore the cycles up to the finished strengthening and thereafter describe the viscoelasticity mathematically. However, this method is doubtful as the strains during the strengthening may be high in relation to the failure strain.

Model Joint Sinusoidal Cycling

As previously mentioned, for the first time, the bondline failure strain was measured on low and fast sinusoidal cycles. As an example, Figure 9 shows some strain cycles recorded just before

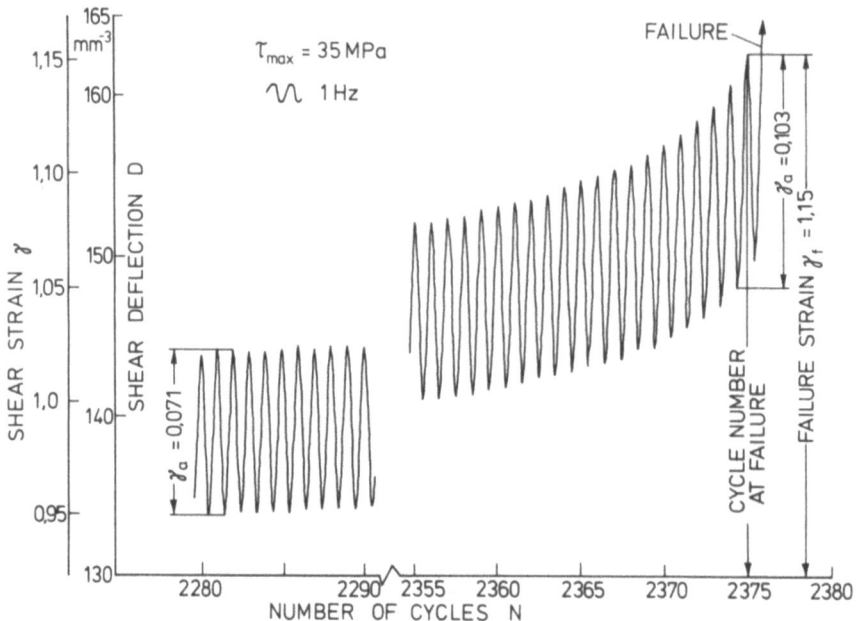

Figure 9. Shear strain vs. numbers of cycles during 1 Hz – sinusoidal cycling to determine the failure strain.

the bondline failed. On the left the first 10 cycles of the recorded last 100 cycles are to be seen, on the right the last 20 cycles are shown with the last complete strain amplitude from which the bondline failure strain was determined.

Figure 10 shows the envelopes of 1 Hz – shear strain amplitudes recorded at different shear stresses. Although the semilogarithmic scale distorts the diagram, the following can be inferred from the test results:

At low shear stresses ($\tau > 25$ MPa) the bondline strains are approximately constant up to a high number of cycles. The bondline deforms quasi-elastically. At higher shear stresses the strains increase and the characteristic of the strain behavior is similar to

EFFECTS OF LOW CYCLE LOADING

the static creep.[5] The creep is caused by the effective mean shear stress. During the initial period of cycling the creep rate is high, later it is low and a short time before the bondline fails the rate increases again.

The amplitudes of the strain are constant for a long time and increase just before the bondline fails. With increasing stresses, the number of cycles at failure decreases. Furthermore, the failure strain decreases with decreasing shear stresses and increasing number of cycles. This phenomenon was observed for the first time.

During some sinusoidal tests the stress-strain loops were also recorded, an example is presented in Figure 11. The loops recorded at different shear stresses indicate following more important facts.

At high stresses with an initial elastic-plastic bondline displacement, the strain does not reach its maximum value at the maximum stress. Moreover, the loop is not closed after the unloading.

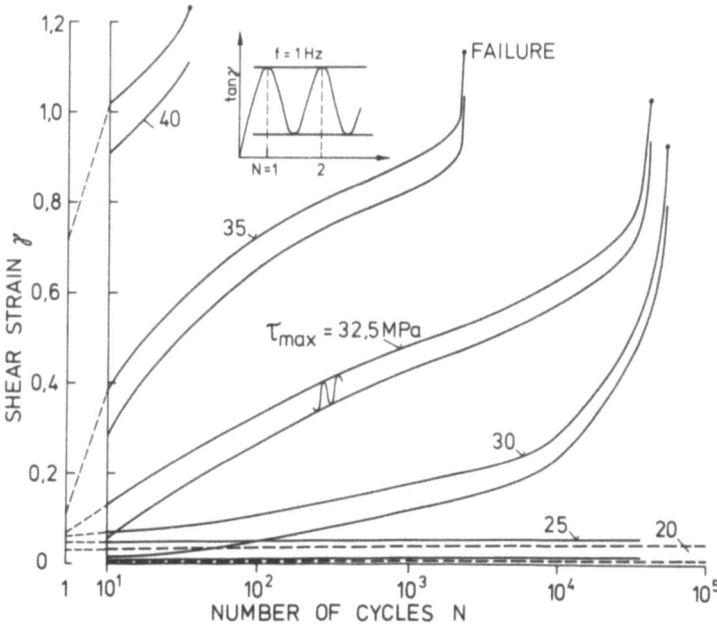

Figure 10. Envelopes of shear strain amplitudes measured on 1 Hz-sinusoidal cycling at different shear stresses.

This means the strain is not recovered. Such a time-delayed nonreversible strain behavior with a time lag in phase of stress and strain is typical of nonlinear-visoelastic materials such as polymers. With an increase in cycles the lag in phase disappears and the bondline deforms quasi-elastically up to a short time before the failure. In this short time nonreversible strains appear again.

The change of the strain behavior from nonlinear-viscoelastic to elastic is probably due to the effect of the strengthening in the adhesive similar to that at squarewave cyling. The longer the bondline is cyclically loaded the more the adhesive strengthens. After the strengthening the adhesive becomes brittle. This effect probably causes the observed decrease of the failure strain during sinusoidal loading at medium shear stresses and high number of cycles, see Figure 10. The results of tests at the frequency of 10 Hz seem to confirm this hypothesis, see Table III which includes all the measured failure strains and numbers of cycles up to failure at different shear stresses and cycle frequencies.

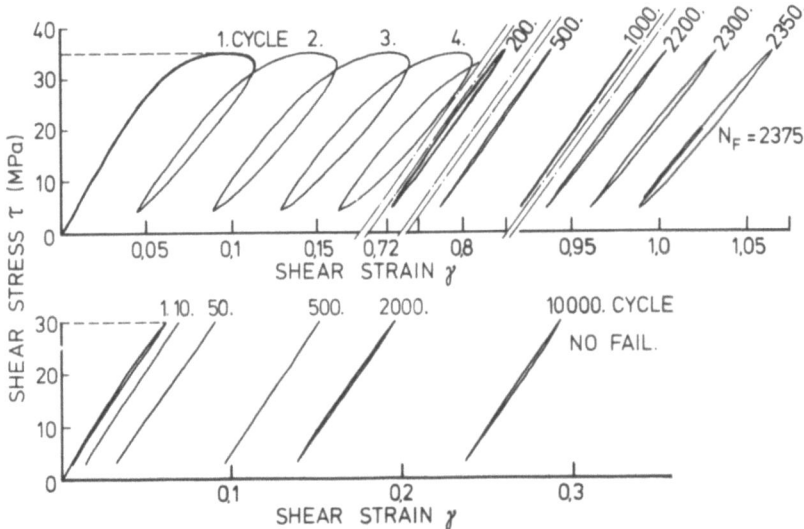

Figure 11. Shear stress-shear strain loops on sinusoidal cycling (1 Hz) at different shear stresses.

Table III. Test Results of FM-73-Adhesive Bondlines under Sinusoidal Cycling at Different Frequencies and Different Shear Stresses.

Shear Stress τ_{max}	f = 1 Hz		f = 10 Hz	
	N_f	γ_f	N_f	γ_f
40 MPa	3.3×10^1	1.24		
35	2.4×10^3	1.13	2.5×10^4	0.86
	7.2×10^3	1.13	2.9×10^4	0.89
	7.3×10^3	1.11	2.1×10^4	0.87
32.5	1.2×10^4	1.12	1.0×10^5	0.58
	1.3×10^4	1.06	1.5×10^5	0.89
	4.0×10^4	1.03		
30	5.0×10^4	0.93	1.9×10^5	0.80
			2.1×10^5	0.70
			3.2×10^5	0.68
25	1.0×10^5	no failure	1.6×10^6	0.73
20	2.8×10^5	no failure		

N_f = numbers of cycle at failure, γ_f = shear strain at failure

Post-Test Examination of Bondlines

The failure surfaces of the bondlines failed on squarewave and sinusoidal cycling did not differ significantly from the faces of bondlines failed after static loading or creep. The bondlines failed in the middle of the adhesive layer. However, a "finger nail crack front" on the end of the lap was not identified. It has been reported that such crack front extend in the bondline more and more on succesive cycles.

Actual Lap Joint Squarewave Cycling

The test results described were obtained on specimens with short bondlines. It is assumed that they are representative of adhesive sections in longer bondlines. To prove results concerning shear stresses causing negligible residual strains, standardized thin adherend lap joints were cyclically loaded on squarewave mode with load and unload periods of 1 hour. During cycling the bondline strain was measured in the middle of the lap length in the previous manner by the use of a shear strain extensometer with measuring points placed at shorter distances than in Figure 2.

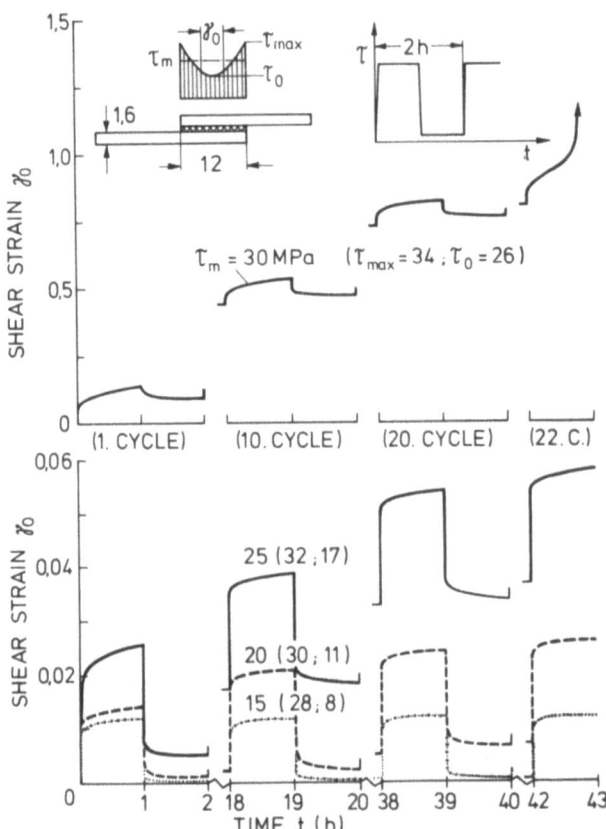

Figure 12. Shear strain vs. time for a bondline section in the middle of an actual lap joint on squarewave cycling at different mean shear stresses.

The joints were loaded at the mean shear stresses τ_m = 15, 20, 25, and 30 MPa. The respective maximum shear stresses at the end of the overlap were calculated to be τ_{max} = 28, 30, 32, and 34 MPa; the minimum stresses in the middle of the lap to be τ_0 = 8, 11, 17, and 26 MPa, respectively.[4]

The results are presented in Figure 12. At low mean shear stresses the bondline deforms quasi-elastically, at medium mean stresses the strains are negligible because the shear stresses in the middle of the lap length are lower than the estimated limit of linear-viscoelasticity of the adhesive tested ($\tau \approx 15$ MPa).

At high mean stresses and stresses in the middle, which are higher than the mentioned limit, the bondline deforms nonreversibly and residual strains remain.

At very high mean stresses and stresses in the middle, which exceed considerably the critical limit, the number of cycles at failure is low.

The example confirms the requirement that in long-time loaded bondlines, at the bondline section with the lowest stress, stresses or strains of the adhesive should not exceed the limit of the linear viscoelasticity. Otherwise, high nonreversible bondline strains must be expected.

CONCLUSIONS

Adhesively bonded thick adherend metal joints (model joints) were low cyclically loaded on squarewave and sinusoidal modes with different period times. During cycling the bondline shear strain as well as the failure strain were quantitatively measured. The tests confirm the qualitative wellknown time-dependent strain behavior of adhesives. Also the following new information was obtained with regard to engineering design of structural adhesive bonded joints.

Only at low stresses the adhesives deform linear-viscoelastically and the time-dependence of the strains can be described mathematically. For the adhesive tested the stress level is about 15 MPa. At medium stresses (τ = 15 - 25 MPa) the strains contain plastic portions. For cycles of short periods, e.g. 1 Hz-sinusoidal cycles, the time is not adequate to cause disadvantageous residual strains. However, on squarewave cycles with long periods of constant loads residual strains build up in the first few cycles; later the bondline deforms quasi-elastically. This change of the strain behavior may be a result of the strengthening of the adhesive. At high stresses ($\tau > 25$ MPa), more plastic strains appear which are not reversible on squarewave cycles. The residual strains build up to a critical strain at which the bondlines fail. The failure strain corresponds to the failure strain at static short-time loading or at creep. Therefore the exceeding of the strain capacity may be a failure criterion for bondlines under squarewave cycling.

The exceeding of the strain capacity may also be a failure criterion for bondlines under sinusoidal cycling on low and medium frequencies. This hypothesis is a consequence of the observed shear strain behavior at medium and high shear stresses. The constant load amplitudes cause constant strain amplitudes, but the effective mean load causes an increase of the total strain which is analogous to the strain behavior at static creep tests under sustained loads. In sight of these facts, the failure criterion cannot be a fatigue failure caused by a progressive crack front in the bondline as observed in other investigations.[7] A progressive crack front, of course, must continually reduce the bond area. At constant load amplitudes the stress will also continually increase and therefore the strain amplitudes too.

The observed strain behavior is strictly valid only for short bondlines with a homogenous shear stress distribution in the bondline. In bonded structures, however, the stresses in the bondlines are mostly shear stresses combined with normal stresses. Therefore the adhesive strain behavior is very complex and also affected by its dependence on time and strengthening. However, at the recommended low stresses the computation methods of the structural mechanics are applicable, providing a linear-elasticity or a linear-viscoelasticity of stresses and strains. Furthermore, at low stresses, the strain capacity of the adhesive is not fully exploited, so that there are strain reserves in the case of local bondline stress concentrations.

The tests described and the measuring method allow determination of adhesive material properties for shear loading required in engineering design of bonded joints. Most probably the observed effects of the shear strain behavior are transferable to tension loaded joints.

Finally, the allowable shear stress requirements determined were successfully proved to be applicable to actual bondlines by tests of cyclically loaded thin adherend lap joints.

ACKNOWLEDGEMENT

Grateful acknowledgement is made to Messrs. G. Klinger and G. Neumann who carefully carried out the experiments.

REFERENCES

1. J. Romanko, in "AGARD-Lecture, Ser. No. 102," p. 4-1, March 1979.
2. V. H. Kenner, W. G. Knauss, and H. Chai, Experimental Mechanics, $\underline{22}$, 75 (Feb. 1982).
3. W. Althof and W. Brockmann, in "Bicentennial of Material Progress, SAMPE Symp. Proc. Vol. 21," p. 581, April 1976.
4. W. Althof, in "New Horizons - Material and Processes for the Eighties, SAMPE NCTS 11," p. 309, Nov. 1979.
5. W. Althof, J. Reinforced Plastics and Composites, $\underline{1}$, 29 (Jan. 1982).
6. A. Matting and G. Ulmer, Kautschuk und Gummi - Kunststoffe, $\underline{16}$, 334 (1963).
7. J. Romanko and W. G. Knauss, in "Developments in Adhesives - 2," A. J. Kinloch, Editor, p. 173, Applied Science Publishers, London, 1981.

EFFECT OF SCRIM CLOTH ON ADHESIVELY BONDED JOINTS

E.C. Francis and D. Gutierrez-Lemini

United Technologies
Chemical Systems Division
Sunnyvale, California 94086

This report discusses the effect of the scrim cloth on the fracture behavior of adhesively bonded joints. The thick-adherend model joint specimen was analyzed with the TEXGAP finite element computer code. Three situations were considered: (1) uniform adhesive without inclusion (scrim), (2) adhesive with bonded inclusions, and (3) adhesive with debonded inclusions. The material properties and specimen geometry were those of a model joint previously analyzed without inclusions by General Dynamics/Fort Worth Division (GD/FWD).

Significant differences in the K_I and K_{II} values were noted between the joints with bonded and unbonded inclusions. The results gave insight into why the scrim weakens the bond and provides a crack-growth path. Scanning electron microscope (SEM) photographs were obtained which show that the scrim cloth is either poorly bonded or not chemically bonded to the FM-73 epoxy.

INTRODUCTION

The use of adhesives for bonding aircraft components has increased greatly in the last two decades. In fact, the high strength-to-weight ratios of adhesively bonded joints has led recently to their application in primary fuselage structures[1] and thus to the development of suitable criteria for evaluating the integrity and structural durability of adhesively bonded joints[2].

To control the adhesive thickness in a bonded joint, a scrim cloth is generally utilized in the adhesive. Published analyses have neglected the presence of the scrim or mat carrier (references 2, 3 and 4), which, as described in the following section, may weaken a bonded joint. To investigate the structural influence of the scrim cloth on the fracture behavior of adhesively bonded joints, we idealized the fibers of the scrim as cylindrical Dacron inclusions located near the tip of a crack at an adhesive-adherend interface. The results obtained are supported by experiment, as described below.

Experimentally Observed Effect of the Cloth Carrier

There is evidence that the scrim or mat cloth carrier influences the response and fracture behavior of structural adhesives. The cloth in FM-73 is most likely not bonded to the epoxy resin, or if it is bonded, the bond is very weak, as inferred from Figure 1. This figure shows SEM photographs of fracture surfaces of FM-73 adhesive that were taken from slow static tests of thick-adherend model joints. The photographs show a smooth surface between the cloth and the epoxy resin. One would also suspect that the scrim or mat provides a pathway for preferential moisture ingression. Consequently, the mat may represent a built-in defect. The neat adhesive is, in fact, similar to a foam, even though reported values of Poisson's ratio ranging from 0.32 to 0.40 (Reference 5) are much higher than those typical of foams (0.10 to 0.16).

Most response and fracture tests on the neat adhesive are performed so that the scrim either reinforces or weakens the epoxy and the fracture plane extends through the Dacron mat. In bonded-joint fracture tests, on the other hand, the fracture plane is parallel to the mat. It is not clear how one can meaningfully relate these two fracture processes, or even the fracture data. Fracture data from the neat-neat adhesive (FM-73U) may perhaps be more relevant to fracture in a bonded joint.

EFFECT OF SCRIM CLOTH

Figure 1. SEM photographs of fracture surfaces of FM-73 adhesive.

Recently, GD/FWD[2] found that fatigue life increased by a factor of four (25,000 cycles to 100,000 cycles) and that the aluminum adherend failed, rather than the adhesive, in tests of thick-adherend model joint specimens bonded with FM-73U. These results suggest again that the scrim may indeed be detrimental to the structural durability of bonded joints.

Finite Element Analysis of the Model Joint

The thick-adherend model joint shown in Figure 2 was analyzed by us at Chemical Systems Division. The corresponding finite element idealization is depicted in Figures 3A and 3B. The geometry, loading, and material properties of the structure correspond to those used previously by the GD/FWD (see, e.g., QPR 5 of Reference 2), except for the fact that we included only one crack instead of the two antisymmetrically placed cracks present in the GD/FWD model.

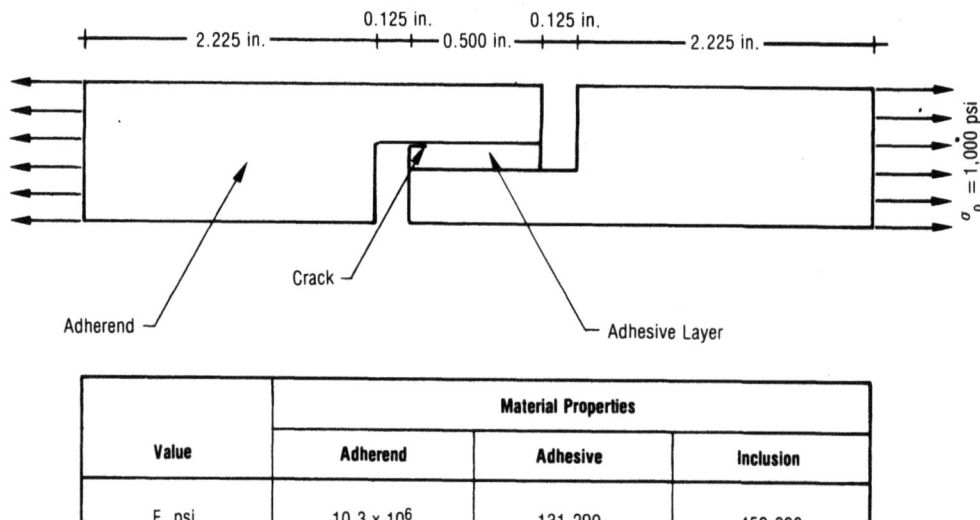

Figure 2. Geometry, loading, and material properties used in the analyses of the model joint.

The behavior of the model joint with a single crack near the adhesive-adherend interface was examined under the following conditions:

1. By considering a homogeneous adhesive without inclusions

2. By placing a couple of inclusions of modulus 450,000 psi and Poisson's ratio 0.32, bonded to the homogeneous adhesive

3. By assuming these inclusions to have a very low modulus (i.e., 10 psi).

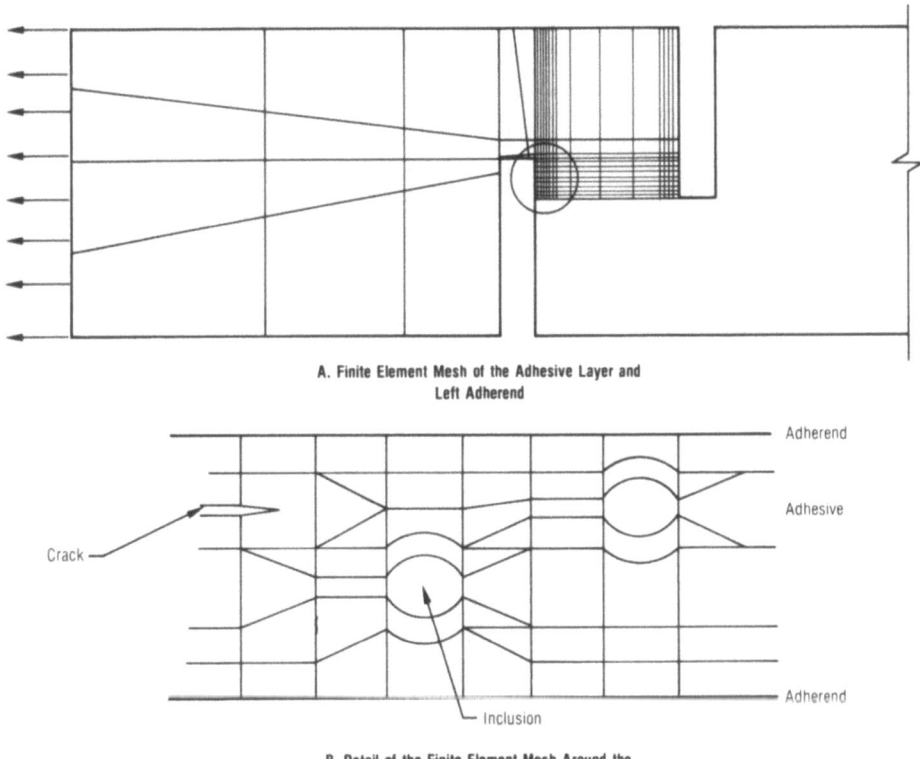

Figure 3. Finite element model of the bonded joint.

The region around the crack tip was modeled with the crack element of TEXGAP-2.5, which is based on a displacement hybrid formulation and yields the stress intensity factors directly. General Dynamics, on the otherhand, used vectorial crack opening-displacement data to obtain the stress intensities. The results of these analyses are summarized in Table I.

While the model employed is a crude approximation to the scrim cloth, significant differences were noted in the magnitude of K_I if the scrim is unbonded or perhaps only weakly bonded. This observation supports our contention of the importance of the scrim and gives some insight into why the scrim may weaken the bond and provide a crack-growth path. The discrepancy with GD/FWD results for K_{II} is not known, but it may be related to the fact that we used a singular element, whereas GD/FWD did not.

Table I. Stress Intensity Factors for the Model Joint with and Without Inclusions, and with a 10-Mil Crack

Value	Without Inclusion		With Inclusion (CSD)	Unbonded Inclusion (CSD)
	CSD	GD/FW		
K_I, psi $\sqrt{in.}$	9.2	8.0	8.9	18.5
K_{II}, psi $\sqrt{in.}$	45.2	66.5	45.4	39.9

DISCUSSION

The analyses reported herein, as well as those carried out by GD/FWD, bring out several important aspects concerning the behavior of adhesively bonded joints:

1. The thickness of the adhesive layer is so small (only a few mils) that the size of the yield zone may be of critical importance. If this is the case, the stress intensity factors may not adequately characterize the response fields at the crack tip. Tests at GD/FWD on adhesive layers containing a scrim show whitening around crack tips on a scale that is comparable to the thickness, thus suggesting that a stress field exists associated with the stress intensity factor concept used in the analyses.

2. The adhesive itself is probably not an isotropic material, as was assumed in the analyses, since the mat carrier introduces an in-plane reinforcement that is quite different from the through-the-thickness reinforcement. Even without the scrim, the material is probably orthotropic because of the thin layer; the boundaries introduce orientation effects in the molecules and give a directional effect to the curing process.

3. Structural adhesives may respond in a complicated, nonlinear viscoelastic or viscoplastic manner which depends on the history of loading, temperature, moisture, and state of "damage" induced by these environments. Thus, their idealization as linear elastic materials may very well be an oversimplified assumption.

4. The mat carrier is clearly not a set of isolated inclusions in a matrix of neat-neat adhesive, but rather a mesh of fibers interacting in a complicated manner.

In view of these peculiar characteristics of structural adhesives, the results obtained so far for adhesively bonded joints must be regarded carefully in practical design.

CONCLUSIONS

The analyses reported herein show, qualitatively, the detrimental effect of the scrim cloth when it is not bonded to the surrounding adhesive. The results also show that when the scrim cloth is strongly bonded to the adhesive, the system behaves essentially as if the carrier were not present. Experimental evidence indicates the existence of weak bonds between the mat carrier and the adhesive, thus increasing the possibility of crack propagation in the system.

REFERENCES

1. E. W. Thrall, Jr., "PABST Program Test Results", Adhesives Age, pp. 22-23, Oct. 1979.
2. J. Romanko, K.M. Liechti and W.G. Knauss, "Integrated Methodology for Adhesive Bonded Joint Life Predictions", AFWAL-TR-82-4139, Nov. 1982*.
3. K. J. Sen and R. M. Jones, "Stresses in Double-Lap Joints Bonded with a Viscoelastic Adhesive: Part I", AIAA J., $\underline{18}$, No. 10, 1237 (1980).
4. T. R. Brussat, S. T. Chiu and S. Mostovoy, "Fracture Mechanics for Structural Adhesive Bonds - Final Report", Technical Report AFML-TR-77-163, Lockheed - California Company, October 1977*.
5. J. Romanko and W. G. Knauss, "Fatigue Behavior of Adhesively Bonded Joints", Final Technical Report, AFWAL-TR-80-4037, April 1980*.

*Reports available from Air Force Wright Aeronautical Laboratory - Material Laboratory - Air Force Command - Wright Patterson AFB, Ohio 45433.

Part VI
Fracture Aspects

MECHANICAL MEASUREMENT OF INTERATOMIC BONDING ENERGIES AT INTERFACES

E.H. Andrews

Materials Department
Queen Mary College
Mile End Road, London, E1 4NS, England

Until recently it has not been possible to relate, quantitatively, the mechanical strength of adhesive joints to the surface chemistry of the system. This situation has been changed dramatically by use of the Generalized Theory of Fracture Mechanics (GFM) which states that the measured work of debonding, θ, per unit area is given by

$$\theta = \theta_o \, \Phi \, (\dot{c}, T, \varepsilon)$$

where θ_o is the energy to break interatomic bonds across the fracture plane and Φ is a loss function dependent on crack velocity, temperature and strain. Using this approach it is possible to extract the value of θ_o from direct mechanical measurements of θ. Results are presented for rubberlike and structural adhesives and for the effects of water environments on θ_o.

INTRODUCTION

While it is obvious that the strength of interatomic bonds across the interface must affect the mechanical strength of an adhesive joint, a quantitative correlation between these parameters has proved difficult to establish. Thus Dahlquist[1] found a correlation between thermodynamic work of adhesion (w_A) and joint strength in peel tests but Kaelble[2] failed to do so in his studies of an alkyl acrylate copolymer. Mittal[3,4] has reviewed the relationship of w_A to joint strength in various systems. Even when correlations are obtained, of course, the absolute magnitude of joint strength far exceeds w_A. A second problem is that the atomic interactions across an interface are difficult to determine quantitatively, even measurements of w_A depending heavily upon theoretical assumptions. Thirdly, most mechanical tests of adhesive strength determine the strength of the joint rather than of the interface, and thus depend upon joint configuration, method of loading and so on.

These problems have been largely overcome in recent studies by the author and his co-workers at Queen Mary College and the purpose of this paper is to review the results of these studies and assess the progress made to date.

THEORETICAL BASIS OF THE STUDIES

The key to all the work summarized below lies in a new formulation of fracture mechanics called 'Generalized fracture mechanics' (GFM)[5]. The use of fracture mechanics in adhesive testing is well established and has the recognized capability of eliminating the effects of joint geometry or configuration. Essentially, fracture mechanics measures the resistance of a glue-line or adhesive interface to the propagation of failure or de-bonding. GFM is essentially the same in this respect, i.e. it measures the total energy required to produce unit area of fracture or debonding. This "failure energy" is denoted by $2\mathcal{T}$ (two identical surfaces) for cohesive fracture or θ (one fracture plane) for adhesive or interfacial failure. Where GFM goes beyond normal fracture mechanics is in its ability to separate the total energy requirement into two factors; thus for cohesive fracture

$$2\mathcal{T} = 2\mathcal{T}_0 \, \Phi \, (\dot{c}, T, \varepsilon_0) \qquad (1)$$

where $\mathcal{T}_0$ is the intrinsic failure energy or the energy actually used to create unit area of new surface and Φ is the 'loss function' dependent upon crack velocity, temperature and the overall strain level in the specimen. In the simplest cast, $2\mathcal{T}_0$ corresponds to the energy needed to break unit area of interatomic bonds across the fracture plane. The loss function has the form[5]

$$\Phi = \left\{ \frac{k_1}{k_1 - \Sigma \beta g \delta v} \right\} \quad (2)$$

where v is the reduced volume, k_1 and g are explicit distribution functions of input energy density in the specimen and β is the hysteresis ratio of the solid (the fractional energy loss in a strain cycle); β is generally dependent on crack velocity, temperature and strain intensity and varies between zero for a perfectly elastic material and unity for a perfectly plastic one. Clearly $\Phi = 1$ for perfectly elastic behaviour and (it turns out) achieves an infinite value before $\beta = 1$.

The same equation can be applied to adhesive failure (see Figure 1). If we consider an infinite block of adhesive adhering to an infinite block of a <u>rigid</u> substrate, we obtain, for crack propagation along the interface

$$\theta = \theta_o \Phi \quad (3)$$

where Φ is the loss function for the adhesive only because the substrate, being rigid, does not deform; θ_o is now the energy to break unit area of interatomic bonds across the interface. If the interface is secondarily bonded, θ_o is identical to the thermodynamic

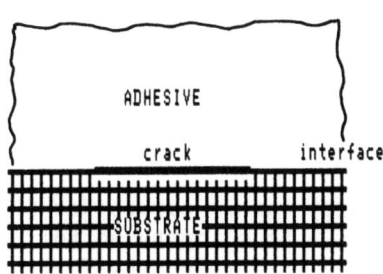

Figure 1. Crack at an adhesive-substrate interface.

work of adhesion. Otherwise, θ_o contains contributions from the severance of primary bonds. If failure is not interfacial but occurs through the adhesive or substrate, the cohesive failure Equation (1) can be used. Indeed, for mixed-mode failure Equations (1) and (3) can be combined mathematically to give a weighted average failure energy.

APPLICATION OF THEORY

Taking logarithms we have,

$$\text{Cohesive: } \log 2\mathcal{T} = \log 2\mathcal{T}_0 + \log \Phi (\dot{c}, T, \varepsilon_o) \qquad (4)$$

$$\text{Adhesive: } \log \theta = \log \theta_o + \log \Phi (\dot{c}, T, \varepsilon_o) \qquad (5)$$

For a given adhesive, a plot of $\log 2\mathcal{T}$ or $\log \theta$ against any parameter (e.g. crack velocity) on which Φ depends should thus produce a family of parallel curves displaced vertically by the factor $\log (2\mathcal{T}_0/\theta_{on})$ where θ_{on} is the θ_o value for the n th

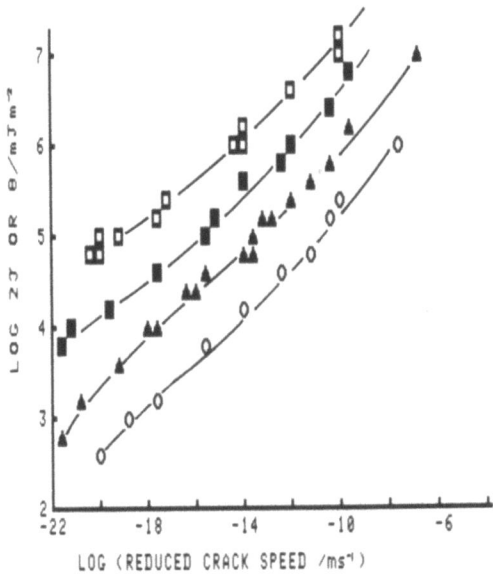

Figure 2. Failure energy master curves for SBR adhesive bonded to various substrate films. From top to bottom, cohesive failure energy for SBR; adhesive failure energy for the following substrates: PET; PCTFE; Nylon 11; fluorinated ethylene-propylene copolymer. Curves are superimposable by vertical shift.

different substrate. This is regardless of the particular shape of the curve, so that is is not necessary to know in advance how Φ behaves. Provided such a family of curves is obtained, they may be superimposed by vertical shift to form a master curve, and the vertical shifts used to determine θ_o from a known value of $\mathcal{J}_o$. That is, provided the cohesive bonding term $\mathcal{J}_o$ can be determined or calculated by other means (as it generally can[6,7]), the interfacial bonding energy θ_o may be precisely deduced.

ELASTOMERIC ADHESIVES

The overall failure energy $2\mathcal{J}$ or θ can be measured experimentally in terms of the applied constraints and specimen geometry as has been fully documented elsewhere[8]. For a single-edge notch specimen, for example

$$2\mathcal{J} \text{ or } \theta = k_1 c W_{o\,crit} \qquad (6)$$

where $W_{o\,crit}$ is the input energy density at which the crack propagates and c is the crack length, whilst for a 90° peel test,

$$2\mathcal{J} \text{ or } \theta = P/b \qquad (7)$$

where P is the peel force and b the breadth of the peeled strip. Other test configurations (with their own formulae) may be employed for this determination. Figure 2 shows the definitive log $2\mathcal{J}$ or log θ plot against crack velocity for an SBR adhesive bonded to a variety of polymer film substrates. The substrates were made effectively rigid by sticking a thin film of the substrate to a metal base-block. For SBR, also, the Williams, Landel, Ferry transform can be applied and the plot made against reduced velocity, incorporating data recorded at different temperatures.

Table I. Values of θ_o and thermodynamic work of adhesion, w_A, for styrene-butadiene rubber bonded to various plastic substrates.

Substrate	θ_o/mJ m^{-2}	w_A/mJ m^{-2}
Fluorinated ethylene propylene copolymer A	21.9	48.4
Polychlortrifluorethylene	74.9	62.5
Nylon 11	70.8	71.4
Polyethylene terephthalate	79.4	72.3
Fluorinated ethylene propylene copolymer B	68.5	56.8

The expected set of parallel curves is obtained. The value of $\mathcal{J}_o$ can be measured in limiting fatigue tests[9,10] or else calculated[11] and θ_o derived for each substrate. Tables I and II show the results, comparing θ_o with w_A, the latter being determined from contact angle measurements[6]. For substrates where debonding was purely adhesive, θ_o corresponds closely with w_A (see Table I), but where primary bonds were present (Table II), $\theta_o \gg w_A$ and failure is partly cohesive[12].

Table II. Values of θ_o and w_A for styrene-butadiene rubber bonded to fluorinated ethylene propylene copolymer film etched with sodium naphthalene to provide primary bonding sites on the substrate.

Etching time (s)	θ_o/mJ m^{-2}	w_A/mJ m^{-2}
10	851	68.0
20	1170	70.2
60	1290	69.8
90	1620	71.1
120	1780	71.1
500	2420	72.7
1000	1990	71.8

EFFECTS OF WATER ON EPOXY/GLASS BONDS

If θ_o is determined as a function of time-of-immersion of the bonded specimen in water, extremely valuable information can be obtained concerning the mechanism of attachment and the influence of water on the joint. Such studies have been carried out on epoxy/glass joints.

The adhesive bonding of polymeric resins to glass assumes major importance in the glass reinforced composite materials used increasingly in structural and aerospace application. Good bonding is essential, not only for the structural integrity of the composite but also to ensure adequate stress transfer from the polymer matrix to the reinforcing fibres.

Epoxy resins are capable of providing an excellent adhesive bond to glass, but the adhesion deteriorates progressively in the presence of moisture. This process is slowed, but not eliminated, if the glass fibres are 'sized' with coupling agents to promote adhesion and durability.

A number of studies have been carried out on the deterioration of glass-to-epoxy bonds in water[13,14] but quantitative conclusions are not easy to obtain because of the complex stress patterns produced by differential thermal contraction between the resin and the glass as the system cools from its curing temperature. Furthermore, conventional adhesion tests measure parameters which are only indirectly related to the actual interatomic bonding energies at the interface[8]. It is therefore difficult to monitor changes in these bonding energies with exposure to water or with the addition of coupling agents.

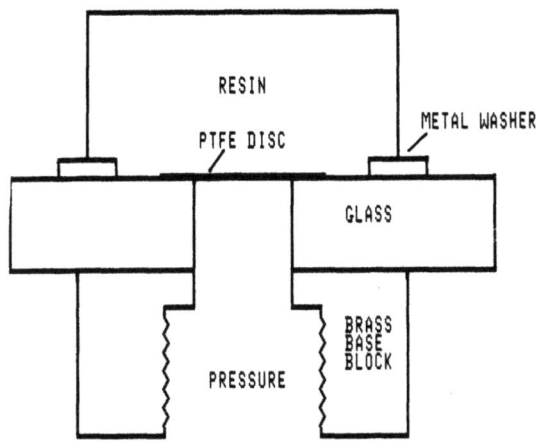

Figure 3. Test specimen for the Andrews-Stevenson fracture test applied to epoxy-resin/glass bonds.

Many of these problems can be overcome by combining the experimental technique of Andrews and Stevenson[15] with the analytical method of GFM. The epoxy resin used throughout this work was "Shell Epikote 828" a diglicidyl ether of bis-phenol A of molecular mass ∿370. The hardener was "Shell Epikure 114", a blend of two cycloaliphatic amines with added benzyl alcohol as an accelerator. These components were mixed in the stoichiometric mass ratio of 5:2, cast and allowed to gel for 24 hours at room temperature, before post curing at 130°C for 1.5 h. Specimens were cooled at 30°C/h. The resulting resin has a glass transition temperature at 72°C.

Three different silane coupling agents were employed (supplied by Union Carbide Limited) as follows:

<u>A 1120</u> N-β-aminoethyl-γ-aminopropyl trimethoxy silane

<u>A 187</u> γ-glycidoxypropyl trimethoxysilane

<u>A 1100</u> γ-aminopropyltriethoxysilane

The silanes were added to the epoxy resin, before casting, in various weight proportions.

The basic specimen employed is illustrated in Figure 3 and consists of a 2" x 2" square 'coupon' of 'Pyrex' glass on to which is cast a squat cylindrical block of epoxy resin approximately 30 mm in radius and 8 ∿ 11 mm in height.

The drilled central hole is covered by a thin circular disc of PTFE before the resin is cast, the disc thus creating a non-adhering region which acts as an enclosed circular crack which can be pressurized internally through the hole. This system provides a totally plane-strain fracture mechanics test in which, moreover, stresses due to differential thermal contraction between the resin and the glass cannot assist crack propagation as long as the latter is interfacial[15]. From the critical pressure for failure and the specimen dimensions, the fracture energy $2\mathcal{J}$ (cohesive) or θ (adhesive) may be evaluated as discussed elsewhere[15]. Crack velocity was measured by high speed photography, again as detailed in the earlier paper.

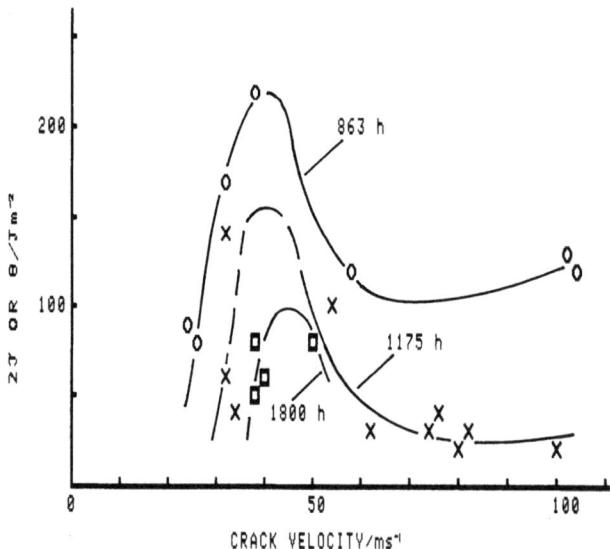

Figure 4. Fracture energy versus crack velocity data for epoxy resin bonded to glass (with silane A187 at 0.1% concentration). Specimens immersed in water at 80°C for the times shown. Note that although data shown here are not numerous, the general shape of the curves is common to a very large number of tests using different substrates, coupling agents and immersion conditions.

INTERATOMIC BONDING ENERGIES AT INTERFACES

Results

The basic data obtained form the fracture tests take the form shown in Figure 4 where cohesive ($2\mathfrak{I}$) or adhesive (θ) failure energy is plotted as ordinate against crack velocity $\dot{c}$ as abcissa for various times of immersion in water at 80°C and 7.8 pH. Immersion causes a steady overall decrease in failure energy without any marked change in the velocity dependence. Resin incorporating coupling agent suffers a much slower degradation of interfacial strength. For example, a given system containing coupling agent deteriorates after 860 h to about the same degree as obtained after 40 h with unprotected resin. Data for all resin mixes can be displayed in the same form as Figure 4. Following the super-position procedures outlined earlier results may be obtained of the kind displayed in Figure 5, which is a 'master curve' obtained by vertical super-position of the logarithmic data for resin containing A1120 coupling agent in various proportions and various immersion times exceeding 100 h. The shape of the master curve does not appear to depend upon the coupling agent employed, its concentration, the time of immersion (after the first 100 h), or the pH of the immersion medium.

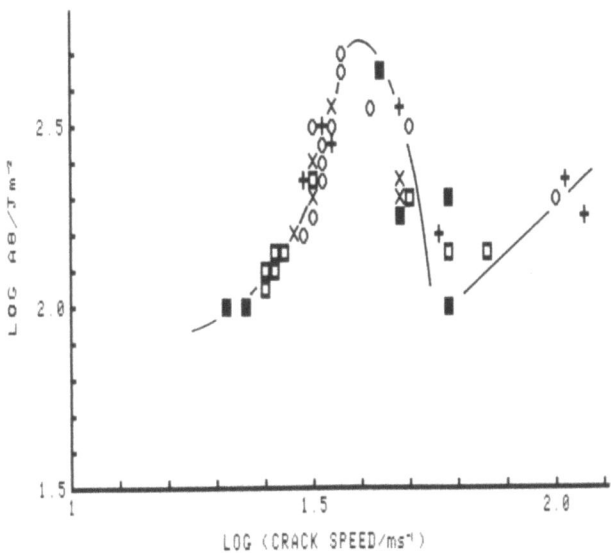

Figure 5. Master curve for failure energy formed by super-position of data such as shown in Figure 4. Data shown are for 0.05% of A 1120 silane in epoxy resin bonded to glass, immersed in water for various periods of time from 500 to 2000 h at 80°C.

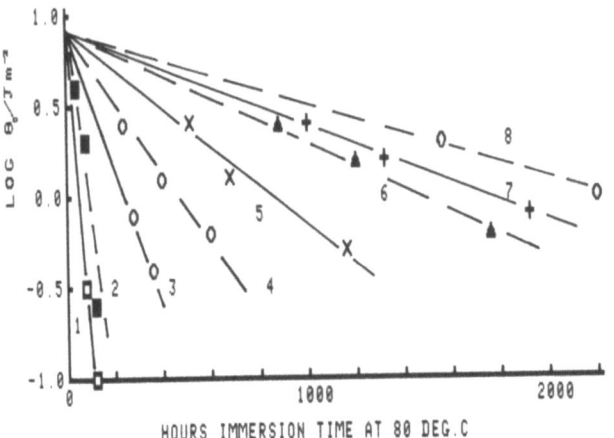

Figure 6. Collected data showing variation of θ_o with time of immersion. (1) 0.05% A187, pH = 13 (2) no silane (3) 0.05% A187 pH = 2 (4) 0.05% A187 (5) 0.05% A1120 (6) 0.1% A187 (7) 0.1% A1120 (8) 1.0% A1100.

Reaction Rate Constants

As outlined earlier, the vertical shifts required to give a master curve give θ_o provided $\mathcal{J}_o$ is known. Comprehensive results for log θ_o are plotted in Figure 6 against time of immersion for a range of coupling agents, concentrations and pH values. A linear dependence of log θ_o upon time is characteristic of a first order chemical reaction, the negative slope of such a plot being what we shall call the overall rate constant, k':

$$k' = [A][B] \, k_{80}/2.303 \qquad (9)$$

where [A] and [B] are respectively the concentrations of hydrolysable bonds and water at the interface, and k_{80} is the true rate constant at 80°C. For comparative purposes it is sufficient to talk in terms of k' since the water concentration should be sensibly constant at its equilibrium uptake value (some 8% at 80°C).

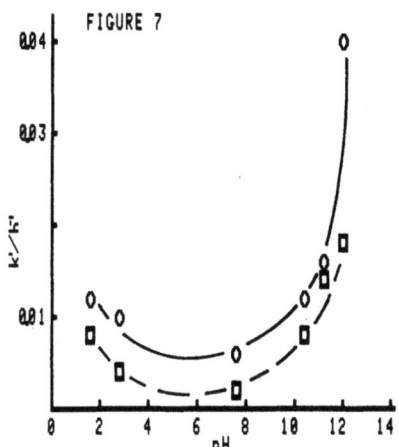

Figure 7. Hydrolysis rate constant as a function of pH for (circles) resin without silane and (squares) resin with 0.05% A187.

Variation of Rate Constants

Finally, we may seek information on the interfacial hydrolysis process by examining the variation of k' with the type and concentration of coupling agent and with pH.

Figure 7 shows k' as a function of pH. Hydrolytic activity is at a minimum for neutral pH, rising dramatically for both acidic and alkaline conditions. This is in distinct contrast to the results obtained for the same resin using titanium as a substrate[17] There, acidic conditions accelerated the hydrolysis reaction but alkalinity retarded it. This difference constitutes strong circumstantial evidence that the hydrolysing bonds are genuinely interfacial. If the locus of hydrolysis were the bulk resin, no difference should be observed between glass and metal substrates.

Dry Value of θ_o

All the data, regardless of resin composition and pH, converge within experimental error to a single value of θ_o at $t = 0$, i.e. for the dry resin/glass bond. This point is, of course, experimentally inaccessible since without immersion all specimens fracture cohesively and no information on the strength of the interface can be deduced (except that $\theta_o > 2\mathcal{J}_o$). The extrapolation of data for specimens at different pH values should, of course, give a single intercept, since all these joints are identical before immersion. Nevertheless, the fact that they do converge to a single value of θ_o is strong evidence that this value can be taken as the actual interfacial bonding energy of the freshly made interface. The point of major interest is that even specimens differing widely in silane content also yield the same θ_o (dry).

The value in question is 7.25 Jm^{-2}, some twenty-four times larger than the expected van der Waals interaction energy for such a system[16]. This indicates that at least some of the interfacial bonds are primary, but a precise proportion is more difficult to specify. A figure of 25 to 50% of primary bonds would appear to be reasonable, based on the relative dissociation energies of primary and secondary bonds. However, as Ahagon and Gent[10] point out in their work on elastomeric adhesives bonded to glass, a single primary interfacial bond may require many times the energy for rupture of a single C-C bond because it is attached to a length of network chain. Taking this factor into account, the magnitude of θ_o (dry) can be explained if between 30% and 60% of all interfacial bonds are primary in nature. The lower figure applies if all network chains terminating on the surface are chemically bonded to it and if the cross-link density is the same at the interface as in the bulk. The higher figure would apply if the network chain effect were totally suppressed by chain immobilisation at the interface.

The most surprising feature of the θ_o (dry) result is that it appears insensitive to the quantity of silane added to the resin, even though the rate constant for hydrolysis is greatly enhanced by the coupling agent. The obvious explanation is that, with the addition of silane, the total number of interfacial primary bonds remains the same but that the type of bond changes. That is, primary bonds which are established in the absence of silane (possibly R-C-O-Si) are progressively replaced by different bonds (possibly R-Si-O-Si) until the interface is saturated with silane-related bonds. Under dry conditions these two types of bond would have very similar strength, but under hydrolytic action the silane-related bond would degrade much more slowly.

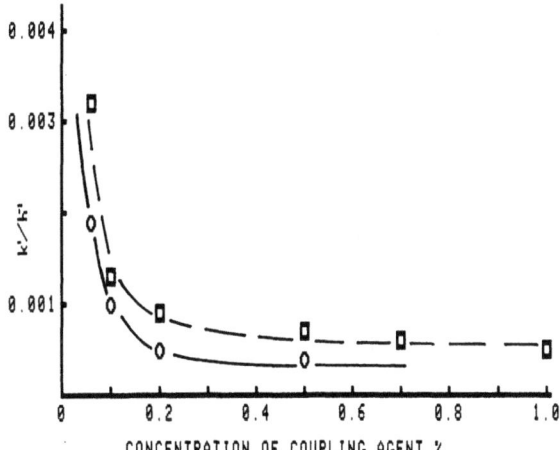

Figure 8. Hydrolysis rate constant as a function of coupling agent concentration. Circles A1120, squares A187.

The variation of k' with silane concentration is shown in Figure 8 for silanes A187 and A1120. There is an almost linear decrease of k' with concentration up to 0.05% w/w but at concentrations as low as 0.1%, saturation is beginning to occur. Although k' continues to fall slowly, saturation is largely achieved at concentrations of 0.5% for both silanes.

CONCLUSION

Fracture mechanics experiments, combined with the GFM analysis of the data, have made it possible to measure directly the interfacial bonding energy, θ_o, for various adhesive/substrate systems. The role of this bonding in determining the mechanical strength of joints is also made clear and quantitative in such studies. Once θ_o is known, its dependence upon such important variables as surface pre-treatments, coupling agent type and concentration and environmental exposure can be determined. The physical and chemical mechanisms involved may then become clear.

REFERENCES

1. C.A. Dahlquist, in "Aspects of Adhesion", D.J. Alner, Editor, Vol. 5, p. 183, University of London Press, 1969.
2. D.H. Kaelble, J. Adhesion, $\underline{1}$, 102 (1969).
3. K.L. Mittal, in "Adhesion Science and Technology", L.H. Lee, Editor, p. 129, Plenum Press, 1975.
4. K.L. Mittal, Polym. Eng. Sci., $\underline{17}$, 467 (1977).
5. E.H. Andrews, J. Materials Sci., $\underline{9}$, 887 (1974).
6. E.H. Andrews and Y. Fukahori, J. Materials Sci., $\underline{13}$, 1307 (1977).
7. N.E. King and E.H. Andrews, J. Materials Sci., $\underline{13}$, 1291 (1978).
8. E.H. Andrews and A.J. Kinloch, Proc. Roy. Soc. (Lond.) A, $\underline{332}$, 385 (1973).
9. G.J. Lake and P.B. Lindley, J. Appl. Polym. Sci., $\underline{9}$, 1233 (1965).
10. A. Ahagon and A.N. Gent, J. Polym. Sci., (Polym. Phys. Ed.), $\underline{13}$, 1903 (1975).
11. G.J. Lake and A.G. Thomas, Proc. Roy. Soc. (Lond.) A, $\underline{300}$, 108 (1967).
12. E.H. Andrews and A.J. Kinloch, Proc. Roy. Soc. (Lond.) A, $\underline{332}$, 401 (1973).
13. J. Comyn, D.M. Brewis, R.J. Shalash and J.L. Tegg, in "Adhesion 3", K.W. Allen, Editor, P. 13, Applied Science Publishers, Barking, Essex, 1979.
14. J.P. Sargent and K.H.G. Ashbee, J. Adhesion, $\underline{11}$, 175 (1980).
15. E.H. Andrews and A. Stevenson, J. Materials Sci., $\underline{13}$, 1680 (1978).
16. E.H. Andrews and N.E. King, J. Materials Sci., $\underline{11}$, 2004 (1976).
17. E.H. Andrews and A. Stevenson, J. Adhesion, $\underline{11}$, 17 (1980).

REVIEW OF CONTINUUM MECHANICS FACTORS IN ADHESIVE FRACTURE

M. L. Williams

School of Engineering
University of Pittsburgh
Pittsburgh, Pennsylvania 15261

The possibility of treating adhesive fracture as an engineering analysis problem in continuum mechanics has been emphasized in previous reviews, with applications to designs which utilize bonded surfaces and/or various composite materials. Cost-effective use of this analysis ability will ultimately depend, of course, upon the accuracy of fracture and structural life prediction, as well as the availability of methods to conduct non-destructive examination.

This review summarizes some of the principal implications from continuum mechanics which influence the character of adhesive fracture, including some new experimental data for the rod pull-out problem. The intent is to furnish an overview of major mechanics factors to those chemists whose interest lies in making direct contributions to the engineering design of adhesively bonded structures.

INTRODUCTION

Continuum mechanics has been used to analyze various adhesive bonding configurations since at least the first part of this century. As the use of bonded components became more widely recognized and reliable as a design option, the subject of adhesion has correspondingly attracted increased attention from both scientists and engineers. Efforts by the former have tended to focus upon a basic understanding of the chemical structure of the mating components such as reflected in the nature of the interfacial forces and the relative contact or wetting angles. On the other hand efforts of engineers have gravitated toward producing acceptably accurate assessments of various aspects of structural integrity, including the development of representative and reliable test methods for quality control.

Since the author presented his first continuum mechanics review to the American Chemical Society approximately ten years ago,[1] followed by one with Anderson to the International Congress on Fracture in 1977,[2] an excellent review, dealing with both the scientific and engineering approaches, has recently (1980-82) been presented by Kinloch.[3] In addition the monograph by Anderson, Bennett and DeVries[4] has provided considerable amplification, including many important aspects of testing methods. With these recent references readily available, it is considered sufficient for the present purposes to call attention to this review literature, reiterate major (continuum mechanics) points, and provide supplementation in a relevant manner.

Because of this ready availability of the recent reviews on the subject of continuum mechanics and adhesive fracture, and particularly the full exposition presented so recently by Kinloch,[3] the present coverage will be directed to chemists with the intent of transmitting an insight into the type of conclusions which might reasonably be expected from co-workers in mechanics. Consequently, and despite its importance, there will be little said of the importance of a Griffith energy balance type approach to cohesive or adhesive fracture in elastic, plastic, or viscoelastic materials.

Instead, after some initial remarks relating to the types of configurations presently used in the community, the first part will be directed toward the utility of continuum mechanics in interpreting fracture through the mechanical properties of the materials involved, the effect of local or global geometry of the part, and the importance of loading orientation(s) with respect to the line of fracture. The second part will be devoted to a discussion of identifying potential adhesive fracture locations primarily through a readily definable knowledge of singular stresses. The third part will supplement the Kinloch review by discussing certain aspects of the rod or fiber pull-out problem of interest in composite materials, civil engineering, and automobile tire design, with special emphasis on some of our continuing experiments and numerical analysis of the partial debonding and adhesive crack growth from, along, around and into the glass-rubber interface of the embedded rod. The concluding section deals with some aspects of adhesive fracture which appear to the author to be a collective responsibility among chemists, physicists, and mechanicists if a more rapid and enhanced understanding of the phenomenon is to be attained.

CONFIGURATIONS

Generally speaking the various geometries discussed in the literature seem to fall into two general categories: practical design components and engineering data specimens. Neither of course is completely exclusive. Very practical single and double shear lap joints have been used as test specimens to assess the import of normal stress in the adhesive perpendicular to the angle of pull; on the other hand, the pressurized blister test configuration used to measure the fundamental specific adhesive fracture energy (γ_a) also yields stress and deformation formulas useful in analyzing the (blister) debonding of paint and thin films or the welding efficiency for explosively bonded steel plates.[1] (Figure 1.)

The principal adhesive configurations are as follows: stress analyses for all of them are available to varying degrees of accuracy. The butt-end joint, including its tubular relative called the "napkin ring" butt joint, had its early history in connection with welded joints in metals and, in rubber, with bridge bearings and shock mounts. The group at the British Rubber Producers Association, stimulated by the early work of Rivlin, Thomas, Andrews, Gent and co-workers, analyzed the effective modulus in tension and compression of thin rubber disks bonded to rigid mountings. This configuration was subsequently adopted in the American rocket industry to study triaxial tension failure and flaw generation in the rubber liner which was bonded to the rocket motor case.

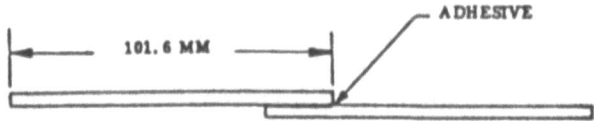

Figure 1-A. *Single lap shear test.*

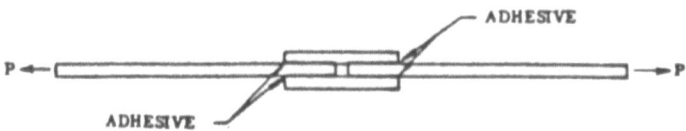

Figure 1-B. *Double lap shear test.*

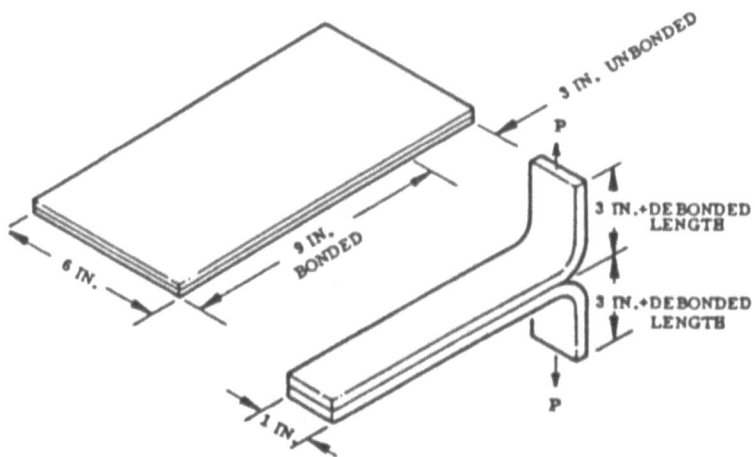

Figure 1-C. *T-peel test.*

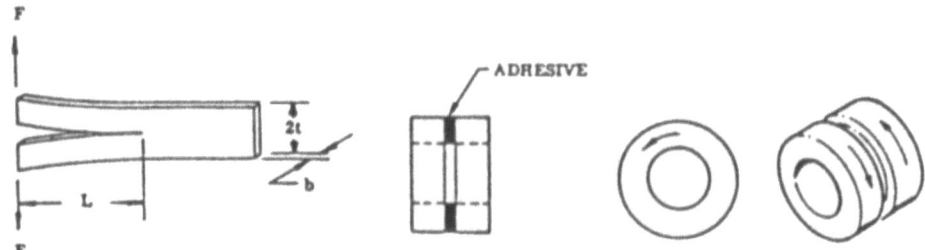

Figure 1-D. Double cantilever cleavage specimen

Figure 1-E. *Torsional shear stress specimen.*

Figure 1. Several typical adhesive test specimens[2,4].

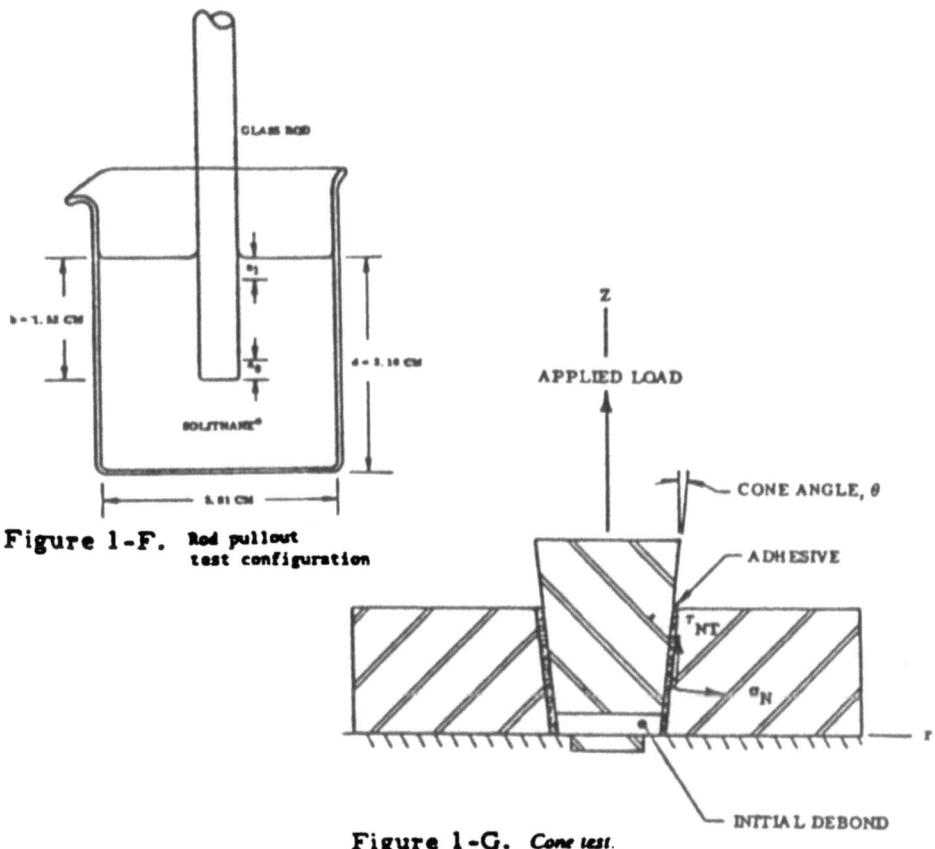

Figure 1-F. Rod pullout test configuration

Figure 1-G. Cone test.

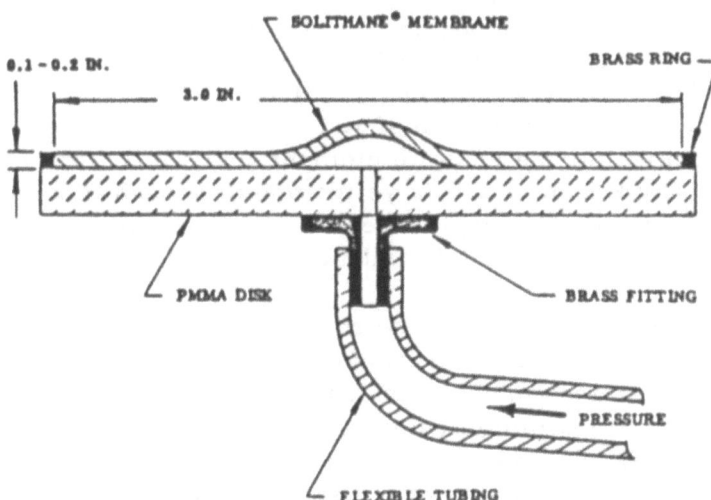

Figure 1-H. Blister test specimen.

Figure 1 (continued)

The stresses in this "poker-chip" specimen and its variations, including consideration of the stress singularities at points of the interface, have been used to examine the propensity for nominally normal stress debonding. (Substantial portions of the limited distribution Messner work has now been reproduced.[4]) The split double cantilever beam (DCB), adopted from the Obreimoff test for the mica fracture energy and Gilman's test for the splitting forces between sheets of mica, and the pressurized or point-loaded blister test used by Williams and Jones and by Malyshev and Salganik,[5] also measure primarily the normal stress or opening mode of fracture (Mode I), but contain important shear components.

On the other hand, single and double shear lap joints with and without scarffed ends deal primarily with the shear mode of fracture (Mode II) as described in the Kinloch review. The conical plug pull-out test, originally introduced by Anderson[9] in conjunction with dental research dealing with gold fillings, has the virtue of varying the proportion of the Mode I (normal stress) and Mode II (shear stress) contribution to fracture. It may also be used for torsion tests (Mode III) by twisting the plug. In one of its limit forms--the straight pull out--it represents the very practical problem faced by automobile tire manufacturers who wire-reinforce the tread and wall. Alternately, and usually in shorter lengths, but not necessarily much different length-to-diameter ratios, the pull-out or "tooth-pick" problem is the elemental problem in fiber-reinforced materials. Our experimental and analytical work, growing out of the initial experiments of W. B. Jones and Masahisa Takashi, was reported earlier[2] and has been extended by Atkinson, Betz, Smelser, and Avila[6] to include definitive experimental data and limited parametric variation of rod geometry and debond length and location making use of finite element analysis. None of our analyses, nor any known to us, is yet complete enough to associate the reinforcement of a single fiber in an infinite medium with that of randomly oriented and distributed multi-fibers.

The DCB and blister specimens were originally introduced to measure Mode I specific adhesive fracture energy (γ_a^*), which is the continuum mechanics adhesive analog of the specific cohesive fracture energy (γ_c^*) commonly denoted in metals as "fracture toughness." The latter specimen has the virtue that, being circular, it has no "sides" to become environmentally contaminated, although the test environment can be changed at the crack front by changing the pressurizing medium. As remarked earlier, "membrane" thickness blister tests with combinations of pressure and temperature have certain applications to thin film adhesion. Further extensions of the basic blister configuration have been used to assess the basic influence of adhesive layer thickness and, by a double sided blister, the stiffening effects of different adherends.

Final mention should include various partial debonding analyses of cylindrical and spherical inclusions, and the related cases of elliptical rod and ellipsoidal geometries, which are important in granular composite media such as cement-aggregate concrete, or in the potentially important, fairly new sulfur-aggregate construction material. (Interestingly enough, one of the major drawbacks is the apparent lack of adherence of the smoother, larger particles which suggests that present continuum models should include interfacial layers of finite thickness, as introduced into the blister test, and from the chemical side the introduction of appropriate surfactants.*) In more generality, granular composite material adhesives technology may well apply to models of powdered metals, and even the increasingly interesting metal matrix composites.

CONTINUUM MECHANICS

A mechanics analysis can be expected to yield information on the influence of mechanical properties of the materials involved, geometric configuration of the design, and the type of loading.

The first--mechanical properties--is reflected mainly in the stress-strain law and includes the effects of time and temperature, separately or together. Separately one may assess thermal strain through a combination of the coefficient of thermal expansion (α) and the temperature change (ΔT), or the magnitude of creep deformation at essentially constant temperature. Together, time (rate) and temperature are important for many adhesives and appear in the temperature-reduced time, WLF[7] shift factor (a_T). Discovered experimentally, it comes about through a semi-empirical correlation between viscosity and free volume of viscoelastic materials incorporating another material property, the glass temperature (T_g). The fundamental material descriptors of stiffness (modulus of elasticity, E or G) and degree of incompressibility ($\Delta V/V$, Poisson's ratio ν) are typical quantities that form

---

*Private communication Dr. D. Saylak, Texas A&M Univ., 1982.

the matrix of material constants which for most materials relate the six components of stress (three shear and three normal) to similar components of strain.

The second--geometric configuration--has already been implied in the reference to the categories of geometries discussed in the literature, e.g., design components such as single and double shear lap joints and specimens used to gather engineering data on the properties of the materials involved. For example, the overall distinction between single and double shear specimens in terms of the presence or absence of a moment imposed upon the joint is commonly accepted. Other perhaps less obvious geometric changes however can also be assessed. One of the more important geometric effects pertinent to adhesion is the thickness (h) of the adhesive interlayer.[4,8] Another is the presence or absence of fillets at sharp corners, usually interpreted by stress analysts in terms of a stress concentration factor, and by chemists, in terms of a wetting angle between the adhesive and the adherend. Many of the geometric considerations, however, relate mainly to the arrangement of material, say for low weight of the assembly. As such they are of major concern to the stress analyst who frequently finds it necessary to employ rather sophisticated and powerful numerical techniques, such as the "finite-element method," to deal with these design configurations, although such methods are occasionally used even in analyzing engineering data specimens.

The third--type of loading--deals with the manner in which loads are oriented and applied to the entire design or the individual component. It may be either mechanical or thermal in nature, imposed in a time dependent way (shock, steady, repeated or oscillatory fatigue), and from one or more directions. For our present purposes it is sufficient to think in terms of a simple specimen subjected to steady mechanical loads imposed from one, two, or three directions at right angles to each other, along with one, two, or three sets of shear forces. Specializing to simple geometries subject to adhesive fracture, the mechanics analyst refers to a crack normal to the primary direction of applied loading as subjected to "Mode I" (normal stress only) loading. Similarly if shear loads are applied parallel to the crack, it is subjected to Mode II (shear) or Mode III (shear) loading. Figure 2 illustrates the three modes for cohesive fracture. As will be seen later, whereas for an isotropic homogeneous cracked specimen, as in Figure 3a, the mode descriptions are unambiguous, the same

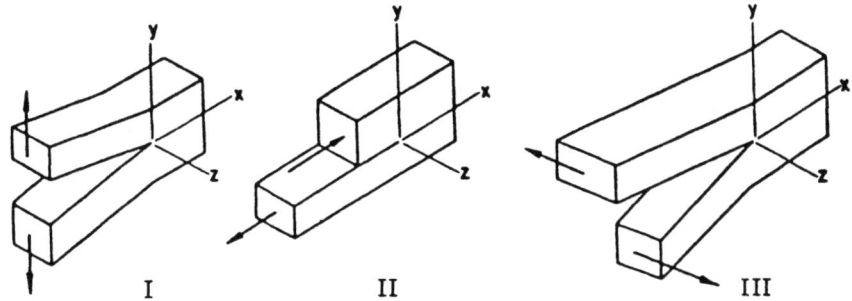

Figure 2. The basic modes of Crack Surface Displacements.

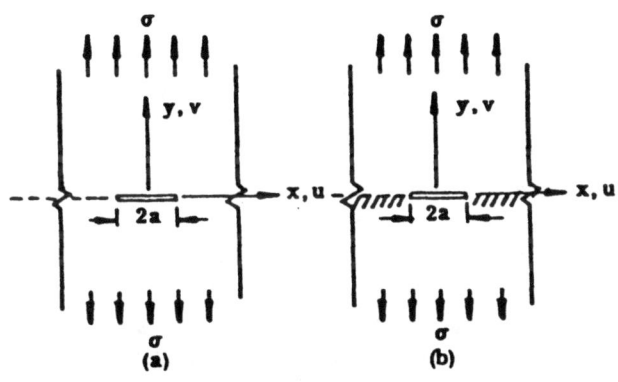

Boundary Conditions: $|x| > a$

(1) $v(x,o) = o$
(2) $\tau(x,o) = G[\partial u(x,o)/\partial y + \partial v(x,o)/\partial x] = o$
or because of (1), equivalently

(2a) $\dfrac{\partial u(x,o)}{\partial y} = o$

Figure 3a. Cohesive Fracture

$$\sigma_{cr} = \sqrt{\dfrac{2}{\pi}\dfrac{E\gamma_c}{a}}$$

Boundary Conditions: $|x| > a$

(1) $v(x,o) = o$
(2) $u(x,o) = o$

Figure 3b. Adhesive Fracture

$$\sigma_{cr} = \sqrt{\dfrac{2}{\pi}\dfrac{E\gamma_a}{a}}$$

Figure 3. Comparison of essential boundary conditions for cohesive and adhesive fracture. The difference is slight and concerns only whether the lateral displacement u(x,o), or its normal derivative, ∂u(x,o)/∂y, is prescribed. Both sets of boundary conditions lead to singular stresses[2].

can not be said for a bi-material case, say cracked along the bond line. If a simple one material center-cracked specimen (Figure 3a) is loaded perpendicular to the crack it will be subject solely to the opening mode (Mode I), i.e., there is no shear stress along the line of crack prolongation. If, on the other hand, the material is different in the half-planes above and below the crack (Figure 3b), a load applied in the same direction as before will still create an "opening mode" but there <u>will</u> be shear along the line of crack prolongation. Hence the common intuitive usage, which implies for bi-material adhesive specimens that an "opening mode" produces only a local normal stress (and no shear stress) at the crack tip, or, similarly, a "sliding mode" produces only a local shear stress (and no normal stress), is incorrect. As will be shown later, the intuitive conclusion is, however, substantially correct most of the time for the usual material combinations. Nevertheless, it is important to be aware of the true situation so that improper extrapolations will not be made.

Indeed a preliminary experimental study was conducted by Anderson[9,4] using a cone pull-out test in which he could vary the proportion of Mode I: Mode II: Mode III. The initial evidence suggests that a different amount of new fracture area (ΔS) is generated in each of the three modes of loading. This experimental observation is in contrast to the usual assumption in the analysis using <u>projected</u> new fracture area for each mode, but allowing for different specific adhesive fracture energy (γ_a, in-lbs/in^2) for each mode. Hence this preliminary conclusion suggests that the physical fracture process could be incorrectly modelled by the continuum mechanics analyst and a reassessment in conjunction with chemists and physicists might be very worthwhile.

SINGULAR STRESSES IN MULTI-MEDIA*

In mathematics it is commonly accepted that a knowledge of singular solutions is essential for representing complete solutions in the large. In this sense there is an important parallel in fracture, particularly to the extent that conclusions based upon a linearly elastic material and small deformations may represent guidelines for the actual behavior. For this reason it is important

---

*The examples given are confined essentially to two materials, although in principle the approach can readily be extended.

to emphasize the existence of mathematical singular solutions in both cohesive and adhesive fracture analysis, as well as simultaneously recognizing their limitations. Clearly the infinite stresses predicted mathematically at certain sharp corners and at crack tips in an interface can not exist physically--but the stress magnitudes may be near enough to it to cause practical trouble!

First for cohesive fracture, e.g., Figure 2 (I) - Mode I. Assuming the crack is in a thin sheet of unit thickness, the material mathematically occupies a region 2π about the crack tip and the crack which has two unloaded (free) faces encloses zero space between them. Starting from a more general case of a V-shaped crack having two free sides and an opening angle ω between them, the character of the stress variation in the vicinity of the crack in terms of the radial distance (r) from its tip can be obtained[10] as*

$$\sigma \sim r^{\lambda(\alpha)-1} f(\theta) \: : \: \alpha = 2\pi - \omega \tag{1}$$

where $\lambda(\alpha)$ has been extracted from Reference 10 and presented in Figure 4. For the special case of a crack, $\omega \to 0$ or $\alpha = 2\pi$, $\lambda(2\pi)$ is found to be 1/2. In such a case the exponent of the radius becomes negative, and

$$\sigma \sim \frac{f(\theta)}{\sqrt{r}} \tag{2}$$

and the stress becomes unbounded at the origin with its characteristic inverse square root singularity.

Equation (1) reveals another interesting feature for thin sheets. At the particular angle, α^*, at which $\lambda(\alpha^*) = 1$, there will be no singular or infinite stress at the origin. Reading from the Figure 4 for free-free edges, one finds $\alpha^* = \pi$. (For smaller included angles, the stress will be even less severe at the origin.) More generally, depending upon the boundary conditions on the faces, the stress at any notch tip will remain finite whenever an opening angle exists for which $\lambda(\alpha) \geq 1$.

For a case of interest to adhesion, consider a rigid adherend upon which an adhesive material having an (included) wetting angle (α) is laid down, or a scarffed joint with an (included) bond angle

---

*Note that the λ used here[10], say λ_W, is different from that used in Reference 4, say λ_A, such that $\lambda_A = 1 - \lambda_W$.

to the base plate (α). In this situation the material would have an included angle (α) and rigid-free boundary conditions. From Figure 4, one finds $\lambda(\alpha^*) = 1$ at $\alpha^* = 62^{0+}$. Hence one concludes that if the included angle $\lambda^* \leq 55^0$,[+] there will be no high stress at the V-tip. Note that the more the "adhesive" does not "wet" the base, the larger α is, and the stronger the stress singularity and the higher the stress at the V-tip, consistent with experience.

The above results were deduced using the data for thin sheets (plane stress).[10] A similar curve can be obtained for clamped-free boundary conditions and plane strain, i.e., infinitely thick sheet on an infinite rigid base, and yields 45 degrees as the largest (included) angle without inviting large stresses at the tip of an incompressible ($\nu = 1/2$) adhesive, i.e., $\sin^2\lambda\alpha = 1 - \lambda^2 \sin^2\alpha$.

Other pertinent deductions can be quickly made. For example as will be used later, consider the cross-section through the geometry of the circular rod pull-out test (Figure 5). Assuming the rod to be rigid compared to the incompressible rubber matrix in which it is bonded, one can easily identify two more "corner problems." Between the free surface and the rigid (fixed) side along the rod, the matrix material occupies an included angle of 90 degrees. Is the stress (mathematically) infinite at the free surface corner? And then if it is, how strong is the singularity compared to the other possible singular location at the corner of the rod tip? Here, with the rod tip flat, the matrix material is rigid (fixed) along the bottom of the rod face and its side, thus subtending an included angle of 270 degrees. Assuming for the moment the cross section of Figure 5 represents indeed a thin flat sheet (plane stress) instead of the actual axially symmetric geometry it really is, one reads from Figure 4

$$\lambda(\pi/2) = 0.76 \quad\Rightarrow\quad \sigma \sim \mathcal{r}^{-0.24}$$
clamped-
free

$$\lambda(3\pi/2) = 0.35 \quad\Rightarrow\quad \sigma \sim \mathcal{r}^{-0.65}$$
clamped-
clamped

Thus there are stress singularities at both locations and the one at

---
[+] The curve in Figure 4 prepared for $\nu = 0.3$ shows $\alpha^* = 62^0$. A similar curve prepared for $\nu = 0.5$ would actually give 55^0. (Ref. 10).

CONTINUUM MECHANICS FACTORS IN ADHESIVE FRACTURE

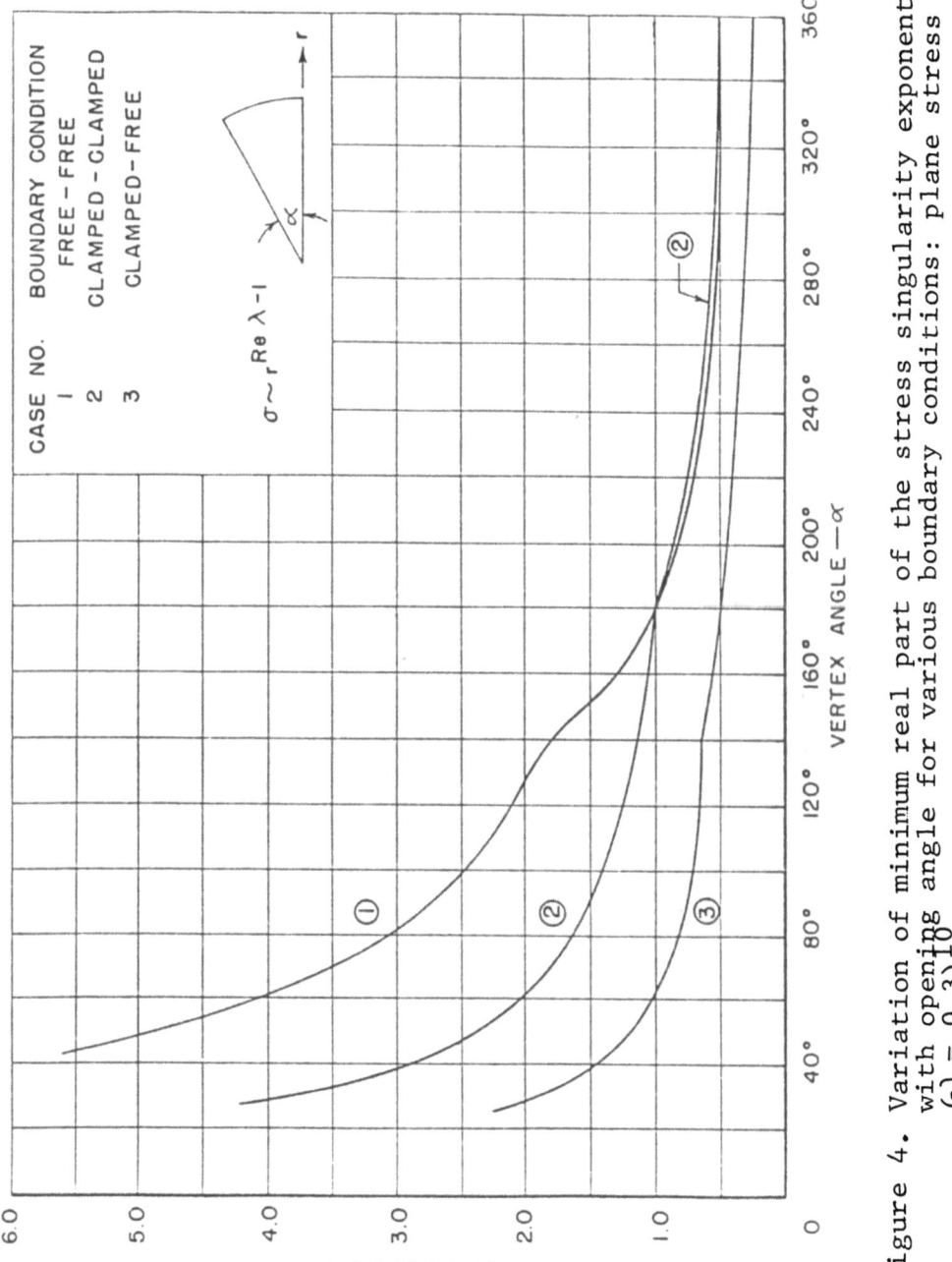

Figure 4. Variation of minimum real part of the stress singularity exponent with opening angle for various boundary conditions: plane stress ($\nu = 0.3$).

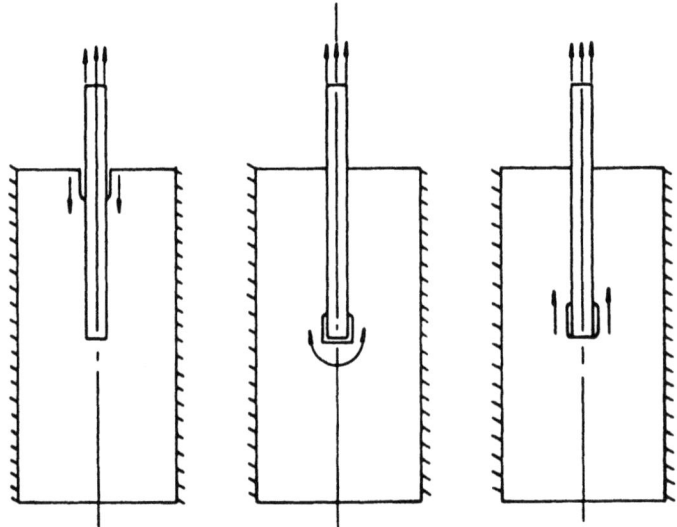

Figure 5. Profile of hypothesized unbonds of axisymmetric body in rod pull-out test[2].

the corner of the rod tip is stronger. (Actually the behavior at the interface of the rod is better associated with plane strain,[11]) although the conclusions are qualitatively similar. This situation, in the limit, can be understood by thinking of the rod as having a very large radius so that the material approaches an infinite thickness as the rod radius becomes infinite. The cross-section in Figure 5 would thus be the same for either plane stress or plane strain, but Figure 4 for the critical eigenvalue $\lambda(\alpha)$ is determined for plane strain from an equation containing Poisson's ratio which is slightly different from the one for plane stress.[10]

As a further example, while pursuing this limit analogy of axial symmetry to plane strain, return to a previous plane strain conclusion that for a 45 degree included angle between rigid and free faces there would be no stress singularity. This result would suggest that a cylindrical butt-end adhesive joint, or the "poker-chip" test of incompressible rubber as used in our early Caltech work[12] and analyzed numerically by Messner to have high stresses[4] at the periphery, would have finite or negligible stresses there if 45 degree indentation fillets were incorporated instead of the 90 degree (straight sided) configuration we used. Such a design would give increased assurance and reliability of more regularly obtaining the desired hydrostatic tensile condition at the center of the specimen.

As many bi-material configurations do not involve the convenient rigid-incompressible material combination, a general formulation was set-up to determine the character of stress singularity to be expected at the crack tip along the interface between two materials with different properties.[13] While originally prepared to investigate layered geophysical fault lines, the analysis is easily adapted to adhesive fracture. The most intriguing discovery was a stress singularity whose order was still the characteristic inverse square root type, but it was modulated by an oscillating trigonometric function such that instead of Equation (2), one obtained behavior such as

$$\sigma \sim \frac{1}{\sqrt{r}} \sin(b \ln r) \qquad (3)$$

Also the shear stress was of the same character and did not vanish unless the two materials were identical, which was reflected in the material dependency of $b = b(E_1, E_2, \nu_1, \nu_2)$, i.e., $b = 0$ if $E_1 = E_2 = E$ and $\nu_1 = \nu_2 = \nu$, thus recovering the result Equation (2) for cohesive fracture.

This result, namely that both normal and shear stress coexisted along a bi-material interface near the crack tip, became clear physically after a bit of reflection, but produced the aforementioned result that a Mode I opening mode would have not only the expected normal stress across the crack, but also an accompanying shear stress. (By extension to Mode II, the converse is evident.) Because of the existing tendency for the uninitiated to speak of "shear" or "normal stress debond" adhesive tests, it is important to make some assessment of the relative contributions of each in, say, nominally Mode I or II loadings. Fortunately this can be done through a relatively simple, but mathematically sophisticated, example reproduced[4] in Appendix A. The principal result presented in this appendix permits the interested reader to determine the exact percentage contents of shear to normal stress in Mode I, i.e. $(\tau/\sigma)_I$, and normal to shear stress in Mode II, i.e., $(\sigma/\tau)_{II}$, for the desired material combination

$$-(\tau/\sigma)_I = (\sigma/\tau)_{II} = \tan[\zeta \ln(\varepsilon/2a)] \tag{4}$$

where ζ is material dependent and ε is distance from the tip.

The oscillatory character of the predicted stress singularity should also be understood. From physical grounds, there must be a shear stress along the bond line if the materials above and below it are different (Figure 3b). Their lateral contraction is different; for them to maintain continuity they must be displaced relative to each other, which requires the shear stress. But why oscillatory? The net shear stress, say on each side of the (symmetric specimen) centerline must vanish, so the shear stress must reverse direction, which it does, in between each node of the trignometric cycle. Finally it was shown by Malyshev and Salganik[5] and Erdogan[14] that the oscillatory character was confined very, very closely to the crack tip, well within any local distortion due to, say, plasticity, so that the basic solution still represented the interface stresses quite well.

As a final comment on the utility of critical stress singularity evaluation in bi-material combinations at various included angle combinations, the tabulation completed by Hein and Erdogan[15] should be particularly noted. Some of the cases are quoted by Anderson et al[4], and one of them is of special interest: plane stress or plane strain of two quarter planes bonded along a common edge with the other two faces unloaded and continuous. In general there is a weak singular stress where the bond line intersects the free surface and varies from $r^{-0.22}$ to $r^{-0.00}$ (plane stress) as the modulus ratio of the two materials varies from several orders of magnitude to unity, respectively.

THE ROD FIBER PULL-OUT PROBLEM

The rod or fiber pull-out problem has stimulated considerable and long standing interest, perhaps from an early civil engineering need to analyze the stability of pilings by friction and length, or the strengthening of concrete by reinforcing bars. Another source of early interest was in aeronautics due to the drive toward low weight design which led to integral or mechanical attached stiffeners to thin sheet. Thus the transfer of load between the strong, stiff member to the surrounding material became of considerable practical importance. Inasmuch as many of the previous analyses have a direct bearing upon the stress analysis of a rod or fiber wholly or partially embedded in a surrounding softer or harder matrix, and furthermore, current composites materials development is properly concerned with maintaining the internal interfacial integrity, it is appropriate to supplement the Kinloch review in this regard. A selected limited bibliography biased toward recent contributions has also been prepared and is appended.

Because of the relative scarcity of precise experimental data for the pull-out problem in an adhesive debonding mode, and stimulated by the somewhat unusual crack growth behavior in the related situation of thermally driven debonding of rubbery rocket fuel from the surrounding motor case, the major emphasis was placed upon experiments to discover the general nature and progression of the failure. In the rocket problem the fuel essentially filled a rigid cylindrical case when positioned and cured vertically. The top surface was free of external loading, flat, and bonded to the case with very little fillet. In view of the foregoing remarks, the material occupied a space subtending 90 degrees with its faces about the vertex being rigid and free. When the temperature changed in a way to exert a thermal stress across the bond line, there was a tendency for fracture to initiate at the 90 degree corner and quickly proceed to propagate along the case interface. What was surprising however was the fact that the crack surface only proceeded a short way, i.e., much less than a case radius, before coming to a stop.

This stable propagation is rather unusual, though observed early by Benbow in the indented punch problem.[16] In that case the broadening conical "mushroom-ring" collar-like fracture was not catastropic or unstable, but required increasing load to extend it. Similarly in the bonded rocket fuel, the crack front only proceeded along the interface while there was sufficient strain energy being released to drive it, as subsequently confirmed by stress analysis.[17]

It developed that the fracture surface in a rod pull-out test behaved in somewhat the same manner. In Figure 5, for example, there are several possible fracture modes shown, depending upon the assumed point of initiation, but it was not a priori clear from analysis as to the origin or propagation of the failure. After considerable exploratory work by W. B. Jones and G. P. Anderson while at Utah, and Dr. Masahisa Takashi from Aoyama University, while a visiting professor at Pittsburgh, a satisfactory experimental procedure evolved. It was then refined and standardized at Pittsburgh by Dr. Eric Betz of the University of Newcastle, Australia. His carefully accomplished experiments led to the data reported earlier.[2] It was subsequently analyzed analytically and numerically by Atkinson, Avila, Betz and Smelser. These results have just been published[6] which reinforce the visual phenomenological description. The suspected high (singular) stresses at the free surface and the rod tip corners were confirmed and, after conducting the requisite theoretical energy balance, seemed to show a general consistency between the theory and experiments. For the size of rod used in the experiments, the initial fracture generally occurred at the center of the flat tip of the rod rather like the origin of a penny-shape crack between the glass rod end and the rubber matrix. It then propagated across the flat end and turned round the corner and crawled up the rod until--like the rocket motor behavior--stopping upon encountering a compressive normal stress state across the interface.

The next step was a second initiation, this time at the corner by the surface, after which the fracture surface then propagated downward to join the previous, stopped crack and complete the debonding. This behavior did not always occur, especially when the specimen was precracked without load to see if the initial crack location could be forced toward a different origin.

Figures 6-7 show the applied force and displacement history for the tip and surface debonds, respectively, as associated with the debonding progression along the interface. For our present purposes it is sufficient to state that a numerical analysis was carried out from which the change in strain energy with crack position (strain energy release rate) could be accurately calculated and correlated with the experimental data. For example the compressive stress field at the lower end of the rod was found, and the relative energy release rates were such that the order of crack initiation and growth were properly predicted. The maximum displacements for the two cases (Figures 6-7) have been consolidated on one curve for comparative purposes (Figure 8).

The practical conclusion from this work is that we now have sufficient confidence in the analytical work, which leaned heavily upon the use of the characteristic stress singularities, to proceed with a parametric variation of the length/diameter ratio of the rod, adjustments in the tip configuration, and the degree of partial embedment in the matrix. Aside from producing a measure of the matrix stiffening by the rod, we presume it will now be possible to evaluate the critical loading at which internal deterioration of the interface affects the load carrying ability of the combination.

CONCLUSION

While there are many potential areas for further work in adhesive fracture from the point of view of continuum mechanics alone, I would like to reiterate my earlier[1] remarks relating to two areas where collaboration with chemists and physicists appears especially profitable.

The first area, referred to earlier in this work, is to establist any consistency, if indeed it is present, in a fundamental specific adhesive fracture energy (γ_a) independent of the loading mode. Three fracture modes have been defined (Figure 2). As found by Anderson[9,4] and reported and discussed by Kinloch,[3] the amount of new surface area (ΔS_k) created seems to be sensitive to the loading mode (k = I, II, III) imposed. The analytical treatment assumes that the total surface energy (Γ) created is split into three separate specific fracture energies ($\gamma_a \gamma_a^{(k)}$) by mode, each multiplied by the <u>projected</u> new fracture area which of course does not recognize the existence of different height/depth of hills/valleys.

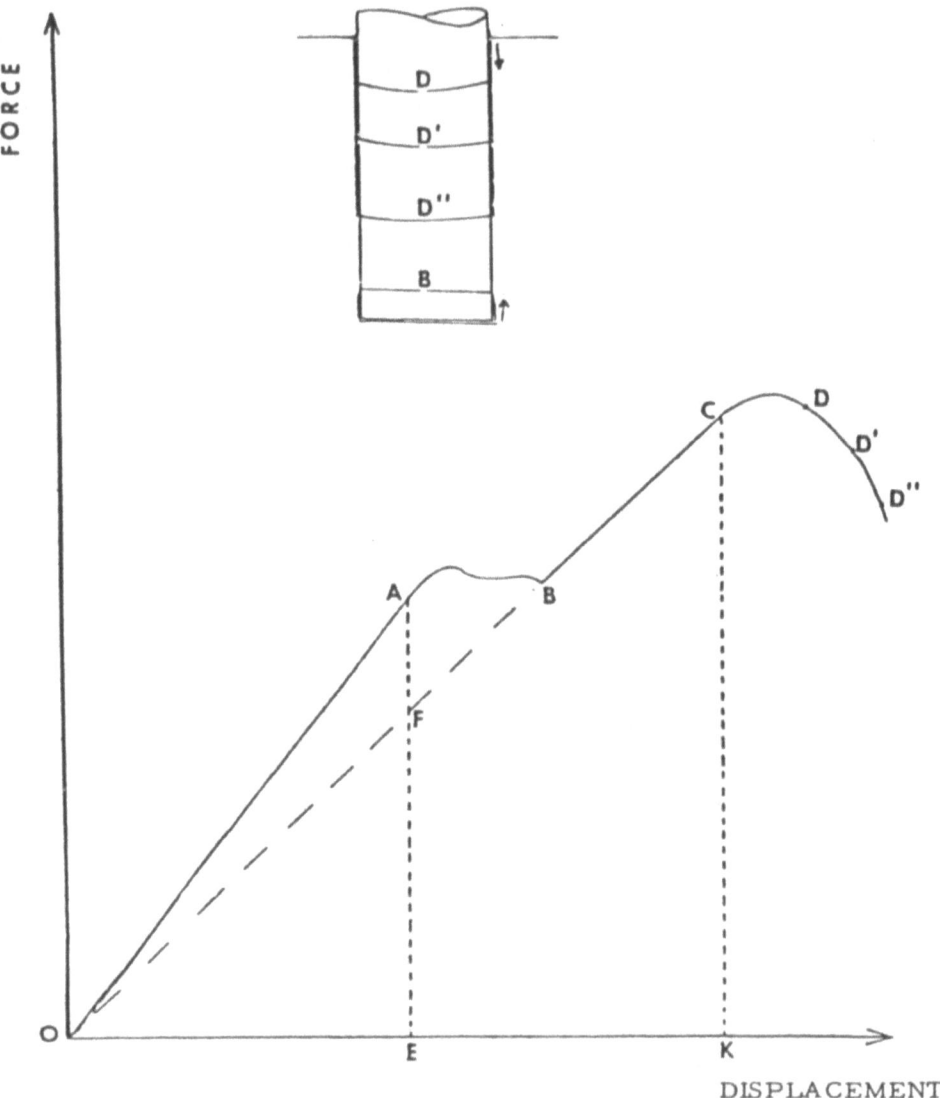

Figure 6. Typical force-displacement diagram for tip debond[6].

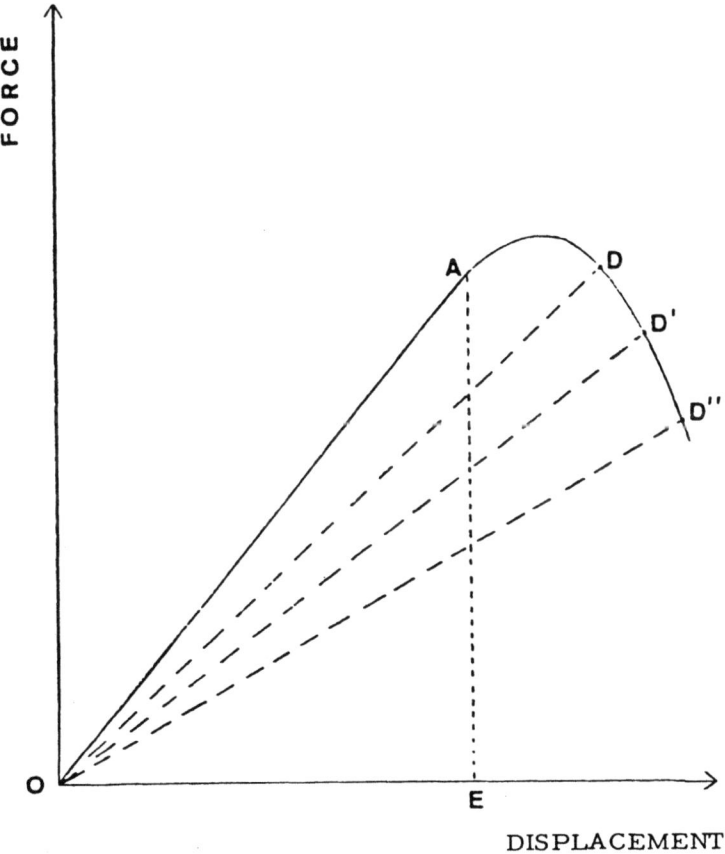

Figure 7. Typical force-displacement diagram for surface debond[6].

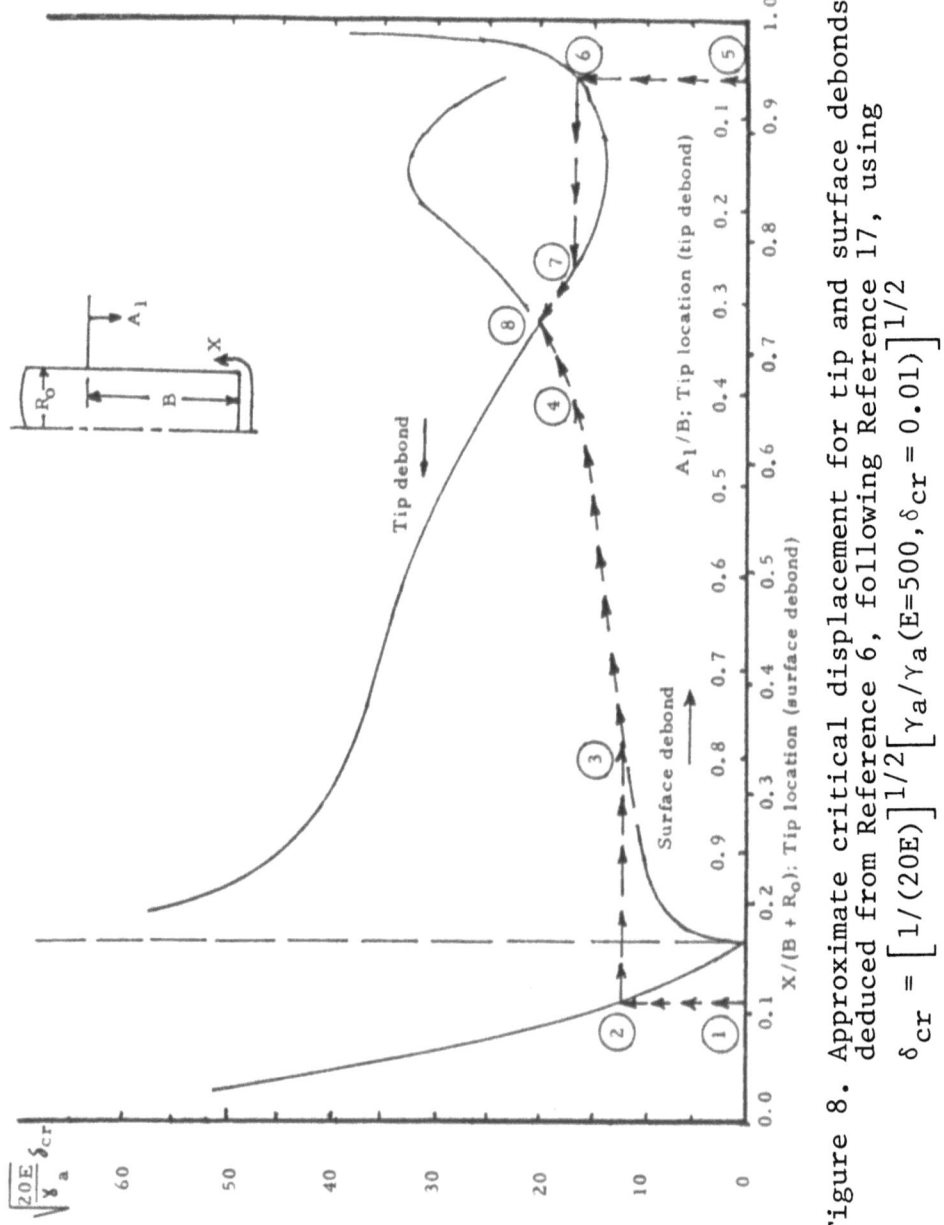

Figure 8. Approximate critical displacement for tip and surface debonds deduced from Reference 6, following Reference 17, using
$$\delta_{cr} = \left[1/(20E)\right]^{1/2}\left[\gamma_a/\gamma_a(E{=}500, \delta_{cr}=0.01)\right]^{1/2}$$

Molecular characteristics	Symbol	Modified power law parameters				
		E_g	E_e	T_o	n	T_g
Cross-link density	ν_e					
Chain stiffness	N_s					
Monomeric friction coefficient	ζ_0					
Solubility parameter	δ_p					
Molecular weight	M					
Heterogeneity index	M_w/M_n					
Molecular weight between entanglements	M_e					
Degree of crystallinity	Λ					
Volume fraction of filler	$\emptyset$					
Volume fraction of plasticizer	V_p					

Figure 9. Basic Interaction Matrix arrangement showing chemical structure and mechanical properties parameters[18].

Hence the assumed change in energy to create new fracture area is
$\Delta \Gamma = [\gamma_a^{(I)} + \gamma_a^{(II)} + \gamma_a^{(III)}] \Delta S$, whereas there may be physicochemical grounds to boldly postulate that there is one unique specific adhesive (or cohesive) fracture energy ($\gamma_a^{(0)}$) which is perhaps identifiable with a chemical structure parameter, e.g., bond strength independent of fracture mode, to give instead
$\Delta \Gamma = \gamma_a^{(0)} [\Delta S_I + \Delta S_{II} + \Delta S_{III}]$.

The second area of collaboration is the convenience as well as significance in developing an Interaction Matrix[18] to provide an explicit guide to each other's terminology and phenomenology. (Figure 9). The column elements could be mechanical descriptors such as rubbery modulus (E_e), glassy modulus (E_g), etc., representing the relaxation modulus through the modified power law

$$E_{rel} = E_e + (E_g - E_e)/[1 + (t_r/\tau_o)]^n \tag{5}$$

The row elements could then be the chemical descriptors such as cross link density (ν_e), molecular weight (M_w), etc. The object is to obtain quantitative mechanical-chemical associations for all row-column intersections in the matrix. The simplest one of course, from rubber elasticity theory, is $E_e = 3\nu_e kT$; the complicated ones may need to be empirical. In any event a joint effort to develop such a scheme even in terms of only first order correlation would seem to benefit communication between the fields--and eventually technical progress.

ACKNOWLEDGEMENT

Major portions of this work were completed under a grant from the USAF Office of Aerospace Research whose support is gratefully acknowledged. The author also is indebted to M. C. Williams, Assistant Editor of the International Journal of Fracture, for providing the selected rod pull-out bibliography.

REFERENCES

1. M. L. Williams, J. Adhesion, 4, 307 (1972).
2. M. L. Williams and G. P. Anderson, in "Fracture 1977," 4th Intl. Congr. Fract., 1, 643 (1977), University of Waterloo Press, 1978.
3. A. J. Kinloch, J. Matl. Sci., Part I, 15, 2141 (1980); Part II 17, 617 (1982).

4. G. P. Anderson, S. J. Bennett, and K. L. DeVries, "Analysis and Testing of Adhesive Bonds," Academic Press, New York, 1977.
5. B. M. Malyshev and R. L. Salganik, Int. J. Fract. Mech., 1, 114 (1965).
6. C. Atkinson, J. Avila, E. Betz, and R. E. Smelser, J. Mech. Phys. Solids, 30, 97 (1982).
7. M. L. Williams, R. F. Landel, and J. D. Ferry, J. Am. Chem. Soc., 77, 3701 (1955).
8. M. L. Williams, J. Appl. Polym. Sci., 14, 1121 (1970).
9. G. P. Anderson, "Applied Adhesive Fracture Mechanics," Ph.D. Dissertation, Univ. of Utah, Salt Lake City, 1973.
10. M. L. Williams, J. Appl. Mech., 19, 526 (1952); 20, 590 (1953).
11. A. R. Zak, J. Appl. Mech., 31, 150 (1964).
12. M. L. Williams, Int. J. Fract. Mech., 1, 292 (1965).
13. M. L. Williams, Bull. Seismol. Soc. Am., 49, 199 (1959).
14. F. Erdogan, J. Appl. Mech., 32, 403 (1965).
15. V. L. Hein and F. Erdogan, Int. J. Fract., 7, 317 (1971).
16. J. J. Benbow, Proc. Phys. Soc., B75, 697 (1960).
17. M. L. Williams, in "Proc. 4th ICRPG Working Group," San Diego, Appl. Phys. Lab., Johns Hopkins Univ., November 1965; abr. in Proc. 5th U.S. Nat. Congr. Appl. Mech., 460 (1966).
18. M. L. Williams and F. N. Kelley, Rubber Chem. Technol., 42, 1175 (1969).

SELECTED BIBLIOGRAPHY - ROD PULL OUT PROBLEM

Adams, D. F., see Wolrath, D. E.
Allen, G., see Williams, T.
Allred, R. E., and Schuster, D. M. 1973 J. Mater. Sci. 8, 245
Archangelssha, I. N., and Mileiko, S. T. 1976 J. Mater. Sci. 11, 356
Arridge, R. G. C., see Takaku, A.
ASTM 1971 ASTM, D2138-67, Book ASTM Stand. Part 28, 913
ASTM 1971 ASTM, D2630-70, Book ASTM Stand. Part 28, 1009
Atkins, A. G. and Mai, Y. W. 1976 J. Mater. Sci. 11, 2297
Atkins, A. G., see Marston, T. U.

Atkinson, C.	1972	Int. J. Eng. Sci. <u>10</u>, 45
Aucouturier, M., see Rebout, D.		
Aveston, J., and Sillwood, J.M.	1976	J. Mater. Sci. <u>11</u>, 1877
Avila, J.	1980	M.S. Thesis. Analytical and numerical study of the fiber pull-out problem. University of Pittsburgh, Pittsburgh, PA
Bader, M.G., Bailey, J.E., and Bell, I.	1973	J. Phys. D <u>6</u>, 572
Bailey, J.E., see Bader, M.G.		
Barry, P.W.	1978	J. Mater. Sci. <u>13</u>, 2177
Batson, G.	1976	Mat. Sci. Eng. <u>25</u>, 53
Beaumont, P.W.R., and Harris, B.	1972	J. Mater. Sci. <u>7</u>, 1265
Beaumont, P.W.R., and Plimpton, B.	1977	J. Mater. Sci. <u>12</u>, 1853
Beaumont, P.W.R., see Kirk, J.N.		
Beaumont, P.W.R., see Harris, B.		
Becker, E., and Brisbane, J.	1965	Report No. S76. Application of the finite element method to stress and analysis of solid propellant rocket grain. Rohm & Haas Co., Huntsville, AL
Bel'chenko, G.L., and Gubenko, S.L.	1978	Met. Sci. & Heat Treat. <u>20</u>, 551
Bell, I., see Bader, M.G.		
Betz, E.	1978	Report No. SETEC-ME 79-34. Experimental studies of the fiber pull-out problem. University of Pittsburgh, Pittsburgh, PA
Blumentritt, B.F., Vu, B.T., and Cooper, S.L.	1974	Polymer Eng. Sci. <u>14</u>, 633
Blumentritt, B.F., Vu, B.T., and Cooper, S.S.	1975	Composites, <u>6</u>, p. 105
Blumentritt, B.F., see Eagles, D.B.		
Bowden, P.B.	1970	J. Mater. Sci. <u>5</u>, 517

Bowling, J., and Groves, G.W. 1979 J. Mater. Sci. <u>14</u>, 431

Bowling, J., and Groves, G.W. 1979 J. Mater. Sci. <u>14</u>, 443

Brisbane, J., see Becker, E.

Broutman, L.J., see Mallick, P.K.

Broutman, L.J., see Gaggar, S.

Brown, C.B. 1966 J. Franklin Inst. <u>282</u>, 271

Capiati, N.J., and Porter, R.S. 1975 J. Mater. Sci. <u>10</u>, 671

Compton, J., see Lessig, E.T.

Cooper, G.A., and Piggott, M.R. 1977 Advances in Research on the Strength and Fracture of Materials, <u>1</u>, 557, Pergamon Press

Cooper, S.L., see Eagles, D.B.

Cooper, S.L., see Blumentritt, B.F.

Cottrell, A.H. 1964 Proc. Roy. Soc. <u>A282</u>, 2

de Ferran, E.M., see Harris, B.

Eagles, D.B., Blumentritt, B.F., and Cooper, S.L. 1976 J. Appl. Polymer Sci. <u>20</u>, 435

Ebert, L.J., see Kim, H.C.

Ellis, J.H., see Loveless, H.S.

Erdogan, F., and Pacella, A.H. 1974 Int. J. Solids Struct. <u>10</u>, 785

Felbeck, D.K., see Marston, T.U.

Field, J.E., see Gorham, D.A.

Fielding-Russell, G.S., see Nicholson, D.W.

Fielding-Russell, G.S., Nicholson, D.W., and Livingston, D.I. 1979 ASTM STP 694, p. 153

Finello, D., see Marcus, H.L.

Gagger, S., and Broutman, L.J. 1976 Flaw Growth and Fracture, ASTM, 10th Nat. Symp. on Fracture Mechanics, p. 310

Gent, A.N., see Nicholson, D.W.

Gershon, B., and Marom, G. 1975 J. Mater. Sci. <u>10</u>, 1549
Ghesquiere, A. 1976 J. Appl. Polymer Sci. <u>20</u>, 891
Goodier, J.N., and Hsu, C.S. 1956 J. Appl. Mech. <u>76</u>, 271
Gorham, D.A., and Field, J.E. 1976 J. Phys. D<u>9</u>, 1529
Groves, G.W., see Bowling, J.
Groves, G.W., see Morton, J.
Gubenko, S.L., see Bel'chenko, G.L.
Hadjis, N., and Piggott, M.R. 1977 J. Mater. Sci. <u>12</u>, 358
Harris, B., see Beaumont, P.W.R.
Harris, B., Beaumont, P.W.R., and de Ferran, E.M. 1971 J. Mater. Sci. <u>6</u>, 238
Harris, B., Morley, J., and Phillips, D.C. 1975 J. Mater. Sci. <u>10</u>, 2050
Harris, B., see McGuire, M.A.
Hastings, G.W. 1978 Composites <u>9</u>, 193
Henriksen, M., and Thornton, H.R. 1979 SAMPE Qtly. <u>10</u>, 15
Hoover, W.R. 1977 J. Compos. Mater. <u>11</u>, 17
Hsu, C.S., see Goodier, J.N.
Hull, D., see Jones, M.L.C.
Huntsberger, J.R. 1978 Adhesives Age <u>21</u>, 23
Ishikawa, T., Koyama, K., and Kobayashi, S. 1977 J. Compos. Mater. <u>11</u>, 332
Jones, M.L.C., and Hull, D. 1979 J. Mater. Sci. <u>14</u>, 165
Kanninen, M., see Rybicki, E.
Kaufman, M.S., see Williams, T.
Keer, L.M., see Luk, V.K.
Kelly, A., and Zweben, C. 1976 J. Mater. Sci. <u>11</u>, 582
Kendall, K. 1975 J. Mater. Sci. <u>10</u>, 1011
Kim, H.C., and Ebert, L.J. 1978 J. Compos. Mater. <u>12</u>, 139
Kirk, J.N., Monro, M., and Beaumont, P.W.R. 1978 J. Mater. Sci. <u>13</u>, 2197
Kobayashi, S., see Ishikawa, T.

Koyama, K., see Ishikawa, T.
Lawrence, P.	1972	J. Mater. Sci. 7, 1
Lessign, E. T., and Compton, J.	1946	Rubber Chem. Technol. 19, 223

Livingston, D. I., see Nicholson, D. W.
Livingston, D. I., see Fielding-Russell, G. S.

Loveless, H. S., and Ellis, J. H.	1977	J. Test. Eval. 5, 369
Luk, V. K., and Keer, L. M.	1979	Int. J. Solids Struct. 5, 587

Mai, Y. W., see Atkins, A. G.

Majumdar, A. J.	1974	Cement Concrete Res. 4, 247

Majumdar, A. J., see Stucke, M. S.

Mallick, P. K., and Broutman, L. J.	1974	J. Mater. Sci. 9, 1420
Marcus, H. L., and Finello, D.	1978	Analytic Methods for Studying Fiber/Matrix Interface. Texas University at Austin, Mat. Sci. Lab., Cont. N00014-78-C-0094

Marom, G., see Gershon, B.

Marston, T. U., Atkins, A. G., and Felbeck, D. K.	1974	J. Mater. Sci. 9, 447
Maslov, B. V.	1978	Sov. Appl. Mechs. 14, 95
McGuire, M. A., and Harris, B.	1974	J. Phys. D 7, 1788

Mileiko, S. T., see Archangelssha, I. N.

Millman, R. S., and Morley, J. G.	1975	J. Phys. D 8, 1065
Millman, R. S., and Morley, J. G.	1976	Mater. Sci. Eng. 23, 1
Mindlin, R. D.	1936	Physics 7, 195
Miyase, A., and Piekarski, K.	1977	J. Compos. Mater. 11, 33

Morley, J., see Harris, B.

Morley, J. G.	1976	Physics Reports 27, 245

Morley, J. G., see Millman, R. S.

Morton, J., and Groves, G. W.	1974	J. Mater. Sci. 9, 1436
Morton, J., and Groves, G. W.	1975	J. Mater. Sci. 10, 170

Morton, J., and Groves, G.W. 1976 J. Mater. Sci. 11, 617
Muki, R., and Sternberg, E. 1969 Int. J. Solids Struct. 5, 587
Muki, R., and Sternberg, E. 1970 Int. J. Solids Struct. 6, 69
Munro, M., see Kirk, J.N.
Nicholson, D.W., Livingston, 1979 Tire Sci. Technol. 6, 114
    D.I., and Fielding-
    Russell, G.S.
Nicholson, D.W., Livingston, 1978 Tire Sci. Technol. 6, 71
    D.I., Fielding-Russell,
    G.S., and Gent, A.N.
O'Brien, T.K., and 1977 J. Test. Eval. 5, 384
    Reifsnider, K.L.
Pacella, A.H., see
    Erdogan, F.
Phillips, D.C., see Harris, B.
Phillips, D.C. 1974 J. Mater. Sci. 9, 1847
Piekarski, K., see
    Miyase, A.
Piggott, M.R., see
    Cooper, G.A.
Piggott, M.R. 1970 J. Mater. Sci. 5, 669
Piggott, M.R. 1974 J. Mater. Sci. 9, 494
Piggott, M.R., see Hadjis, N.
Pinchin, D.J. 1976 J. Mater. Sci. 11, 1578
Pinchin, D.J., and 1974 J. Mater. Sci. 9, 300
    Woodhams, R.T.
Plimpton, B., see
    Beaumont, P.W.R.
Porter, R.S., see
    Capiati, N.J.
Rebout, D., Stohr, J.F., 1978 J. Mater. Sci. 13, 2333
    and Aucouturier, M.
Reifsnider, K.L., see
    O'Brien, T.K.
Rybicki, E., and 1977 Hybrid and Select Metal
    Kanninen, M.   Matrix Composites. AIAA
        Journal, 53
Schuster, D.M., see
    Allred, R.E.
Short, D., see
    Summerscales, J.
Sillwood, J.M., see
    Aveston, J.

Skolnik, L.	1974	Rubber Chem. Technol. $\underline{47}$, 434
Sternberg, E., see Muki, R.		
Stohr, J.F., see Rebout, D.		
Stucke, M.S., and Majumdar, A.J.	1976	J. Mater. Sci. $\underline{11}$, 1019
Summerscales, J., and Short, D.	1978	Composites $\underline{9}$, 157
Takaku, A., and Arridge, R.G.C.	1973	J. Phys. D $\underline{6}$, 2038
Thornton, H.R., see Henriksen, M.		
Tuler, F.R., see Wagner, H.E.		
Vu, B.T., see Blumentritt, B.F.		
Wagner, H.D., and Tuler, F.R.	1979	J. Mater. Sci. $\underline{14}$, 500
Weetsman, J.	1977	J. Compos. Mater. $\underline{11}$, 378
Williams, T., Allen, G., and Kaufman, M.S.	1973	J. Mater. Sci. $\underline{8}$, 1765
Wolrath, D.E., and Adams, D.F.	1976	J. Compos. Mater. $\underline{10}$, 44
Woodhams, R.T., see Pinchin, D.J.		
Yamaki, J.	1976	J. Phys. D $\underline{9}$, 115
Zak, A.R.	1964	J. Appl. Mech. $\underline{31}$, 150
Zweben, C., see Kelly, A.		

APPENDIX A

The Stress Distribution along a Bi-material Interface of Finite Length

When the character of the stress singularity near the crack tip in a bi-material body was determined,[13] the region over which the major oscillations extended was indeterminate because there was no need for a characteristic length parameter in the analysis. This part of the singularity behavior was studied by Malyshev and Salganik[5] and Erdogan[14] who showed that the violent part of the oscillations was confined to a very small region near the crack tip. This feature, and the co-existence of shear and normal stresses along the interface, can be illustrated by using a simple example.

The Muskhelishvili complex function approach was chosen by Anderson et al[4] to illustrate the stresses for the thin plate geometry subjected to a combined shear (P) and normal load (Q) as shown in the insert.

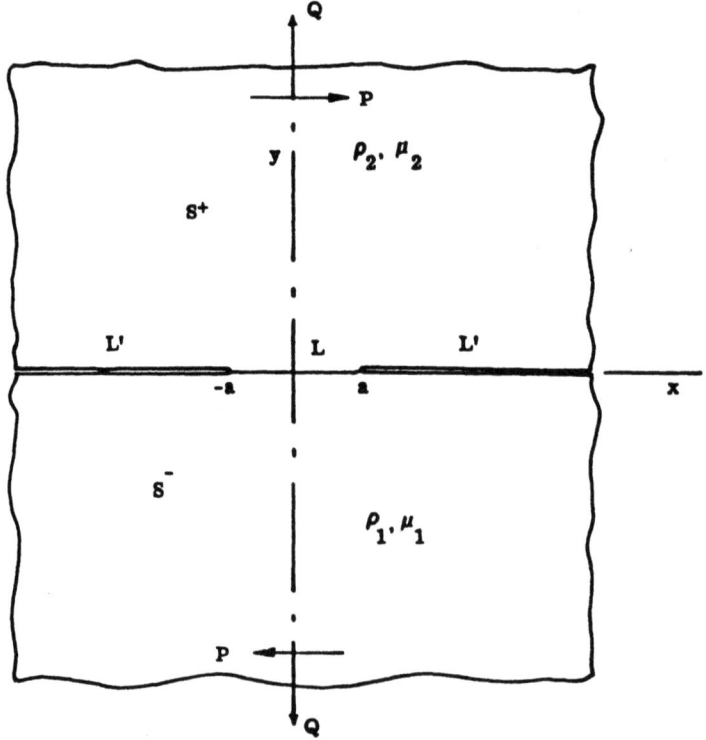

Two semi-infinite crack geometry and loads analyzed.

CONTINUUM MECHANICS FACTORS IN ADHESIVE FRACTURE 735

The parameters which reflect the different material properties of the upper (S^+) and lower (S^-) regions are defined as

$$\alpha = \frac{(\rho_1/\mu_1) + (1/\mu_2)}{(\rho_2/\mu_2) + (1/\mu_1)}$$

$$\zeta = (1/2\pi) \ln \alpha$$

$$\rho_k = 3 - 4\nu_k \quad : \text{plane strain}$$

$$= \frac{3 - \nu_k}{1 + \nu_k} \quad : \text{plane stress}$$

in which μ_k is the shear modulus in the material ($k = 1, 2$) and ν_k is the appropriate Poisson's ratio.

Without elaborating upon the entire solution,[4] the shear stress (τ_{xy}) and normal stress (σ_y) on the bond line near the crack tip can be found, in the vicinity ($y = 0$, $x = a - \xi$) of the bond line, to be

$$\sigma_y(\xi, 0) = \frac{(1 + \alpha^{-1}) \exp(\pi\zeta)}{2\pi\sqrt{2a\xi}} \left[Q \cos\left(\zeta \ln \frac{\xi}{2a}\right) + P \sin\left(\zeta \ln \frac{\xi}{2a}\right) \right] \quad (A-1)$$

$$\tau_{xy}(\xi, 0) = \frac{(1 + \alpha^{-1}) \exp(\pi\zeta)}{2\pi\sqrt{2a\xi}} \left[P \cos\left(\zeta \ln \frac{\xi}{2a}\right) - Q \sin\left(\zeta \ln \frac{\xi}{2a}\right) \right] \quad (A-2)$$

For a Mode I loading, i.e., $P = 0$,

$$\left.\frac{\tau_{xy}}{\sigma_y}\right|_I = -\tan\left(\zeta \ln \frac{\xi}{2a}\right) \quad (A-3)$$

For a Mode II loading, i.e., $Q = 0$,

$$\left.\frac{\sigma_y}{\tau_{xy}}\right|_{II} = \tan\left(\zeta \ln \frac{\xi}{2a}\right) \quad (A-4)$$

The first of these, (A-3), represents the contribution of shear stress for Mode I loading, which in the case of identical materials ($\alpha = 1$, $\zeta = 0$) would be zero. Similarly, the second, (A-4), gives the normal to shear stress ratio for Mode II. Note that

$$(-\tau_{xy}/\sigma_y)_{I\,(P=0)} = (\sigma_y/\tau_{xy})_{II\,(Q=0)} = \tan\left(\zeta \ln \frac{\xi}{2a}\right) \quad (A-5)$$

A maximum of this function, (A-5), occurs at

$$\frac{1}{2\pi} \ln \alpha \ln \varepsilon/2a = \pm \pi/2$$

at which value the "negligible" component becomes infinitely large! The question arises: For what material properties or location can this situation occur? For example, if one of the plane strain media has an infinite shear modulus, say $\mu_2 \to \infty$, then $\alpha = \rho_1 = 3 - 4\nu_1$, or

$$\frac{1}{2\pi} [\ln(3 - 4\nu_1)] \ln[\varepsilon/2a] = -\frac{\pi}{2}$$

Now if the Poisson's ratio of the material bonded to it is very nearly incompressible, say $\nu_1 = 0.49$, then

$$\ln \frac{\varepsilon}{2a} = -\frac{\pi^2}{\ln 1.04} \approx -\frac{\pi^2}{0.04} \approx -250$$

which would be predicted at $\varepsilon/2a \sim \exp(-250)$ which is too small a distance for that magnitude to hold much significance in continuum mechanics. Or, alternatively, selecting a small distance from the crack tip, say $\varepsilon/a = 0.1$,

$$\tan\left[\frac{1}{2\pi} \ln(3 - 4\nu_1) \ln(\varepsilon/2a)\right] = \tan\left[\frac{0.04}{2\pi}(-3.00)\right] \sim 0.02$$

and the shear stress is a reasonably small proportion of the normal stress that close to the crack. (Note that if the material is ideally incompressible, then $\nu_1 = 1/2$ and the tangent function is identically zero.)

The important point to note is that a simple calculation of (A-5) applicable to specific materials of immediate interest provides a guideline as to the relative importance of the shear-normal interaction in the vicinity of a crack tip along the interface.

The following table shows the variation with distance for a plane strain case in which $E_1/E_2 = 10$, $\nu_1 = 0.3$, $\nu_2 = 0.4$.

ε/a	$\tan[\xi \ln(\varepsilon/2a)]$
10^{-1}	0.12
10^{-2}	0.21
10^{-3}	0.30
10^{-4}	0.40

which for this particular case would as a practical matter probably correspond to a maximum shear stress which was of the order of 15 percent of the normal stress.

FRACTURE ENERGETICS OF ADHESIVE JOINTS

Lieng-Huang Lee

Webster Research Center
Xerox Corporation
Webster, New York   14580

In this review, we relate various fracture energetics theories to the types of polymers commonly used in the adhesive joints. The energy-balance concept of Griffith's fracture criterion is emphasized. The same concept led to the development of various methods of determining the fracture energies (or the failure energies) of both rigid and flexible adhesive joints.

I. INTRODUCTION

Fracture mechanics has been applied successfully[1] during the last decade to assessing the strengths of adhesive joints, either by the energy-balance or by the stress-intensity factor approach. The former uses fracture energy, G_c, as the fracture criterion; the latter uses fracture toughness, K_c. Though G_c and K_c are related for linear elastic systems through the Irwin's treatment[2], G_c is still generally preferred in expressing the adhesive strenth. One of the major difficulties in obtaining K_c directly and accurately is that significant differences[3] may be observed in the stress field from those estimated for a homogeneous material outside a very short distance in front of the crack tip.

Many papers in the literature emphasize the energy-balance approach derived from the Griffith's criterion;[4] however, they are sometimes contradictory. Thus, one of the purposes of this paper is to define the application of each approach with respect to the type of polymer, e.g., brittle, ductile, or viscoelastic. Subsequently, we intend to show how the fracture energetics of adhesive joints are derived from the cohesive fracture properties of the three types of polymers.

Fracture energetics of brittle, glassy polymers[5] have been discussed in terms of effective fracture surface energy, Γ, fracture energy, G_c, and fracture toughness, K_c. Fracture energetics of ductile and viscoelastic polymers have been treated in a separate paper.[6] For ductile polymers, the critical J-integral[7] (or ductile fracture energy), J_c, is a better measure than G_c. For viscoelastic polymers, T, the tear energy,[8] is frequently determined in terms of G_c. In fact, the actual equivalence for G_c should be T_c, the critical tear energy (or elastomeric fracture energy). In the recent generalized theory of fracture mechanics,[26] the failure energy T is claimed to embrace G_c, T_c and J_c.

In practice, it is difficult to classify an adhesive joint as brittle, ductile, or viscoelastic. For technologists, it is easier to differentiate rigid from flexible joints. Thus, in the following discussion, we shall treat adhesive joints according to the latter classification but with the three types of polymers in mind.

II. FRACTURE ENERGETICS OF RIGID ADHESIVE JOINTS

1. Measurement of Fracture Energy

Early work by Ripling, Mostovoy and Patrick[9] involved the application of linear elastic fracture mechanics (LEFM) for the determination of crack growth of rigid adhesive joints. Originally, the double-cantilever beam was used to measure fracture energies,

G_{Ic} (Mode I) and G_{IIc} (Mode II). Since then, the tapered double-cantilever beam has been more widely used.[10] Assuming that an adhesive behaves linearly, its total fracture energy (Mode I) can be expressed as:

$$G_{Ic} = \frac{P^2}{2B}\left(\frac{\partial C}{\partial a}\right)_p , \qquad (1)$$

where p is the wedge load; B (originally b) is the thickness of the specimen; C is the compliance, and $(\partial C/\partial a)_p$ can be determined from a specific specimen geometry.

Bascom's results[11] on fracture energies of an epoxy adhesive are shown in Table I. The calculated values of Γ and K_c are also included for comparison. The fracture energy of the bulk is slightly higher than that of the resin in the joint when Mode I is concerned. However, they are nearly equal when the mixed mode of fracture is taken into account.

2. Intrinsic Fracture Energy

When the crack speed is very low, or the polymer is truly linear-elastic, the fracture energy, G_c, should approach the intrinsic fracture energy, $(G_c)_o$. Under normal conditions, G_c consists chiefly of the work of plastic deformation for each surface, W_d:

$$G_c = 2\Gamma = 2(\gamma_s + W_d). \qquad (2)$$

However, at the threshold condition, $W_d \to 0$, and $G_c \to (G_c)_o$, so,

$$(G_c)_o = 2\gamma_s = W_a. \qquad (3)$$

In other words, under the limiting conditions, the fracture energy approaches the thermodynamic work of adhesion, especially for truly viscoelastic solids. For plastically deforming solids, some residual plastic work must be done to sever intermolecular bonds and create surface. This often takes the form of craze when homogeneous plastics flow is no longer present. Crazing energies are much greater than W_a.

Under normal conditions, the adhesive fracture energy, $(G_c)^{ad}$, is affected by the adhesive composition, crack speed, temperature, and environmental variations. One of the major factors influencing $(G_c)^{ad}$ is reinforcement with elastomers.

3. Toughening Mechanism

Reinforcement of a brittle resin (or adhesive) with an elastomer greatly increases the fracture energy of the adhesive. This

Table I. Effective Fracture Surface Energy, Fracture Energy, and Fracture Toughness of Epoxy Polymer and Adhesive*.

Epoxy Resin (HHPA-DGEBA)	Modulus (10^3MN/M^2)	Modulus (10^{10}dyne/cm^2)	Γ (Calc) (J/M^2)	Γ (Calc) (10^5erg/cm^2)	G_c (J/M^2)	G_c (10^5erg/cm^2)	K_c (Calc) (10^8N/M$^{3/2}$)	K_c (Calc) (10^{10}dyne/cm$^{3/2}$)
Bulk	3.86	3.86	68.0	0.68	136.0	1.36	5.3	5.3
Adhesive Mode I	3.86	3.86	58.0	0.58	116.0	1.16	4.5	4.5
Mixed Modes (I and II)	3.86	3.86	70.0	0.70	140.0	1.40	5.4**	5.4

* Ref: Modulus and fracture energy data were obtained from the paper by W.D. Bascom, C.O. Timmons and R.L. Jones, J. Mat. Sci., 10, 1037 (1975).

** The calculation for $K_{I,IIC}$ was assumed to be that for K_{IC}.

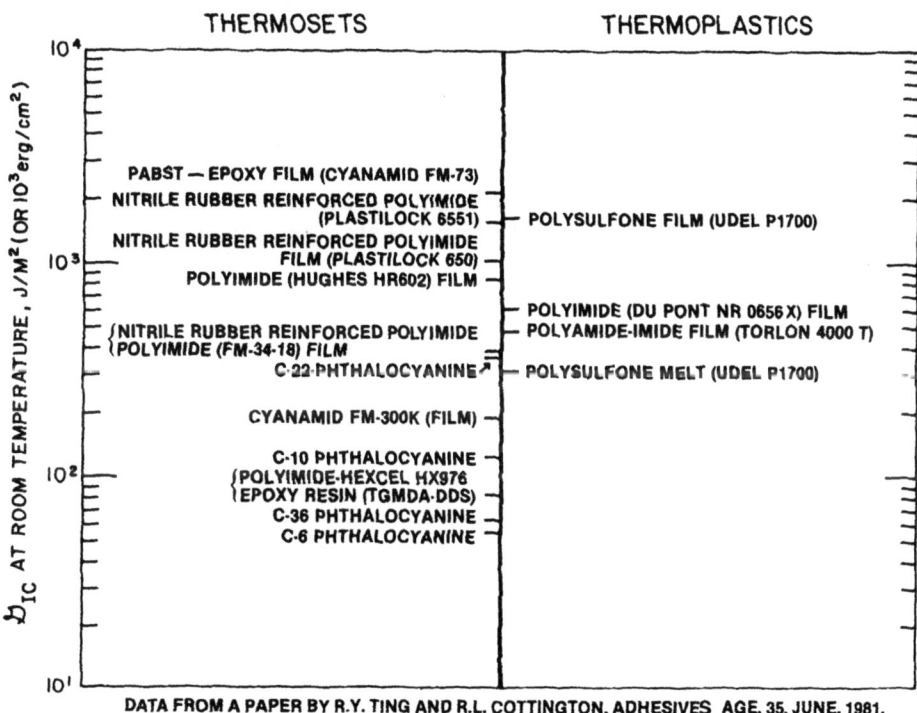

Figure 1. Fracture energies of thermosets and thermoplastics. (Data taken from a paper by R.Y. Ting and R.L. Cottington, Adhesives Age, 24, 35, June, 1981).

is illustrated by Bascom's results on the reinforcement of an epoxy resin with a CTBN nitrile rubber (Table II).

Recently, Ting and Cottington[13] published a series of G_c values for high performance adhesives (Fig. 1). In Fig. 1, we note that the reinforcement of polyimide with an elastomer also significantly increases the fracture energy.

Bascom, et al., theorize[11] that the mechanism of reinforcement is the increase of the plastic zone size at the crack tip with the presence of the rubber particles. Since most polymers are more or less viscoelastic, a quantitative analysis of the plastic deformation zone can involve the time-temperature dependent viscoelastic properties. The results, if attainable, would be too complex for illustration. Thus, an elastic-plastic model has been used with the assumption that within the deformation zone there is a wedge of material at yield stress, σ_y, but at the failure strain, ε_f. Surrounding this wedge, the material is at the yield stress, σ_y, and strain, ε_y outside the elastic-plastic boundary which envelopes a volume with a diameter of $2r_y$ (Fig. 2).

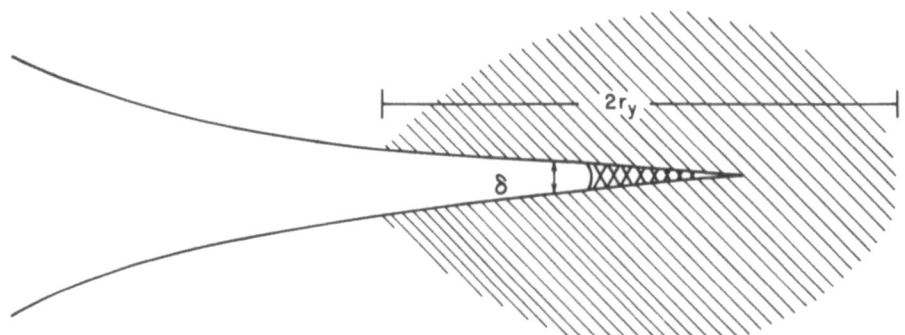

Figure 2. Crack opening displacement model of an elastic-plastic material. W.D. Bascom et al., J. Appl. Polym. Sci., <u>19</u>, 2545 (1975).

From Griffith's equation, Bascom[11] derived the relation between G_{Ic} and $(r_y)_c$:

$$G_{Ic} \simeq 6\pi\sigma_y\varepsilon_y(r_y)_c . \qquad (4)$$

Table II. Effective Fracture Surface Energies and Fracture Energies of Elastomer-modified Epoxy Resins.

Adhesive Resin	Γ (Calc)				G_c			
	Cleavage (I) (J/M²)	Cleavage (I) (10⁵erg/cm²)	Cleavage+Shear(I,II) (J/M²)	Cleavage+Shear(I,II) (10⁵erg/cm²)	Cleavage (I) (J/M²)	Cleavage (I) (10⁵erg/cm²)	Cleavage+Shear(I,II) (J/M²)	Cleavage+Shear(I,II) (10⁵erg/cm²)
Unmodified Epoxy (HHPA-DGEBA)	58.0	0.58	68.0	0.68	116.0	1.16	136.0	1.36
CTBN-Epoxy (Piperidine-DGEBA)								
10% Elastomer	1750	17.5	55.0	0.55	3500	35.0	110.0	1.1
30% Elastomer	1100	11.0	55.0	0.55	2200	22.0	110.0	1.1
Commercial								
Elastomer-Epoxy	1150	11.5	435	4.35	2300	23.0	870	8.7
Nylon-Epoxy	3050	30.5	375	3.75	6100	61.0	750	7.5

Ref: W.D. Bascom, R.L. Cottington and C.O. Timmons, Appl. Poly. Sym. 32, 165 (1977).

Strictly speaking, neither G_{Ic} nor $(r_y)_c$ is a true material constant because both of them are time-temperature dependent. Furthermore, G_{Ic} is also affected by the thickness of the specimen. It should also be emphasized that Griffith equation is based on linear elastic assumptions which are not valid for a material which yields. Equations such as (4) and (5), therefore, are approximations for <u>small</u> amounts of plastic deformation and are invalid if significant plastic zones are formed.

4. Prediction of Long-Term Stability

Since G_{Ic} depends on many variables, it cannot be used as a single parameter for predicting the long-term stability of an adhesive joint. The plastic zone size r_y discussed in the preceding paragraph should be considered an important parameter for predicting long-term stability.

Under plane strain condition, at fracture, the plastic zone size is:

$$(r_y)_c = \frac{1}{6\pi(1-\nu^2)\varepsilon_y^2} \left[\frac{G_{Ic}}{E(t)}\right], \qquad (5)$$

where $E(t)$ is time-dependent modulus. According to Eq. (5), $(r_y)_c$ is a function of $G_{Ic}/E(t)$. Thus, $(r_y)_c$ can only become stable or constant when $G_{Ic}/E(t)$ is constant. Therefore, $(r_y)_c$ is a failure criterion for the long-term stability of a structural adhesive. This criterion has been verified by Williams et al.,[15,16] for polymethyl methacrylate and polycarbonate. A value of $(r_y)_c$ for an epoxy adhesive has been calculated to be 16 μm.

The latter study is based on a Dugdale line-plastic zone model.[17] Based on this model, the crack opening displacement (COD) at fracture, δ_c, may be related to G_{Ic} and $E(t)$ as follows:

$$\delta_c = \frac{1}{(1-\nu^2)\varepsilon_y} \left[\frac{G_{Ic}}{E(t)}\right] \qquad (6)$$

For an epoxy resin, δ_c has been calculated to be 4.5 μm.

5. The Emerging Acceptance of the J-Integral

The importance of the plastic deformation at the crack-tip could signify the gradual acceptance of using Rice's J-Integral[7] for determining fracture energies of ductile polymers as well as adhesives. The J-integral[7] can be determined experimentally according to the following equation:

FRACTURE ENERGETICS OF ADHESIVE JOINTS

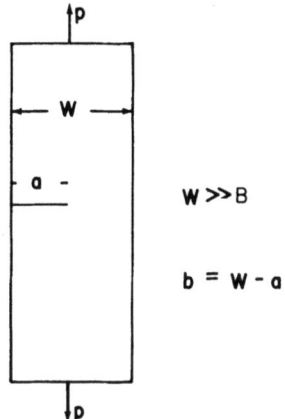

Figure 3. Tensile specimen for the J-integral

$$J = \frac{2U_p}{B(w-a)} = \frac{2U_p}{Bb}, \quad (7)$$

where U_p is the total potential energy; a is the crack length at the notch edge; B is thickness of the test piece, w is the width and b is the ligand length (Fig. 3).

At the crack initiation point, J takes the critical value, J_c, which is called the ductile fracture energy. Its value is:

$$J_c = \frac{2(U_p)_{max}}{Bb}, \quad (8)$$

where $(U_p)_{max}$ is the maximum potential energy corresponding to the initiation of crack growth. In general, $J_c > G_c$ for ductile polymers because J_c more inclusively accounts for the plasticity component than does G_c.[18] A series of J_c values for polymers has been determined by Ferguson et al.[14] (Fig. 4). It is likely that J_c values for more adhesives[20] will be made available in the future.

III. FRACTURE ENERGETICS OF FLEXIBLE ADHESIVE JOINTS

Fracture energetics of elastomers have been developed by Rivlin and Thomas.[8] Their work was originally derived from Griffith's fracture criterion.[4] Thus, the fracture mechanics

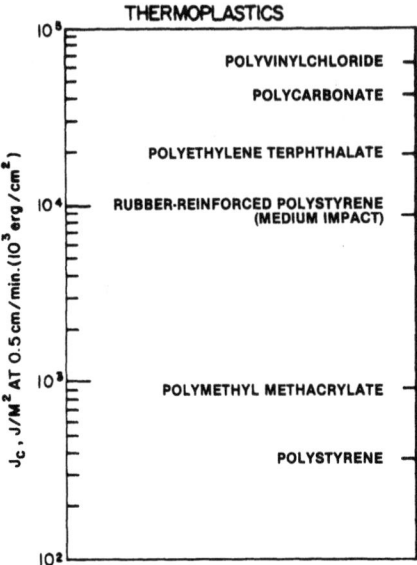

Figure 4. J_c values for thermoplastics. (Data taken from a paper by R.J. Ferguson, G.P. Marshall and J.G. Williams, Polymer 14, 451 (1973)).

of flexible adhesive joints is basically similar to that of rigid adhesive joints.

In general, the adhesive failure energy, θ, described by Gent and Kinloch[21] is essentially equivalent to the cohesive fracture energy, G_c or T_c, discussed previously.[5,6]

The measurements of the adhesive failure energy can be carried out in the three forms of Eqs (9), (10), and (11).[21]

Simple extension (Fig. 5):

$$(G_c)^{ad} = \theta = kaU_b, \qquad (9)$$

Pure shear (Fig. 6):

$$(G_c)^{ad} = \theta = h_o U_b, \qquad (10)$$

and

Peeling (Fig. 6):

$$(G_c)^{ad} = \theta = P'/B_o, \qquad (11)$$

where U_b is the strain energy density at break (or strain energy per unit volume); h_o is the unstrained height; B_o (originally t_o) is the unstrained thickness, and P' is the peel force per unit

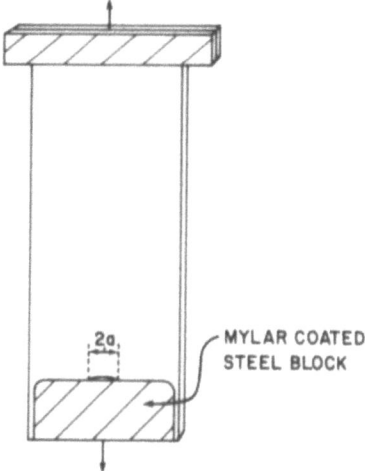

Figure 5. Sketch of simple extension testpiece. (A.N. Gent and A.J. Kinloch, J. Polym. Sci., A-2, 9, 659 (1971)).

width of the adhesive layer. For simple extension, only one side of the crack plane, i.e., the elastomer side, is considered; for the cohesive fracture, two sides are used for the summation.

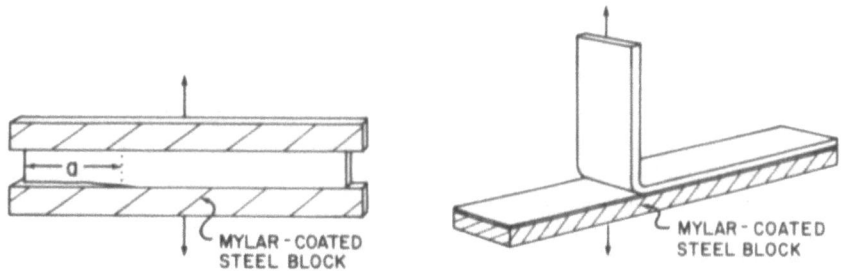

SKETCH OF PURE SHEAR TESTPIECE.   SKETCH OF PEEL TESTPIECE.

Figure 6. Sketches of shear and peel test pieces. (A.N. Gent and A.J. Kinloch, J. Polym. Sci., A-2, 9, 659 (1971)).

For flexible joints, the viscoelastic response of the adhesive is manifested by the dependence of the failure energy upon crack speed and temperature as indicated by the WLF superposition principle.[21]

For simplicity, the adhesive failure energy, θ, has been assumed[22,23] to consist chiefly of the viscoelastic dissipation energy ψ for each surface and the intrinsic failure energy θ_o:

$$\theta = \theta_o + 2\psi. \qquad (12)$$

Thus, when the crack speed is very low or the temperature is high, $\psi \to 0$, and $\theta \to \theta_o$.

The generalized theory of fracture mechanics shows theoretically that θ can also be related to θ_o by a loss function ϕ:[22-26]

$$\theta = \theta_o \phi\,(\dot{a},\,t,\varepsilon_o), \qquad (13)$$

where $\dot{a}$ is the crack speed; t is the temperature, and ε_o is the strain. For cross-linked elastomers, in the absence of a chemical reaction at the interface, θ_o approaches $2\gamma_s$, which is the thermodynamic work of adhesion, W_a. Under limiting conditions, the adhesive failure energy (or adhesive strenth) is equal to the thermodynamic work of adhesion. The data obtained by Andrews and King[25] further illustrates the validity of the relationship indicated in Eq. (13).

IV. CONCLUSIONS

Adhesive fracture energies for both rigid and flexible adhesive joints were discussed separately. For rigid joints, fracture energy G_c is a useful quantity for comparison. However, for predicting long-term stability, we should also consider the plastic zone size, r_y. For ductile polymers (or adhesives) it is more appropriate to measure the ductile fracture energy, J_c.

For flexible adhesive joints, the failure energy θ can be measured by three methods: simple tension, pure shear, and peel. The failure energy can be related to the intrinsic failure energy θ_o by a loss function ϕ. At a very low speed, or at high temperature, $\phi \to 1$, and $\theta \to \theta_o$. If there is no chemical bonding at the interface, the failure energy equals the thermodynamic work of adhesion, W_a. In other words, for an adhesive joint, when the viscoelastic dissipation vanishes, the strength relies solely on the thermodynamic work of adhesion. At the limiting stage, the theory that governs is no longer fracture energetics but surface energetics.

FRACTURE ENERGETICS OF ADHESIVE JOINTS

NOMENCLATURE

a = 1/2 crack length (in the center), or full crack length (at the edge), (cm)

$\dot{a}$ = crack speed (mm/sec)

b = ligand length (cm)

B = thickness (cm)

C = compliance (erg-cm^3)

E = elastic modulus (dyn-cm^{-2})

G_c = fracture energy (erg-cm^{-2})

$(G_c)^{ad}$ = adhesive fracture energy (erg-cm^{-2})

J_c = ductile fracture energy (erg-cm^{-2})

K_c = fracture toughness (dyn-cm$^{-3/2}$)

P = applied wedge load (erg-cm^{-2})

P' = peel force (dyn)

r = radius of the plastic zone (cm)

T = tear energy (erg-cm^{-2})

T_c = elastomeric fracture energy (erg-cm^{-2})

t = temperature (°C)

U_p = total potential energy (erg-cm^{-2})

U_b = strain energy per unit volume (erg-cm^{-2})

W_d = work of deformation for each surface (erg-cm^{-2})

w = width (cm)

γ_s = surface free energy of solid per unit area (erg-cm^{-2})

Γ = effective fracture surface energy (erg-cm^{-2})

δ = crack opening displacement (cm)

ε_o = strain

ν = Poisson's ratio

σ = applied stress (dyn-cm^{-2})

ψ = viscoelastic dissipative work for each surface (erg-cm^{-2})

ϕ = loss function

REFERENCES

1. A.J. Kinloch, J. Materials Sci., 17, 617 (1982).
2. G.R. Irwin, "Fracture" in "Handbuch der Physik," Vol. VI, pp. 551-590, Springer-Verlag, Berlin, 1958.
3. S.S. Wang, J.F. Mandel and F.J. McGarry, Int. J. Fract., 14, 39 (1978).
4. A.A. Griffith, Phil. Trans. Roy. Soc. A221, 163 (1920).
5. L.H. Lee, in Physicochemical Aspects of Polymer Surfaces,"p.523 K.L. Mittal, Editor, Vol. 1, Plenum press, New York (1983).
6. L.H. Lee, in "Adhesive Chemistry - Developments and Trends," L.H. Lee, Editor, Vol. 1, Plenum press, New York (1984).
7. J.R. Rice, J. Appl. Mech. 35, 379 (1968).
8. R.S. Rivlin and A.G. Thomas, J. Polym. Sci., 10, 291 (1953).
9. E.J. Ripling, S. Mostovoy and R.L. Patrick, Mat. Res. Stds., 4, 129 (1964).
10. S. Mostovoy and E.J. Ripling, J. Appl. Polym. Sci., 10, 1351 (1966).
11. W.D. Bascom, C.O. Timmons and R.L. Jones, J. Materials Sci., 10, 1037 (1975).
12. W.D. Bascom, R.L. Cottington and C.O. Timmons, Appl. Polym. Sym., 32, 165 (1977).
13. R.H. Ting and R.L. Cottington, Adhesives Age, 24, 35, June 1981.
14. R.A. Gledhill and A.J. Kinloch, Polym. Eng. Sci., 19, 82 (1979).
15. G.P. Marshall, L.H. Coutts, and J.G. Williams, J. Materials Sci., 9, 1409 (1974).
16. M. Marvin and J.G. Williams, J. Materials Sci., 10, 1883 (1975).
17. D.S. Dugdale, J. Mech. Phys. Solids, 8, 100 (1960).
18. J.D.G. Sumpter and C.E. Turner, "Cracks and Fracture," ASTM STP601, pp. 3-18, American Society for Testing and Materials (1976).
19. R.J. Ferguson, G.P. Marshall, and J.G. Williams, Polymer 14, 451 (1973).
20. S.S. Wang, D.L. Hunston, and A.J. Kinloch, to be published.
21. A.N. Gent and A.J. Kinloch, J. Polym. Sci., A-2, 9, 659 (1971).

22. E.H. Andrews and A.J. Kinloch, Proc. Roy. Soc. London A332, 385 (1973).
23. E.H. Andrews and A.J. Kinloch, Proc. Roy. Soc. London A332, 401 (1973).
24. E.H. Andrews, these proceedings, pp. 689-702.
25. E.H. Andrews and N.E. King, J. Materials Sci., 11, 2004 (1976).
26. E.H. Andrews, J. Materials Sci., 9, 887 (1974).

FRACTURE OF COMPOSITE-ADHESIVE-COMPOSITE SYSTEMS

E. J. Ripling, J. S. Santner and P. B. Crosley

Materials Research Laboratory, Inc.
One Science Road
Glenwood, IL 60425

This program was undertaken to initiate the development of a test method for testing adhesive joints in metal-adhesive-composite systems. The uniform double cantilever beam (UDCB) and the width tapered beam (WTB) specimen geometries were evaluated for measuring Mode I fracture toughness in these systems. The WTB specimen is the preferred geometry in spite of the fact that it is more costly to machine than the UDCB specimen.

The use of loading tabs attached to thin sheets of composites proved to be experimentally unsatisfactory. Consequently, a new system was developed to load thin sheets of adherends. This system allows for the direct measurement of displacement along the load line.

In well made joints separation occurred between the plies rather than in the adhesive.

INTRODUCTION

Adhesive joints fracture by the propagation of crack-like defects, such as unbonded areas, that occur in the glue-line. Fractures of this kind have been studied for more than a dozen years at the Materials Research Laboratory, primarily by Mostovoy and Ripling. These studies described the behavior of metal-adhesive-metal (MAM) systems under a number of loading conditions including continuously increasing loads, alternating load and sustained loads[1]. The behavior of a large number of adhesives was catalogued as a function of joint geometry, test temperature, and environment[2], and it was demonstrated that laboratory test data could be used to predict the fracturing characteristics of complex structures[3]. An ASTM Test Method for measuring the fracture mechanics parameters of adhesive bonds was developed on the basis of these studies.

Progressive separations of the type found in MAM systems are also expected in systems in which one or both of the adherends is a composite. Hence, the study of joint fractures has been extended to include composite adherends. One major reason for carrying out this study was to determine whether or not the test methods developed for metal adherend systems could be extended to include composites, either as composite-adhesive-composite (CAC) or as metal-adhesive-composite (MAC) hybrid systems.

Metal and composite adherends are expected to act differently, and, in general, metal adherend systems are probably easier to classify than composite adherend systems. In most adhesive tests, cracking is produced by elastically bending the adherends so that the elastic modulus of the adherend must be known to measure the toughness of the adhesive. The elastic modulus is essentially constant for a single type of homogeneous metals, such as aluminum alloys. For composites it is not only a function of the reinforcement and matrix species, but also of the lay-up pattern. Second, the apparent bending modulus of metals is only dependent on the beam span-to-height ratio when it is small, while in composites the apparent bending modulus is constant only when the ratio is very large. Finally, and most important, the through-the-thickness toughness of composites is much lower than it is for metals. Hence, for metal adherends, the crack that is initiated in the adhesive is not able to wander into the adherends; this is not the case for composite adherends[8]. These differences place restrictions on specimens designed to measure the fracture mechanics parameters in composite-adherend systems.

TEST MATERIALS AND SPECIMEN ANALYSIS

Test Materials

Two composites were used in this study: both were graphite/epoxy sheets, nominally 0.125 inch (3 mm) thick, having the same number of cross-plies. The ply lay-up, the number of fiber orientations, and total number of plies are given in Table I. Composite A was fabricated to have the 0° plies near the sheet center, where bending stresses are at a minimum, while for composite B the 0° plies were near the surface, where bending stresses are greatest. The two different lay-up patterns were selected so that the two composites would have different flexure moduli.

Table I. Design of Composite Adherends

Designation	Total No. of Plies	No. of Cross Plies	Lay-up Sequence
A	21	4	$[+45/-45/0/90/0/-45/+45]_3$
B	21	4	$[0_3/+45_3/-45_3/90_3/-45_3/+45_3/0_3]$

The adhesive AF-163-2k (0.06 oz/ft^2) utilized in this program is manufactured by the Minnesota Mining and Manufacturing Company. It is a 250F (120C) curing adhesive reported to have excellent water resistance. This lower temperature curing adhesive was selected rather than the 350F (175C) curing ones normally used with composites so that the aluminum adherends could be used without significantly altering their yield strength. The adhesive is generically a modified epoxy.

Specimen Type and Analysis

As was the case for the MAM specimens in Method D3433-75[4], double cantilever beam specimens were selected for this program. Both the uniform height double cantilever beam (UDCB) specimen and the width tapered beam (WTB) specimen were used. Preliminary experiments were conducted on the UDCB specimen. Its geometry is simpler than the WTB specimen geometry and allows the efficient utilization of composite material. The WTB specimen, unlike the UDCB, has the advantage that $\mathcal{L}$ is independent of crack length, and hence its measurement only requires that load be monitored. Unfortunately, it is somewhat more expensive than a UDCB specimen, but its advantage in stress corrosion cracking and fatigue are so great that this additional expense is justified. Both specimen types can be produced by "wide-area" bonding.

Analysis of DCB Specimens

Both the UDCB and WTB specimens have been analyzed earlier[5-8]. However, in using composite adherends it is advantageous to select the adherend thickness so that neither the rotation correction given in Reference 6 nor the large displacement expression given in (7) are needed. Hence, the simplified expression for the crack extension force, $\mathcal{G}$, as given below, is adequate.

For any specimen or structure the compliance, C, and load, P, are related to $\mathcal{G}$ by the expression

$$\mathcal{G} = P^2 C'/(2b) \qquad (2\text{-}1)$$

where $C' = dC/da$, the rate of change of compliance with crack length, and b is the specimen thickness at the crack front position. The compliance at the loading point is the displacement per unit load, i.e., $C = V/P$ where V is the load point displacement. The present concern is the relation between P and V for a crack extending at constant $\mathcal{G}$. The desired relation is obtained by substituting the quantity $P = V/C$ for P in Equation (2-1). This gives

$$P = (2bC\mathcal{G}/C') \cdot (1/V) \qquad (2\text{-}2)$$

For a DCB specimen analyzed as a pair of built-in beams which deform by bending, the compliance change is given by

$$C' = \frac{2a^2}{EI} \qquad (2\text{-}3)$$

where a is the crack length measured from the load line, E is Young's modulus, and I is the moment of inertia of the beam cross section at the crack front position.* For rectangular cross sections

$$C' = 24 \, a^2/(Ebh^3) \qquad (2\text{-}4)$$

The compliance can be obtained by carrying out the integration

$$C = C_o + \int_0^a C' \, da \qquad (2\text{-}5)$$

where C_o can be formally identified as the compliance corresponding to $a = 0$. The integration will depend on the variation with crack

---

* This result is derived in Appendix A.

length of the beam cross section and of Young's modulus.* The simplest case is the uniform DCB specimen where $\underline{b}$ and $\underline{h}$ are independent of $\underline{a}$, and the integration gives

$$C = 8a^3/(Ebh^3) \quad (2\text{-}6)$$

assuming that C_o makes a negligible contribution to the compliance. Substituting Equations (2-4) and (2-6) into Equation (2-2) gives the load-displacement relationship for the uniform DCB specimen:

$$P = (2ab\mathcal{L}/3) \cdot (1/V) \quad (2\text{-}7)$$

A second type of specimen used on this program was the width-tapered-beam (WTB) specimen in which the thickness, $\underline{b}$, increases linearly with the crack length, $\underline{a}$; i.e., $b = a/k$ where $\underline{k}$ is constant. Carrying out the integration indication in Equation (2-5) gives

$$C = C_o + [12k/(Eh^3)] \cdot a^2 \quad (2\text{-}8)$$

Since beam theory is not exact, measured compliance values may differ from Equation (2-8). The procedure then is to fit the experimental compliance data to an equation of the form

$$C = \beta + \alpha a^2 \quad (2\text{-}9)$$

In Equation (2-9) β can be identified with C_o and α is simply a parameter fitted to experimental compliance data. If the beam model were precisely applicable, α would have a value corresponding to $12k/(Eh^3)$; in fact, it might be somewhat different. The compliance derivative, C', obtained from Equation (2-9) is

$$C' = 2\alpha\, a = 2\alpha\, kb \quad (2\text{-}10)$$

and substitution of this value into Equation (2-1) gives

$$\mathcal{L} = P^2 \alpha k \quad (2\text{-}11)$$

To the extent that a single value of α provides a good fit to the compliance date over a range of crack lengths, the relation between $\mathcal{L}$ and P is independent of crack length in that range.

---

* In the composite materials tested on this program the apparent elastic modulus in bending depends on beam length. This situation is addressed in Appendix B.

TEST RESULTS

Three-Point Bend Measurement of Tensile Modulus for Composites A and B

One difficulty encountered in testing DCB specimens with composite adherends, is that the apparent elastic modulus of composites loaded in bending can depend on the adherend's span-to-depth ratio. The results of a recent survey[9] suggest that a span-to-depth ratio of at least 30/1 may be necessary to avoid this effect. As the design and analysis of DCB specimens must take this modulus variation into account, measurements were made on the materials used in this program. Three-point bend specimen tests were conducted according to the ASTM Standard D790[10]. Two series of tests were run to determine whether the modulus was affected by specimen

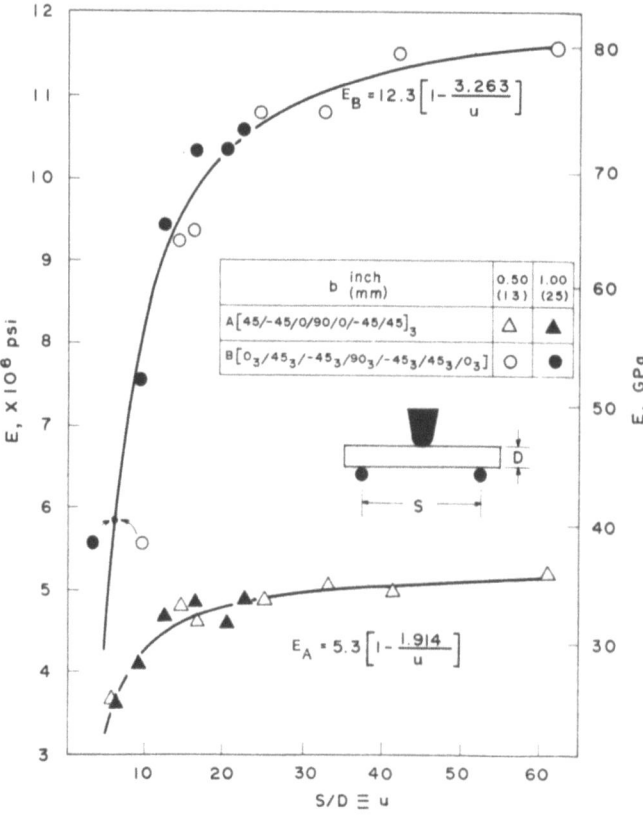

Figure 1. Apparent modulus as a function of span-to-depth ratio, u.

width, b. In the first series, specimens were 0.5 inch (12.7 mm) wide, and in the second, the specimen width was increased to 1.0 inch (25.4 mm). Figure 1 is a plot of the apparent tensile modulus measured as a function of the span-to-depth (S/D) ratio for the specimens with two different widths. The modulus is seen to be independent of width for both composites tested. As expected, placing the 0° plies near the surface of composite B significantly increased its stiffness over composite A, where the 0° plies were located more uniformly throughout the thickness. In addition, the apparent modulus reached a plateau with 90 percent of the saturated value, E_o, near the S/D ratio of 30/1. Composite A reached its 90 percent of the plateau value at a smaller S/D ratio (20/1).

In order to model these effects in the UDCB and WTB specimen geometries, the apparent modulus must be expressed as a function of the crack length, i.e., $E = E(a)$. Several empirical fitting functions having the general shape of Figure 1 were investigated. The best fit was obtained with:

$$E = E_o (1 - u_o/u), \quad u = S/D \tag{3-1}$$

which has a correlation coefficient of better than 0.98 for both composites. In order to relate the crack length, a, in a double cantilever beam to the span in a three-point bend test, the following identifications are made:

$$\frac{S}{2} \rightarrow a, \quad \frac{S_o}{2} \rightarrow a_o, \quad u = \frac{S}{D}, \quad \text{and} \quad u_o = \frac{S_o}{D} \tag{3-2}$$

so that,

$$a_o = \frac{u_o D}{2}, \quad \text{and} \quad \frac{u_o}{u} \rightarrow \frac{a_o}{a} \tag{3-3}$$

Based on the observations for two extremes of composite structure and the correlations between a three-point bend specimen and a double cantilever beam specimen, the requirements for initial crack length were established. A span-to-depth ratio of 30/1 for a three-point beam is equivalent to a ratio of 15/1 for a cantilever beam. Since the composites used in the program were 0.125 inch (3 mm) thick, the initial crack length, a, should be almost two inches (50 mm) long. The initial blunt starter flaw was made shorter than this to allow room for it to jump to form a natural crack on loading. While data is collected for a/h < 15, it is not considered valid until it extends beyond this region. In designing the test specimens for this program, the starter cracks were made to equal about one-half the required value of a/h. In this way, both the specimen size and magnitude of displacements that had to be measured were reduced to manageable sizes. The critical value of a/h = 15 is marked on the test records (Figures 3, 4, 12 and 13).

UDCB Specimen

Specimen shape. Uniform double cantilever beam specimens with the geometry in Figure 2 were fabricated for testing. Individual oversized specimen blanks, approximately 7/8 x 4 inch (22 x 100 mm), were bonded together following the manufacturer's recommended procedure, with the exception of pressure. This value was decreased by half to minimize excessive adhesive run-out experienced during gluing trials with unidirectionally reinforced composites. The purpose of these trials was to establish the characteristic failure mode with composite adherends, and determine the feasibility of measuring adhesive toughness with the UDCB specimen.

Test results of UDCB specimen. Typical load-deflection test records from UDCB specimens with composite A and composite B adherends are shown in Figures 3 and 4, respectively. Both figures show non-linearity associated with the initial loading. In addition, whenever the tests were interrupted to optically measure the crack length, the load decreased by 7-10 percent. This might be due to plastic flow at the crack tip, or slow, subcritical crack extension. No arrest marks or similar features were observed at

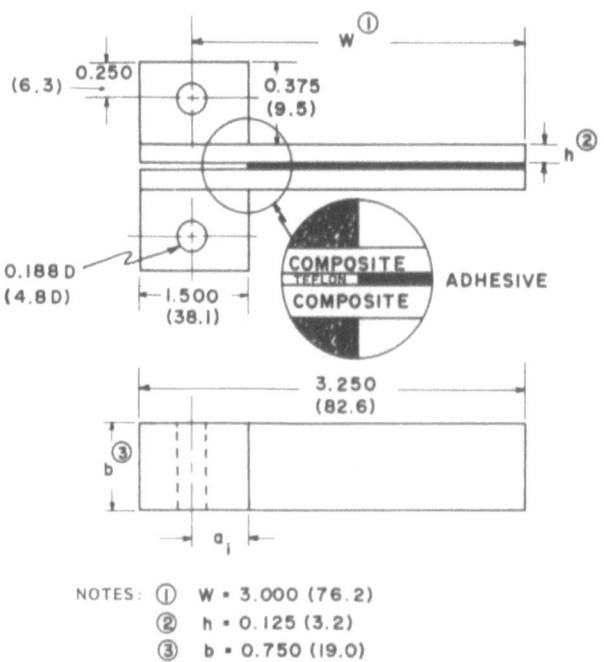

Figure 2. UDCB specimen geometry with W/h = 24. All dimensions are inches (mm).

FRACTURE OF COMPOSITE-ADHESIVE-COMPOSITE SYSTEMS 763

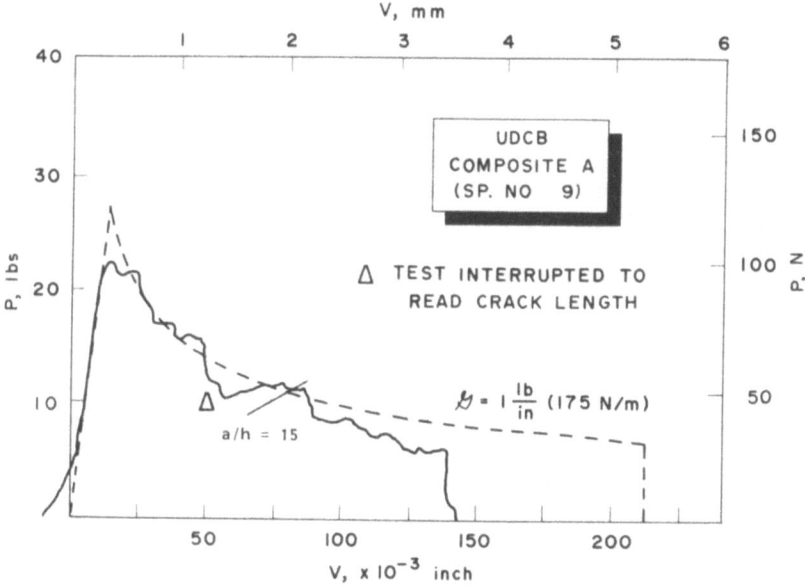

Figure 3. Typical load-deflection curve for similar composite adherends whose failure mode is 100% adhesive failure, i.e., separation at the adhesive-composite interface.

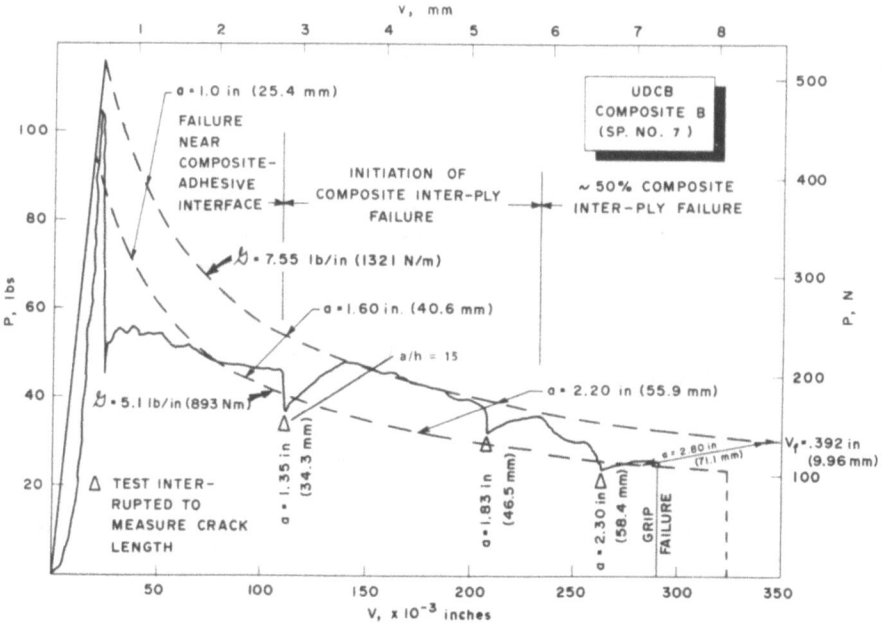

Figure 4. Experimental load-deflection curve for similar composite adherends whose failure mode gradually changes.

those locations using optical fractography up to 50X magnification. Upon continuing the test, the load monotonically increased until the pre-interrupted loads were reached. Further work is needed to understand how sensitive $\mathcal{L}$ is to loading rates.

The broken line in Figure 3 is the P-V curve for $\mathcal{L}$ = 1.0 lb/in (174 N/m) calculated from Equation (2-7). The theoretical curve beyond the initial non-linearity closely matches the experimental line, if the interrupted portion of the test is neglected. The low value of $\mathcal{L}$ is due to the fact that this specimen failed by adhesive failure, i.e., separation at the adhesive-composite interface.

Following the same procedure for the curve in Fig. 4, which is a specimen with B composite adherend, such good agreement was not found. This is due to the fact that the fracture mode changed after the test was stopped to measure the crack length. Failure near the adhesive-composite interface was the primary failure mode until the crack grew to 1.35 inches (34.3 mm). At this point the test was interrupted to measure the crack length. After this interruption, the failure mode changed to include interply separation along one edge of the crack front. The test was interrupted again at a crack length of 1.83 inches (46.5 mm). Approximately half of the crack front failed by interply separation, and the remainder of the crack front continued to fail near the composite-adhesive interface when the crack length reached two inches. Since interrupting the test to measure the crack length is associated with a change in the fracture mode, this procedure appeared to be a questionable practice. Figure 5 shows the macro fracture surfaces of the specimens, which are summarized in Table II.

WTB Specimen

Evaluating the fracture toughness using the UDCB specimen requires the simultaneous measurement of the load and crack length. As noted in the previous section, interrupting a UDCB specimen test to measure the crack length may cause transients to occur. This raises questions of interpreting the test record. The geometry of the WTB specimen is designed so that only knowledge of the load is required to measure $\mathcal{L}$.

<u>Loading thin adherend panels.</u> Loading thin (0.125 inch, i.e., 3.2 mm) adherends with externally attached tabs such as shown in Figure 2, was not a satisfactory experimental arrangement. Separately attaching tabs is an expensive, time consuming process, in which frequent tab failure was experienced. In addition, while several procedures for assuring pin hole alignment were tried, none were completely satisfactory. Therefore, a new approach for loading thin adherend specimens, illustrated in Figure 6, was devised

FRACTURE OF COMPOSITE-ADHESIVE-COMPOSITE SYSTEMS

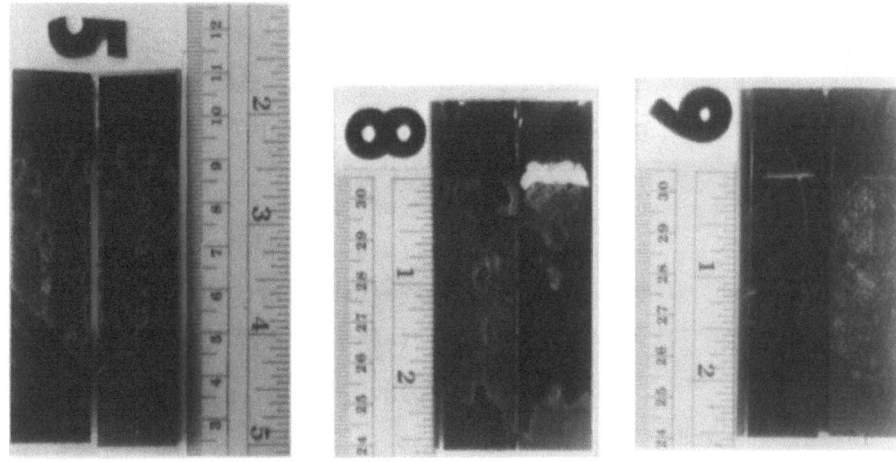

Figure 5a. Fractographs of UDCB specimens with matching composite A adherends.

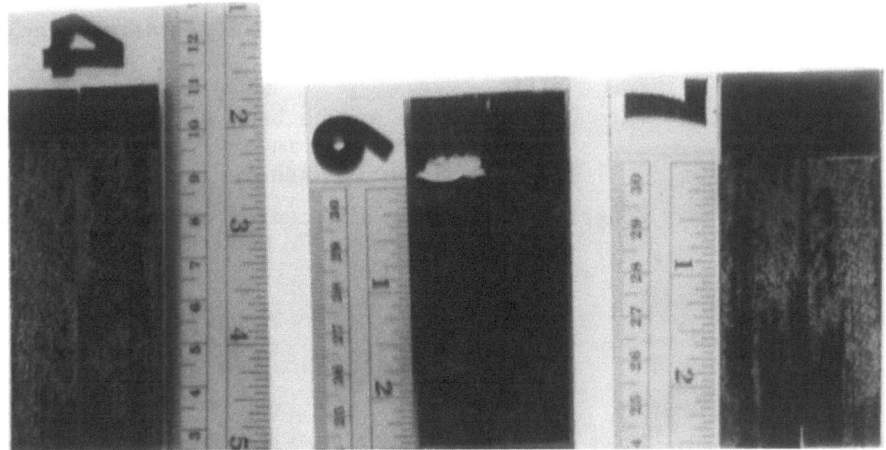

Figure 5b. Fractographs of UDCB specimens with matching composite B adherends.

Table II. UDCB Specimen Test Results

Spec. No.	Adherend	Bond Line Thickness mils (mm)	Failure Mode*
4	B	9(0.23)	CF(a/h<24)/BF(0°)[on edges for (a/h<24)]
6	B	6(0.15)	AF(a/h<24)
7	B	4(0.10)	CF(a/h<16)/BF(0°)[a/h>16]
5	A	5(0.13)	AF(a/h≲16)/BF(0°)[a/h>16]
8	A	5(0.13)	AF(a/h<24)/BF(0°)[7.2<a/h<9.1]
9	A	6(0.15)	AF(a/h<24)

* Failure modes in adhesive tests.

Failure Mode	Description
AF	Adhesive failure: separation at the adhesive-adherend interface.
BF (x°)	Beam failure: failure of the adherend.
	(x = 90°: fracture of the beam arm perpendicular to the bond line.
	(x = 0°: fracture of the beam arm parallel to the bond line (interply separation for composite adherends).
CF	Cohesive failure in the center of the bond line.

which could be applied to either a UDCB specimen or a WTB specimen. However, in the present program, this approach was used only on WTB specimens with the geometry illustrated in Figure 7.

The unique feature of this loading arrangement was the use of the same member for applying the load and measuring displacement. In this way no special effort is required to be certain the displacements being measured are those from the load line. The details of this loading arrangement are shown in Appendix C.

FRACTURE OF COMPOSITE-ADHESIVE-COMPOSITE SYSTEMS

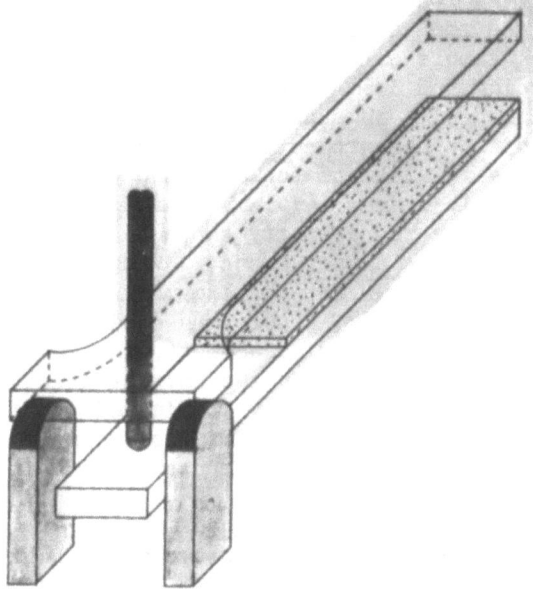

Figure 6. Loading concept for thin adherend specimens.

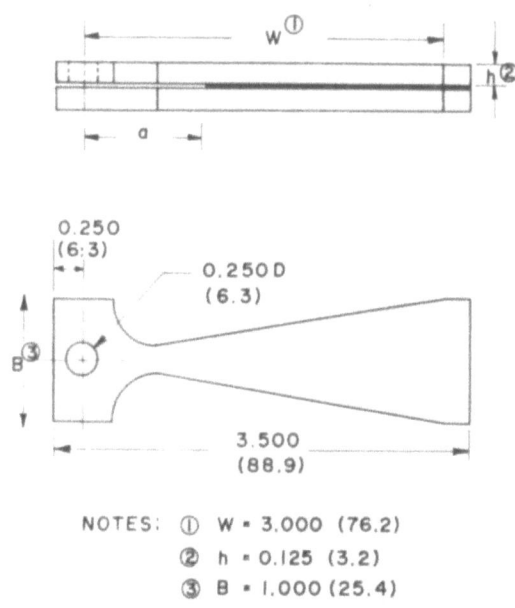

NOTES: ① W = 3.000 (76.2)
② h = 0.125 (3.2)
③ B = 1.000 (25.4)

Figure 7. WTB specimen geometry. All dimensions are inches (mm).

To prevent the load rod from coming into contact with the access hole in the top adherend as the specimen is opened, its nose was tapered. A simple calculation (see Figure 8) of geometrically wedging the loading rod in the top adherend for the worst case specimen (WTB specimen with an adherend modulus of 5.2×10^6 psi, i.e., 35.8 GPa) showed that the projected diameter, d_p, decreases by 95 percent, while the rotation of the adherends causes an offset, e, of approximately 0.17 inches (4.4 mm) at a displacement of 1.0 inch (25 mm). The loading rod geometry shown in Figure C-2 would physically wedge at this displacement. Wedging effects were not noticed during this program for displacements less than about 0.5 inches (13 mm). Since the WTB specimen equilibrium load is well established well before this displacement, the wedging problem was of little practical importance in the current study. However, the potential problem can be completely eliminated by replacing the hole in the top adherend with a slot. The slot geometry would also allow easier loading of the specimen into the test fixture.

<u>Compliance measurement of WTB specimens.</u> Aluminum alloy 2024-T73 was used as the adherend for the compliance measurement. Six specimens were prepared with initial crack lengths ranging

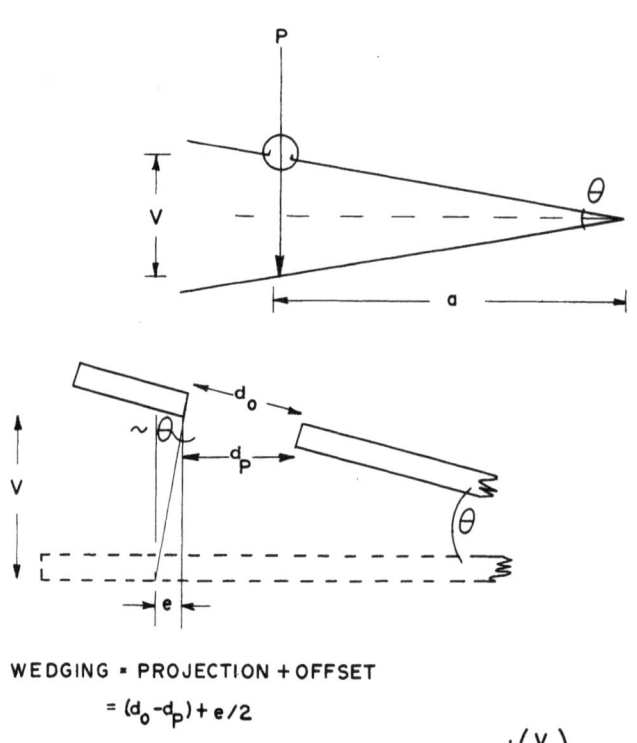

WEDGING = PROJECTION + OFFSET

$= (d_0 - d_p) + e/2$

$= d_0(1-\cos\theta) + 1/2\, V \tan\theta, \quad \theta = \tan^{-1}\left(\dfrac{V}{2a}\right)$

Figure 8. Geometrical wedging of the loading rod during the opening of a cantilever beam specimen.

from 3/8 inch to 2¼ inch (9.5 mm to 57 mm) following the procedure outlined in Appendix D. Each specimen was loaded, removed from the test fixture and re-loaded six times. The reported compliance is the average of these trials. A standard deviation of five percent was observed at each crack length tested. As the WTB specimen compliance is proportional to the crack length squared [see Equation (2-8)], the results were plotted on this scale in Figure 9. A linear least squares regression was used to determine the coefficients in an equation of the form:

$$C = \beta + \alpha a^2 \qquad (2-9)$$

A correlation coefficient of greater than 0.99 statistically confirms what is obvious from the plot. It is noteworthy that the intercept, β, is nearly zero, and that the fitted slope $\alpha = 1.74 \times 10^{-3}$ (in-lb)$^{-1}$ is identical to the theoretical value, $\alpha = 12k/Eh^3$.

<u>WTB test results and fractography.</u> Test results for the WTB specimen composite adherends are summarized in Table III. Fractographs of the first four specimens are shown in Figure 10. These

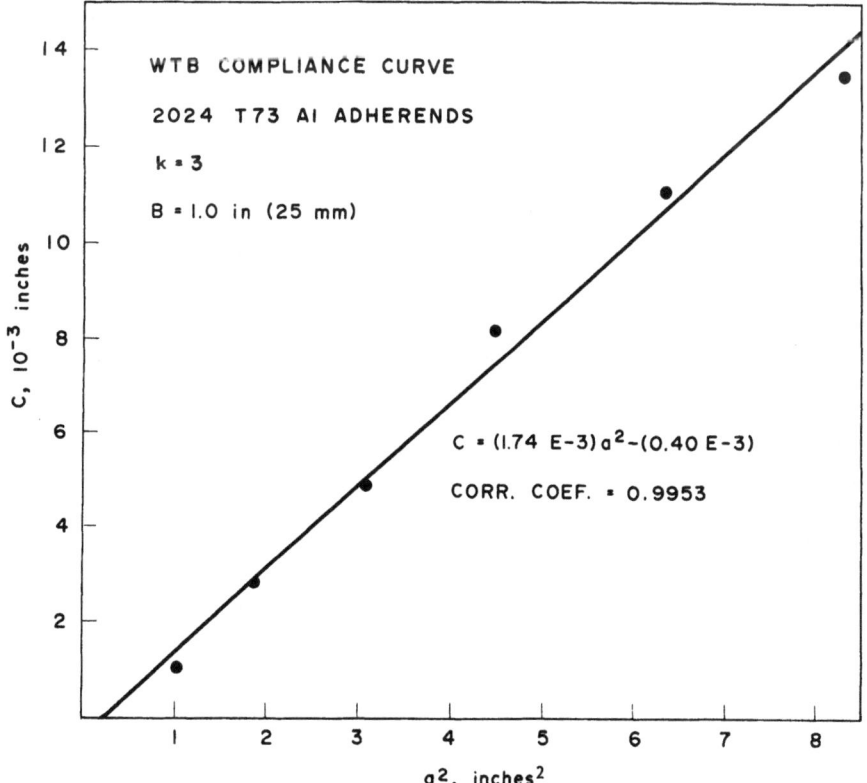

Figure 9. WTB specimen [k = 3, B = 1.0 inch (25 mm)] compliance.

Table III. Composite-Adhesive-Composite Test Results for WTB Specimens.

Outer Ply Orientation	Spec. No.	E 10^6 psi (GPa)	P lbs, (N)	$\mathcal{R}$ lb/in (N/m)	Primary Failure Mode
45°	A1	5.3 (36.6)	11 (49)	0.5 (88)	Adhesive
45°	A2	5.3 (36.6)	10 (44)	0.4 (70)	Adhesive
0°	B1	12.3 (84.7)	31 (138)	1.6 (280)	Mixed*
0°	B2	12.3 (84.7)	*	--	Mixed*
45°	A3	5.3 (36.6)	20 (89)	1.6 (280)	Interply
45°	A4	5.3 (36.6)	20 (89)	1.6 (280)	Interply
0°	B3	12.3 (84.7)	23 (103)	0.9 (158)	Interply
0°	B4	12.3 (84.7)	29 (129)	1.4 (245)	Interply

* Initially adhesive failures transforms into interlaminar ply failure. Specimen No. B1 is primarily adhesive failure, while specimen No. B2 has a larger fraction of its surface with interlaminar separation. There was no region on the load displacement curve where the load remained constant.

four specimens were made with adherends that were cleaned, but not roughened. As was the case for the UDCB specimens, specimens made from composite A showed a greater tendency to separate at the adhesive-composite interface than the B composite.

Subsequently, the composite specimen surfaces were roughened with 400 grit SiC paper to help prevent adhesive failure. Figure 11, which contains the fractographs for this series of tests, shows that the failure mode changed from adhesive failure to wholly interply separation. The fracture in all of these specimens initiated between plies even though an initial unbonded area in the adhesive layer was made with "Teflon" tape. It must be noted that all these bonds were thinner than recommended, however. Interestingly, the load-deflection curves remained remarkably flat, even when the fracture plane changed through the adherend thickness from one interply layer to another, as illustrated in Figure 12 for specimen A-4. Clearly, in this example, the plies in question are not oriented to bear a significant portion of the load in the beam. However, this same phenomenon persists for composite B, where the interply separation removes the 0° fiber from one adherend and effectively adds them to the other (see Figure 13).

FRACTURE OF COMPOSITE-ADHESIVE-COMPOSITE SYSTEMS

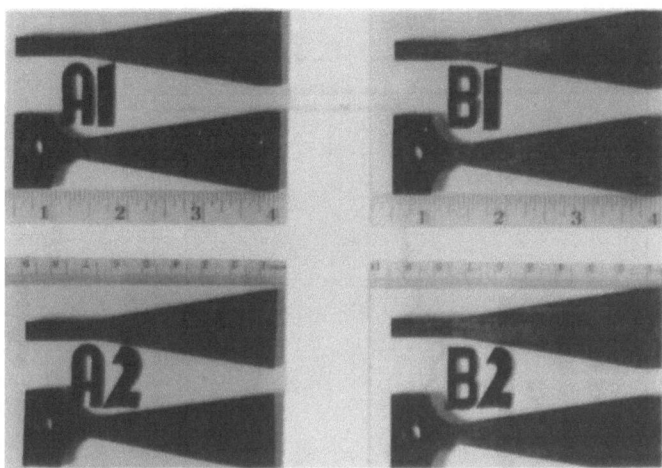

Figure 10. Fractographs of WTB specimens with similar composite adherends prepared without surface roughening prior to bonding. Prefix of specimen number is composite type.

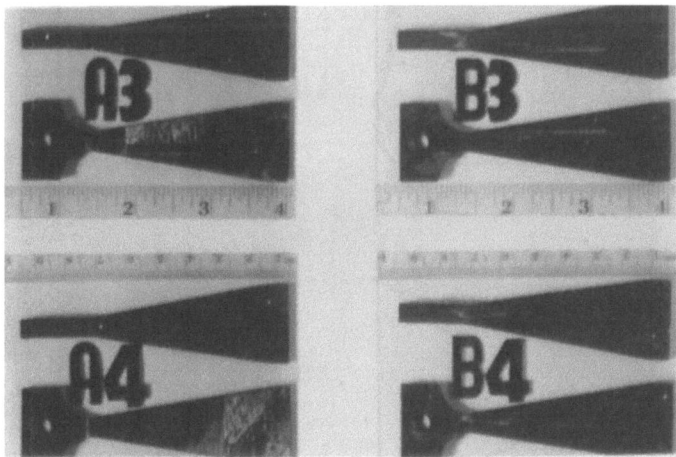

Figure 11. Fractographs of WTB specimens with similar composite adherends prepared with surface roughening prior to bonding. Prefix of specimen number is composite type.

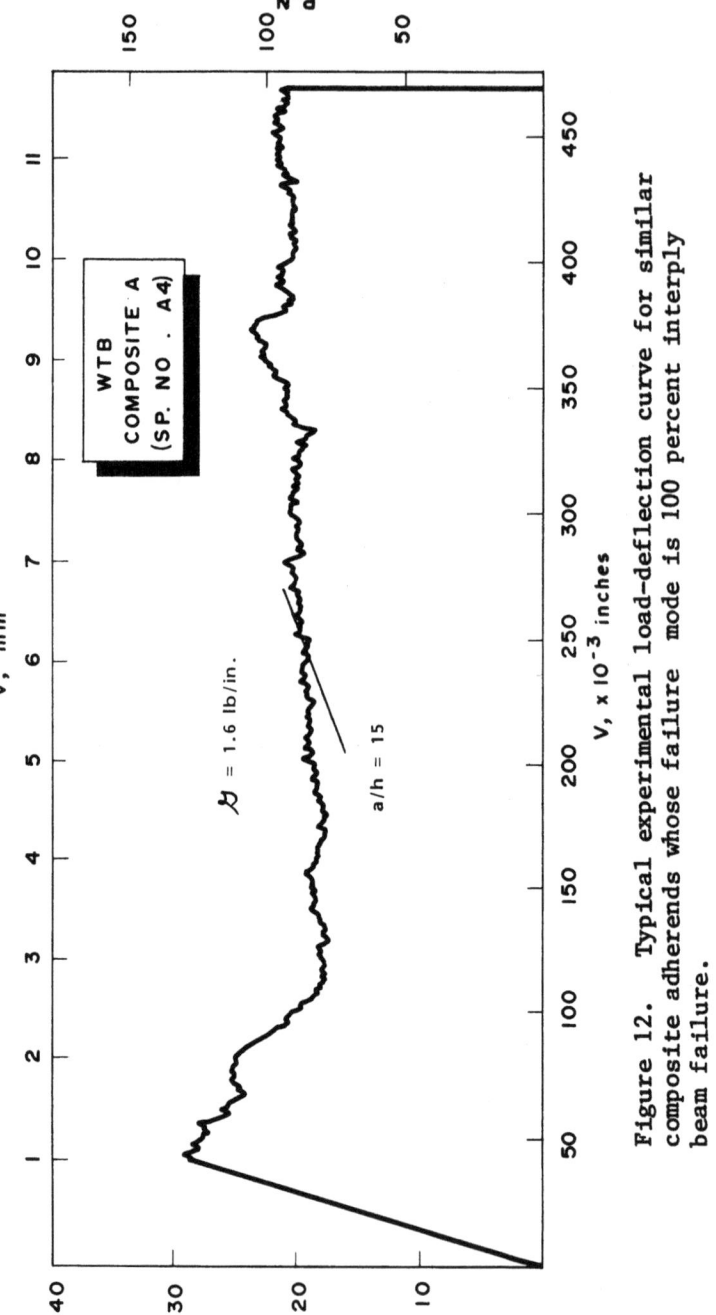

Figure 12. Typical experimental load-deflection curve for similar composite adherends whose failure mode is 100 percent interply beam failure.

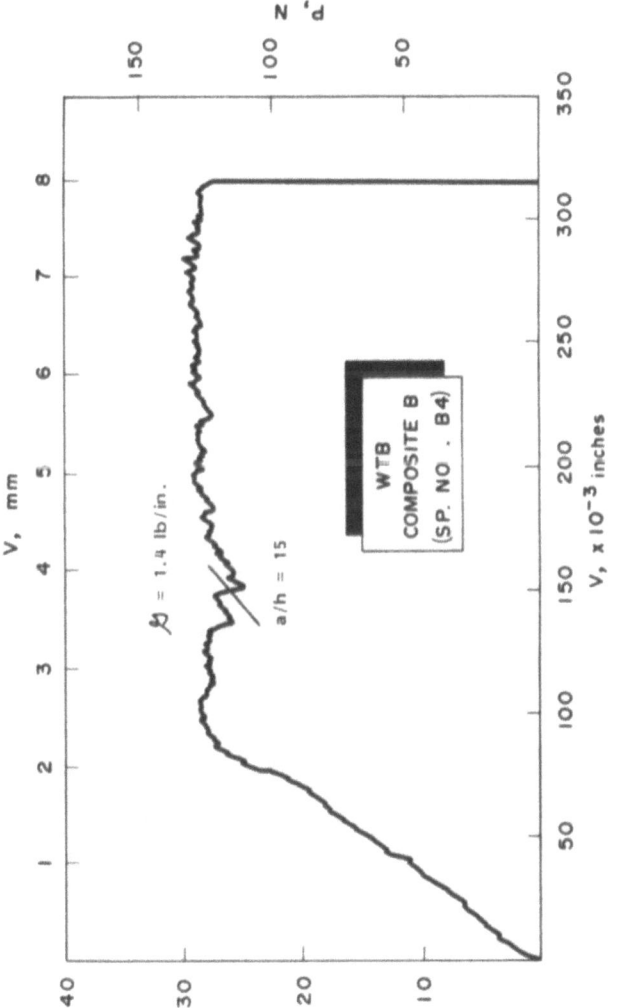

Figure 13. Typical experimental load-deflection curve for similar composite adherends whose failure mode is 100 percent interply beam failure.

It might also be noted that the E values fit within the same scatter band when a/h was more than or less than 15. Obviously, the scatter in toughness overshadowed the variation in modulus.

Separation between the plies near the adhesive was also the fracturing behavior found by Han and Koutsky[8] for adhesives using a glass fiber reinforced polyester composite. Bascom, et al.[11] have pointed out that the interply toughness of composites is always far lower than the bulk toughness of the matrix resin. They suggest that this behavior occurs because the interply layer is a very thin joint which does not allow the formation of a full deformation zone at the crack tip. If this is the case, the fact that these joints were thinner than recommended did not contribute to the fracture morphology.

CONCLUSIONS AND RECOMMENDATIONS

Uniform and width tapered double cantilever beam specimens were used to measure the fracture toughness of adhesive joints made with composite adherends. Both types of test specimen were designed to take into account the fact that the bending elastic modulus of composites is a function of span-to-depth ratio. A novel method for loading thin-adherend DCB specimens was introduced in which the load-point displacement measuring device was also the loading member.

For well made joints, the crack appeared to initiate and propagate between the plies near the adhesive joint for monotonically increasing loads even though a starter flaw was placed in the adhesive. This behavior was also reported for adherends made of glass fiber reinforced polyester composite[8]. It has been suggested that a resin matrix in a composite will always have a lower value of toughness than its bulk value because the interply layer is a thin joint that does not allow for the development of a full deformation zone at the crack tip[11].

Additional Mode I opening load tests will be conducted with other than monotonically increasing loads to determine whether or not the interplies always have lower crack propagation resistance than the adhesive layer under opening mode conditions. Furthermore, joint configurations that do not allow interply separation will also be examined.

ACKNOWLEDGMENT

The study reported in this program was conducted by the Materials Research Laboratory, Inc., of Glenwood, Illinois, and was sponsored by the National Aeronautics and Space Administration through the Lewis Research Center under Contract No. NAS3-21824.

REFERENCES

1a. E. J. Ripling, H. T. Corten and S. Mostovoy, J. Adhesion, 3, 107 (1971). (Also published in SAMPE, 2, 259 (1970).
1b. S. Mostovoy, C. R. Bersch and E. J. Ripling, J. Adhesion, 3, 125 (1971). (Also published in SAMPE, 2, 273 (1970).
1c. E. J. Ripling, S. Mostovoy and C. Bersch, J. Adhesion, 3, 145, (1971). (Also published in SAMPE, 2, 287 (1970).
2. S. Mostovoy and E. J. Ripling in "Adhesion Science and Technology," L. H. Lee, Editor, 9B, 513, Plenum Press, New York, (1975).
3. T. R. Brussat and S. Mostovoy, "Fracture Mechanics for Structural Adhesive Bonds," Final Report for Contract No. F33615-75-C-5224 (15 June 1975 - 15 July 1977), Technical Report No. AFML-TR-77-163, Air Force Materials Laboratory, Wright-Patterson AFB, OH 45433.
4. "Standard Recommended Practice for Fracture Strength in Cleavage of Adhesives in Bonded Joints,"ASTM Standard D3433-75, 1979 Annual Book of ASTM Standards, Part 22: Wood; Adhesives, American Society for Testing and Materials, Philadelphia, PA.
5. S. Mostovoy, P. B. Crosley and E. J. Ripling, J. Materials, 2, 661, (1967).
6. T. R. Brussat, S. T. Chiu and S. Mostovoy, "Fracture Mechanics for Structural Adhesive Bonds," Tenth Interim Report, Contract No. F33615-75-C-5224, Air Force Materials Laboratory, Wright-Patterson AFB, OH 45433.
7. D. F. Devitt, R. A. Schapery and W. L. Bradley, J. Composite Materials, 14, 270 (1980).
8. K. S. Han and J. Koutsky, J. Composite Materials, 15, 371, (1980).
9. C. Zweben, W. S. Smith and M. W. Wardle, "Composite Materials: Testing and Design (Fifth Conference)," ASTM STP 674, S. W. Tsai, editor, 228, American Society for Testing and Materials, Philadelphia, PA, (1979).
10. "Standard Test Method for Flexural Properties of Plastics and Electrical Insulating Material," ASTM Standard D790-71 (Reapproved 1978), 1979 Annual Book of ASTM Standards, Part 35: Plastics - General Test Methods, Nomenclature, American Society for Testing and Materials, Philadelphia, PA.
11. W. D. Bascom, J. L. Bitner, R. J. Moulton and A. R. Siebert, Composites, 11, 9 (1980).

APPENDIX A

Demonstration that $\mathcal{G}$ Depends on Beam Cross Section at the Crack Tip

Double cantilever beam (DCB) specimens can be analyzed as a pair of opposed built-in beams. Consider the bending of the beam shown in Figure A-1. The curvature is given by

$$y'' = \frac{M}{EI} = \frac{Px}{EI} \tag{A-1}$$

the slope by,

$$y' = \int_x^a \frac{Px}{EI} dx \tag{A-2}$$

and the deflection by,

$$y = \int_x^a \int_x^a \frac{Px}{EI} dx\, dx \tag{A-3}$$

where,

M = moment (= Px)
E = Young's modulus
I = moment of inertia

A condition of zero slope at $x = \underline{a}$ has been incorporated in the limits of integration. In general, EI may be considered to be a function of x.

The compliance of a DCB specimen composed of two such beams is taken as

$$C = \frac{2y_o}{P} = \int_o^a \int_x^a \frac{2x}{EI} dx\, dx \tag{A-4}$$

where
y_o = y evaluated at $x = 0$

Let F be a function such that

$$F(a) - F(x) = \int_x^a \frac{2x}{EI} dx \tag{A-5}$$

or,

$$\frac{dF}{dx} = \frac{2x}{EI} \tag{A-6}$$

The compliance can then be written

$$C = aF(a) - \int_0^a F(x)\,dx \tag{A-7}$$

and, upon differentiation,

$$\frac{dC}{da} = aF'(a) \tag{A-8}$$

or

$$\frac{dC}{da} = \frac{2a^2}{EI}\bigg|_{x=a} \tag{A-9}$$

The noteworthy feature of this result is that, within the limits of beam theory, dC/da, which is the crucial factor in determining $\mathcal{L}$, depends only on the beam dimensions (specifically, on EI) evaluated at the crack front.

A more complete description takes into account shear deformation of the beam. In this case

$$\frac{dC}{da} = \frac{2a^2}{EI} + \frac{6(1+\nu)}{EA} \tag{A-10}$$

where the second term represents the shear contribution to dC/da, and where

ν = Poisson's ratio
A = beam cross-section area

Again, all the quantities are evaluated at the crack front position. For a rectangular beam of height $\underline{h}$, $A = 12\,I/h^2$ and

$$\frac{dC}{da} = \frac{2a^2}{EI}\left[1 + \frac{(1+\nu)}{4}\left(\frac{h}{a}\right)^2\right] \tag{A-11}$$

Thus, so long as h/a < 1/8 and ν = 1/3, the contribution of a shear is less than 1/2 percent.

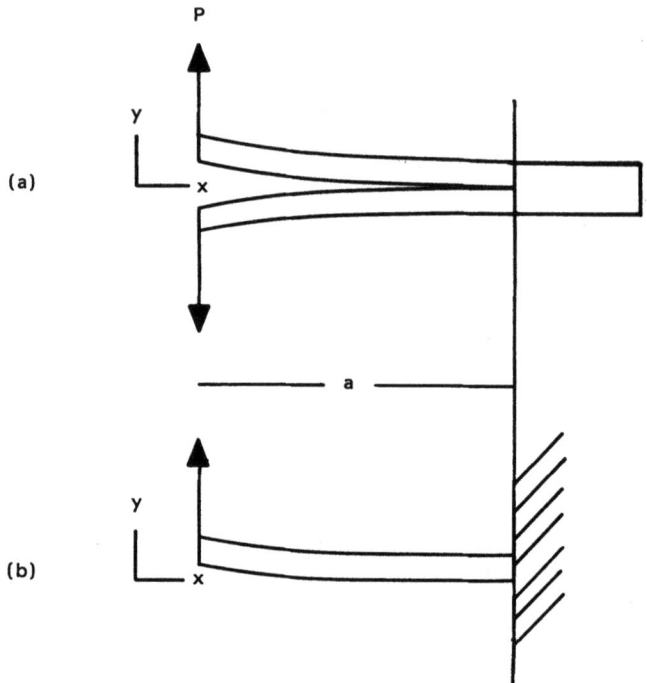

Figure A1. Nomenclature used to describe the DCB specimen, (a), as a pair of opposed built-in beams (b).

APPENDIX B

Effect of Crack Length Dependent Elastic Modulus on DCB Specimen Behavior

For a DCB specimen, the load-displacement relation for a crack propagating at constant $\mathscr{G}$ is given by Equation (2-2) as

$$PV = (2b \mathscr{G}) \cdot (C/C') \qquad \text{(B-1)}$$

For a specimen in which the apparent tensile modulus, E, depends on crack length, the expression for C is unchanged as long as E is interpreted as the apparent modulus. Because the apparent modulus depends on $\underline{a}$, the expression for C' (derivative of C with respect to $\underline{a}$) is different.

<u>UDCB Specimen.</u> For the UDCB specimen

$$C = \frac{8a^3}{Ebh^3} \qquad \text{(B-2)}$$

as given by Equation (2-6), but evaluation of C' gives

FRACTURE OF COMPOSITE-ADHESIVE-COMPOSITE SYSTEMS

$$C' = \frac{dC}{da} = \frac{24a^2}{EBh^3}\left[1 - \frac{aE'}{3E}\right] \tag{B-3}$$

where E is the apparent modulus, and $E' = dE/da$. Substituting for C and C' in Equation (B-1) gives

$$P\dot{v} = (2b\dot{\ell})\cdot(a/3)\cdot\phi \tag{B-4}$$

where

$$\phi = [1 - aE'/(3E)]^{-1} \tag{B-5}$$

The factor ϕ can be recognized as a correction to the expression, Equation (2-7) obtained for constant E. Substituting the experimentally fitted modulus expression, Equation (3-1) into Equation (B-5) gives

$$\phi = \frac{3(a - a_o)}{3a - 4a_o} \tag{B-6}$$

WTB Specimen. For the WTB specimen the experimental compliance was adequately represented by

$$C = \frac{12\ ka^2}{Eh^3} \tag{B-7}$$

Differentiation with respect to $\underline{a}$ gives

$$C' = \frac{24ka}{Eh^3}\left(1 - \frac{aE'}{2E}\right) \tag{B-8}$$

The relationship between P and $\dot{\ell}$ as given in Equation (2-1) is

$$P^2 = (2b\dot{\ell})\cdot(1/C') \tag{B-9}$$

Substituting Equation (B-8) for C' gives

$$P^2 = (2b\dot{\ell})\cdot\frac{Eh^3}{24ka}\cdot\left(1 - \frac{aE'}{2E}\right)^{-1} \tag{B-10}$$

Noting that $k = a/b$ gives

$$P^2 = \frac{\dot{\ell} Eh^3}{12k^2}\left(1 - \frac{aE'}{2E}\right)^{-1} \tag{B-11}$$

The dependence of P on $\underline{a}$ arises, of course, from the dependence of E on $\underline{a}$. Using the E-a relationship given in the test, Equation (3-1) gives

$$P^2 = \frac{2E_o h^3}{12k^2} \left\{ \frac{2(a-a_o)^2}{a(2a-3a_o)} \right\} \tag{B-12}$$

The term in curly brackets represents the correction due to the variable Young's modulus.

APPENDIX C

Press Design for Loading Double Cantilever Beam Specimens

Figure A-1 illustrates the press designed for use in a conventional machanical test machine. The purpose of the fixture is to provide a reliable means of applying a load to the bottom adherend of a double cantilever beam concentric with the dowel on which the top adherend rests. The loading rod is machined to a hemisphere of the same diameter of the dowel pins in the fixture. The rod is also tapered from the hemisphere, as shown in Figure C-2, to prevent it from rubbing the circular opening in the top adherend, which rotates as the specimen is opened. Thus, both halves of the double cantilever beam specimen are free to rotate with this design.
Figure C-3 shows the fixture with an LVDT coil attached to the top. The magnetic core is an integral part of the load rod (Figure C-2) and provides a direct measure of the load line displacement.
Figure C-3(b) illustrates that the loading rod shoulder drops to the top adherend after the specimen fracture. The LVDT was chosen to have a one inch (25 mm) range based on the calculated WTB deflections for aluminum adherends (10.5×10^6 psi, i.e., 72 GPa modulus).

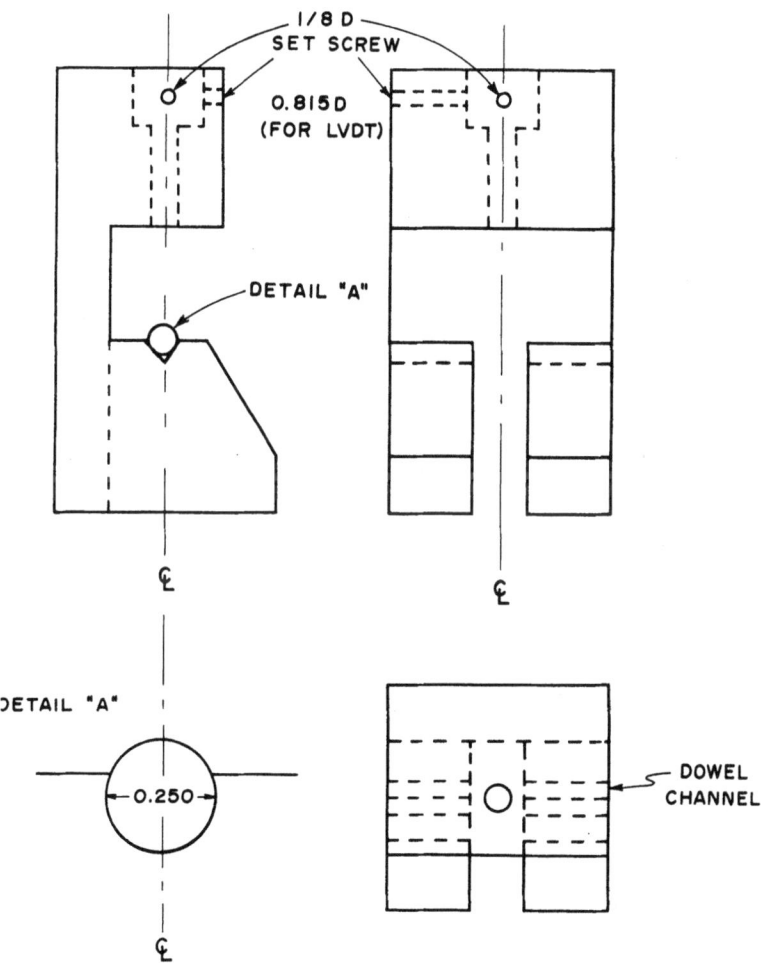

Figure C1. Point loading adhesive testing fixture.

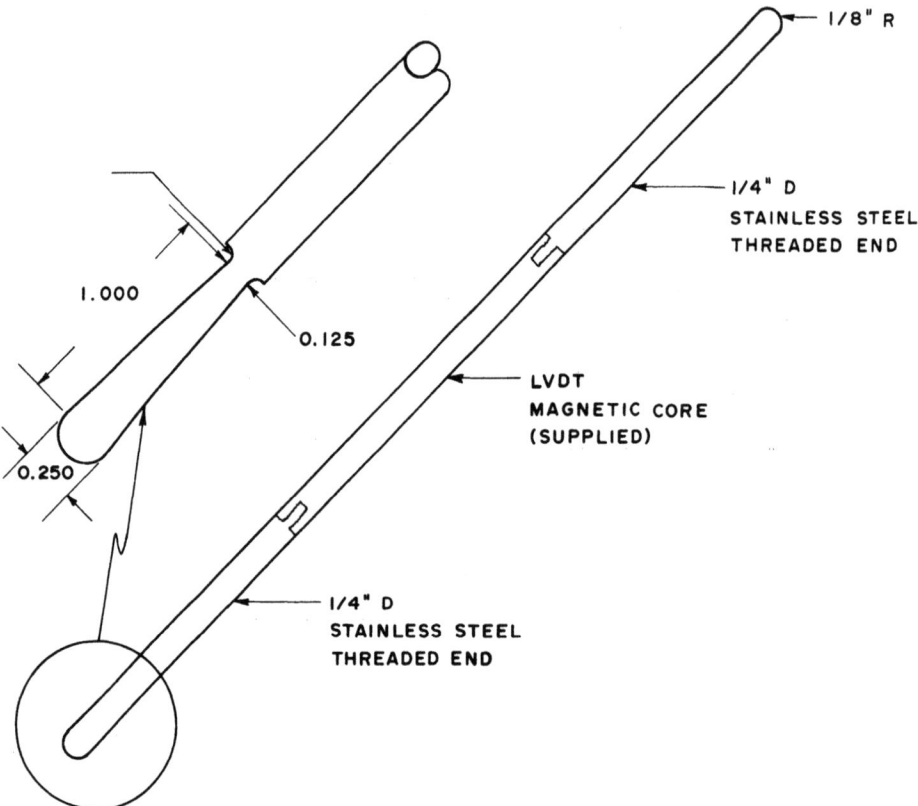

Figure C2. Loading rod for point loading adhesive double cantilever beam specimen.

(a)   (b)

Figure C3. Point loading adhesive testing fixture with LVDT attached to top (a) side view (b) front view.

APPENDIX D

Procedure for Preparing Width Tapered Beam (WTB) Adhesively Bonded Specimens

1. Prepare pre-adhesive blanks with the geometry shown in drawing D-1.

2. Ultrasonically clean the pre-adhesive blanks by immersion in a non-aqueous solvent bath. For non-metallic adherends, follow the standard recommended practice for preparing surfaces of plastics prior to adhesive bonding: ANSI/ASTM D2093-69 (Reapproved 1976). For titanium adherends, the stabilized phosphate-fluoride treatment (method I-S) is used. For aluminum adherends, the phosphoric acid anodizing treatment is used.

3. The bonding conditions prescribed by the adhesive manufacturer are followed on the WTB pre-adhesive blanks with the exception of pressure which is reduced by half to minimize run-out. Two alignment holes are provided at the end of the specimen to prevent the bottom adherend from slipping relative to the top during the cure cycle. It is not necessary to use this precaution when a press is used.

4. The specimen is machined to the final configuration shown in drawing D-2. Sawing the taper angle into the previously bonded top and bottom pre-adhesive blanks provides a uniform, smooth surface between the adherends and the adhesive. The machining conditions can be adjusted to accommodate the adherend material; plastic, non-metallic composite, metallic composite, or metal.

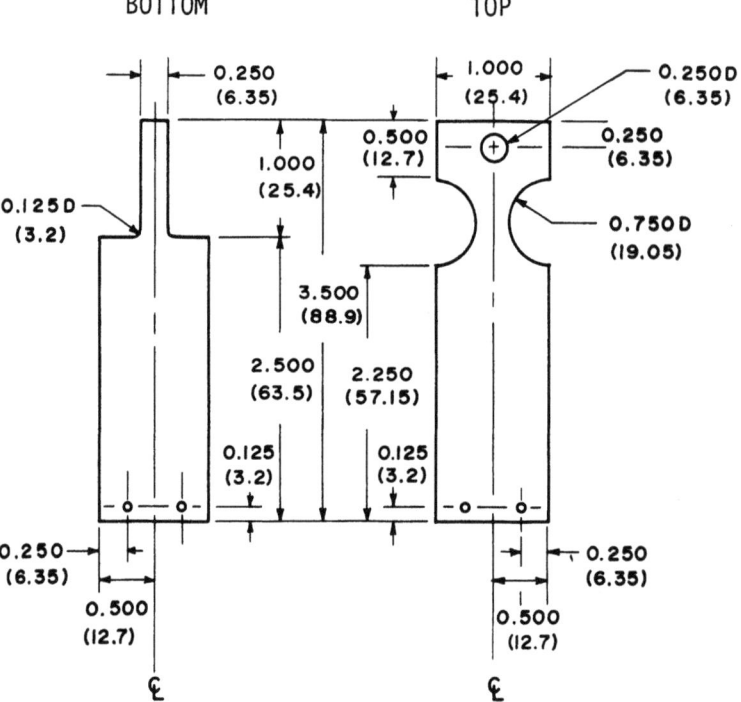

Figure. D1. WTB specimen pre-adhesive blanks. Dimensions are inches (mm).

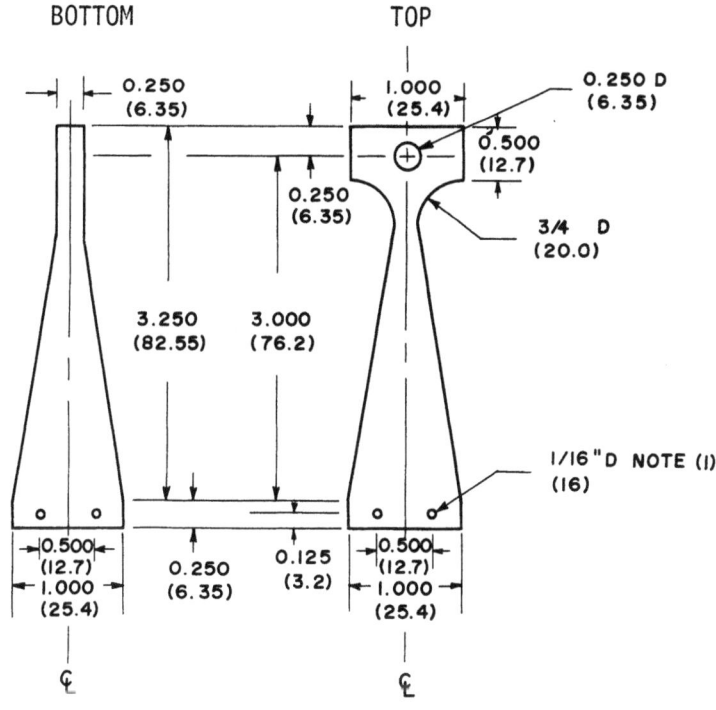

Figure D2. WTB specimen design.

NOMENCLATURE

A   beam cross section area.

a   crack length for a double cantilever beam (DCB) fracture specimen measured from the point of loading to the end of the cracked region.

a_i initial starter flaw length in a DCB specimen.

a_o coefficient used to describe the change of apparent tensile modulus as a crack grows in a DCB specimen.

b   crack front length.

B   maximum specimen width for a width tapered beam specimen.

C   specimen compliance (displacement per unit load).

C'   rate of change of specimen compliance with crack length, dC/da.

D   depth of a 3-point bend specimen (same as h in a DCB fracture specimen).

DCB   double cantilever beam.

E   tensile modulus.

E_o coefficient used to describe the change of apparent tensile modulus as the span-to-depth increases for a three-point bend specimen.

$\mathscr{G}$ crack extension force.

h   height of one beam for a double cantilever beam fracture specimen.

I   moment of inertia.

k   ratio of a/b for a width tapered beam (WTB) specimen.

M   bending moment (Px).

P   load applied to specimen.

S   distance between support for a three-point bend specimen.

V   opening measured at the load line for a double cantilever beam fracture specimen.

W   length of a double cantilever beam fracture specimen from the center of loading to the end of the specimen in the direction of the crack growth.

u   span-to-depth ratio (S/D) for a three-point bend specimen.

x   spatial coodinate in appendix

UDCB   uniform double cantilever beam fracture specimen.

WTB   width tapered beam fracture specimen.

y   spatial coordinate in appendix.

α, β coefficients for a linear least squares regression relating compliance and crack length.

θ	coefficient relating load to crack extension force for a width tapered beam fracture specimen.
$\varkappa$	constant of integration.
ρ	coefficient relating crack length to load for a uniform double cantilever beam fracture specimen.
ν	Poisson's ratio.
ϕ	correction factor to the DCB specimen behavior when the crack length dependent tensile modulus is considered.

CHARACTERIZATION OF THE FRACTURE BEHAVIOR OF ADHESIVE JOINTS

D. L. Hunston

Polymer Division
National Bureau of Standards
Washington, D.C. 20234, USA

A. J. Kinloch and S. J. Shaw

Ministry of Defence (PE), P.E.R.M.E.
Waltham Abbey, Essex, U.K.

S. S. Wang

University of Illinois
Urbana, IL. 61801, USA

The desire to use adhesives and composites in structural applications has led to a need for a failure prediction capability for the polymers used in such systems. Unfortunately, this task is greatly complicated by the failure load being dependent not only upon the specimen geometry but also on the previous history of loading, temperature, environment, etc. For the tough, rubber-modified polymers that are of most interest for structural applications the effects of previous history can be dramatic. As a result, predictions based on measurements at a single set of conditions can lead to dangerous over or under estimates of the fracture behavior. In an effort to understand this problem the present work has studied the fracture behavior of various polymer formulations using bulk and adhesive joint specimens tested over a variety of different thermal and loading histories.

INTRODUCTION

The utilization of composites and structural adhesives in aerospace, marine and automotive systems has increased markedly in the last decade and this dramatic growth rate shows every sign of continuing in the future. Since it is often desirable to join composites with structural adhesives, this growth has placed a heavy burden on the present state of adhesive bonding in terms of both producing new materials that meet the rigorous demands of these applications and developing the failure models required for efficient design with such materials. Predicting the behavior of structural adhesives is a difficult task because they are formulated as complex mixtures of polymers and they are subjected to constraints resulting from the fact that they are restricted to a thin layer between relatively high-modulus substrates. The formulations are designed to give tough materials and do so by the introduction of complex viscoelastic and plastic energy absorption processes. This makes it difficult to develop a basic understanding and predictive capability for failure. The situation is further complicated by the fact that the stress-strain field in a thin adhesive bond layer is difficult to analyse even for linear-elastic materials, and structural adhesives are clearly not linear elastic.

In an effort to address this highly interdisciplinary problem, a cooperative program has been established among scientists at the National Bureau of Standards, Washington, D.C.; Ministry of Defence (PERME) U.K.; Naval Research Laboratories, Washington, D.C.; and the University of Illinois, Urbana. The objective in this effort is to combine expertise in the areas of materials science, polymer mechanics, engineering and stress analysis in a unified effort to gain a better understanding of adhesive joint failure. To accomplish this, the program is identifying and studying the major variables that influence adhesive joint failure. This provides useful information for design, identifies problem areas in failure prediction and helps to build the data base of information that must be established if true design criteria are ever to be successfully formulated. This program has been underway for two years and the purpose of this paper is to briefly outline some of the areas that have been studied and to indicate future research directions.

BACKGROUND

The polymers used in structural adhesives are generally high-modulus glassy polymers that are highly-crosslinked to give good elevated temperature properties and to minimize creep. Unfortunately, high crosslinking can produce brittle materials and thus

Figure 1. Replica transmission electron micrograph of fracture surface of rubber-modified epoxy material (bar indicates 2μm).

it is necessary to toughen them if they are to be used in structural applications. One of the most successful methods used to toughen crosslinked epoxy systems is the addition of a rubber which phase-separates. The curing reactions are controlled so that the final material is a matrix of epoxy with small rubber particles dispersed in and bonded to the matrix, as shown in Figure 1. The reason for utilizing this type of two-phase system can be understood as follows.[1-3] It is not difficult to generate materials that are highly resistant to crack growth but in doing so it is often necessary to make considerable sacrifices in other properties such as modulus, elevated temperature properties and creep resistance. The advantage of the rubber-modified material is that the two-phase nature of the system makes possible toughening mechanisms that do not occur in a single-phase material. These mechanisms can greatly increase the resistance to crack growth in the material. Equally important is the feature that these toughening processes are restricted to localised regions at the tips of flaws, cracks or other stress concentrations. Thus, since the matrix itself contains little or no rubber, the bulk properties such as modulus, heat distortion temperature, and creep resistance are not greatly reduced compared to those of the unmodified epoxy. Consequently, the resistance to crack growth is increased with a minimum sacrifice in other properties. For this reason most commercial structural adhesives are based on a technology of this type.

Although the exact details of the toughening mechanisms are not completely understood, it is clear that non-linear viscoelastic/plastic processes are involved.[1,4,5] This means that only a stress-strain analysis involving a sophisticated constitutive equation can hope to provide an adequate characterization of the system. Moreover, the region in which these processes occur is not small compared to the dimensions of the adhesive bond layer thickness.[6] Consequently, the behavior of the bulk material is modified by the constraining effect of the substrates when such polymers are employed as adhesives. To address this situation the present program is divided into two parts. Part one considers the behavior of bulk specimens of both rubber-modified and unmodified epoxy to provide base-line data. Part two then seeks to understand how the constraints imposed by restricting the adhesive between the substrates alter the bulk behavior. Only mode-I loading will be discussed here; however, many of the conclusions would be applicable to other types of loading as well.

EXPERIMENTAL

Materials

Experiments were performed using an unmodified epoxy and two different types of rubber-modified epoxies. All materials were cured with piperidine (5 parts per hundred resin, phr) at 120°C for 16 hours. Specimens of the first type of modified epoxy were formulated by adding various amounts (0, 5, 15 and 17.5 phr) of a liquid rubber (carboxyl-terminated poly(butadiene-acrylonitrile), CTBN) to the epoxy resin (diglycidyl ether of bisphenol-A) before adding the piperidine. For the second type of modified system 24 phr of bisphenol-A (BPA) was added to the epoxy together with 5 phr of CTBN, prior to the addition of piperidine. With this second rubber-modified epoxy system the cure chemistry is such that a different final project is obtained. Unlike the former rubber-modified systems where rubber particles of about 1 to 5 μm diameter are obtained (see Figure 1), the CTBN/BPA modified epoxy not only results in particles of this diameter but also many small particles with diameters of 0.2 μm and less. This microstructure has been termed "bimodal". To minimize any effects of physical and/or chemical aging, all specimens were given the same thermal history prior to testing, i.e., a very slow cool (>12 hrs.) from 120°C to room temperature.

Bulk Specimens

Bulk fracture tests were conducted using single-edge notched, compact tension and tapered-double-cantilever-beam specimens.[1,2] After inserting a sharp precrack the specimens were loaded to failure using crosshead-speeds ranging from 5mm/s to 0.0005mm/s and temperatures from -60°C to +60°C. The results of these tests were converted to fracture energies using standard procedures.[1-3] Although the bending modulus is relatively insensitive to time and temperature over the range of conditions employed for the fracture tests, care was taken to use modulus values in the fracture energy calculation that were measured at approximately the same temperatures and time scales as the corresponding fracture experiments. All tests employ mode-I type loading.

In addition to fracture tests, the basic mechanical properties of the materials were also determined. Linear viscoelastic properties, such as the shear storage modulus and shear loss modulus, were measured using a specially designed and computerized test device which measures the dynamic mechanical and stress relaxation properties of polymers.[7] Non-linear properties were also determined in compression and tension tests; for example, the true compressive yield stress was determined from uniaxial compression tests. This information provides data on basic behavior as well as input for the stress analysis calculations.

Joint Specimens

The specimen geometry employed for the adhesive joint fracture tests was a tapered-double-cantilever-beam joint, as shown schematically in Figure 2. The substrate material was either aluminium-alloy or mild-steel. As for the bulk fracture experiments, tests were conducted over a wide range of rates and temperatures[1,5] and standard procedures were used to calculate fracture energies.[1-3] All tests employ mode-I type loading.

RESULTS AND DISCUSSIONS

Bulk Behavior

<u>Bulk mechanical properties</u>. The dynamic shear modulus

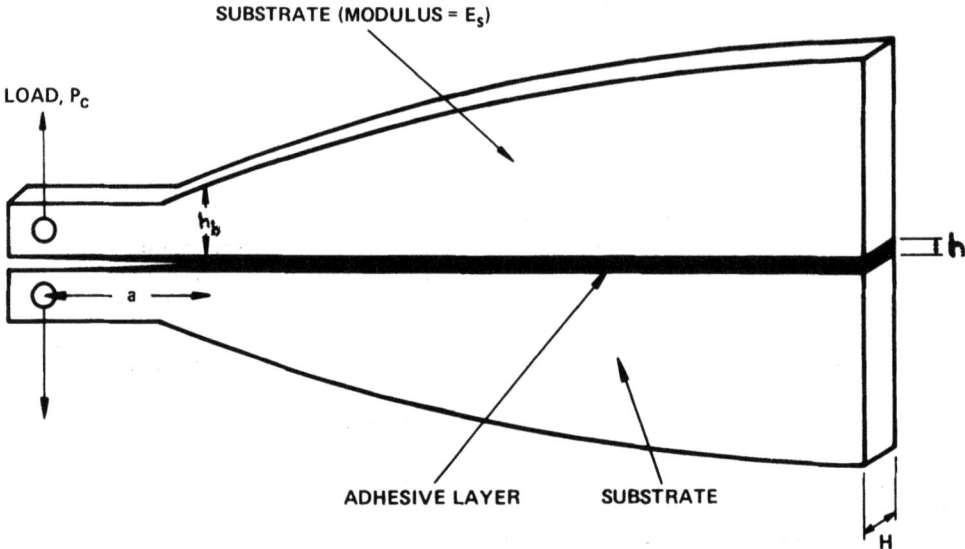

Figure 2. Tapered-double-cantilever-beam adhesive joint specimen.

in the linear viscoelastic range was measured over 4 to 5 decades of frequency and a temperature range of 80°C. The results were then examined for the applicability of time-temperature superposition. Despite the rather complex nature of the modified epoxies, it was found that data for both the modified and unmodified materials superimposed quite well in the range from room temperature to somewhat above the glass transition temperature. The shift factors, a_T, needed for the superposition are plotted against temperature in Figure 3. Also shown in this figure are results from shear stress relaxation experiments performed on the same materials. These experiments in the linear viscoelastic range indicate that reasonably good superposition can be obtained with only a horizontal shift; however, slightly better results are produced when a small vertical shift is included.

Figure 3 also contains the results from non-linear viscoelastic data obtained by applying time-temperature superposition to curves for true compressive yield stress vs. time to yield (time required to go from zero load to the yield point). In this experiment the experimental scatter and the small curvature in the isothermal curves made it impossible to test for a vertical shift. The results for linear viscoelasticity suggest that a vertical shift is present but that it is small. Consequently, the data here were treated using only a horizontal shift. Since the objective in this work was to investigate general trends

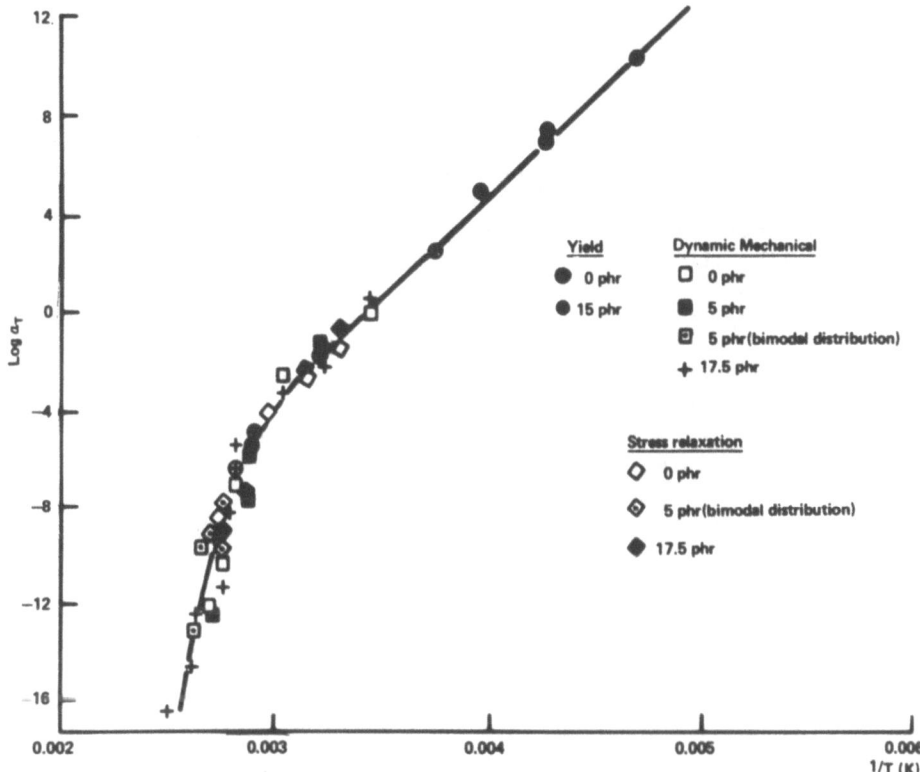

Figure 3. Value of the shift factor, a_T, needed for superpositioning of bulk mechanical data as a function of one over temperature.

rather than to make highly accurate extrapolations, this procedure was thought to be a reasonable first step.

As shown in Figure 3, the shift factors deduced from all three experiments on a variety of different materials agree quite well, particularly when the difficulties involved in maintaining similar thermal histories in the different samples are considered. Consequently, while the absolute magnitudes of the various mechanical properties depend on composition, the shift factors do not.

In the temperature range that is of interest for the fracture behavior, below 60°C, the shift factors can be fitted to a simple Arrhenius type relationship

$$a_T = B \, e^{\Delta E/RT} \quad (1)$$

where B is a constant and ΔE is an activation energy. Values for ΔE are given in Table I.

Bulk fracture properties. It is well known that the energy required to fracture even brittle, unmodified epoxy materials is significantly greater than that required to break the chemical bonds involved. Most of the energy expended in fracture goes to viscoelastic and plastic deformation processes at the crack tip. These deformation processes are rate and temperature dependent and thus the fracture energy, $\mathcal{G}_{Ic}$, also varies with rate and temperature. However, in unmodified epoxies the magnitude of these effects, like the size of the deformation zone at the crack tip, is small. On the other hand, for the rubber-modified epoxies the fracture energy may exceed that of the unmodified material by more than an order of magnitude[8,9]. This arises because the deformation zone is considerably larger and this results in the rate and temperature dependence of the fracture energy being considerably greater. Figure 4, for example, shows the values of $\mathcal{G}_{Ic}$ for two modified epoxy formulations as a function of temperature measured at three crosshead speeds: 2mm/s (F), 0.08 mm/s (M) and 0.0008 mm/s (S). These data demonstrate not only that the fracture

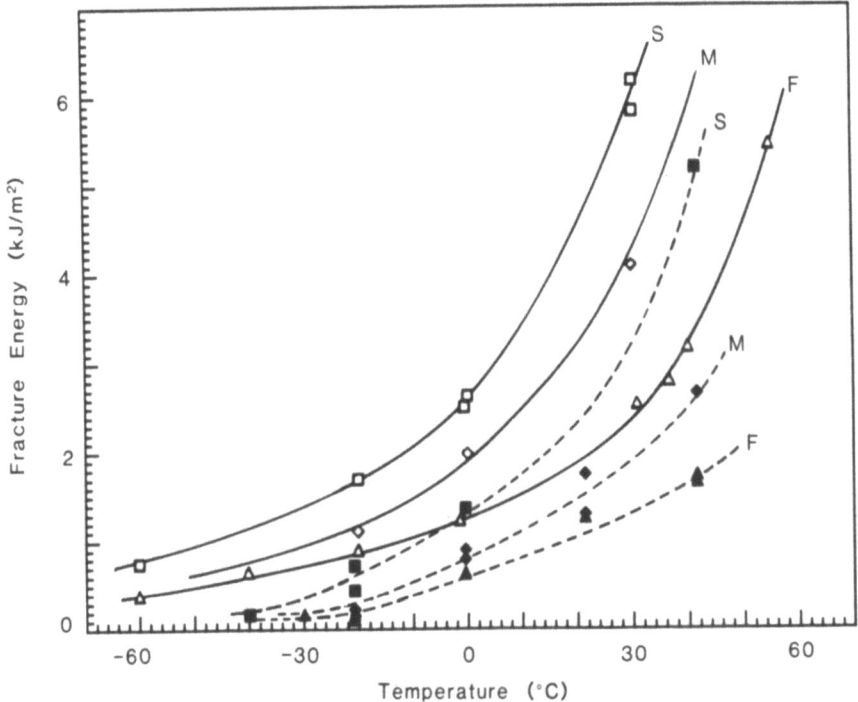

Figure 4. Fracture energy versus temperature at three crosshead speeds for bulk specimens of 17.5 phr CTBN-epoxy: open symbols, solid lines; and of 5 phr CTBN-BPA-epoxy: filled symbols, dotted lines.

energy varies dramatically with temperature and loading rate but also that different formulations can have very different rate and temperature dependences. Consequently, the conventional method of testing materials at one set of conditions and presenting the resulting values in tabular form can be very misleading. For a given formulation the data obtained in one test may provide very little information about how the material will respond in a test under different set of conditions. Moreover, comparisons between formulations can be totally incorrect if the temperature and/or loading rate is changed. Consequently, only when a complete characterization of $\mathcal{D}_{Ic}$ is performed can various formulations be evaluated and compared.

One method we have developed to characterize the loading rate and temperature dependence is to perform a series of constant crosshead speed experiments at various temperatures and speeds. The data may then be presented in the form shown in Figure 4, but may also be modelled by the expression:

$$\mathcal{D}_{Ic} = \mathcal{D}_{Ics} + A \left(\frac{t_f}{a_T}\right)^m \qquad (2)$$

where $\mathcal{D}_{Ic}$ = the measured value of the mode-I fracture energy,

$\mathcal{D}_{Ics}$ = the minimum value of $\mathcal{D}_{Ic}$ at low temperatures and high speeds (see Figure 5),

A,m = constants,

t_f = the time taken for the load to increase from zero to that required for the onset of crack growth,

a_T = the shift factor taken from time-temperature superpositioning of the bulk mechanical data (see Figure 3).

Typical plots of $\mathcal{D}_{Ic}$ versus Log (t_f/a_T) are shown in Figure 5 and within the limits of the experimental uncertainty, the data fall close to a single curve. Consequently, the time-temperature shift factors obtained from measurements of viscoelastic properties can be used to reduce the fracture data. It should be noted that with the fracture data to an even greater extent than with the yield data, the experimental scatter makes it difficult to judge the accuracy of extrapolations based on time-temperature superposition; i.e., it is impossible to estimate the uncertainty in values of $\mathcal{D}_{Ic}$ in Figure 5 for times to failure outside the range directly measured. What is important here however is that the general relationship between temperature and time for the fracture data is roughly equivalent to that for the basic viscoelastic properties. Now the shift factor, a_T, may in turn be expressed as in Equation

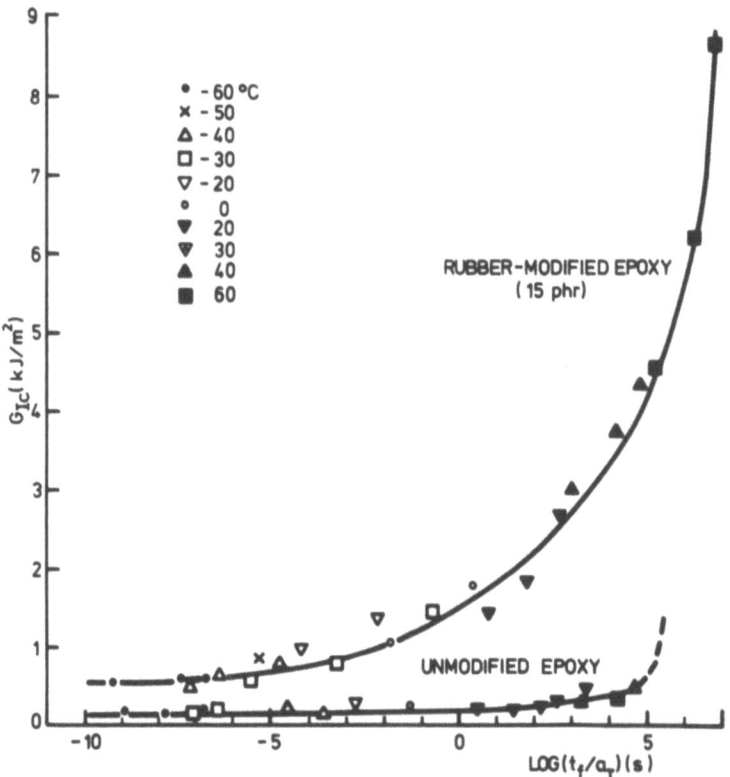

Figure 5. Fracture energy versus time-to-failure, t_f/shift factor, a_T. Data shown for 15 phr CTBN-epoxy and unmodified epoxy.

(1) to give:

$$\mathcal{G}_{Ic} = \mathcal{G}_{Ics} + C\, t_f^m\, e^{-\Delta E_m/RT} \qquad (3)$$

where the parameters C, m and ΔE largely characterise the magnitude, rate dependence and temperature dependence of $\mathcal{G}_{Ic}$. As an example, when the different formulations shown in Figure 4 and 5 are analysed in this way, they give similar values of ΔE but very different values for m, c, and $\mathcal{G}_{Ics}$ as may be seen from Table I. Consequently, an expression such as Equation (3) provides a useful way to evaluate the bulk mode-I fracture behavior of different formulations.

Table I. Fracture Parameters.

Formulation	$\mathcal{G}_{Ics}$ (kJ/m^2)	C	m	ΔE(kJ/mol)
5 phr CTBN + 24 phr BPA (bimodal)	0.2	21650	0.15	160.5
5.0 phr CTBN	0.2	1236	0.12	160.5
15.0 phr CTBN	0.6	2935	0.12	160.5
17.5 phr CTBN	0.6	4064	0.12	160.5

Adhesive Joint Behavior

Effect of rate, temperature and joint geometry. The mode-I fracture behavior of unmodified epoxy in an adhesive joint is in most ways a predictable extension of the failure behavior of bulk specimens: the fracture energy has the same value and is not significantly dependent upon the detailed geometry of the adhesive joint. However, with rubber-modified epoxies the size of the deformation zone is often similar to the thickness of the adhesive layer and thus interactions must be considered.

For the rubber-modified epoxy, the mode-I adhesive fracture energy, $\mathcal{G}_{Ic}$ (joint), is shown as a function of adhesive bond thickness, h, at various temperatures (specimen width, H=12mm; crosshead speed, $\dot{y}$=0.002mm/s) in Figure 6, at various rates (temperature=22°C; specimen width=12mm) in Figure 7 and at various specimen width (temperature=22°C; rate=0.02mm/s) in Figure 8. In all cases for the joints the adhesive fracture behavior is dominated by a strong dependence upon the adhesive bond thickness, h, employed, as previously found by Bascom and co-workers[6,10] for tough adhesive systems. The adhesive fracture energy passes through a maximum, $\mathcal{G}_{Icm}$ (joint), at a certain bond thickness, h_m. The value of this maximum, $\mathcal{G}_{Icm}$ (joint), is however dependent upon the temperature, rate and specimen width, as may be seen from Figures 6, 7 and 8 respectively. Finally, in discussing the experimental results, it is noteworthy that throughout these tests the locus of joint failure was cohesive in the adhesive layer.

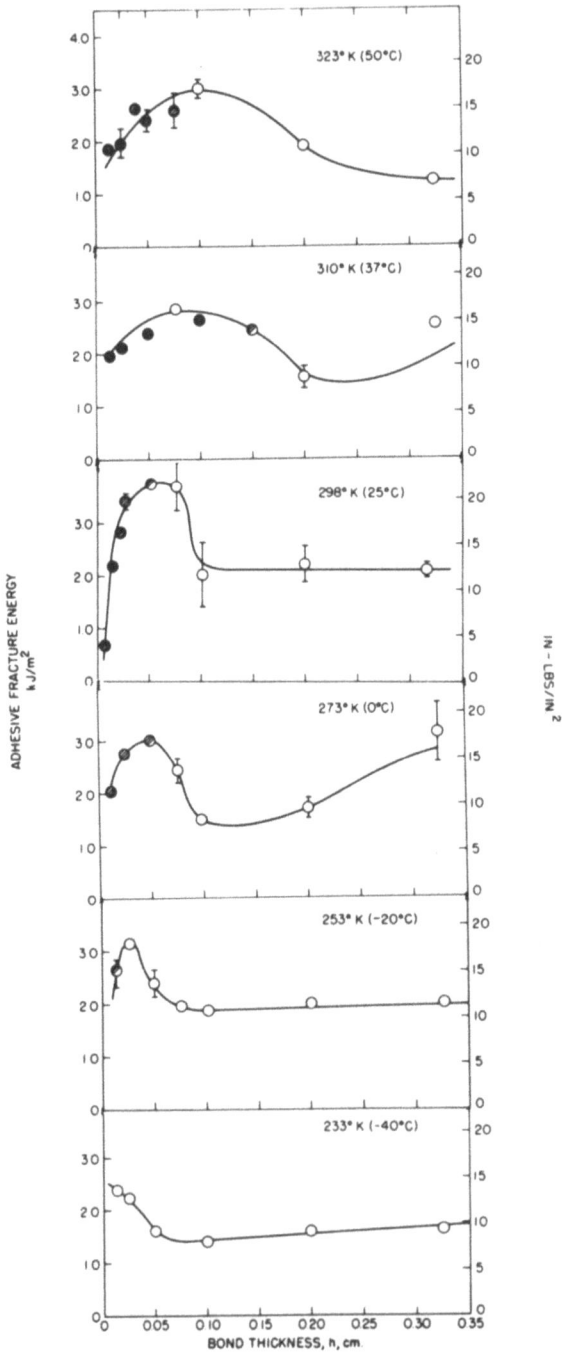

Figure 6. Adhesive fracture energy versus bond thickness at various temperatures (H = 12mm; $\dot{y}$ = 0.02mm/s).

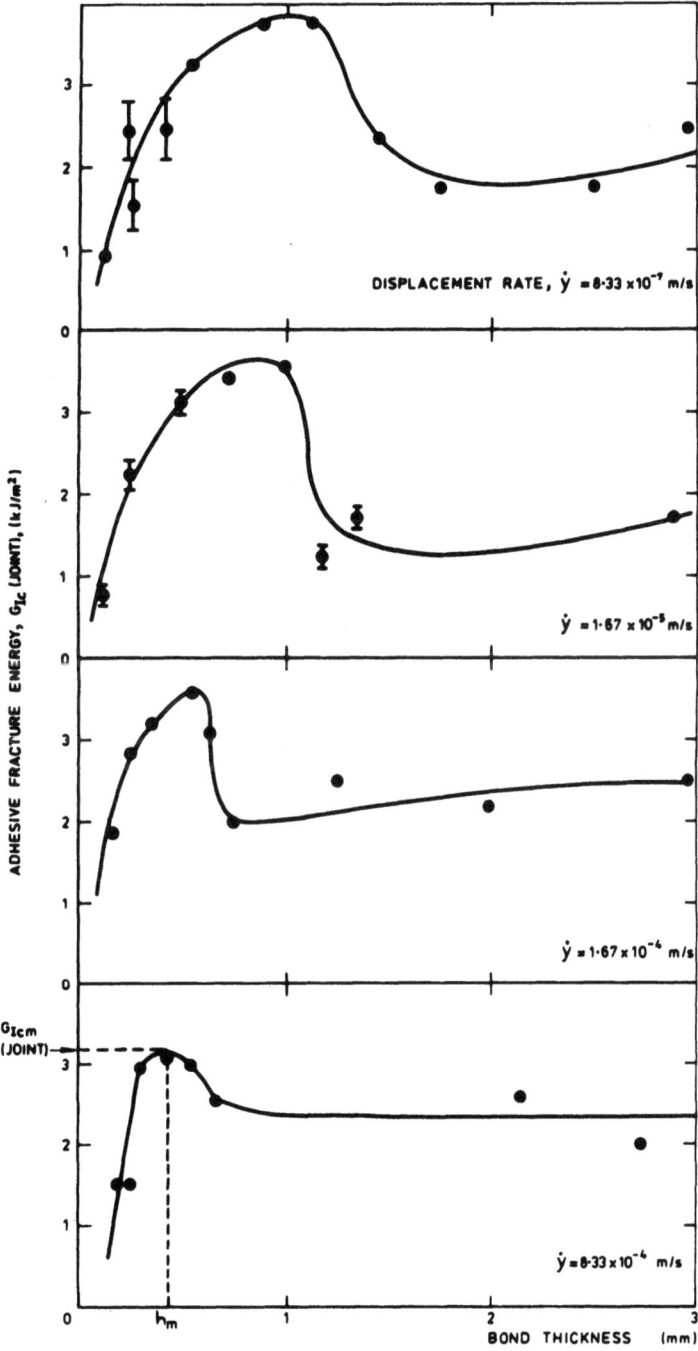

Figure 7. Adhesive fracture energy versus bond thickness at various crosshead speeds (T = 20°C; H = 12 mm).

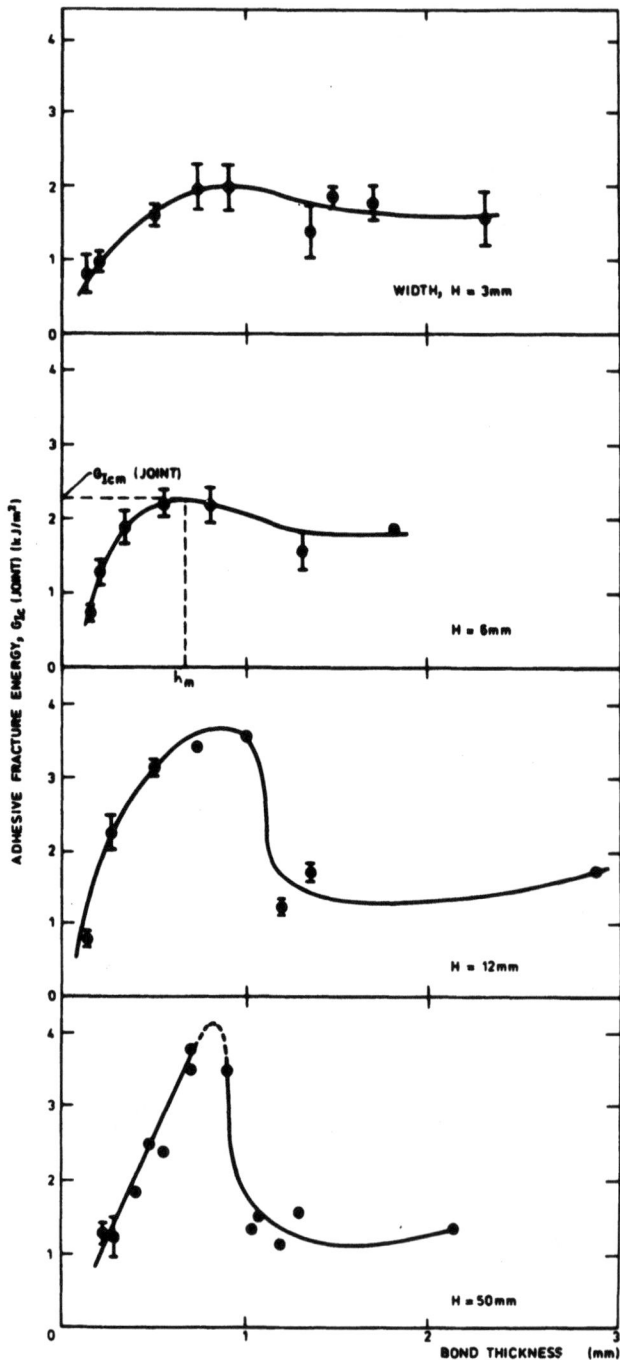

Figure 8. Adhesive fracture energy versus bond thickness at various joint widths (T = 20°C; $\dot{y}$ = 0.02mm/s).

Theoretical analysis. Detailed calculations have shown[11] that it is possible to estimate the optimum bond thickness, h_m, (for obtaining $\mathcal{G}_{Icm}$ (joint)) for a wide range of test rates and temperatures by using a plane-stress elastic-plastic model of the crack tip deformation zone (Figure 9). Based on the conventional "quasi-elastic" approach to estimate small scale crack-tip yielding, the radius of the deformation zone, r_{Iyc}, can be calculated. The optimum bond thickness is then predicted by assuming that the maximum adhesive fracture energy is obtained when the size of the zone, $2r_{Iyc}$, matches the bond thickness. The parameters in the model are taken to be time dependent and the time to failure, t_f, in the fracture test is used to select the appropriate values:

$$h_m = 2r_{Iyc} = \frac{\mathcal{G}_{Ic}(t_f)}{\pi \sigma_y(t_y) e_y(t_f)} \qquad (4)$$

where σ_y is the yield stress, and e_y is the yield strain.

Values of h_m calculated from Equation (4) are shown together with the experimental values in Table II. As may be seen, the correlation is very good. The decline in $\mathcal{G}_{Ic}$ (joint) values at adhesive bond thicknesses, h, less than h_m may now be readily understood. Namely, the presence of the high-modulus substrates restricts the full volume of the plastic zone from developing; and, since the toughness is mainly controlled by the energy dissipated in forming the deformation zone, the adhesive fracture energy is steadily reduced as the bond thickness is decreased.[6,10]

Beyond the conclusions discussed above, there are three questions concerning the adhesive fracture behavior which can only be answered by considering the details of the stress-strain field that is present in the adhesive bond layer. The questions are: (i) why does the value of $\mathcal{G}_{Ic}$ (joint) decrease at bond thicknesses greater than h_m, (ii) why is $\mathcal{G}_{Icm}$ (joint) > $\mathcal{G}_{Ic}$ (Bulk), and (iii) why does the value of $\mathcal{G}_{Icm}$ (joint) depend upon specimen width when the value of $\mathcal{G}_{Ic}$ (bulk) is <u>not</u> significantly dependent upon specimen width? Although complete answers to these questions are not yet possible, some interesting qualitative notions can be offered. As implied in the opening statement of this paragraph, the stress-strain field ahead of a crack in an adhesive layer is

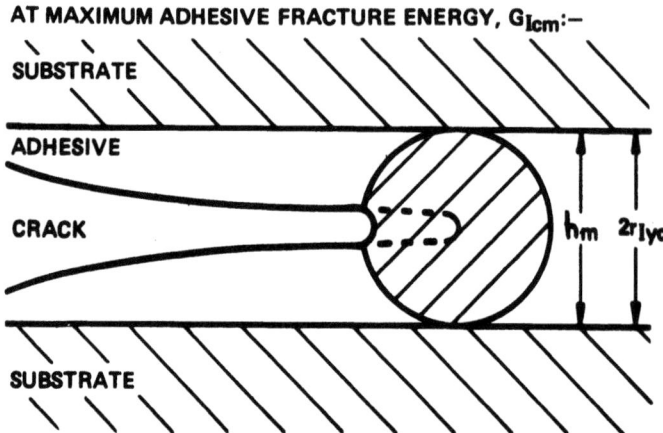

Figure 9. Elastic-plastic model for deformation zone at the crack tip. As drawn, $h = h_m$.

greatly influenced by the constraints imposed by the presence of the high-modulus substrates. Wang and co-workers have conducted such a stress analysis using both an elastic[12] and an elastic-plastic[13] finite element analysis. These analyses demonstrate that as the adhesive bond thickness decreases, the high crack-tip stress levels extend farther down the adhesive layer than they do in bulk specimens or thick adhesive bonds. This results in a change in the shape of the deformation zone in the thin adhesive bonds; i.e., the zone extends farther down the bond line.

From the discussion above it is clear that there are two effects that must be considered when estimating the size of the deformation zone and the expected value of fracture energy: the dimensions of the zone in the adhesive bond thickness direction and its length down the adhesive bond layer. Very thick bonds give results similar to bulk specimens. However, as the bond thickness is decreased, the first effect is the increasing degree of constraint which causes an extension of the high stress levels further down the bond layer and in response to this the deformation zone extends further down the bond. This results in an

Table II. Comparison of Calculated and Experimental Values of Adhesive Bond Thickness, h_m, at Maximum Adhesive Fracture Energy, $\mathcal{J}_{Icm}$ (joint).

Temp, T (°C)	Crosshead speed, $\dot{y}$ (mm/s)	Joint width H(mm)	Value of h_m (mm) Theoretical (from eqn (4))	Experimental
20	0.0008	12	0.85	1.0
20	0.02	12	0.70	0.8
20	0.2	12	0.49	0.55
20	1.2	12	0.43	0.4
20	0.02	3	0.75	0.85
20	0.02	6	0.72	0.65
20	0.02	12	0.70	0.8
20	0.02	25	0.65	0.7
20	0.02	50	0.61	0.8
50	0.02	12	1.6	1.1
37	0.02	12	1.16	0.9
25	0.02	12	0.57	0.6
0	0.02	12	0.39	0.5
-20	0.02	12	0.15	0.25
-40	0.02	12	0.05	<0.1

increase in the zone size and the adhesive fracture energy. As the bond thickness continues to decrease eventually a point is reached where the extension down the bond layer becomes secondary to the restricting effect of the decreasing space available in the thickness direction. The zone size and fracture energy then begin to decrease after passing through a maximum. Experimental observations[14] of the growth of the crack-tip deformation zone made with high-speed movies are in good qualitative agreement with the predictions of the elastic finite-element analysis[12] and a more detailed quantitative analysis that is to be published by Wang et.al.[13]. Finally, the degree of constraint is obviously also a function of joint width, H. For example, as H increases the degree of constraint increases and thus $\mathcal{J}_{Icm}$ (joint would be predicted to increase in value compared to $\mathcal{J}_{Ic}$ (bulk), and this is indeed observed as may be seen from Figure 10. Thus, the same arguments may be used to explain the effects of both adhesive bond thickness and specimen width upon the adhesive fracture energy. Although a quantitative explanation of these effects will probably require a detailed stress-strain analysis with a more realistic constitutive equation, the simple idea presented here is qualitatively consistent with all of the experimentally observed results.

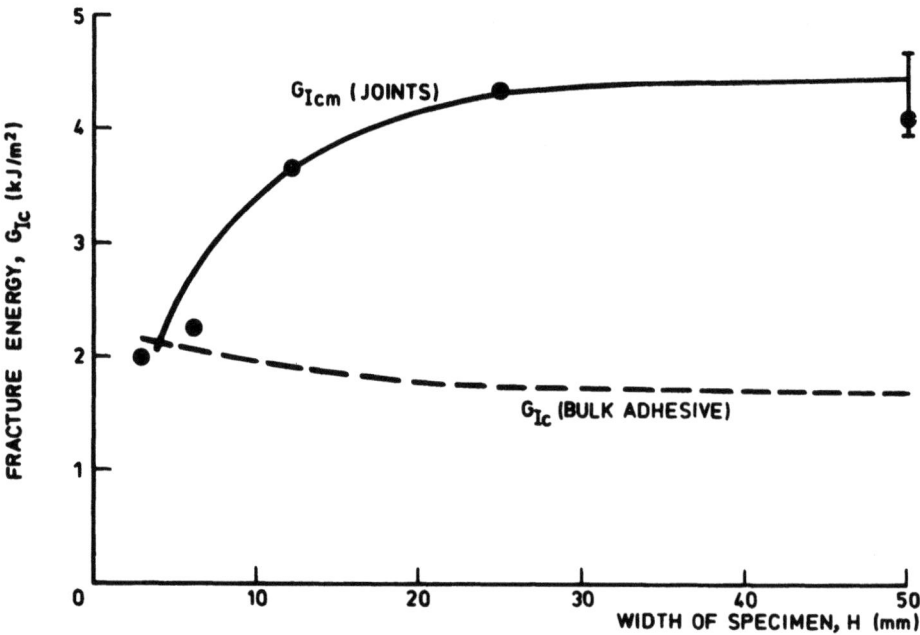

Figure 10. Fracture energy versus specimen width, H (T = 20°C; $\dot{y}$ = 0.02mm/s).

CONCLUSIONS

The studies on the bulk epoxy materials have demonstrated that there is a correlation between temperature and time-to-failure that makes it possible to apply time-temperature superposition techniques to help characterize the fracture behavior of these materials. This characterization has enabled a semi-empirical equation to be developed which provides a means of modelling the properties of the materials over a wide range of test rates and temperatures. Moreover, since the behavior of rubber-modified epoxies changes substantially when the formulation is altered, these techniques should provide a basis on which to establish structure-property relationships that will assist in the development of optimized materials.

The studies on the adhesive joints fabricated with rubber-modified epoxies have shown that their fracture behavior is dependent upon (i) the fracture energy of the bulk epoxy adhesive and (ii) the degree to which the geometry of the bonded joint enhances and/or restricts the toughening processes normally present in the bulk adhesive. The importance of this second parameter has been demonstrated experimentally and qualitatively modelled using both elastic and elastic-plastic stress-analyses.

REFERENCES

1. J.L. Bitner, J.L. Rushford, S. Rose, D.L. Hunston and C.K. Riew, J. Adhesion, 13, 3 (1981).
2. A.J. Kinloch and R.J. Young, "Fracture Behavior of Polymers", Applied Science Publishers, London, 1983.
3. D.L. Hunston, J.L. Bitner, J.L. Rushford and J. Oroshnik, J. Elast. Plastics, 12, 133 (1980).
4. A.J. Kinloch and J.G. Williams, J.Mater.Sci., 15, 987 (1980).
5. A.J. Kinloch, S.J. Shaw and D.L. Hunston, in "Proceedings Int. Conf. on Yield, Deformation and Fracture", held in Cambridge, Plastics Rubber Inst., London, 1982.
6. W.D. Bascom and R.L. Cottington, J. Adhesion, 7, 333 (1976).
7. D.L. Hunston, W.D. Bascom, E.E. Wells, J.D. Fahey and J.L. Bitner, in "Adhesion and Adsorption of Polymers", L.H. Lee, Editor, Pt. A. p.321, Plenum Press, New York, 1980.
8. E.H. Rose, A.R. Siebert and R.S. Drake, Mod. Plastics, 47, 110 (1970).
9. J.N. Sulton, R.C. Laible and F.J. McGarry, J. Appl. Polym. Sci., 6, 127 (1971).
10. W.D. Bascom, R.L. Cottington, R.L. Jones, and P. Peyser, J. Appl. Polym. Sci., 19, 2545 (1975).
11. A.J. Kinloch and S.J. Shaw, J. Adhesion, 12, 59 (1981).
12. S.S. Wang, J.F. Mandell and F.J. McGarry, Intern. J. Fracture, 14, 39 (1978).
13. S.S. Wang, D.L. Hunston and A.J. Kinloch, to be published.
14. D.L. Hunston, A.J. Kinloch and J.L. Rushford, to be published.

STRUCTURAL PRECURSORS TO FRACTURE IN ADHESIVE JOINTS

Wartan A. Jemian

Materials Engineering
Department of Mechanical Engineering
Auburn University, AL 36849

Sructure-property relations in adhesive joints are reviewed for the purpose of explaining the general nature of conditions and structural processes leading to fracture. Reference is made to selected homogeneous materials and adhesive joints to illustrate specific features. The adhesive joint is described as a composite structure composed of adherend and adhesive layers and various chemical and structural transition layers. The nature and dimensions of materials structural features are considered. Traditional and modern methods of analysis are summarized with special emphasis placed on the use of materials characteristic curves to indicate the existence of structural precursors to fracture. These features, in relatively small quantities, severely alter the mechanical properties of the adhesive joint. In only a few instances have structural precursors to fracture been directly observed and identified due to their small dimensions, distribution in the system and limited quantity. It is concluded that identifying the presence of structural precursors to fracture by the use of materials characteristic curves may aid the avoidance of adhesive joint failure.

INTRODUCTION

The general study of structural precursors to fracture is important in adhesive joints as well as any structural materials system because it promises the basis for a method to anticipate and hence avoid fracture. Even before this objective is fully realized, it is advantageous to better understand the nature of events leading to fracture. This study requires a definition of the significant features of the original structure and a review of structural changes that are known to occur in the separate, homogeneous materials of the adhesive joints.

An adhesive joint is a layered structure in which strong, elastically stiff materials are chemically bonded (covalent, ionic, metallic bonding) through a second substance, which has a lower strength and stiffness. These are called the "adherend" and "adhesive", respectively. The principles governing properties of the joint relate also to more general cases of adhesion, such as plating, in which the term, "joint", is not appropriate. The adhesive joint usually is a combination of any basic type of material such as metals and alloys, glassy or crystalline ceramics, and any of the many organic plastics and rubbers or other high molecular weight compounds. Therefore, this review draws upon common ideas and similarities in the special fields touched upon.

In a symmetric loading system which is frequently the case for lap-shear specimens with a center of inversion symmetry at the center of the overlap, deformation and fracture are usually also symmetric. Figure 1 depicts the nature and sequence of progressive changes that take place in aluminum-epoxy-aluminum lap shear samples under load, as reported by Wilcox and Jemian.[1,2] Initially, crack nucleation begins by a void condensation process, followed by slow crack growth toward the center of the overlap until a critical condition develops for rapid crack extension. The initial processes start at the ends of the overlap close to the center of the adhesive and rapid crack growth occurs close to the adherend at the interface with the higher shear stress. The final separation at the center of the overlap is a tearing process under overload conditions after the joint has effectively lost all of its original strength. Bending, initial separation, crack extension and final decoupling of the two alloy strips occur symmetrically, within the precision of observation. There are similar, progressive changes that take place in a non-symmetric joint, where separation occurs along one surface only. Therefore, in this analysis a single adherend-adhesive composite is used to represent an entire assembly or joint.

Materials structure and phenomenology represent two principal considerations in understanding the characteristics of materials. Materials are articulated systems where the nature of the response to any action depends on the details of physical arrangement at all

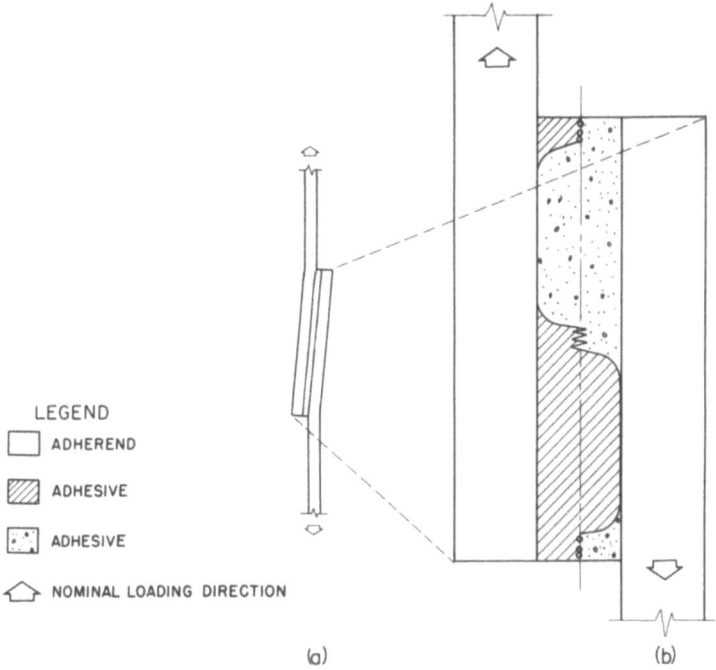

Figure 1. Configuration of the lap shear joint.
(a) Macro-structure under load
(b) Profile of the fracture path

structural levels. This aspect is emphasized in this presentation. However design requires numerical values which are obtained by the application of phenomenological relations, which are identified with each deformation process.

Geometry, dimensions and configuration are important. The mechanical actions transferred through the bulk structural members to the region of the interface produce components of normal and shear strain. Table I lists the basic elements of the terminology of solid mechanics that are referred to. Figure 2 illustrates the relationship of these variables to a volume element of the material. The volume element represents any portion of the adhesive joint, or other materials system, under load. Loading configuration and sample deformation control the proportions and distribution of stress components, including the interfacial region. These change with loading and sample shape. In addition to their general stress distribution, local features such as cracks and interphase boundaries between phases of markedly different stiffness produce stress fields that are locally complex and include concentrated stress components.[3,4]

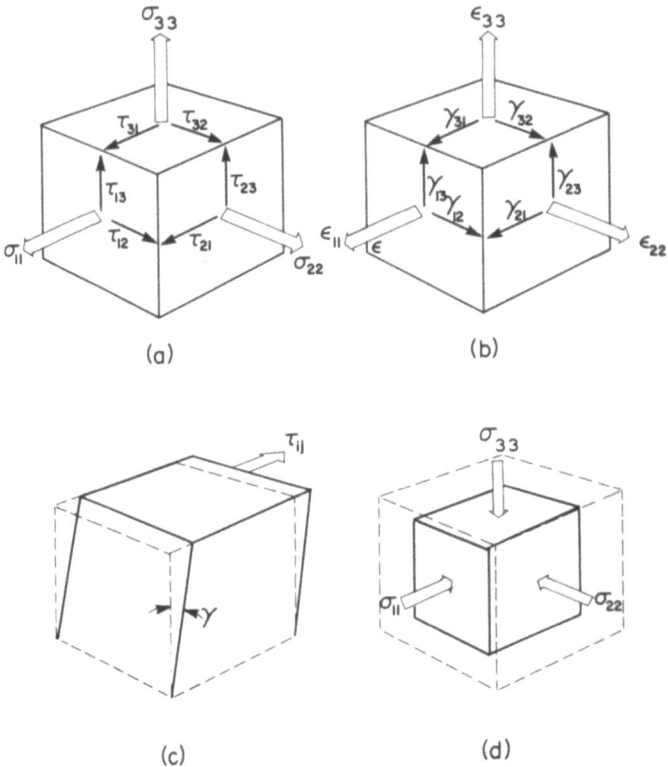

Figure 2. Mechanical Actions and Materials Response.
(a) Stress components   (c) Shear, showing the shear angle, γ
(b) Strain components   (d) Dilatation under hydrostatic pressure

Adhesion testing is no longer limited to the measurement of a few, arbitrary sample properties. Adhesion scientists recognize structural similarities in bulk constituents and transition zones of adhesive joints although terminology varies considerably. At different stages in the loading a variety of changes take place. Some are the same as those that occur in the bulk components, separately loaded. Others have their origin in the unique chemical and structural conditions of the adhesive joint.

Mechanical Characteristics

Flow (stress-strain), creep and stress relaxation curves provide comprehensive information about loading response and sample behavior.[3] Certain features are expected for each type of sample by material and configuration. In the lap-shear configuration the separation surface has features of the form shown in Figure 1. Different combinations of materials and dimensions produce different combinations of these features and others not shown in this

Table I. Forces, Stresses, Strains, Strain Rate and Deformations in Materials. (Refer to Figure 2)

FORCE	STRESS	STRAIN	STRAIN RATE	DILATATION	SHEAR
Normal	σ_{11}	ε_{11}	$\dot{\varepsilon}_{11}$	*	
Normal	σ_{22}	ε_{22}	$\dot{\varepsilon}_{22}$	*	
Normal	σ_{33}	ε_{33}	$\dot{\varepsilon}_{33}$	*	
Shear	τ_{12}	γ_{12}	$\dot{\gamma}_{12}$		*
Shear	τ_{23}	γ_{23}	$\dot{\gamma}_{23}$		*
Shear	τ_{31}	γ_{31}	$\dot{\gamma}_{31}$		*

*Deformation associated with the listed variables.
NOTE: The subscripts, i, j, in σ_{ij} etc. refer to force or displacement in the j-direction on a surface normal to the i-direction.

sketch. There is greater variety possible. It is frequently possible to interpret these differences in terms of the relating magnitudes of stress and strain components of Table 1 and Figure 2, and their time derivatives. At a crack at the edge of the overlap, at the beginning of adhesive failure, the opening stress is maximum and the shear stress is zero, as shown schematically in Figure 3. This distribution was described by Masubuchi and Keith[4], in a review of deformation mechanisms, confirmed by finite element structural analysis by Grimes[5], photoelastically by Jemian and Ventrice[6], and supported by a closed-form mathematical solution procedure by Renton and Vinson[7]. The photoelastic analysis showed that the stress distribution form is constant and follows the crack tip as it advances in the fracture process.[6]

In a long joint, a transient "steady state" may be established until terminal conditions prevail. The conditions of the symmetric, tapered double cantilever fracture toughness test sample and the peel test sample represent more clear examples of the action of steady state stress distributions during the fracture process.

The materials characteristic is the sequential response of the sample to a programmed action under controlled conditions. Stress-strain curves are one form of mechanical materials characteristic. Other materials characteristics depict response to thermal exposure, magnetic or electric fields, etc. There are three related types of materials properties; (i) slopes and parameters of the continuous portions of the characteristic curve, (ii) critical values of either the action or reaction variable where the features of the curve change, and (iii) a critical value of some condition, e.g. temperature or pressure, that causes the characteristic curve to have a different shape or show some new feature.

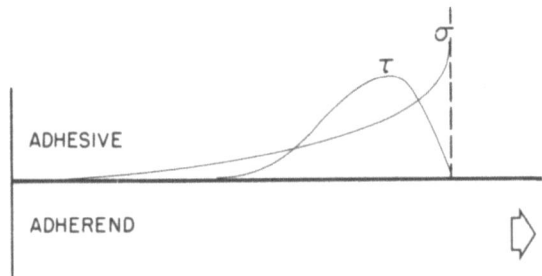

Figure 3. Stress distribution in the lap shear sample.

Well known examples of these properties are provided by the standard test procedures for homogeneous alloy samples.[3] The typical tensile materials characteristic for a ductile alloy sample is shown in two ways in Figure 4. The engineering stress-strain curve is the more common representation and is equivalent (it presents the same information in a different form) to the true stress-natural strain curve which relates better to phenomenological representations. The two curves coincide closely in the range of elastic deformation, the first smooth region below the elastic limit, σ_e. The other regions, uniform plastic flow and non-uniform plastic flow differ in form and deformation process. Phenomenological relations are identified with each region in Figure 4.

Three transitions are also shown with critical value properties identified. The elastic limit marks the onset of slip, the ultimate tensile strength, which is the engineering measure of the maximum load that can be supported, marks the onset of significant production and accumulation of voids. Breaking, the last transition involving crack nucleation and growth, is usually overlooked in testing alloy samples. This terminal detail is frequently displayed in tests of adhesive joints because response times are different. This final portion of the characteristic in the alloy sample involves crack extension under the action of essentially constant displacement, a stress relaxation process.

In the alloy sample, the first deformation process is the distortion of the crystal structure in the grains without any changes between nearest neighbor atoms. Uniform plastic deformation, typified phenomenologically by a constant strain hardening exponent, n, is principally a constant volume process with a small increase in elastic distortion proportional to the level of

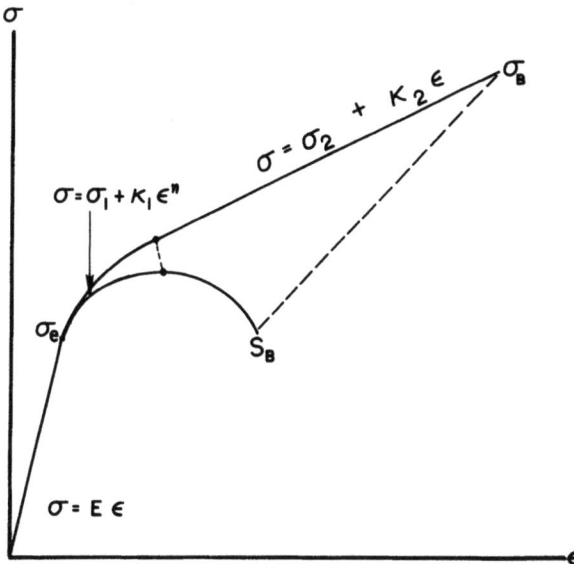

Figure 4. Mechanical Characteristic of ductile alloy. σ_e, Elastic limit; UTS, ultimate tensile strength; S_B, σ_B, breaking strength; K_1, K_2, constants.

loading. Other processes such as vacancy production occur but are only important in the later stages of deformation. Vacancy production leading eventually to void formation results in the typical fibrous fracture surfaces observed in ductile materials. The UTS marks the onset of significant void production as the yield strength marks the onset of slip.

The ductile-brittle transition temperature of the Charpy test or dropping weight impact test is an example of the third type of related materials property. In this procedure, a separate sample is tested at different temperatures spanning the temperature range where the fracture work, as well as the dynamic stress-strain characteristic curve change. The temperature associated with the changing characteristic is reported as a property of the material.[3]

A constant linear slope or smooth constant form of non-linear characteristic indicates that structural changes within the system are of the same kind. Changes in curvature or line continuity indicate the action of some new internal structural process. "Pop in" in a fracture toughness test is such a feature as is the more subtle deviation from linearity observed in deformation curves in the region of the elastic limit or in the rate of change of specific volume with temperature at the glass transition temperature of a glassy phase. The observation of such a feature indicates a significant change in the material.

Phase Structure Model

The aluminum-epoxy system is referred to as a typical adhesive joint system. The epoxy is a thermosetting polymer with a multiply connected network of primary bonds along aliphatic and aromatic molecular segments. In addition to the strong covalent bonds, distributed secondary bonds also stabilize the phase structure. The epoxy is an amorphous materials system. The adherend is polycrystalline. Experience shows that surface preparation of the adherends is as important as formulation of the adhesive to the ultimate joint properties. Degreasing, followed by chromic acid etching and finished by controlled washing and drying are critical to the final result. The objective is to form a coherent layer of hexagonal Bayerite (β-$Al_2O_3 \cdot 3H_2O$), which strongly absorbs the epoxy, rather than pseudo-Boehmite ($Al_2O_3 \cdot X\ H_2O$, $X \approx 2$) that is formed in boiling water.[8,9]

The final joint involves the transition from alloy to adhesive through an hydrated oxide and an internal surface of attachment where epoxy molecular segments are chemically bonded to the oxide. This attachment is highly organized in comparison to the bulk epoxy phase. The regular termination of the epoxy branches at the oxide surface produces a region in the epoxy with a non-typical molecular organization. In this region the molecular segments are closely aligned and positioned to match the oxide surface.

The attachment density at the interface is estimated to be of the order of $10^{18} m^{-2}$, taking the molecular spacing to be approximately 1 nm. A typical hydrated alumina surface provides $3.2 \times 10^{18} m^{-2}$ sites. By comparison there are $1.4 \times 10^{19} m^{-2}$ attachments across (111) in aluminum.

These features are represented schematically in Figure 5. The alloy and coherent oxide layer of the adherend are represented by arrays of dots. It is assumed that this region is within a single grain of the alloy. The epoxy is represented by lines, intended to indicate molecular conformation in a general way. As a natural consequence of the structures of the phases and their relationships, a number of layered regions can be identified. These are:

    M   metallic substrate
    OMI oxide-metal interface
    AOI Adhesive-oxide interface, referred to above as the internal plane of attachment.

OA Ordered Adhesive, where the effects of the structural regularity of the adhesive molecule and the plane of attachment persist.

BA The remaining adhesive phase with a structural organization not influenced by the interfacial attachment.

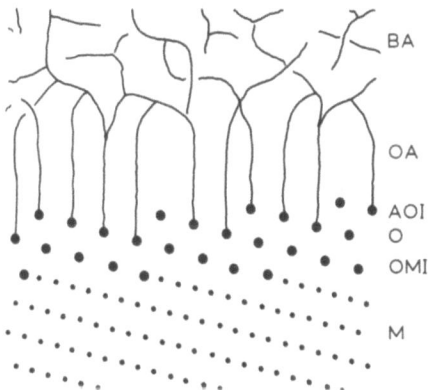

Figure 5. Phase structure in the regions of the aluminum-epoxy adhesive joint (schematic). BA, bulk adhesive; OA, ordered adhesive; AOI, adhesive-oxide interface; O, oxide layer; OMI, Oxide-Metal Interface; and M, metallic substrate within the adherend.

If the adhesive layer is thin the effects of surface organization affect the structure of the entire adhesive layer. This, in turn, produces changes in adhesive joint characteristics, which are typically an increase in measured strength and a shift in fracture locus toward or into the adherend or oxide.

Interfacial contaminants, such as the monolayers or partial monolayers of Langmuir[10,11], Blodgett[12], Langmuir and Schaeffer[13], and Jacquet [14] interrupt the interface bonding and lower measured adhesive strengths, as well as lead to apparently clean interface separations. Jacquet used this effect to initiate a peel test ("T" test) of an electrodeposit. Jemian[15] found that interfacial contamination controllably reduced dislocation interactions with metal-metal interfaces. These effects, however, are in a separate category from the structural features that develop out of a well formed joint under load.

Stress Effects

The application of forces to the adhesive joint invariably results in complex stress distributions including both general stress field details and local variations due to the presence of structural features. There are many examples of the effects of simplified, homogeneous stress fields on various materials. In some situations these effects are attributed to the stress field alone and in other cases the effects are complicated by ambient conditions or external factors. The stress effects include enhancement or development of the following processes and structural features:

Elastic deformation,
Flow,
Void formation,
Impurity production,
Transformations, and
Transition macrostructures.

These processes are described in a manner consistent with the different phase structures involved in homogeneous materials and corresponding regions of the adhesive joint. Table II is an approximate size scale for structural features in adhesive joints.

Elastic deformation, in reference to alloys and other engineering materials, is extension and bending within the yield strength or elastic limit, and is usually in direct proportion to the stress. Elastomers, rubber modified adhesives, and a variety of other materials undergo initial deformations, when the load is gradually applied, that are continuous, non-linear functions of the stress and may result in pronounced shape changes within the applicable range. These processes are recoverable distortions of intercomponent spacing in the phase structure and changes in component shape. The energy content of the elastically deformed phase is increased due to both potential energy and entropy factors associated with spacing and shape changes, respectively. Phenomenologically, the possible deformations expand into the realm of finite elasticity[16] which includes the effects of major shape changes and other nonlinearities.

In the adhesive joint the adherend and oxide layer, which are the stiffest constituents, undergo linear elastic deformation. The bulk polymer is capable of some recoverable extension by rotations about carbon-carbon bonds within molecular branches. This process also involves some reversible intercomponent shear. The polymer at the oxide surface is considerably restricted and has a preferred orientation and structure. These factors increase the elastic stiffness in homogeneous samples and have the same effect in the adhesive joint.

Table II. Size scale for features of materials structure in adhesive joints

SIZE	FEATURE	
	ELEMENTAL SOLID	MACROMOLECULAR PHASE
1 mm		
	CRACK	
0.1 mm	GRAIN TRANSFORMATION NEEDLE DISLOCATION CLUSTER	SPHERULITE
10 μm	SLIP LINE	
		CRAZE
1 μm		
0.1 μm		SHEAR BAND MICROCRAZE VOIDS
	VACANCY CLUSTER	
10 nm		FIBRILS
1 nm	BURGERS VECTORS ANIONS	MOLECULAR SPACINGS
0.1 nm	ATOMS CATIONS LATTICE VACANCIES	

Flow processes differ in mechanism in solid and glassy phases due to crystal compatibility requirements in the former and lack of such requirements in the latter. Flow includes any structural process involving a sequence of phase structure equilibrium states. When the load is removed the system retains its immediate configuration. Thus, flow results in permanent set. Flow is enhanced by an increase in temperature, especially if the increase involves the glass transition. Gent[17] finds that the effect of an applied stress is to enhance flow mechanically, thus effectively reducing the glass transition temperature. Crystalline materials flow by slip, twinning, and stress induced martensite transformation. This is true of both metallic and non-metallic crystals. These changes occur by progressive action and involve dislocation motions which mark the extent of the crystallographic shear.

When flow occurs in either a crystalline or glassy phase, rotations are usually induced. Molecular branches tend to align with the principal normal stress direction and crystallographic deformation directions align with the direction of maximum shear.

Flow in a glassy macromolecular substance involves local relaxations between neighbor branches. Permanent set is achieved if the new configurations are similar to the original. This is viscous flow in which the strain rate, $\dot{\varepsilon}$ or $\dot{\gamma}$, is a function of the stress. The relation is linear for Newtonian liquids. The tensile viscosity, ϕ, and shear viscosity, μ, are defined as $\phi_{ii} = \sigma_{ii}/\dot{\varepsilon}_{ii}$ and $\mu_{ii} = \dot{\gamma}_{ii}/\tau_{ii}$, where $\phi = 3\mu$.[3]

Alfrey discusses general relations between bulk polymer mechanical properties and adhesive joint properties, including yield strengths.[18] The principal deformation processes in glassy polymers are shear band formation and crazing.[19-22] Microshear bands, which generally form at about 40° to the compression direction are illustrated in Figure 6. Diffuse shear bands which also form under compression, but closer to 45°, are illustrated in Figure 7. Crazing, which forms under tension is illustrated in Figure 8. The craze surface is perpendicular to the loading direction as shown. Oxborough and Bowden found that $\sigma_{11} - \nu(\sigma_{21} + \sigma_{23}) = E\varepsilon_{11}$, where E is Young's modulus and ν is Poisson's ratio, provides a meaningful criterion for crazing.[23] Figure 9 shows the criteria for polymer deformation in two dimensions. The elliptical boundary for shear band formation is similar to the Von Mises criterion for yielding in metals. Its center, however, is displaced towards biaxial compression. Craze formation occurs at a much lower stress magnitude and in the tensile region.

STRUCTURAL PRECURSORS TO FRACTURE 821

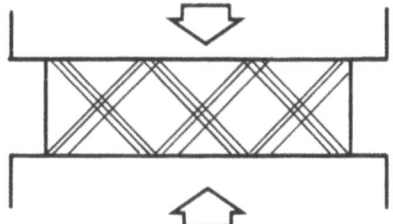

Figure 6. Microshear band formation in bulk plastic under compression. Typical bands are inclined approximately 40° to the compression surface.

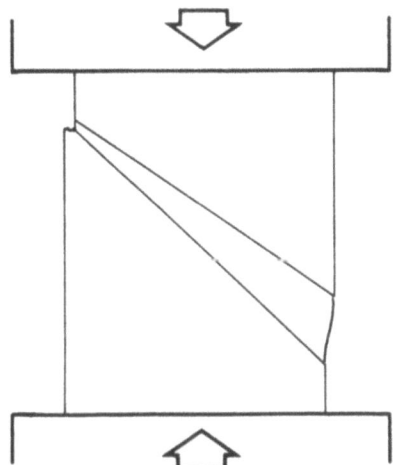

Figure 7. Diffuse shear band formation by compression of bulk plastic. Shear band is inclined approximately 45° to the compression surface.

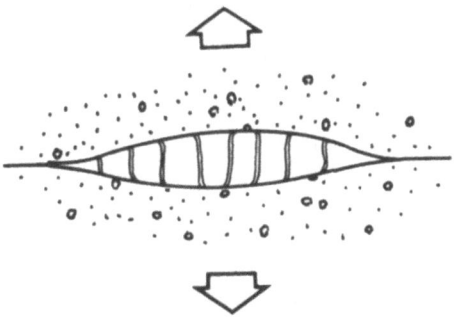

Figure 8. Crazing. The craze is an opening crossed by oriented fibrils in the plastic.

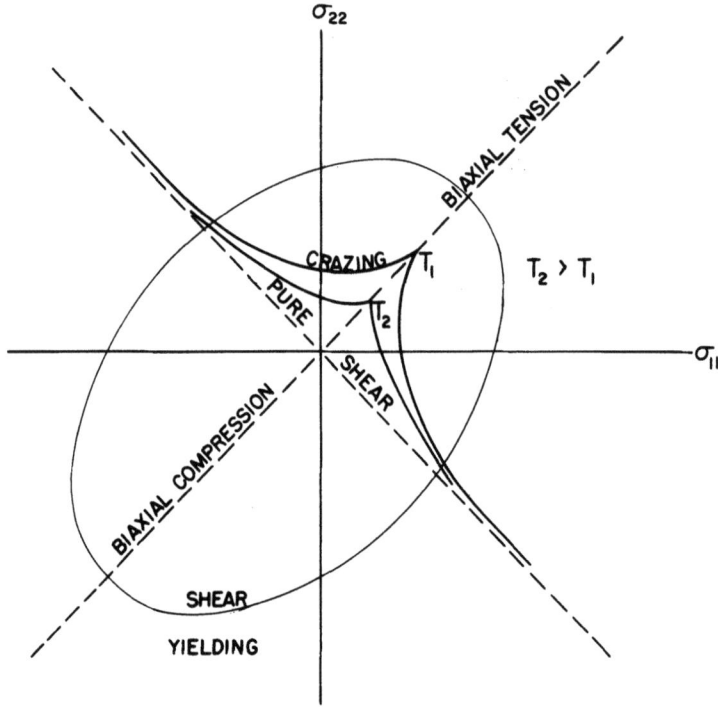

Figure 9. Criteria for shear yielding and crazing in polymers. The effect of temperature on crazing is indicated.

A further distinction between the two deformations is that crazing is dilatational and associated with a decrease in density. There is also a loss in cohesion and strength but not to the same extent as in crack formation. The craze contains substance and is capable of supporting loads to a reduced extent. Crazing develops from voids. The craze volume is marked by many fibrils oriented across the surface. The fibrils have dimensions of the order of 10 nm thickness and the spacing between is of the same order.

Crazing is a deformation mode observed in thermoplastic polymers or weakly crosslinked thermosets. Void formation, which is a precursor to crazing has been observed in thermosetting adhesive systems.

Flow in the ordered adhesive is restricted. A number of investigators have reported an increase in fracture strength with a decrease in adhesive thickness.[24-27] Bascom and Cottington report deviations in this relationship and provide a more detailed explanation.[28] The increase in strength with decreasing joint thickness that is generally observed is attributed to a different specific mechanism in each system. However, the trend is general and the fracture mode changes from ductile to brittle. Possible explanations are that the flow processes are eliminated in these adhesive layers due to a boundary effect which is an apparent manifestation of the Bikerman weak boundary layer theory.[29] Another contributing factor is the ability of the ordered adhesive to support a higher stress due to the increased elastic limit in the ordered structure (flow is severely restricted). A third possible reason is that since the thickness of the adhesive layer is reduced and the principal deformation occurs in this layer, the effective strain rate is increased for any given loading rate. This favors brittle fracture.

Void formation is an included process in the deformation and fracture of both alloys and polymers.[3,19,30] The increase of entropy of mixing accounts for an equilibrium vacancy content in crystals that have an Arrhenius temperature dependency. Dislocation interactions and diffusion provide mechanisms for the attainment of these concentrations. Continued vacancy production by dislocation interactions in the loaded member cause vacancy super-saturations that are relieved by diffusion and clustering. These structural features are usually of the order of 15 nm in small dimension. Macromolecular glassy phases include structural features that provide for analogous effects. Void production is accompanied by decreases in specific gravity and strength.

As a point of interest, vacancies are generally found to have a strengthening effect in metals and dilute solid solutions.[30] They are associated with local elastic distortion in the crystal so that the volume of the missing particle is partially occupied. These lattice distortions cause dislocation interactions which reduce dislocation mobility. On the other hand in a tensile test void production opposes work hardening, leading to the pronounced necking observed beyond the ultimate load. The center gage thinning reported by Miklowitz,[31,32] MacGreggor,[33] and others is attributed to other factors.

Impurity production is more a result of external factors, such as atmospheric humidity and the absorption of ultra-violet radiation than the application of stresses alone. The degradation of plastics is comprehensively reviewed by Pinner.[34] In addition to these factors the dilatation of a portion of the joint due to the stress field favors the accommodation of molecules and molecular fragments that alter the local bonding, phase stability and mechanical properties. Impurities are expected to be of the order of molecular spacings as represented in Table II.

Transformations, including crystallization, stress-induced martensite transformations of crystalline phases and chemical reactions, occur spontaneously to reduce the energy of a substance with stored elastic energy. Under stress the system is in a higher energy state. If a new configuration is possible at a lower energy, it will form. Polymers react to form new molecular groups by mechanisms involving free radicals or chemical agents already in the system. These changes, as is the case of impurity production are usually eliminated from adhesive joints by joint design and materials selection. It is possible, however, to develop an undesirable layer and lose strength and toughness. The transformation product may be large enough to be included in the next category.

Transition macrostructures are features in the macroscopic size range that form out of the elements of the original phase structure under the conditions that prevail. They are distinguished by their structure, size, and shape and are transitional since they lead naturally to other features connected with the fracture process. Of these features listed in Table II it is possible to single out

    shear bands,
    crazing,
    voids, and
    crystallites or other transformation products.

These are are the structural precursors to fracture and are detectable. Critical dislocation clusters are also precursors to fracture but not in adhesive joints. The failure envelope for the adherend in the well designed adhesive joint is much larger than that of the adhesive so that adhesive failure shields against adherend failure.

The actual development of a craze or other macrofeature makes a profound change in terms of stress concentrations and system behavior. Beardmore and Rabinowitz show that the stress for crazing in PMMA is approximately 80% of the fracture stress.[35] This is analogous to experience in metal fatigue failure in that 90% of the life of the item elapses before the fatigue crack is nucleated. The appearance of the precursor is a positive indication of imminent fracture.

METHODS OF CHARACTERIZATION

The most notable development for adhesion science is the availability of a large variety of sophisticated, reliable instruments, that provide information well beyond the capabilities of the classical microscope, radiograph and universal test machine. The older instruments, never-the-less, retain their importance in basic analytical procedures. The new instruments provide sensitivity and method out of the range of the earlier models. The functions that are now available are in the following categories:

    chemical and structural analysis
    observation
    process monitoring
    materials characteristic determination.

Chemical and structural analyses may be applied to the _in situ_ determination of submicroscopic sample regions. This includes the electron microprobe, EDAX attachments to scanning electron microscopes, Auger electron spectroscopy, selected area electron diffraction, ellipsometry and other surface characterization techniques. To this list can be added three facilities that require removal of portions of the sample for analysis but that can be used to identify the existence of molecular fragments and transition structures. These include the scanning calorimeter, NMR spectrometer, chromatograph and IR spectrometer.

Modern observational techniques include the transmission and scanning electron microscopes, polarized light, phase contrast and other interference techniques for the enhancement of contrast. There are many such methods and they all contribute information.

Process monitoring may now be accomplished by such methods as acoustic emission in which characteristic mechanical waves are emitted in conjunction with specific processes. These sounds can be isolated from a noisy environment to provide a reliable signal of the occurrence of some specific event in the sample. Alternate methods, in which the sample interacts with radiation, which is monitored, are also available. Ultrasonics provide reflection techniques, useful if new surfaces are formed and attenuation methods that have proven useful in grain size determination and similar applicatons. A variety of physical properties may now be employed along with sensitive transducers and signal processing equipment. In many cases signal data can be computer processed to isolate and enhance the desired information.

This same signal enhancement can be used to record the fine structure of a materials characteristic. Although last on this list, the materials characteristic is the most direct analytical test and provides the most useful information to identify separate deformation processes and to recognize the formation of structural precursors to fracture.

CONCLUSION

The science of materials has progressed to such an extent that a number of transition structures are recognized in materials under load. Techniques are available to systematically identify their presence and formation. The first indication of the presence of new structural entities is provided by discontinuities in materials characteristic curves or changes under different conditions even though direct confirmation may not be possible. Other, more specific information may be gained by modern methods of fracture surface observation, microchemical analysis and macroscopic structural determination. Eventually in-process monitoring can be used to detect the presence of structural precursors to the fracture process and thus avoid adhesive joint failure.

ACKNOWLEDGEMENT

The author is grateful for helpful discussions with B.Z. Jang, who very generously provided information related to this subject from his PhD dissertation at Massachusetts Institute of Technology.

REFERENCES

1. R.C. Wilcox and W.A. Jemian, Metall. Rev., 1,13 (1972).
2. R.C. Wilcox and W.A. Jemian, Polym. Eng. Sci., 13, 40 (1973).
3. A.H. Cottrell, "The Mechanical Properties of Matter," John Wiley & Sons, New York, 1964.
4. K. Masubuchi and R.E. Keith, Final Report on "Fundamentals of Deformation Characteristics of Adhesive-Bonded Joints and Metal-Adhesive Interfaces," Contract No. DA-01-021-AMC-14693(Z), Battelle Memorial Institute, Columbus, 1967.
5. G.C. Grimes, Aero. Adhes. and Elast., 249-258 (1970).
6. W.A. Jemian and M.B. Ventrice, J. Adhesion, 1,, 190 (1969).
7. W.J. Renton and J.R. Vinson, "The Analysis and Design of Composite Material Bonded Joints under Static and Fatigue Loadings,"Contract No. AFOSR-1760-72, University of Delaware, 1973.
8. F.J. Boerio, C.A. Gosselin, R.G. Dillingham, and H.W. Liu, J. Adhesion, 13,159 (1981).
9. W.T. McCarville and J.P. Bell, J. Appl. Polymer Sci., 18,335 (1974).
10. I. Langmuir, J. Amer. Chem. Soc., 38,2221 (1916).
11. I. Langmuir, J. Amer. Chem. Soc., 39,1848 (1917).
12. K. Blodgett, J. Amer. Chem. Soc., 56,495 (1934).
13. I. Langmuir and V. Schaeffer, Chem. Rev., 24, 181 (1939).
14. P. Jacquet, Trans. Electrochem. Soc., 66, 393 (1934).

15. W.A. Jemian, "Adhesion and Bonding of Coatings on Metals," PhD Thesis, Rensselaer Poly. Inst., Troy, 1956.
16. Y.C. Fung, "Foundations of Solid Mechanics," Prentice Hall, 1965.
17. A.N. Gent, J. Mater. Sci., $\underline{5}$, 925 (1970).
18. T. Alfrey, in "Treatise on Adhesion and Adhesives," R.L. Patrick, Editor, Vol. 1, Marcel Dekker, New York, 1966.
19. C.B. Buchnall, "Toughened Plastics," Applied Science Publishers, London, 1977.
20. R.P. Kambour, J. Polymer Sci., $\underline{D7}$,1 (1973).
21. S. Rabinowitz and P. Beardmore, CRC Crit. Revs. Macromol. Sci., $\underline{1}$,1 (1972).
22. A.S. Argon, J.G. Hannoosh and M.M. Salama, Fracture 1977, $\underline{1}$, 445 (1977).
23. R.J. Oxborough and P.B. Bowden, Phil. Mag., $\underline{28}$, 547 (1973).
24. J.J. Bikerman, "The Science of Adhesive Joints," Academic Press, New York, 1961.
25. J.L. Gardon, in "Treatise on Adhesion and Adhesives," R.L. Patrick, Editor, Vol. 1, Marcel Dekker, New York, 1966.
26. R.W. Bryant and W.A. Dukes, in "Structural Adhesives Bonding," M.J. Bodnar, Editor, p 81, Interscience, New York, 1966.
27. W.C. Wake, in "Adhesion," D.D. Eley, Editor, Oxford University Press, p.191, 1961.
28. W.D. Bascom and R.L. Cottington, J. Adhesion, $\underline{7}$, 333 (1976).
29. J.J. Bikerman, in "Adhesion and Adhesives," R. Houwink and G. Salomon, Editors, Vol. 1, 2nd Ed., Elsevier, 1965.
30. J. Friedel, "Dislocations," Pergamon Press, 1964.
31. J. Miklowitz, J. Appl. Mech., Trans, ASME, $\underline{70}$, 274 (1948).
32. J. Miklowitz, J. Appl. Mech., Trans. ASME, $\underline{72}$, 159 (1950).
33. C.W. MacGreggor, Proc. ASTM, $\underline{40}$, 508 (1940).
34. S.H. Pinner, "Weathering and Degradation of Plastics," Gordon and Breach, 1966.
35. P. Beardmore and S. Rabinowitz, J. Mater. Sci., $\underline{10}$, 1763 (1975).

FRACTURE TOUGHNESS OF ELASTOMER MODIFIED EPOXY ADHESIVES

A.A. Donatelli[*], C.T. Mooney[*] and J.C. Bolger[**]

[*]University of Lowell, Lowell, MA 01854
[**]Amicon Corporation, Lexington, MA 02173

Elastomer modification is an effective way of toughening glassy epoxy resins. The toughening mechanisms of phase separation and gellation have been studied extensively. This paper reports the effects of carboxyl terminated acrylonitrile rubber on several single component epoxy adhesives, based on Epon 828 and cured with dicyandiamide. Variables include accelerator type and level, cure temperature and elastomer level. There is an optimum amount of rubber which may be added to each system to yield maximum shear strength. Peel strength is maximized in two regions. Rapid gellation of these adhesives may be detrimental to the phase separation and chain extension mechanism and thereby cause lowering of fracture toughness and strength properties.

INTRODUCTION

Elastomeric modifiers have been used extensively in the adhesive industry to improve the mechanical properties of glassy epoxy adhesives. Many structural adhesives have benefitted from the addition of small amounts, ca. 5-20 parts per hundred resin (phr) of elastomer.

Toughening is achieved by introducing an elastomer such as a carboxyl terminated copolymer of butadiene and acrylonitrile ("HYCAR CTBN") into a glassy epoxy resin. The elastomer which is initially dissolved in the resin precipitates to form a second phase during cure. This second phase appears as small domains (5000-30000 Å) within the glassy matrix. The formation of this second phase is dependent on such factors as concentration, cure temperature and rate of reaction.[1,2]

The importance of carboxyl termination in the toughening mechanism stems from the fact that the precipitated phase is not pure rubber but rather an epoxy terminated liquid rubber. The following reactions have been proposed to account for this phenomenon.[3]

$$\sim C(=O)-OH + CH_2-CH\sim \rightarrow \sim C(=O)-O-CH_2-CH(OH)\sim \quad (1)$$

$$\sim CH_2(OH)\sim + CH_2-CH\sim \rightarrow \sim CH(O-CH_2-CH(OH)\sim)\sim \quad (2)$$

Reaction 1 is the chain extension reaction which occurs between the carboxyl terminated elastomer and the epoxy resin. Reaction 2 is a cross linking reaction which competes with reaction 1 for epoxide groups.

Although carboxyl termination is an important step in the toughening mechanism, the molecular weight of the starting resin also plays an important role. Toughening has been increased by adding Bisphenol A into liquid epoxy resins which are then modified with an elastomer.[4,5] In theory, the addition of Bisphenol A causes chain extension which further increases toughness.

In practice, mixtures containing Bisphenol A have unusually high viscosities and are difficult to handle. The increased strength generally does not justify the handling disadvantages.

EXPERIMENTAL

Several single component epoxy systems were prepared. All systems studied contained 100 parts of a diglycidyl ether of Bisphenol A (Epon 828) plus 10 phr dicyandiamide and 2 phr CAB-O-SIL.

The first variable studied was 0 to 100 phr elastomer modifier (HYCAR CTBN x 13). An "optimum" system was then selected and two catalysts were added. The catalysts studied were 3-(p-chlorophenyl)-1,1-dimethylurea and nonyl phenol. Both were varied from 0-8 phr. Systems were cured at temperatures ranging from 120°C to 180°C.

Performance properties measured were Shore D Hardness, 160°C hot plate gel time ("stroke cure") and two strength tests to measure adhesive joint fracture; i.e; tensile shear strength per ASTM D1002 and "T" - peel strength per ASTM D1876.

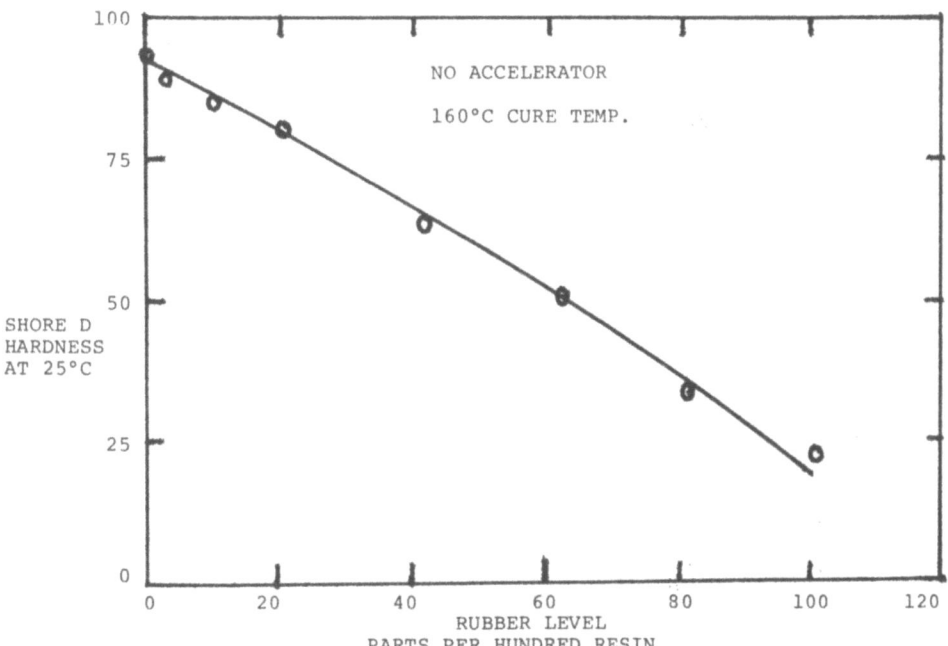

Figure 1. Shore D Hardness versus rubber level.

RESULTS AND DISCUSSION

The addition of small amounts of elastomer modifier increases the toughness of glassy epoxy resins. In addition, elastomer modifiers also have a softening or plasticizing effect. Figure 1 shows Shore D hardness measured at 25°C versus rubber level in an unaccelerated epoxy resin matrix. A nearly linear relationship exists between rubber level and hardness. Maximum softening occurs at addition levels above 20 phr. Between 0-20 phr of elastomer a 11.8% decrease in hardness is observed, but between 20-40 phr of elastomer a 24.4% decrease in hardness occurs.

This toughening versus softening effect is best illustrated in Figures 2 and 3. Figure 2 plots tensile shear strength versus rubber level for an unaccelerated epoxy resin matrix. These systems were cured for one hour at 180°C and one hour at 160°C. Both exhibit the same behavior. A bell shaped curve is observed with shear strength increasing sharply between 0 and 20 phr rubber. This supports the concept of softening rather than toughening at addition levels above 20 phr. Above this level the materials lose their rigidity, which is almost as important to shear strength as the crack stopping imparted by the small rubbery domains present in cured system with low addition levels of rubber.

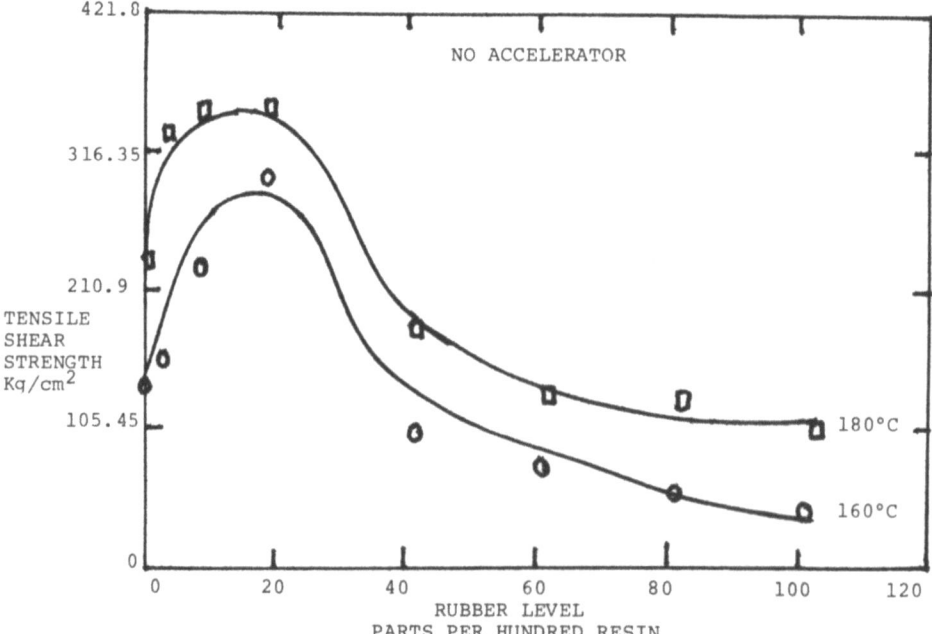

Figure 2. Tensile shear strength versus rubber level.

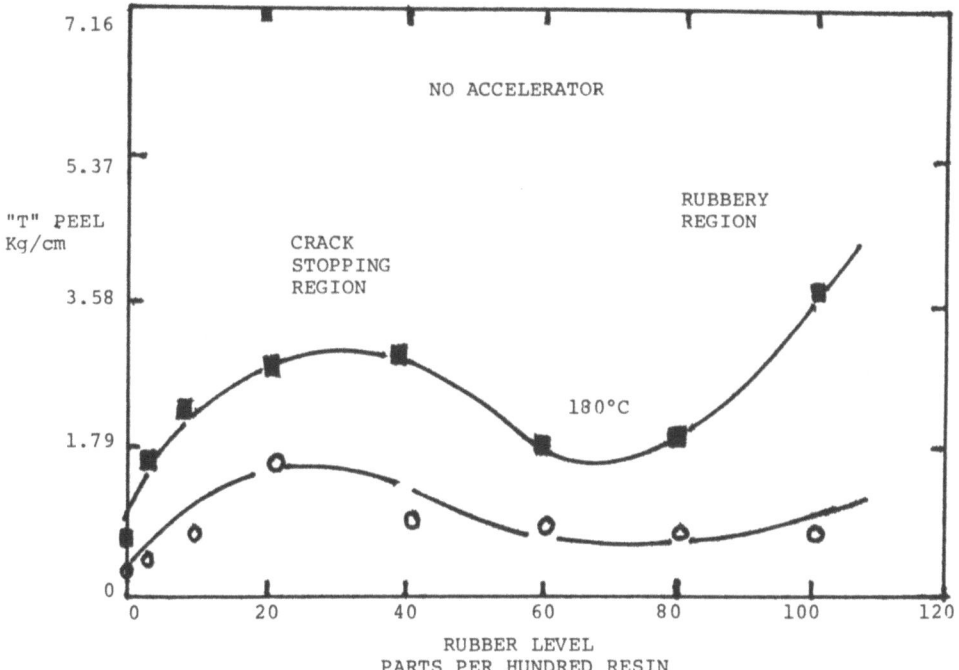

Figure 3. "T" peel strength versus rubber level.

Figure 3 presents "T" peel strength plotted aginst rubber level for the same systems presented in Figure 2. Peel strength increases between 0 and 20 phr rubber, levels off, decreases to 60 phr then starts to increase again. In the first region at low addition levels, peel strength is being increased by the crack stopping effect of the precipitated rubbery domains. Unlike shear strength, peel strength is not as adversely affected by a soft material. This is supported by the fact that peel strength increases after the 60 phr addition level. The fracture mode indicates that rigidity is somewhat of a disadvantage and the same effect can be achieved through the internal strength of the material. These data indicate that there is a transition period from a discontinuous precipitated rubber phase which depends on crack stopping to achieve its strength, to a continuous rubbery phase which depends on its internal strength.

Several adhesives possess a good balance of strength properties. One disadvantage of dicyandiamide is the high temperatures at which it must be cured. Dicyandiamide has a minimum activation temperature of approximately 150°C. To obtain a realistic cure time, 160°C must be used.

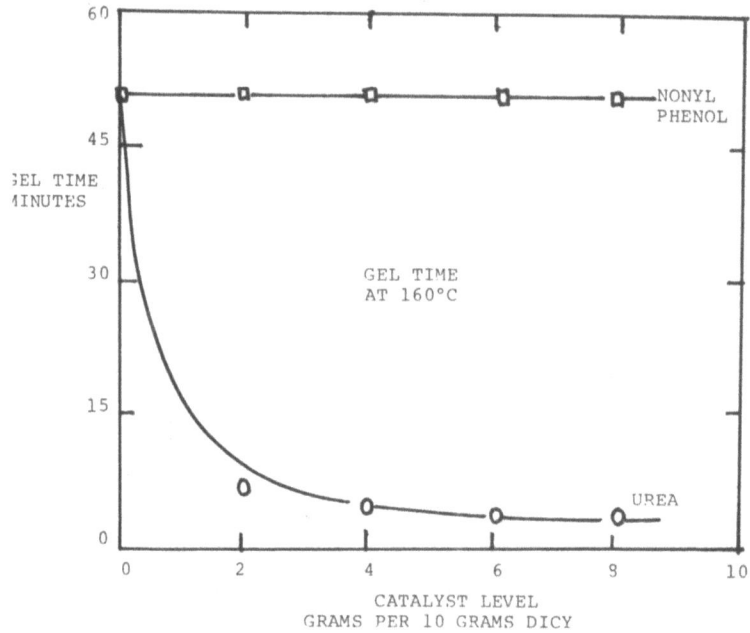

Figure 4. 160°C hot plate gel time versus catalyst level.

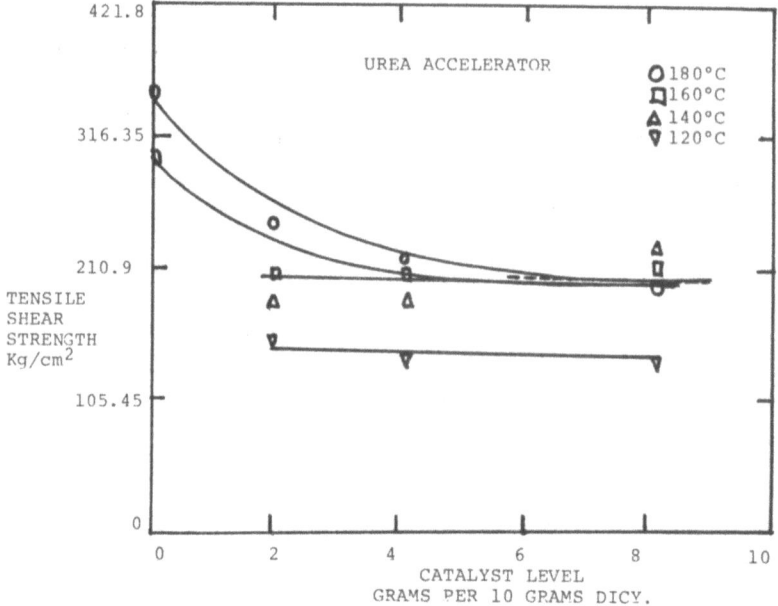

Figure 5. Tensile shear strength versus urea type accelerator level, at constant 20 phr rubber level.

FRACTURE TOUGHNESS OF EPOXY ADHESIVES

Two accelerators were used in this study, 3-(p-chlorophenyl)-1,1-dimethylurea and nonyl phenol. Figure 4 shows the effect of the addition of these accelerators to a system which contained 20 phr of elastomer modifier. The addition of nonyl phenol had no effect on the 160°C hot plate gel time. The urea accelerator significantly reduced the gel time between 0 and 2 parts of urea per 10 parts of dicyandiamide, then heading towards a plateau at the 8 parts addition level indicating that additions of urea above this level would yield no advantage. It is evident that the acid catalyst (nonyl phenol) has no accelerating effect whereas the basic catalyst (urea type) has a significant effect on the dicyandiamide cure rate.

The accelerators have a detrimental effect on tensile shear strength. Figure 5 shows the effect of the urea accelerator level on shear strength at four different cure temperatures. There is a large decrease in strength between the control which contains no accelerator and the adhesive which contains two parts of urea accelerator. Although there is a large decrease in strength initially there is no appreciable strength difference between the 2 and 8 parts addition level of the urea catalyst at 180°C and 140°C cure temperatures. There is a significant decrease in strength in the systems when the cure temperature is decreased to 120°C. As with the higher cure temperatures, there is no appre-

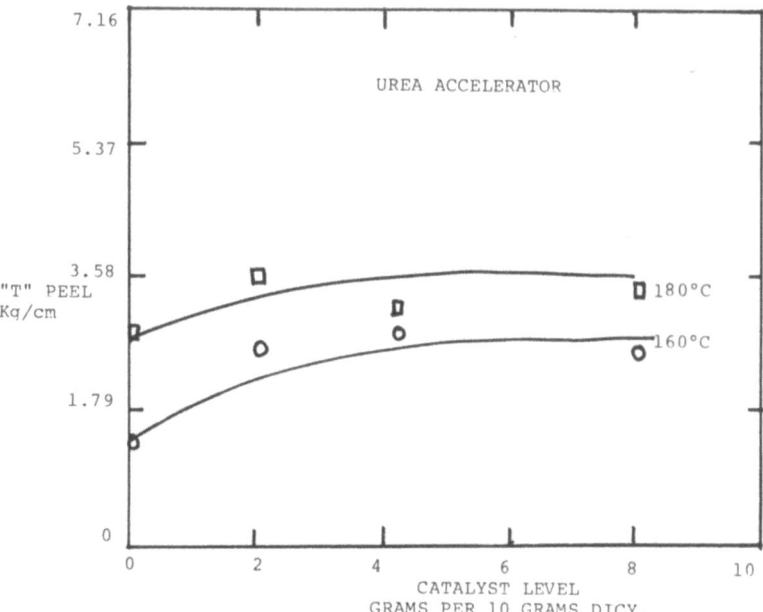

Figure 6. "T" peel strength versus urea type accelerator level, at constant 20 phr rubber level.

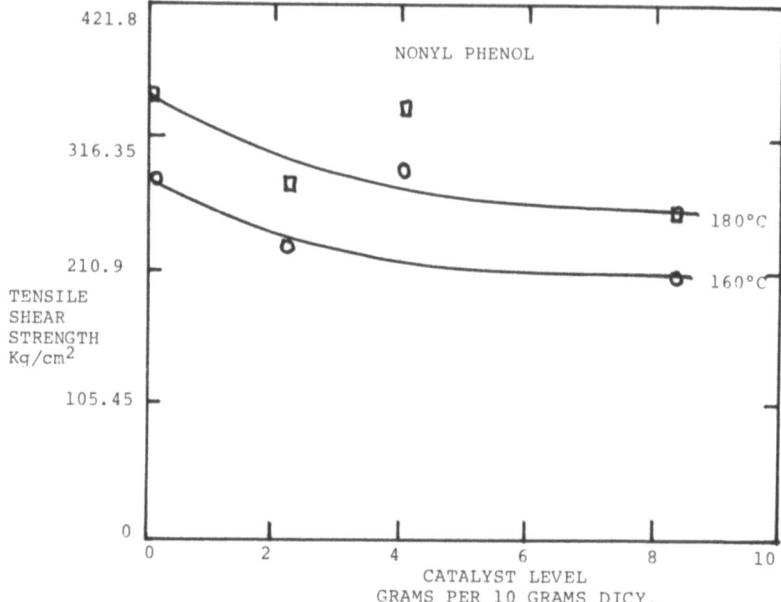

Figure 7. Tensile shear strength versus nonyl phenol level, at constant 20 phr rubber level.

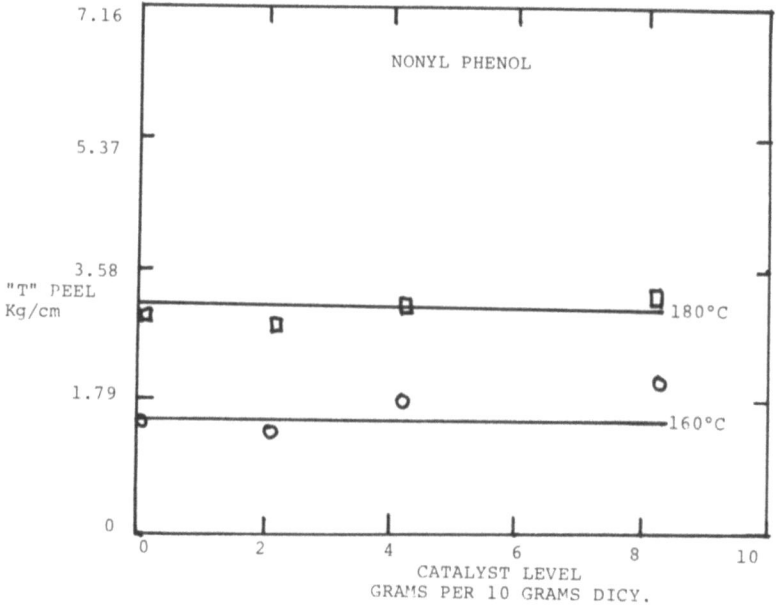

Figure 8. "T" peel strength versus nonyl phenol level, at constant 20 phr rubber level.

ciable strength difference observed between the 2 and 8 parts addition level of the urea catalyst. The accelerator may interfere with the phase separation and hence the toughening mechanism.

The strength loss effect of the urea accelerator is not observed with "T" - peel strength. Figure 6 shows a slight but not significant increase in peel strength with urea level. As was shown previously in Figure 3 there are two regions where peel strength can be maximized, the crack stopping region and the rubbery region. If the phase separation mechanism is inhibited by the presence of an accelerator than a rubbery phase would be present in the cured system and the peel strength can be maintained.

Nonyl phenol does not function as an accelerator but it does have an effect on the strength properties of the system. Figure 7 plots tensile shear strength versus nonyl phenol level. There is a decrease in strength as the level of nonyl phenol is increased. It is interesting to note that the data points at 4 phr are well above the respective curves. It would appear from this, plus the n on accelerating effect observed in Figure 4, that the nonyl phenol functions as a plasticizer.

Figure 8 shows the effect of the nonyl phenol on "T" peel strength. Nonyl phenol has no effect on the peel strength.

CONCLUSIONS

1. Rubber addition reduces hardness, increases flexibility.

2. Tensile shear strength is maximized at ca 20% rubber addition level.

3. Peel strength is maximized in two regions, the crack stopping region and the rubbery region.

4. Increased cure temperature (up to 180°C) increases all strength properties.

5. Accelerators may be used to speed cure rate and to permit cure at as low as 120°C to 140°C.

    a. Urea is a more effective accelerator than nonyl phenol.

    b. Nonyl phenol appears to function as a plasticizer.

    c. Both accelerators cause a reduction in tensile shear strength.

    d. Accelerators have a negligible effect on peel strength.

REFERENCES

1. F. J. McGarry and J. N. Sultan, Polym. Eng. Sci., 13 (1), 29 (1973).
2. R. S. Drake, E. H. Rowe and A. R. Siebert, Mod. Plast., 47, 110 (1970).
3. A. R. Siebert and C. K. Riew, "The Chemistry of Rubber Toughened Epoxy Resin I", B. F. Goodrich Research Center, Brecksville, Ohio, Paper presented at the 161st American Chemical Society Meeting, (1971).
4. L. T. Manzione and J. K. Gillham, J. Appl. Polym. Sci., 26, 907 (1981).
5. C. B. Bucknall and T. Yoshii, Br. Polym. J., 10 (1), 53 (1978).
6. W. D. Bascom, R. L. Cottington, R. L. Jones and P. Peyser, J. Appl. Polym. Sci., 19, 2545 (1975).

A THREE-DIMENSIONAL ANALYSIS OF A BUTT JOINT WITH A FLAW

R. S. Alwar* and K. N. Ramachandran Nambisan**

*Applied Mechanics Department, IIT, Madras
 600036, India
**Civil Engineering Department, R.E.C., Calicut
 673601, India

A three-dimensional finite element method is employed to analyze an adhesive butt joint with a through-the-thickness crack in the center of the adhesive layer. The finite element model used accommodates both the extremely small dimensions of the adhesive layer and the larger dimensions of the rest of the region. The crack tip region is modelled by isoparametric quarter-point degenerate brick elements which give the required square root singularity for stresses at the crack tip. The results from a fracture mechanics approach corroborate the earlier experimental results that the strength of an adhesive joint increases with decrease in adhesive layer thickness. It is also observed that the stress intensity factor varies in a non-linear fashion along the crack front.

INTRODUCTION

Significant savings in weight and fabrication cost can be realized through the use of adhesive bonding in aircraft structures. The definition of an adhesive bond encompasses a number of situations from bonded structural components to fibers bonded together by the matrix in a fibrous composite. When a composite is made with an excellent bond between the fibers and the matrix, it is found to be brittle and have very low toughness value. Hence the application of the principle of linear elastic fracture mechanics in the design of bonded structures becomes very important. Extensive studies have been conducted by Mostovoy and Ripling[1] on flaw tolerance and crack growth in adhesives under opening mode. Their studies showed that the cracks tend to grow in a center of bond (CoB) fashion.

Trantina[2] employed a two-dimensional analysis, using conventional six noded triangular elements, and analyzed a single edge notched (SEN) specimen. He calculated the compliance rate for various adhesive layer thicknesses and compared the results with experimental values. He showed that compliance rate decreased as the adhesive layer thickness became thinner. This corroborates the experimental finding from a fracture mechanics point of view that the bond strength increases with decrease in adhesive layer thickness. Wang et al.[3] analyzed a Double Cantilever Beam (DCB) Specimen with an adhesive interlayer by a two-dimensional hybrid finite element method. They found the stress field close to the crack tip in the adhesive to be similar to that of a monolithic system. However, they found no influence of the adhesive layer thickness on the stress intensity factor in the case of the DCB specimen. But, for a stretching field, an increase in adhesive strength with decrease of adhesive layer thickness was noted by Gardon[4] and Williams[5]. The same result was reported by many others who worked with adhesive structural joining. In fact, Meissner and Baldauf[6] had demonstrated that at very low adhesive layer thickness, the ultimate tensile strength of a butt joint can exceed the ultimate tensile strength of the adhesive. A discussion of the factors which may contribute to the strength of the adhesive joint with decrease in adhesive thickness was given by Anderson et al.[7]. Trantina[8], using a two-dimensional finite element analysis, studied the effect of the location of the crack tip for an inclined crack in a butt joint.

Two-dimensional analysis for an adhesive joint may not be adequate, especially for large thickness to crack length ratios and with sudden change of material properties in a region close to the crack front as in the case of bonded joints. The object of the present investigation is to analyze a two material (aluminum-epoxy-aluminum) system in the form of a butt joint using three-dimensional finite element method and to apply the fracture mechanics concept to the strength of butt joints.

METHOD OF ANALYSIS

The geometry and dimensions of the problem analyzed are shown in Figure (1). Edge-cracked and center-cracked bonded joints are analyzed with the crack situated in the middle of the adhesive layer thickness. In the multi-material body, the stress intensification near the crack tip is produced from the geometric discontinuity (crack) and the material discontinuity. The numerical analysis becomes complicated by the vast disparity in dimensions between the adherend boundaries and the region across the adhesive layer thickness where the stress distribution must be accurately described. The finite element mesh must accommodate both the small dimensions of the adhesive thickness and the larger dimensions of the rest of the specimen. It is essential to model the adhesive layer and the crack tip region accurately with a number of elements across the adhesive thickness to discern the effect of the material discontinuity.

Using the symmetry about x-z and x-y planes, only 1/4 of the plate is considered for solution in the case of edge crack; and using the additional symmetry about y-z plane, only 1/8 of the plate is taken for solution in the case of center-cracked plate.

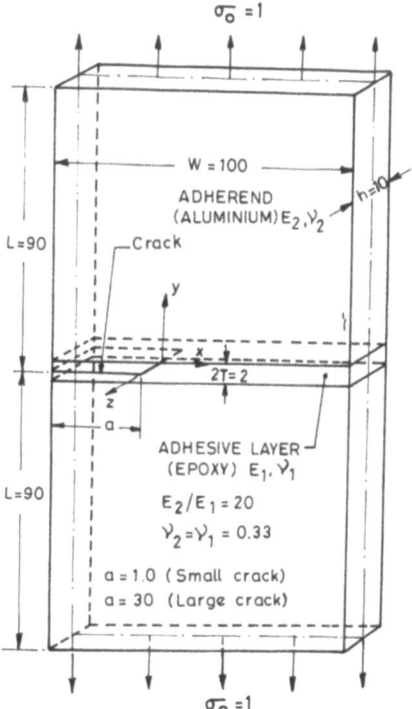

Figure 1. Geometry of the bonded joint and the coordinate system at the crack tip.

The adhesive layer thickness (=2T) has been represented by eight
layers of elements in the crack tip region and by four layers in
the rest of the region. The crack tip element size is taken as
0.1 with the crack tip element size to crack length ratio of
0.00333 for large crack and 0.1 for small crack. Crack tip region
is modelled by singular[9] and transition elements[10] with a proper
grading of the element sizes to avoid large distortions of the
elements. Because of the interaction of the singularity with the
material discontinuity in the immediate neighborhood of the crack
tip, a more refined mesh grading has been adopted in the region
between the crack tip and adherend-adhesive interface. The mesh
pattern adopted for the analysis is shown in Figure (2).

The ratio of the modulus of elasticity of the adherend
(aluminum) to that of the adhesive (epoxy) is taken to be 20 and
the Poisson's ratio ν for both the materials is taken as 0.33.
The analysis assumes both adherend and adhesive to be linear
elastic materials.

EXTRACTION OF STRESS INTENSITY FACTORS

The three-dimensional stress and displacement fields in the
close neighborhood of the crack front have the following form.
(See Figure (3) for the crack tip coordinates).

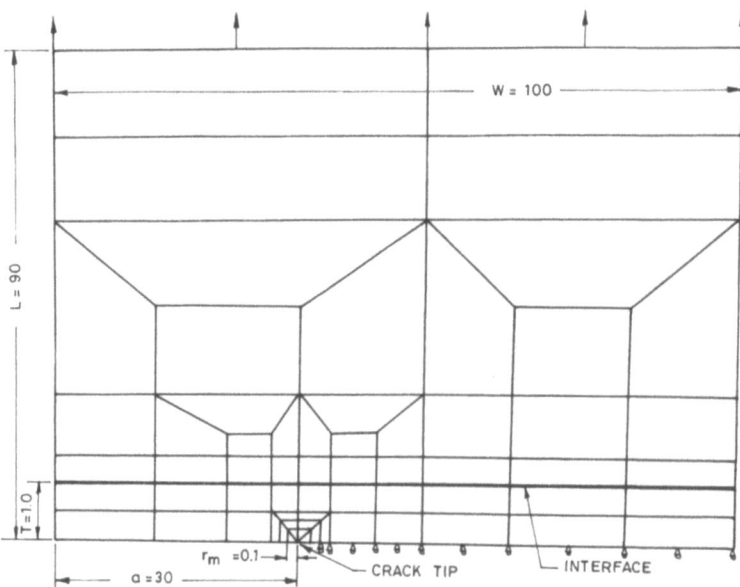

Figure 2. Finite element model of the bonded joint (not to
scale).

$$\sigma_x = \frac{K_1(z)}{(2r)^{1/2}} \cos\frac{\theta}{2} (1 - \sin\frac{\theta}{2} \sin\frac{3\theta}{2})$$

$$- \frac{K_2(z)}{(2r)^{1/2}} \sin\frac{\theta}{2} (2+\cos\frac{\theta}{2} \cos\frac{3\theta}{2}) + 0(1) \qquad ..(1)$$

$$\sigma_y = \frac{K_1(z)}{(2r)^{1/2}} \cos\frac{\theta}{2} (1+\sin\frac{\theta}{2} \sin\frac{3\theta}{2}$$

$$+ \frac{K_2(z)}{(2r)^{1/2}} \sin\frac{\theta}{2} \cos\frac{\theta}{2} \cos\frac{3\theta}{2} + 0(1) \qquad ..(2)$$

$$\sigma_z = \nu (\sigma_x + \sigma_y) \qquad ..(3)$$

$$\tau_{xy} = \frac{K_1(z)}{(2r)^{1/2}} \sin\frac{\theta}{2} \cos\frac{\theta}{2} \cos\frac{3\theta}{2}$$

$$+ \frac{K_2(z)}{(2r)^{1/2}} \cos\frac{\theta}{2} (1-\sin\frac{\theta}{2} \sin\frac{3\theta}{2}) + 0(1) \qquad ..(4)$$

$$\tau_{xz} = -\frac{K_3(z)}{(2r)^{1/2}} \sin\frac{\theta}{2} + 0(1) \qquad ..(5)$$

$$\tau_{yz} = -\frac{K_3(z)}{(2r)^{1/2}} \cos\frac{\theta}{2} + 0(1) \qquad ..(6)$$

$$4Gu = K_1(z)(2r)^{1/2} \cos\frac{\theta}{2} (\kappa - 1+2\sin^2\frac{\theta}{2})$$

$$+ K_2(z)(2r)^{1/2} \sin\frac{\theta}{2} (\kappa +1+2\cos^2\frac{\theta}{2}) + 0(r) \qquad ..(7)$$

$$4Gv = K_1(z)(2r)^{1/2} \sin\frac{\theta}{2} (\kappa +1-2\cos^2\frac{\theta}{2})$$

$$- K_2(z)(2r)^{1/2} \cos\frac{\theta}{2} (\kappa -1-2\sin^2\frac{\theta}{2}) + 0(r) \qquad ..(8)$$

$$Gw = K_3(z)(2r)^{1/2} \sin\frac{\theta}{2} + 0(r) \qquad ..(9)$$

where $\kappa = 3-4\nu$ for plane strain

$\qquad = \frac{3-\nu}{1+\nu}$ for generalized plane stress

and G = Modulus of rigidity.

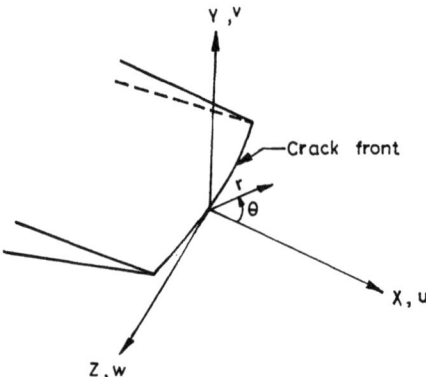

Figure 3. Coordinate system near the crack front.

An exact 3-D solution[11] incorporates $\kappa = 3-4\nu$ and hence this value has been used in the present analysis.

The finite element values of displacements or stresses are matched with the elastic singular solutions in the vicinity of the crack tip (Equations 1-9), yielding the stress intensity factors (SIF) as a function of distance from the crack tip. Extrapolation to the crack tip, using the exact stress or displacement expressions, yields accurate values for the stress intensity factors. Stresses are calculated at two consecutive gauss points very close to the crack tip and lying on a radial line from the crack front, using a 10x10x10 gauss point rule. Substituting the computed values of σ_y in Equation (2) and extrapolating using square root singular behavior of the stresses, the value of SIF at $r = 0$ is obtained. Alternatively the crack opening displacements (COD) at the quarter-node and end-node of the crack tip element may be used to calculate the stress intensity factors $K_{1,1}$ and $K_{1,2}$ respectively using Equation (8) and the extrapolated value of the stress intensity factors K_1 at $r=0$ is given by $K_1 = 2K_{1,1} - K_{1,2}$. The evaluation of K_1 in this manner incorporates the $\sqrt{r}$ displacement variation near the crack tip and eliminates the rigid body motion and linear terms from the displacements. The K_1 variation along the crack front is obtained by calculating the continuous displacement field on the traction-free surface of the crack tip element from the element nodal displacement values and the shape functions of the element.

RESULTS AND DISCUSSION

A study of the accuracy of the present analysis has been made by comparing the results with those of the two-dimensional analysis by Trantina[2]. The comparison has been made for the case of very large crack length for which 2-D analysis may be accurate enough. For small crack length with large h/a value, the problem should be treated as a three-dimensional one[12]. Though the two-dimensional analysis by Trantina was done using conventional elements to model the crack tip region without singular and transition elements, the application of energy principle to get the compliance derivative is expected to give accurate results. Moreover, experimental verification of the numerical results had also been obtained in the above analysis. Figure (4) shows the comparison of the present 3-D results with those obtained by Trantina. The present analysis gives SIF values which are not uniform along the crack front. The maximum value of the SIF is obtained in the mid-plane of the plate and the value decreases as we proceed towards the free surface of the plate. It may be observed from the figure that there is reasonable agreement between the 3-D results and the 2-D results by Trantina. The 3-D mid-plane values are found to be higher than the 2-D values by 8-12 percent. It may be observed that 3-D values are higher than the corresponding 2-D values in the case of monolithic plates.

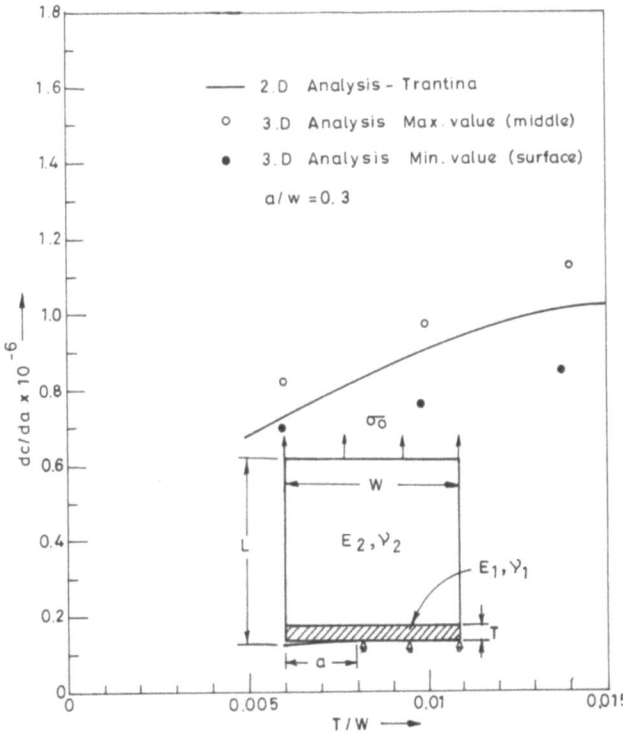

Figure 4. Compliance derivative dc/da as a function of adhesive thickness.

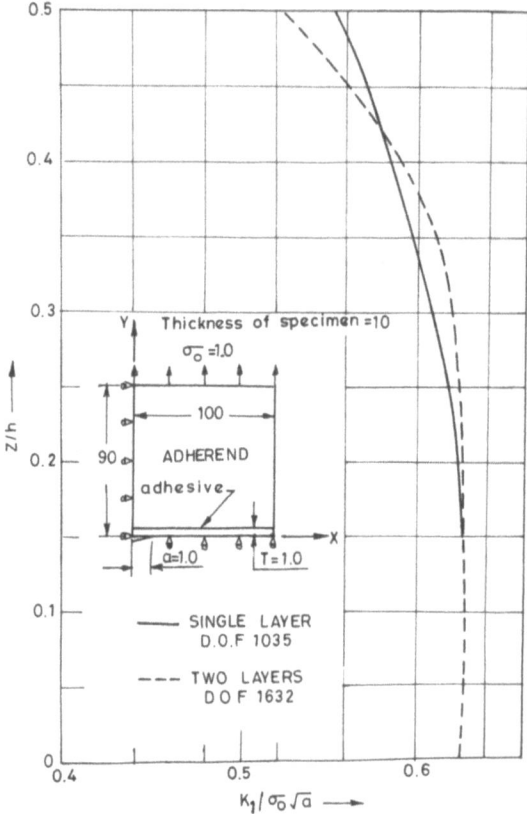

Figure 5. Convergence study for SIF variation along the crack front.

Figure (5) shows the convergence of SIF variation in the interior of a center-cracked plate for single and two layer solutions for a/W = T/W - 0.01. Since 'a' is of the same order of magnitude as the adhesive layer thickness and less compared to the specimen thickness, the crack-tip stress distribution will be three-dimensional in nature. Convergence study could not be done with more number of layers of elements across the specimen thickness because of the prohibitive computer time involved in such an analysis. The experience in analyzing the monolithic plates with a center crack has shown that two and three layer solutions agree well for the SIF variation throughout the thickness of the plate. It is seen from the figure that in the interior of the plate up to z/h = 0.2 from the mid-plane, the agreement between one and two layer solutions is excellent. The reduction towards the surface for the two layer solution compared to one layer solution is about 6 percent. The results presented in this paper are corresponding to single layer solutions and hence the values in the mid-plane region are expected to be accurate, whereas the surface values are expected to be higher than the actual by about 6-10 percent. From the two-layer solution, the decrease of SIF at the surface compared to the mid-plane value is about 16 percent. A decrease of SIF value at the surface, of the order of 15-30 percent, is reported in the case of monolithic plates[13,14].

Figure (6) gives the 3-D results for the stress intensity factor as a function of adhesive layer thickness for a large crack length (a = 30.0). It can be seen from the figure that the value of SIF decreases with the decrease in adhesive layer thickness which corroborates the earlier experimental results that the strength of an adhesive joint increases with the decrease in adhesive layer thickness.

Figure (7) shows the mid-plane SIF variation as a function of adhesive layer thickness for the case of a very small crack length (a = 1.0). It is observed that the stress intensity factor reduces with decrease in adhesive layer thickness for this case also. Comparing both Figures (6) and (7), it can be seen that as h/a increases $K_1/\sigma_0\sqrt{a}$ also increases which is true for the case of a monolithic plate[15].

Figures (8-11) show the stress distribution across the plate thickness in the neighborhood of the crack front for points lying around the crack tip. It can be seen from these figures that the transverse shear stresses are extremely small compared to the other stresses. It may be observed that for bending and stretching loads, these shear stresses are bounded and since the points selected are close to the axis of symmetry, the transverse shears have to be very small. Figure (8) shows that $\sigma_x \simeq \sigma_y$ at points very close to the line of crack extension (ie. at $\theta = 0.74°$) and that the condition of pseudo plane strain, i.e., $\sigma_z = \nu(\sigma_x + \sigma_y)$ is also satisfied. The pseudo plane strain condition is found to exist at all the points in the neighborhood of the crack front. This is in accordance with the 3-D analytical solution[11] that $\sigma_x = \sigma_y$ along the line of crack extension and that a pseudo plane strain condition exists in the close vicinity of the crack front. The figures also show that the values of σ_y increases from $\theta = 0°$ to approximately $\theta = 60°$ and then gradually decreases to zero value at $\theta = 180°$. The value of σ_x decreases from $\theta = 0°$ to approximately $\theta = 75°$ and then shows an upward trend up to $\theta = 135°$ and then drops to zero at $\theta = 180°$. The value of τ_{xy} increases up to $\theta = 30°$ and then decreases to assume a maximum negative value at $\theta = 100°$ and then drops to zero at $\theta = 180°$. The above θ-variation of stresses follow the theoretical θ-variation as represented in reference 16. This indicates that a correct θ-variation of stresses in the neighborhood of the crack front is obtained with the finite element mesh pattern adopted in the analysis.

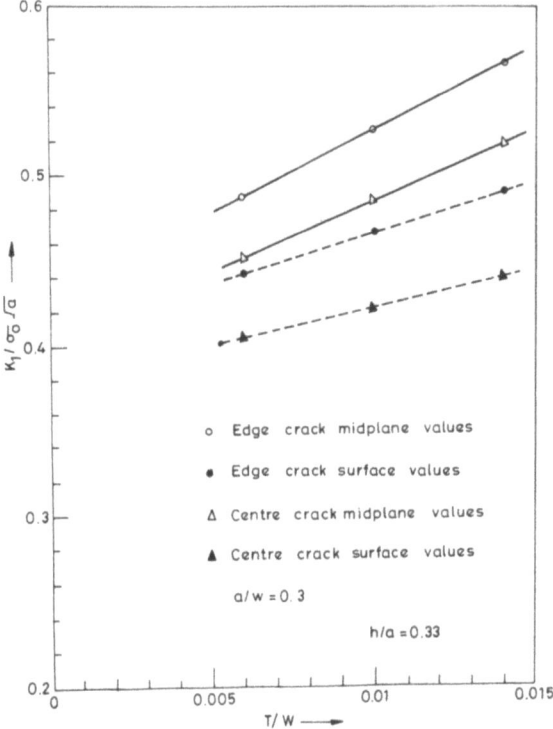

Figure 6. $K_1/\sigma_0\sqrt{a}$ as a function of adhesive thickness for a/W = 0.3.

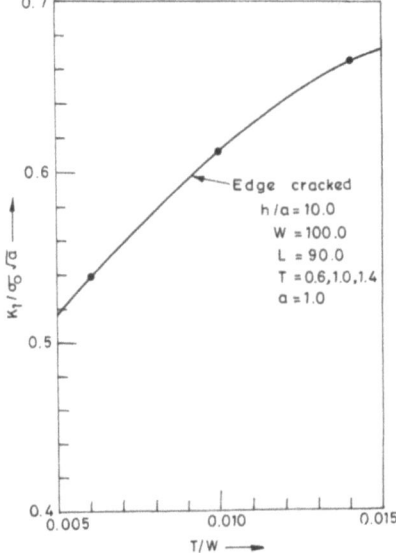

Figure 7. $K_1/\sigma_0\sqrt{a}$ (mid-plane) as a function of adhesive thickness for a/W = 0.01.

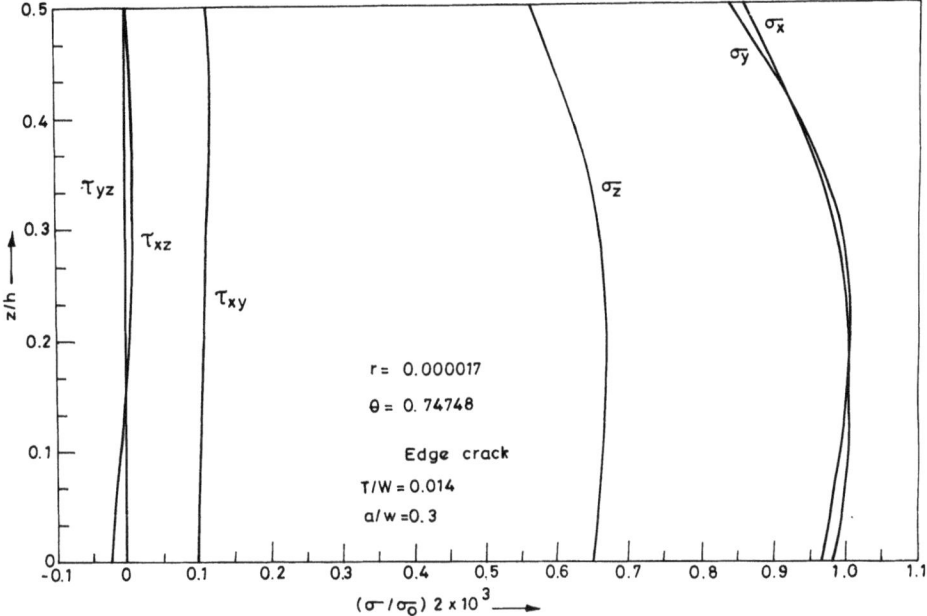

Figure 8. Stress distribution σ/σ_0 near the crack front at $r = 0.000017$ and $\theta = 0.74748°$ - edge cracked plate.

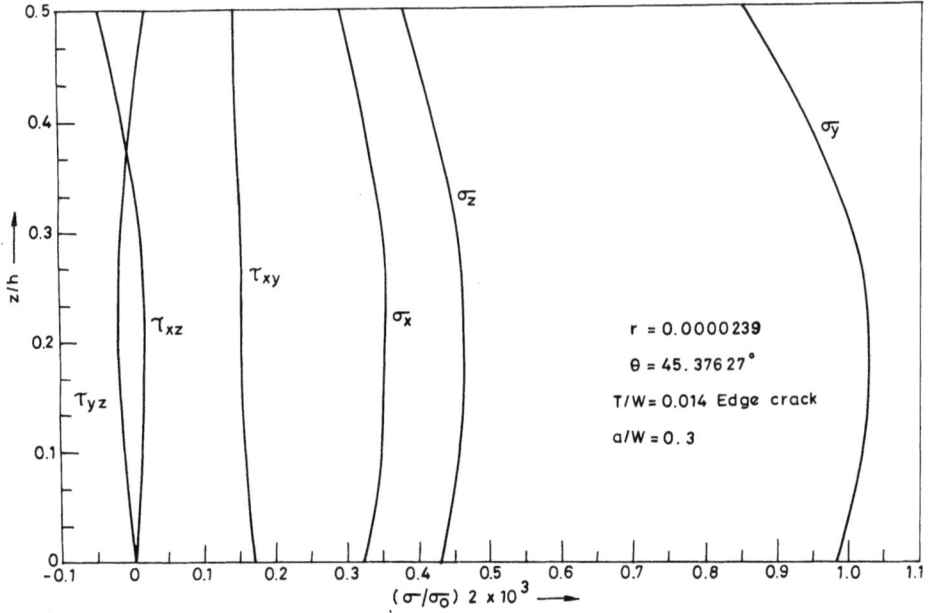

Figure 9. Stress distribution σ/σ_0 near the crack front at $r = 0.0000239$ and $\theta = 45.376°$ - edge cracked plate.

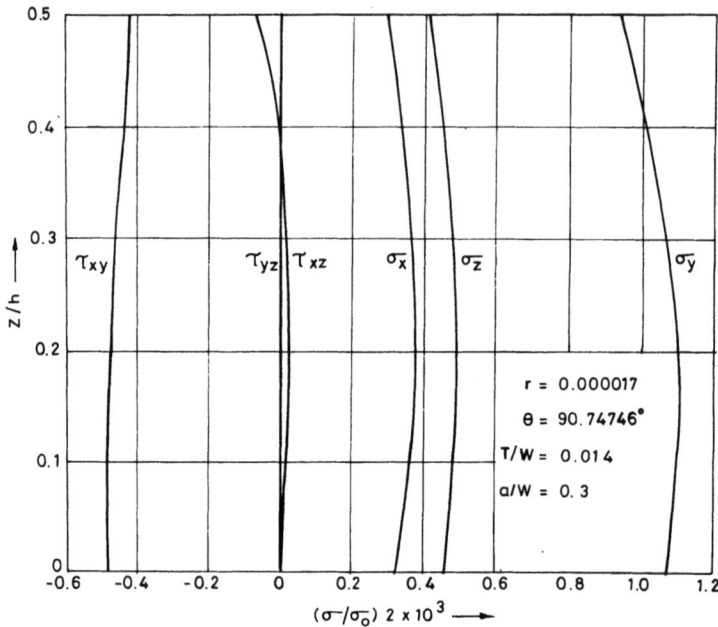

Figure 10. Stress distribution σ/σ_o near the crack front at $r = 0.000017$ and $\theta = 90.74746°$ - edge cracked plate.

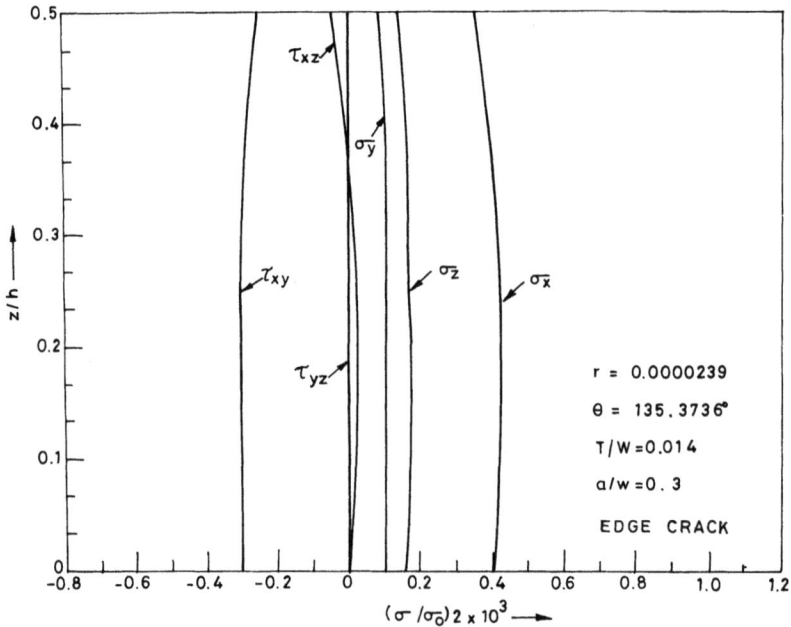

Figure 11. Stress distribution σ/σ_o near the crack front at $r = 0.0000239$ and $\theta = 135.3736°$ - edge cracked plate.

ANALYSIS OF A BUTT JOINT WITH A FLAW

Figure (12) compares the stress-based and COD-based values of stress intensity factors across the plate thickness and it is seen that there is good agreement between the two results, the maximum difference being about 3 percent.

CONCLUSIONS

The 3-D analysis gives results which are 8-12 percent higher compared to 2-D values in the interior portion of the adhesive.

The SIF is uniform in the central region of the adhesive and decreases towards the free surface. This reduction is about 16 percent for the case studied.

For an adhesively bonded joint the value of SIF decreases with decrease in adhesive layer thickness. This corroborates the experimental observations that the strength of a bonded joint increases with decrease in adhesive layer thickness.

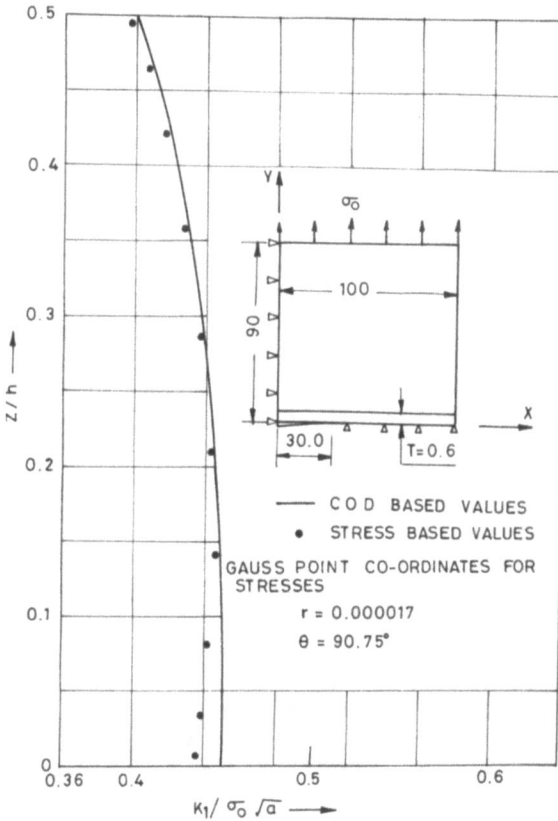

Figure 12. Comparison of stress-based and COD-based SIF values across plate thickness.

APPENDIX

NOTATION

$2a$	−	Crack length
E	−	Modulus of elasticity
G	−	Shear modulus
h	−	Plate thickness
K_1, K_{11}, K_{111}	−	Stress intensity factors
$2W$	−	Width of Plate
$2T$	−	Thickness of adhesive layer
u,v,w	−	Displacement components in X,Y,Z direction
r,θ	−	Polar coordinates
ν	−	Poisson's ratio

REFERENCES

1. S. Mostovoy and E. J. Ripling, J. Appl. Polymer Sci., 10, 1351 (1966).
2. G. G. Trantina, J. Composite Mater., 6, 192 (1972).
3. S. S. Wang, J. F. Mandell and F. J. McGarry, Int. J. Fracture, 14, 39 (1978).
4. J. L. Gardon, in "Treatise on Adhesion and Adhesives", R. L. Patrick, Editor, Vol. 1, pp. 269-324, Marcel Dekker, New York, 1967.
5. M. L. Williams, J. Appl. Polym. Sci., 14, 1121 (1970).
6. H. P. Meissner and G. H. Baldauf, Trans. Soc. Amer. Mech. Engr., 697 (1951).
7. G. P. Anderson, K. L. DeVries and M. L. Williams, Int. J. Fracture, 10, 565 (1974).
8. G. G. Trantina, J. Composite Mater., 6, 371 (1972).
9. R. S. Barsoum, Int. J. Numerical Meth. Eng., 11, 85 (1977).
10. P. P. Lynn and A. R. Ingraffea, Int. J. Numerical Meth. Eng., 12, 1031 (1978).

11. G. C. Sih, M. L. Williams and J. L. Swedlow, "Three-dimensional stress distribution near a sharp crack in a plate of finite thickness", Air Force Materials Lab., Wright-Patterson Air Force Base, AFML-TR-66-242 (1966).
12. G. C. Sih and R. J. Hartranft, Int. J. Fract., 7, 39 (1971).
13. P. D. Hibitt, in "Mechanics of Fracture, Plates and Shells with Cracks", G. C. Sih, Editor, Noordhoff Int. Pub., Leyden (1977).
14. I. S. Raju and J. C. Newman, Jr., "Three-dimensional finite element analysis of finite thickness fracture specimens", NASA-TN-D 8414, May (1977).
15. G. C. Sih and R. J. Hartranft, Int. J. Fract., 9, 75 (1973).
16. R. S. Barsoum, Int. J. Numerical Meth. Eng., 10, 551 (1976).

THE INFLUENCE OF LAYER THICKNESS AND INTERNAL STRESSES ON THE BOND STRENGTH OF METAL-TO-CERAMIC JOINTS

W. Diem, G. Elssner, T. Suga and G. Petzow

Max-Planck-Institut für Metallforschung
D-7000 Stuttgart, W. Germany

Sandwich-like joints between Zr, Zr-Nb-Zr or Hf foils of different thickness and rectangular bodies of hot-pressed silicon nitride (HPSN) were fabricated by solid-state bonding at 1150 °C or 1200 °C. Bend bars of 30x5x2 mm were cut from the welded pieces and notches of variable depth a were introduced at the metal-to-ceramic interface. The bond strength was described by critical energy release rates using experimentally determined correction functions and correction functions calculated by means of the FEM method. The effects of layer thickness and crack position on the correction functions of materials joints are discussed from a more general point of view by a comparison with the correction functions of isotropic materials and bimaterials. Measurements of the critical energy release rates of the metal-to-ceramic joints and microstructural investigations indicate that the reduction of internal stresses by microcrack formation and by allotropic transformation of the metal component control together with the metal layer thickness the bond strength of the sandwich-like joints. The bend test method is also applicable to adhesively bonded joints. The influence of the adhesive thickness and of the crack position on the correction function can be evaluated according to the principles described in this paper.

INTRODUCTION

In most of the materials joints thermal stresses are produced during the manufacturing process. They may lead to a reduction in bond quality. Thermal stresses promote slow crack growth and the formation of microcracks in the interface region. Their magnitude is not only affected by the differences in the elastic and thermal properties of the bonded materials but depend also on the thickness of intermediate layers.

Solid-state bonded metal-to-ceramic joints consisting of two bodies of hot-pressed silicon nitride (HPSN) and an intermediate foil of a refractory metal were chosen to study the influence of layer thickness and of an allotropic phase transformation on the bond strength[1].

The bond strength of the metal-to-ceramic joints was characterized by the critical energy release rate G_C obtained by a fracture mechanics testing method. The critical release rate G_C and the fracture resistance K_C are closely related to each other. Both can be taken as a measure of the bond strength of metal-to-ceramic joints and of semi-brittle joints. Generally, the energy release rate G and the stress intensity factor K are described by the relations[2]

$$G = (1-\beta^2)K^2/E^* \qquad (1)$$

$$G = K_o^2 Y_G/E^* \qquad (2)$$

$$K^2 = K_o^2 Y_G/(1-\beta^2) \qquad (3)$$

where β is a Dundur's composite parameter given by the elastic properties of the two materials adjacent to the crack front, E^* the effective modulus of elasticity, and K_o a constant proportional to the applied stresses. Y_G is a so-called correction function which depends on the crack length at the interface, the elastic properties of the bulk materials of the joint, and the specimen geometry.

For bend test specimens the correction function Y_G can be obtained experimentally by compliance measurements using the equation[2]

$$Y_G = dC^*/d(a/W) \qquad (4)$$

where C^* is a normalized compliance, a the crack or notch depth, and W the height of the specimen.

To get true data of the critical energy release rate G_C the dependence of the correction function Y_G on both the ratio a/W and the layer thickness d must be determined. If different materials combinations are investigated this procedure will be rather tedious. However, some combinations as refractory metal/ceramic joints show relatively small differences in the elastic properties of their constituents when compared to others as for example adhesive joints. Therefore, it may be useful to estimate beforehand the maximal deviation of the correction function Y_G from the wellknown correction function Y_{iso} of isotropic materials. This feature is discussed from a more general point of view in the next section.

THE CORRECTION FUNCTION AND LAYER THICKNESS EFFECTS

The correction function Y_G for a crack at the interface of a materials joint depends on the elastic properties of all materials of the joint. Consequently for a joint consisting only of the two materials 1 and 2 the stresses under in-plane deformations and the correction function Y_G can be described by the Dundurs' parameters[2] α_{12} and β_{12}.

Consider the case that a crack is situated at an interface between the two materials A and B of an intermediate multilayer of thickness d_m between the materials 1 and 2 (Figure 1). The thickness d_m is negligibly small compared to the specimen size. If the elastic properties of the materials A and B are known, a simple relation can be derived from considerations of the continuity of the energy release rate during continuous changes of the materials geometry and the crack tip position.

$$Y_{AB} = \frac{(1-\alpha_{AB})(1+\alpha_{2B})}{(1-\alpha_{12})(1-\alpha_{2B})} Y_{12}, \quad d_m/W \ll 1 \qquad (5)$$

where α_{ij} is a Dundurs' parameter related to the materials i and j and Y_{ij} the correction function Y_G for an interfacial crack specified by the materials i and j. This relation stands also for a diffusion layer, if its thickness d_m is small compared to the specimen size and its elastic properties at the crack tip are known.

In cases where the crack runs irregularly through a thin intermediate multilayer of thickness d_m (Figure 2) and a specification of the elastic properties at the crack tip is not possible the correction function Y_{12} can only be calculated. In Figure 2 two other possible fracture paths are shown where the cracks are running parallel to the interface into the materials 1 and 2. They may be designated as cohesive failures. When the layer thickness d is small compared with the specimen height W the correction function Y_G of interfacial cracks and cracks near the interface can be related to the correction Y_{iso} for isotropic specimens using one composite parameter α_{12} for the description of the elastic properties of materials 1 and 2.

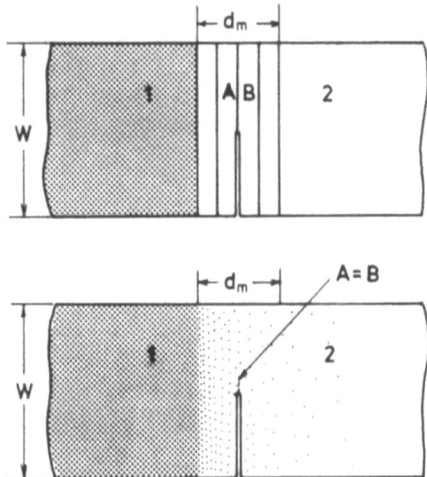

Figure 1. Cracks in intermediate layers of thickness d_m between the two materials 1 and 2.

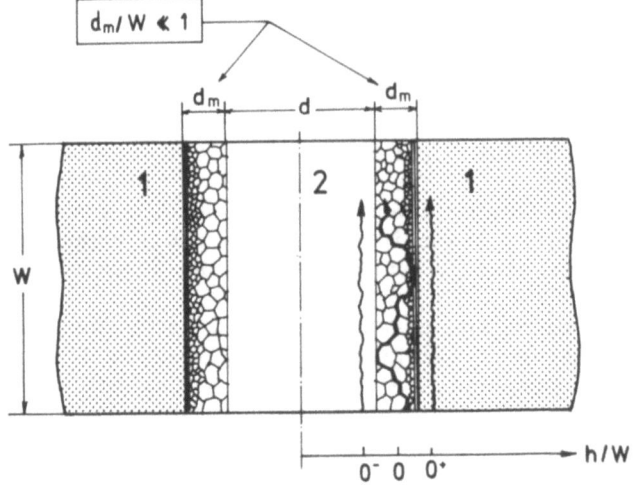

Figure 2. Fracture paths within an intermediate layer and parallel to the interfaces in the materials 1 and 2.

A more general view of the relationship between the isotropic correction function Y_{iso} and the correction functions of materials joints is given in Figure 3. Thin intermediate layers which may exist between the materials 1 and 2 are omitted in the sketches of the specimens. Y_{bi}, Y_1, and Y_2 denote the correction functions Y_G for an interfacial crack in a simple bimaterial or for cracks in the materials 1 or 2, respectively. As can be seen, Y_G for interfacial cracks at variable ratios d/W is bounded by the correction functions $Y_{bi}(d/W \to \infty)$ and $(1+\alpha)Y_{iso}(d/W \to 0)$. If the influence of stress concentrations at the bonded edges is not large the value of Y_{bi} will be slightly smaller than the value of Y_{iso}. For a center crack in an infinite bonded plane the relation $Y_{bi} > 0.842\ Y_{iso}$ holds, e. g. $Y_{bi} = 0.974\ Y_{iso}$ for glass/epoxy and $Y_{bi} = 0.994\ Y_{iso}$ for HPSN/Hf joints. At constant layer thickness and variable ratios h/W (h = distance of the crack from the interface) the correction function Y_G changes its value discontinuously from $Y_{12}/(1+\alpha)$ over Y_{12} to $Y_{12}/(1-\alpha)$ when the crack position shifts from material 1 to material 2. With increasing distance of the crack from the interface the correction function $Y_{12}/(1+\alpha)$ will be substituted by Y_1 and finally by Y_{iso} if material 1 is considered or $Y_{12}/(1-\alpha)$ will be substituted by Y_2 if the crack is situated in an intermediate layer material 2 between two bodies of material 1.

EXPERIMENTAL

Sandwich-like joints consisting of two hot-pressed silicon nitride (HPSN) bodies and an intermediate layer of a refractory metal were manufactured by solid-state bonding[3]. Zr and Hf foils and Zr-Nb-Zr triple layers so different thickness ranging from 0.125 to 2.0 mm were chosen as intermediate metal layers. The HPSN/Zr/HPSN and the HPSN/Zr,Nb,Zr/HPSN joints were solid-state bonded at 1150 °C and the HPSN/Hf/HPSN joints at 1200 °C. The metal and ceramic parts were welded toghether under a pressure of 10 N/mm² in a high vacuum. Some properties and the composition of the materials used for the solid-state bonded joints are given in Table I.

Three-point bend test specimens of the dimensions 30x5x2 mm were cut from the solid-state bonded materials. Notches of width 50 to 100 µm and of depth a were introduced into the metal-to-ceramic interface (Figure 4) by a precision cutting device. The bend tests were carried out with a model 1474 Zwick machine at a cross-head speed of 1 µm/min. In the set-up load and deflection signals were treated in real time by a directly connected microprocessor. To obtain true load-deflection curves and compliance data, the measured deflections were corrected for the deformation of the test device during loading.

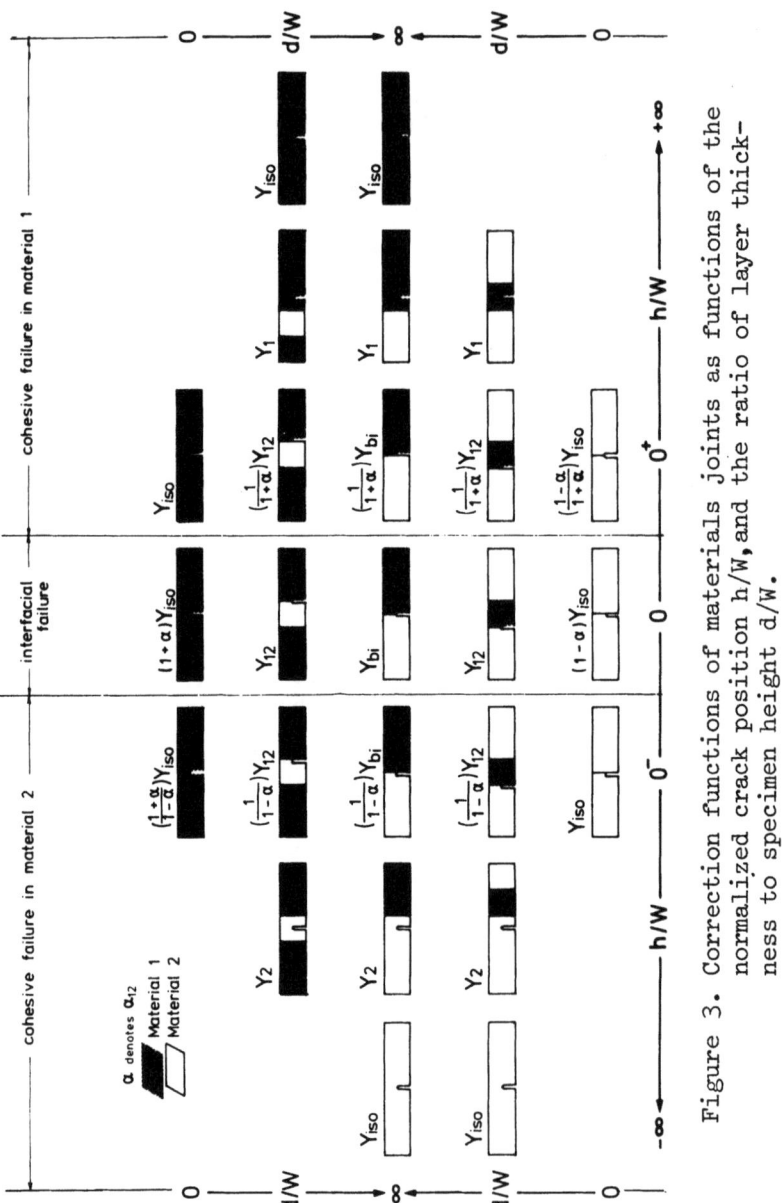

Figure 3. Correction functions of materials joints as functions of the normalized crack position h/W, and the ratio of layer thickness to specimen height d/W.

Table I. Materials used for solid-state bonded parts.

Material	Composition (w/o)	E(GPa)	ν	$\alpha_T (10^{-6}/K)$
Zr	99.8 Zr, 0.1 O, 0.06 Fe, 0.01 Cr,N,H	99	0.33	7.3
Hf	97.0 Hf, 2.8 Zr, 0.015 O, 0.01 Fe,Ta,C,N	141	0.30	6.0
Nb	99.8 Nb, 0.05 Ta, 0.03 W, 0.01 Zr,Ti,Fe	108	0.38	8.4
HPSN	97.0 Si_3N_4, 0.98Mg, 0.14 Ca, Y_2O_3, Si_2ON_2	314	0.28	2.7

E is the Young's modulus, ν the Poisson ratio, and α_T the thermal expansion coefficient.

NUMERICAL ANALYSIS

The stress analysis of a materials joint with thin intermediate metal layers requires a fine mesh of finite elements for the specimen. The present calculations were performed with about 800 quadratic isoparametric elements under the condition of plane strain. For the crack tip eight quarter-point singular elements of length 0.005 W were employed. The correction functions were calculated by means of the virtual extension and the J-integral method. The calculated results for four-point bend test specimens of an isotropic material differ within ± 1 % from the wellknown analytical values[4] (Figure 5). Figure 6 shows a mesh of finite elements used for the stress analysis of a HPSN/Hf/HPSN three-point bend test specimen.

RESULTS AND DISCUSSION

In Figure 7 the normalized compliance C* and the corresponding correction function Y_G of HPSN/Hf/HPSN joints are plotted against the ratio of crack depth to specimen height a/W for different ratios of metal layer thickness to specimen height d/W. The curves are calculated by means of the finite elements method. The dots represent experimental data. Minor deviations between the calculated and the experimental values seem to be due to the small specimen size and the assumption of a sharp crack in the FEM calculations. For HPSN/Hf/HPSN joints the composite parameters α and β are -0.366 and 0.098 respectively. As previously mentioned, the correction functions are bounded by two extreme curves: the curve of the correction function Y_{bi} near the curve for the isotropic case and the curve of $Y_G = (1+\alpha)Y_{iso}^{bi}$ for d/W = 0.

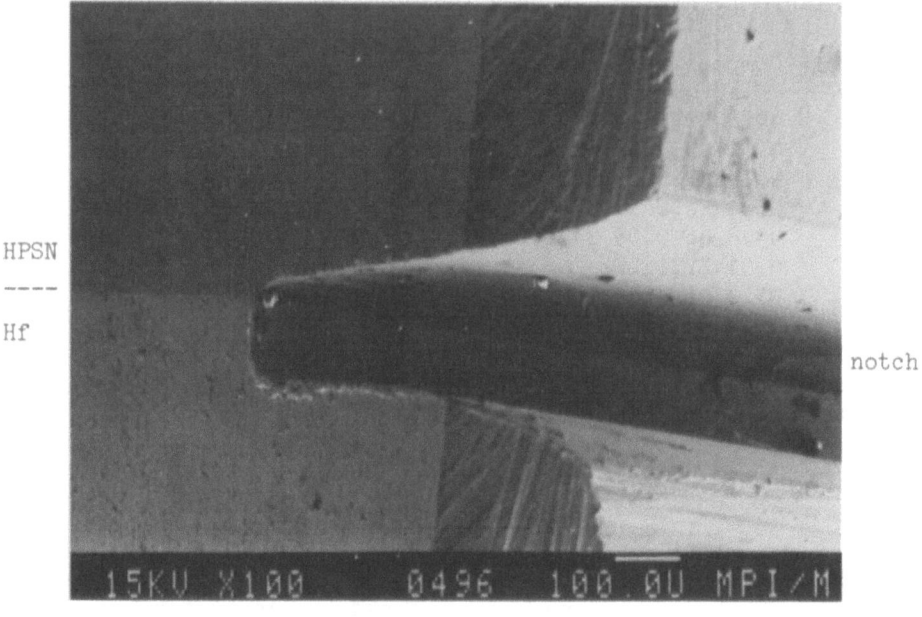

Figure 4. SEM picture of a notch cut into the interface of a HPSN/Hf/HPSN joint.

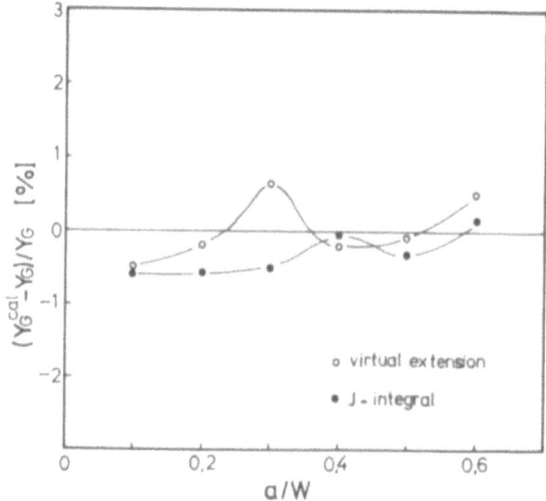

Figure 5. Comparison of the calculated correction function Y_G (calc) for an isotropic four-point bend test specimen with the correction function Y_G obtained by Gross and Srawley[4].

INFLUENCE OF LAYER THICKNESS AND INTERNAL STRESSES 863

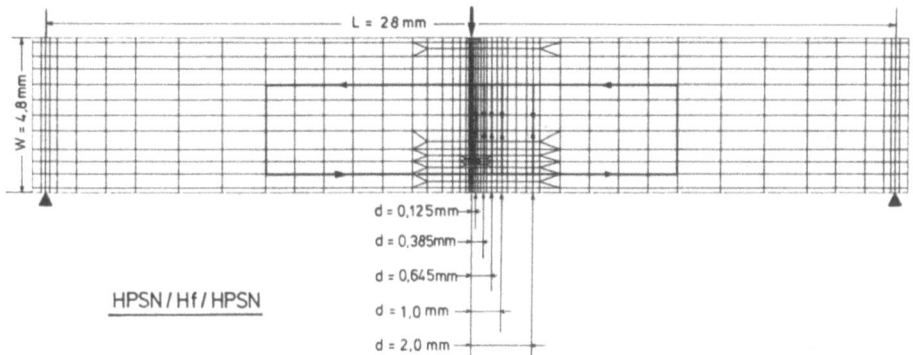

Figure 6. FEM mesh used for the stress analysis of a HPSN/Hf/HPSN three-point bend test specimen. Heavy lines indicate the integral paths.

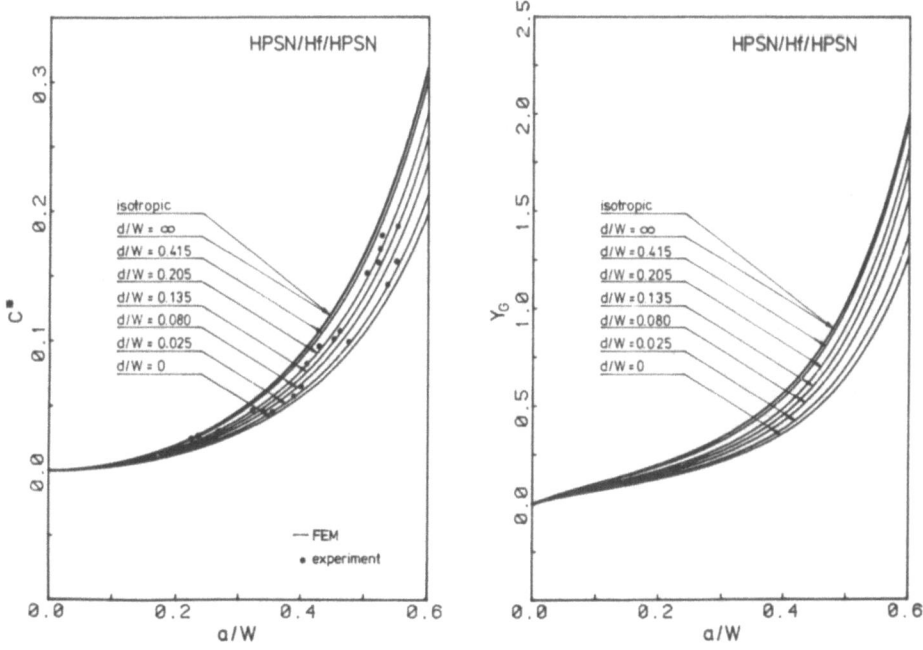

Figure 7. Normalized compliance C* and correction function Y_G versus ratio a/W for HPSN/Hf/HPSN joints. Three-point bend test. Dots are measured and lines calculated values.

Figure 8 shows experimental data of the critical energy release rate G_c of a HPSN/Hf/HPSN joint with a Hf layer of 0.125 mm thickness as a function of the ratio a/W. The observed independence of the G_c data on the ratio a/W demonstrates the applicability of the correction function Y_G.

Metal-to-ceramic joints fabricated by solid-state bonding at 1150 to 1200 °C from bulk HPSN pieces with an intermediate Zr or Hf layer or a triple layer Zr-Nb-Zr are characterized by a thin reaction layer of about 5 μm thickness which is composed of silicides, nitrides, and ternary metal-silicon-nitrogen compounds of the metals Zr of Hf[2]. Furthermore, a very thin Ca- and Si-oxide rich glassy layer develops during the welding process on the HPSN interface which may lead to a reduction of bond strength. The formation of these glassy layers is caused by the hot-pressing aids added to the HPSN material. To separate the influence of layer thickness and allotropic transformation of the metal component different model combinations were chosen. Triple intermediate layers Zr-Nb-Zr consist of a rigid niobium foil of variable thickness coated on both sides by Zr foils of 0.125 mm thickness. They cause a thickness-independent intimate contact of the mating ceramic and metal interfaces during welding which is only achieved by the plastic deformation of the Zr layers without any visible deformation of the Nb foil or the HPSN parts. HPSN/Zr,Hf,Zr/HPSN joints are characterized by the same type of reaction layers independent of the total metal layer thickness. Nb and Zr form solid solutions at 1150 °C. Therefore, no reaction layers develop at the Zr/Nb interfaces. HPSN/Zr/HPSN joints of different metal layer thickness and, for comparison, HPSN/Hf/HPSN joints were used to study the influence of an allotropic transformation of the metal on the bond strength of the joint. Zr transforms during cooling down from the welding temperature from a cubic body-centered β-structure to a hexagonal close-packed α-structure. The transformation of the pure metal at 862 °C is accompanied by an elongation of about 0.1 %. Hafnium as another metal of the group IV A shows a β/α-transformation at 1750 °C. Therefore welding at 1200 °C and cooling down is not influenced by a phase transformation. However, both metals form the same types of reaction layers because of their chemical similarity.

G_c data of HPSN/Zr,Nb,Hf/HPSN, HPSN/Hf/HPSN, and HPSN/Zr/HPSN joints were evaluated from the fracture load F_c, the specimen geometry, and the correction functions Y_G according to the relation[2]

$$G_C = F_C^2 \cdot \frac{9 e^2 \pi Y_G}{B^2 W^3 E^*} \tag{6}$$

where 2e is the lower span length in the case of three-point bend testing, B the width, and W the height of the specimen.

INFLUENCE OF LAYER THICKNESS AND INTERNAL STRESSES   865

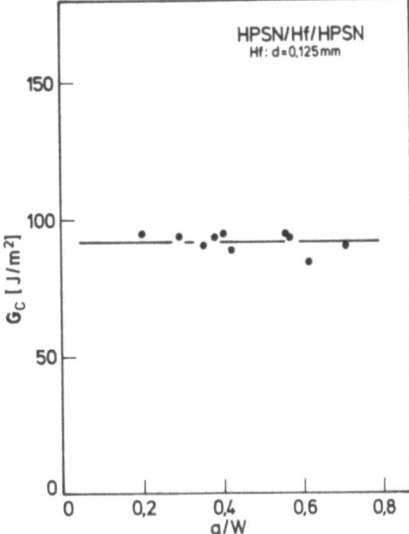

Figure 8. Critical energy release rate G_C of a HPSN/Hf/HPSN joint with a Hf layer of 0.125 mm thickness as a function of the ratio a/W.

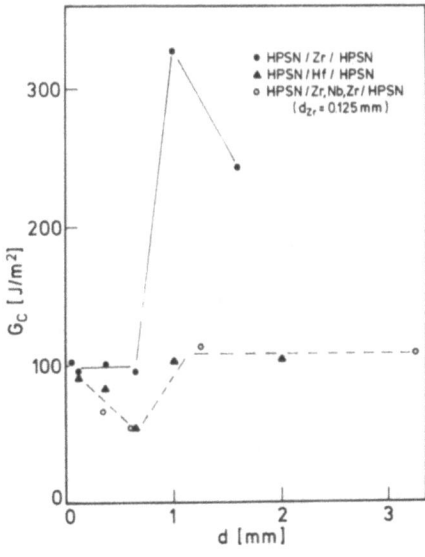

Figure 9. G_C data of refractory metal/HPSN joints as a function of the intermediate metal layer thickness d.

Figure 9 shows the experimentally determined G_C data of the joints as a function of the layer thickness d. The plotted data represent mean values obtained from 10 to 15 measurements per layer thickness. The scatter is ± 10 %. For HPSN/Hf/HPSN and HPSN/Zr,Nb,Zr/HPSN joints the critical energy release rate G_C decreases from 94 J/m² at d = 0.125 mm to 54 J/m² at d = 0.6 mm and increases again to a value of 107 J/m² at d = 1 mm. For layer thickness of d > 1 mm up to d = 3.25 mm nearly the same G_C values were observed. HPSN/Zr/HPSN joints show for Zr layers of the thickness 0.055, 0.385, and 0.64 mm a G_C value of 100 J/m². From d_{Zr} = 0.64 mm to d_{Zr} = 1.0 mm a steep increase to a G_C value of 328 J/m² occurs and at a thickness d_{Zr} = 1.56 mm the G_C value is reduced to 244 J/m².

During cooling down from the welding temperature considerable thermal stresses are built up in the interface region due to the differences in the thermal expansion coefficients of the ceramic material and the refractory metals (Table I). They lead to the formation of microcracks perpendicular to the interface. Figures 10 and 11 show two examples of microcracks which terminate at the α-Zr/reaction layer interface (HPSN/Zr/HPSN, d_{Zr} = 0.125 mm) or at the reaction layer/HPSN interface (HPSN/Zr,Nb,Zr/HPSN, d = 0.6 mm). For HPSN/Zr,Nb,Zr/HPSN and HPSN/Hf/HPSN joints the lengths of the cracks increase with increasing metal layer thickness from 1 μm at d = 0.125 mm to about 30 μm at d = 3.25 mm and the tips of the microcracks shift with increasing metal layer thickness from the metal/reaction layer interface over the reaction layer/HPSN interface into the adjacent regions of the ceramic materials. Similar observations were made with HPSN/Zr/HPSN joints. The length of the microcracks increase from about 1 μm at d_{Zr} = 0.125 mm to about 4 μm at d_{Zr} = 1.0 mm. However, the tips of the microcracks were always situated at the metal/reaction layer interface independently of the thickness of the Zr layer. Fractographic investigations showed that the failure of all three types of materials joints occured predominantly in the interfacial regions where the microcracks terminate. This observation and the dependence of the microcrack position on the metal layer thickness can be used to explain the dependence of the bond strength data on the layer thickness. For the HPSN/Hf/HPSN and the HPSN/Zr,Nb,Zr/HPSN joints the minimum in bond strength at a metal layer thickness of about 0.6 mm (Figure 9) is due to a fracture at the interface between the reaction layer and the HPSN. This interface is rather weak caused by the already mentioned formation of a thin glassy layer. At a metal layer thickness of 1.0 mm microcracks surmount the reaction layer/HPSN interface and enter the ceramic part of the joint. Thereby a stress relief of the interface region occurs which leads to an increase in bond strength. HPSN/Zr/HPSN joints show up to a metal layer thickness of 0.645 mm a constant G_C value of 100 J/m² and a constant fracture behaviour. The fracture of the specimens occurs in the metal/reaction layer interface. The steep increase in bond strength for specimens with a Zr layer thickness of 1.0 mm can be

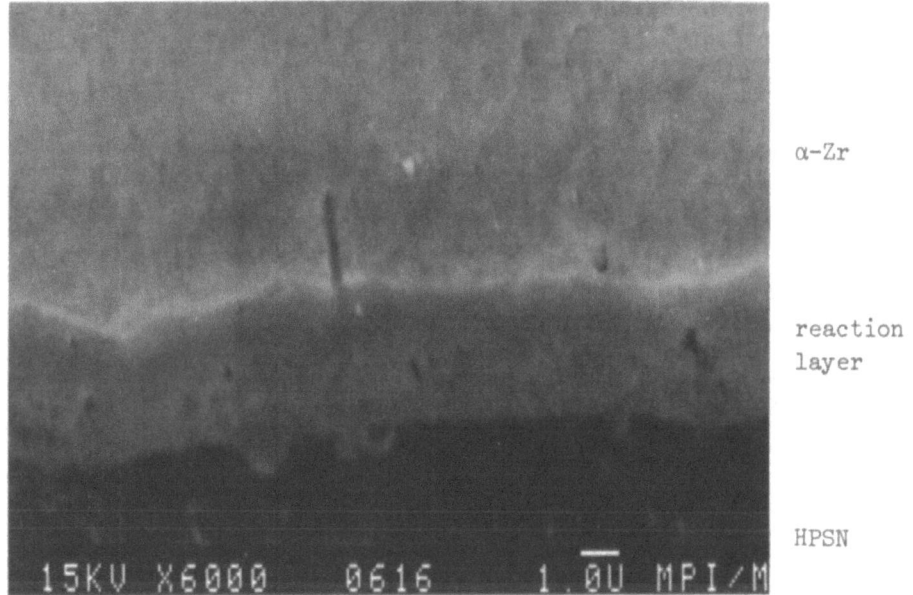

Figure 10. Microcrack terminating at the interface between α-Zr and reaction layer. HPSN/Zr/HPSN joint with d_{Zr}=0.125 mm. SEM picture.

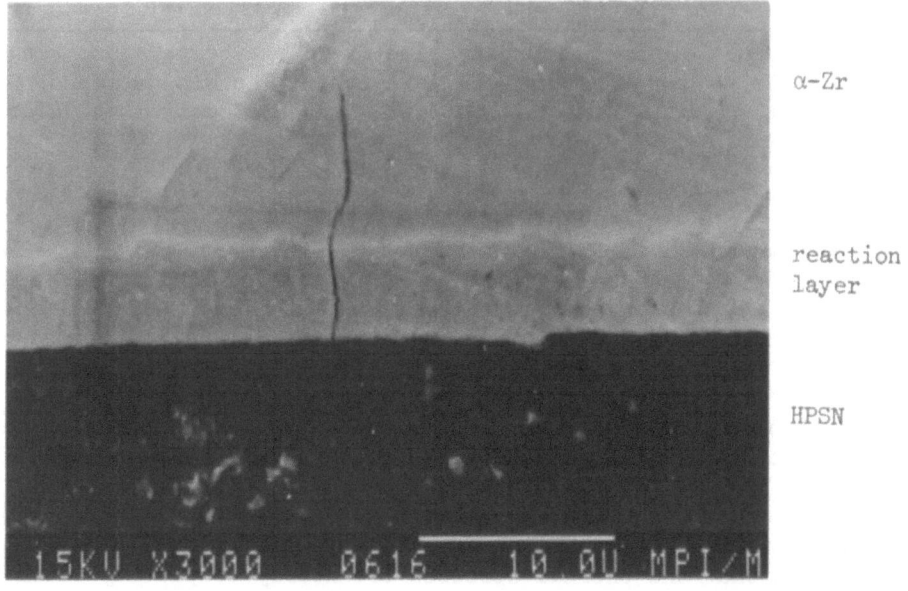

Figure 11. Microcrack terminating at the reaction layer/HPSN interface. HPSN/Zr,Nb,Zr/HPSN joint with a metal layer thickness of 0.6 mm. SEM picture.

explained by the existence of rather short microcracks terminating again at the metal/reaction layer interface. The rather weak reaction layer/HPSN interface is not influenced by microcracks. At the critical load the crack propagates into the reaction layer alternating its fracture path from the reaction layer/HPSN interface to the reaction layer/Zr interface. The high G_C value of 328 J/m² seems to be due to these irregularities in the fracture behaviour.

Table II. Microcrack densities N in solid-state bonded HPSN/metal joints.

Intermediate metal layer	d (nm)	N (1/mm)	A (µm)	N·A (10^{-3})	$\Delta T \Delta \alpha_T$ (10^{-3})	ΔT (°C)
Zr,Nb,Zr	0.35	17	0.15	2.6	4.0	
	0.6	31	0.15	4.7	5.0	1120
	1.25	35	0.15	5.3	5.5	
Hf	0.125	36	0.12	4.3	4.0	
	0.385	34	0.12	4.1	4.0	
	0.64	33	0.12	4.0	4.0	1170
	1.0	30	0.12	3.6	4.0	
	2.0	32	0.12	3.8	4.0	
Zr	0.385	13	0.12	1.5	1.7*	830*
	1.0	14	0.12	1.7	1.7*	

** Because of the plasticity of ß-Zr in the temperature range 1150 to 862 °C ΔT = 830 K is assumed. The elongation of 0.1% during the ß/α transformation is subtracted from the $\Delta T \Delta \alpha_T$ value.

The magnitude of thermal stresses in materials joints can be estimated by the strain mismatch $\Delta \alpha_T \cdot \Delta T$ where $\Delta \alpha_T$ is the difference between the expansion coefficients of the metal and the ceramic and ΔT is the difference between welding temperature and room temperature. On the other hand, the reduction of internal stresses by microcrack formation is given by the product N·A where N is the number of microcrack per unit length and A the mean opening of the microcracks. A was determined by SEM measurements at a magnification of 30 000:1. Table II shows a comparison between N·A and $\Delta T \Delta \alpha_T$. As can be seen the agreement between the two values is good for all materials combinations investigated. This leads to the conclusion that the thermal stresses are nearly completely eliminated by microcrack formation. The reduction of internal stesses by the β/α transformation of the Zr layer in a HPSN/Zr/HPSN joint is also demonstrated. The measureded N•A values of HPSN/Zr/HPSN joints are smaller by a factor of more than 2 than the N•A values of HPSN/Hf/HPSN and HPSN/Zr,Nb,Zr/HPSN joints.

CONCLUDING REMARKS

G_C measurements by means of single edge notched bend test specimens require a precise determination of correction functions Y_G. If the elastic properties of the materials adjacent to an interfacial crack in a metal-to-ceramic or in an adhesively bonded joint are known the somewhat tedious experimental determination can be replaced by a calculation of the correction function. Following this procedure accurate bond strength determinations are feasible both for metal-to-ceramic and adhesively bonded joints. Some relations between the correction function Y_{iso} for isotropic materials and the correction function Y_{bi} for a bimaterial can be derived, if the dependence of the correction function Y_G on the thickness of intermediate layers and on the crack position is considered.

The results obtained here on the influence of layer thickness and phase transformation on the bond strength of joints between hot-pressed silicon nitride and refractory metals indicate that the critical energy release rate G_C is controlled by the position of the cracks in the interfacial region and the reduction of internal stresses. Rather high bond strength values can be obtained by the applications of metal layers which undergo a stress reducing phase transformation.

ACKNOWLEDGEMENT

Support of this work by the Deutsche Forschungsgemeinschaft is gratefully acknowledged.

REFERENCES

1. G. Elssner, W. Diem, and J.S. Wallace in "Surface and Interfaces in Ceramic and Ceramic-Metal Systems", J. Pask and A. Evans, Editors, p. 629, Plenum Publishing Corporation, New York, 1981
2. W. Diem, G. Elssner, T. Suga, and G. Petzow, these proceedings these proceedings, pp. 871-882
3. G. Elssner and U. Krohn, Z. Metallkde. 70, 71 (1979)
4. B. Gross and J.E. Srawley, NASA TN D-2603 (1965)
5. H. Nowotny, B. Lux, and H. Kudielka, Mh. Chemie 87, 447 (1956)

BOND STRENGTH CHARACTERIZATION OF METAL-TO-CERAMIC AND ADHESIVE JOINTS BY CRITICAL ENERGY RELEASE RATES

W. Diem, G. Elssner, T. Suga, and G. Petzow

Max-Planck-Institut für Metallforschung
D-7000 Stuttgart, W. Germany

Small three- or four-point bend test specimens of 30 to 60 mm length were used to characterize the bond strength of metal-to-ceramic and adhesive joints by critical energy release rates. The testing procedure involves the determination of correction functions which describe the influence of the geometry and the elastic properties of the joint. The correction functions are obtained by compliance measurements or can be calculated by means of the finite element method. A fracture resistance parameter K_C derived from the critical energy release rate is introduced. K_C allows a comparison of the bond strength of different joints without considering differences in their elastic properties. Examples of a bond strength determination of glass/epoxy joints, and of the influence of crystal orientation on the strength of sapphire/Cu joints serve to illustrate the utility of this bend test method to determine critical energy release rates of semi-brittle materials joints.

INTRODUCTION

Modern technology has placed increasing demands on the strength properties of metal-to-ceramic and adhesive joints. The soundness of these types of joints must be assured to guarantee the structure in which they are incorporated. Tests to determine the bond strength properties are, therefore, of major importance for the quality assessment and development of composite structures.

Conventional bend and tensile tests or peel tests provide bond strength data which are functions of the specimen size and the test method employed. However, in practice and for a better understanding of the bonding behaviour of joints, strength data are needed as a function of fabrication parameters and parameters such as chemical and mechanical compatibility, intrinsic materials properties and local characteristics of the interface region such as flaws and unbonded areas. This may be provided by fracture mechanics test methods yielding values which describe the fracture resistance or the critical energy release rate of the joints. The present paper reports on the determination of the bond quality of specimens notched exactly at the metal-to-ceramic or the adherend-adhesive interface in terms of critical energy release rates using correction functions experimentally or numerically determined in three- or four-point bend tests.

CRITICAL ENERGY RELEASE RATE

One general method of studying semi-brittle fracture employs energy rate considerations which are based on a global energy balance for creating new surfaces under the assumption of an autonomy of the fracture process ($\dot{\gamma}_F = 0$)

$$\dot{U}_E + \dot{U}_K + 2\gamma_F \dot{A} = Q + \dot{W}_S \qquad (1)$$

where U_E is the internal energy, U_K the kinetic energy, γ_F the fracture surface (or interfacial) energy incorporating dissipated energy associated with crack extension, A the crack surface area, Q the heat flux, and W_S the work due to applied surface and body forces. The dots denote differentiation with respect to time. The energy release rate G is defined by

$$G = \frac{d}{dA}(W_S - U_E) \qquad (2)$$

The critical energy release rate G_C corresponds to the onset of unstable crack extension. It represents the fracture surface or interfacial energy

$$G_C = 2\gamma_F \qquad (3)$$

when the load is applied slowly and kinetic and thermal effects are negligible.

Some continuum mechanics models for an interfacial crack in ideally bonded elastic materials have been developed[1-2] which allows us to establish a correlation between G and the stress intensity faktor K. Choosing the definition[3]

$$K = K_I - i K_{II} = \sqrt{2\pi} \lim_{r \to 0} r^{-(\bar{\lambda}_{(1)}-1)} (\sigma_\theta - i \tau_{r\theta})_{\theta=0} \quad (4)$$

where $\lambda_{(1)}$ is the first eigenvalue of the equation for the local stress field corresponding to the chosen crack tip model. K_I and K_{II} are the stress intensity factors for model I and II, the local stress field can be described independently of the elastic properties by the stress intensity factor K:

$$(\sigma_\theta - i \tau_{r\theta})_{\theta=0} = (K/\sqrt{a\pi}) r^{\bar{\lambda}_{(1)}-1} \quad (5)$$

In the models mentioned above the energy release G is related to the stress intensity factor $K^2 = K_I^2 + K_{II}^2$ by

$$G = \frac{1-\beta^2}{E^*} K^2 \quad (6)$$

where ß is a composite parameter (see equation 8) and E^* an effective modulus of elasticity given by

$$\frac{1}{E^*} = \frac{1}{16} \left(\frac{\kappa_1+1}{\mu_1} + \frac{\kappa_2+1}{\mu_2} \right) \quad (7)$$

where μ_j (j=1,2) are the shear moduli of the materials j adjacent to the interface, $\kappa_j = 3-4\nu_j$ (plane strain), $\kappa_j=(3-\nu_j)/(1+\nu_j)$ (plane stress) and ν_j the Poisson ratios.

The elastic behaviour of the two bonded materials can be expressed by the Dundurs' parameters α and β[4],

$$\alpha = \frac{k(\kappa_1+1)-(\kappa_2+1)}{k(\kappa_1+1)+(\kappa_2+1)}$$

$$\beta = \frac{k(\kappa_1-1)-(\kappa_2-1)}{k(\kappa_1+1)+(\kappa_2+1)} \quad (8)$$

where $k = \mu_2/\mu_1$.

Although up to now the mechanisms of interfacial separation have not been completely understood and the description of this type of failure by a continuum mechanics model is not completely warranted, it seems appropriate to utilize both parameters G and K for the characterization of the bond quality of materials joints.

For engineering applications the specification of bond quality by critical stress intensity factors or fracture resistance values K_C may be preferred because these parameters allow a comparison of strength between different materials joints and even between joints and bulk materials. However, for joints which do not fracture exactly at the interface, or for materials combinations with no distinct interface, it is necessary to specify the chosen Dundurs' parameter α and β and the effective modulus E^* to avoid misinterpretations of the given K_C data.

We used for the investigation of the bond strength of metal-to-ceramic joints three- and four-point bend test specimens with lengths of 30 to 60 mm. The sandwich-like specimen consist of two rectangular ceramic bodies connected by a thin layer of metal and were notched exactly at the metal/ceramic interface. These relatively small specimens of simple geometry[5] were chosen because of the scarcity of material, their better fabricability and their adaptability for high temperature tests. For comparison, glass/epoxy specimens of similar size were tested.

THE CORRECTION FUNCTIONS

The energy release rate and the stress intensity factor depend not only on the applied load and the elastic properties but also on the specimen geometry. For material joints the influence of both the specimen geometry and the elastic properties of the constituents is described by the so-called correction functions Y_G. Consequently, the energy release rate G and the stress intensity factor K can be expressed by

$$G = (K_o^2/E^*)Y_G \tag{9}$$

and

$$K^2 = K_o^2 Y_G /(1-\beta^2) \tag{10}$$

where K_o is a constant for a given specimen proportional to the applied stress.

The energy rate G for three- and four-point bend test specimens is described (see Figure 1) by[3]

$$G = F^2 \frac{9\,e^2\pi}{B^2 W^3} \frac{Y_G(a/W)}{E^*} \tag{11}$$

for a given materials combination with constant thickness d of the intermediate layer. A precise determination of G_C requires an accurate correction function. Y_G can be experimentally obtained by measurements of the compliance C according to the well known relation

$$G = \frac{F^2}{2B} \frac{dC}{da} \qquad (12)$$

By comparison of Equation (11) and (12), it is seen that Y_G can be described by

$$Y_G\left(\frac{a}{W}\right) = \frac{dC^*}{d(a/W)} \qquad (13)$$

where C^* is a normalized compliance

$$C^* = (C-C_o)BW^2E^*/18e^2\pi \qquad (14)$$

and C_o is the compliance of the unnotched specimens. The correction functions Y_G can also be determined by means of the finite element method.

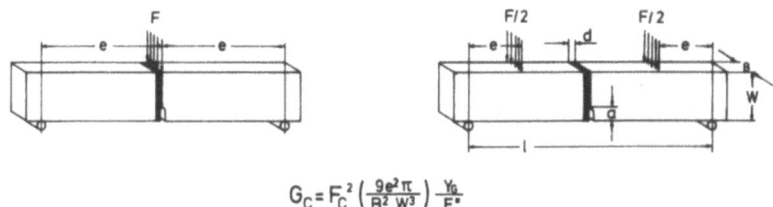

$$G_C = F_C^2 \left(\frac{9e^2\pi}{B^2W^3}\right) \frac{Y_G}{E^*}$$

Figure 1 Bend test configurations for G_C measurements.

EXPERIMENTAL

Soda-lime glass/epoxy/soda-lime glass joints of adhesive thickness 0.1 to 3 mm were prepared. The adhesive employed was an unmodified epoxy resin of DGEBA type (DER 332) mixed with 10 phr TETA curing agent cured 18 h at room temperature and postcured 4 h at 60 °C. Some glass and epoxy specimens were also prepared for investigating their bulk properties. Sapphire/Cu and Al_2O_3/Cu joints were fabricated by an eutectic bonding process[6]. Polycrystalline Al_2O_3 platelets and sapphire platelets of the orientation $[0001]$ and $[0\bar{1}10]$ and of 1.4 mm thickness were joined via Cu layers of 0.26 mm thickness to two pieces of 14 mm length. Some properties and the composition of the materials are given in Table I.

Table I. Materials properties and composition.

Material	Composition (w/o)	E (GPa)	ν	α_T (10^{-6} K^{-1})
soda-lime glass	72 SiO_2, 14.5 Na_2O, 8.5 CaO, 3.5 MgO, 1.5 Al_2O_3	70.3	0.24	9
unmodified epoxy resin DGEBA	epoxy equiv. 172-178, 10 phr TETA curing: 18h/RT+ 4h/60 °C	3.9	0.34	40
sapphire[0001] [0$\bar{1}$10]	99.98 Al_2O_3 + Cr,Ni,Sn,Mn	461 425	-- --	8.8 7.9
polycryst. Al_2O_3	97.0 Al_2O_3, 2.9 SiO_2	350	0.25	8.2

E is the Young's modulus, ν the Poisson ratio, and α_T the thermal expansion coefficient.

Three- and four-point bend test specimens of the dimensions 60x10x10 mm (glass/epoxy) and 30x5x2 mm (alumina/Cu) were cut from the as-received composite materials. Notches of width 50 to 100 μm and of depth a were introduced into the interface or into the glass and epoxy parts of the adhesive joints by a precision cutting device. The notch in the epoxy was sharpened by tapping with a razor blade. The bend tests were carried out with a Model 1474 Zwick machine at a cross-head speed of 1 μm/min. In the set-up load and deflection signals were treated in real time by a directly connected microprocessor. To obtain time load-deflection curves and compliance data, the measured deflections are corrected for the deformation of the test device during loading. Calculations of the correction functions by the finite element method were performed using the virtual extension[7] and the J-integral methods. The results obtained by these two methods differ only by ± 1 %.

RESULTS AND DISCUSSION

Figure 2 shows the dependence of the normalized compliance C^* and the correction function Y_G on the ratio a/W for different layer thickness to specimen height ratios d/W of glass/epoxy/glass four-point bend test specimens. As can be seen, experimental data agree well with the calculated results.

In Figure 3 experimental data for the fracture resistance K_C or K_{IC} for a glass/epoxy joint with d = 1 mm are plotted against the distance h of the notch from the interface. The filled circles denote apparent values obtained by means of the isotropic correction function Y_{iso}[8] and the open circles are the corrected data obtained by means of the correction function Y_G for the adhesive joint tested. The corrected K_{IC} data for the epoxy and glass regions agree very well with the data obtained by fracture tests on the individual bulk materials. The interfacial failure of glass/epoxy joint was characterized by stable crack extension. K_C data of 0.24 MN($m^{3/2}$) at a crack velocity of $1.7 \cdot 10^{-3}$ m/s were obtained by using the correction function shown in Figure 2. As demonstrated in Figure 3 it is generally not advisable to use isotropic correction functions for the determination of K_C data for joints. In the given example the difference between the apparent and the corrected value of the fracture resistance of the joint is about 54 %.

G_C measurements were used to study the influence of the crystal orientation on the bond strength of Al_2O_3/Cu joints. Specimens with a polycrystalline alumina, a sapphire [0001], and a sapphire [0$\bar{1}$10] interface in contact with polycrystalline copper were tested. After the determination of the correction functions Y_G, bond strength measurements were carried out at 77 K to minimize plastic deformation of the metal layers. However, the sapphire/Cu joints showed only an interfacial fracture if the ratio a/W was larger than 0.4. For these rather deeply notched specimens, a plastic deformation of copper could not be avoided as can be seen in Figure 4 by the steep increase of the measured G_C values with a/W ratio. G_C data free from contributions of plastic deformation were obtained by an extrapolation of the measured G_C values to a a/W ratio of 0.39 following the trace of curve for polycrystalline alumina-copper joints as shown in Figure 4 by the two dotted lines. Typical fracture surfaces of the sapphire/Cu specimens are given in Figures 5 and 6. The copper side (Figure 5) is characterized by numerous semispherical pores whose borderlines

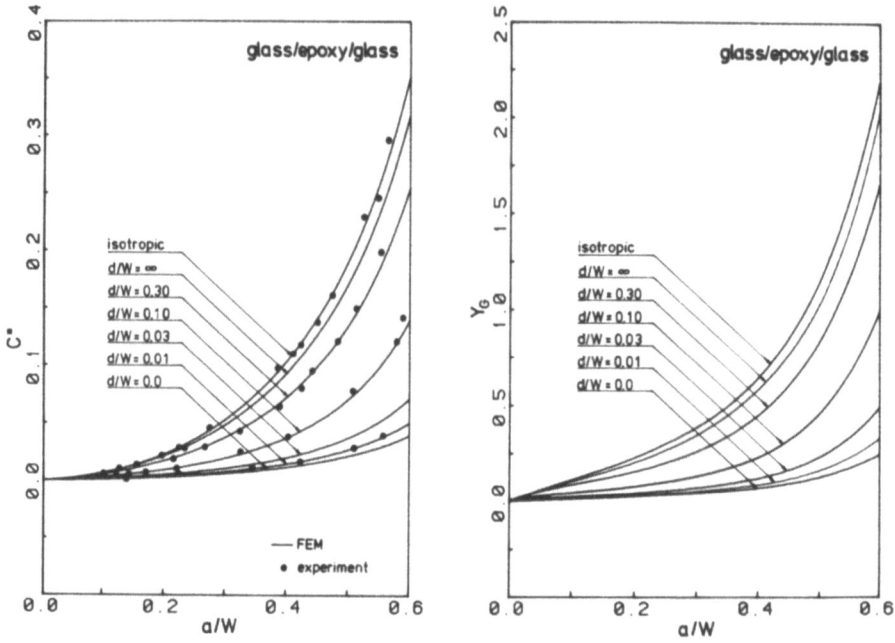

Figure 2. Normalized compliance C^* and correction function Y_G versus ratio a/W for glass/epoxy joints. Four-point bend test. Dots are measured and lines calculated values.

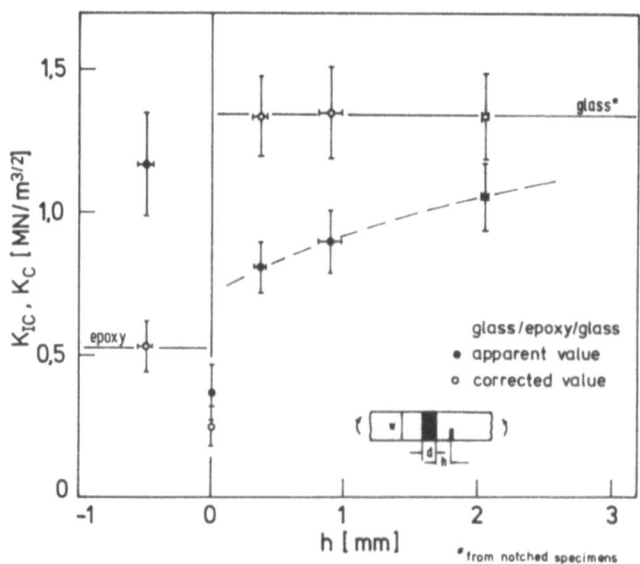

Figure 3. Comparison of apparent and corrected values of K_C and K_{IC} as a function of notch location h for a glass/epoxy joint with d = 1 mm.

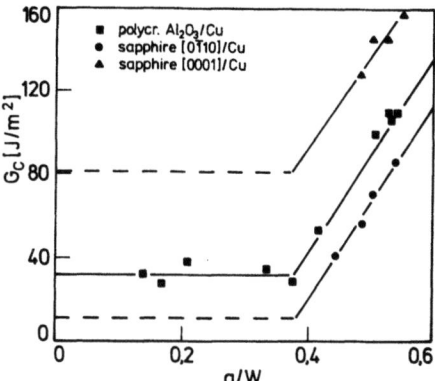

Figure 4. G_C data for Al_2O_3/Cu joints measured at 77 K as a function of a/W ratio.

are covered with copper oxide particles. More detailed investigations of the fracture surfaces lead to the assumption that a $CuAlO_2$ spinel of 0.1 to 1µm thickness is also formed between copper and sapphire. Table II shows the extrapolated G_C data for the joints corrected for the interfacial porosity in comparison to the fracture surface energies $2\gamma_F$ of single crystal and polycrystalline alumina. The experimentally estimated data indicate a strong influence of the crystal orientation on the bond strength of the joints. Furthermore, they compare well with the fracture surface energy data for bulk alumina and sapphire. Therefore, it can be concluded that the interfacial bonding forces of copper/alumina joints are similar to those in bulk alumina. This may be explained by an intermediate thin spinel-like layer between alumina and copper.

Table II. Critical energy release rates G_C of copper-alumina joints and fracture surface energies $2\gamma_F$ of alumina.

Alumina	G_C (J/m²)	Interfacial Porosity(%)	$2\gamma_F$ (J/m²)	References
polycrystalline	35	9	43	10
sapphire [0110]	12	4	13.8	11, 12
sapphire [0001]	93	13	102	11, 12, 13

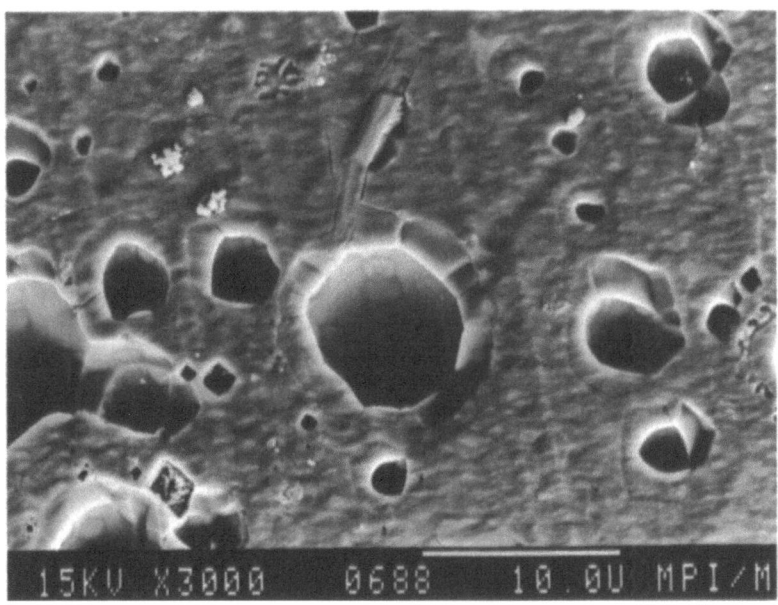

Figure 5. Copper side of an interfacial fracture of a sapphire/Cu joint tested at 77 K. SEM picture.

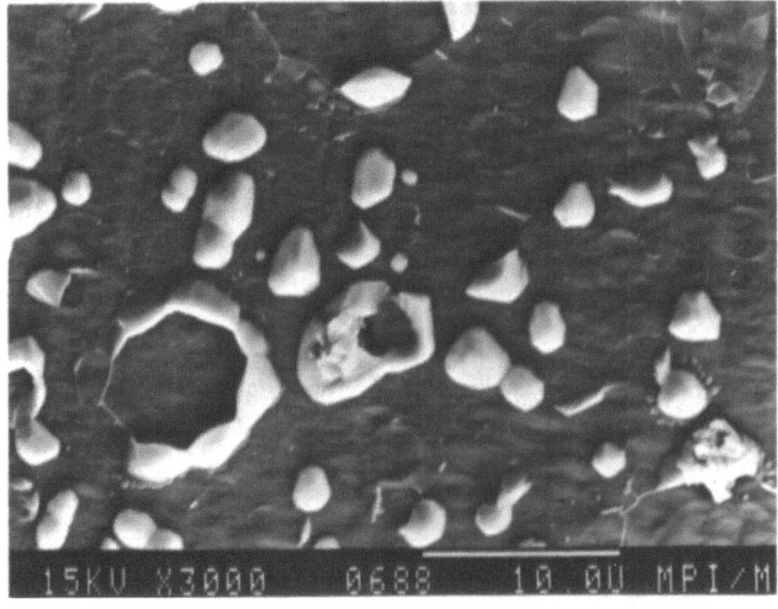

Figure 6. Sapphire side of an interfacial fracture of a sapphire/Cu joint tested at 77 K. SEM picture.

Table III. Critical energy release rates G_{IC} and G_C and fracture resistance data K_{IC} and K_C for some bulk materials and materials joints.

Materials	G_{IC} [J/m²]	K_{IC} [MN/m³/²]	E* [GPa]		
Al_2O_3 (2.9% SiO_2)	33	3.5	373.3		
sapphire [0110]	13.8	2.6 [2]	501.2 [2]		
HPSN	89	5.5	340.7		
soda-lime glass	5.2	0.62	74.6		
amine cured epoxy (DER 332 + 10 phr TETA)	61	0.52	4.4		
plasma-sprayed ZrO_2 coating	26	1.2	52.8		

Materials Joints	G_C [J/m²]	K_C [MN/m³/²]	E* [GPa]	α	β
Al_2O_3(2.9% SiO_2)/Nb	32.8	3.5	183.4	-0.509	-0.064
sapphire [0110] / Cu	12 [1]	1.6 [2]	227.1 [2]	-0.546	-0.144
HPSN / Hf	91	4.5	216.1	-0.366	-0.098
soda-lime glass / amine cured epoxy	6.6 [3]	0.24	8.33	-0.888	-0.210
plasma-sprayed NiCrAlY/Cr/ZrO_2 coating on steel	16	1.1	76.8	-0.458	-0.081

1) extrapolated value
2) treated as an isotropic materials joint
3) stable crack extension

Small-sized specimens of a variety of materials combinations and bulk materials were tested. Table III summarizes some measured data. The G_C data for adhesive joints and metal-to-ceramic joints are of the order of 10 to 100 J/m². The K_C data rank from rather low values below 1 MN/m³/² for glass/epoxy joints to relatively high values above 4 MN/m³/² for special metal-to-ceramic joints.

CONCLUDING REMARKS

Although the micromechanical processes in the interfacial region of a materials joint are highly complex, its critical energy release rate G_C can be measured in a similar manner as for isotropic materials. The necessary correction functions can be computed if the elastic properties are known or if only thin intermediate layers of unknown properties are present; they could also be derived from experimental compliance measurements.

The determination of critical energy release rates gives a more scientific footing for bond strength measurements than conventional testing methods. Small-sized specimens cut from brittle or semi-brittle materials joints can be used to characterize local variations of bond quality and to describe the influence of microstructure in the interfacial region, the influence of fabrication parameters, intermediate layer effects, and crystal orientation effects.

ACKNOWLEDGEMENT

We thank the Deutsche Forschungsgemeinschaft for the financial support of some portions of the work described here.

REFERENCES

1. J.R. Rice and G.C. Sih, J. Appl. Mech. 32, 418 (1965)
2. M. Comninou, J. Appl. Mech. 44, 631 (1977)
3. T. Suga and G. Elssner, unpublished results
4. J. Dundurs, J. Appl. Mech. 36, 650 (1969)
5. R.F. Pabst and G. Elssner, J. Materials Sci. 15, 188 (1980)
6. J.C. Driscoll and Y.E. Sun, Proc. Int. Microelectronics Symp., Oct. 1976, p. 44, Vancouver, Canada
7. T.K. Hellen, Int. J. Num. Mech. Engng. 9, 187 (1975)
8. B. Gross and J.E. Srawley, NASA TN D-2603 (1965)
9. W. Diem, Doctoral Thesis, University of Stuttgart (1982)
10. W. Kromp and R.F. Pabst, Z. Materialprüfung 22, 241 (1980)
11. P.F. Becher, J. Am. Ceram. Soc. 59, 59 (1976)
12. A. Krell, M. Kikuchi and T. Nishino, phys. stat. sol. (a) 52, K45 (1979)
13. S. Morozumi, M. Kikuchi and T. Nishino, J. Mat. Sci. 16, 2137 (1981)

INTERFACIAL PROPERTIES OF FILLED EPOXIDE RESINS

A.C. Moloney*, H.H. Kausch* and H.R. Stieger**

*Laboratoire de Polymères
Swiss Federal Institute of Technology
CH-1007 Lausanne, Switzerland
**Brown Boveri & Cie.
CH-8050 Zurich-Oerlikon, Switzerland

The fracture properties of two commercial epoxide resins both unfilled and filled with varying volume fraction of silica, alumina and dolomite particles have been investigated. The variation in the stress intensity factor with the crack velocity was measured using the double torsion test technique. In order to examine the influence of the resin-filler adhesion on the fracture toughness, alumina particles were treated with three silane compounds. In addition the yield stresses and the flexural strengths were measured. In an attempt to simulate resin-filler interactions and to measure the fracture energy for interfacial separation a scarf joint was used with a bond angle of 45°. This geometry produces a mixture of mode I and mode II loading. The combined mode strain energy release rate ($G_{(I,II)_c}$) was determined.

INTRODUCTION

Particulate fillers are incorporated into plastic components principally because of their lower cost. However, in certain cases mechanical, thermal and electrical properties may be improved, notably the elastic modulus, heat deflection temperature and the arc resistance.

Considerable work has been reported in the literature[1] on the effect of particulate fillers on the mechanical properties of thermoplastic and thermosetting polymers but the underlying mechanisms are still unclear. Since the cost of plastics continues to rise at a rate faster than that of mineral fillers, it is of great importance that fillers and surface treatments be chosen on a more rational basis for specific applications.

In this work we have investigated the fracture properties of two commercial epoxide resins filled with silica, alumina and dolomite particles. The principal parameters which have been varied are volume fraction, filler surface treatment, mechanical resistance of the filler and particle size.

MATERIALS AND EXPERIMENTAL

Fracture of Unfilled Resins and Composites

Two commercial epoxide resins were used in this investigation. Restin A (solid at room temperature) was based on the diglycidyl ether of bisphenol-A, cured with phthalic anhydride. Resin B was an epoxide resin based on dimethyl hydantoin and was also anhydride cured. After thorough mixing and degassing, the resins were poured into pre-heated steel moulds which were rotated to prevent sedimentation of the filler. The curing conditions were followed in accordance with the manufacturers instructions (Ciba-Geigy). Alumina, silica and dolomite particles were used as filler materials, the salient physical properties are compared in Table I.

Silica particles pre-treated with γ-glycidoxypropyltrimethoxysilane (A187 from Union Carbide) are available from Quarzwerke. Alumina particles, however, were treated in our laboratories. Three silane compounds were used in order to change the adhesion between the phases. The first hexamethyldisilazane (HMDS) would be expected to reduce the adhesion[2]. The other two compounds were γ-aminopropyltriethoxysilane (A1100) and A187 from Union Carbide.

Table I Properties of Filler Materials.

Filler	Specific gravity gcm^{-3}	Form of particles	Mean particle size	Young's modulus GPa	Fracture toughness K_{Ic} MNm$^{-3/2}$	Work of fracture G_{Ic} (Jm^{-2})
Alumina	3.97	irregular-rounded edges	~ 6 μm	320 (Ref.4)	5.3 (Ref.7)	40 (Ref.8)
Silica	2.65	irregular-sharp edges	W 10 ~ 60 μm W 6 ~ 100 μm W 4 ~ 160 μm W 1 ~ 300 μm	94 (Ref.5)	0.8 (Ref.6)	4.4 (Ref.6)
Dolomite	2.85	irregular-sharp edges	20 μm	78 (Ref.5)	—	—

These would be expected to improve the resin-filler adhesion. The filler particles were treated using a method based on that suggested by Trachte and DiBenedetto[3].

The double torsion test technique was used to measure the stress intensity factors of these composites with varying volume fraction of filler. Certain authors[9-11] have questioned the validity of this method due to the curved crack front which is produced. It has been shown recently[10] that depending on the dimensions of the double torsion specimen a correction factor may be necessary to give an accurate value for K_{Ic}. The dimensions chosen in this work minimised the correction to within experimental scatter. The difficulty of machining these highly filled resins prevented the use of another geometry. (In a few selected cases, comparative values were obtained by using single edge notch specimens). The stress intensity factors were measured using both the constant displacement rate and the load relaxation methods[7].

The relationship between K_{Ic} and the applied load (P) is shown below:

$$K_{Ic} = W_n P \left[\frac{1 + \nu}{W t^3 t_n k_1} \right]^{\frac{1}{2}} \tag{1}$$

where W_n is the moment arm, ν Poisson's ratio, W the specimen width, t the specimen thickness, t_n the thickness in the plane of the crack and k_1 a geometric factor.

The crack velocities were measured by a method developed in our laboratories by Stalder et al[12] employing the change in the inverse of resistance with time of a thin layer of graphite applied to the specimen by means of a spray.

The two sections of the double torsion specimen after failure were used to measure the flexural strength by a three point bending test. Three faces of the specimen were polished to eliminate surface flaws. For resin B filled with 30 % untreated alumina and 30 % alumina treated with A187, the tensile strengths were also measured.

The yield stresses of the composites (σ_y) were determined using the plane strain compression test developed by Williams and Ford[13]. Employment of this test permitted the use of the failed halves of the double torsion specimens thereby enabling K_{Ic} and σ_y to be compared for the same sample.

Combined Mode Adhesion Test

In order to simulate the resin-filler interactions and measure the fracture energy for interfacial separation, a modified version of the combined mode adhesive test specimen, analysed by Trantina[14], was employed. This geometry has been used to study epoxide-aluminium adhesive bonds[15,16]. However, for our purposes it was desirable to investigate the interactions between alumina ceramic and epoxide resin. It was impractical to fabricate such a specimen from alumina so instead a block of the ceramic was inserted into a steel support as shown in Figure 1.

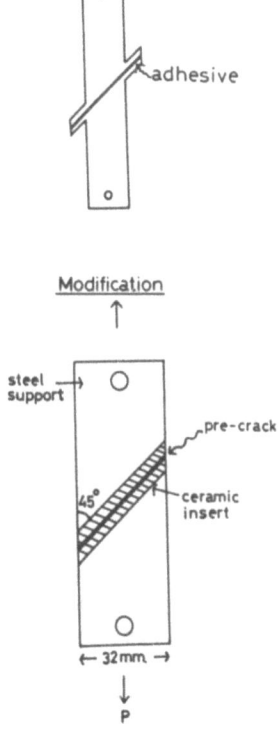

Figure 1. Scarf joint for mixed mode adhesive fracture.

Blocks of alumina ceramic type AL 23 (obtained from Degussa S.A.) were machined with diamond cutting tools to the required dimensions. The alumina was treated by degreasing with trichloroethylene in a Soxhlet apparatus. The steel supports were grit blasted and the

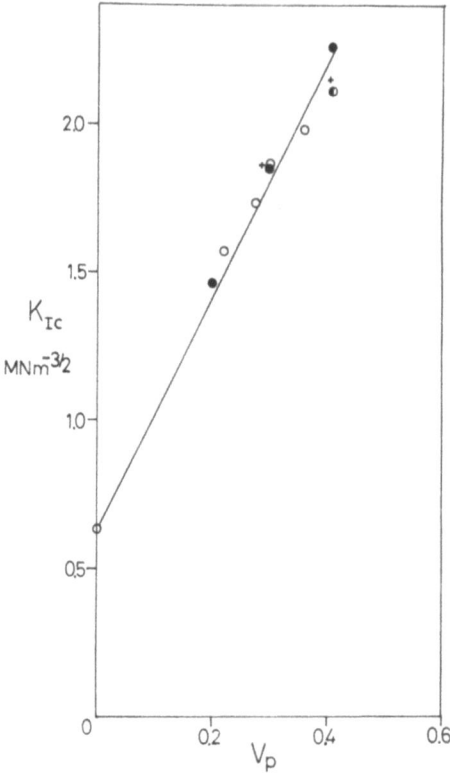

Figure 2. The variation in the stress intensity factor with the volume fraction of filler.
Resin A oAl_2O_3(double torsion DT); •SiO_2(DT); ⊕SiO_2+A187; + SiO_2 (single edge notch SEN).

treated alumina bonded into the support using a secondary adhesive Ciba Geigy AV138/HV998. The specimens were assembled in aluminium jigs to ensure correct alignment of the adherends. Resin A was used as the adhesive in these tests. After curing, a razor blade was used as a wedge to initiate a sharp pre-crack into the specimen. Since resin A is transparent, the initial crack length could easily be measured prior to testing. The specimens were loaded in tension in an Instron testing machine at a constant crosshead displacement rate of 0.1 mm/min, the temperature being controlled to 23 °C ± 2 °C. The fracture surfaces were analysed by the scanning electron microscope.

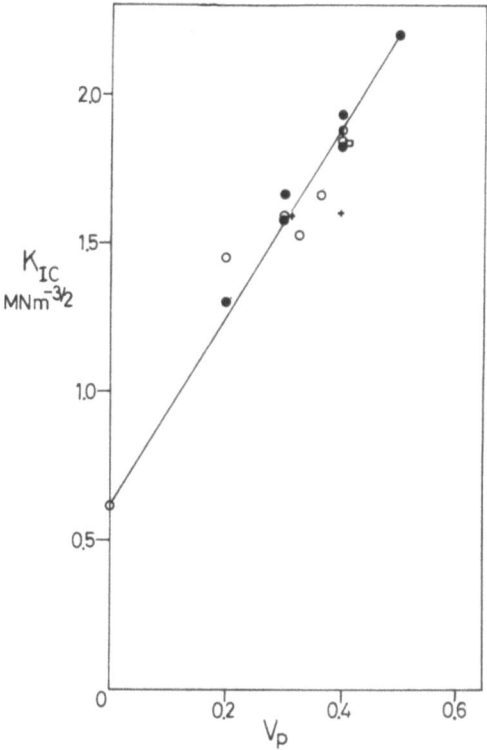

Figure 3. The variation in the stress intensity factor with the volume fraction of filler.
Resin B oAl_2O_3(DT); •SiO_2(DT); ⊙SiO_2+A187; +SiO_2(single edge notch SEN).

RESULTS AND DISCUSSION

The relationship between K_{IC} and the volume fraction of filler was found to be linear for both resin systems as shown in Figures 2 and 3. For a given resin the data for silica and alumina could be superimposed. This result is surprising firstly, because of the dissimilarity in particle size of the two fillers and secondly, because of the discrepancy in elastic moduli of these two filler materials (see Table I). Thirdly, the appearance of the fracture surfaces under the SEM was completely different. The silica particles were well-bonded to the resin, whereas each alumina particle was clearly de-bonded from the resin (see Figures 4 and 5). These results strongly suggest that for these irregularly shaped and relatively "strong" particles neither the particle size at a constant volume fraction of filler nor the filler-particle adhesion greatly affects the toughness.

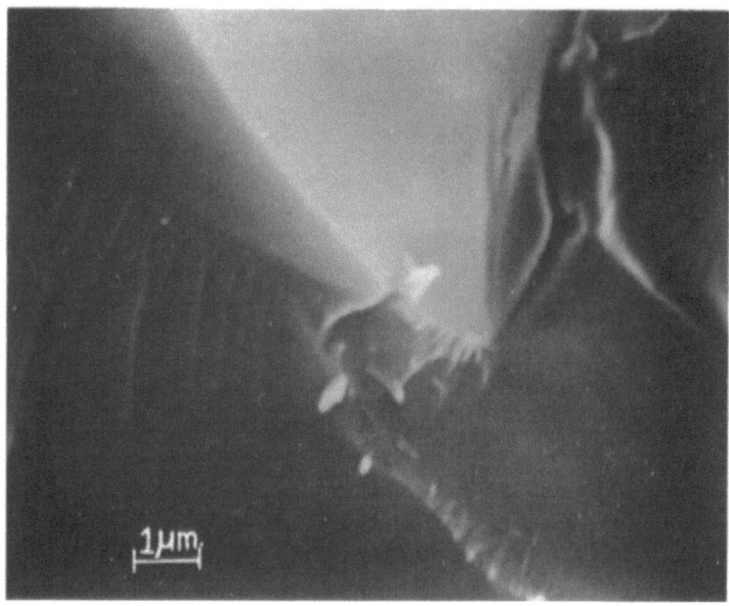

Figure 4. Scanning electron micrograph of the fracture surface of silica filled resin B.

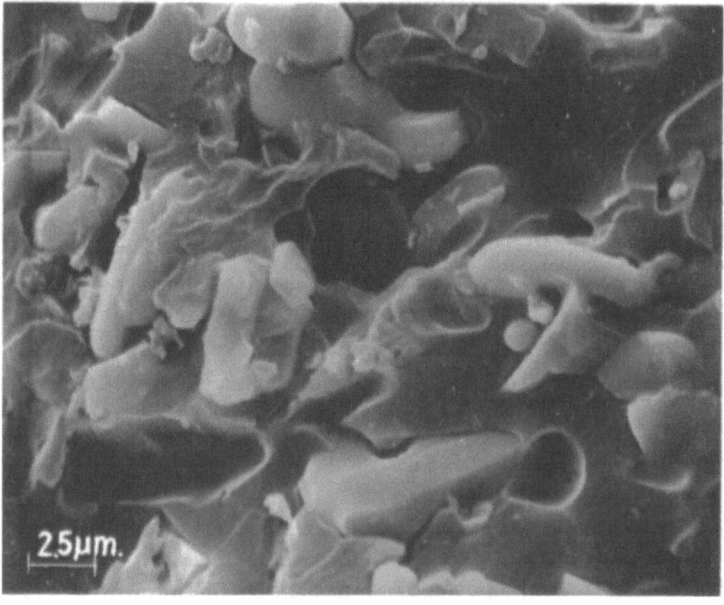

Figure 5. Scanning electron micrograph of the fracture surface of alumina filled resin B.

In order to investigate the former parameter, silica was chosen since this is available in a wide range of particle sizes. The mean particle size was varied between 300 μm and 60 μm. As can be seen from Table II the stress intensity factor is little affected at a constant volume fraction of filler.

Table II The Influence of Filler Particle Size on the Stress Intensity Factor.

Resin B + 40 % SiO_2

Grade of Quartz (Quarzwerke)	Mean particle size μm	Mean value of K_{Ic} (MN m$^{-3/2}$)
W 1	∿ 300	1.76 ± 0.01
W 4	∿ 160	1.74 ± 0.02
W 6	∿ 100	1.87 ± 0.01
W 10	∿ 60	1.83 ± 0.02

To determine the effect of resin-particle adhesion on the fracture toughness, alumina filled resins were chosen. For both resin systems, treatment with the silanes A1100 and A187 increased the toughness as compared with the untreated material and treatment with HMDS slightly decreased the values. However, the differences were not large and compared with the unfilled resin are negligible. For example, for unfilled resin B the value of K_{Ic} was 0.6 NM m$^{-3/2}$ and with 30 % by volume of filler this increased to between 1.46 and 1.67 irrespective of the filler surface treatment (Table III). Scanning electron micrographs of the fracture surfaces are shown in Figures 6 to 8. Treatment of the filler with HMDS resulted in a fracture surface very similar to that of the untreated filler (see Figure 5). Treatment of the alumina with A187 improved markedly the adhesion between the resin and the filler, as demonstrated in Figure 7. For the alumina treated with the silane A1100 some particles were well-bonded and others poorly bonded, as shown in Figure 8.

Broutman and Sahu[17] and Hammond and Quayle[18] have also investigated the effects of surface treatment of glass bead filled polyester resins. In spite of the fact that they used similar resins and fillers, their results were contradictory as to which

Table III Stress Intensity Factor of Composites Filled with Treated and Untreated Fillers.

Resin A

Filler	Treatment	Mean value K_{Ic} MN m$^{-3/2}$
30 % Alumina	none	1.86 ± 0.01
	HMDS	1.80 ± 0.01
	A187	1.97 ± 0.04
	A1100	1.99 ± 0.04

Resin B

Filler	Treatment	Mean value K_{Ic} MN m$^{-3/2}$
30 % Alumina	none	1.53 ± 0.02
	HMDS	1.46 ± 0.04
	A187	1.64 ± 0.01
	A1100	1.67 ± 0.02

system was the toughest. We have also conducted some tests on glass bead filled composites[19] and found the properties to be somewhat different to those of the irregularly shaped silica and alumina particles.

For the silica and alumina filled resins, the presence of filler particles provokes a considerable increase in toughness which is independent of the resin-filler adhesion. There are a number of possible mechanisms for this phenomenon. Firstly, the toughness may be increased by the filler diverting the crack and causing a larger surface area of fracture. However, the increase in area is insufficient to account for the increase in toughness. Secondly, in the case of some filled systems, using for example metallic or rubber particles[20], energy may be absorbed by deformation of the filler, thus causing increases in toughness. This mechanism is not feasible for brittle glass and ceramic fillers. Thirdly, the increase in toughness may arise from an increased plastic deformation

INTERFACIAL PROPERTIES OF FILLED EPOXIDE RESINS 893

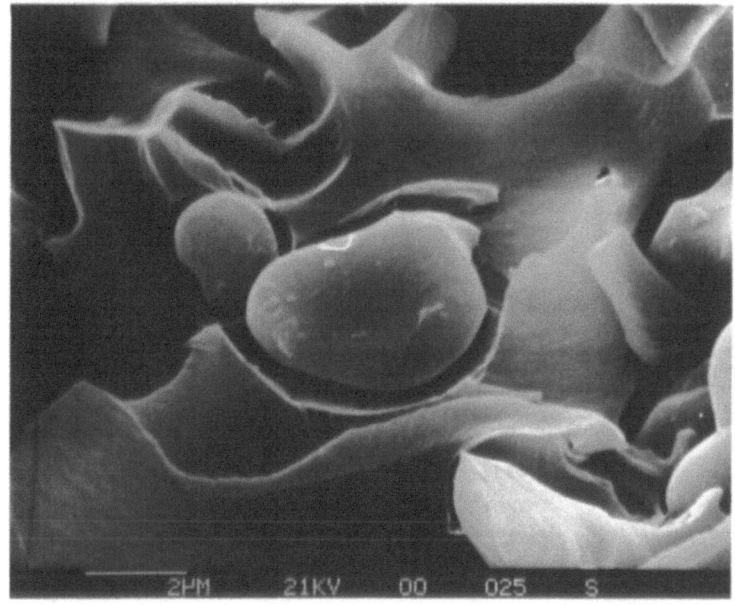

Figure 6. Scanning electron micrograph of the fracture surface of resin A filled with alumina treated with HMDS.

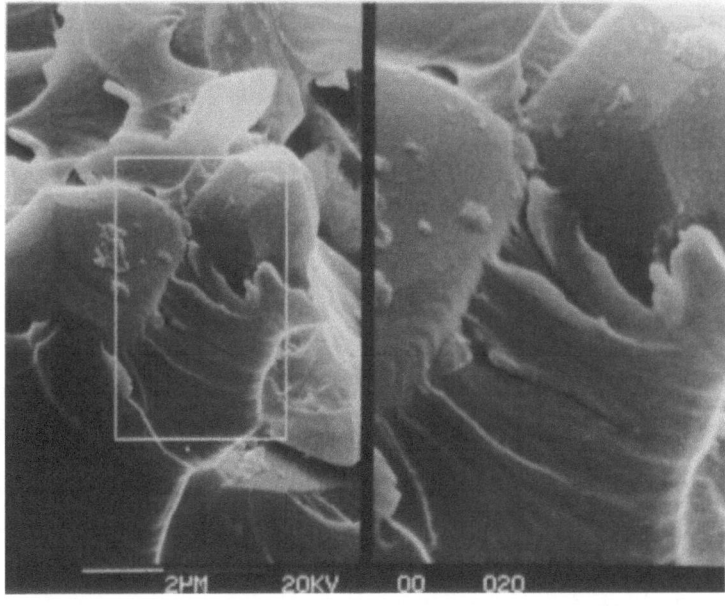

Figure 7. Scanning electron micrograph of the fracture surface of resin B filled with alumina treated with A187.

Figure 8. Scanning electron micrograph of the fracture surface of resin A filled with alumina treated with A1100.

of the matrix. Fourthly, the augmentation may be due to the obstacles pinning the crack and causing the crack front to bow out between the particles.

This latter explanation was put forward by Lange[21] who proposed that the increase in fracture energy was proportional to the inverse of the interparticle spacing (2c). He proposed an equation of the form:

$$\gamma_{comp} = \gamma_{res} + \frac{T}{2c} \qquad (2)$$

where γ_{comp} and γ_{res} are the fracture energies of the composite and pure resin respectively and T the line tension effect. However, this analysis is not directly applicable as T is a function of the particle size (2r). For a penny-shaped crack, Lange deduced that the line tension was:

$$T = \frac{2r}{3} (\gamma_{res}) \qquad (3)$$

The ratio of the particle size to the particle spacing is proportional to the volume fraction V_p by $V_p(1-V_p)^{-1}$. Thus, Lange's equation predicts a linear relationship between the fracture energy and the ratio of r to c. However, this linear relationship has not been observed in practice except for the case of a ceramic composite, glass filled alumina[22]. For particulate filled plastics such a linear dependence is not found. The strain energy release rate (G_{Ic}) is related to the stress intensity factor as follows:

$$G_{Ic} = \frac{K_{Ic}^2(1-\nu^2)}{E} \qquad (4)$$

where E is Young's modulus.

The increase in E with volume fraction is well-known to be a complex function of the moduli of the two phases and a parabolic relationship is found[23-25]. The effect that E rises at a faster rate than K_{Ic} with increasing volume fraction of filler means that even for "strong" fillers the value of G_{Ic} will go through a maximum at a certain volume fraction. An example of the relationship between G_{Ic} and V_p is shown in Figure 9, a very similar result has been obtained by Young and Beaumont[26].

The analysis of Lange was carried further by Evans[27] who calculated the increase in strain energy necessary to bow out the crack between the particles. Evans also calculated the ratio of G_{Ic} of the composite to that of the resin or the ratio of the tensile strength of the composite to that of the resin. Attempts to predict changes in tensile strength in this manner are problematic because the presence of the second phase inclusions increases the size of the "inherent flaws" even if the adhesion between the resin and filler is good. The flexural and tensile strengths of the alumina filled resins are shown in Table IV. The strengths were greatly improved by treatment with A187 and A1100 and impaired by the HMDS treatment. The tensile strength and the stress intensity factor are related as follows:

$$K_{Ic} = Y \sigma a^{\frac{1}{2}} \qquad (5)$$

where σ is the applied stress, Y a geometric factor and a the effective "inherent flaw" size.

The tensile strength must be used here since the flexural strengths are too close to the yield point. Inserting the relevant values into this equation gives a flaw size of 400 μm for resin B filled with 30 % untreated alumina and 100 μm for this resin with

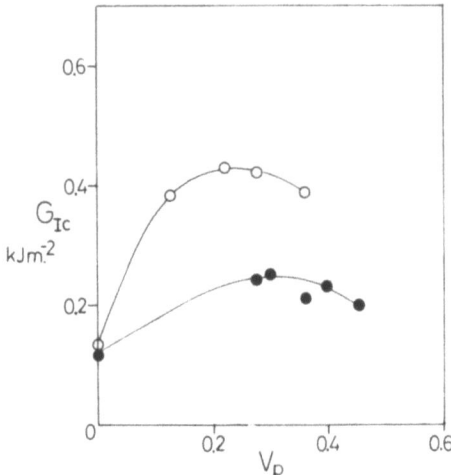

Figure 9. The variation in the strain energy release rate with volume fraction of filler. o resin A + alumina; ● resin B + alumina.

30 % A187 treated alumina. Thus, even for a composite with well-bonded particles the effective flaw size is greater than that of the pure resin which is about 50 μm. The very large flaw sizes for composites prepared from untreated alumina probably arises from the linking of poorly bonded particles and the existence of agglomerates[23].

Evans[27] has calculated the relative tensile strength in the presence of inclusions, $\sigma_{composite}/\sigma_{resin}$, assuming that the flaw size is constant. This is clearly invalid for these composites since we have seen that changing the adhesion between the phases alters the flaw size. However, this ratio may be converted into $K_{Ic composite}/K_{Ic resin}$. The increase in strain energy (U) due to the crack bending is:

$$U = \frac{16(1-\nu^3)c^3}{\pi E}(\sigma_A)^2 \int_0^{\pi/2} \{\int_0^{\pi/2}(1+\frac{c'}{2r_o+2c\sin\sin})^{\frac{1}{2}}d\beta\}^2 \sin\theta d\theta - U_o \quad (6)$$

where σ_A is the stress required to propagate a crack through a series of obstacles. Thus, differentiating U with respect to c and

Table IV Strengths of Composites Made from Treated and Untreated Fillers.

Resin	Filler	Treatment	Flexural strength MPa	Tensile strength MPa
A	30% Alumina	HDMS	106.8	-
		A1100	144.9	-
		A187	154.7	-
B	30% Alumina	none	95.7	45
		HMDS	82.2	-
		A187	131.9	87.8
		A1100	136.2	-

carrying out a numerical integration of the double integral, the ratio of $K_{Iccomp}/K_{Icresin}$ may be obtained. These calculations have been carried out by Green et al[28]. The values derived from this analysis are shown in Figure 10, and compared with the experimental values. At low filler contents the theoretical values are in very good agreement; at higher volume fractions there is some scatter. This may arise because the crack front bows out in a semi-elliptical manner rather than a semi-circular[28]. However, this analysis shows that the increase in the stress intensity factor of these resins may be explained by a mechanism of crack pinning.

The mechanism proposed by Lange[21] and Evans[27] is clearly invalid in the case of weak particles. For the dolomite filled resins it was found that above a certain critical volume fraction (∼20%) a plateau is reached and the stress intensity factor no longer increases (see Figure 11). This appears to be due to the propagation of cracks through the particles. Some evidence in favour of this explanation was provided from scanning electron micrograph of the fracture surface (see Figure 12). Lange and Radford[29] have observed a similar plateau with an aluminium hydroxide filled epoxide resin.

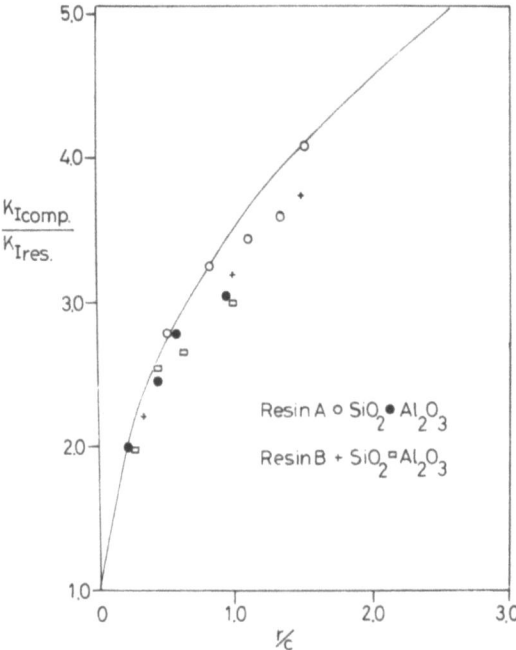

Figure 10. The increase in the stress intensity factor required to move a semi-circular crack through a series of obstacles ___ theoretical curve.

It is interesting to note that for the dolomite composites the relationship between the yield stress and the volume fraction of filler is completely different from that for the silica and alumina composites. As demonstrated in Figure 13 the yield stress is constant with increasing volume fraction of dolomite whereas for silica and alumina the yield stress is considerably increased. There appears to be a correlation between the dependences of K_{Ic} on volume fraction on one hand and σ_y on volume fraction on the other hand but the precise nature of this correlation is unclear at present.

INTERFACIAL PROPERTIES OF FILLED EPOXIDE RESINS 899

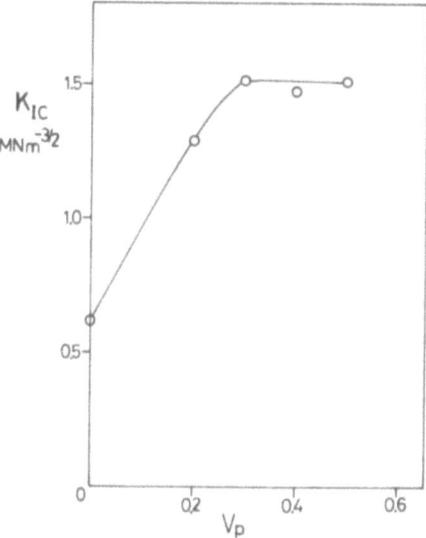

Figure 11. The variation in the stress intensity factor with the volume fraction of filler. Resin B + domolmite.

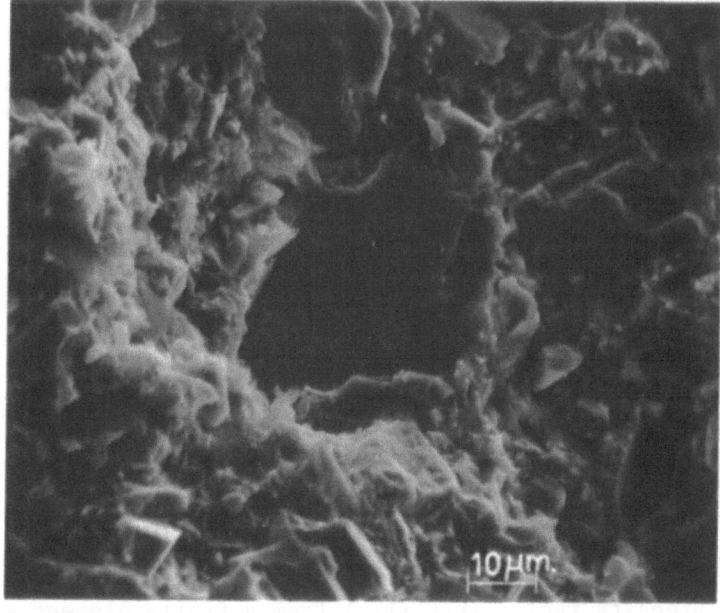

Figure 12. Scanning electron micrograph of the fracture surface of dolomite filled resin B.

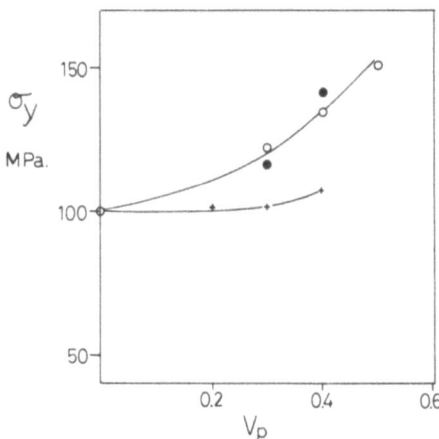

Figure 13. The variation in yield stress with volume fraction of filler. Resin B o alumina; o silica; + dolomite.

Combined Mode Adhesion Tests

The fracture of carefully prepared adhesive joints occurs normally in the centre of the bondline[30]. In the case of aluminium adherends bonded with an epoxide adhesive in the tapered double cantilever beam geometry, for example, failure was within the adhesive and the value of K_{Ic} measured was in good agreement with that of the pure bulk resin[31]. However, for the purposes of this work centre of bond failure was undesirable and a geometry was sought that produced fracture at the interface. With the scarf joint Trantina[14] and Bascom et al[15] found that failure was at, or very close to, the adherend surface and with a bond angle of 45° there was little tendency for the crack to jump from one interface to the other. This result was also confirmed in our work. The scanning electron micrograph in Figure 14 clearly shows the grains of alumina and, in some cases, resin is left adhering between the grains.

Trantina[14] and Bascom et al[15] calculated the combined mode stress intensity factor ($K_{(I,II)c}$) from a finite element analysis of a homogeneous sheet containing a slant edge crack when subjected

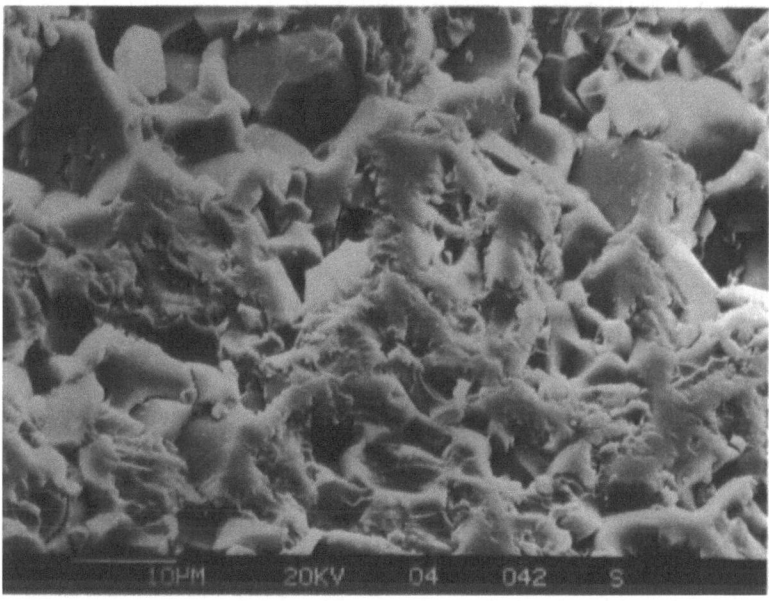

Figure 14. Scanning electron micrograph of the fracture surface of a scarf joint between alumina and resin B.

to a uniform tensile stress. They justified this approach since the compliance of the scarf joint was found to be the same as that of a solid aluminium specimen with a slant edge crack. However, this result would be expected since the glue line is very thin with respect to the rest of the aluminium specimen and even though the modulus of aluminium is ~ 25 x that of the resin, it is principally the aluminium that contributes to the compliance of the whole specimen. There are certain problems in determining the stress intensity factor at the interface between two materials since it is a crack tip parameter and its value depends on the properties of the material near the crack tip. It is extremely difficult to define the mechanical parameters of an interfacial zone. For example, it is unclear how the interfacial modulus may be defined. Bascom et al[15] used the modulus of aluminium in order to calculate the value of $G_{(I,II)c}$ from:

$$G_{(I,II)c} = \frac{K_{(I,II)c}^2}{E} \qquad (7)$$

However, the validity of this may be questioned, especially when the moduli of the two phases are so different. Other authors[32] have used a combination of the moduli of the two materials (E_1 and E_2) as shown below:

$$G_{(I,II)_c} = \frac{1}{2}(\frac{1}{E_1} + \frac{1}{E_2}) K_{(I,II)_c}^{2} \qquad (8)$$

To avoid these problems of interfacial moduli, we decided to calculate $G_{(I,II)_c}$ using a compliance calibration. An extensometer was connected to the ceramic inserts and the compliance (C) was measured for specimens with varying crack lengths (a). $G_{(I,II)_c}$ was then calculated from the following relationship

$$G_{(I,II)_c} = \frac{P^2}{2B} \frac{dC}{da} \qquad (9)$$

where P is the load at break and B the thickness of the specimen. The combined mode strain energy release rate for interfacial fracture between degreased alumina and epoxide resin A was found to be 320 Jm^{-2}. Preliminary results have indicated that this value varies with the surface treatment of the alumina.

There are little data in the literature with which to compare this value. The limited work reported has concerned the mixed mode loading of aluminium-epoxide bonds. Mulville et al[33,34] used a bending specimen where one half was aluminium and the other epoxide. For unmodified epoxide resins they found values of $G_{(I,II)_c}$ of up to 300 Jm^{-2} depending on the surface treatment of the aluminium. For modified epoxide resins the values were considerably higher, between 1300 and 3000 Jm^{-2} depending on the angle of loading[33]. Saxena[32] used compact tension and centre cracked tension specimens which had a ratio of K_{II}/K_I of $\sim$ 10 % and found values of $G_{(I,II)_c}$ for aluminium-epoxide bonds of the order of 270 Jm^{-2}. As described above, Bascom and co-workers[35] used the modulus of aluminium in their calculations of $G_{(I,II)_c}$, however, they obtained values between 80 and 300 Jm^{-2} depending on the surface treatment of the aluminium. This was for an unmodified epoxide resin. For commercial epoxide resins adhesives modified with nylon or an elastomer, they found values up to 900 Jm^{-2}.

For the unmodified epoxide bonds the strain energy release rate for interfacial fracture is generally higher than that of the pure resin (G_{Ic} for bulk resin A is 130 Jm^{-2}). This is probably because there is a restraint on the deformation at the crack tip as propagation is forced to the adherend-resin interface[35].

It appears from these limited results that the scarf joint is a promising method for modelling the fracture of particulate-filled epoxide resins. The interfacial interactions between epoxides and a wide range of substrates may be investigated. It is proposed to extend this work by testing other surface treatments of alumina and then to use quartz glass.

ACKNOWLEDGEMENTS

The authors would like to thank Professor J.G. Williams and Dr. A.J. Kinloch for helpful discussions.

REFERENCES

1. D.C. Phillips and B. Harris, in "Polymer Engineering Composites", M.O.W. Richardson Editor, Ch. 2, Applied Science, London 1977.
2. K.L. Mittal, Solid State Technol. (May 1979) 89.
3. K.L. Trachte and A.T. DiBenedetto, Int. J. Polym. Mater., $\underline{1}$, 75 (1971).
4. R.W. Davidge and G. Tappin, J. Mater. Sci., $\underline{3}$, 165 (1968).
5. C. Tourenq and F. Gragger, Naturstein Ind., $\underline{11}$, 19 (1965).
6. S.M. Wiederhorn, J. Amer. Ceram. Soc., $\underline{52}$, 99 (1969).
7. A.G. Evans, J. Mater. Sci., $\underline{7}$, 1137 (1971).
8. H.G. Tattersall and G. Tappin, J. Mater. Sci., $\underline{1}$, 296 (1966).
9. B.J. Pletka, E.R. Fuller and B.G. Loepke, in "Fracture Mechanics Applied to Brittle Materials", STP No. 678, p. 3, 19, ASTM, Philadelphia, PA, 1979.
10. B. Stalder and H.H. Kausch, J. Mater. Sci., $\underline{17}$, 2481 (1982).
11. P. Leevers, J. Mater. Sci., $\underline{17}$, 2469 (1982).
12. B. Stalder, Ph. Béguelin and H.H. Kausch, Int. J. Fract., in press (1982).
13. J.G. Williams and H. Ford, J. Mech. Eng. Sci., $\underline{6}$, 7 (1964).
14. G.C. Trantina, J. Comp. Mater., $\underline{6}$, 371 (1972).
15. W.D. Bascom, C.O. Timmons and R.L. Jones, J. Mater. Sci., $\underline{10}$, 1037 (1975).
16. W.D. Bascom and J. Oroshnik, J. Mater. Sci., $\underline{13}$, 1411 (1978).
17. L.J. Broutman and S. Sahu, Mater. Eng. Sci., $\underline{8}$, 98 (1971).
18. J.C. Hammond and D.C. Quayle, Second Int. Conf. on Deformation, Yield and Fracture of Polymers, P.R.I., March 1973, Cambridge, UK.

19. A.C. Moloney, H.H. Kausch and H.R. Stieger, manuscript in preparation (1983).
20. S. Kunz-Douglass, P.W.R. Beaumont and M.F. Ashby, J. Mater. Sci., 15, 1109 (1980).
21. F.F. Lange, Phil. Mag., 22, 983 (1970).
22. F.F. Lange, J. Amer. Ceram. Soc., 54, 614 (1971).
23. A.C. Moloney, H.H. Kausch and H.R. Stieger, J. Mater. Sci., 18, 208 (1983).
24. J.A. Manson and L.H. Sperling, "Polymer Blends and Composites", Plenum Press, New York, 1976.
25. O. Ishai and L.J. Cohen, Int. J. Mech. Sci., 9, 539 (1967).
26. R.J. Young and P.W.R. Beaumont, J. Mater. Sci., 10, 1343 (1975) and 12, 684 (1977).
27. A.G. Evans, Phil. Mag., 26/6, 1327 (1972).
28. D.J. Green, P.S. Nicholson and J.D. Embery, J. Mater. Sci., 14, 1657 (1979).
29. F.F. Lange and K.C. Radford, J. Mater. Sci., 6, 1197 (1971).
30. A.J. Kinloch, J. Mater. Sci., 15, 2141 (1980).
31. R.A. Gledhill, A.J. Kinloch, S. Yamini and R.J. Young, Polymer, 19, 574 (1978).
32. A.K. Saxena, Fibre Sci. Technol., 12, 111 (1979).
33. D.R. Mulville, P.W. Mast and R.N. Vaishnav, J. Eng. Mater. Technol., 100, 27 (1978).
34. D.R. Mulville and R.N. Vaishnav, J. Adhesion, 7, 215 (1975).
35. W.D. Bascom, R.L. Cottington and C.O. Timmons, J. Appl. Polym. Sci. Appl. Polym. Symp., 32, 165 (1977).

ABOUT THE CONTRIBUTORS

R. D. Adams is Reader in Mechanical Engineering, University of Bristol, Bristol, England. He received his Ph.D. degree in 1967 from Cambridge University. He and his team have been working on the stress and failure analysis of adhesive joints since 1970 and much of their work has involved the use of finite element techniques. His other interests are in composite materials, nondestructive testing, and vibration damping. He is a Fellow of the Institution of Mechanical Engineers and the Institute of Physics.

Walter Althof is Chief Scientist of the working group "Adhesive Bonded Joints" at the DFVLR - Institute for Structural Mechanics, Braunschweig, W. Germany which he jointed in 1955. His research interest is in the elastomechanic properties of structural aircraft adhesives, and has numerous publications dealing with the engineering aspects of adhesive bonding. Because of his experience in testing adhesive joints, he has participated in the standardization of test methods of adhesively bonded joints in Germany.

R. S. Alwar is presently Professor in the Department of Applied Mechanics at the Indian Institute of Technology, Madras, India which he joined in 1963. He received his Ph.D. degree in 1963 from the Indian Institute of Science, Bangalore. He has guided the research of 9 Ph.D. scholars and 2 Master's candidates and has 40 publications to his credit. He is a consultant to various organizations like Bharat Heavy Electricals, Indian Space Research Organization, Ministry of Railways, Defense Research and Development Organization, and several other private industries.

Garron P. Anderson is presently on the staff at Thiokol Corp. in Brigham City, UT. He received his Ph.D. degree in Mechanical Engineering from the University of Utah in 1973. His current activities include development of analytical and experimental techniques for evaluation of adhesives and solid propellants, development of service life evaluation techniques, and development of numerical stress analysis techniques. He has been a consultant to a number of industrial and government establishments, and his publications include 30 technical papers and one book written mainly in the areas of adhesion testing, analysis of bond stresses, and the mechanics of solid propellants.

Edgar Andrews has been Professor of Materials at Queen Mary College, University of London since 1968 when he established the Department of Materials at that college. He was Dean of Engineering at Queen Mary College, 1970-1973. His publications include <u>Fracture in Polymers</u>, a monograph published in 1968 and over 90 research papers and review articles. He has also published several books of a theological nature. He was awarded the A. A. Griffith Silver Medal by the Materials Science Club of Great Britain in 1977 for his work on the fracture of materials.

W. L. Baun is with the Air Force Materials Laboratory, Wright-Patterson Air Force Base, Ohio. He carried out graduate studies at the Air Force Institute of Technology and Ohio State University. After five years of commissioned service in the Air Force, he accepted a position as a research chemist in the Air Force Materials Laboratory where he has carried out research on bulk and surface characterization of solid state materials.

A. Beevers is at the Oxford Polytechnic, Oxford, England where in 1968 he established a specialist facility in joining and fastening technology. In addition to providing a teaching laboratory for engineering students, this facility undertakes a variety of R&D programs. He carried out postgraduate work in resistance welding at British Welding Research Association and was awarded an M.Sc. for his research on spot welding of hardenable steels.

J. P. Bell is Professor at the University of Connecticut, Storrs, which he joined in 1969 and has been very active in both the Chemical Engineering Department and the Institute of Materials Science. Prior to the University of Connecticut, he worked at Monsanto's Chemstrand Research Center for 3-1/2 years and the duPont Plastics Department for 6-1/2 years. He received his D.Sc. degree in Chemical Engineering from MIT. His publications in textile fibers and epoxy resin structure-property relationships are rather well known, and he has held extended professional appointments in West Germany, Italy and Israel.

Elisabeth J. Berger is currently an advanced scientist at the Owens-Corning Fiberglas Technical Center in Granville, OH. She received her Ph.D. degree in Physical Chemistry from the Pennsylvania State University in 1980.

F. James Boerio is presently Professor of Materials Science at the University of Cincinnati which he joined in 1970. He received his Ph.D. degree in Macromolecular Science from Case

Western Reserve University in 1971. His current research interests include adhesive bonding of metals and polymers and surface analysis, especially using infrared and Raman spectroscopy. He has published about 50 papers describing his research results.

Justin C. Bolger is currently managing the division which produces epoxy and other formulated products at the Amicon Corporation, Lexington, MA which he cofounded in 1962. He received his Sc.D. degree in Chemical Engineering from MIT.

Joseph Boutilier is a Research Associate with Singer-Kearfott in Little Falls, NJ where he has been a materials and processing technician for fourteen years. He has worked with the development of experimental instrumentation for the testing of materials. He has a certificate in Electronic Technology from Newark College of Engineering, and an Associates Degree in Applied Science, Electronic Technology, from the County College of Morris, Randolph, NJ.

Walter Brockmann has been the Head of the Structures and Composites Department in the Fraunhofer-Institut fur angewandte Materialforschung in Bremen, W. Germany since 1970. He graduated as a Diplom-Ingenieur in 1966 followed by work as a scientist with Prof. Matting in the area of adhesion and adhesives, and obtained his Ph.D. degree in 1969 in this field. He has published a book and more than 100 papers dealing with adhesion, bonding techniques and composites.

P. B. Crosley is a consulting engineer at the Materials Research Laboratory, Inc., Glenwood, IL. He received his Ph.D. degree in Metallurgical Engineering from Illinois Institute of Technology, Chicago, in 1966. He has over a dozen publications dealing with the measurement of fracture toughness in a variety of material systems and the prediction of failure in structures exposed to different service conditions.

R. G. Davidson is currently leader of the optical spectroscopy section, Polymer Research Group, Materials Research Laboratories, Ascot Vale, Victoria, Australia. He graduated in Applied Chemistry in 1962 from the Royal Melbourne Institute of Technology and his interests have included dyes and pigments, explosives, and synthetic rubber. He is presently interested in the application of UV-VIS-IR spectroscopy to the characterization of polymeric materials, and to the study of polymer degradation.

K. L. DeVries is presently Professor at the University of Utah, Salt Lake City and was Department Chairman during 1969-1981. He received his Ph.D. degree in Mechanical Engineering from the University of Utah in 1962. During 1975-1976, he headed the Polymer Program at the National Science Foundation, and has served on several NSF, NBS, and NRC advisory committees. He served as the

Chairman of Gordon Research Conference on Science of Adhesion in 1976, as well as Gordon Conference on Basic Failure Mechanisms in Polymers and Composites in 1980. He is the coauthor of the book, Analysis and Testing of Adhesive Bonds (1979), and his publications include approximately 150 papers, chapters in books, proceedings, etc. on fracture, polymers, adhesives, biomaterials, and dental materials.

Wolfgang Diem is with Degussa, Hanau, W. Germany ad was a research assistant at the Max-Planck-Institute for Metals Research in Stuttgart, W. Germany. He obtained his doctorate at the University of Stuttgart in 1982. His interests include physical metallurgy, mechanical engineering, and fabrication and testing of metal-to-ceramic joints and adhesively bonded joints.

R. G. Dillingham is currently a candidate for a Ph.D. degree in Materials Science at the University of Cincinnati where he received his M.S. degree in Materials Science in 1982. His research interests include adhesive bonding of metals and the influence of metal oxides on the curing reactions of thermosetting adhesives.

A. A. Donatelli is Associate Professor of Chemical Engineering at the University of Lowell, Lowell, MA. He received his Ph.D. degree in Chemical Engineering from Lehigh University. His research interests are in the areas of polymer science and mass transfer.

Yona Eckstein is currently an advanced scientist at the Owens-Corning Fiberglas Technical Center in Granville, OH. Before her current position, she spent two years as a Research Associate at the Institute of Polymer Science, University of Akron. She received her Ph.D. degree from Hebrew University, Jerusalem.

Gerhard Elssner is a research staff member at Max-Planck-Institute for Metals Research, Stuttgart, W. Germany. He obtained his doctorate in physical metallurgy at the University of Stuttgart in 1967. His group is involved in research on the mechanical properties and the microstructure of metal-to-ceramic transitions. His interests include materials science, composite materials, ceramography and metallography. He has published about 50 papers.

R. A. Everett, Jr. is currently involved in fatigue and fracture research of adhesively bonded structures at the U. S. Army Structures Laboratory located at NASA Langley Research Center, Hampton, VA. He received an M.S. in Engineering Mechanics from Old Dominion University in 1980. Prior to his current position, he worked at the Army Missile Command, the Army Aviation Laboratory, Lockheed-Georgia Company, and Lockheed-California Company. He has authored or coauthored about 15 technical papers.

ABOUT THE CONTRIBUTORS

R. Exalto is with Fokker B.V. Technological Centre, Schiphol-Oost, The Netherlands.

Eugene C. Francis is Chief Structural Research Engineer, Chemical Systems Division, United Technologies, Sunnyvale, CA. He received B.A. in Physics from the University of California. He has published about 50 papers involving finite and nonlinear analysis and experimentsl testing of filled and unfilled polymer systems, viscoelastic fracture mechanics, nonlinear constitutive theories, experimental stress analysis, service life analysis, adhesive bonding, etc.

Francois Gaillard is a Research Associate in the Applied Chemistry and Chemical Engineering Department, University Claude Bernard, Lyon, France where he received his M.S. in Materials Science in 1980. He is involved in the characterization of stainless steels submitted to various passivation treatments.

Yvon Gilibert is currently Assistant Professor in Civil Engineering at Rheims University. He graduated in Engineering from C.N.A.M., France in 1968 and obtained a Science Doctorate from Rheims University in 1978. From 1965-1972, he worked in far infrared and Raman effect spectroscopy at Rheims University, and has worked in the field of experimental mechanics of adhesive joints since 1973. He is a consultant in Civil Engineering.

D. Gutierrez-Lemini is Senior Structural Research Engineer, Chemical Systems Division, United Technologies, Sunnyvale, CA. Received Ph.D. degree from the University of Utah.

J. A. Harris is Research Assistant in Mechanical Engineering, University of Bristol, England, where he received B.Sc. degree (First Class Honours) in 1977. He has several years experience in adhesives research with particular interest in impact performance and failure of bonded joints.

Otto-Diedrich Hennemann has been since 1978 in the Department of Nonmetallic Materials and Adhesive Techniques in the Fraunhofer-Institut fur angewandte Materialforschung in Bremen, W. Germany. He was graduated as a Diplom-Physicist in 1972 followed by work in the area of spinwaves with Prof. Urban and received his Ph.D. in 1975 in this field from the University of Bonn.

Edward J. Hughes is Principal Scientist with Singer-Kearfott in Little Falls, N.J. where he has been for 17 years and has conducted extensive research into the microstrain behavior of metals, adhesives and composite materials. Prior to that he was a Senior Scientist at the Materials Research Corp., Orangeburg, N.Y. He received his Ph.D. degree in Physical Metallurgy from Birmingham University, England.

D. L. Hunston has been with the National Bureau of Standards since 1980 where he is currently helping to establish a program in adhesives and composites. Prior to coming to NBS, he was (1971-1980) at the Naval Research Laboratory, Washington, D.C. where he studied polymeric materials and between 1978 and 1980 directed the Adhesives and Polymer Composites Section. He received his Ph.D. degree in Physical Chemistry from Kent State University, followed by 2 years at Northwestern University on an NIH Postdoctoral Fellowship.

R. W. Hylands has since March 1980 been Principal of Portadown Technical College, one of the largest further educational establishments in N. Ireland. Prior to his current position, he has had a number of academic appointments including Principal Lecturer and Director of Studies and Head of School of Building at the new Ulster Polytechnic. At the Polytechnic he carried out a program of research associated with prestressed timber beams which led to the award of Ph.D. He has published papers and booklets dealing with strength characteristics of adhesives, home grown timber, deflection of non-uniform beams and buckling of non-uniform struts.

T. Igarashi is Associate Professor at Gunma University, Kiryu, Japan where he has been since 1974. He graduated from Tokyo Institute of Technology in 1953 and holds a Ph.D. degree. Before coming to Gunma University, he was with the Research Laboratory of Hitachi Cable Ltd. (1961-1974). He is an Associate Editor of Journal of Adhesion, and his publications include: <u>Theory of Pressure Sensitive Adhesion and its Applications</u>, 1978, Keieikaihatsu Center, Tokyo; <u>Testing Methods and Estimation of Polymeric Materials, Mechanical Test Methods</u>, 1980, Baihukan, Tokyo; and <u>Ultrasonic Wave and Nondestructive Testing</u>, Keieikaihatsu Center, Nagoya, to be published.

Ravi B. Jathar is currently Quality Control and Product Development Manager at Elmendorf Board Corp., Claremont, N.H. Prior to his present position, he was Technical Manager, Forest Industries Pvt. Ltd. (India), and Research Assistant/Trainee at the Research Center of Bison-Werke, Springe, W. Germany. He received his M.S. degree in Wood Technology from the University of Minnesota.

J. P. Jeandrau is an ingenieur at the Centre Mecanique-Chimie-Materiaux, Etablissement Technique Central de l'Armement, Arcueil Cedex, France where he is responsible for Laboratoire "Collages et Adhesifs". He received Doctorat de 3eme Cycle (Chimie-Physique) and the subject of the thesis was kinetic and synthetic aspects of the photoreduction of aromatic ketones.

Wartan Jemian is Professor of Mechanical Engineering and Materials Engineering at Auburn University. He received his Ph.D.

degree in 1956 from Rensselaer Polytechnic Institute. He was Senior Fellow and Head of the Power Rectifier Fellowship of Mellon Institute with major interests in materials and junction structure-property relationship until 1962. He is active in research in adhesion, dynamic characteristics of materials and solidification morphology.

W. S. Johnson is currently the group leader of the Adhesively Bonded Structures Group of the Fatigue and Fracture Branch at NASA Langley Research Center, Hampton, VA. He received a Ph.D. in Solid and Structural Mechanics from Duke University. He has previously worked as an engineer at the U.S. Naval Ship R&D Center and at General Dynamics/Fort Worth. He has authored or coauthored approximately 30 technical papers in the area of fatigue and fracture of materials. He is currently chairman of an ASTM task group concerned with standardizing durability testing of adhesively bonded joints.

Hans-Henning Kausch has been Director of the Polymer Laboratory at the Swiss Federal Institute of Technology, Lausanne, Switzerland, since 1976. Prior to his current position, he was associated with the Battelle Institute eV, Frankfurt am Main and the Deutsches Kunststoff-Institute, Darmstadt. He received his Dr. rer. nat. in Nuclear Physics from the University of Gottingen, and his interests turned to polymeric materials with his affiliation with the Mannesmann-Research Institute in 1960. He is predominantly concerned with the various interrelations between molecular properties, physical modifications of polymers through orientation, reinforcement and processing and macroscopic mechanical properties.

A. C. P. Kho is currently carrying out research on the dynamic performance of bonded box-section structures. He graduated in Engineering from Oxford Polytechnic in 1981.

A. J. Kinloch is with the Ministry of Defense at Waltham Abbey, U.K., where he is currently leader of a group engaged in research and development in the field of adhesives, adhesion and polymer fracture. He received his Ph.D. degree in Materials Science in 1972 from the Queen Mary College, London University.

Ronald A. Kline is Assistant Professor at the University of Oklahoma, Norman, OK. Before his current position, he was Senior Research Engineer with General Motors Research Laboratories, Warren, MI (1979-1982) and during 1978-1979 he was Senior Research Scientist with General Dynamics in Fort Worth, TX. He received his Ph.D. degree in Mechanics and Materials Science from Johns Hopkins University in 1978. His research interests are nondestructive testing, mechanics of composite materials, and adhesive bonding.

Wolfgang G. Knauss is presently Professor at California Institute of Technology which he joined in 1965. He is very well known in the field of fracture of polymers and viscoelasticity. He has been a consultant to many industrial organizations, and has also served as a consultant on the AFML/MBC sponsored program on "Fatigue Behavior of Adhesively Bonded Joints" and on the AFML/MBM sponsored program on "Time-Dependent Environmental Behavior of Graphite/Epoxy Composites". He is currently associate investigator in the Integrated Methodology for Adhesive Bonded Joint Life Predictions program. He received his Ph.D. degree in 1963 from Caltech.

Hansgeorg Kollek has been since 1976 with the Fraunhofer-Institut fur angewandte Materialforschung in Bremen, W. Germany. He received his degreees in Analytical Chemistry fom the University of Goettingen, W. Germany. He is currently working in the area of chemistry of adhesion and adhesives, especially the aging behavior of metal bonds, and has published several papers dealing with this subject.

Dimiter L. Kotzev is Research Scientist at the Scientific-Industrial Center for Special Polymers, Sofia, Bulgaria. Hold doctorate degree.

A. Kwakernaak is with Fokker B.V. Technological Centre, Schiphol-Oost, The Netherlands.

Lieng-Huang Lee is a Senior Scientist of Xerox Corp., Webster, N.Y. He is the author of over 50 technical papers and 25 patents and editor of six books on adhesion, friction and wear of polymers. Recently, he was selected as the distinguished scholar to participate in the U.S.-China Exchange Program by the National Academy of Sciences and the Chinese Academy of Sciences.

K. M. Liechti is currently Associate Professor, Department of Aerospace Engineering and Engineering Mechanics, University of Texas at Austin. Before this he was with the Materials Research Laboratory of General Dynamics, Fort Worth, TX which he joined in March 1980 to work on the Integrated Methodology for Adhesive Bonded Life Prediction Program. He has been interested, inter alia, in the development of failure criteria to explain the failure modes observed in the model lap shear joints. He received his Ph.D. degree in Aeronautics from California Institute of Technology in 1980.

Yu. S. Lipatov is with the Institute of Macromolecular Chemistry, The Ukrainian SSR Academy of Sciences, Kiev, USSR.

S. Mall is presently engaged in fatigue and fracture research of adhesively bonded structures at NASA Langley Research Center,

Hampton, VA. He received his Ph.D. in 1977 from the University of Washington, Seattle, and thereafter joined the faculty of the University of Maine, Orono and became Associate Professor in 1981. He obtained M.Sc. in Mechanical Engineering from Banaras Hindu University, India in 1966. He has authored and coauthored more than 25 technical papers in fatigue and fracture mechanics.

N. T. McDevitt is with the Materials Laboratory, Wright-Patterson Air Force Base, Ohio. He received his graduate degree in Chemistry from Lehigh University. After graduation, he accepted a commission in the United States Air Force. Following his military tour in the Air Force Materials Laboratory, he remained in the laboratory as a civilian research chemist. During his career in the Materials Laboratory, he has specialized in molecular spectroscopy (IR and Raman) and more recently research on adhesive bonding.

J. Dean Minford is a Scientific Associate in the Product and Process Engineering Division at Alcoa Laboratories, Alcoa Center, PA. He received his Ph.D. degree in Bio-Organic Chemistry from the University of Pittsburgh in 1951. He has focussed his attention for over 25 years on the study of joining aluminum to itself and other materials through the use of adhesives; and has published more than 50 technical papers dealing with the durability of aluminum bonded joints. He is the author of the Alcoa Handbook on Bonding and has written a number of chapters for various books. He is the recipient of the 1982 Adhesives Age Award for outstanding contributions to the adhesives industry.

Kashmiri Lal Mittal* is presently employed at the IBM Corporation at Hopewell Junction, N.Y. He received his M.Sc. (First Class First) in 1966 from Indian Institute of Technology, New Delhi, and Ph.D. in Colloid Chemistry in 1970 from the University of Southern California. In the last ten years, he has organized and chaired a number of very successful international symposia and in addition to this volume, he has edited nineteen more volumes as follows: Adsorption at Interfaces, and Colloidal Dispersions and Micellar Behavior (1975); Micellization, Solubilization, and Microemulsions, Volumes 1 & 2 (1977); Adhesion Measurement of Thin Films, Thick Films and Bulk Coatings (1978); Surface Contamination: Genesis, Detection, and Control, Volumes 1 & 2 (1979); Solution Chemistry of Surfactants, Volumes 1 & 2 (1979); Solution Behavior of Surfactants - Theoretical and Applied Aspects, Volumes 1 & 2 (1982); and Physicochemical Aspects of Polymer Surfaces, Volumes 1 & 2 (1983); Adhesion Aspects of Polymeric Coatings, (1983); Surfactants in Solution, Volumes 1, 2 & 3 (1984); and Polyimides: Synthesis, Characterization and Applications, Volumes 1 & 2 (1984). Also he is Editor of the series, Treatise on Clean Surface Technology, a multi-volume work in progress. In addition to these

---
* As editor of this volume.

books he has published more than 50 papers in the areas of surface and colloid chemistry, adhesion, polymers, etc. He has given many invited talks on the multifarious facets of surface science, particularly adhesion, on the invitation of various societies and organizations in many countries all over the world, and is always a sought-after speaker. He is a Fellow of the American Institute of Chemists and Indian Chemical Society, is listed in American Men and Women of Science, Who's Who in the East, Men of Achievement and many other reference works. He is or has been a member of the Editorial Boards of a number of scientific and technical journals. Currently, he is Vice-President of the India Chemists and Chemical Engineers Club (USA).

Anne C. Moloney has been at the Swiss Federal Institute of Technology, Lausanne, Switzerland since 1979 where she is studying the fracture behavior of particulate filled epoxide resins. She obtained her Ph.D. degree in 1979 for work conducted on the effect of carriers on the environmental stability of adhesive joints. From 1976 to 1979 she worked at the Leicester Polytechnic, England.

Charles T. Mooney is R&D Manager for the Electronic Materials Group at Amicon Corp., Polymer Products Division, Lexington, MA. He received his B.S. degree in Chemistry/Biology from Boston College and is currently a candidate for the M.S. degree in Chemical Engineering at the University of Lowell, Lowell, MA.

C. E. M. Morris currently heads the structural adhesives work at the Materials Research Laboratories, Ascot Vale, Victoria, Australia. Before joining MRL, she spent three years at the Corporate Laboratory of I.C.I. Ltd. in Cheshire, U.K. She received her Ph.D. degree in Physical Chemistry in 1967 from the University of Sydney. she has conducted research programmes on the kinetics and mechanisms of free radical polymerization, surface modification of polyolefins for improved adhesion and characterization procedures for polymers, and is currently interested in epoxy-based systems for aerospace applications.

William J. O'Brien is Professor of Dental Materials at the University of Michigan and Director of the Surface Science Laboratory in the Dental Research Institute. He received his Ph.D. from the University of Michigan in 1967 and has been on its faculty since 1970. He is coeditor of Outline of Dental Materials and Their Selection with Gunnar Ryge and coauthor of Dental Materials: Properties and Manipulation with Drs. Craig and Powers. He has served as a consultant to the American Dental Association and World Health Organization, and has been very active in A.D.A. and I.A.D.R. He is currently a representative of the American Association of Dental Schools to the American National Standards Institute Committee on Dental Materials. His current research is in the optical properties and adhesion of dental materials and has published in the areas of precious metals, sealants, surface phenomena, ceramics, and polymers.

Jae M. Park has recently completed his Ph.D. degree in Chemical Engineering at the University of Connecticut, Storrs with thesis research on epoxy resin bonding to copper. Prior to coming to the United States, he worked for three years as a Scientist at the Korean Institute for Science and Technology. His undergraduate work was in chemical engineering at Seoul National University.

P. J. Pearce is affiliated with the Materials Research Laboratories , Ascot Vale, Victoria, Australia which he joined in 1978. Before joining MRL, he was at the Explosives Research and Development Establishment, Waltham Abbey, U.K. and later at Joint Tropical Research Unit, Innisfail, Australia carrying out research which included the effects of a tropical environment on a wide variety of polymeric materials. He received his M.S. degree in Polymer Chemistry from Salford University in 1974, and his current studies involve the characterization and cure kinetics on a variety of structural adhesives and composites. He recently completed a 15 month attachment to the Army Materials and Mechanics Research Center, USA.

S. S. Pesetskii is a Senior Researcher in the Institute of Mechanics of Metal-Polymer Systems, Byelorussian SSR Academy of Sciences, USSR, where he has been since 1972. He defended his candidate thesis in 1980. He is interested in technological problems of polymers and metal-adhesive bonds, and has more than 50 papers and 25 USSR patents to his credits.

Gunter Petzow is a member of the board of directors of the Max-Planck-Institute for Metals Research in Stuttgart and also Professor at the Universities of Stuttgart and Berlin. He received his doctorate in 1959 from the University of Stuttgart. His research activities have been concerned with problems in the field of physical metallurgy, powder metallurgy, special ceramics and with phase diagrams of metallic and ceramic materials. He has received several awards among them the Kuczynski Diploma, the honorary membership of the Korean Institute of Metals, doctor honoris causa of the Tokyo Institute of Technology, the Hume-Rothery Prize of the Metals Society, London, and the ASM Fellowship. He has published more than 250 research papers and four books and holds eight patents.

Roscoe A. Pike is a Senior Materials Scientist at the United Technologies Research Center, E. Hartford, CT which he joined in 1967. He received his Ph.D. degree in Organic Chemistry from MIT in 1953. His fields of activity include high temperature resin based composites, adhesive technology, thermoplastic resins and microelectronic packaging. His areas of publications include silicone chemistry, grinding fluids, advanced composites and adhesives; and he has been awarded 35 U.S. patents.

A. Pizzi is with the National Timber Research Institute, Pretoria, South Africa. He had his education at the University of Rome where he obtained a Doctorate in Physical Chemistry of Polymers. He emigrated to South Africa in 1969 and continued his studies at the University of the Orange Free State in Bloemfontain and obtained a Ph.D. degree in Applied Organic Chemistry. He was awarded the 1980 Forest Products Research Society (USA) Bork Award for his work on wood adhesives based on bark tannin extracts, and was nominated for the 1982 Sir Stuart Mollingson Forest Research Medal (U.K.). He is on the editorial boards of Intl. J. Adhesion Adhesives and J.Wood Chemistry Technology, and is the editor and principal author of a book entitled Wood Adhesives Chemistry and Technology to appear in 1983. His research interests are wood adhesives, chemistry of wood preservation, wood chemistry, polymers and conformational analysis, and has published 150 papers.

K. N. Ramachandran Nambissan is presently Assistant Professor in the Civil Engineering Department of Regional Engineering College, Calicut, India. He completed his Ph.D. degree at the Indian Institute of Technology, Madras, India. His dissertation was in the field of fracture mechanics and the present paper forms a part of the research undertaken for his doctoral degree. He is the author of several publications.

R. Ramaswamy is a Senior Scientist in the Polymers and Special Chemicals Division of Vikram Sarabhai Space Centre (Indian Space Research Organization) at Trivandrum. He received his Ph.D. in 1971 from Poona University, India. He and his group are engaged in the development of specialty adhesives and sealants for space and aircraft applications. His interests include adhesion mechanisms and structure-property relationships of adhesives. He has published papers in the fields of ion exchange resins and adhesion.

Stephen T. Rasmussen is Assistant Professor of Basic Dental Technology at the School of Dentistry, Case Western Reserve University, Cleveland, OH which he joined in 1979. He received his Ph.D. degree in Materials Science and Engineering in 1970 from the University of Utah. His current research deals with the fracture properties of human tooth structure and dental materials, and the adhesion of dental materials and has published in these areas.

E. J. Ripling is President of the Materials Research Laboratory, Inc., Glenwood, IL. He received his Ph.D. degree in Metallurgy in 1952 from the Case Institute of Technology. Work on the application of fracture mechanics to adhesive joints has been conducted at the Materials Research Laboratory over the past two decades and has resulted in over a dozen publications in the open literature. In recognition of his work on fracture mechanics, he was elected a Fellow of the American Society of Metals in 1970. He is a member of several NMAB Committees, and chairman of an ASTM Task Group on Crack Arrest.

Alain A. Roche is with the Applied Chemistry and Chemical Engineering Department, University Claude Bernard, Lyon, France and received his thesis degree in 1978 from the same university. During 1979-1980, he was Research Scientist at the Air Force Materials Laboratory, Wright-Patterson Air Force Base, Ohio; and he was appointed "Attache de Recherche" at the CNRS in 1980. He is the author of over 30 papers dealing with the characterization of solid surfaces and the mechanical properties of metal/adhesive systems.

Stanislav I. Rokhlin is with the Materials Engineering Department of the Ben-Gurion University of the Negev, Beer-Sheva, Israel. He received his Ph.D. degree in Physics in 1972. He has published over 30 papers and holds some patents. His research interests include wave propagation and diffraction and application of ultrasonic waves for materials evaluation especially thin films and composite materials.

Maurice J. Romand is Head of the Materials Science Division of the Applied Chemistry and Chemical Engineering Department, University Claude Bernard, Lyon, France where he received his Ph.D. in Physical Chemistry in 1970. His present research activities are mainly devoted to the characterization of passive films formed on various substrates, and to the development of low-energy electron-induced x-ray spectroscopy (LEEIXS) for characterizing solid surfaces and thin films, and has published some 70 papers, most of them dealing with applications of x-ray emission spectroscopy and surface characterization.

John Romanko is currently Engineering Specialist Sr., Materials Research Laboratory, General Dynamics/Fort Worth Division, TX. He received his Ph.D. in Molecular Physics in 1954 from the University of Toronto. He has conducted studies in full field displacement (strain) analysis in materials and composites using laser holographic and interferometry techniques. He is currently involved in adhesive bonding and is program manager on three USAF programs on adhesive bonding. He has published on holography and adhesive bonding.

Klaus Ruhsland is a certified engineer for plastic material engineering and head of a research team for adhesive bonding of metals and welding of plastics at the Central Institute for Welding Engineering of the GDR, Halle (Saale), E. Germany. In 1977 he was awarded the Goethe Prize for Science and Technology of the City Council of Berlin for developing the adhesive Epasol FV/ZIS 939. He has published about 60 technical papers and is coauthor of the technical manual <u>Plastwerkstoffe in der Feingeratetechnik</u> published in 1973.

John L. Rutherford is Director, Materials and Process Laboratory, Singer-Kearfott in Little Falls, N.J. where he has been for 20 years. Prior to that he was a Research Physicist at the Franklin Institute Laboratories in Philadelphia where he helped develop floating zone-refining techniques for iron and titanium. He has a Ph.D. degree in Metallurgy from the University of Pennsylvania. He is listed in American Men of Science.

J. S. Santner is a Research Engineer at the Materials Research Laboratory, Inc. Glenwood, IL. He received his Ph.D. degree in Materials Science and Engineering in 1975 from Northwestern University. His correlating the fatigue and fracture properties of structural materials to their microstructures characterized by light, scanning and transmission electron microscopy has resulted in over a dozen publications in the open literature.

J. P. Sargent is currently doing postdoctoral research at Bristol University, England. He received his Ph.D. degree studying the linear thermal expansion coefficients of carbon fiber reinforced plastic. He is presently interested in the application of interferometry to the study of the behavior of adhesive joints when exposed to humid environments and also has a keen interest in microprocessors.

P. Sasidharan Achary is a Scientist in the Polymers and Special Chemicals Division of Vikram Sarabhai Space Centre (Indian Space Research Organisation), Trivandrum. He received B.Sc. (Special) in 1970 from Kerala University, India. He is engaged in the research and development of various adhesives and sealants suitable for aerospace applications. He has about 10 years experience in the development of various adhesives and sealants based on epoxies, phenolics, acrylates and polysulfides.

L. M. Sergeeva is with the Institute of Macromolecular Chemistry, The Ukrainian SSR Academy of Sciences, Kiev, USSR.

Gedalia Sharon is currently on leave at the University of Utah from Rafael Armament Division, Haifa, Israel. He received B.S. in 1967 in Chemical Engineering from Technion, Israel and M.S. in Industrial Engineering in 1981. He has conducted studies in the areas of adhesives and joining.

S. J. Shaw has been with the Propellants, Explosives and Rocket Motor Establishment, Ministry of Defense, Waltham Abbey, U.K. since 1976 where he is currently engaged in research and development in the field of structural adhesives. He obtained his M.Sc. in Polymer Technology in 1975 from the Institute of Polymer

Technology, Loughborough University, Loughborough, U.K.

S. V. Shcherbakov is with the Institute of Mechanics of Metal-Polymer Systems, Byelorussian SSR Academy of Science, USSR where he heads the Design Office. He has a Candidate of Science degree. He is interested in designing and producing machine elements from polymer-based composite materials, and in improving the strength of reinforced systems. He has over 50 USSR patents to his credit and has coauthored three monographs: <u>Plastic Gearings</u> (1965); <u>Metallopolymeric Materials and Items</u> (1978); and <u>Metallopolymeric Gearings</u> (1981).

Margaret Sheridan is currently a graduate student in Ph.D. program in Chemistry at VPI&SU, Blacksburg, VA. She received her B.S. degree from Gannon College.

Francois P. Sidoroff is Head of the Solid Mechanics Laboratory, Ecole Centrale de Lyon, Ecully, France which he joined in 1978. He received his Ph.D. degree in Mechanics in 1976 from the Ecole Nationale Superieure des Arts et Metiers (ENSAM) in Paris, and was appointed "Charge de Recherche" at the CNRS. He is concerned with wide-ranging research programs including solid material viscoelasticity, plasticity, viscoplasticity and behavior modeling, and has published over 40 papers in these areas.

V. E. Starzhynskii has been with the Institute of Mechanics of Metal-Polymer systems, Byelorussian SSR Academy of Sciences, USSR since 1959. He defended his Candidate Thesis in 1968, and since 1971 has been the Head of the Laboratory of Metallopolymeric Structures. He is interested in the development of engineering composite materials, designing of metallopolymeric machine parts, and increasing the load bearing capacity of drives. He has published numerous papers and has coauthored two monographs <u>Metallopolymeric Materials and Items</u> (1978), and <u>Metallopolymeric Gearings</u> (1981).

H. R. Stieger is presently Head of the Polymeric Materials Department, Brown Boveri & Cie, Zurich, Switzerland which he joined in 1975. He received his Dr. phil. II in Chemistry from the University of Zurich in 1973.

Tadatomo Suga is research assistant at the Max-Planck-Institute for Metals Research in Stuttgart, W. Germany. He obtained his master of engineering at the University of Tokyo in 1979. Recently he has been working on fracture mechanics problems of metal-to-ceramic and adhesively bonded joints.

Robert Y. Ting is presently Head, Materials Research, Underwater Sound Reference Division of the Naval Research Laboratory in Orlando, FL. In his current position he is responsible for R&D work on materials used in the Navy's underwater acoustic systems. He received his Ph.D. degree from the University of California, La Jolla. He has published extensively on the various subjects such as drag reduction, cavitation, polymer rheology, composite materials, fracture mechanics and adhesion. He is listed in American Men and Women of Science, Who's Who in the East, and Who's Who of Sino-Americans.

T. T. Todosiychuk is with the Institute of Macromolecular Chemistry, The Ukrainian SSR Academy of Sciences, Kiev, USSR.

P. K. Tsarev is with the Institute of Macromolecular Chemistry, The Ukrainian SSR Academy of Sciences, Kiev, USSR.

A. M. Usmani is a Senior Research Staff, Research Institute, at the University of Petroleum and Minerals, Dharan, Saudi Arabia. Prior to his current position, he was Research Professor and Associate Director of Polymer Research at the University of Dayton. He has more than 20 years of experience in polymer science and technology and has been associated with a number of industrial concerns. He received his Ph.D. degree in Chemistry from North Dakota State University and has more than 100 publications and many issued U.S. patents to his credit.

H. A. van Hoof is with Fokker B.V., Technological Centre, Schiphol-Oost, The Netherlands.

John D. Venables is the Manager of the Materials and Surface Science Department at Martin Marietta Laboratories, Baltimore, MD. He has been with the Laboratories since 1964, and is responsible for studying properties of the transition metal carbides, nitrogen ceramics, and adhesive bonded structures. He was a Research Fellow at the University of Warwick in 1968-1969 and obtained his Ph.D. degree there in 1971. He has been a consultant on the Viking Mars Lander program and the Space Shuttle program.

Georges Verchery is Professor of Materials Science at Ecole des Mines, Saint-Etienne, France. He graduated in Engineering from Ecole Polytechnique, Paris in 1967, and obtained a Science Doctorate from the University of Paris in 1973. He has worked in the field of mechanics of composite materials and interfaces since 1967. He is a consultant in Mechanical Engineering and Composite Materials.

S. S. Wang is Associate Professor at the University of Illinois which he joined in 1977. He received his Sc.D. degree from MIT in 1975 and stayed there for an additional two years as a Research Associate and Lecturer. His work involves experimental and analytical studies of adhesives and composites with special emphasis on advanced analytical and finite element analysis techniques.

In 1982, he received the Xerox Award for Outstanding Faculty Research.

Thomas C. Ward is Professor of Chemistry at VPI&SU, Blacksburg, VA which he joined in 1968. He received his Ph.D. degree in Physical Chemistry in 1966 from Princeton University and spent two years in England with Prof. Manfred Gordon. He has received several teaching honors, including the Wine Award and is a member of the Academy of Teaching Excellence at VPI&SU. He has also been a major contributor to the ACS polymer short course program since its inception in 1976. His publications have been principally in the areas of polymer characterization, testing, adhesion and long term property evaluation of polymeric materials.

M. L. Williams is presently Dean of the School of Engineering, University of Pittsburgh having come there in 1973 from the University of Utah where he was Dean of the College of Engineering and Distinguished Professor of Engineering. Prior to these 8 years in Salt Lake City, he had been at Caltech as Professor of Aeronautics since receiving his doctorate there in 1950. His main professional research activities have been concentrated in the field of general materials behavior and structural mechanics and design, with a specific interest in both cohesive and adhesive fracture which is reflected in his being Editor-in-Chief of the International Journal of Fracture. He has lectured nationally and internationally, for example as a Sigma Xi National Lecturer and as a Visiting Professor at the Imperial College. He has served on many governmental technical advisory committees such as the National Institute of Dental Research, NSF, NASA and National Materials Advisory Board.

Richard S. Williams is a Senior Research Engineer at the United Technologies Research Center, E. Hartford, CT. He received his Ph.D. degree in Materials Engineering Science from Virginia Tech in 1975. His research has spanned all areas of NDT and experimental mechanics, with special emphasis on ultrasonics and acoustic emission, and has over 60 publications and presentations. He most recently served as Editor of the ASNT Symposium on Ultrasonic Reliability. He is an active member of ASNT. In addition to his Associate Technical Editor position with Materials Evaluation, he has served or is serving on a number of committees.

INDEX

Accelerated Procedures for Glue-Wood Bonds, 395-417
Acoustic Method for Adhesion Estimation in Filled Systems, 440-445
Adhering Systems (Single Adherend/Adhesive), 19-30, 85-102
  effect of surface conditioning of the adherend on the mechanical behavior of, 85-102
  practical adhesion measurement in, 19-30
Adhesion Estimation Methods in Filled Polymer Systems, 433-450
  acoustic method, 440-445
  compensation method, 435-438
  volumetric method, 438-440
Adhesive
  curing and viscoelastic properties by interface waves, 328-337
Adhesive Bonded Joints (see also Adhesive Joints of)
  characterization of
    by ultrasonic surface and interface waves, 307-345
  dimensional stability of, 137-150, 151-164
    effect of moisture on, 137-150
  durability of, 453-467, 469-484, 485-501, 503-521, 523-540, 541-553, 555-564, 694-701
  effect of scrim cloth on, 679-685
  evaluation of

Adhesive Bonded Joints (cont.)
  role of surface and bulk characterization, 3-17
  fracture aspects of, 689-702, 703-737, 739-753, 755-787, 789-807, 809-827, 839-853
  influence of aging of film adhesive on, 231-246
  influence of surface roughness on, 69-84
  life prediction methodology for, 567-586
  model of the interfaces formed in, 6
  nondestructive evaluation of, 347-367
  stress analysis of, 587-610
  three-point bend test for, 381-394
Adhesive Bonded Thin Gauge Sheet Metal Box-Section Beams
  performance of, 627-637
Adhesive Bond Lines (Shear Stressed)
  effects of low cycle loading, 659-677
Adhesive Bonding
  of contaminated surfaces, 259-266
  of oily surfaces, 259-266
  vibrational, 262-266
Adhesive Bonding Process
  integrating surface treatment of bonding parts in, 257-266
Adhesive Failure Energy, 748-750

Adhesive Failure in Dentistry, 295-300
Adhesive Fracture
  continuum mechanics factors in, 703-737
  structural precursors to, 809-827
Adhesive Fracture Energy of
  Epoxy Adhesive Joints, 799-803
  effect of bond thickness on, 800-803
Adhesive Joints of
  aluminum (alloy), 195-212, 233, 240, 245, 383, 453-467, 479, 485-501, 503-521, 616, 793, 817
  box-section beams, 627-637
  composites, 639-658, 755-787
  copper, 523-540
  glass, 694-701
  graphite/epoxy, 639-658, 757
  honeycomb structures, 369-380
  iron, 549
  microscope cover slip, 152
  oily surfaces, 259- 266
  polyamide 11, 133
  polyamide 6.6, 133
  polyethylene (low density), 130
  polyethylene (high density), 130
  polymers, 121-136, 692-694
  polyoxymethylene, 132
  polypropylene, 132
  polytetrafluoroethylene, 129
  steel, 7, 15, 69, 85-102, 165-193, 195-212, 351, 429, 564, 793
  titanium, 8, 138, 158, 453-467, 541-553
Adhesive Lap Joints
  impact strength of, 611-626
Adhesive Test Methods, Evaluation of, 269-287
  blister, 296, 304, 707
  cone, 706
  double cantilever cleavage, 706, 762, 794
  lap shear, 271-283
    double lap shear, 706
    single lap shear, 706

Adhesive Test Methods, Evaluation of, (cont.)
  rod pull-out, 707
  tensile bond, 283-286
  torsional shear stress, 706
  T-peel, 706
Adhesive Test Specimens, 706-707
Adhesively Bonded Composites, 639-658, 755-787
Adhesively Bonded Honeycomb Structures
  ultrasonic assessment of cure rate effects in, 369-380
Ageing of Structural Film Adhesives
  and effect on joint strength, 231-246
Aliphatic Polyamides-Metal Bonds, 195-212
Alumina
  fracture surface energies of, 879
Aluminum Adhesive Bonding (see Adhesive Joint of)
Aluminum Alloys, Oxides on
  and characterization nomograms, 113-115
  morphology of, 104-106, 107-113, 455-456, 472
  polymer identification in, 473
  surface characterization of, 103-120
  surface impedance measurements of, 107
  surface potential difference of, 106-113
  ternary phase diagram of, 462
Aluminum Hydroxide
  morphology of, 460
Aluminum Polymer Bonds
  durability of, 453-467, 469-484, 485-501, 503-521
Aminosilane Treatment
  of titanium, effect on Ti/epoxy adhesive joints, 541-553.
  of iron, effect on iron/epoxy adhesive joints, 549
Auger Electron Spectroscopy, 9

INDEX

Benzotriazole (BTA) Coupling Agents for Copper, 523-540
Blister Technique, 296, 304, 707
Bond Length
 effect on monowire joints, 178-179
 effect on multiwire joints, 183-190
Bondability of Low Surface Energy Materials
 effect of surface treatment on, 121-236
Bonded Honeycomb Structures
 ultrasonic assessment of cure rate effects in, 369-380
Bonding Parts Surface Treatment
 integration of, in adhesive bonding process, 257-266
Bondline Stiffness, 387
Box-Section Beams (Adhesively Bonded)
 performance of, 627-637
Bulk Characterization
 and evaluation of adhesive joints, 3-17
Butt Joint with a Flaw
 three dimensional analysis of, 839-853

Ceramic-Metal Crown, 290
Ceramic-Metal Joints, 855-869, 871-882
Characterization Methods for Locus of Failure, 7-15
Characterization Nomograms for Representing Oxide Film Properties, 113-115
Chromic-Sulfuric Treatment of Polymers, 124
Composite-Composite Bond Joints
 cycling debonding of, 639-658
 fracture of, 755-787
Contact Angles, Measurement of, 125
 effect of surface treatment of polymer substrates on, table, 128
Continuum Mechanics Factors in Adhesive Fracture, 703-737

Copper Adhesion to Epoxy, 523-540
Copper-Alumina Joints
 critical energy release rates of, 879
Copper-BTA Complex, 524
Crazing in Polymers, 821-824
Creep of Epoxy Adhesives, 140-146
Critical Energy Release Rates
 in Metal-Ceramic Joints, 855-869, 871-882
Critical Fracture Energy, 62
Critical Surface Tension of Wetting, 122
 for various polymers, table, 123
Cure Rate Effects, Ultrasonic Assessment of, 369-380
Curing of Adhesives
 investigation by interface waves, 328-337
Cyanamid FM73 Adhesive
 shear stress-shear strain diagram of, 663
Cycle (low) Loading
 effect of, on shear stressed adhesive bond lines, 659-677
Cyclic Debonding of Adhesively Bonded Composites, 639-658

Deformation of Adhesive Joint
 study of, by Moire patterns, 151
Dental Adhesion Testing
 a critical appraisal of, 289-305
Dentistry
 adhesive failure in, 295-300
Diallyl Phthalate (DAP)
 epoxy IPN adhesives, 252-254
 structural adhesives based on, 247-256
Dimensional Stability of Adhesively Bonded Joints, 137-150, 151-164
 effect of moisture on, 137-150
Double Cantilever Beam Specimen
 fractographs of, 765
 geometry of, 706, 762, 794

Ductile Fracture Energy, 747-748
Durability Factor, 505
Durability of Metal-Polymer
  Bonds, 453-467, 469-484,
  485-501, 503-521, 523-540,
  541-553, 555-564, 694-701
    in industrial atmosphere, 495-497
    in salt fog cycle environment, 510, 513-514
    in seacoast atmosphere, 497-500
Dynamic Mechanical Analysis
    and polyurethane coating curing, 46-48
    of aged epoxy adhesive, 239
Dynamic Mechanical Relative Shear
  Modulus and Damping
    for epoxy bonded joint, 356
    for poly(allyl 2-cyanoacrylate) bonded joint, 359
    for poly(ethyl 2-cyanoacrylate) bonded joint, 358
    for polyester/polysulfone block copolymer bonded joint, 357
Dynamic Mechanical Storage
  Modulus and Damping
    for epoxy free film, 355
    for polyester/polysulfone block copolymer film, 357
    for poly(ethyl 2-cyanoacrylate) film, 358

Effective Fracture Surface Energy
    of epoxy adhesive, 742, 745
Effective Shear Modulus, 337
Elastic Interface Waves, 321-328
Elastomer-Metal Bonding, 555-564
Electrochemical Pretreament of Copper
    and its effect on joint strength, 523-540
E-Glass Single Filaments
    epoxy resin wetting of, 51-65
    silane treatment of, 51, 54-65
    surface free energy of, 62
    work of adhesion on, 58
Electron Energy Loss Spectroscopy
    and polymer identification in aluminum oxide, 473

Electron Spectroscopy for Chemical Analysis (ESCA), 10
    investigation of adhesively bonded Al joints by, 479
Enamel-Composite Failures, 300
Enamel-Resin System, SEM of, 293
Energy Dissipation
    effect on peel strength, 419-432
Epoxide Resins (Filled)
    interfacial properties of, 883-904
Epoxy Adhesion to Copper, 523-540
Epoxy Adhesive
    aging of, effect on joint strength, 231-246
    creep of, 140-146
    dynamic mechanical analysis of aged, 239
    effective surface energy, fracture energy and fracture toughness of, 742
    HPLC analysis of fresh and aged, 237
    modified (Epasol FV/ZIS939), 259-262
      and bonding of oily surfaces, 259-260
    moisture absorption in, 148
Epoxy Adhesive (Rubber or Elastomer Modified)
    effect of loading rate on, 613, 618-619
    effective fracture surface energies and fracture energies of, 745
    fracture energy of, 796
    fracture toughness of, 829-838
    replica TEM of fracture surface of, 791
Epoxy Adhesive or Bonded Joints
  (see also Adhesive Bonded Joints), 8, 69, 121-136, 138, 151-164, 165-193, 231-246, 259-260, 356, 383, 485-501, 503-521, 523-540, 541-553, 549, 639-655, 694-701, 799-803, 817

INDEX

Epoxy Bonded Joints of
  aluminum, 233, 240, 245, 383, 616, 793, 817
  copper, 523-540
  glass, 694-701
  graphite/epoxy, 639-658, 757
  iron, 549
  microscope cover slip, 152
  polymers, 121-136
  steel, 69, 85-102, 165-193, 351, 429, 793
  titanium, 8, 138, 158, 459, 541-553
Epoxy Free Film
  dynamic mechanical storage modulus and damping for, 355
Epoxy Resin Wetting of E-Glass Single Filaments, 51-65

Failure, Locus of, 4-7, 10-15
Filled Polymer Systems
  estimation of adhesion in, 433-450
Film Adhesives (Structural)
  aging of, and effect on joint strength, 231-246
Finger Joints (Exterior Grade), 213-230
Finite Element Analysis of
  adhesive bonded joints, 648-652
  box-section structure, 635-637
  butt joint with a flaw, 839-853
  single lap joint, 617-618, 622-625
  thick adherend model joint, 682-684
Flexibilized Diallyl Phthalate Adhesives, 250-251
Fracture Aspects of Adhesive Joints, 689-702, 703-737, 739-753, 759-787, 789-807, 809-827, 839-853
Fracture Energy
  measurement of, 740-741
  of adhesive joints, 739-753
  of epoxy adhesives, 742, 745
  of glass-epoxy composite, 62
  of thermoplastics and thermosets, 743

Fracture of Metal-Ceramic Joints, 855-869, 871-882
Fracture Properties of Filled Epoxide Resins, 883-904
Fracture Toughness, 740
  of epoxy adhesive, 742
  of epoxy (rubber modified), 829-838

Generalized Theory of Fracture Mechanics, 689
Glass-Epoxy Bonds
  effects of silanes on, 695-701
  effects of water on, 694-701
Glue Line Thickness
  effect of, on monowire joints, 176-177, 181
Glue-Wood Bonds
  accelerated aging procedures for, 395-417
Gold-Aluminum Diffusion Couple
  surface acoustic wave phase shift in, 315
GPC Analysis of Nitrile Epoxy Adhesive, 242
Griffith Fracture Energy for Glass-Epoxy Composite, 62
Groups in a Resin and Property Obtained, Table, 249

Hand Pulled Peel Test, 44, 49
Honeycomb Structures
  ultrasonic assessment of cure rate effects in, 369-380
"Honeymoon" Adhesives, 213-230
Hot-Pressed Silicon Nitride Joints, 856
HPLC Analysis of
  fresh and aged epoxy adhesive, 237
  nitrile-epoxy adhesive, 244
Hydrothermal Stability of Titanium-Epoxy Adhesive Joints, 541-553

Impact Strength of Adhesive Lap Joints, 611-627
Impact Torsion Test Rig, 615

Industrial Atmosphere
   bond durability in, 495-497
Infra-red Spectrum of Amino-
   silane on Titanium, 544-547
Inhibitors and the Durability of
   Adhesive Bonds to Al, 463-464
Integrated Methodology for
   Adhesive Bonded Joint Life
   Predictions, 567-586
Interatomic Bonding Energies at
   Interfaces
   mechanical measurement of,
     689-702
Interface in a Single Two Com-
   ponent System, 5
Interface Waves, Application of
   in investigating curing and
   viscoelastic properties of
   adhesives, 328-337
Interfaces in an Adhesive Bond-
   ing System, 6
Interferometric Technique
   and study of deformation of
     adhesive joint, 151
Intermediate Adhesive Layers, 196
   effect on adhesive strength,
     197-212
Interpenetrating Polymer Network,
   249, 252-255
   DAP/epoxy, 252-255
Interphase, 5
Intrinsic Fracture Energy, 741
Ion Scattering Spectrometry, 9
Iron-Epoxy Adhesive Bonds and
   Effect of Aminosilane, 549

J-Integral, 746-747

Lap Shear Test
   for adhesive joints, 271-283,
     706
   stress analysis in, 273-283
   stress intensity factors for,
     277
Leaky Surface Waves, 316-317
Life Prediction Methodology for
   Adhesive Bonded Joints, 567-586
Locus of Failure, 4-7, 10-15
   characterization methods for,
     7-15

Low Energy Electron-Induced
   X-ray Spectroscopy (LEEIXS)
   and surface characterization,
     85
   schematic diagram of, 87
Low Surface Energy Materials
   effect of surface treatment
     on the wettability and
     bondability of, 121-136
   suitable surface treatments
     and adhesives for, 134

Mechanical Measurement of
   Interatomic Bonding Energies
   at Interfaces, 689-702
Mechanical Properties of Joints
   influence of surface rough-
     ness on, 69-84
Metal-Ceramic Crown, 290
Metal-Ceramic Joints, 855-869,
   871-882
   influence of layer thickness
     and internal stress on,
     855-869
Metal-Polymer Bonds (see
   Adhesive Joints of)
   durability of, 453-467, 469-
     484, 485-501, 503-521, 523-
     540, 541-553, 555-564, 694-
     701
Metal Surfaces
   interaction with phenol for-
     maldehyde resin, 35
Modified Epoxy Adhesive (Epasol
   FV/ZIS939)
   and bonding to oily surfaces,
     259-260
Moire Fringes, 154
   and study of deformation of
     an adhesive joint, 151
Moisture (see also Durability
   of Metal-Polymer Bonds)
   effect of, on the dimensional
     stability of adhesive bond-
     ed joints, 137-150
Mono and Multiple-Wire Steel to
   Steel Joints, 165-193

Naphthalene-Sodium Etching of
   Polymers, 125

INDEX

Nitrile Rubber Modified Epoxy
  aging of, 232, 240-245
  and tensile shear strength of Al-Al lap joints, 245
  GPC analysis of, 242
  HPLC analysis of, 244
Nondestructive Evaluation of Bonded Joints, 347-367

Oily Surfaces
  bonding by modified epoxy adhesive (Epasol FV/ZIS939), 259-260
Optical Micrographs of Fracture Surfaces in Ti/Epoxy Bonds, 551-552

Particle Filled Epoxide Resin
  fracture properties of, 883-904
Peel Strength and Energy Dissipation, 419-432
Peel Test, Hand Pulled, 44, 49
Phase Boundary Sensitive Test, 19
Phenol Formaldehyde Resin
  and interactions with metal surfaces, 35
  effect of, on the adhesion of polysulfide sealant, 31-40
Phenolic ("Honeymoon") Fast Setting Adhesives
  and exterior grade finger joints, 213-230
Plasma Treatment of Polymers, 125
Polyamides
  adhesive joint strength with metals (Al, steel), 195-212
Polybutadiene-Al Powder System
  adhesion estimation in, 444
Polymer-Metal Bonds (see Metal-Polymer Bonds)
Polymeric Films (Thin)
  and metal-polyamide joints, 195-212
Polymers
  bondability of, 121-136, 692-694
  contact angles of, effect of surface treatment on, table, 128

Polymers (cont.)
  critical surface tension of wetting of, table, 123
  work of adhesion for SBR bonded to, 693-694
Polysulfide Sealant
  adhesion of, 31-40
    effect of phenol formaldehyde resin, 31-40
  crosslink density of, 33, 37-38
  cure rate of, 34
  polyurethane coating interfacial aspects, 41-50
    SEM study of, 44-46
  pot life of, 34, 38
  water absorption in, 34, 39
Polyurethane
  chemistry, 42-43
  coating curing by DMA, 46-49
  coating-polysulfide sealant interfacial aspects, 41-50
Porcelain-Metal
  failures, 299
  interface, test of, 302
  specimens, 296
Practical Adhesion, 20, 32

Quality Control (In-Plant) of Glue-Wood Bonds
  accelerated aging procedures for, 395-417

Rayleigh-Angle Method, 317-321
Recursive Structural Identification Method, 370
Resin-Enamel/Dentin Interface
  test of, 302
Resin Properties vs. Groups in, table, 249
Rod Pull-out Test, 707, 716, 719-721
Roughness (Surface)
  influence of, on mechanical properties of joints, 69-84
  parameters, 72
Rubber-Metal Adhesion, 555-564

Salt Fog Cycle Environment
  bondability in, 510, 513-514

Sapphire/Copper Joints, 871-882
Scrim Cloth
   effect of, on adhesively bonded joints, 679-685
Seacoast Atmosphere
   bond durability in, 497-500
Secondary Ion Mass Spectrometry, 9
SEM of
   CrO_3 anodized Al surface, 105
   enamel-resin system, 293
   fracture of sapphire joint, 880
   fracture surfaces of filled epoxy resin, 890-894
   fracture surfaces of FM-73 adhesive, 681
   fracture surfaces of Ti/epoxy lap joints, 551
   sealant-coating interface, 44-46
Shear Strength of Glass Fiber/Epoxy Resin, 51
Short Beam Shear Test, 384
Silane Treatment of
   alumina particles, 884
   E-glass filaments, 51, 54-65
      wetting of, 58
   glass, 695-701
   silica particles, 884
   titanium adherend, 541-553
Stainless Steel (AISI-304)
   and epoxy joint, 85-102
      effect of surface conditioning upon, 85-102
   effect of surface conditioning on oxide thickness of, 93
   LEEIXS spectra of, 95
Steel, Adhesive Bonding of, 7, 15, 69, 85-102, 165-193, 351, 429, 793
Steel to Steel (Mono and Multiple-Wire) Joints, 165-193
Stoneley Wave, 321
Stress Analysis of Adhesive Joints, 587-610
Stress Energy Release Rates and cyclic debonding, 639-658
Stress Intensity Factors for Lap Shear Tests, 277
Structural Adhesives
   based on diallyl phthalate, 247-256
   form and composition of, 248
Structural Film Adhesives
   aging of, and effect on joint strength, 231-246
Styrene/Isoprene/Styrene Triblock Copolymers, 353
   dynamic mechanical damping of, free films and bonded joints, 362-364
Surface Acoustic Waves, 310-316
   phase shift in the Au-Al diffusion couple, 315
Surface Behavior Diagram, 461-462
Surface Characterization
   and evaluation of adhesive bonds, 3-17
   methods for locus of failure, 5-15
   of anodic oxides on Al alloys, 103-120
Surface Free Energy of Silane Coated Fibers, 62
Surface Impedance Measurements of anodic oxides, 107, 116-117
Surface Potential Diffeerence of anodic oxides, 106-113
Surface Roughness
   influence of, on mechanical properties of joints, 69-84
   parameters, 72
Surface Treatment of Bonding Parts
   integration in the adhesive bonding process, 257-266
Surface Treatments of Low Surface Energy Materials, 124-125
   effect of, on contact angles, table, 128

Tapered Double Cantilever Beam Adhesive Joint Specimen, 794
Tensile Bond Tests, 283-285
   stresses in, 284

Theory of Attachment Site, 35
Thermoplastics
    ductile fracture energy of, 748
    fracture energies of, 743
Thermosets
    fracture energies of, 743
Thin Films
    application of surface wave method to, 315
Thin-Gauge Sheet Metal Box-Section Beams
    adhesive bonding of, 627-637
Three Point Bend Test for Adhesive Joints, 381-394
Three Point Flexural Test for an Adhering System, 19-30, 85
Titanium-Adhesive Bonding, 8, 138, 158, 453-467, 541-553
Titanium Pretreatment Processes, 458

Ultrasonic Assessment of Cure Rate Effects in Bonded Honeycomb Structures, 369-380
Ultrasonic Surface and Interface Waves
    and adhesive joint characterization, 307-345
Underwater Sonar Systems
    elastomer/metal bonds applicable in, 555

Vibrational Adhesive Bonding, 262-266

Viscoelastic Properties of Adhesives
    investigation by interface waves, 328-337

Weak Boundary Layer, 4, 6, 11, 35, 337, 448, 480

Wedge Test, 12-14, 459, 463-464, 473

Wettability of Low Energy Surface Materials
    effect of surface treatment on, 121-136

Width Tapered Beam Specimen
    fractograph of, 771
    geometry of, 767

Wire (Steel Mono and Multiple) Joints, 165-193

Wood-Glue Bonds
    accelerated aging procedures for, 395-417

Work of Adhesion
    for SBR bonded to various plastic substrates, 693
    on glass filaments, 58

X-ray Photoelectron Spectroscopy (see ESCA)

Yield Strength of the Interface, 387

MIX
Papier aus verantwortungsvollen Quellen
Paper from responsible sources
FSC® C105338

If you have any concerns about our products,
you can contact us on
ProductSafety@springernature.com

In case Publisher is established outside the EU,
the EU authorized representative is:
**Springer Nature Customer Service Center GmbH
Europaplatz 3, 69115 Heidelberg, Germany**

Printed by Libri Plureos GmbH
in Hamburg, Germany